Principles
of Cloning

EDITORIAL BOARD

PRINCIPLES OF CLONING

Edited by

JOSE CIBELLI

Advanced Cell Technology
Worcester, Massachusetts

ROBERT P. LANZA

Advanced Cell Technology
Worcester, Massachusetts

KEITH H. S. CAMPBELL

Department of Animal Physiology
University of Nottingham
Loughborough
Leicestershire, United Kingdom

MICHAEL D. WEST

Advanced Cell Technology
Worcester, Massachusetts

ACADEMIC PRESS

An imprint of Elsevier Science

Amsterdam Boston London New York Oxford Paris San Diego San Francisco Singapore Sydney Tokyo

Academic Press
An Elsevier Science Imprint
525 B Street, Suite 1900, San Diego, California 92101-4495, USA
http://www.academicpress.com

Academic Press
84 Theobalds Road, London WC1X 8RR, UK
http://www.academicpress.com

Library of Congress Catalog Card Number: 2002101649

International Standard Book Number: 0-12-174597-X

PRINTED IN THE UNITED STATES OF AMERICA
02   03   04   05   06   07   MM   9   8   7   6   5   4   3   2   1

We dedicate this book to all the pioneers who contributed, consciously or not, to this new field of science.

In particular, we want to mention those no longer with us: Alan Wolffe, Jacques Loeb, Hans Spemann, Robert Briggs, Thomas King, and Stephen Jay Gould.

CONTENTS

PART IV: CURRENTLY SOUGHT AFTER SPECIES

PART V: NUCLEAR TRANSFER IN PRIMATES

PART VI: APPLICATIONS

CONTRIBUTORS

Jennifer D. Ambroggio
IVF Clinic
University of Southern California
Los Angeles, California 90033
Chapter 19

Michael Bader
Department of Pediatrics
Northwestern University Medical School
Children's Memorial Institute for Education and Research
Chicago, Illinois 60614
Chapter 21

Esmail Behboodi
Genzyme Transgenics Corporation
Framingham, Massachusetts 01701
Chapter 25

Zeki Beyhan
Department of Animal Sciences
University of Wisconsin—Madison
Madison, Wisconsin 53706
Chapter 19

Michele Boiani
Germline Development Group at the Center for Animal
    Transgenesis and Germ Cell Research
University of Pennsylvania
Kennett Square, Pennsylvania 19348
Chapter 6

J. A. Byrne
Wellcome CRUK Institute
and Department of Zoology
University of Cambridge
Cambridge CB2 1QR, United Kingdom
Chapter 14

Keith H. S. Campbell
School of Biosciences
University of Nottingham
Sutton Bonington, Loughborough
Leicestershire LE12 5RD, United Kingdom
Chapter 20

Pascale Chavatte-Palmer
Biologie du Développement et Biotechnologies
Unité Mixte de Recherche Institut National de la
    Recherche Agronomique/Ecole Nationale Vétérinaire
    d'Alfort
78352 Jouy en Josas, France
Chapter 12

LiHow Chen
Genzyme Transgenics Corporation
Framingham, Massachusetts 01701
Chapter 25

Philippe Collas
Institute of Medical Biochemistry
University of Oslo
Oslo 317, Norway
Chapter 2

Margaret M. Destrempes
Genzyme Transgenics Corporation
Framingham, Massachusetts 01701
Chapter 25

András Dinnyés*
Department of Animal Science/Biotechnology Center
University of Connecticut
Storrs, Connecticut 06269
Chapter 17

Tanja Dominko
Advanced Cell Technology
Worcester, Massachusetts 01605
Chapter 22

Patrick W. Dunne
Department of Veterinary Anatomy and Public Health
Center for Animal Biotechnology and Genomics
Texas A&M University
College Station, Texas 77843
Chapter 11

Yann Echelard
Genzyme Transgenics Corporation
Framingham, Massachusetts 01701
Chapter 25

Kevin Eggan
Whitehead Institute for Biomedical Research
and Department of Biology
Massachusetts Institute of Technology
Cambridge, Massachusetts 02142
Chapter 4

Neal L. First
Department of Animal Sciences
University of Wisconsin—Madison
Madison, Wisconsin 53706
Chapter 19

*Current affiliations: Roslin Institute, Department of Gene Expression and Development, Roslin, United Kingdom; and
Research Group for Animal Breeding, Hungarian Academy of Sciences and Szent, Istvan University, Godollo, Hungary.

Rafael A. Fissore
Paige Laboratories
Department of Veterinary and Animal Sciences
University of Massachusetts
Amherst, Massachusetts 01003
Chapter 2

Robert H. Foote
Department of Animal Science
Cornell University
Ithaca, New York 14853
Chapter 1

Vasiliy Galat
Max Delbrück Center for Molecular Medicine
D-13092 Berlin, Germany
Chapter 21

David K. Gardner
Colorado Center for Reproductive Medicine
Englewood, Colorado 80110
Chapter 9

R. L. Gardner
Department of Zoology
University of Oxford
Oxford OX1 3PS, United Kingdom
Chapter 27

Ronald M. Green
Ethics Institute
Dartmouth College
Hanover, New Hampshire 03755
Chapter 26

J. B. Gurdon
Wellcome CRUK Institute
and Department of Zoology
University of Cambridge
Cambridge CB2 1QR, United Kingdom
Chapter 14

Jonathan R. Hill
College of Veterinary Medicine
Cornell University
Ithaca, New York 14853
Chapter 12

Philip Iannaccone
Department of Pediatrics
Northwestern University Medical School
Children's Memorial Institute for Education and Research
Chicago, Illinois 60614
Chapter 21

Kimiko Inoue
RIKEN Bioresource Center
Ibaraki 305-0074, Japan
Chapter 8

Rudolf Jaenisch
Whitehead Institute for Biomedical Research
and Department of Biology
Massachusetts Institute of Technology
Cambridge, Massachusetts 02142
Chapter 4

Y. Kato
Laboratory of Animal Reproduction
College of Agriculture
Kinki University
Nara 631-8505, Japan
Chapter 13

Manabu Kurokawa
Paige Laboratories
Department of Veterinary and Animal Sciences
University of Massachusetts
Amherst, Massachusetts 01003
Chapter 2

Michelle Lane
Colorado Center for Reproductive Medicine
Englewood, Colorado 80110
Chapter 9

Crista Martinovich
Departments of Obstetrics-Gynecology and Reproductive
    Sciences and Pittsburgh Development Center of
    Magee—Women's Research Institute
University of Pittsburgh School of Medicine
Pittsburgh, Pennsylvania 15213
Chapter 22

Harry M. Meade
Genzyme Transgenics Corporation
Framingham, Massachusetts 01701
Chapter 25

Narumi Ogonuki
RIKEN Bioresource Center
Ibaraki 305-0074, Japan
Chapter 8

Atsuo Ogura
RIKEN Bioresource Center
Ibaraki 305-0074, Japan
Chapter 8

Kenjiro Ozato
Bioscience Center
Nagoya University
Nagoya 464-8601, Japan
Chapter 15

Raymond L. Page
Advanced Cell Technology
Worcester, Massachusetts 01605
Chapter 7

Anthony C. F. Perry[†]
Advanced Cell Technology
Worcester, Massachusetts 01605
Chapter 16

Jorge A. Piedrahita
Center for Animal Biotechnology and Genomics
College of Veterinary Medicine
Texas A&M University
College Station, Texas 77843
Chapter 11

[†]Current affiliation: Laboratory of Mammalian Molecular Embryology, RIKEN Center for Developmental Biology, Chuou-ku, Kobe 650-0047, Japan.

Randall S. Prather
Department of Animal Sciences
University of Missouri—Columbia
Columbia, Missouri 65211
Chapter 18

Gerald Schatten
Departments of Obstetrics-Gynecology and Reproductive
 Sciences and Pittsburgh Development Center of
 Magee—Women's Research Institute
University of Pittsburgh School of Medicine
Pittsburgh, Pennsylvania 15213
Chapter 22

Hans R. Schöler
Germline Development Group at the Center for Animal
 Transgenesis and Germ Cell Research
University of Pennsylvania
Kennett Square, Pennsylvania 19348
Chapter 6

George E. Seidel, Jr.
Animal Reproduction and Biotechnology Laboratory
Department of Physiology
Colorado State University
Fort Collins, Colorado 80523
Chapter 10

Calvin Simerly
Departments of Obstetrics-Gynecology and Reproductive
 Sciences and Pittsburgh Development Center of
 Magee—Women's Research Institute
University of Pittsburgh School of Medicine
Pittsburgh, Pennsylvania 15213
Chapter 22

Jeremy Smyth
Paige Laboratories
Department of Veterinary and Animal Sciences
University of Massachusetts
Amherst, Massachusetts 01003
Chapter 2

Steven L. Stice
Department of Animal and Dairy Science
University of Georgia
Athens, Georgia 30602
Chapter 24

M. Azim Surani
Wellcome CRUK Institute
Physiological Laboratory
University of Cambridge
Cambridge CB2 1QR, United Kingdom
Chapter 5

Masako Tada
Department of Development and Differentiation
Institute for Frontier Medical Sciences
Kyoto University
Kyoto 606-8507, Japan
Chapter 5

Takashi Tada
Department of Development and Differentiation
Institute for Frontier Medical Sciences
Kyoto University
Kyoto 606-8507, Japan
Chapter 5

X. Cindy Tian
Department of Animal Science/Biotechnology Center
University of Connecticut
Storrs, Connecticut 06269
Chapter 17

Alan Trounson
Institute of Reproduction and Development
Monash University
Clayton, Victoria 3168, Australia
Chapter 23

Y. Tsunoda
Laboratory of Animal Reproduction
College of Agriculture
Kinki University
Nara 631-8505, Japan
Chapter 13

Fyodor D. Urnov
Sangamo BioSciences, Inc.
Pt. Richmond Tech Centre
Richmond, California 94804
Chapter 3

Yuko Wakamatsu
Bioscience Center
Nagoya University
Nagoya 464-8601, Japan
Chapter 15

Teruhiko Wakayama[‡]
Advanced Cell Technology
Worcester, Massachusetts 01605
Chapter 16

Alan P. Wolffe[§]
Sangamo BioSciences, Inc.
Pt. Richmond Tech Centre
Richmond, California 94804
Chapter 3

Xiangzhong Yang
Department of Animal Science/Biotechnology Center
University of Connecticut
Storrs, Connecticut 06269
Chapter 17

[‡]Current affiliation: RIKEN Center for Developmental Biology, Chuou-ku, Kobe 650-0047, Japan.
[§]Deceased.

PREFACE

In 1997, Dolly the cloned sheep began a biological revolution. She showed us the way to physiologically dedifferentiate already committed somatic cells, opening the gate to a whole new world of possibilities in the study of basic biological mechanisms that touch the origins of life at its core. It is clear that cloning vertebrates using somatic cells is here to stay. Unfortunately, at this time, we know little about how this process works, and it is difficult to imagine all the applications this technology will bring to fruition. This book tries to address some of these questions.

Part I of *Principles of Cloning* explores the basic known biological processes and lays out critical points to consider when interpreting cloning experiments that are now being conducted. The nucleus itself, chromatin structure, epigenetic changes, and the role of the egg are emphasized. Part II describes the methods utilized during cloning in an attempt to make them available to students and professionals who are interested in nuclear transfer techniques in vertebrates. Parts III–V contain analyses of species that already have been cloned and species that are still being studied, including nonhuman primates. We are delighted to say that the contributors in these sections published the first articles on most of these species. Experts in the area have also contributed to the applications of cloning covered in Part VI; we expect this part of the book to communicate the benefits of cloning to society in the medical and agricultural fields. Part VII addresses the ethical concerns evoked by this technology and is followed by an analysis of what the future may bring.

We hope this book challenges readers, encouraging them to think beyond the cloning procedure itself to the creation of new individuals or pluripotent stem cells. Interesting questions regarding the mechanisms of dedifferentiation must now be addressed; the answers will perhaps allow scientists to recreate the same phenomenon in a more consistent and simpler way.

Jose Cibelli

FOREWORD

Every living mammal originates from the transplantation of a sperm nucleus into an egg. Biparental inheritance and genetic diversity have been the rewards of this natural form of nuclear transplantation. Modification of this process for laboratory use was first accomplished in the middle of the past century. The 27 chapters in *Principles of Cloning* not only provide a fitting 50th anniversary tribute to this pioneering achievement, but also serve as a record of the notable scientific advances that have emanated from cloning experiments.

Nuclear transplantation has had a much more profound impact on our understanding of the processes of development and differentiation than is often appreciated. It has demonstrated not only that cell differentiation takes place without the loss or permanent inactivation of genes, but has also provided the basis for the new field of genomic imprinting. In addition, elegant studies based on the transfer of nuclei between somatic cells have yielded important insights into the mechanisms of differentiation and have contributed to gene mapping and the identification of tumor suppressor genes. If the past achievements of nuclear transplantation have been to extend significantly our knowledge of nuclear plasticity and differentiation, what problems and progress can be expected from the cloning experiments of the future?

Many investigators have reported that fully differentiated cells in amphibia and mammals revert to a totipotent state when exposed to the cytoplasm of oocytes and eggs. Although widely acclaimed, this important advance is still beset by major difficulties and uncertainties arising from the exceedingly small percentage of differentiated nuclei that develop into viable young after transplantation. The current emphasis on cloning an ever-increasing range of mammals has demonstrated the universality of the problem; irrespective of the species studied, only 1 to 2% of cloned embryos survive to birth. In my view, the repetitious and disappointing nature of these results argues strongly for a redirection of effort. Would it not be more rewarding to move away from the cloning of ever more animals and focus heavily instead on the molecular and cellular events associated with nuclear reprogramming? Many fundamental issues associated with cloning deserve investigation. For example, does the high rate of failure indicate that the capacity to undergo full reprogramming resides in a small and specialized subset of adult nuclei only? Alternatively, are all differentiated nuclei equally capable of undergoing reprogramming but prevented from doing so due to imperfections in the cloning procedures? Likewise, what is the nature of the epigenetic and higher order events that are associated with nuclear reprogramming? Priority given to these studies will not only be of importance to cloning but will also contribute to a more comprehensive understanding of developmental processes in both embryos and stem cells.

A sharper focus on the mechanisms of reprogramming will be significant both in restoring cloning to its original role of answering important biological questions and also in redressing the balance between the academic and the commercial aspects of cloning.

Although the identification of nuclear changes underpinning reprogramming is essential, this advance may well be eclipsed by studies on cytoplasmic proteins involved in the initiation and regulation of nuclear reprogramming. The identification of reprogramming proteins will be important for both scientific and medical reasons. In addition to a likely role in the preconditioning of somatic nuclei before transplantation, cytoplasmic reprogramming proteins may well have an even more vital future role in the preparation of somatic cells for direct use in tissue repair procedures. These considerations, together with the intellectual challenge of discovering how the reprogramming proteins act at the level of the nucleus, provide adequate incentives for a major focus on this emerging area of research. However, it would be unhelpful to minimize the problems that face investigators working on either the cytoplasmic or the nuclear aspects of reprogramming of cloned embryos. These difficulties include working with very limited amounts of material and having an inadequate set of markers with which to delineate the progress of nuclear reprogramming. Although these difficulties are substantial, the present advances in molecular technology are sufficiently great to ensure the problems will be overcome.

Although I believe fundamental studies offer the best long-term solutions to the problems of cloning, it is nevertheless apparent that the major interest at present is focused firmly on the potential of cloning for medical and biotechnological use. To be of value, the debate on the commercialization of cloning must be realistic, flexible, broadly based, and comparative in nature. In addition, it is imperative that the deep concerns felt by a substantial proportion of the public about some aspects of cloning are addressed.

Before looking toward future commercial developments, it is instructive to recall some of the existing biomedical successes achieved using nuclear transplantation techniques. Perhaps the least controversial but most important of its biotechnological successes has been in the production of monoclonal antibodies for therapeutic and experimental purposes. A comparable success has been made in the transplantation of nuclei for the treatment of male infertility; the so-called intracytoplasmic sperm injection procedure. These two examples demonstrate that nuclear transplantation can have a significant but acceptable impact on widely disparate areas of human well-being. Controversy emerged a few years ago when sheep were cloned by somatic nuclear transplantation amidst a fanfare of publicity. There followed a period of largely unfavorable press comment and unrealistic commercial expectations. Now a new and balanced approach is emerging in which technical problems, ethical considerations, and alternative technologies are being carefully weighed. The debate about therapeutic cloning exemplifies this more measured approach and has led to a consensus in which embryonic stem cells are adjudged to offer the best short-term prospects for tissue repair, while adult stem cells seem destined to predominate in the longer term. The need for flexibility is, however, paramount because rapid developments in adult stem cell biology are likely to challenge the prevailing consensus and may well point to the advantages of making more immediate use of the patient's own cells for tissue repair. A development of this nature would have major implications for the role of therapeutic cloning.

In conclusion, cloning is in transition from a distinguished past to a new future. If used in an imaginative manner, this move will ensure that nuclear transplantation continues to serve as a beacon of scientific excellence and as an important contributor to human health.

Robert M. Moor
The Babraham Institute
Cambridge CB2 4AT, United Kingdom

INTRODUCTION

CHAPTER 1

HISTORICAL PERSPECTIVE

Robert H. Foote

As we reflect on the transition from nature's ancient unassisted cloning to the modern era of attempting to understand the basic mechanisms of arrays of gene action during development in a complex genome (Oliver *et al.*, 2001; Venter *et al.*, 2001), we should not forget basic ideas and paths that pioneers laid out before us, stimulating modern efforts to clone mammals. Who directed attention to the basic molecules responsible for hereditary characteristics? Avery *et al.* (1944) are often overlooked as the group that discovered DNA as the molecule of life. Soon after, Beadle (1946) described how genes regulate biology. Spemann (1938), and others, established a model in amphibians that served as an excellent guideline to build on to clone mammals. The study of sperm, eggs, and embryos, the development of successful procedures for artificial insemination, and embryo manipulation, culture, and transfer provided a background for cloning research (Heape, 1897; Brackett *et al.*, 1981; Foote, 1998, 1999).

ANCIENT CLONING (ASEXUAL REPRODUCTION)—IN THE BEGINNING

Cloning, or asexual reproduction, a time-honored method of reproduction, is the method by which many extinct and existing species of organisms reproduced. "Clone" comes from the Greek word "klon," meaning twig. In horticulture clones are defined as descendants of a single plant. This clearly indicates an expected genetic identity within a clone. When a particular genotype is well-adapted to a particular environment, cloning is a simple and favorable way to reproduce.

Many plants can propagate by both asexual (vegetative) and sexual reproduction. Many simple one-celled organisms, such as *Escherichia coli*, propagate primarily by a type of asexual reproduction called binary fission. A relatively simple single circular chromosome containing DNA replicates, and by binary fission one bacterium becomes two identical progeny cells. However, even *E. coli* have sex factors and occasionally conjugate. Yeasts have various stages in which they exist with haploid, diploid, or a mixture of one or two copies of DNA. Simple protozoa, such as paramecia, reproduce both asexually and sexually. Thus, in the evolution of plant and animal types, many variations of sexual and asexual reproduction have arisen.

Sexual reproduction permits a continuum of genetic diversity whereby an organism can adapt to survive and thrive in many environments. Even earthworms, which are hermaphrodites, do not self-fertilize. Within a group of organisms, the diversity of genotypes promoted by sexual reproduction likely is responsible for the multitude of types that have adapted to fill various biological niches throughout the evolutionary process.

SPONTANEOUS REGENERATION

Simple multicellular organisms such as *Planaria* (flatworms) can regenerate parts extensively. When cut in half each remaining half of the flatworm can regenerate the missing half to form a complete organism. Starfish can regenerate an arm that has been broken off. Various amphibians can regenerate limbs. However, in more complex animal forms regeneration generally is limited to repair of injury to tissues or organs, such as cut skin and broken bones. Damaged whole organs are not replaced spontaneously. Such repair would require a complex turning on and off of an array of genes in the appropriate sequence, as happens during embryological development, but without triggering action of other developmental genes. This beautiful process of development is poorly understood, but the advances in cloning and related technologies provide a golden opportunity to design studies to reveal more about how the intricacies of nature work.

TYPES OF CLONING

In this brief historical overview the term "cloning" will be used to reflect propagation of replicated biological material in many forms. Emphasis will be on studies designed to produce functional cloned individuals through a controlled series of mitotic divisions and cellular differentiation leading to formation of the fetus and placenta, and finally to the birth of complex individuals. Most of the millions of cells in the individual presumably contain the same genetic material in their nuclei. A cell that can give rise to a complete individual would be totipotent. Many early embryonic cells have the potential to produce cloned individuals. Other cells, such as embryonic stem (ES) cells, have the ability to produce the three germ layers, but, as currently manipulated *in vitro*, they may not result in formation of a normal trophoblast, placenta, and fetus. These cells are pluripotent, unless the nuclei are placed in the cytoplasm of oocytes, wherein they may be totipotent. Restrictions on development, as we currently describe them, may change rapidly as research moves our understanding forward. The manipulation, implications, and applications of pluripotent and totipotent cells will be found in various specialized chapters that follow in this book. With the unfolding of the information and technology that allow us to explore as never before the biology of life and its regeneration, it is an exciting time to live.

EXAMPLES OF NATURAL CLONING IN HUMANS AND ANIMALS

Identical twin embryos occur spontaneously at low frequency in human pregnancies, by mechanisms that are not well understood. One possibility is that the zygote divides into two blastomeres that ultimately form two identical individuals instead of a single individual, but duplication can also occur at somewhat later stages of early embryonic cell divisions. Identical multiplets beyond twins are relatively rare in nature, but it is known that the fertilized egg in the armadillo species divides mitotically, producing multiple blastomeres. These develop individually to produce four or more progeny, all genetically identical.

EXPERIMENTAL CLONING OF NONMAMMALS

AMPHIBIANS

Important discoveries that were made in early studies of amphibians provided guidelines for research with mammals, but not all were heeded. Spemann (1938) demonstrated that nuclei of newt salamanders are pluripotent up to the eight-cell stage.

This eventually led to intensive studies with *Rana pipiens* and *Xenopus laevis*, using nuclear transfer to determine whether a somatic cell underwent stable changes in the genome during morphogenesis, or whether the nuclei in differentiated somatic cells were capable of directing formation of new individuals (for references, see McKinnel, 1981; Gurdon, 1986; Di Berardino, 1997). It was recognized that oocyte cytoplasm had a major effect on activity of somatic cell nuclei. Therefore, nuclear transfer to enucleated oocytes was the method of choice to test pluripotency or totipotency of these nuclei.

Studies by Briggs and King (1952) and King and Briggs (1955) showed that oocytes receiving blastula nuclei in *R. pipiens* could be reared to maturity, but gastrula nuclei from the mesoderm or endoderm injected into oocytes were unable to develop into mature frogs (Briggs and King, 1960). Results with *Xenopus* were somewhat more successful (Fischberg *et al.*, 1958)—nuclei from the endoderm of tail-bud larvae were totipotent. Gurdon (1962b) also produced fertile adult frogs following transfer of nuclei from the intestines of tadpoles.

These studies revealed several significant points. First, there was a difference between the genetic reprogramming in *Xenopus* versus *Rana* under the laboratory conditions used to rear frogs to maturity. Second, some somatic cells could reach a significant stage in development, i.e., the nuclei did not require reprogramming beyond any that might occur immediately following injection into oocytes (Gurdon, 1962b). However, attempts to use nuclei from more advanced cell types demonstrated only pluripotency (see McKinnell, 1981; Gurdon, 1986). Even diploid spermatogenic cells, which might be expected to retain totipotency, were only pluripotent (Di Berardino and Hoffner, 1971). Thus, the view was held by some that nuclei in highly specialized somatic cells had undergone irreversible changes in DNA that precluded them from becoming totipotent.

One of the problems noted was the persistent prevalence of chromosomal anomalies in embryos derived from pluripotent nuclei (Di Berardino and Hoffner, 1970). In some experiments it was shown that this was an experimentally produced artifact because the recipient eggs were not completely enucleated. However, chromosomal anomalies were more likely associated with the differential rates of division between the recipient egg and the slower dividing nuclei of somatic cells (Di Berardino and Hoffner, 1970).

It was known that enucleated amphibian ovarian oocytes as recipients had an advantage over ovulated and fertilized oocytes (eggs). Oocytes do not divide in culture without some stimulus. Gurdon (1986) stated that in amphibians "a single nucleus transplanted into an egg will have increased to 10^4 nuclei in 7 hours, whereas 500 somatic nuclei injected into an oocyte will still be 500 nuclei after several days," because the oocyte does not respond spontaneously to injected nuclei. The oocyte can be maintained in what appears to be an unchanged state for up to 3 weeks.

The results with nuclear transfer in amphibians provided substantial information on likely successful techniques for cloning mammalian embryos, particularly because genetic material important for development has been so highly conserved during evolution. McKinnell (1981) stated "that procedures are available with mammals that are entirely analagous to the methods that have resulted in successful cloning with amphibians." To what extent did we take advantage of all this information in attempting to clone mammals?

CLONING CARROTS

During the time that intensive studies on amphibian development using nuclear transfer were taking place, successful cloning of carrots received little attention (Steward *et al.*, 1958; Steward, 1970). The carrot studies illustrated the importance of providing the appropriate culture conditions for the proper expression of the genetic material (nature vs. nurture). The paper by Steward (1970) is extraordinary, and worth reading by anyone who is a student of cloning. In this paper Steward traces the research from the time that Haberlandt predicted that one should be able to grow artificial embryos from somatic cells until they were cloned. He documents in detail the development of special conditions required to isolate viable cells in quantity and to nurture them in an ordered way so that they will recapitulate embryogenesis instead of undergoing random cell proliferation. He pointed out the comparative totipotency of somatic cells nuclei in plants and amphibians, but noted the importance of oocyte cytoplasm in promoting embryogenesis in amphibians. As an example of the scope and vision of this paper the concluding paragraph is quoted:

> Thus, the subject has developed greatly since Haberlandt's bold, but unsupported, prophecy of 1902. However, there is still an outstanding gap to be filled. This involves the precise understanding of the means by which the totipotent cells of the angiosperm plant body (whose nuclei and cytoplasm clearly contain all the necessary information to become a whole plant and also the machinery to transcribe it) are constrained in the orderly course of development to do what they should, when they should and where they are. Thus, the free cell cultures of angiosperms, and the morphogenesis which they may be induced to exhibit, now furnish systems upon which these important problems may be investigated. No problems, not even those of outer space, have greater challenge than the marvelously coordinated journey through time and space of a fertilized egg, or of suitably stimulated somatic cells, as each may develop into a complete organism.

CLONING MAMMALS

Several overviews of the status of cloning have been published in the modern, rapidly moving field of mammalian cloning (Campbell, 1999; Colman, 1999; Chan, 1999; Dominko *et al.*, 1999; Fissore *et al.*, 1999; Polejaeva and Campbell, 2000; Foote, 2001; Peura *et al.*, 2001; Westhusin *et al.*, 2001; Renard *et al.*, 2002). These publications, in addition to the following chapters, should be considered for additional references.

PRODUCING HOMOZYGOTES EXPERIMENTALLY

There are several ways that homozygous offspring may be produced, some with nearly identical genetics. Highly inbred mice are used routinely to provide uniformity among test animals, although they are not completely homozygous. A variety of methods have been used to diploidize haploid embryos, but developmental failures are high and they do not lead to multiple clones. These procedures contributed minimally to development of somatic cell cloning by nuclear transfer and will not be outlined here. They were reviewed by Markert (1982) and by Seidel (1982, 1983).

PRODUCING MULTIPLE CLONES WITHOUT NUCLEAR TRANSFER TO OOCYTES

The principle that one blastomere is totipotent in the two-cell embryo was demonstrated by Nicholas and Hall (1942). They derived progeny in rats resulting from half of a two-cell embryo following destruction of one blastomere.

Moustafa and Hahn (1978) divided morulae into halves and succeeded in producing eight sets of twin mice. This principle was independently applied by Ozil *et al.* (1982) and Williams *et al.* (1982) to produce identical twins in cattle. This procedure has also been applied commercially to increase the progeny of embryos collected from superovulated cattle.

Willadsen (1979, 1982) and Willadsen and Fehilly (1983) separated the blastomeres from sheep embryos at the two-, four-, and eight-cell stages, and placed them into surrogate pig zonae pellucidae. Substantial pregnancy rates were obtained with blastomeres from the two- and four-cell embryos, but they were very low when using single blastomeres from eight-cell embryos. The procedure may have been successful because Willadsen cleverly kept the blastomeres in close contact in agar chips, and an *in vivo* culture system was used. These studies also demonstrated that heterologous zonae pellucidae could be used. Also, the procedure has been successful in cattle, horses, and pigs. This advance in cloning demonstrated that individual blastomeres retained totipotency up to the eight-cell stage in sheep.

These researchers found that when combining blastomeres from four-cell and eight-cell embryos, the blastomeres from the four-cell embryos tended to form trophoblast and those from the eight-cell embryos tended to form the inner cell mass in these chimeras. Markert and Petters (1978) had reported that in triply chimeric mice the fetus may be derived from as few as three inner cell mass cells. Although chimeras are a mosaic of cloned cells, they are mentioned here because they contributed to the search for producing true clones.

Chimeras have been useful in understanding normal and abnormal embryogenesis by including appropriate genetic markers. Chimerism also has been used to produce interspecific pregnancies by production of the trophoblast and eventually the placenta of the species serving as a surrogate mother, and a fetus of another closely related species. These various applications have been reviewed by Anderson (1987). Because chimeric animals are characterized by variability in the proportion of cells that comprise the mosaic composition of different tissues, including the germ line, these animals cannot be copied with reasonable fidelity.

MOUSE CLONING LIMITATIONS AND CONTROVERSY

Further stimulation of research, seeking to resolve the limited success of cloning mammals, was provided by Illmensee and Hoppe (1981). These researchers reported that three genetically identical (cloned) mice were produced by injecting inner cell mass (ICM) cells into enucleated zygotes. The zygote has already been triggered to undergo further development as a result of fertilization. Although this was a major achievement, most researchers thought at the time that the DNA of adult cells was irreversibly modified, and the report was met with skepticism. Other researchers were unable to repeat these results.

McGrath and Solter (1984b) had reported that both maternal and paternal genomes were required for mouse embryogenesis. A decade later Kimura and Yanagimachi (1995) reported that mouse secondary spermatocyte nuclei injected into oocytes would produce young, indicating "that gametic imprinting of mouse spermatogenic cells is completed either in the testis before the second meiotic division or within the cytoplasm of a mature oocyte after artificial nuclear transfer."

The methylation and demethylation of chromosome proteins and other changes during spermatogenesis were considered to be part of the problem associated with the low efficiency of nuclear transfer as a practical method of producing clones. The roles played by these non-DNA changes in chromosome structure still are not understood. In addition, McGrath and Solter (1984a), from carefully done research, reported that nuclear transfer in mice from one-cell and two-cell stages could support blastocyst formation, but the nuclei from four-cell or later stages did not support blastocyst formation. Based on this collective information these researchers concluded that the differential functioning of the maternal and paternal genomes might be required for early development, and that this condition would not exist in nuclei of adult cells. Consequently, the prevailing view then was that cloning in mammals using somatic cell nuclei was not likely to be successful.

Many years later Cheong et al. (1993) were able to obtain newborn mice following injection of nuclei from eight-cell stage embryos into enucleated oocytes, provided the donor nuclei were obtained from the early stage (G1 stage) of the cell cycle. This indicated that complete reprogramming was possible after the embryonic genome was activated, but that chromatin structure changes in later stages of the cycle reduced the totipotentiality of the nuclei. However, somatic cells have since been used successfully as donor nuclei (Wakayami and Yanagimachi, 2001). It is not surprising that different results were obtained by different groups working with different strains of mice, and making small but not so subtle changes in procedures with the different selection of donor nuclei, or oocyte recipients, and varying micromanipulation and other steps through fusion, culture, and transfer. A remarkable beneficial effect of heterozygosity of the donor genome was reported by Eggan et al. (2001). All but one of the six strains of F_1 hybrid mice used to form ES cell lines were capable of producing live adult pups by nuclear transfer. Only one pup was produced by similar procedures using four inbred (homozygous) donor strains. This opens new avenues to explore gene expression during development. Most domestic animal donors used to date likely are highly heterozygous.

CLONING DOMESTIC ANIMALS USING EMBRYONIC CELLS

Despite the early failures of attempts to clone mice there were reasons to be optimistic about cloning domestic animals. Large-scale attempts to produce cattle clones by using multiple blastomeres of early embryos and recycling cloned embryos (Bondioli et al., 1990) offered an opportunity to disseminate the genes of prize male–female combinations. Also, much could be learned about gene imprinting, cell lineage, etc., to fill large gaps in our understanding of early development. Now genetic programming at early stages of embryonic development in mammals could be studied in vitro.

Many techniques needed for successful cloning have been improved in recent years. Extensive research has been done and is continuing on embryo culture media and procedures (Daniel, 1978; Biggers, 1987; Gordon, 1994; Bavister, 1995; Foote, 1999, 2001). Embryo transfer became a standard practice, especially in cattle, and embryos could be preserved by cryopreservation following the work by Whittingham et al. (1972), with possible multiplication after some of the clones had been transferred and grown to maturity for phenotypic evaluation (Westhusin et al., 1991). These successful procedures represented major advances in knowledge and application since the first successful embryo transfer in cattle by Willett et al. (1951).

In this brief overview on past development of cloning in mammals, examples of problems and progress are given for different species. However, emphasis is on cattle, being the species researched most extensively because of its agricultural importance (Massey, 1990). However, sheep often have served as a less costly ruminant model (Willadsen, 1986). Among domestic animals, pigs have been particularly difficult to clone (Prather et al., 1989; Li et al., 2000; Onishi et al., 2000; Polejaeva et al., 2000), although genetically engineered pigs have been cloned in limited numbers (Bondioli et al., 2001).

Rabbits have been a useful model for nuclear cloning (Yang et al., 1990; Collas and Robl, 1990; Yang, 1991). As indicated previously, the mouse has been difficult to clone (Tsunoda and Kato, 1993; Wakayama et al., 1998; Wakayama and Yanagimachi, 2001), although Tsunoda et al. (1987) produced offspring following blastomere transfer to enucleated two-cell embryos.

Tools and Skills for Micromanipulation

Anyone who researched embryo micromanipulation in the 1970s and 1980s will remember the necessary emphasis on producing the appropriate micropipettes and micromanipulators, and learning how to handle embryos gently and deftly. One of the pioneers in this field, T. P. Lin, described equipment used to inject mouse eggs (Lin, 1971), and for intracytoplasmic sperm injection (ICSI) of mouse sperm. Lin described the consistency and size of the egg, and the proper equipment needed to penetrate the zona pellucida, and the vitellus. During microinjection, mouse eggs often shrank without appreciable damage, leaving a perivitelline space. Later Yang and co-workers (Yang *et al.*, 1990) expanded this space in rabbit oocytes by using hypertonic media to facilitate nuclear transfer without mechanically disturbing the cytoplast.

Gardner (1978) used microinjection techniques extensively to produce chimeras. He described the equipment needed in considerable detail, advantages and limitations, and stepwise procedures to follow for one to be successful. He emphasized the need for patience as much as manual dexterity. Finally, Gardner emphasized the need to have spare needles because as he stated "trying to continue with broken needles or partly occluded pipettes leads to a vicious circle of abortive operations and increasing exasperation." Practitioners of the art of micromanipulation can relate to these words of wisdom.

Embryonic Stages of Donor DNA and Totipotency

Based on the earlier amphibian experiments, it seemed clear that during embryogenesis individual cells lost their totipotency during differentiation. Consequently, mammalian cloning was focused on using early embryonic cells as totipotent stem cells. From an agricultural standpoint it was important to devise ways to use the most advanced embryos possible (maximal number of cells) so that the potential yield of clones would be maximal.

Why had the mouse been a poor model? What could be learned from this fact? The mouse genome is active at the two-cell stage. This might be a reason that cloning was so unsuccessful in mice, because there may not have been enough time to reprogram nuclei from more advanced embryonic cells, when they were transferred. Embryonic genomes of other animals normally are activated one to three cycles later than is the case for mice (Table 1). This may have facilitated cloning by nuclear transfer with more advanced embryonic cells from other species. Also, we know now that heterozygous mice can be cloned much more easily than inbred strains, and domestic animals are highly heterozygous.

During development the embryo undergoes a multide of synchronous morphologic changes. These include membrane polarity, cytoplasmic polarity, and changes in the microvilli over the surface of the blastomeres. These changes were reviewed by Ziomek (1987), with permanent cellular polarity established in the mouse by the eight-cell stage. As blastomeres continue to divide, the outer polar cells preferentially form the trophectoderm (TE) layer of the blastocyst, the precursor of the placenta. The inner nonpolar cells preferentially form the inner cell mass, giving rise to the fetus. How advanced could an embryo become before cells were no longer directly totipotent?

Table 1 Species Comparison of Stages of Genome Activation and Maintenance of Nuclear Totipotency before Dolly

	Stages in development by species[a]				
Event	Mouse	Pig	Rabbit	Cow	Sheep
Genome activation	Two cell	Two cell	Two cell	Eight to sixteen cell	Eight to sixteen cell
Nuclear totipotency	Two cell	Two cell	Thirty-two cell	Thirty-two cell	Sixty-four cell

[a]From the literature. Also, *Xenopus laevis* contains about 4000 cells before embryonic genome transcription, and tadpole intestinal epithelial cells were totipotent when transferred to enucleated oocytes.

Based on this background of events in the mouse, Koyama *et al.* (1994) examined the polarization of microvilli on the blastomeres of rabbit and cattle embryos through six cell cycles with the aid of scanning electron microscopy. In cattle, extensive polarization of the microvilli occurred at the 16-cell stage, whereas in the rabbit it occurred at the 32-cell stage. These polarity results provided further support for the hypothesis that loss of totipotency was a gradual phenomenon, that species differed, and that some cells might retain totipotency after multiple mitoses.

Stice and Robl (1988) were the first to clone rabbits using blastomeres obtained from 8-cell embryos. Later Collas *et al.* (1992) studied the stage of the cell cycle of 8- and 16-cell rabbit embryos and reported that 59, 32, and 3% of the nuclear transfers from donor blastomeres at the early, middle, and late stages of cell division, respectively, resulted in nuclear transplant blastocysts.

A contrast between pluripotency and totipotency is demonstrated in the difficult to clone pig. Prather *et al.* (1989) prepared 88 nuclear transfer embryos using two-, four-, or eight-cell nuclei, and one piglet resulted from the transfer of a four-cell nucleus to an enucleated, activated oocyte in metaphase II. This was the only cloned pig reported for more than 10 years. Recently Li *et al.* (2000) reported that two piglets were born following blastomere transfer from four-cell embryos produced *in vivo* to enucleated oocytes matured *in vivo*. Onishi *et al.* (2000) produced one piglet and Polejaeva *et al.* (2000) produced five piglets following injection of somatic cells into enucleated oocytes. In addition, freshly obtained pig ICM cells transferred to blastocysts have resulted in live chimeric progeny (Anderson *et al.*, 1994), and ICM-derived embryonic stem cells have contributed to a chimeric pig (Wheeler, 1994).

Research on nuclear transfer in cattle was facilitated by using slaughterhouse ovaries as a source of viable cytoplasts. This inexpensive source of oocytes was utilized by Lu *et al.* (1987), and reviewed by Gordon (1994) to produce progeny following fertilization *in vitro* and embryo transfer. Prather *et al.* (1987) studied development *in vitro* of blastomeres obtained from 2- to 32-cell bovine embryos following injection into enucleated oocytes. Development in culture tended to decrease when blastomeres from the more advanced-stage embryos were used. When 23 morula- or blastocyst-stage embryos were transferred to 13 heifers, two calves were produced. It is worth noting that the embryo culture system used at that time caused a block in embryo development at the 8- or 16-cell stage, so the embryos were cultured in ligated sheep oviducts. In contrast, when blastomeres from two-, four-, and eight-cell embryos were transplanted into pronuclear recipient embryos and placed in sheep oviducts for 5 days, no development was observed (Robl *et al.*, 1987). However, control nuclei from pronuclear embryos removed, reinserted, and cultured in sheep oviducts resulted in development to morulae or blastocysts, and normal offspring following transfer.

Many questions remained on the source of nuclei, the recipient cytoplasts, and the techniques to use. With advances in methodology, Keefer *et al.* (1994) produced calves following transfer of bovine ICM cells to enucleated oocytes. Collas and Barnes (1994) also produced progeny using ICM donor nuclei, but no progeny resulted when using granulosa cell donor nuclei. Thus totipotency of ICM cells was demonstrated as improved laboratory procedures became available, but nuclei from differentiated somatic cells were not totipotent under these conditions, or at least at an efficiency that was detected by the number of transfers.

Variability in Donor Nuclei, Recipient Oocytes, and Cell Cycle Synchrony

Because of the variability in results as experimental cloning progressed, it is valuable to examine some of the sources of variability. McGrath and Solter (1984a) used zygotes and Tsunoda *et al.* (1987) used early-stage embryos as recipient cytoplasts, but the earlier work with amphibians (Gurdon, 1962a,b, 1986) indicated that oocytes not yet activated to initiate mitosis were potentially more useful recipients of transferred nuclei. Willadsen (1986) was the first to demonstrate that nuclear transfer to mature oocytes in sheep could lead to successful reprogramming and the production of young.

Considerable advances appeared to be in the making when Granada Genetics, Inc. mounted a major effort to clone cattle commercially (Bondioli *et al.*, 1990; Massey, 1990). This group used both fresh embryos and frozen-thawed embryos that were obtained 5 to 6 days after superovulating donor cows (average number of blastomeres ranged from 28 to 48). There was no difference in development rate using blastomeres from embryos at different stages of development. However, the pregnancy rate was higher when fresh embryos were used as blastomere donors.

In addition to the stage of embryonic development used as a source of donor nuclei, the importance of donor–recipient cell cycle synchrony was recognized, and much early work focused on determining the optimal stage of the cell cycle for the recipient cell at the time of transfer (Robl *et al.*, 1986; Fissore *et al.*, 1999). Initially bovine oocytes were in short supply and were expensive because they came from superovulated cattle (Bondioli *et al.*, 1990). This was a labor-intensive and costly procedure. However, based partly on the success of Lu *et al.* (1987) with slaughterhouse ovaries, oocytes were matured *in vitro* (IVM), and these IVM oocytes were enucleated and used as the recipients for nuclear transfer. Massey (1990) reported pregnancy results with both sources of oocytes. Prather *et al.* (1987) had used oocytes from slaughtered cattle, but the IVM oocytes did not fuse readily with the transferred nucleus. Robl and Stice (1989) noted that further improvement in IVM procedures was needed for these inexpensive oocytes to be used as the cytoplast for accommodating any reprogramming of early embryo-stage donor nuclei. Also, it was important to develop procedures to enucleate completely the recipient oocytes (Westhusin *et al.*, 1992; Bondioli, 1993).

Bondioli *et al.* (1990) attempted to increase the number of clones by making cloned embryos from the previous generation of clones. However, the success rate decreased with repeated generations. Subsequently, others have attempted to increase the number of clones using methodologies that may have improved with time. Stice and Keefer (1993) obtained only 10, 2, and 3% calves with clones produced from one, two, and three recycled donor clones, respectively. Peura *et al.* (2001) compared one, four, and seven rounds of donor nuclei produced by nuclear transfer, and calving percentages resulting from the selected embryos transferred were 25, 4, and 0%, respectively. Whether the reduction in totipotent recycled nuclei is a result of intrinsic genomic modification during reprogramming and/or epigenetic factors during repeated manipulation is unknown.

Stice *et al.* (1994) obtained calves following nuclear transfer of blastomeres obtained from day 5 or 6 embryos that had been produced *in vivo* in superovulated cows as well as from blastomeres isolated from day 5 or 6 embryos produced by *in vitro* maturation, fertilization, and culture. Results were best when the aged recipient oocytes were activated prior to the time of fusion.

Similar studies were in progress in sheep (Smith and Wilmut, 1989). Reconstituted sheep embryos derived from either 16-cell (day 4) embryos or from the ICM of day 6 embryos resulted in 18% becoming lambs. Collas and Robl (1990) tested a variety of procedures in rabbits, including oocyte age, which increased the efficiency of the nuclear transfer procedure. With the best procedures, 23/110 (21%) of the cloned embryos derived from 8- to 16-cell stage donor nuclei developed into young, compared with 4/41 (10%) of the control nonmanipulated embryos.

Many procedures were used to obtain highly viable donor nuclei and recipient cytoplasts, as well as to ensure appropriate synchrony between the cell-cycle stage of the donor and recipient cells (Prather *et al.*, 1987; Robl and Stice, 1989; Smith and Wilmut, 1990; Foote and Yang, 1992; Yang and Anderson, 1992; Campbell *et al.*, 1996b; Heyman and Renard, 1996). Donor nuclei and recipient cytoplasts respond differently to varying culture conditions. It is amazing to note how many published procedures were successful to some extent, and it is not clear what is optimal. Certainly improvements have been made in preparing the donor nuclei (blastomeres) and recipient oocytes, in ensuring complete enucleation of the oocytes with minimal damage during injection of the donor nucleus into the recipient oocyte, in fusing and activating the oocyte more appropriately, and in improved culture and transfer of the manufactured embryos. The development of a more compatible system has allowed nuclei from more advanced-stage embryos to express their totipotency, limited in the beginning by suboptimal handling procedures.

Advances have been made in understanding the biology of cell-cycle regulation. For example, changes in substances critical in cell-cycle regulation, such as meiosis–mitosis phase (maturation)-promoting factor (MPF, $p34^{cdc2}$/cyclin B) and cytostatic factor (CSF), have been identified. Oocyte maturation, germinal vesicle breakdown, and chromosome condensation are promoted by MPF. Oocytes are prevented from exiting metaphase II by CSF, which is inactivated after fertilization or parthenogenetic activation. Addition of various biochemical criteria associated with detailed morphologic changes (Kaňka *et al.*, 1991) will assist in defining oocytes that are competent to perform the role to which they are uniquely fitted for mammalian cloning.

Embryonic Stem Cells

Brief comment on ES cells is in order because they were intensively investigated as a possible means of establishing pluripotent cell lines for domestic animals following earlier success with certain strains of mice (see Yang and Anderson, 1992; Anderson, 1999; Piedrahita *et al.*, 1999). These cells can be multiplied in an undifferentiated state under precise culture conditions. These cell lines were valuable for developmental studies, use of genetic markers to study cell lineage in chimeras, and possible genetic engineering. However, few lines from domestic animals were successfully developed, they were difficult to maintain, and they were of limited value in producing clones because they were not totipotent. With the successful cloning from somatic cells (Wilmut *et al.*, 1997), and the ability to maintain totipotent lines of fibroblasts, the value of ES cells for experimental or practical ends diminished.

Cloning Domestic Animals from Somatic Cells

In 1996 there was a signal (Campbell *et al.*, 1996a,b) that the world of cloning and stem cell biotechnology would never be the same. This was followed with the paper by Wilmut *et al.* (1997) of the production of Dolly, a viable lamb resulting from a cultured cell derived from somatic mammary gland cells harvested from a 6-year-old ewe. Some questions were raised concerning this "impossible" event, but this carefully conducted research has stood the test of subsequent experimentation by others. Only one viable lamb was produced from the original fusion of 277 donor nuclei and enucleated oocytes. However, one was enough to stimulate the scientific community to explore the mechanisms that made this remarkable event possible, and to utilize this knowledge to improve the efficiency and safety of cloning procedures. Needless to say, the world news media flooded the print press and airwaves with fanciful stories; "rubber stamping" and "ditto machine" cartoons spread fear and some thoughtful concern among the public and legislative bodies. My own 1997 files are filled with daily communiques, and requests for mass media interviews, because Dr. Yang in our laboratory had cloned rabbits and cattle from embryonic cells many years earlier (Yang, 1991).

What Wilmut, Campbell, and colleagues had done (Campbell *et al.*, 1996a,b; Wilmut *et al.*, 1997) was to culture cells on a minimal medium that caused the cells to exit the growth cycle and become arrested in the quiescent (G0) stage. These G0 nuclei are diploid. They were transferred to metaphase II (MII)-stage enucleated oocytes, avoiding chromosomal damage due to premature chromosome condensation. Cells maintained in this quiescent G0 stage for a period of time can undergo various changes in chromatin structure with alteration of specific factors that could promote or result in reprogramming. Many of these important changes during embryogenesis are dealt with in various chapters in this book. The expansion of knowledge rather than commercial cloning of animals likely will be the first payoff from cloning and allied research.

Colman (1999) summarized the practical results of live births reported from somatic cell (nuclear) transfer in cattle, sheep, goats, and mice. Production of multiple cattle clones has been encouraging (Kato *et al.*, 1998; Wells *et al.*, 1999). Numerous reviews have been written in the past decade discussing the implications, future avenues for research, and applications of cloning for animal agriculture and medicine (Robl and Stice, 1989; First, 1990; Smith and Wilmut, 1990; Yang and Anderson, 1992; Di Berardino, 1997; Wilmut *et al.*, 1998; Stice *et al.*, 1998; Colman, 1999; Seidel, 2000). Improved techniques for somatic cell processing and donor–recipient synchrony (Stice *et al.*, 1998; Campbell, 1999; Colman, 1999; Dominko *et al.*, 1999; Mohamed Nour *et al.*, 2000) are being investigated in many laboratories. These include the ability to maintain totipotency following limited (Cibelli *et al.*, 1998) or multiple passages of cells (Kubota *et al.*, 2000), important for experimental gene targeting and practical genetic engineering (Murray, 1999; Piedrahita *et al.*, 1999; Polejaeva and Campbell, 2000; Piedrahita, 2000; Georges, 2001; Westhusin *et al.*, 2001). Combining genomic information with the ability to detect selected gene expression in embryos will facilitate marker-assisted selection (Chan, 1999; Band *et al.*, 2000; Bredbaka, 2001; Cross, 2001; Georges, 2001; Roberts, 2001). With new knowledge rapid progress is being made when compared to earlier reports based on theory and less efficient procedures (Van Vleck, 1981; Hammer *et al.*, 1985; Pursel *et al.*, 1989).

An area of major interest not only to cloning, but also to aging and cancer research, is the regulation of cell proliferation and senescence, stimulated by the report on telomere length in Dolly (Shiels *et al.*, 1999). However, telomere length does not seem to be a problem in other studies (Lanza *et al.*, 2000a; Tian *et al.*, 2000; Yang and Tian, 2000; Betts *et al.*, 2001; Xu and Yang, 2001), nor is telomere length the only component affecting the inherent ability of cells to continue to proliferate.

Cloned animals as producers of pharmaceutical products (Wilmut *et al.*, 1990; Wall *et al.*, 1997; Ziomek, 1998; Brink *et al.*, 2000), as organ donors (Stice *et al.*, 1998; Colman, 1999; Bondioli *et al.*, 2001), and as models for studying diseases are other examples of useful applications. Cloning also has an application in preserving endangered species (Wildt, 1992; Wells *et al.*, 1998; Lanza *et al.*, 2000b; Solti *et al.*, 2000). However, there are many unsolved problems, such as the high incidence of anomalies (Garry *et al.*, 1996; Hill *et al.*, 1999) and the inefficiency of the current procedures (Colman, 1999; Dominko *et al.*, 1999). These must be improved before they become cost-effective for most applications.

One of the most exciting fringe benefits is the major increase in research on developing procedures to enable many types of differentiated somatic cells to be reprogrammed and guided into forming specific cells and tissues. This has great potential for improving the quality of life for our aging human population afflicted with age-associated diseases.

This has been and will continue to be a wonderful journey on the road to understanding the biology of, and pondering the meaning of, life (Foote, 1992, 1998, 1999). It has been marked by insight, determination, and serendipity. The latter word, coined by Walpole in 1754 (Remer, 1965), combines sagacity with accidental discovery, and is not correctly defined in the dictionary. The largest room in the world is the room for improvement. Exciting times lie ahead as the mysteries of developmental biology are waiting to be unraveled. The opportunities in this area are as exciting for the twenty-first century as landing on the moon was in the twentieth century. These powerful techniques also put powerful emphasis on us to discern how these technologies might best be applied, especially to our own species. We should all ask and ponder the question, "just because we can do it, should we?" Those who blaze trails should also assume some responsibility for the consequences.

ACKNOWLEDGMENTS

Helpful comments were made by Drs. X. Yang and G. E. Seidel, Jr.; D. Bevins assisted with manuscript preparation.

REFERENCES

Anderson, G. B. (1987). Use of chimeras to study development. *J. Reprod. Fertil.* (Suppl.) **34**, 251–259.

Anderson, G. B. (1999). Embryonic stem cells in agricultural species. *In* "Transgenic Animals in Agriculture" (J. D. Murray, G. B. Anderson, A. M. Oberbauer, and M. M. McGloughlin, eds.), pp. 57–66. CABI Publ., New York.

Anderson, G. B., Choi, S. J., and Bondurant, R. H. (1994). Survival of porcine inner cell masses in culture and after injection into blastocysts. *Theriogenology* **42**, 212–221.

Avery, O. T., MacLeod, C. M., and McCarty, M. (1944). Induction of transformation by a deoxyribonucleic acid fraction isolated from pneumococcus type III. *J. Exp. Med.* **79**, 137–158.

Band, M. R., Larson, J. H., Rebiez, M., Green, C. A., Heyen, D. W., Donowan, J., Windish, R., Steining, C., Mahyuddin, P., Womack, J. E., and Lewin, H. A. (2000). An ordered comparative map of the cattle and human genomes. *Genome Res.* **10**, 1359–1368.

Bavister, B. D. (1995). Culture of preimplantation embryos: Facts and artifacts. *Hum. Reprod. Update* **1**, 91–148.

Beadle, G. (1946). Genes and the chemistry of the organism. *Am. Sci.* **34**, 31–53.

Betts, D., Bordignon, V., Hill, J., Winger, Q., Westhusin, M., Smith, L., and King, W. (2001). Reprogramming of telomerase activity and rebuilding of telomere length in cloned cattle. *Proc. Natl. Acad. Sci. U.S.A.* **98**, 1077–1082.

Biggers, J. D. (1987). Pioneering mammalian embryo culture. *In* "The Mammalian Preimplantation Embryo" (B. D. Bavister, ed.), pp. 1–22. Plenum Press, New York.

Bondioli, K., Ramsoondar, J., Williams, B., Costa, C., and Fodor, W. (2001). Cloned pigs generated from cultured skin fibroblasts derived from a H—Transferase transgenic boar. *Mol. Reprod. Dev.* **60**, 189–195.

Bondioli, K. R. (1993). Nuclear transfer in cattle. *Mol. Reprod. Dev.* **36**, 274–275.

Bondioli, K. R., Westhusin, M. E., and Looney, C. R. (1990). Production of identical offspring by nuclear transfer. *Theriogenology* **33**, 165–174.

Brackett, B. G., Seidel, G. E., Jr., and Seidel, S. M., eds. (1981). "New Technologies in Animal Breeding." Academic Press, New York.

Bredbacka, P. (2001). Progress on methods of gene detection in preimplantation embryos. *Theriogenology* **55**, 23–34.

Briggs, R., and King, T. J. (1952). Transplantation of living nuclei from blastula cells into enucleated frogs' eggs. *Proc. Natl. Acad. Sci. U.S.A.* **38**, 455–463.

Briggs, R., and King, T. J. (1960). Nuclear transplantation studies on the early gastrula (*Rana pipiens*). *Dev. Biol.* **2**, 252–270.

Brink, M. F., Bishop, M. D., and Pieper, F. R. (2000). Developing efficient strategies for the generation of transgenic cattle which produce biopharmaceuticals in milk. *Theriogenology* **53**, 139–148.

Campbell, K. H. S. (1999). Nuclear equivalence, nuclear transfer, and the cell cycle. *Cloning* **1**, 3–15.

Campbell, K. H., McWhir, J., Ritchie, W. A., and Wilmut, I. (1996a). Sheep cloned by nuclear transfer from a cultured cell line. *Nature* **380**, 64–66.

Campbell, K. H. S., Ritchie, W. A., McWhir, J., and Wilmut, I. (1996b). Cloning farm animals by nuclear transfer: From cell cycles to cells. *Int. Embryo Soc. Trans. Newsl.* **14**(1), 12–17.

Chan, A. W. S. (1999). Transgenic animals: Current and alternative strategies. *Cloning* **1**, 25–46.

Cheong, H. T., Takahashi, Y., and Kanagawa, H. (1993). Birth of mice after transplantation of early cell-cycle-stage embryonic nuclei into enucleated oocytes. *Biol. Reprod.* **48**, 958–963.

Cibelli, J. B., Stice, S. L., Golueke, P. J., Kane, J. J., Jerry, J., Blackwell, C., Ponce de León, F. A., and Robl, J. L. (1998). Cloned transgenic calves produced from nonquiescent fetal fibroblasts. *Science* **280**, 1256–1258.

Collas, P., and Barnes, F. L. (1994). Nuclear transplantation by microinjection of inner cell mass and granulosa cell nuclei. *Mol. Reprod. Dev.* **38**, 264–267.

Collas, P., and Robl, J. M. (1990). Factors affecting the efficiency of nuclear transplantation in the rabbit embryo. *Biol. Reprod.* **43**, 877–884.

Collas, P., Balise, J. J., and Robl, J. M. (1992). Influence of the cell cycle stage of the donor nucleus on development of nuclear transplant rabbit embryos. *Biol. Reprod.* **46**, 492–500.

Colman, A. (1999). Somatic cell nuclear transfer in mammals: Progress and applications. *Cloning* **1**, 185–200.

Cross, J. C. (2001). Genes regulating embryonic and fetal survival. *Theriogenology* **55**, 193–207.

Daniel, J. C., Jr., ed. (1978). "Methods in Mammalian Reproduction." Academic Press, New York.

Di Berardino, M. A. (1997). "Genomic Potential of Differentiated Cells." Columbia Univ. Press, New York.

Di Berardino, M. A., and Hoffner, N. (1970). Original of chromosomal abnormalities in nuclear transplants—A reevaluation of nuclear differentiation and nuclear equivalence in amphibians. *Dev. Biol.* **23**, 185–209.

Di Berardino, M. A., and Hoffner, N. (1971). Development and chromosomal constitution of nuclear-transplants derived from male germ cells. *J. Exp. Zool.* **176**, 61–72.

Dominko, T., Ramalho-Santos, J., Chan, A., Moreno, R. D., Luetjens, C. M., Simerly, C., Hewitson, L., Takahashi, D., Martinovich, C., White, J. M., and Schatten, G. (1999). Optimization strategies for production of mammalian embryos by nuclear transfer. *Cloning* **3**, 143–152.

Eggan, K., Akutsu, H., Loring, J., Jackson-Grusby, L., Klemm, M., Rideout, W. M., 3rd, Yanagimachi, R., and Jaenisch, R. (2001). Hybrid vigor, fetal overgrowth, and viability of mice derived by nculear cloning and tetraploid embryo complementation. *Proc. Natl. Acad. Sci. U.S.A.* **98**, 6205–6214.

First, N. L. (1990). New animal breeding techniques and their application. *J. Reprod. Fertil.* (Suppl.) **41**, 3–14.

Fischberg, M., Gurdon, J. B., and Elsdale, T. R. (1958). Nuclear transplantation in *Xenopus laevis.* *Nature (London)* **181**, 424.

Fissore, R. A., Long, C. R., Duncan, R. P., and Robl, J. M. (1999). Initiation and organization of events during the first cell cycle in mammals: Applications in cloning. *Cloning* **1**, 89–100.

Foote, R. H. (1992). Ethical concerns of new animal biotechnologies. *Proc. Symp., Cloning Mammals by Nuclear Transplantation* (G. E. Seidel, Jr., ed.), pp. 45–48. Colorado State University, Ft. Collins, Colorado.

Foote, R. H. (1998). "Artificial Insemination to Cloning. Tracing 50 Years of Research." Published by the author, Ithaca, New York.

Foote, R. H. (1999). Development of reproductive biotechnologies in domestic animals from artificial insemination to cloning: A perspective. *Cloning* **1**, 133–142.

Foote, R. H. (2001). Developments in animal reproductive biotechnology. *In* "Assisted Fertilization and Nuclear Transfer in Mammals" (D. P. Wolf and M. Zelinski-Wooten, eds.), pp. 3–20. Humana Press, Totowa, New York.

Foote, R. H., and Yang, X. (1992). Cloning bovine embryos. *Reprod. Domest. Anim.* **27**, 13–21.

Gardner, R. L. (1978). Production of chimeras by injecting cells or tissue into the blastocyst. *In* "Methods in Mammalian Reproduction" (J. C. Daniel, Jr., ed.), pp. 137–165. Academic Press, New York.

Garry, F. B., Adams, R., McCann, J. P., and Odde, K. G. (1996). Postnatal characteristics of calves produced by nuclear transfer cloning. *Theriogenology* **45**, 141–152.

Georges, M. (2001). Recent progress in livestock genomics and potential impact on breeding programs. *Theriogenology* **55**, 15–21.

Gordon, I. (1994). "Laboratory Production of Cattle Embryos." CAB Int., Wallingford, Oxon, UK.

Gurdon, J. B. (1962a). Adult frogs derived from the nuclei of single somatic cells. *Dev. Biol.* **5**, 68–83.

Gurdon, J. B. (1962b). The developmental capacity of nuclei taken from intestinal epithelium cells of feeding tadpoles. *J. Embryol. Exp. Morphol.* **10**, 622–640.

Gurdon, J. B. (1986). Nuclear transplantation in eggs and oocytes. *J. Cell Sci.* (Suppl.) **4**, 287–318.

Hammer, R. E., Pursel, V., Rexroad, C. E., Jr., Wall, R. J., Boldt, D. J., Ebert, K. M., Palmiter, R. D., and Brinster, R. L. (1985). Production of transgenic rabbits, sheep and pigs by microinjection. *Nature* **315**, 680–683.

Heape, W. (1897). The artificial insemination of mammals and subsequent possible fertilization or impregnation of their ova. *Proc. R. Soc. Lond. [B] Biol. Sci.* **61**, 52–63.

Heyman, Y., and Renard, J. P. (1996). Cloning of domestic species. *Anim. Reprod. Sci.* **42**, 427–436.

Hill, J. R., Roussel, A. J., Cibelli, J. B., Edwards, J. F., Hooper, N. L., Miller, M. W., Thompson, J. A., Looney, C. R., Westhusin, M. E., Robl, J. M., and Stice, S. L. (1999). Clinical and pathologic features of cloned transgenic calves and fetuses (13 case studies). *Theriogenology* **51**, 1451–1465.

Illmensee, K., and Hoppe, P. C. (1981). Nuclear transplantation in *Mus musculus*: Developmental potential of nuclei from preimplantation embryos. *Cell* **23**, 9–18.

Kaňka, J., Fulka, J., Jr., Fulka, J., and Petr, J. (1991). Nuclear transplantation in bovine embryo: Fine structural and autoradiographic studies. *Mol. Reprod. Dev.* **29**, 110–116.

Kato, Y., Tani, T., Sotomaru, Y., Kurokawa, K., Kato, J. Y., Doguchi, H., Yasue, H., and Tsunoda, Y. (1998). Eight calves cloned from somatic cells of a single adult. *Science* **282**, 2095–2098.

Keefer, C. L., Stice, S. L., and Matthews, D. L. (1994). Bovine inner cell mass cells and donor nuclei in the production of nuclear transfer embryos and calves. *Biol. Reprod.* **50**, 935–939.

Kimura, Y., and Yanagimachi, R. (1995). Development of normal mice from oocytes injected with secondary spermatocyte nuclei. *Biol. Reprod.* **55**, 855–862.

King, T. J., and Briggs, R. (1955). Changes in the nuclei of differentiating gastrula cells, as demonstrated by nuclear transplantation. *Proc. Natl. Acad. Sci. U.S.A.* **41**, 321–325.

Koyama, H., Suzuki, H., Yang, X., Jiang, S., and Foote, R. H. (1994). Analysis of polarity of bovine and rabbit embryos by scanning electron microscopy. *Biol. Reprod.* **50**, 163–170.

Kubota, C., Yamakuchi, H., Todoroki, J., Mizoshita, K., Tabara, N., Barber, M., and Yang, X. (2000). Six cloned calves produced from adult fibroblast cells after long-term culture. *Proc. Natl. Acad. Sci. U.S.A.* **97**, 990–995.

Lanza, R. P., Cibelli, J. B., Blackwell, C., Cristofalo, V. J., Francis, M. K., Baerlocher, G. M., Mak, J., Schertzer, M., Chavez, E. A., Sawyer, N., Lansdorp, P. M., and West, M. D. (2000a). Extension of cell life-span and telomere length in animals cloned from senescent somatic cells. *Science* **288**, 665–669.

Lanza, R. P., Cibelli, J. B., Diaz, F., Moraes, C. T., Farin, P. W., Farin, C. E., Hammer, C. J., West, M. D., and Damiani, P. (2000b). Cloning of an endangered species (*Bos gaurus*) using interspecies nuclear transfer. *Cloning* **2**, 79–90.

Li, G.-P., Tan, J.-H., Sun, Q.-Y., Meng, Q.-G., Yue, K.-Z., Sun, X.-S., Li, Z.-Y., Wang, H.-B., and Xu, L.-B. (2000). Cloned piglets born after nuclear transplantation of embryonic blastomeres into porcine oocytes matured *in vivo*. *Cloning* **1**, 45–52.

Lin, T. P. (1971). Egg micromanipulation. *In* "Methods in Mammalian Embryology" (J. C. Daniel, Jr., ed.), pp. 157–171. W. H. Freeman and Co., San Francisco.

Lu, K. H., Gordon, I., Gallaher, M., and McGovern, H. (1987). Pregnancy established in cattle by transfer of embryos derived from *in vitro* fertilization of oocytes matured *in vitro*. *Vet. Rec.* **121**, 259–260.

Markert, C. L. (1982). Parthenogenesis, homozygosity, and cloning in mammals. *J. Hered.* **73**, 390–397.

Markert, C. L., and Petters, R. M. (1978). Manufactured hexaparental mice show that adults are derived from three embryonic cells. *Science* **202**, 56–58.

Massey, J. M. (1990). Animal production industry in the year 2000 A.D. *J. Reprod Fertil.* (Suppl.) **41**, 199–208.

McGrath, J. S., and Solter, D. (1984a). Inability of mouse blastomere nuclei transferred to enucleated zygotes to support development *in vitro*. *Science* **226**, 1317–1319.

McGrath, J. S., and Solter, D. (1984b). Completion of mouse embryogenesis requires both the maternal and paternal genomes. *Cell* **37**, 179–183.

McKinnell, R. G. (1981). Amphibian nuclear transplantation: State of the art. *In* "New Technologies in Animal Breeding" (B. G. Brackett, G. E. Seidel, Jr., and S. M. Seidel, eds.), pp. 163–180. Academic Press, New York.

Mohamed Nour, M. S., Ikeda, K., and Takahashi, Y. (2000). Bovine nuclear transfer using cumulus cells derived from serum-starved and confluent cultures. *J. Reprod. Dev.* **46**, 86–92.

Moustafa, L., and Hahn, J. (1978). Experimentelle Erzeugung von identischen Mäusezwillingen. *Dtsch. Tierärztl. Wochenschr.* **85**, 242–244.

Murray, J. D. (1999). Genetic modification of animals in the next century. *Theriogenology* **51**, 149–159.

Nicholas, J. S., and Hall, B. V. (1942). Experiments on developing rats. II. The development of isolated blastomeres and fused eggs. *J. Exp. Zool.* **90**, 441–459.

Oliver, M., Aggarwal, A., Allen, J., *et al.* (2001). A high-resolution radiation hybrid map of the human genome draft sequence. *Science* **291**, 1298–1302.

Onishi, A., Masaki, I., Akita, T., Makawa, S., Takeda, K., Awata, T., Hanada, H., and Perry, A. C. F. (2000). Pig cloning by microinjection of fetal fibroblast nuclei. *Science* **289**, 1188–1190.

Ozil, J. P., Heyman, Y., and Renard, J. P. (1982). Production of monozygotic twins by micromanipulation and cervical transfer in the cow. *Vet. Rec.* **110**, 126–127.

Peura, T. T., Lane, M. W., Lewis, I. M., and Trounson, A. O. (2001). Development of bovine embryo-derived clones after increasing rounds of nuclear recycling. *Mol. Reprod. Dev.* **58**, 384–389.

Piedrahita, J. A. (2000). Targeted modification of the domestic animal genome. *Theriogenology* **53**, 105–116.

Piedrahita, J. A., Dunne, P., Lee, C.-K., Moore, K., Rucker, E., and Vazquez, J. C. (1999). Use of embryonic and somatic cells for production of transgenic domestic animals. *Cloning* **1**, 73–87.

Polejaeva, I. A., and Campbell, K. H. S. (2000). New advances in somatic cell nuclear transfer: Application in transgenesis. *Theriogenology* **53**, 117–126.

Polejaeva, I. A., Chen, S. H., Vaught, T. D., Page, R. L., Mullins, J., Ball, S., Dai, Y. F., Boone, J., Walter, S., Ayares, D. L., Colman, A., and Campbell, K. H. S. (2000). Cloned pigs produced by nuclear transfer from adult somatic cells. *Nature* **407**, 86–90.

Prather, R. S., Barnes, F. L., Sims, M. M., Robl, J. M., and First, N. L. (1987). Nuclear transplantation in the bovine embryo: Assessment of donor nuclei and recipient oocyte. *Biol. Reprod.* **37**, 859–866.

Prather, R. S., Sims, M. M., and First, N. L. (1989). Nuclear transplantation in early pig embryos. *Biol. Reprod.* **41**, 414–418.

Pursel, V. G., Pinkert, C. A., Miller, K. F., Boldt, R. L., and Hammer, R. E. (1989). Genetic engineering of livestock. *Science* **244**, 1281–1288.

Remer, T. G., ed. (1965). "Serendipity and the Three Princes." Univ. Oklahoma Press, Norman, OK.

Renard, J. P., Zhou, Q., LeBouris, D., Charatte-Palmer, P., Hue, I., and Vignon, X. (2002). Nuclear transfer technologies: Between successes and doubts. *Theriogenology* **57**, 203–222.

Roberts, R. M. (2001). The place of farm animal species in the new genomics world of reproductive biology. *Biol. Reprod.* **64**, 409–417.

Robl, J. M., and Stice, S. L. (1989). Prospects for the commercial cloning of animals by nuclear transplantation. *Theriogenology* **31**, 75–84.

Robl, J. M., Gilligan, B., Critser, E. S., and First, N. L. (1986). Nuclear transplantation in mouse embryos: Assessment of recipient cell stage. *Biol. Reprod.* **34**, 733–739.

Robl, J. M, Prather, R., Barnes, F., Eyestone, W., Northey, D., Gilligan, B., and First, N. (1987). Nuclear transplantation in bovine embryos. *J. Anim Sci.* **64**, 642–647.

Seidel, G. E., Jr. (1982). Applications of microsurgery to mammalian embryos. *Theriogenology* **17**, 23–34.

Seidel, G. E., Jr. (1983). Production of genetically identical sets of mammals: Cloning? *J. Exp. Zool.* **228**, 347–354.

Seidel, G. E., Jr. (2000). Reproductive biotechnology and "big" biological questions. *Theriogenology* **53**, 187–194.

Shiels, P. G., Kind, A. J., Campbell, K. H. S., Waddington, D., Wilmut, I., Colman, A., and Schnieke, A. E. (1999). Analysis of telomere lengths in cloned sheep. *Nature* **398**, 316–317.

Smith, L. C., and Wilmut, I. (1989). Influence of nuclear and cytoplasmic activity on the development *in vivo* of sheep embryos after nuclear transplantation. *Biol. Reprod.* **40**, 1027–1035.

Smith, L. C., and Wilmut, I. (1990). Factors affecting the viability of nuclear transplanted embryos. *Theriogenology* **33**, 153–164.

Solti, L., Crichton, E. G., Loskutoff, N. M., and Cseh, S. (2000). Economical and ecological importance of indigenous livestock and the application of assisted reproduction to their preservation. *Theriogenology* **53**, 149–162.

Spemann, H. (1938). "Embryonic Development and Induction." Yale Univ. Press, New Haven, Conn.

Steward, F. C. (1970). From cultured cells to whole plants: The induction and control of their growth and morphogenesis. *Proc. R. Soc. Lond. [B] Biol. Sci.* **175**, 1–30.

Steward, F. C., Mapes, M. O., and Mears, K. (1958). Growth and organized development of cultured cells. *Am. J. Bot.* **45**, 705–713.

Stice, S. L., and Keefer, C. L. (1993). Multiple generational bovine embryo cloning. *Biol. Reprod.* **48**, 715–719.

Stice, S. L., and Robl, J. M. (1988). Nuclear reprogramming in nuclear transplant rabbit embryos. *Biol. Reprod.* **39**, 657–664.

Stice, S. L., Keefer, C. L., and Matthews, L. (1994). Bovine nuclear transfer embryos: Oocyte activation prior to blastomere fusion. *Mol. Reprod. Dev.* **38**, 61–68.

Stice, S. L., Robl, J. M., Ponce de Leon, F. A., Jerry, J., Golaeke, P. G., Cibelli, J. B., and Kane, J. J. (1998). Cloning: New breakthroughs leading to commercial opportunities. *Theriogenology* **49**, 129–138.

Tian, X., Xu, J., and Yang, X. (2000). Normal telomere lengths found in cloned cattle. *Nat. Genet.* **26**, 272–273.

Tsunoda, Y., and Kato, Y. (1993). Nuclear transplantation of embryonic stem cells in mice. *J. Reprod. Fertil.* **98**, 537–540.

Tsunoda, Y., Yasui, T., Shioda, Y., Nakamura, K., Uchida, T., and Sugie, T. (1987). Full term development of mouse blastomere nuclei transplanted into enucleated two cell embryos. *J. Exp. Zool.* **242**, 147–151.

Van Vleck, L. D. (1981). Potential genetic impact of artificial insemination, sex selection, embryo transfer, cloning and selfing in dairy cattle. *In* "New Technologies in Animal Breeding" (B. C. Brackett, G. E. Seidel, Jr., and S. M. Seidel, eds.), pp. 221–242. Academic Press, New York.

Venter, J. C., Adams, M. D., Myers, E. W., *et al.* (2001). The sequence of the human genome. *Science* **291**, 1304–1351.

Wakayama, T., and Yanagimachi, R. (2001). Mouse cloning with nucleus donor cells of different age and types. *Mol. Reprod. Dev.* **58**, 376–383.

Wakayama, T., Perry, A. C. F., Zuccotti, M., Johnson, K. R., and Yanagimachi, R. (1998). Full term development of mice from enucleated oocytes injected from cumulus cell nuclei. *Nature* **394**, 369–374.

Wall, R. J., Kerr, D. E., and Bondioli, K. R. (1997). Transgenic dairy cattle: Genetic engineering on a large scale. *J. Dairy Sci.* **80**, 2213–2224.

Wells, D. N., Misica, P. M., Tervit, H. R., and Vivanco, W. H. (1998). Adult somatic cell nuclear transfer is used to preserve the last surviving cow of the Enderby Island cattle breed. *Reprod. Fertil. Dev.* **10**, 369–378.

Wells, D. N., Misica, P. M., and Tervit, H. R. (1999). Production of cloned calves following nuclear transfer with cultured adult mural granulosa cells. *Biol. Reprod.* **60**, 996–1005.

Westhusin, M. E., Pryor, J. H., and Bondioli, K. R. (1991). Nuclear transfer in the bovine embryo: A comparison of 5-day, 6-day, frozen-thawed, and nuclear transfer donor embryos. *Mol. Reprod. Dev.* **28**, 119–123.

Westhusin, M. E., Levanduski, M. J., Scarborough, R., Looney, C. R., and Bondioli, K. R. (1992). Viable embryos and normal calves after nuclear transfer into Hoechst stained demi-oocytes of cows. *J. Reprod. Fertil.* **95**, 475–480.

Westhusin, M. E., Long, C. R., Shin, T., Hill, J. R., Looney, C. R., Pryor, J. H., and Piedrahita, J. A. (2001). Cloning to reproduce desired genotypes. *Theriogenology* **55**, 35–49.

Wheeler, M. B. (1994). Development and validation of swine embryonic stem cells—A review. *Reprod. Fertil. Dev.* **6**, 563–568.

Whittingham, D. G., Leibo, S. P., and Mazur, P. (1972). Survival of mouse embryos frozen to −196°C and −269°C. *Science* **178**, 411–414.

Wildt, D. E. (1992). Genetic resource banks for conserving wildlife species: Justification, examples and becoming organized on a global basis. *Anim. Reprod. Sci.* **28**, 247–257.

Willadsen, S. M. (1979). A method for culture of micromanipulated sheep embryos and its use to produce monozygotic twins. *Nature* **277**, 298–300.

Willadsen, S. M. (1982). Manipulation of eggs. *In* "Mammalian Egg Transfer" (C. E. Adams, ed.), pp. 185–210. CRC Press, Boca Raton, Florida.

Willadsen, S. M. (1986). Nuclear transplantation in sheep embryos. *Nature (London)* **20**, 63–65.

Willadsen, S. M., and Fehilly, C. B. (1983). The developmental potential and regulatory capacity of blastomeres from 2-, 4-, and 8-cell sheep embryos. *In* "Fertilization of the Human Egg *in Vitro*—Biological Basis and Clinical Application" (H. M. Beier and H. R. Lindner, eds.), pp. 353–357. Springer Verlag, Berlin.

Willett, E. L., Black, W. G., Casida, L. E., Stone, W. H., and Buckner, P. G. (1951). Successful transfer of a fertilized bovine ovum. *Science* **113**, 247.

Williams, T. J., Elsden, R. P., and Seidel, G. E., Jr. (1982). Identical twin bovine pregnancies derived from bisected embryos. *Theriogenology* **17**, 114 (abstr.).

Wilmut, I., Archibald, A. L., Harris, S., McClenaghan, M., Simons, J. P., Whitelaw, C. B. A., and Clark, A. J. (1990). Modification of milk composition. *J. Reprod. Fertil.* (Suppl.) **41**, 199–208.

Wilmut, I., Schnieke, A. E., McWhir, J., Kind, A. J., and Campbell, K. H. S. (1997). Viable offspring derived from fetal and adult mammalian cells. *Nature* **385**, 810–813.

Wilmut, I., Young, L., and Campbell, K. H. S. (1998). Embryonic and somatic cell cloning. Reprod. Fertil. Dev. **10**, 639–643.

Xu, J., and Yang, X. (2001). Telomerase activity in early bovine embryos derived from parthenogenetic activation and nuclear transfer. *Biol. Reprod.* **64**, 770–774.

Yang, X. (1991). Embryo cloning by nuclear transfer in cattle and rabbits. *Int. Embryo Trans. Soc. Newsl.* **9**(4), 10–22.

Yang, X., and Anderson, G. B. (1992). Manipulation of mammalian embryos: principle, progress and future possibilities. *Theriogenology* **38**, 315–335.

Yang, X., and Tian, X. C. (2000). Cloning adult animals—What is the genetic age of the clones? *Cloning* **3**, 123–128.

Yang, X., Zhang, L., Kovacs, A., Tobback, C., and Foote, R. H. (1990). Potential of hypertonic medium treatment for embryo micromanipulation: II. Assessment of nuclear transplantation methodology, isolation, subzona insertion, and electrofusion of blastomeres to intact or functionally enucleated oocytes in rabbits. *Mol. Reprod. Dev.* **27**, 118–129.

Ziomek, C. A. (1987). Cell polarity in the mouse preimplantation embryo. *In* "The Mammalian Preimplantation Embryo" (B. D. Bavister, ed.), pp. 23–41. Plenum Publ. Corp., New York.

Ziomek, C. A. (1998). Commercialization of proteins produced in the mammary gland. *Theriogenology* **49**, 139–144.

PART I

BASIC BIOLOGICAL PROCESSES

CHAPTER 2

ACTIVATION OF MAMMALIAN OOCYTES

Rafael A. Fissore,[1] Jeremy Smyth, Manabu Kurokawa, and Philippe Collas

INTRODUCTION

Mammalian oocytes are ovulated arrested at the metaphase stage of the second meiotic division (MII), from which, if fertilized in a timely manner, they will initiate embryonic development. Prior to the MII arrest, oocytes undergo maturation, during which they progress from the G2/M stage of the first meiotic division to MII. During this process (described in detail in Chapter 1), which is completed in a short period of time compared to the total life span of the female gamete, oocytes undergo significant reorganization and redistribution of organelles and acquire a full complement of signaling molecules (for review seeMiyazaki *et al.*, 1993; Carroll *et al.*, 1996). These changes render mammalian oocytes competent to exit meiosis and initiate embryonic development.

Exit from the MII arrest is accomplished by fertilization and is commonly referred to as "oocyte activation." Oocyte activation comprises a sequence of cellular changes, all of which must be faithfully completed to assure development to term. These events can be arbitrarily organized as early and late events (Schultz and Kopf, 1995). Early events include the initiation of intracellular calcium ($[Ca^{2+}]_i$) oscillations, which triggers all other events of activation, including cortical granule exocytosis to prevent polyspermy, recruitment of maternal mRNAs, and resumption of meiosis with extrusion of the second polar body. Late events include formation of male and female pronuclei, DNA synthesis, and first mitotic cleavage.

In all species studied to date, oocyte activation is triggered by Ca^{2+} release (Whitaker and Patel, 1990). In mammalian oocytes, unlike those of echinoderms and *Xenopus*, multiple $[Ca^{2+}]_i$ oscillations are needed to achieve full activation (Stricker, 1999). It is thought that production of inositol 1,4,5-trisphosphate (IP3) plays a significant role in stimulating Ca^{2+} release during fertilization. IP3 is produced by the hydrolysis of phosphatidylinositol 4,5-bisphosphate by a phospholipase C (PLC). IP3 then binds to its ligand-gated receptor, the IP3R [located in the endoplasmic reticulum (ER), the Ca^{2+} store of the cell], promoting Ca^{2+} release (Patel *et al.*, 1999). Neither the signaling mechanism by which the sperm initiates production of IP3 nor the pathway by which it maintains high levels of IP3 for 4 to 24 hours is known (Jaffe *et al.*, 2001). In this chapter we discuss the Ca^{2+} requirements of mammalian oocytes necessary to initiate activation and development and compare the mechanisms of action of the different parthenogenetic procedures currently in use. We discuss the probable pathway(s) by which the sperm may initiate Ca^{2+} release and the mechanism(s) that may control the persistence and termination of oscillations. The cellular and molecular events required for pronuclear assembly as

---

[1]To whom correspondence should be addressed.

well as likely differences in the assembly and composition of nuclear envelope membranes formed following the transfer of a somatic nucleus, and the impact that this may have on embryo development, are also discussed. Finally, recent evidence suggesting a role for $[Ca^{2+}]_i$ oscillations as an apoptotic-inducing agent, rather than an activating agent, will also be examined.

REQUIREMENT OF Ca^{2+} FOR OOCYTE ACTIVATION

It has long been known that $[Ca^{2+}]_i$ elevations play a universal role in the initiation of oocyte/egg activation (Steinhardt *et al.*, 1974; Tarkowski, 1975; Whittingham and Siracusa, 1980). In the early studies, which were carried out in oocytes or eggs of several invertebrate and vertebrate species, increases in $[Ca^{2+}]_i$ were generated by exposure to Ca^{2+} ionophores (Steinhardt *et al.*, 1974), by direct injection of Ca^{2+} (Fulton and Whittingham, 1978), or by exposure to electrical pulses, which are known to generate transitory pores in the plasma membrane of cells, allowing Ca^{2+} influx into the oocytes (Tarkowski, 1975; Ozil, 1990). All these treatments initiated morphological changes that were consistent with sperm-induced oocyte activation. Further, experiments in which the changes in $[Ca^{2+}]_i$ were blocked by addition of BAPTA-AM, a Ca^{2+} chelator, resulted in inhibition of sperm and Ca^{2+}-induced oocyte activation (Kline and Kline, 1992). Collectively, these results established the role of Ca^{2+} release as the trigger of development.

EARLY EVENTS OF ACTIVATION REQUIRE LESS $[Ca^{2+}]_i$

Although the aforementioned parthenogenetic treatments induced initiation of development, there was a clear age-dependent effect on the activation triggered by these agents. Importantly, it was observed that oocytes stimulated several hours postovulation approximately 16–20 hours following injection of human chorionic gonadotropin in the mouse (Xu *et al.*, 1997; Abbot *et al.*, 1998), or matured for more than 24 hours and then activated (in the case of bovine oocytes), were easily activated by a single rise in $[Ca^{2+}]_i$, which is the typical response induced by the agents mentioned above (Nagai, 1987; Ware *et al.*, 1989; Collas *et al.*, 1989). However, when recently ovulated oocytes were exposed to these treatments, not only were fewer oocytes activated, but also the activation that was initiated was abortive. For example, in response to a single rise in $[Ca^{2+}]_i$, recently ovulated oocytes could exhibit one or several of the following events: cortical granule exocytosis, recruitment of mRNAs, and, in some cases, resumption of meiosis, although in most studies they failed to progress into interphase (Ozil, 1990; Xu *et al.*, 1996). Similarly, it was shown that inhibition of early events of activation requires higher concentrations of BAPTA-AM than is required to block late events of activation (Kline and Kline, 1992), further supporting the differential Ca^{2+} requirement for events of activation.

LATE EVENTS OF ACTIVATION REQUIRE $[Ca^{2+}]_i$ OSCILLATIONS

Fertilization can initiate activation of ovulated mammalian oocytes regardless of their age. Therefore, it was assumed and then tested that the robustness of the sperm-induced activation signal may be related to the sperm's ability to induce $[Ca^{2+}]_i$ oscillations. For instance, it was shown by Ozil (1990) and Collas *et al.* (1989) that the administration of repetitive electrical pulses resulted in higher rates of extrusion of the second polar body and pronuclear formation than in those cases in which a single pulse was administered. Furthermore, the higher the magnitude of the stimulation, and presumably of the corresponding intracellular Ca^{2+} change induced by the pulse, the greater the rate of pronuclear formation (Ozil, 1990, 1998). Interestingly, as the intensity of the Ca^{2+} stimulation increased, the time to pronuclear formation decreased, suggesting that the molecules that control MII arrest may also be involved in controlling the dynamics of pronuclear formation.

It is well known that the MII arrest of mammalian oocytes is mediated by a cytostatic factor (CSF); CSF contains several molecules, including maturation-promoting factor (MPF), a complex formed by cdk1 and cyclin B1 (Draetta and Beach, 1988); MAPK; c-mos; and, probably, p90^{rsk} (Sagata, 1997; Ferrell, 1999). Inactivation of MPF, which is mediated by the degradation of cyclin B by the proteasome, makes possible the exit from meiosis, and this is followed significantly later by inactivation of MAPK, which correlates with pronuclear formation (Moos *et al.*, 1995; Liu *et al.*, 1998). Although the molecules responsible for degradation/inactivation of the components of CSF in mammals are not clearly known, in frogs, Ca^{2+}-calmodulin-dependent protein kinase II (CaMKII), a $[Ca^{2+}]_i$ decoding molecule (De Koninck and Schulman, 1998; Dupont and Goldbeter, 1998), appears to play a role in the inactivation of MPF. For example, addition of a constitutively active form of CaMKII to frog egg extracts with high CSF activity triggers degradation of cyclin B and loss of MPF activity, even in the absence of Ca^{2+}. Furthermore, injection of these treated extracts into embryonic blastomeres fails to induce cleavage arrest (Lorca *et al.*, 1993), a clear indication that CSF has been inactivated. In mammalian eggs, the role of CaMKII in meiosis exit has not been thoroughly tested, although the addition of CaMKII inhibitors blocks/delays resumption of meiosis, supporting a similar role for this molecule in mammals (Tatone *et al.*, 1999). Furthermore, $[Ca^{2+}]_i$ elevations in mouse oocytes induced by exposure to ethanol or ionomycin trigger transient activation of CaMKII activity (Winston and Maro, 1995; Johnson *et al.*, 1998). Additional support for the role of Ca^{2+} in the regulation of MPF activity was provided by experiments in which the generation of $[Ca^{2+}]_i$ increases induced inactivation of MPF activity (Collas *et al.*, 1993), which was assumed to be due to degradation of cyclin B. This inactivation of MPF was temporary, and a few hours after the rise in $[Ca^{2+}]_i$ the activity of MPF rebounded. These results were interpreted to mean that because mammalian MII oocytes are translationally active, a single rise tends to produce a partial cyclin B degradation and, in the absence of persistent Ca^{2+} signaling, cyclin B reaccumulates, reestablishing MPF activity. Regarding MAPK, although it is known that MAPK activity is down-regulated by phosphatases (Charles *et al.*, 1992; Keyse and Emslie, 1992), how these phosphatases are activated during early fertilization is not known and remains an area in need of additional investigation. Therefore, it appears that the requirement for prolonged oscillations is intended, at least in part, to ensure exit from meiosis and progression into interphase.

Given the realization of the need for multiple rises to accomplish full activation, parthenogenetic treatments were designed to replicate such Ca^{2+} responses. In the mouse system, these efforts were rewarded—the addition of $SrCl_2$ into a $CaCl_2$-free medium stimulated long-term Ca^{2+} oscillations that produced high rates of development (Kline and Kline, 1992; Bos-Mikich *et al.*, 1997). $SrCl_2$, however, does not induce Ca^{2+} responses in bovine or porcine oocytes, although in oocytes of these species injection of several agonists triggers oscillations and acceptable rates of activation (see Fissore *et al.*, 1999, for review). Nonetheless, injection procedures are time consuming and they incur greater oocyte losses, which has limited the use of these agents. Thus, new parthenogenetic procedures were developed that combined the need for a rise in $[Ca^{2+}]_i$, induced by a brief exposure to a Ca^{2+} ionophore, with persistent kinase inactivation, which was provided by 4–6 hours of exposure either to a kinase inhibitor such as 6-DMAP (Susko-Parrish *et al.*, 1994) or to a protein synthesis inhibitor such as cycloheximide (Presicce and Yang, 1994). Although these treatments have proved very successful and have resulted in the birth of young following nuclear transfer procedures, the broad-spectrum nature of these compounds may have unintended detrimental effects on development that need to be investigated. It is possible, however, that the development of more specific kinase inhibitors such as roscovitine or butyrolactone, which are specifically targeted to inhibit the active component(s) of cdk1, the catalytic component of MPF (Meijer *et al.*, 1997; Motlik *et al.*, 1998), may be equally effective in inducing activation and may lack unwanted side effects.

EFFECT OF Ca^{2+} STIMULATION ON DEVELOPMENT

Besides affecting the initiation and completion of activation, the magnitude of the Ca^{2+} stimulation may also have an impact on later stages of development. For instance, it was shown that besides the number of $[Ca^{2+}]_i$ spikes administered by electrical pulses, the amplitude (mainly) and interval of the spikes had a significant impact on the morphology and organization of the early fetuses that resulted from the transfer of those embryos into recipients (Ozil and Huneau, 2001). This report showed that embryos produced by exposure to the optimal activation treatment, which mimicked fertilization-associated $[Ca^{2+}]_i$ rises in amplitude and interval, resulted in the highest pregnancy and implantation rates. This concurs with previous data demonstrating that additional $[Ca^{2+}]_i$ oscillations produced by longer exposures to $SrCl_2$ produced parthenogenetic embryos that exhibited higher ratios of inner cell mass to trophectoderm cells and, presumably, that this may result in embryos with better capacity to develop and implant (Bos-Mikich *et al.*, 1997). The question that then arises is how can a Ca^{2+} signal that takes place within 2 hours of fertilization have an impact on events that occur 10 days later? The functional significance of $[Ca^{2+}]_i$ oscillations as a mechanism to encode precise information remains to be fully elucidated. In somatic cells, the frequency and/or amplitude of the $[Ca^{2+}]_i$ rises can be altered by exposure to a cell-permeant caged IP3 capable of inducing controlled, long-term oscillations (Li *et al.*, 1998). Studies using this system have revealed that modifications of these parameters result in altered gene expression, confirming that Ca^{2+} may be used to encode differential patterns of gene expression (Li *et al.*, 1998). Furthermore, it was shown that Ca^{2+} might affect the function of a transcriptional repressor, Dream, suggesting that $[Ca^{2+}]_i$ oscillations regulate gene expression by operating through a variety of different mechanisms (Carrion *et al.*, 1999). Interestingly, the target of $[Ca^{2+}]_i$ rises in mammalian eggs is unlikely to be the immediate regulation of gene transcription, because, in these cells, transcription is maintained at insignificant levels from the time of germinal vesicle breakdown (GVBD) to the time of the maternal–zygotic transition, which occurs 36–96 hours postfertilization, depending on the species. Therefore, $[Ca^{2+}]_i$ oscillations at fertilization are likely to influence the levels of protein translation or, alternatively, they may activate Ca^{2+}-sensitive proteins already present in the oocyte. One possible target protein is the already discussed CaMKII, which is specially suited to decode the information encoded in these oscillations, given its special properties that allow it to remain active for periods longer than the time during which $[Ca^{2+}]_i$ is elevated (De Koninck and Schulman, 1998; Dupont and Goldbeter, 1998). Finally, $[Ca^{2+}]_i$ elevations have been shown to modify proteins that bind DNA and, in this manner, they may induce conformational changes in the chromatin that may alter gene expression later on in development, as was proposed by Ozil and Huneau (2001).

Despite these beneficial effects of Ca^{2+} on development, in a study using nuclear transfer-generated embryos, it was shown that the activation method, whether $[Ca^{2+}]_i$ oscillations or a single $[Ca^{2+}]_i$ rise, had no impact on the rates of development to term (Kishikawa *et al.*, 1999). It is important to note that these embryos were generated from slightly aged oocytes and this might have negated the beneficial effects of $[Ca^{2+}]_i$ oscillations on development (Tarin *et al.*, 1999; Gordo *et al.*, 2000). However, it remains to be carefully tested whether the beneficial effects of multiple $[Ca^{2+}]_i$ oscillations reported in several studies are exclusively due to the persistent down-regulation of MPF and MAPK activity that they induce, or, on the other hand, if $[Ca^{2+}]_i$ oscillations may differentially modify the zygote's chromatin structure to optimize gene expression and development. These possibilities can be easily tested. For example, if the latter is true, the rates of development to term of embryos generated by nuclear transfer should be maximal under conditions that elicit multiple oscillations. Conversely, if the former alternative better explains the mechanisms of action of $[Ca^{2+}]_i$ oscillations, then activation by specific kinase inhibitors may produce similar developmental rates. Also, the effects of different activation procedures on the expression of specific genes at different stages of development should be tested.

HOW DOES THE SPERM TRIGGER Ca²⁺ RELEASE?

The mechanism by which the sperm initiates Ca^{2+} release has not been elucidated, although several hypotheses that have been proposed are discussed here. We concentrate on two of the possible mechanisms triggered by the sperm, the "receptor theory" and the "fusion/sperm factor (SF) theory" (for additional reviews on the topic see Schultz and Kopf, 1995; Swann and Lai, 1997). Another hypothesis, "the conduit theory," which proposes that the sperm, during fertilization, promotes Ca^{2+} influx and, in this manner, makes possible the generation of $[Ca^{2+}]_i$ oscillations, will not be discussed because a series of experiments have demonstrated that Ca^{2+} influx alone cannot replicate/sustain the generation of oscillations (Igusa and Miyazaki, 1983; Fissore and Robl, 1992; Swann and Ozil, 1994; Jones *et al.*, 1998a).

RECEPTOR THEORY

The receptor theory proposes that the sperm acts as a ligand for a receptor in the plasma membrane of oocyte/egg (A and B, Fig. 1). This theory is based on the assumption that oocytes may behave as somatic cells, in which the addition of hormones/ligands initiates the generation of $[Ca^{2+}]_i$ oscillations following stimulation of specific surface receptors (Whitaker and Swann, 1993; Shultz and Kopf, 1995). Engagement of these receptors results in the activation of PLCs, most likely PLCβ and/or γ, and subsequent IP3 production. Activation of PLCβ isoforms is likely mediated by G-proteins; that this pathway is active in mammalian oocytes was demonstrated by the finding that injection of GTPγ[S], a nonhydrolazable activator of G-proteins, induced long-lasting Ca^{2+} responses (Miyazaki, 1988). Likewise, expression and stimulation of exogenous muscarinic receptors in oocytes resulted in activation and generation of $[Ca^{2+}]_i$ oscillations (Miyazaki, 1991; Williams *et al.*, 1992; Moore *et al.*, 1993). Also, when the activation of G-protein signaling in mammalian oocytes was inhibited by injection of GDPβ[S], a nondegradable inhibitor of G-proteins, fertilization-induced oscillations were blocked/inhibited (Miyazaki, 1988; Fissore and Robl, 1994; Moore *et al.*, 1994). Nevertheless, when this pathway was inhibited by injection of a function-blocking antibody against Gαq, which is the Gα protein most likely to stimulate PLCβ, fertilization-initiated oscillations were not blocked (Williams *et al.*, 1998), and these results were interpreted to mean that G-proteins are not involved in the fertilization-induced oscillations. Despite these results, the role of G-proteins in mammalian fertilization cannot be ruled out. It is important to note, for instance, that homozygous embryos generated from the crossing of PLCβ3⁻/⁺ mice, PLCβ3 being an isoform widely present in MII mouse oocytes (Wang *et al.*, 1998), were developmentally incompetent and the arrest seemed to occur at the two-cell stage, which is the earliest arrest stage of any of the PLC knockouts so far reported. This early cessation of cleavage may very well reflect inadequate activation, suggesting the involvement of the G-protein pathway in mammalian fertilization.

A G-protein linked receptor hypothesis

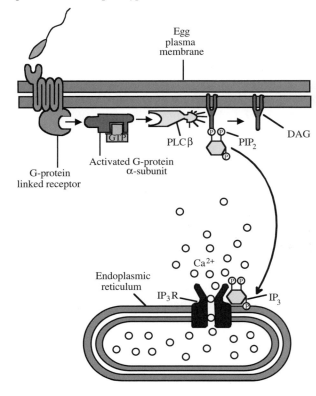

B Receptor tyrosine kinase hypothesis

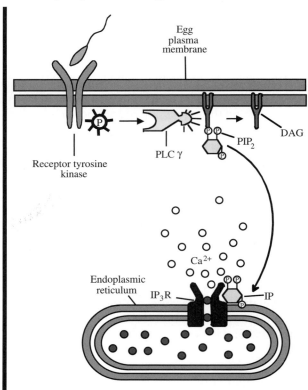

C Fusion hypothesis

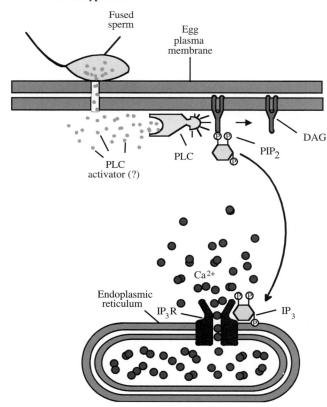

Figure 1 Possible mechanism(s) by which the sperm may initiate $[Ca^{2+}]_i$ oscillations in mammalian oocytes. The receptor model presumes the binding and activation of a surface receptor coupled to either a G-protein (A) or a tyrosine kinase (B) signal transduction pathway, which culminates in the activation of a phospolipase C (PLC) and subsequent production of inositol 1,4,5-trisphosphate (IP3). PIP_2, Phosphatidylinositol 4,5-bisphosphate; DAG, diacylglycerot; IP3R, Ip3 receptor. (C) Possible fusion mechanism: the fused sperm releases a factor into the egg cytoplasm. The nature of the sperm factor is unknown, but it may be a PLC isoform or egg PLC activator.

Another group of cell surface receptors, those linked to protein tyrosine kinases (PTKs), can also trigger IP3 production in oocytes/eggs, although, in this case, PLCγs are the isoforms involved and are most likely activated by tyrosine phosphorylation (Carroll *et al.*, 1997; Jaffe *et al.*, 2001). The findings that tyrosine kinase activity increases immediately/soon after fertilization in echinoderm and mammalian eggs/oocytes (Ciapa and Epel, 1991; Ben-Yosef *et al.*, 1998) suggested the participation of this pathway in sperm-induced oscillations. The increase in PTK activity likely serves to stimulate the enzymatic activity of PLCγ and this was shown to be the case in recently fertilized sea urchin eggs (Rongish *et al.*, 1999). Whether PLCγ activity is required to mediate the fertilization-associated $[Ca^{2+}]_i$ rise in sea urchin eggs was tested by preinjection of SH2 domains corresponding to PLCγ, which act as a competitive inhibitor of PLCγ by blocking the activation of the endogenous enzyme. Inhibition of PLCγ activity resulted in inhibition of the sperm-associated Ca^{2+} release, although injection of IP3 was able to elicit Ca^{2+} release, indicating that the inhibition is upstream of IP3 (Carroll *et al.*, 1997). A PTK that has been shown to participate in the activation of PLCγ is the SRC-family kinase. In echinoderm eggs, this kinase was shown to be activated immediately after fertilization and to associate with the SH2 domains of PLCγ (Giusti *et al.*, 1999). Furthermore, inhibition of the SRC-family kinase by injection of SH2 domains homologous to the SRC protein blocked Ca^{2+} release and activation (Abassi *et al.*, 2000; Kinsey and Shen, 2000). Also, in support of this view, injection of constitutively active SRC kinase induced Ca^{2+} release and activation (Giusti *et al.*, 2000).

In oocytes of vertebrates, such as in *Xenopus* and mammals, the role of PTK/SRC-family kinase/PLCγ remains to be fully demonstrated, although data from *Xenopus* studies offer tantalizing evidence for participation in fertilization. In these oocytes, the presence of a SRC-like kinase, Xyc, has been demonstrated (Sato *et al.*, 1999; 2000). Further, inhibition of this pathway by exposure of eggs/extracts to PP1, an inhibitor of SRC-family kinases, or injection of a variety of specific tyrosine kinase inhibitors (Glahn *et al.*, 1999; Sato *et al.*, 2000), or culture with PLC inhibitors such as U-73122 (Sato *et al.*, 2000), inhibited sperm-induced Ca^{2+} release and activation. Remarkably, the activation of PLCγ does not appear to be mediated by binding of SH2 domains because injection of an excess of these proteins failed to block sperm-associated Ca^{2+} release and activation (Runft *et al.*, 1999). Likewise, in mammalian oocytes, injection of SH2 domain proteins corresponding to PLCγ failed to abrogate fertilization-induced oscillations (Mehlmann *et al.*, 1998). Nevertheless, the role of PTK in mammalian fertilization is suggested by the findings that PTK inhibitors such genistein blocked/delayed the generation of $[Ca^{2+}]_i$ oscillations, and addition of U-73122 also blocked the initiation/persistence of sperm- and sperm factor-triggered $[Ca^{2+}]_i$ oscillations (Dupont *et al.*, 1996; Wu *et al.*, 2001). Thus, it appears that the sperm may activate a tyrosine kinase during mammalian fertilization. However, the identity of the molecules involved as well as the significance of this step in the initiation and persistence of oscillations require additional investigation.

Despite that the aforementioned results implicate G-proteins and PTK in the activation of PLCs, a significant gap remains in our understanding of which cell surface receptor mediates the sperm's stimulation of the phosphoinositide (PI) pathway. In mammalian oocytes, integrins appear to be the proteins involved in the sperm–oocyte interaction. Interestingly, in frog eggs and in bovine oocytes, extracellular addition of peptides that bind integrin receptors induced Ca^{2+} release (Schilling *et al.*, 1998; Campbell *et al.*, 2000). Nevertheless, *in vitro* fertilization studies using oocytes of mice lacking α6β1 integrin exhibited normal fertilization and activation of development (Miller *et al.*, 2000). Further, as was the case with integrin-deficient mice, *in vitro* fertilization studies using sperm lacking fertilin β, the molecule complementary to integrin, showed that despite deficient binding these sperm were able to initiate Ca^{2+} responses and induce activation (Cho *et al.*, 1998).

Remarkably, it was clearly demonstrated that unless the sperm can fuse and be introduced into the oocyte cytoplasm, it is unable to induce Ca^{2+} responses and activation (Le Naour *et al.*, 2000; Miyado *et al.*, 2000). These seminal observations were made utilizing a knockout mouse in which the expression of the CD9 protein had been eliminated. CD9 is a member of the tetraspan-membrane proteins and is reported to play a role in adhesion and cell motility. It was observed that females lacking CD9 exhibited severe infertility and that the oocytes of these females could bind sperm, although sperm were unable to fuse. In the absence of fusion, activation or $[Ca^{2+}]_i$ oscillations were not observed, although injection of sperm circumvented the inability of these oocytes to be activated. Collectively, studies supporting the receptor theory have demonstrated that the intracellular machinery necessary to generate IP3 is present in mammalian oocytes. Nonetheless, it appears that during fertilization, production of IP3 is not stimulated by a signal from a surface receptor but may be activated by a novel factor/mechanism delivered by the sperm that still remains to be elucidated.

FUSION HYPOTHESIS

The lack of a candidate receptor for the sperm has led to the search for alternative hypotheses. One theory that has gained support is the "fusion hypothesis." This hypothesis proposes that a sperm product is delivered into the oocyte cytoplasm after gamete fusion and is responsible for initiating $[Ca^{2+}]_i$ oscillations (Fig. 1C). Evidence from several laboratories supports the role of this mechanism in mammalian fertilization. For instance, it has been demonstrated by monitoring the transfer of fluorescent dyes between sperm and oocytes that the initiation of oscillations occurs after fusion of the gametes has taken place (Lawrence *et al.*, 1997; Jones *et al.*, 1998a). Second, injection of sperm extracts from a variety of sperm sources into mammalian and nonmammalian oocytes generates Ca^{2+} responses that closely mimic those initiated by autologous fertilization (Fig. 2B) (Swann and Lai, 1997; Stricker, 1999). Further, injection of these extracts initiates parthenogenetic development *in*

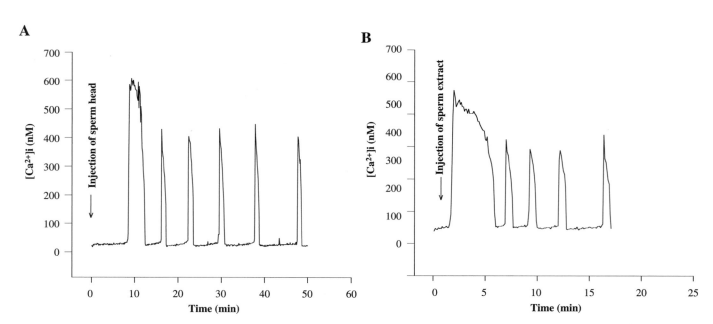

Figure 2 *Injection of sperm and sperm extract initiates physiological Ca^{2+} responses in mouse oocytes. A mouse sperm head was injected into a mouse oocyte using a pipette driven by a piezoelectric unit (A). Sperm extracts (0.5 mg/ml, concentration in the pipette) were delivered into oocytes using a 2-μm glass pipette and pneumatic pressure (B).*

vitro (Stice and Robl, 1990; Wu *et al.*, 1998a) and, when combined with injection of spermatogenic cells without Ca^{2+}-activating ability, they support development to term and birth of live young (Sakurai *et al.*, 1999). Finally, the advent and success of intracytoplasmic sperm injection (ICSI) (Palermo *et al.*, 1992), a technique during which the membranes of the sperm do not come in contact with the oocytes' membranes, has strengthened the validity of the fusion hypothesis. In this technique, whole sperm are injected into oocytes, and sperm delivered in this manner initiate Ca^{2+} responses that closely resemble those initiated by normal fertilization (Fig. 2A) (Tesarik and Souza, 1994; Nakano *et al.*, 1997) and support high rates of development to term. Hence, these results collectively support the hypothesis that a sperm-associated product activates the oocytes' Ca^{2+}, releasing machinery from inside the oocytes' cytoplasm, favoring an inside-out sequence of events to activate Ca^{2+} release.

The nature and composition of the Ca^{2+}-active molecule in sperm extracts remain to be elucidated. It was originally proposed that IP3 may be the active molecule in sperm (Dale *et al.*, 1985; Tosti *et al.*, 1993). However, this theory is at odds with the suggested proteinaceous nature of the active molecule (Swann, 1990; Wu *et al.*, 1997; Stricker, 1997), as well as with the likely amounts of this molecule required to maintain oscillations for over 20 hours in some species. Two different protein molecules were then suggested to represent the active molecule in sperm— oscillin, which was later shown to be glucosamine 6-phosphate deaminase, and tr-kit, a truncated version of c-kit (Parrington *et al.*, 1996; Sette *et al.*, 1997). Although these molecules were demonstrated to be present in crude sperm and cellular extracts and their injection into oocytes induced oscillations and/or activation, on further purification or injection of recombinant proteins they were unable to elicit Ca^{2+} responses (Wolosker *et al.*, 1998; Wu *et al.*, 1998b; Wolny *et al.*, 1999). Therefore, the nature of the active component(s) of the sperm remains to be elucidated.

SPERM EXTRACTS STIMULATE THE PHOSPHOINOSITIDE PATHWAY

What, then, is the nature of the active molecule in the sperm? Some important clues came from reports detailing the signaling pathway stimulated by sperm extracts. It has long been known that during fertilization the sperm induces an increase in polyphosphoinositide metabolism and IP3 production (Turner *et al.*, 1984; Stith *et al.*, 1994). These studies, however, were carried out using mass assays in groups of oocytes, which may not offer a highly synchronous population, complicating the interpretation of the events that lead to the activation of the pathway. Taking advantage of an *in vitro* system, egg extracts, it was shown that addition of sperm factor (SF) induces IP3 production (Jones *et al.*, 1998b). Further, it was demonstrated at the single-cell level that, simultaneously with Ca^{2+} release, injection of SF induces IP3 production (Wu *et al.*, 2001). These results confirm the strong stimulation of the PI pathway and IP3 production by sperm extracts and, most likely, by the sperm. What now remains to be determined is how the active component of the sperm/SF stimulates this pathway. It has been proposed that SF is a PLC capable of initiating and sustaining IP3 production during fertilization (Jones *et al.*, 2000). The sperm PLC has been suggested to be active at very low concentrations of Ca^{2+}, although Ca^{2+} can stimulate it, and it appears to have at least twice the specific activity of the PLCs in other tissues (Rice *et al.*, 2000). Interestingly, injections of recombinant PLCs, which have significantly higher PLC activity, failed to initiate $[Ca^{2+}]_i$ oscillations in mammalian oocytes. This, together with the fact that mammalian oocytes contain significant amounts of PLCs, raises questions as to whether the active component of SF is a PLC or an activator of the oocytes' PLCs.

Additional points of contention regarding the active component(s) in sperm extracts is whether the active component is present in different sperm compartments, and whether a single molecule is responsible for the Ca^{2+}-releasing activity or whether there are several active molecules (Perry *et al.*, 1999; 2000). Regarding the first topic, it is important to note that most of the studies that addressed the location of the active component in sperm used mouse sperm, which possess significantly less amounts of activity than does sperm from other mammalian species. Second, it is entirely possible that the active component of SF may be in an insoluble association with a protein in the sperm and, as time postpenetration increases, the molecule becomes progressively solubilized by the oocyte cytoplasm. This is supported by the findings that female pronuclei that form after fertilization acquire Ca^{2+}-releasing ability, as demonstrated by transfer of these pronuclei into MII oocytes (Kono *et al.*, 1995). Remarkably, female pronuclei that form after parthenogenetic activation do not posses this ability (Kono *et al.*, 1995). These results can be interpreted to mean that a protein brought in by the sperm, following sperm head decondensation, becomes part of the cytoplasm, and it or a protein activated by it is targeted to the pronuclear envelope region. Whether this redistribution of the active molecule takes place in all mammalian species, and the mechanisms responsible for targeting it to the nucleus, remain unknown and are areas in need of research. Finally, whether the sperm contains several different molecules or whether the active molecule is active in a complex is not known. However, our preliminary data following chromatographic fractionation of soluble and less soluble sperm compartments suggest that a single molecule may be responsible for the Ca^{2+} activity of mammalian sperm (M. Kurokawa and H. Wu, unpublished results). Collectively, the published information demonstrates that a sperm-derived protein, on fusion, initiates Ca^{2+} release by stimulating the PI pathway, and that the distribution of the factor in several sperm compartments with potentially different solubilization properties, may be specially suited to support the long-lasting generation of oscillations, which are necessary for successful mammalian fertilization (see Fig. 3 for model of sperm factor location and release).

REGULATION OF FERTILIZATION-ASSOCIATED [Ca²⁺]ᵢ OSCILLATIONS

Even though fertilization in mammals initiates long-lasting [Ca^{2+}]$_i$ oscillations, species-specific differences in duration and amplitude of the oscillations are apparent on closer scrutiny. For instance, in mouse zygotes, oscillations cease at approximately the time of pronuclear formation (Jones *et al.*, 1996; Deguchi *et al.*, 1998). In contrast, in rabbit zygotes, the oscillations persist well into the pronuclear stage, and in bovine zygotes, oscillations are observed into the first mitosis (Nakada *et al.*, 1995). Remarkably, the amplitudes of the oscillations in rabbit and bovine zygotes appear to steadily decrease as time postfertilization increases (Fissore and Robl, 1993; Nakada *et al.*, 1995). It is assumed that the long persistence of oscillations in mammalian oocytes is required to ensure exit from meiosis, but why in some species these oscillations are terminated prematurely, whereas in others there is a gradual decline in the amplitude of the rises, is not known. Moreover, the mechanisms implicated in this modification of the zygote's Ca^{2+} responsiveness remain to be elucidated.

Oocytes have several mechanisms to control the pattern of oscillations and whether each of these mechanisms operates independently or concomitantly needs to be investigated. One way in which oocytes may control Ca^{2+} release is by downregulation of the IP3Rs. The well-known isoform 1 of this receptor (IP3R1) represents the most abundant Ca^{2+} channel in mammalian oocytes and is likely to mediate most of the sperm-associated Ca^{2+} release (Mehlmann *et al.*, 1996; He *et al.*, 1997, 1999; Parrington *et al.*, 1998). It is well-documented that fertilization triggers progressive down-regulation of the receptor, with a decrease of nearly 50% of the mass of the receptor by 4 hours postinsemination, a time at which in mouse zygotes oscillations appear to cease (Parrington *et al.*, 1998; Jellerette *et al.*, 2000). This degra-

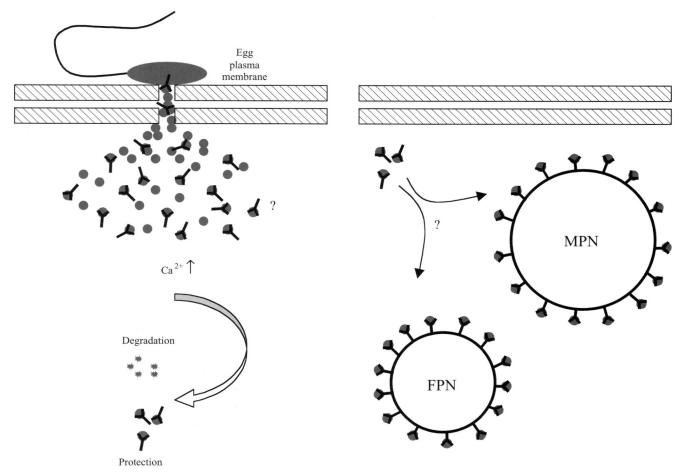

Figure 3 *Model by which the Ca²⁺-active component(s) of the sperm may be released into the cytoplasm of oocytes, and how, after 4–6 hours, it appears to colocalize to the male pronucleus (MPN) and to the female pronucleus (FPN). The gray circles and oval represent the active Ca²⁺ components of the sperm. The black "Y" shapes depict hypothetical sperm molecules that bind the Ca²⁺-active component, protect it from degradation, and may be responsible for targeting it to the pronuclear envelope region. Whether the proposed model is accurate remains to be determined along with the nature of the molecules, suggested to be either the active Ca²⁺ molecule or the protector/targeting molecule.*

dation of the receptor, which is mediated by the proteasome, appears specific to fertilization because it is not induced by oscillations generated by exposure to $SrCl_2$, which do not rely on IP3 production (Jellerette *et al.*, 2000; Brind *et al.*, 2000). A second mechanism that may modulate Ca^{2+} release is the changes in stage of the cell cycle. As previously mentioned, in mouse zygotes fertilization-associated oscillations cease at the time of pronuclear formation, a stage at which both MPF and MAPK are low, although some Ca^{2+} responsiveness appears to be regained after pronuclear envelope breakdown (Kono *et al.*, 1996). Additional support for the decreased Ca^{2+} responsiveness at the pronuclear stage is provided by experiments showing that injection of IP3 or sperm extracts at this stage is unable to induce oscillations (Tang *et al.*, 2000), and in another report culture of oocytes with roscovitine, an inhibitor of MPF activity, inhibited fertilization-associated oscillations, although roscovitine may have some undesired effects on the Ca^{2+}-ATPase pump (Deng and Shen, 2000). Nonetheless, the above results suggest an association between MPF activity and $[Ca^{2+}]_i$ oscillations. Likewise, in ascidian eggs, persistently high levels of MPF activity by expression of nondegradable cyclins significantly prolonged the duration of oscillations (Levasseur and McDougall, 2000). Remarkably, in mammals other than the mouse, the persistence of $[Ca^{2+}]_i$ oscillations does not correlate as closely with the inactivation profile of MPF, suggesting that other kinases, such as MAPK and CaMKII, may modulate the Ca^{2+} responsiveness of the zygotes.

How these kinases modulate Ca^{2+} release is not known, although several potential targets exist. First, the IP3R, which possesses several phosphorylation consensus sites, could be influenced by changes in the stage of the cell cycle (Jayaraman *et al.*, 1996). Alternatively, the content of the Ca^{2+} stores could be modified as time postfertilization increases and approaches the first mitosis (Tang *et al.*, 2000). Whether this is the result of a direct modification of the activity of the Ca^{2+}-ATPase pump (SERCA), which regulates Ca^{2+} uptake into the ER, or is a consequence of the cell-cycle dependence of the store-operated Ca^{2+} channel (Machaca and Haun, 2000), which contributes to the refilling of the stores after IP3 stimulation, remains to be determined. Despite that the precise mechanism responsible for the progressive down-regulation of the Ca^{2+} signal is not known, its elucidation may answer a significant biological question and may as well contribute to the development of better parthenogenetic activation procedures, which will have an impact on the success of cloning. It is interesting to point out that both the cessation of oscillations and the decrease in amplitude of the $[Ca^{2+}]_i$ responses appear to occur around the time of pronuclear formation, and this suggests a possible regulatory function of Ca^{2+} release in pronuclear assembly and/or initiation of DNA synthesis.

REQUIREMENTS FOR ASSEMBLY OF COMPETENT PRONUCLEI

Proper assembly of a functional pronucleus is vital for many cellular functions in the zygote and early embryo. Pronuclei contain either female or male chromosome complements in a decondensed form and enclosed by a double nuclear membrane fenestrated by nuclear pore complexes (NPCs). Between the chromatin and the inner nuclear membrane, a network of intermediate filament proteins called "lamins" constitutes the nuclear lamina. The lamina contributes to the nuclear architecture but is also implicated in DNA replication and transcription. The nucleus is a highly dynamic organelle, as illustrated, for example, by the extensive morphological and biochemical changes the sperm nucleus undergoes at fertilization. In the egg cytoplasm, the inactive, compact, fertilizing sperm nucleus is disassembled and remodeled into a large, replication-competent, male pronucleus. Concomitantly, the female pronucleus assembles from a condensed chromatin structure (telophase II chromosomes in most mammals) on oocyte activation. Another example of pronuclear assembly lies in the remodeling of a somatic nucleus into a pronuclear-like structure following transfer into an oocyte.

Morphological changes associated with male pronuclear formation at fertilization have been documented in several species, including mammals (e.g., Longo and Anderson, 1968; Perreault, 1990). However, insights into molecular processes governing pronuclear assembly have emanated from the use of cell-free systems derived from, notably, *Xenopus* and sea urchin gametes (Lohka and Masui, 1984; Cameron and Poccia, 1994; Collas and Poccia, 1995a).

DISASSEMBLY OF THE SPERM NUCLEAR ENVELOPE

The first event of male pronuclear development is likely the disassembly of sperm nuclear envelope components. Little is known about the solubilization of sperm nuclear membranes, most likely due to the lack of an appropriate *in vitro* system to study this phenomenon (membrane-intact sperm nuclei do not breakdown *in vitro*) (P. Collas, unpublished observations). Nonetheless, disassembly of the sperm nuclear lamina is better understood. The lamina of most somatic nuclei consists of

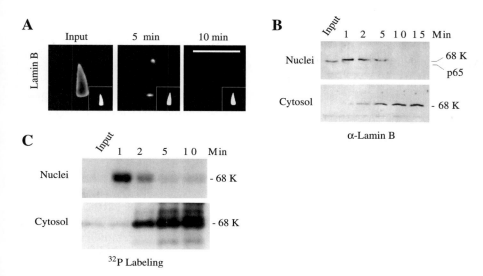

Figure 4 *Phosphorylation and solubilization of the sea urchin sperm nuclear lamina in vitro. (A) Immunofluorescence localization of lamin B in nuclei incubated in egg cytosolic extract. The lamin is removed first from the lateral aspects of the nucleus then from the nuclear poles. Insets, DNA; bar, 5 μm. (B) Immunoblotting analysis of lamin B from nuclei incubated in lamin B-depleted cytosol. Solubilized sperm lamin B is gradually detected in the cytosol in a 68-kDa form. (C) Autoradiograms of nuclear and cytosolic proteins during incubation of nuclei in cytosol as in (B). Lamin B is rapidly phosphorylated (1 minute).*

A- and B-type lamins, whereas that of sperm usually consists of B-type lamins and in some instances (e.g., rat) sperm-specific lamin variants. Sea urchin sperm contain a single B-type lamin of 65 kDa (Collas *et al.*, 1995). In egg cytosolic extract, sperm lamina disassembly occurs within minutes (A and B, Fig. 4) (Collas *et al.*, 1997). Concomitantly, the solubilized lamin appears in the cytosol in a phosphorylated form, as shown by rapid ^{32}P incorporation into sperm lamin B prior to its solubilization (Fig. 4). Lamin B phosphorylation *in vitro* is inhibited by BAPTA, indicating that the reaction requires Ca^{2+}.

Several lamin kinases have been identified, including CDK1, the Ca^{2+}-dependent kinase C (PKC), and the cAMP-dependent protein kinase A (PKA). However, specific inhibitors of PKC and immunodepletion of egg-soluble PKC are the most effective inhibitors of lamin B phosphorylation, and biochemical studies have shown that the sea urchin sperm lamin kinase is an egg PKC (Collas *et al.*, 1997). Interestingly, inhibition of sperm lamin solubilization systematically prevents decondensation of the sperm chromatin, suggesting that both events are related. Nevertheless, although PKC-mediated sperm lamina disassembly precedes chromatin decondensation, it is not sufficient to induce decondensation.

ASSEMBLY OF THE MALE PRONUCLEUS

Molecules from the oocyte cytoplasm are equally important for the formation of the male pronucleus. Indeed, development of the pronucleus from sperm DNA requires maternal histones, nuclear lamins, and nucleoporins (Yanagimachi, 1994). Furthermore, the zygotic centrosome is assembled around the sperm centriole from tubulin and centrosomal proteins of maternal origin in most species (e.g., Schatten, 1994).

Chromatin Decondensation

In mammalian oocytes, a pool of glutathione, synthesized during the first meiosis, seems to ensure the reduction of disulfide bonds in the sperm nucleus and thereby promotes chromatin decondensation during fertilization. Indeed, disulfide cross-linking of the sperm protamines is responsible for the tight packaging and hypercondensation of the DNA in the sperm nucleus. After destabilization of the disulfide bonds, the protamines are replaced by oocyte-derived histones, and the sperm nucleus develops into the male pronucleus.

Two stages of sea urchin sperm chromatin decondensation that have been identified *in vitro* are reminiscent of those occurring *in vivo*: a membrane-independent decondensation phase and a membrane-dependent swelling phase (Collas and Poccia, 1995b). Decondensation requires cytosolic factors and ATP hydrolysis. It is promoted in activated egg cytosol and at alkaline pHs. Nuclear swelling necessitates the previous assembly of a nuclear envelope and requires ATP, soluble lamins, Ca^{2+}, and cytosolic factors. The final nuclear diameter is limited by suboptimal amounts of vesicles in the assembly reaction, amount of ATP in the extract, or depletion of lamins from the extract. From these observations, chromatin decondensation appears to be a property of the chromatin and cytosol, whereas nuclear swelling requires a nuclear envelope with a lamina. Sea urchin sperm chromatin decondensation is associated with replacement of sperm-specific histone variants by maternal histones (reviewed in Poccia and Green, 1992).

Assembly of the Pronuclear Envelope

Targeting of the bulk of nuclear envelope vesicles to chromatin is probably mediated by integral membrane proteins (Collas *et al.*, 1996). Evidence suggests that pronuclear envelope assembly requires a cooperation between distinct cytoplasmic vesicle populations, or distinct fractions of a continuous endoplasmic reticulum (Collas and Poccia, 1996). The various vesicle populations required for nuclear vesicle binding and fusion suggest the involvement of multiple interactions. *In vitro*, fusion of chromatin-bound vesicles is promoted by cytosolic components, Ca^{2+}, ATP, and GTP hydrolysis. GTP hydrolysis may be mediated by vesicle-associated monomeric G-proteins, although their function in nuclear envelope assembly remains unclear. A role of lumenal Ca^{2+} has also been suggested but clearly remains to be elucidated.

GETTING READY FOR DNA SYNTHESIS: ASSEMBLY OF NUCLEAR PORES AND OF THE LAMINA

Assembly of NPCs

Nuclear pore complexes are essential components of the nuclear envelope because they mediate the exchange of macromolecules between the nucleus and the cytoplasm. Little is known about how and when NPCs assemble in the pronuclear envelope. In the sea urchin *in vitro* pronuclear assembly system, the lamina assembles primarily by nuclear import of soluble lamins (see below), thus the first NPCs must form prior to initiation of lamina assembly. Data from somatic and embryonic systems suggest that a number of NPC components are targeted from the cytoplasm to the chromatin surface during early stages of nuclear assembly, whereas others, perhaps of greater structural importance for the NPC architecture, are incorporated into the nuclear envelope during the last stages of assembly to anchor the NPC to the nuclear membrane.

Assembly of the Nuclear Lamina

Assembly of the nuclear envelope *in vitro* appears to be a lamin-independent process. Nevertheless, lamina-free nuclei are small and their nuclear envelope is devoid of NPCs (Collas *et al.*, 1996). The lamina assembles during a later stage of pronuclear development and is associated with swelling of the nucleus. *In vitro* experiments have shown that the lamina assembles by import of soluble B-type lamins from the egg cytoplasm, and subsequent polymerization of the imported lamins. A limited proportion of the new nuclear lamina may also be supplied by B-type lamins attached to nuclear precursor vesicles, however, this fraction is not strictly required for lamina assembly. Nuclear swelling accompanies lamina assembly and necessitates incorporation of additional membrane vesicles into the pronuclear envelope (Collas *et al.*, 1996).

Nuclei devoid of lamina and of nuclear pores are incapable of replicating DNA, whereas assembly of these structures confers DNA replication ability at least *in vitro* (see, e.g., Goldberg *et al.*, 1995). This has led to the suggestion that lamina assembly is a prerequisite for replication in *in vitro*-assembled pronuclei. This view has been supported by increasing evidence that the nuclear lamina is implicated in replication in somatic cells (e.g., Moir *et al.*, 2000).

PRONUCLEAR ASSEMBLY AFTER NUCLEAR TRANSPLANTATION INTO OOCYTES: REQUIREMENTS AND IMPACT ON DEVELOPMENTAL COMPETENCE OF THE RESULTING EMBRYO

The study of nuclear dedifferentiation through the transplantation of nuclei into oocytes has provided insights on the remodeling of a nucleus at various stages of its differentiation into a "pronucleus." With the aim of cloning animals, nuclear transfers have been performed mostly into (1) enucleated MII oocytes that were subsequently activated to initiate embryo development or (2) into oocytes that have been activated prior to nuclear transfer. Nuclei from undifferentiated embryos direct development to term of reconstituted embryos in both cases. However, nuclei from later stage embryos (Sims and First, 1993; Collas and Barnes, 1994), cultured fetal cells (Campbell *et al.*, 1996), or differentiated somatic cells (e.g., Wilmut *et al.*, 1997) direct development to term only after transfer into metaphase II oocytes. The most likely explanation for this difference is the ability of the metaphase II oocyte to remodel extensively, and perhaps thereby reprogram, the transplanted nucleus.

Remodeling of an exogenous nucleus in a metaphase II oocyte seems very similar to remodeling of the sperm nucleus at fertilization or in cell-free extracts. Remodeling of the transplanted nucleus and *de novo* formation of a pronucleus entails disassembly of the transplanted nucleus and reconstitution of a new nucleus. Nuclear disassembly in meiotic cytoplasm involves nuclear envelope breakdown and chromatin condensation (Szöllösi *et al.*, 1986, 1988; Collas and Robl, 1991; Barnes *et al.*, 1993). The extent of nuclear envelope disassembly is related to the level of maturation-promoting factor activity in the oocyte, as illustrated in Fig. 5. Electron microscopic examination of bovine morula nuclei transplanted into MII oocytes that were activated either at the time of fusion or several hours prior to fusion reveals that various degrees of nuclear envelope disassembly, partial (Fig. 5C) or complete (Fig. 5D), take place. As a result of nuclear envelope disassembly and chromatin condensation, DNA synthesis in transplanted nuclei is inhibited or severely repressed (Barnes *et al.*, 1993). Transcription is also decreased or inhibited, a phenomenon associated with changes in nucleolar ultrastructure (Kanka *et al.*, 1996). In contrast, nuclei transplanted into S-phase cytoplasm retain their interphase characteristics, as judged by decondensed chromatin, an intact nuclear envelope (Fig. 5A), and virtually uninterrupted DNA synthesis (Szöllösi *et al.*, 1986). Following transfer between telophase II and the pronuclear stage, compromised nuclear structures are detected, notably blebbing nuclear envelopes (Fig. 5C) (Szöllösi *et al.*, 1988), and the chromatin does not significantly condense or decondense under these conditions.

Following activation of the recipient oocyte, a new nucleus assembles from the exogenous chromatin. Nuclear reassembly involves decondensation of the chromatin, formation of a new nuclear envelope, polymerization of a new lamina, formation of nucleoli, and expansion of the remodeled nucleus (Szöllösi *et al.*, 1986; Stice and Robl, 1988). DNA replication resumes as the nucleus swells, presumably

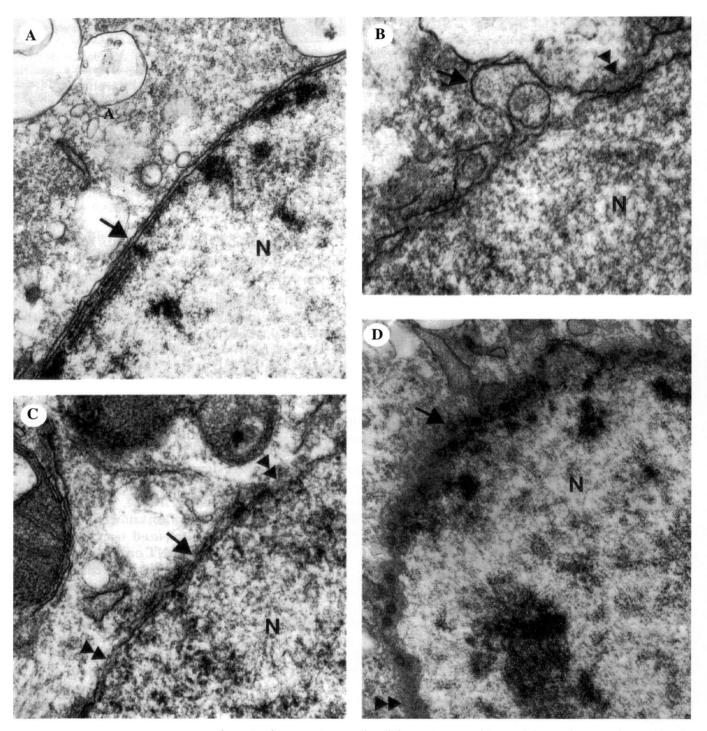

Figure 5 *Electron micrographs of the various conditions of the nuclear envelope of bovine embryo nuclei (N) following transplantation into oocytes. (A) Intact nuclear envelope (arrow) with distinct inner and outer membranes (×25,000). (B) Partial nuclear envelope; the outer membrane exhibits breaks (double arrow) and blebs off (arrow) (×50,000). (C) Fragmented nuclear envelope; both membranes display breaks (arrow and double arrows) (×40,000). (D) Absence of nuclear envelope—the envelope has broken down; heterochromatin (double arrow) and putative nuclear pore complex remnants (arrow) indicate the nuclear boundary (×25,000). From Barnes et al. (1993), Molecular and Reproductive Development, Copyright 1993 Wiley, with permission.*

after the reformation of a functional nuclear envelope concomitantly with initiation of DNA synthesis in the female pronucleus (Barnes *et al.*, 1993). The degree of remodeling of transplanted nuclei, however, is related to the cell-cycle stage of the host oocyte on transplantation, and thereby to the extent of previous disassembly. It is well established that nuclei that undergo complete disassembly in MII cytoplasm are remodeled into complete and swollen nuclei, whereas nuclei that do not completely disassemble after transfer into preactivated oocytes do not significantly swell.

From these observations, it is clear that the oocyte cytoplasmic environment directs the extent of remodeling of transplanted nuclei. Direct contact of the chromatin with the oocyte cytoplasm (as a consequence of nuclear envelope breakdown) seems necessary for complete remodeling to occur. This is also true for fertilizing sperm, because sperm nuclei containing intact membranes microinjected into eggs or incubated in egg extract fail to develop into pronuclei.

$[Ca^{2+}]_i$ OSCILLATIONS AND APOPTOSIS IN MAMMALIAN OOCYTES

As emphasized in this chapter, Ca^{2+} oscillations are a hallmark of mammalian fertilization and are required to trigger activation of recently ovulated eggs. Also, as previously noted, parthenogenetic agents that trigger a single Ca^{2+} rise are effective only in activating aged eggs. Additional studies have revealed that a minimum number of Ca^{2+} increases, approximately nine, is required to activate newly ovulated eggs (Swann and Parrington, 1999). However, excessive Ca^{2+} stimulation has also been shown to have negative effects on cell metabolism and Ca^{2+} has been associated with programmed cell death (apoptosis) and necrosis (Trump and Berezesky, 1995). Interestingly, mammalian eggs, which use sophisticated methods to prevent polyspermy, occasionally fail to block the penetration of multiple sperm, and when this occurs the frequency of oscillations is significantly increased (Faure *et al.*, 1999). The physiological impact of such altered Ca^{2+} responses is unclear, as is the effect of normal Ca^{2+} stimulation on aged oocytes, and these topics are discussed in more detail in the next section.

$[Ca^{2+}]_i$ OSCILLATIONS SIGNAL APOPTOSIS IN AGED EGGS

Although Ca^{2+} responses have been associated with necrosis and apoptosis in numerous cell types, in mammalian eggs the role of Ca^{2+} has been exclusively associated with activation of development. Notably, apoptosis has been shown to occur in the female gonad and gametes (Morita and Tilly, 1999). Furthermore, eggs either recovered from older females, aged *in vitro*, or exposed to nonphysiological stimuli and suboptimal culture conditions exhibit typical signs of apoptosis such as cell and DNA fragmentation (Perez *et al.*, 1999). Fertilization is also associated with fragmentation. In an early study, Marston and Chang (1964) demonstrated that delayed inseminations resulted in increased rates of embryo fragmentation and decreased rates of *in vitro* development. These data suggested that the fertilizing Ca^{2+} signal might, if imposed into aged or developmentally impaired eggs, induce premature termination of development rather than normal activation. This possibility was tested by injecting sperm extracts eliciting physiological Ca^{2+} responses into newly ovulated and *in vitro*-aged mouse eggs. Injection of sperm extracts induced Ca^{2+} signals that initiated normal activation in newly ovulated eggs, but in aged eggs they induced fragmentation and caspase activation (Gordo *et al.*, 2000) (Fig. 6). Interestingly, in the same study, induction of oscillations in recently ovulated oocytes at exceedingly high frequency and for a prolonged period of time also induced abnormal activation that resulted in cell-cycle arrest. Thus, it appears that fertilization can initiate two developmental programs, normal development or cell-cycle arrest/apoptosis, according to the frequency of the oscillations and age of the oocytes.

Ca²⁺ RELEASE AND ACTIVATION OF THE APOPTOTIC PATHWAY IN MAMMALIAN EGGS

The molecular mechanisms responsible for the increased susceptibility of aged eggs, or eggs from aged females, to undergo apoptosis are not fully elucidated. Importantly, mammalian eggs express a variety of proteins of the Bcl-2 family that are known to regulate apoptosis. The Bcl-2 family contains antiapoptotic members such as Bcl-2 and Bcl-xl, and proapoptotic members such as Bax and Bad (Exley *et al.*, 1999; Jurisicova *et al.*, 1998a,b). Moreover, caspases are also expressed in eggs (Malcov *et al.*, 1997) and are activated in mammalian embryos undergoing fragmentation (Perez *et al.*, 1999; Gordo *et al.*, 2000) (Fig. 6). Thus, it is possible that proteins and mRNA levels of antiapoptotic members of the Bcl-2 family may be degraded or inactivated as eggs age, rendering these eggs susceptible to apoptotic stimuli. Findings consistent with this hypothesis have been reported. Furthermore, the same studies also demonstrated that the mRNA levels of proapoptotic genes remain at relatively constant levels during aging in eggs (Exley *et al.*, 1999; Jurisicova *et al.*, 1998a,b), tilting the delicate balance between antiapoptotic and proapoptotic genes in favor of the latter. There are several possibilities to explain how $[Ca^{2+}]_i$ oscillations may trigger death in aged eggs. First, it has been reported that a close relationship exists between IP3-mediated Ca^{2+} release and apoptosis in cells exposed to proapoptotic stimuli (Szalai *et al.*, 1999). In this report, mitochondria of mildly injured cells responded to IP3-induced Ca^{2+} release with release of cytochrome *c*, an activator of caspases, and loss of mitochondrial membrane

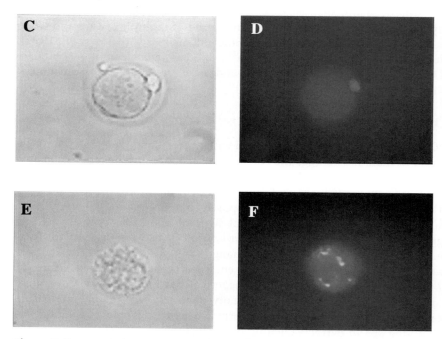

Figure 6 *Fragmented, sperm extract-injected MII aged oocytes exhibit widespread caspase activity. Phase-contrast images of uninjected MII aged oocyte (40 hours post-hCG) and freshly ovulated oocyte (16 hours post-hCG) injected with 1 mg/ml SF show no signs of cytoplasmic fragmentation (C, D) and caspase activity is solely present in the polar body (D). Aged oocytes (40 hours post-hCG) injected with 1 mg/ml sperm extract show cytoplasmic fragmentation (E) and exhibit clear evidence of extensive caspase activity (F). From Gordo et al. (2000) (Biology of Reproduction by Gorde, Wu, He, Fissore. Copyright 2000 by Soc. for the Study of Reproduction. Reproduced with permission of Soc. for the Study of Reproduction).*

potential. This led to cellular fragmentation rather than to normal ATP production. In aged eggs, the aging process, associated with a decrease in levels of antiapoptotic proteins, may serve as the proapoptotic stimulus that sensitizes eggs to activate caspases following Ca^{2+} release. Alternatively, as Bcl-2 and Bcl-xl modulate Ca^{2+} homeostasis in the ER, they may also modulate expression and function of the Ca^{2+}-ATPase pump, which maintains Ca^{2+} levels in the ER (Kuo *et al.*, 1998). Interestingly, emptying of the cell's Ca^{2+} stores can also serve as an equally effective stimulus of cell death (Jiang *et al.*, 1994; Bian *et al.*, 1997). Thus, it is possible that in aged eggs, which may contain partially depleted ER stores due to a partially dysfunctional Ca^{2+} pump, persistent stimulation of Ca^{2+} release triggered by fertilization may deplete the stores and induce apoptosis. That the ER is altered in aged eggs has been suggested by the finding that injection of IP3 triggers significantly less Ca^{2+} release in aged eggs than in newly ovulated eggs (Takahashi *et al.*, 2000). Thus, fertilization-associated $[Ca^{2+}]_i$ oscillations can serve as a natural trigger of the apoptotic program in aged eggs that are less developmentally competent. Elucidation of the mechanism by which antiapoptotic proteins are lost or inactivated during egg aging may prolong the fertilizable life span of mammalian eggs and may result in higher rates of *in vitro* development of mammalian embryos.

REFERENCES

Abassi, Y. A., Carroll, D. J., Giusti, A. F., Belton, R. J., Jr., and Foltz, K. R. (2000). Evidence that Src-type tyrosine kinase activity is necessary for initiation of calcium release at fertilization in sea urchin eggs. *Dev. Biol.* **218**, 206–219.

Abbott, A. L., Xu, Z., Kopf, G. S., Ducibella, T., and Schultz, R. M. (1998). *In vitro* culture retards spontaneous activation of cell cycle progression and cortical granule exocytosis that normally occur in *in vivo* unfertilized mouse eggs. *Biol. Reprod.* **59**, 1515–1522.

Barnes, F. L., Collas, P., Powell, R., King, W. A., Westhusin, M., and Sheperd, D. (1993). Influence of recipient oocyte cell cycle stage on DNA synthesis, nuclear envelope breakdown, chromosome constitution and development in nuclear transplant bovine embryos. *Mol. Reprod. Dev.* **36**, 33–41.

Ben-Yosef, D., Talmor, A., Shwartz, L., Granot, Y., and Shalgi, R. (1998). Tyrosyl-phosphorylated proteins are involved in regulation of meiosis in the rat egg. *Mol. Reprod. Dev.* **49**, 176–185.

Bian, X., Hughes, F. M. J., Huang, Y., Cidlowski, J. A., and Putney, J. W. J. (1997). Roles of cytoplasmic Ca^{2+} and intracellular Ca^{2+} stores in induction and suppression of apoptosis in S49 cells. *Am. J. Physiol.* **272**, C1241–C1249.

Bos-Mikich, A., Whittingham, D. G., and Jones, K. T. (1997). Meiotic and mitotic Ca^{2+} oscillations affect cell composition in resulting blastocysts. *Dev. Biol.* **182**, 172–179.

Brind, S., Swann, K., and Carroll, J. (2000). Inositol 1,4,5-trisphosphate receptors are downregulated in mouse oocytes in response to sperm or adenophostin A but not to increases in intracellular Ca^{2+} or egg activation. *Dev. Biol.* **223**, 251–265.

Cameron, L. A., and Poccia, D. L. (1994). *In vitro* development of the sea urchin male pronucleus. *Dev. Biol.* **162**, 568–578.

Campbell, K. H. S., McWhir, J., Ritchie, W. A., and Wilmut, I. (1996). Sheep cloned by nuclear transplantation from a cultured cell line. *Nature* **380**, 64–66.

Campbell, K. D., Reed, W. A., and White, K. L. (2000). Ability of integrins to mediate fertilization, intracellular calcium release, and parthenogenetic development in bovine oocytes. *Biol. Reprod.* **62**, 1702–1709.

Carrion, A. M., Link, W. A., Ledo, F., Mellstrom, B., and Naranjo, J. R. (1999). DREAM is a Ca^{2+}-regulated transcriptional repressor. *Nature* **398**, 80–84.

Carroll, J., Jones, K. T., and Whittingham, D. G. (1996). Ca^{2+} release and the development of Ca^{2+} release mechanisms during oocyte maturation: A prelude to fertilization. *Rev. Reprod.* **3**, 137–143.

Carroll, D. J., Ramarao, C. S., Mehlmann, L. M., Roche, S., Terasaki, M., and Jaffe, L. A. (1997). Calcium release at fertilization in starfish eggs is mediated by phospholipase Cγ. *J. Cell Biol.* **138**, 1303–1311.

Charles, C. H., Abler, A. S., and Lau, L. F. (1992). cDNA sequence of a growth factor-inducible immediate early gene and characterization of its encoded protein. *Oncogene* **7**, 187–190.

Cho, C., Bunch, D. O., Faure, J. E., Goulding, E. H., Eddy, E. M., Primakoff, P., and Myles, D. G. (1998). Fertilization defects in sperm from mice lacking fertilin β. *Science* **281**, 1857–1859.

Ciapa, B., and Epel, D. (1991). A rapid change in phosphorylation on tyrosine accompanies fertilization of sea urchin eggs. *FEBS Lett.* **295**, 167–170.

Collas, P., and Barnes, F. L. (1994). Nuclear transplantation by microinjection of inner cell mass and granulosa cell nuclei. *Mol. Reprod. Dev.* **38**, 264–267.

Collas, P., and Poccia, D. L. (1995a). Lipophilic structures of sperm nuclei target membrane vesicle binding and are incorporated into the nuclear envelope. *Dev. Biol.* **169**, 123–135.

Collas, P., and Poccia, D. L. (1995b). Formation of the sea urchin male pronucleus *in vitro*: Membrane-independent chromatin decondensation and nuclear envelope-dependent nuclear swelling. *Mol. Reprod. Dev.* **42**, 106–113.

Collas, P., and Poccia, D. L. (1996). Distinct egg membrane vesicles differing in binding and fusion properties contribute to sea urchin male pronuclear envelopes formed *in vitro*. *J. Cell Sci.* **109**, 1275–1283.

Collas, P., and Robl, J. M. (1991). Relationship between nuclear remodeling and development in nuclear transplant rabbit embryos. *Biol. Reprod.* **45**, 455–465.

Collas, P., Balise, J. J., Hofman, G. A., and Robl, J. M. (1989). Electrical activation of mouse oocytes. *Theriogenology* **32**, 835–844.

Collas, P., Sullivan, E. J., and Barnes, F. L. (1993). Histone H1 kinase activity in bovine oocytes following calcium stimulation. *Mol. Reprod. Dev.* **34**, 224–231.

Collas, P., Pinto-Correia, C., and Poccia, D. L. (1995). Lamin dynamics during sea urchin male pronuclear formation *in vitro*. *Exp. Cell Res.* **219**, 687–698.

Collas, P., Courvalin, J.-C., and Poccia, D. L. (1996). Targeting of membranes to sea urchin sperm chromatin is mediated by a lamin B receptor-like integral membrane protein. *J. Cell Biol.* **135**, 1715–1725.

Collas, P., Thompson, L., Fields, A. P., Poccia, D. L., and Courvalin, J.-C. (1997). Protein kinase C-mediated interphase lamin B phosphorylation and solubilization. *J. Biol. Chem.* **272**, 21274–21280.

Dale, B., DeFelice L. J., and Ehrenstein, G. (1985). Injection of a soluble sperm extract into sea urchin eggs triggers the cortical reaction. *Experientia* **41**, 1068–1070.

Deguchi, R., Shirakawa, H., Oda, S., Mohri, T., and Miyasaki, S. (1998). Spatiotemporal analysis of Ca^{2+} waves in relation to the sperm entry site and animal–vegetal axis during Ca^{2+} oscillations in fertilizing mouse eggs. *Dev. Biol.* **218**, 299–313.

De Koninck, P., and Schulman, H. (1998). Sensitivity of CaM kinase II to the frequency of Ca^{2+} oscillations. *Science* **279**, 227–230.

Deng, M. Q., and Shen, S. S. (2000). A specific inhibitor of p34(cdc20)/cyclin B suppresses fertilization-induced Ca^{2+} oscillations in mouse eggs. *Biol. Reprod.* **62**, 873–878.

Draetta, G., and Beach, D. (1988). Activation of cdc2 protein kinase during mitosis in human cells: Cell cycle-dependent phosphorylation and subunit rearrangement. *Cell* **54**, 17–26.

Dupont, G., and Goldbeter, A. (1998). CaM kinase II as frequency decoder of Ca^{2+} oscillations. *BioEssays* **20**, 607–610.

Dupont, G., McGuinness, O. M., Johnson, M. H., Berridge, M. J., and Borgese, F. (1996). Phospholipase C in mouse oocytes: Characterization of β and γ isoforms and their possible involvement in sperm-induced Ca^{2+} spiking. *Biochem. J.* **316**, 583–591.

Exley, G. E., Tang, C., McElhinny, A. S., and Warner, C. M. (1999). Expression of caspase and BCL-2 apoptotic family members in mouse preimplantation embryos. *Biol. Reprod.* **61**, 231–239.

Faure, J. E., Myles, D. G., and Primakoff, P. (1999). The frequency of calcium oscillations in mouse eggs at fertilization is modulated by the number of fused sperm. *Dev. Biol.* **213**, 370–377.

Ferrel J. E., Jr. (1999). Xenopus oocyte maturation: new lessons from a good egg. *BioEssays* **21**, 833–842.

Fissore, R. A., and Robl, J. M. (1992). Intracellular Ca^{2+} response of rabbit oocytes to electrical stimulation. *Mol. Reprod. Dev.* **32**, 9–16.

Fissore, R. A., and Robl, J. M. (1993). Sperm, inositol trisphosphate, and thimerosal-induced intracellular Ca^{2+} elevations in rabbit eggs. *Dev. Biol.* **159**, 122–130.

Fissore, R. A., and Robl, J. M. (1994). Mechanism of calcium oscillations in fertilized rabbit eggs. *Dev. Biol.* **166**, 634–642.

Fissore, R. A., Long, C. R., Duncan, R. P., and Robl, J. M. (1999). Initiation and organization of the first cell cycle in mammalians: Applications in cloning. *Cloning* **1**, 89–100.

Fulton, B. P., and Whittingham, D. G. (1978). Activation of mammalian oocytes by intracellular injection of calcium. *Nature* **273**, 149–151.

Giusti, A. F., Carroll, D. J., Abassi, Y. A., and Foltz, K. R. (1999). Evidence that a starfish egg Src family tyrosine kinase associates with PLC-γ1 SH2 domains at fertilization. *Dev. Biol.* **208**, 189–199.

Giusti, A. F., Xu, W., Hinkle, B., Terasaki, M., and Jaffe, L. A. (2000). Evidence that fertilization activates starfish eggs by sequential activation of a src-like kinase and phospholipase Cγ. *J. Biol. Chem.* **275**, 16788–16794.

Glahn, D., Mark, S. D., Behr, R. K., and Nucitelli, R. (1999). Tyrosine kinase inhibitors block sperm-induced egg activation in *Xenopus laevis*. *Dev. Biol.* **205**, 171–180.

Goldberg, M., Jenkins, H., Allen, T., Whitfiled, W. G., and Hutchison, C. J. (1995). *Xenopus* lamin B_3 has a direct role in the assembly of a replication competent nucleus: Evidence from cell-free egg extracts. *J. Cell Sci.* **108**, 3451–3461.

Gordo, A. C., Wu, H., He, C. L., and Fissore, R. A. (2000). Injection of sperm cytosolic factor into mouse metaphase II oocytes induces different developmental fates according to the frequency of [Ca^{2+}]i oscillations and oocyte age. *Biol. Reprod.* **62**, 1370–1379.

He, C. L., Damiani, P., Parys, J. B., and Fissore, R. A. (1997). Calcium, calcium release receptors, and meiotic resumption in bovine oocytes. *Biol. Reprod.* **57**, 1245–1255.

He, C. L., Damiani, P., Ducibella, T., Takahashi, M., Tanzawa, K., Parys, J. B., and Fissore, R. A. (1999). Isoforms of the inositol 1,4,5-trisphosphate receptor are expressed in bovine oocytes and ovaries: The type-1 isoform is down-regulated by fertilization and by injection of adenophostin A. *Biol. Reprod.* **61**, 935–943.

Igusa, Y., and Miyazaki, S. (1983). Effects of altered extracellular and intracellular calcium concentration on hyperpolarizing responses of the hamster egg. *J. Physiol.* **340**, 611–632.

Jaffe, L. A., Giusti, A. F., Carroll, D, J., and Foltz K. R. (2001). Ca^{2+} signaling during fertilization in echinoderm eggs. *Cell Dev. Biol.* **12**, 45–51.

Jayaraman, T., Ondrias, K., Ondriasova, E., and Marks, A. R. (1996). Regulation of the inositol 1,4,5-trisphosphate receptor by tyrosine phosphorylation. *Science* **272**, 1492–1494.

Jellerette, T., He, C. L., Wu, H., Parys, J. B., and Fissore, R. A. (2000). Down-regulation of the inositol 1,4,5-trisphosphate receptor in mouse eggs following fertilization or parthenogenetic activation. *Dev. Biol.* **223**, 238–250.

Jiang, S., Chow, S. C., Nicotera, P., and Orrenius, S. (1994). Intracellular Ca^{2+} signals activate apoptosis in thymocytes: Studies using the Ca^{2+}-ATPase inhibitor thapsigargin. *Exp. Cell Res.* **212**, 84–92.

Johnson, J., Bierle, B. M., Gallicano, G. I., and Capco, D. G. (1998). Calcium/calmodulin-dependent protein kinase II and calmodulin: regulators of the meiotic spindle in mouse eggs. *Dev. Biol.* **204**, 464–477.

Jones, K. T., Carroll, J., Merriman, J. A., Whittingham, D. G., and Kono, T. (1995). Repetitive sperm-induced Ca^{2+} transients in mouse oocytes are cell-cycle dependent. *Development* **121**, 3259–3266.

Jones, K. T., Soeller, C., and Cannell, M. B. (1998a). The passage of Ca^{2+} and fluorescent markers between the sperm and egg after fusion in the mouse. *Development* **125**, 4627–4635.

Jones, K. T., Cruttwell, C., Parrington, J., and Swann, K. (1998b). A mammalian sperm cytosolic phospholipase C activity generates inositol trisphosphate and causes Ca^{2+} release in sea urchin egg homogenates. *FEBS Lett.* **437**, 297–300.

Jones, K. T., Matsuda, M., Parrington, J., Katan, M., and Swann, K. (2000). Different Ca^{2+}-releasing abilities of sperm extracts compared with tissue extracts and phospholipase C isoforms in sea urchin egg homogenate and mouse eggs. *Biochem. J.* **346**(Pt. 3), 743–749.

Jurisicova, A., Latham, K. E., Casper, R. F., and Varmuza, S. L. (1998a). Expression and regulation of genes associated with cell death during murine preimplantation embryo development. *Mol. Reprod. Dev.* **51**, 243–253.

Jurisicova, A., Rogers, I., Fasciani, A., Casper, R. F., and Varmuza, S. (1998b). Effect of maternal age and conditions of fertilization on programmed cell death during murine preimplantation embryo development. *Mol. Hum. Reprod.* **4**, 139–145.

Kanka, J., Hozak, P., Heyman, Y., Chesne, P., Renard, J.-P., and Flechon, J. E. (1996). Transcriptional activity and nucleolar ultrastructure of embryonic rabbit nuclei after transplantation to enucleated oocytes. *Mol. Reprod. Dev.* **14**, 135–144.

Keyse, S. M., and Emslie, E. A. (1992). Oxidative stress and heat shock induce a human gene encoding a protein-tyrosine phosphatase. *Nature* **359**, 644–647.

Kinsey, W. H., and Shen, S. (2000). Role of the Fyn Kinase in calcium release during fertilization in the sea urchin egg. *Dev. Biol.* **225**, 254–264.

Kishikawa, H., Wakayama T., and Yanagimachi, R. (1999). Comparison of oocyte-activating agents for mouse cloning. *Cloning* **1**, 153–159.

Kline, D., and Kline, J. T. (1992). Repetitive calcium transients and the role of calcium in exocytosis and cell cycle activation in the mouse egg. *Dev. Biol.* **149**, 80–89.

Kono, T., Carroll, J., Swann, K., and Whittingham, D. G. (1995). Nuclei from fertilized mouse embryos have Ca^{2+} releasing activity. *Development* **121**, 1123–1128.

Kono, T., Jones, K. T., Bos-Mikich, A., Whittingham, D. G., and Carroll, J. (1996). A cell-cycle associated change in Ca^{2+} releasing activity leads to the generation of Ca^{2+} transients in mouse embryos during the first mitosis. *J. Cell Biol.* **132**, 915–923.

Kuo, T. H., Kim, H. R., Zhu, L., Yu, Y., Lin, H. M., and Tsang, W. (1998). Modulation of endoplasmic reticulum calcium pump by Bcl-2. *Oncogene* **17**, 1903–1910.

Lawrence, Y., Whitaker, M., and Swann, K. (1997). Sperm-oocyte fusion is the prelude to the initial Ca^{2+} increase at fertilization in the mouse. *Development* **124**, 223–241.

Le Naour, F., Rubinstein, E., Jasmin, C., Prenant, M., and Boucheix, C. (2000). Severely reduced female fertility in CD9-deficient mice. *Science* **287**, 319–321.

Levasseur, M., and McDougall, A. (2000). Sperm-induced calcium oscillations at fertilization in ascidians are controlled by cyclin B1-depedent kinase. *Development* **127**, 631–641.

Li, W., Llopis, J., Whitney, M., Zlokarnik, G., and Tsien, R. Y. (1998). Cell-permeant caged InsP3 ester shows that Ca^{2+} spike frequency can optimize gene expression. *Nature* **392**, 936–941.

Liu, L., Ju, J. C., and Yang, X. (1998). Differential inactivation of maturation-promoting factor and mitogen-activated protein kinase following parthenogenetic activation of bovine oocytes. *Biol. Reprod.* **59**, 537–545.

Lohka, M. J., and Masui, Y. (1984). Roles of cytosol and cytoplasmic particles in nuclear envelope assembly and sperm pronuclear formation in cell-free preparations from amphibian eggs. *J. Cell Biol.* **98**, 1222–1230.

Longo, F. J., and Anderson, E. (1968). The fine structure of pronuclear development and fusion in the sea urchin, *Arbacia punctulata. J. Cell Biol.* **39**, 339–368.

Lorca, T., Cruzalegui, F. H., Fesquet, D., Cavadore, J. C., Mery, J., Means, A., and Doree, M. (1993). Calmodulin-dependent protein kinase II mediates inactivation of MPF and CSF upon fertilization of *Xenopus* eggs. *Nature* **366**, 270–273.

Machaca, K., and Haun, S. (2000). Store-operated Ca^{2+} entry inactivates at the germinal vesicle breakdown stage of *Xenopus* meiosis. *J. Biol. Chem.* **49**, 38710–38715.

Malcov, M., Ben-Yosef, D., Glaser, T., and Shalgi, R. (1997). Changes in calpain during meiosis in the rat egg. *Mol. Reprod. Dev.* **48**, 119–126.

Marston, J. H., and Chang, M. C. (1964). The fertilizable life of ova and their morphology following delayed insemination in mature and immature mice. *J. Exp. Zool.* **155**, 237–252.

Mehlmann, L. M., Mikoshiba, K., and Kline, D. (1996). Redistribution and increase in cortical inositol 1,4,5-trisphosphate receptors after meiotic maturation of the mouse oocyte. *Dev. Biol.* **180**, 489–498.

Mehlmann, L. M., Carpenter, G., Rhee, S. G., and Jaffe, L. A. (1998). SH2 domain-mediated activation of phospholipase Cγ is not required to initiate Ca^{2+} release at fertilization of mouse eggs. *Dev. Biol.* **203**, 221–232.

Meijer, L., Borgne, A., Mulner, O., Chong, J. P., Blow, J. J., Inagaki, N., Inagaki, M., Delcros, J. G., and Moulinoux, J. P. (1997). Biochemical and cellular effects of roscovitine, a potent and selective inhibitor of the cyclin-dependent kinases cdc2, cdk2 and cdk5. *Eur. J. Biochem.* **243**, 527–536.

Miller, B. J., Georges-Labouesse, E., Primakoff, P., and Myles, D. G. (2000). Normal fertilizationoccurs with eggs lacking the integrin alpha6beta1 and is CD9-dependent. *J. Cell Biol.* **149**, 1289–1296.

Miyado, K., Yamada, G., Yamada, S., Hasuwa, H., Nakamura, Y., Ryu, F., Suzuki, K., Kosai, K., Inoue, K., Ogura, A., Okabe, M., and Mekada, E. (2000). Requirement of CD9 on the egg plasma membrane for fertilization. *Science* **287**, 321–324.

Miyazaki, S. (1988). Inositol 1,4,5-trisphosphate-induced calcium release and guanine nucleotide-binding protein-mediated periodic calcium rises in golden hamster eggs. *J. Cell Biol.* **106**, 345–353.

Miyazaki, S. (1991). Repetitive calcium transients in hamster oocytes. Cell *Calcium* **12**, 205–216.

Miyazaki, S., Shirakawa, H., Nakada, K., and Honda, Y. (1993). Essential role of the inositol 1,4,5-trisphosphate receptor/Ca^{2+} release channel in Ca^{2+} waves and Ca^{2+} oscillations at fertilization of mammalian eggs. *Dev. Biol.* **158**, 62–78.

Moir, R. D., Spann, T. P., Herrmann, H., and Goldman, R. D. (2000). Disruption of nuclear lamin organization blocks the elongation phase of DNA replication. *J. Cell Biol.* **149**, 1179–1191.

Moore, G. D., Kopf, G. S., and Schultz, R. M. (1993). Complete mouse egg activation in the absence of sperm by stimulation of an exogenous G protein-coupled receptor. *Dev. Biol.* **159**, 669–678.

Moore, G. D., Ayabe, T., Visconti, P. E., Schultz, R. M., and Kopf, G. S. (1994). Roles of heterotrimeric and monomeric G proteins in sperm-induced activation of mouse eggs. *Development* **120**, 3313–3323.

Moos, J., Visconti, P. W., Moore, G. D., Schultz, R. M., and Kopf, G. S. (1995). Potential role of mitogen-activated protein kinase (MAP) kinase in pronuclear envelope assembly and disassembly following fertilization in mouse eggs. *Biol. Reprod.* **53**, 692–699.

Morita, Y., and Tilly, J. L. (1999). Oocyte apoptosis: like sand through an hourglass. *Dev. Biol.* **213**, 1–17.

Motlik, J., Pavlok, A., Kubelka, M., Kalous, J., and Kalab, P. (1998). Interplay between cdc2 kinase and MAP kinase pathway during maturation of mammalian oocytes. *Theriogenology* **49**, 461–469.

Nagai, T. (1987). Parthenogenetic activation of cattle follicular oocytes *in vitro* with ethanol. *Gamete Res.* **16**, 243–249.

Nakada, K., Mizuno, J., Shiashi, K., Endo, K., and Miyazaki, S. (1995). Initiation, persistence, and cessation of the series of intracellular Ca^{2+} responses during fertilization in bovine eggs. *J. Reprod. Dev.* **41**, 77–84.

Nakano, Y., Shirakawa, H., Mitsuhashi, N., Kuwabara, Y., and Miyazaki, S. (1997). Spatiotemporal dynamics of intracellular calcium in the mouse egg injected with spermatozoon. *Mol. Hum. Reprod.* **3**, 1087–1093.

Ozil, J. P. (1990). The parthenogenetic development of rabbit oocytes after repetitive pulsatile electrical stimulation. *Development* **109**, 117–127.

Ozil, J. P. (1998). Role of calcium oscillations in mammalian egg activation: Experimental approach. *Biophys. Chem.* **72**, 141–152.

Ozil, J. P., and Huneau, D. (2001). Activation of rabbit oocytes: the impact of the Ca^{2+} signal regime on development. *Development* **128**, 917–928.

Palermo, G. D., Joris, H., Devroey, P., and Van Sterteghem, A. C. (1992). Pregnancies after intracytoplasmic injection of single spermatozoon into an oocyte. *Lancet* **340**, 17–18.

Parrington, J., Swann, K., Shevchenko, V. I., Sesay, A. K., and Lai, F. A. (1996). Calcium oscillations in mammalian eggs triggered by a soluble sperm protein. *Nature* **379**, 364–368.

Parrington, J., Brind, S., De Smedt, H., Gangeswaran, R., Lai, F. A., Wojcikiewicz, R., and Carroll, J. (1998). Expression of inositol 1,4,5-trisphosphate receptors in mouse oocytes and early embryos: The type I isoform is upregulated in oocytes and downregulated after fertilization. *Dev. Biol.* **203**, 451–461.

Patel S., Joseph, S. K., and Thomas, A. P. (1999). Molecular properties of inositol 1,4,5-trisphosphate receptors. *Cell Calcium* **25**, 247–264.

Perez, G. I., Tao, X. J., and Tilly, J. L. (1999). Fragmentation and death (a.k.a. apoptosis) of ovulated oocytes. *Mol. Hum. Reprod.* **5**, 414–420.

Perreault, S. D. (1990). Regulation of sperm nuclear reactivation during fertilization. *In* "Fertilization in Mammals" (B. D. Bavister, J. Cummins, and E. R. S. Roldan, eds.), pp. 285–296. Serono Symposia, Norwell, Mass.

Perry, A. C. F., Wakayama, T., and Yanagimachi, R. (1999). A novel trans-complementation assay suggests full mammalian oocyte activation is coordinately initiated by multiple, submembrane sperm compartments. *Biol. Reprod.* **60**, 747–755.

Perry, A. C., Wakayama, T., Cooke, I. M., and Yanagimachi, R. (2000). Mammalian oocyte activation by the synergistic action of discrete sperm head components: Induction of calcium transients and involvement of proteolysis. *Dev. Biol.* **217**, 386–393.

Poccia, D. L., and Green, G. R. (1992). Packaging and unpackaging the sea urchin sperm genome. *Trends Biochem. Sci.* **17**, 223–227.

Presicce, G. A., and Yang, X. (1994). Parthenogenetic development of bovine oocytes matured *in vitro* for 24 hr and activated by ethanol and cycloheximide. *Mol. Reprod. Dev.* **38**, 380–385.

Rice, A., Parrington, J., Jones, K. T., and Swann, K. (2000). Mammalian sperm contain a Ca^{2+}-sensitive phospholipase C activity that can generate InsP3 from PIP2 associated with intracellular organelles. *Dev. Biol.* **228**, 125–135.

Rongish, B. J., Wu, W., and Kinsey, W. H. (1999). Fertilization-induced activation of phoslopase C in the sea urchin egg. *Dev. Biol.* **215**, 147–154.

Runft, L. L., Watras, J., and Jaffe, L. A. (1999). Calcium release at fertilization of *Xenopus* eggs requires type I IP(3) receptors, but not SH2 domain-mediated activation of PLCgamma or G(q)-mediated activation of PLCbeta. *Dev. Biol.* **214**, 399–411.

Sagata, N. (1997). What does Mos do in oocytes and somatic cells? *BioEssays* **19**, 13–21.

Sakurai, A., Oda, S., Kuwabara, Y., and Miyazaki, S. (1999). Fertilization, embryonic development, and offspring from mouse eggs injected with round spermatids combined with Ca^{2+} oscillation-inducing sperm factor. *Mol. Hum. Reprod.* **5**, 132–138.

Sato, K., Iwao, Y., Fujimura, T., Tamaki, I., Ogawa, K., Iwasaki, T., Tokmakov, A. A., Hatano, O., and Fukami, Y. (1999). Evidence for the involvement of a Src-related tyrosine kinase in *Xenopus* egg activation. *Dev. Biol.* **209**, 308–320.

Sato, K., Tokmakov, A. A., Iwasaki, T., and Fukam, Y. (2000). Tyrosine kinase-dependent activation of phospholipase C gamma is required for calcium transient in *Xenopus* egg fertilization. *Dev. Biol.* **224**, 453–469.

Schatten, G. (1994). The centrosome and its mode of inheritance: the reduction of the centrosome during gametogenesis and its restoration during fertilization. *Dev. Biol.* **165**, 299–335.

Schultz, R. M., and Kopf, G. S. (1995). Molecular basis of mammalian egg activation. *Curr. Top. Dev. Biol.* **30**, 21–62.

Sette, C., Bevilacqua, A., Bianchini, A., Mangia, F., Geremia, R., and Rossi, P. (1997). Parthenogenetic activation of mouse eggs by microinjection of a truncated c-kit tyrosine kinase present in spermatozoa. *Development* **124**, 2267–2274.

Shilling, F. M., Magie, C. R., and Nuccitelli, R. (1998). Voltage-dependent activation of frog eggs by a sperm surface disintegrin peptide. *Dev. Biol.* **202**, 113–124.

Sims, M., and First, N. L. (1993). Production of calves by transfer of nuclei from cultured inner cell mass cells. *Proc. Natl. Acad. Sci. U.S.A.* **90**, 6143–6147.

Steinhardt, R. A., Epel, D., Carroll, E. D., and Yangimachi, R. (1974). Is calcium ionophore a universal activator of unfertilized eggs? *Nature* **252**, 41–43.

Stice, S. L., and Robl, J. M. (1988). Nuclear reprogramming in nuclear transplant rabbit embryos. *Biol. Reprod.* **39**, 657–664.

Stice, S. L., and Robl, J. M. (1990). Activation of mammalian oocytes by a factor obtained from rabbit sperm. *Mol. Reprod. Dev.* **25**, 272–280.

Stricker, S. A. (1997). Intracellular injections of a soluble sperm factor trigger Ca^{2+} oscillations and meiotic maturation in unfertilized oocytes of a marine worm. *Dev. Biol.* **186**, 185–201.

Stricker, S. A. (1999). Comparative biology of calcium signaling during fertilization and egg activation in animals. *Dev. Biol.* **211**, 157–176.

Susko-Parrish, J. L., Leibfried-Rutledge, M. L., Northey, D. L., Schutzkus, V., and First, N. (1994). Inhibition of protein kinases after calcium transient causes transition of bovine oocytes to embryonic cycles without meiotic completion. *Dev. Biol.* **166**, 729–739.

Swann, K. (1990). A cytosolic sperm factor stimulates repetitive calcium increases and mimics fertilization in hamster eggs. *Development* **110**, 1295–1302.

Swann, K., and Lai, A. (1997). A novel signaling mechanisms for generation Ca^{2+} oscillations at fertilization in mammals. *BioEssays* **19**, 371–378.

Swann, K., and Ozil, J. P. (1994). Dynamics of the calcium signal that triggers mammalian egg activation. *Int. Rev. Cytol.* **152**, 183–222.

Swann, K., and Parrington, J. (1999). Mechanism of Ca^{2+} release at fertilization in mammals. *J. Exp. Zool.* **285**, 267–275.

Szalai, G., Krishnamurthy, R., and Hajnoczky, G. (1999). Apoptosis driven by IP_3-linked mitochondrial calcium signals. *EMBO J.* **18**, 6349–6361.

Szöllösi, S., Czolowska, R., Soltynska, M. S., and Tarkowski, A. K. (1986). Remodeling of thymocyte nuclei in activated mouse oocytes: an ultrastructural study. *Eur. J. Cell Biol.* **42**, 140–151.

Szöllösi, S., Czolowska, R., Szöllösi, M. S., and Tarkowski, A. K. (1988). Remodeling of mouse thymocyte nuclei depends on the time of their transfer into activated, homologous oocytes. *J. Cell Sci.* **91**, 603–613.

Takahashi, T., Saito, H., Hiroi, M., Doi, K., and Takahashi, E. (2000). Effects of aging on inositol 1,4,5-triphosphate-induced Ca^{2+} release in unfertilized mouse oocytes. *Mol. Reprod. Dev.* **55**, 299–306.

Tang, T.-S., Dong, J.-B., Huang, X.-Y., and Sun, F.-Z. (2000). Ca^{2+} oscillations induced by a cytosolic sperm factor are mediated by a maternal machinery that functions only once in mammalian eggs. *Development* **127**, 1141–1150.

Tarin, J. J., Albalá-Pérez, S., Aguilar, A., Minarro, J., Hermenegildo, C., and Cano C. (1999). Long-term effects of postovulatory aging of mouse oocytes on offspring: A two-generational study. *Biol. Reprod.* **61**, 1347–1355.

Tarkowski, A. K. (1975). Induced parthenogenesis in the mouse. *Symp. Soc. Dev. Biol.* **33**, 107–129.

Tatone, C., Iorio, R., Francione, A., Gioia, L., and Colonna, R. (1999). Biochemical and biological effects of KN-93, an inhibitor of calmodulin-dependent protein kinase II, on the initial events of mouse egg activation induced by ethanol. *J. Reprod. Fertil.* **115**, 151–157.

Tesarik, J., and Sousa, M. (1994). Comparison of Ca^{2+} responses in human oocytes fertilized by subzonal insemination and by intracytoplasmic sperm injection. *Fertil. Steril.* **62**, 1197–1204.

Tosti, E., Palumbo, A., and Dale, B. (1993). Inositol triphosphate in human and ascidian spermatozoa. *Mol. Reprod. Dev.* **35**, 52–56.

Trump, B. F., and Berezesky, I. K. (1995). Calcium-mediated cell injury and cell death. *FASEB J.* **9**, 219–228.

Turner, P. R., Sheetz, M. P., and Jaffe, L. A. (1984). Fertilization increases the polyphosphoinositide content of sea urchin eggs. *Nature* **310**, 414–415.

Wang, S., Gebre-Medhin, S., Betsholtz, C., Stalberg, P., Zhou, Y., Larsson, C., Weber, G., Feinstein, R., Oberg, K., Gobl, A., and Skogseid, B. (1998). Targeted disruption of the mouse phospholipase C beta3 gene results in early embryonic lethality. *FEBS Lett.* **441**, 261–265.

Ware, C. B., Barnes, F. L., Maiki-laurila, M., and First, N. L. (1989). Age dependence of bovine oocyte activtion. *Gamete Res.* **22**, 265–275.

Whitaker, M. J., and Patel, R. (1990). Calcium and cell cycle control. *Development* **108**, 525–542.

Whitaker, M. J., and Swann, K. (1993). Lighting the fuse at fertilization. *Development* **117**, 1–12.

Whittingham, D. G., and Siracusa, G. (1978). The involvement of calcium in the activation of mammalian oocytes. *Exp. Cell Res.* **113**, 311–317.

Williams, C. J., Schultz, R. M., and Kopf, G. S. (1992). Role of G proteins in mouse egg activation: Stimulatory effects of acetylcholine on the ZP2 to ZP2f conversion and pronuclear formation in eggs expressing a functional m1 muscarinic receptor. *Dev. Biol.* **151**, 288–296.

Williams, C. J., Mehlmann, L. M., Jaffe, L. A., Kopf, G. S., and Schultz, R. M. (1998). Evidence that Gq family G proteins do not function in mouse egg activation at fertilization. *Dev. Biol.* **198**, 116–127.

Wilmut, I., Schnieke, A. E., McWhir, J., Kind, A. J., and Campbell, K. H. S. (1997). Viable offspring derived from fetal and adult mammalian cells. *Nature* **385**, 810–813.

Winston, N. J., and Maro, B. (1995). Calmodulin-dependent protein kinase II is activated transiently in ethanol stimulated mouse oocytes. *Dev. Biol.* **170**, 350–352.

Wolny, Y. M., Fissore, R. A., Wu, H., Reis, M. M., Colombero, L. T., Ergun, B., Rosenwaks, Z., and Palermo, G. D. (1999). Human glucosamine-6-phosphate isomerase, a homologue of hamster oscillin, does not appear to be involved in Ca^{2+} release in mammalian oocytes. *Mol. Reprod. Dev.* **52**, 277–287.

Wolosker, H., Kline, D., Bian, Y., Blackshaw, S., Cameron, A. M., Fralich, T. J., Schnaar, R. L., and Snyder, S. H. (1998). Molecularly cloned mammalian glucosamine-6-phosphate deaminase localizes to transporting epithelium and lacks oscillin activity. *FASEB J.* **12**, 91–99.

Wu, H., He, C. L., and Fissore, R. A. (1997). Injection of a porcine sperm factor triggers calcium oscillations in mouse oocytes and bovine eggs. *Mol. Reprod. Dev.* **46**, 176–189.

Wu, H., He, C. L., and Fissore, R. A. (1998a). Injection of a porcine sperm factor induces activation of mouse eggs. *Mol. Reprod. Dev.* **49**, 37–47.

Wu, H., He, C. L., Jehn, B., Black, S. J., and Fissore, R. A. (1998b). Partial characterization of the calcium-releasing activity of porcine sperm cytosolic extracts. *Dev. Biol.* **203**, 369–381.

Wu, H., Smyth, J., Luzzi, V., Fukami, K., Takenawa, T., Allbritton, N. L., and Fissore, R. (2001). Sperm factor induces $[Ca^{2+}]i$ oscillations by stimulating the phosphoinositide pathway. *Biol. Reprod.* **64**, 1338–1349.

Xu, Z., LeFevre, L., Ducibella, T., Schultz, R. M., and Kopf, G. S. (1996). Effects of calcium-BAPTA buffers and the calmodulin antagonist W-7 on mouse egg activation. *Dev. Biol.* **180,** 594–604.

Xu, Z., Abbott, A., Kopf, G. S., Schultz, R. M., and Ducibella, T. (1997). Spontaneous activation of ovulated mouse eggs: Time-dependent effects on M-phase exit, cortical granule exocytosis, maternal messenger ribonucleic acid recruitment, and inositol 1,4,5,-trisphosphate sensitivity. *Biol. Reprod.* **57,** 743–750.

Yanagimachi, R. (1994). Mammalian fertilization. *In* "The Physiology of Reproduction," 2nd Ed. (E. Knobil and J. D. Neill, eds.), pp. 189–317. Raven Press Inc., New York.

CHAPTER 3

THE NUCLEUS

Fyodor D. Urnov[1] and Alan P. Wolffe

The several months of mammalian embyonic ontogeny from fertilization to birth feature the unfolding of one of the most intricate phenomena in biology—the gradual reprogramming of the zygotic genome into each one of the several hundred cell-type-specific states that it must assume to allow for cell differentiation. With two notable exceptions—cells of the immune system that recombine their immuno-globulin and T cell receptor loci, and gametes—this progressive and ultimately drastic change in gene expression patterns from zygote to terminally differentiated cell in an adult organism occurs without any alteration in primary DNA sequence. Similarly, the process of "reverse nuclear ontogeny," during which a differentiated nucleus is remodeled by the cytoplasm of an enucleated egg into a totipotent state (Kikyo and Wolffe, 2000), also occurs without any concomitant change in the DNA sequence. The extraordinary plasticity of genomic behavior is amply manifested not only in the striking phenotypic differences between genetically equivalent cells (e.g., neurons and myocytes), but in the seeming facility with which some terminally differentiated cells, such as hepatocytes, revert to an undifferentiated state upon appropriate stimuli (e.g., hepatectomy). This is the central biological property of the eukaryotic genome: it does not simply encode the entire proteome of a given species, but also controls its expression in an extraordinarily fluid and dynamic manner, one that simultaneously allows both acutely inducible as well as stable modes of mRNA production by particular genes.

It is clear that prokaryotic paradigms for dynamic gene control do not suffice to explain this property of eukaryotic genomes (Lemon and Tjian, 2000): thus, the engagement of particular stretches of the genome by specific regulatory proteins, such as transcription factors (in a manner analogous to the λ or the *lac* repressor), while very important in eukarya, is only part of a highly intricate regulatory scheme. This additional, very significant, complexity that eukaryotes acquired during evolution stems from their much larger genome size, significantly greater number of genes, the requirement to establish multiple distinct modes of tissue-specific gene expression from the same genome, and, most importantly, evolutionary pressure to effect highly nonbinary, graded gene expression modes responsive to a wide variety of stimuli. It is truly remarkable that the same molecular device yields a solution to all these demands: the union of the genome with histone proteins to form chromatin (Wolffe, 1998). Unfortunately, and in emphatic contradiction to, or in the complete absence of, robust experimental data regarding many of its features, the most pervasive image regarding chromatin is one found in introductory textbooks—a "beads-on-a-string" necklace of nucleosomes, each drawn as a monolithic hockey puck that compacts into an internally homogeneous braidlike entity, and then the braid folds further to form the metaphase chromosome. These drawings—in particular those aspects that pertain to higher order structure—are familiar to people

---

[1]To whom correspondence should be addressed.

with even a passing knowledge of biology, and yet they exclude or obscure many known and essential *in vivo* features of chromatin:

1. Structurally and functionally, the nucleosome is not like a puck and is not a monolithic entity.

2. All nucleosomes are not the same—there is extraordinary idiosyncrasy from locus to locus regarding the chemical state of the histones within nucleosomes, and in the way with which they interact with DNA.

3. Whatever the physical structure of chromatin above the "bead-on-a-string" level, it is not a homogeneous braid, and the genome assumes a wide range of chromatin structure states *in vivo*.

4. At all levels of chromatin compaction, it contains a significant number of nonhistone components that are integral to its structure and function—with only a few exceptions, e.g., the winged-helix transcription factor HNF3 (Cirillo *et al.*, 1998), the high-mobility group proteins (Bustin, 1999), or certain nuclear hormone receptors (Urnov and Wolffe, 2001a,c), we have little structural information on how these proteins become part of chromatin; it is this lack of knowledge—rather than some experimentally established fact—that explains their absence from drawings in reviews and textbooks.

5. Finally, chromosomes do not float in space, but function within the tight spatial constraints of the nucleus, where a complex and poorly understood proteinaceous network (the nuclear matrix) colludes with the nuclear lamina that underlies the inner nuclear membrane to ensure proper chromosome position and behavior.

All of the properties of chromatin and chromosomes described above are essential to effecting dynamic alterations in gene expression; in fact, remarkably, the genome is functionally so diverse precisely *because* it is compacted into chromatin. Thus, transcription factors and other regulatory entities collude with, amend, and are abetted by the nucleoprotein architecture of their template.

In the present chapter, we briefly describe the current state of knowledge, and lack thereof, about the nucleus as the highly functional residence of the genome in the form of chromatin.

DNA AND GENOME STRUCTURE

The nucleus of each human cell contains 8 pg of DNA in the form of 46 separate molecules. Although capable of assuming a wide variety of structural states (Calladine and Drew, 1992; Sinden, 1994), so-called B-form DNA is the predominant one, and will be discussed here. The structure of B-form DNA "immediately suggests" not only its "possible copying mechanism," as very famously noted nearly 50 years ago (Watson and Crick, 1953), but also a number of significant physico-chemical problems that the cell must overcome in order to manage its genome. In aqueous solution of physiological pH and moderate ionic strength, DNA exists as a double-stranded helix with 10.5 base pairs (bp) per helical turn and an axial rise (i.e., distance between two adjacent base pairs) of 0.34×10^{-9} m (Sinden, 1994). Thus, a diploid human genome—which comprises, by latest estimates, 6.4×10^9 bp—is 2.17 m long and yet functions inside a nucleus that is only a few micrometers in diameter. An immediate consequence of such a striking disparity in size between the container (the nucleus) and its contents (the genome) is a requirement for extensive compaction, which reaches its maximum (20,000-fold) during mitosis. The result of this compaction—chromatin—is described later.

As has become apparent from the near-completion of the human genome sequencing projects (Lander *et al.*, 2001; Venter *et al.*, 2001), the primary DNA sequence content of the human genome poses a challenge not only in terms of its sheer length, but also its extraordinary burden of repetitive DNA (more than 50%

by the current estimate), and relative paucity of coding information (less than 5%). This means that a particular nonhistone DNA-binding protein, if challenged with the genome in naked form, would be faced with two copies of its target sequence (assuming the sequence is unique in the haploid genome), and a tremendous excess of "competitor" DNA, much of it repetitive; simple mass-action considerations indicate, therefore, that locus-specific DNA binding would be impossible. In the nucleus, however, a large percentage of repetitive DNA is not easily accessible to binding by nonhistone regulators, because it is assembled into a compact structure termed heterochromatin; thus, in any given cell type, a very significant percentage of the genome is relatively transcriptionally inert and tightly packaged. Chromatin assembly, however, cannot account for the entirety of target specificity displayed by DNA-binding proteins, as evidenced by the remarkable *in vivo* preference of the budding yeast transcription factor Gal4p for only a tiny fraction of its potential binding sites in the yeast genome (Ren *et al.*, 2000). This phenomenon also extends to the SBF/MBF[2] regulators in yeast (Iyer *et al.*, 2001), and cannot be explained by current notions of DNA, chromatin, and nuclear structure.

In addition to problems related to its length and content, the fundamental structure of DNA imposes many requirements on the molecular machines that service it and interpret its contents. At physiological pH, the *outside* of the DNA—its backbone—contains 21 O^- atoms per helical turn; thus, proteins that compact it, the histones, are very rich in amino acids that have positively charged side chains at physiological pH (lysine and arginine). The information content of the DNA, however, resides *inside* the helix—the "base pairs" formed by hydrogen bonds between purines on one strand and pyrimidines on the other are located almost exactly at the geometric center of the helix when viewed in cross-section. The enzymatic machinery, e.g., the polymerases, requires that the hydrogen bonds between the strands be broken and the helix unwound. Many nonhistone DNA-binding proteins can interact with relatively undistorted double-stranded DNA and read its contents by invading one of the two clefts that form continuous grooves on the face of the helix; in many cases, this invasion occurs through the larger of the two clefts, the major groove.

For example, in the most abundant class of DNA-binding proteins in all eukaryotic taxa—the zinc-finger motif—an α-helix arranges itself in the major groove and base-specific recognition is established via electrostatic, hydrogen bond, and hydrophobic interactions between the side chains of the amino acids in the protein helix and particular chemical moieties on the base pairs. Thus, an arginine residue in the protein may establish an electrostatic interaction with a guanine, while an aliphatic side chain of a leucine may engage the methyl group of a thymine in a nonpolar contact (Wolfe *et al.*, 2000).

However a particular protein engages its target sequence in the context of double-stranded DNA, the array of its interaction partners within the nucleic acid molecule is highly asymmetrically distributed; thus, only a particular portion of the surface area on the DNA helix is permissive for binding. The specific manner in which a given DNA stretch is assembled into chomatin will therefore have major consequences for whether a target sequence within this stretch is accessible to a particular regulator *in vivo*—the structural explanation for this phenomenon is provided in the next section, with a description of the structure of chromatin.

DNA-ASSOCIATED PROTEINS IN SOMATIC AND GERM CELLS

As made conspicuous in the famous electron micrographs taken by Ulrich Laemmli of protein-depleted metaphase chromosomes (Paulson and Laemmli, 1977), deproteinized DNA of even a single chromosome brings "tangled web" to a new level of

---

[2]SBF, Swi4/Swi6 cell-cycle box binding factor; MBF, Mlu1 cell-cycle box binding factor.

meaning. Fortunately for our cells, they do not contain naked DNA—all DNA is protein bound at virtually all times. These proteins fall into three major classes: (1) the histones, which includes the four core histones (H2A, H2B, H3, and H4), their variants (such as the specialized centromeric histone H3 called CENP-A and histones H2A.Z and macroH2A), and the linker histones (such as H1 and H5); (2) nonhistone proteins that control chromosome and chromatin structure; among many others, these include architectural components of chromatin, such as the high-mobility group (HMG) proteins (Bustin, 1999), proteins that control the extent of chromatin compaction, such as the condensin SMC proteins (Strunnikov, 1998), and proteins that create a particular structural domain within chromosomes, such as the SIRs and HP1 (Eissenberg and Elgin, 2000; Guarente, 2000); (3) proteins directly related to transcriptional control, such as components of the basal transcriptional machinery and RNA polymerase (at particular loci, such as the rDNA) (Martinez-Balbas et al., 1995; Scheer and Rose, 1984), various transcription factors, such as class II nuclear hormone receptors (Samuels et al., 1982; Urnov and Wolffe, 2001c), and the many chromatin remodeling and modification engines they target, such as histone acetyltransferases (Sterner and Berger, 2000), deacetylases (Khochbin et al., 2001), and ATP-dependent chromatin remodeling complexes (Ahringer, 2000; Sudarsanam and Winston, 2000)—in specific cases, the latter class are thought to remain resident on chromatin after the regulator that targeted them has exited the locus (Cosma et al., 1999).

Rather than descend on the DNA as a chaotic horde, these protein entities assemble into chromatin in an orderly way (Urnov and Wolffe, 2001b), in many cases not well understood. The following discussions briefly review their most salient known features.

THE HISTONES AND THE NUCLEOSOME

The complex and poorly understood transition from viscous and tangled naked DNA to a functioning chromosome inside the nucleus begins with the assembly of each 146 bp of DNA with the eight core histones into a nucleosome core particle (Germond et al., 1975; Kornberg and Thomas, 1974), and continues with the addition of linker histone to form a complex between ~160 bp of DNA and nine molecules of histone protein (this entity is called "the chromatosome"). Approximately 20 bp (this number varies from species to species) will remain in a "linker" between two adjacent nucleosomes. The overwhelming majority of DNA in the nucleus is assembled into such nucleosomes, resulting in ~40 million nucleosomes in each human cell.

Core Histones and the Nucleosome

The core histones, discovered by Albrecht Kossel in avian erythrocytes and first described ~120 years ago in an article with the wonderfully understated title "On a Proteinaceous Component of Cell Nuclei" (Kossel, 1884), share two important things. First, within the family, all core histones share a remarkably lucid common organization of primary and secondary structure: the core histones are small (~110 amino acids), highly basic (~25% lysine and arginine), and organized in two well-defined domains (Fig. 1). On the COOH terminus lies the globular domain: a short α-helix, a loop, a longer α-helix, another loop, and finally another short α-helix. This domain assumes a distinctive secondary and tertiary structure termed "the histone fold": the shorter α-helices lie on top of either end of the longer central helix, approximately perpendicular to its axis. On the NH$_2$ terminus lies the tail domain—its structure is not known, although there are data pointing to some α-helical content (Baneres et al., 1997; Wang et al., 2000). The second thing shared by the histones is the extraordinary across-species conservation between them. For

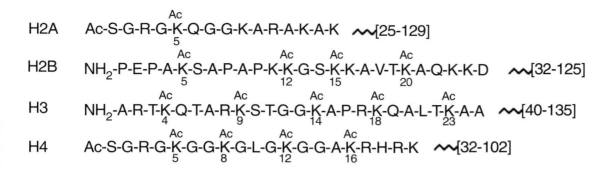

Figure 1 *Bovine (Bos taurus) core histones (Wolffe, 1998). The lower part of the figure shows the overall structural organization of the protein. The amino-terminal tails are shown in the upper half. Some of the lysine residues that are postranslationally modified by acetylation are indicated. From Wolffe (1998).*

example, histones H4 in humans (*Homo sapiens*) and rice (*Oryza sativa*) are identical with the exception of two highly conservative substitutions (valine$_{61}$ → isoleucine and lysine$_{78}$ → arginine). Such extraordinary resistance to mutational pressure doubtless reveals the exceptional functional fit of these proteins.

The unity of structural organization of core histones translated into a precise and invariant sequence of their assembly onto DNA [Fig. 2 presents a simplified version of this process; the reader is encouraged to consult the review by Kaufman and Almouzni (2000) and primary research articles, e.g., Nakagawa *et al.* (2001), for details]. The histone genes present a paradigm of a cell-cycle-regulated gene cluster (Xie *et al.*, 2001); during G$_1$ histone gene transcription is massively up-regulated, large quantities of histone protein are synthesized, and histones H3 and H4 are hyperacetylated on lysine residues within their tail domains by the enzyme histone acetyltransferase (HAT) 1 (Verreault *et al.*, 1998), associate with histone chaperones such as CAF-1 (Verreault, 2000), and then are assembled onto newly synthesized DNA as follows:

1. Histones H3 and H4 heterodimerize via their histone-fold domains—the two longer α-helices at the center of the domain stack against each other, and the shorter helices at the ends abet this interaction in what has been aptly described as a "handshake" (Luger *et al.*, 1997) (Fig. 3).

2. Two such entities then associate into a histone tetramer (only histone H3 mediates this association by forming a four-helix bundle).

3. The histone chaperones then deposit the tetramer onto the DNA, and ~130 bp of DNA is compacted this way, but somewhat loosely, so that the DNA remains permissive for transcription and nonhistone factor binding (Hansen and Wolffe, 1994; Tse *et al.*, 1998a).

4. Chromatin assembly then continues via the chaperone-mediated addition of histones H2A and H2B: these also first heterodimerize with each other via a "handshake" of their central α-helix, and then this heterodimer binds to histone H4 within the H3/H4 tetramer—in striking structural analogy (see item 2 above), a four-helix bundle is formed between histone H2B and histone H4 (Luger *et al.*, 1997). This compacts 146 bp of DNA.

5. Chromatin is "matured" via the action of histone deacetylases (HDACs) that remove the acetyl residues from histone H4 (Ng and Bird, 2000), and the

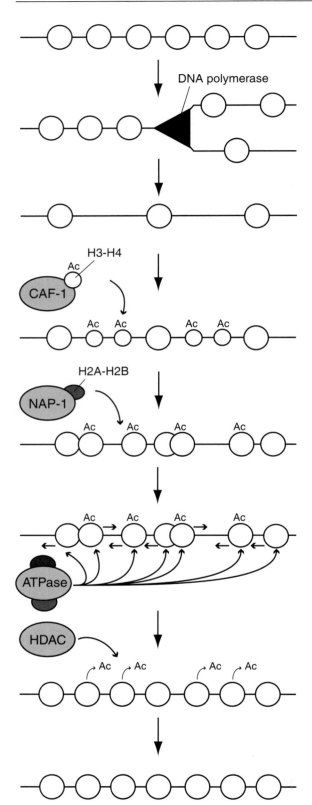

Figure 2 A hypothetical scenario for postreplicative chromatin assembly in vivo [for details, see Kaufman and Almouzni (2000), Nakagawa et al. (2001), and Verreault (2000)].

A

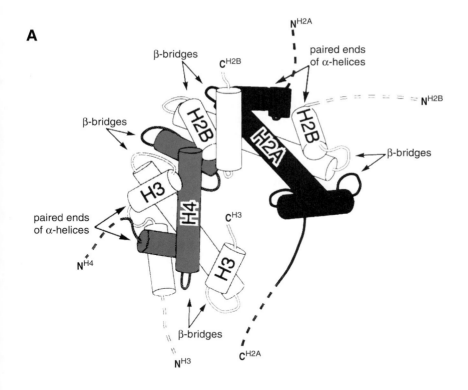

B

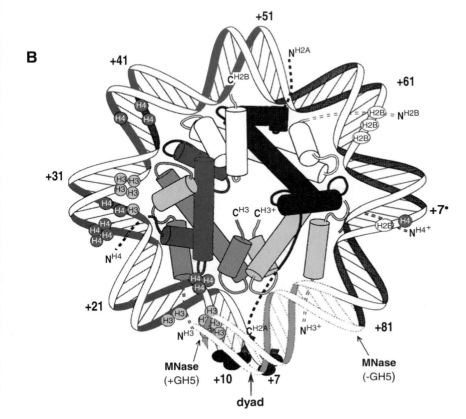

Figure 3 *(A) The histone–histone interface in the nucleosome (Wolffe, 1998). Note the identical "handshake" between histones H3 and H4, and between histones H2A and H2B. Also note that the histones are rotated ~180° relative to the depiction familiar from the X-ray crystal structure (Luger et al., 1997). (B) Core histones within the nucleosome (Luger et al., 1997; Wolffe, 1998). Only one turn of DNA is shown, as is only one copy of histones H4, H2A, and H2B. Two molecules of histone H3 are shown to highlight the four-helix bundle interface in the (H3/H4) tetramer. From Wolffe (1998).*

concomitant nucleosome spacing activity of ATP-dependent chromatin remodeling machines capable of mobilizing nucleosomes (Vignali *et al.*, 2000); the assembly of a single nucleosome by *Drosophila* ACF (Ito *et al.*, 1997) requires ~370 ATP molecules (D. Fyodorov and J. Kadonaga, personal communication). The importance of the spacing activity is particularly apparent because *de novo*-deposited histone octamers must share the DNA with nucleosomes that existed prior to DNA replication; these segregate randomly to the newly synthesized strands and will not assume regular spacing until joined by newly deposited histones and acted on by the spacing ATPase (in the absence of such action, nucleosomes will not form a regular array). Although a great number of HDACs and ATP-dependent chromatin remodeling complexes have been found, the specific enzymes that perform postreplicative chromatin maturation remain to be identified.

6. A molecule of linker histone is added to each nucleosome core particle; the chaperones (if they exist) that mediate this deposition are unknown, and the precise location of the linker histone relative to the histone octamer and the DNA gyres remains a somewhat open issue

It is convenient to consider the "linear" daisy chain of histones in the nucleosome, while remaining acutely aware that this representation only highlights the interaction partners and is completely unrepresentative of the spatial arrangement: H2A—H2B=H4—H3=H3—H4=H2B—H2A. The double lines indicate the four-helix bundle and the single lines denote a "handshake" of the histone fold domains; the heterodimers within which individual histones are engaged in such a handshake are underlined. In terms of ternary structure, the H3/H4 tetramer forms a ∧-shaped core, and the two H2A/H2B heterodimers associate on top of either side of this ∧ to form a wedge. The loops between the α-helices decorate its side and, together with the juxtaposed shorter α-helices at the ends of the histone-fold motifs, form the ramp onto which DNA winds into ~1.7 superhelical left-handed turns. In contrast to the $(H3/H4)_2$–DNA complex, in many cases (see next paragraph) the mature nucleosome core particle presents an obstacle both for binding by most nonhistone factors and for transcription (Wolffe, 1998).

One of the more striking and convincing data sets in the biology of the nucleus comes from experiments with budding yeast strains that contain various mutations in histone genes. Although required for viability, as expected, depletion of histone genes (Wyrick *et al.*, 1999) or point mutations that disrupt the binding of the H2A/H2B heterodimer to the H3/H4 tetramer (Santisteban *et al.*, 1997) have one expected, and one very unexpected, consequence. The former is that transcription driven by a number of genes is increased, which is fully consistent with the repressive role that chromatin assembly plays in gene control (Wolffe, 1998). The latter is that a significant number of genes are *down-regulated* when chromatin is "genetically depleted." There is not a good general explanation for this phenomenon, although in specific cases (e.g., Archer *et al.*, 1992; Schild *et al.*, 1993; Sewack and Hansen, 1997; Urnov and Wolffe, 2001a) the assembly of a highly specific chromatin architecture on a promoter is known to be required for its proper regulation. Thus, histones play a genome-wide and a gene-specific role in transcriptional control. The same holds true of the linker histone as well—once again, the underlying mechanism is not understood.

It is unfortunate that the nucleosome is commonly perceived as a static entity (something like a spool of sewing thread lying in a drawer). Far from it: the DNA is exceedingly structurally distorted in the nucleosome (Hayes *et al.*, 1990)—its helical pitch varies in the structure from 10.0 to 10.7 bp per turn, it is very severely bent, and it spontaneously "breathes" on the surface of the histone octamer, stochastically lifting stretches of itself (Anderson and Widom, 2000; Hayes and Hansen, 2001). In addition, the histone octamer can move ("slide") relative to the

DNA both spontaneously and in an enzyme-facilitated manner (Pazin *et al.*, 1997; Varga-Weisz, 2001). Finally, the interactions between the core histones and between the histones and the DNA can be disrupted in an ATP-dependent manner (Flaus and Owen-Hughes, 2001; Vignali *et al.*, 2000); this yields a highly localized (50–500 bp) disrupted stretch of chromatin termed a "nuclease-hypersensitive site" (Nedospasov and Georgiev, 1980; Wu, 1980), an entity of uncertain structure but one that contains DNA highly accessible to binding by nonhistone factors. All of these phenomena are integral to gene control.

A final disservice done to scholarship in chromatin by the "pucks-on-a-string" drawing is the conspicuous absence from it of core histone NH_2-terminal tails; these project far outside the DNA gyres (Wolffe and Hayes, 1999) but their secondary structure is unclear (Hayes and Hansen, 2001). In their unmodified state, these tails surround the nucleosome with a web of highly positive charge (~44 lysine and arginine residues per nucleosome), and are targets for a large and extraordinarily diverse number of enzymes: acetyltransferases, deacetylases, kinases, phosphatases, methyltransferases, ubiquitin-ligases, etc. (Berger, 2001; Strahl and Allis, 2000). As elaborated below, there is very strong evidence connecting these modifications with structural transitions in the chromatin fiber and with subsequent changes in gene expression levels.

However beautiful and important, the mononucleosome does not exist in nature (except during terminal stages of programmed cell death), in fact, there is very considerable evidence that the functional behavior of the genome as chromatin is greatly affected by levels of compaction above that seen in the nucleosome. This is significantly less understood than the nucleosome, and will be discussed after we describe the other protein components of chromatin.

Male Germ Cell Chromatin: Histones Away

Around the time of his justly celebrated discovery of DNA (Portugal and Cohen, 1977), Friedrich Miescher also discovered a highly basic compound in salmon sperm that he called "protamine" (Miescher, 1874). Some 130 years later, we know this entity to be a complex mixture of small, highly basic proteins essential to one of the more remarkable events in the ontogeny of most metazoa—the remodeling of the nucleus of male gamete precursors into an elongated structure (the sperm head) in which the genome is packaged into a volume that is ~5% of a somatic cell nucleus!

During spermatogenesis, the genome of the haploid gametic precursor is gradually silenced (Steger, 1999), and by the "round spermatid" stage all of the 20 million core histone octamers in the nucleus are physically removed from the DNA and are replaced by two "transition proteins" (TP1 and TP2); these are small (6 and 13 kDa, respectively), are very lysine and arginine rich, and are partly functionally redundant (Yu *et al.*, 2000). Sperm transition proteins and their successors, the protamines, are, arguably, the only proteins related to genome function whose entire primary sequence can be comfortably accommodated in a single line of text. Here, for example, are the sequences of *Mus musculus* transition protein 1 and of *H. sapiens* protamine 1 (PRM1):

TP1:   MSTSRKLKTHGMRRGKNRAPHKGVKRGGSKRKYRKSVLKSRKRGDDASRNYRSHL
PRM1: MARYRCCRSQSRSRYYRQRQRSRRRRRRSCQTRRRAMRCCRPRYRPRCRRH

As can be seen, in TP1 ~40% of amino acids consists of either lysine or arginine, and in PRM1 24 of 51 amino acids are arginine. Although the mechanism of histone replacement by transition proteins is unknown, this event is nothing short of remarkable—in fact, the transition proteins are the only ones known to compete successfully in a nucleus-wide fashion with the histones and the nonhistone DNA-binding proteins (Wolffe, 1998). This competition is particularly notable for its success in the absence of abetting chromatin disruption events such as due to DNA replication or transcription. As such, this is one of the few true nucleus-wide chromatin remodeling events that occur naturally in mammalian ontogeny, and its mechanism, obscure at present, is important for our understanding of other nuclear reprogramming events (Kikyo and Wolffe, 2000).

By the time that transition proteins replace the histones in round spermatids, the nucleus is completely transcriptionally quiescent, and the subsequent chain of events is driven by nontranscriptional regulatory pathways. Specifically, both TP and protamine levels are under translational control: the messages are stored in the cytoplasm in translationally repressed form, and this repression is sequentially released (Zhong et al., 1999). As the gamete precursor progresses from the round spermatid to the elongated spermatid stage (Steger, 1999), action by Ca/calmodulin kinase 4 (Wu et al., 2000)—perhaps the actual phosphorylation of the protamines—leads to the replacement of the transition proteins on sperm chromatin with the protamines [of which there are at least two, both essential (Cho et al., 2001)] and complete genomic quiescence. It is thought that the extraordinary level of compaction of the genome by the protamines is an adaptation that evolved to protect the genome from damage by environmental agents that the sperm may encounter on its voyage toward the female gamete.

It is also possible that the erasure of all parental contribution to the nucleoprotein architecture of the genome has a "resetting" role; because all nonhistone trans-acting factors are removed from the sperm genome, as are all the histones with their attending panel of complex posttranslational modifications, the 23 chromosomes in sperm become something of a regulatory tabula rasa and lose all memory of gene expression states, with only a few loci of persistent epigenetic (nonprotein) modifications (Groudine and Conkin, 1985). This makes the sperm genome fully amenable to assuming a zygotic gene expression state (or any state, for that matter) on fertilization. In this sense, there is a conceptual similarity—although an extraordinary lack of macroscopic and mechanistic likeness!—between totipotency-inducing somatic nuclei remodeling in egg cytoplasm [which is accompanied by the dramatic loss of transcriptional machinery components, such as TBP, from chromatin (Kikyo et al., 2000)] and the removal of all histone and nonhistone regulators from chromatin by the transition proteins and protamines during spermatogenesis: both are nucleus-wide chromatin remodeling events that erase the chromosomes' preexisting nucleoprotein architecture in preparation for embryonic development.

By the same token, the fate of the protamines after fertilization is equally notable: it is thought sperm chromatin decondenses inside the egg and that the protamines are removed by the action of dedicated chaperones, e.g., nucleoplasmin (Philpott and Leno, 1992; Philpott et al., 1991); this event has been successfully modeled in in vitro systems. More importantly, the redeposition of conventional core histones onto the genome provided by the paternal pronucleus occurs concomitant with the reacquisition of transcriptional competence. It is useful to recall that in contrast to zygotic genomes of invertebrates and most vertebrates, where the zygotic genome remains quiescent until a very significant number of embryonic cleavages has occurred (normally, until the so-called midblastula transition), in mammals the genome provided by the paternal pronucleus has to begin transcription after just one round of mitosis. Unfortunately, very little is known about the mechanisms in which the histone redeposition chaperones and the nonhistone regulators (e.g., components of the basal transcriptional machinery) collude in remodeling the previously protamine-laden sperm genome into a conventional haploid chromosome set that is competent for controlling early embryonic development on equal footing with its maternally provided counterpart (which, one might recall, has not had to deal with protamines).

BEYOND THE CORE: ARCHITECTURAL COMPONENTS OF CHROMATIN

Nucleosome assembly converts each 146 bp of DNA (a fiber ~47 nm in length) into a particle that is ~11 nm in diameter, thereby achieving an ~4.3-fold compaction. All subsequent folding of chromatin (essential if it is to function inside the nucleus) occurs on a metanucleosomal level. However poorly understood, it is known to occur through two principal mechanisms: (1) the chemical modification of core histone NH₂-terminal tails that affects the ability of the nucleosomal fiber to condense (Tse *et al.*, 1998b) and (2) the inclusion into, or the action on, the chromatin fiber of additional protein components (in specific cases these become stable components of chromatin and regulate its interphase and metaphase architecture). A surprising finding, elaborated for each class of protein in the cognate section below, is that the genetic ablation of these presumably "constitutive" and "intrinsic" components of chromatin, rather than misregulate the entire transriptome, leads to highly localized lesions in genome behavior (Barra *et al.*, 2000; Calogero *et al.*, 1999; Shen *et al.*, 1995). These data indicate that "general" chromatin components can have locus-specific roles.

Linker Histone

The best understood such additional component is the linker histone, of which histone H1 is most common. In very interesting contrast to the core histones, it is not required for viability of unicellular eukaryotes such as budding yeast (Hellauer *et al.*, 2001) or the ciliate *Tetrahymena* (Shen *et al.*, 1995); instead, its genetic ablation in those organisms leads to defects in the transcriptional regulation of particular genes. The molecular basis of this phenomenon is not understood, but it is unlikely to result from a sequence-specific association of linker histone with DNA, because no sequence preference for chromatin binding has been reported for this molecule.

Several issues in the biology of linker histone are controversial, in particular the physical location of the linker histone molecule relative to the core histone octamer and the DNA gyres (Wolffe, 1998) (Fig. 4); it was originally thought that the linker histone makes relatively symmetrical contacts with DNA entering and exiting the nucleosome, but other data point to an asymmetric location of the protein within the nucleosome (Pruss *et al.*, 1996). Additional support for this model comes from structural analysis of a nonhistone protein, the transcription factor HNF3. In terms of secondary structure, it is almost congruent to the linker histone (both DNA-binding domains assume the "winged helix" motif) (Clark *et al.*, 1993; Ramakrishnan *et al.*, 1993). Interestingly, HNF3 asymmetrically interacts with DNA on just one side of the nucleosome (Cirillo *et al.*, 1998).

Whatever the details of linker histone binding to nucleosomes, it has a local and global effect on chromatin structure. The former is that an additional 15 bp of DNA are compacted, and the mobility of the histone octamer relative to the DNA is significantly restricted (Ura *et al.*, 1995). The latter is that the nucleosomal fiber becomes much more amenable to folding (Wolffe, 1998). There are data indicating that the inclusion of the linker histone into the nucleosomal fiber overcomes a certain component of the intrinsic repulsion that the negatively charged DNA backbone has for other DNA stretches by binding to DNA in a way that shields parts of this charge (Dou and Gorovsky, 2000; Dou *et al.*, 1999).

In marked contrast to core histones, which are known—in the absence of chromatin-disruptive activity—to remain relatively stably bound to the DNA (Kimura and Cook, 2001), linker histone association with chromatin is highly dynamic *in vivo*, and the protein constantly shuttles onto and off of the DNA (Lever *et al.*, 2000; Misteli *et al.*, 2000). These data were generated using *in vivo* laser ablation

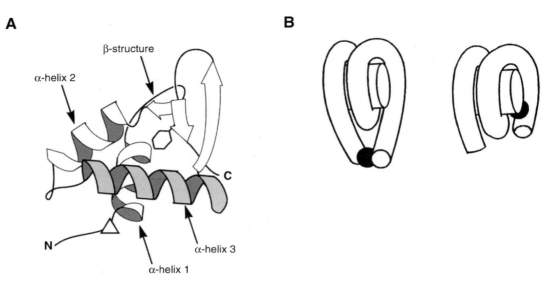

Figure 4 *(A) Schematic representation of the structure of linker histone H5 (Ramakrishnan et al., 1993; Wolffe, 1998). (B) Two alternative models for linker histone location relative to the nucleosome core particle. From Wolffe (1998).*

and imaging techniques known as "fluorescence recovery after photobleaching (FRAP)" (Poo and Cone, 1974) and "fluorescence loss in photobleaching (FLIP)" (Ellenberg *et al.*, 1997). This very surprising finding has been extrapolated to other proteins previously thought to be relatively stably associated with chromatin, such as the glucocorticoid receptor (McNally *et al.*, 2000), estrogen receptor (Shang *et al.*, 2000; Stenoien *et al.*, 2001), and the nucleosome-binding protein HMG-17 (Phair and Misteli, 2000). For their paradigm-rescinding quality, these data rightfully inspired comparisons to Galileo's "In fact, it does move!" (Pederson, 2001). The significance of this mobility is not immediately known, but by way of conjecture, given the earlier mentioned effect of linker histone association on chromatin folding, this suggests that its hit-and-run binding may result in marked instability and fluidity in chromosome structure—at least to the extent that histone H1 is capable of influencing it. Constantly "breathing" chromatin, in turn, may be more amenable than a static structure to receving, interpreting, and rapidly responding to signal transduction stimuli related to genome behavior (Wolffe and Hansen, 2001).

HMG Proteins

One important problem raised by this newly discovered transience in chromatin binding by its "integral" components is the issue of what *bona fide* components of chromatin are. The name of the high-mobility group proteins reveals these proteins' behavior in electrophoretic analysis; the HMGs illustrate this problem quite well. In addition to relatively small size (10–25 kDa), these ubiquitous, abundant proteins (Bustin, 1999) have historically been considered to be architectural components of chromatin (Bustin and Reeves, 1996). Based on structural properties of their DNA/chromatin-binding domain, the HMGs are divided into three subfamilies [in a helpful development to those outside this field, these have been recently renamed according to an internally consistent nomenclature (Bustin, 2001)]: (1) HMG-box proteins (HMGBs), such as HMGB1 and 2 [formerly HMG-1/2 (Bustin, 2001)], that contain a very distinctive motif of three α-helices that fold into an L-shape and bind DNA via the minor groove, leading to its unwinding and bending; (2) nucleosome binding HMG proteins (HMGNs), such as HMGN1 and 2 [formerly HMG-14/17 (Bustin, 2001)], that these contain a 30-amino acid stretch of basic amino acids that allow the protein to prefer binding to nucleosomes rather than to naked DNA (a relatively unique property in the known proteome), which in the context of the chromatin fiber may lead to its unfolding; (3) HMG proteins that contain the "AT-hook" domain (HMGAs), such as HMGA1 and 2 (formerly HMG-I and HMG-Y); the AT-hook consists of a core glycine-arginine-proline motif flanked by multiple arginine residues, invades DNA via the major groove, and in some cases is known to lead to "DNA unbending" (Merika and Thanos, 2001).

Apart from abundance, a property shared by the prototypical HMGs is sequence-nonspecific binding to DNA (HMGB1/2 and HMGA1/2) and chromatin (HMGN1/2); it is important to make the distinction here between the "prototypical" HMG proteins to which this statement applies and the very many sequence-specific DNA-binding proteins that have *an HMG-like DNA-binding domain*). In the case of HMGN1/2, which binds nucleosomes, there is *in vivo* confirmation of such lack of primary sequence specificity (Shirakawa *et al.*, 2000), but HMGA1/2 appears to be broadly distributed through metaphase chromosomes (Saitoh and Laemmli, 1994), where it likely performs some currently unidentified structural role. Furthermore, HMGB1 can replace linker histone in chromatin (Nightingale *et al.*, 1996; Ura *et al.*, 1996) and is sufficiently abundant to engage ~5% of the nucleosomes in the nucleus (Goodwin *et al.*, 1977).

However abundant, the HMGs have proved remarkably resistant to experimental attempts to define their *in vivo* role in chromatin structure and genome behavior. For example, in contradiction to expectation, mice homozygous for a deletion of HMGB1 are able to complete embryonic and fetal development (Calogero *et al.*, 1999; Wolffe, 1999). Thus, contrary to earlier well-justifed predictions based on its abundance and sequence-nonspecific DNA binding, HMGB1 is not required for normal chromosome behavior—in fact, as revealed by their postnatal death due to hypoglycemia, HMGB1-deficient mice are impaired in responding to glucocorticoids. These surprising data indicate that a very abundant, sequence-nonspecific DNA-binding protein is required *in vivo* to drive expression of only a few specific genes (i.e., those that respond to the glucocorticoid receptor)—perhaps by engaging, in cooperation with other proteins, GR-responsive loci to yield a nucleoprotein architecture that is stable over time (Bustin, 1999; Wolffe, 1999). For a different HMG protein, HMGA1, there is evidence that such a structure does, in fact, assemble *in vivo* over the interferon-β promoter (Merika and Thanos, 2001). The mechanism of locus selectivity of HMGB1 action, however, remains unclear, and may involve a "seeing-eye dog" mechanism [the idea, but not the term, is taken from (Bustin, 1999)], where protein–protein interactions with a sequence-specific DNA-binding protein, such as a transcription factor, "guide" the HMG protein to a particular locus, where it then exerts its structural effect on the DNA.

Thus, the chromatin fiber contains a large number of protein components that are not core histones and that by themselves do not appear to possess any primary DNA-binding specificity. These proteins are in principle sufficiently abundant to be general chromatin components, and strong data exist that they affect global chromatin structure; for example, histone H1 deletion in the filamentous fungus *Ascobolus* has a truly spectacular effect and markedly increases overall chromatin accessibility to micrococcal nuclease (Barra *et al.*, 2000). Importantly, these proteins are exceedingly mobile relative to chromatin (Wolffe and Hansen, 2001) and within the nucleus (Pederson, 2001), and some, like certain HMGs, can be removed from chromatin under conditions of relatively moderate ionic strength (~350 mM NaCl), similar to most transcription factors (Bustin, 1999); thus, their liaison with chromatin is transient. In addition, genetic ablation of these proteins leads not to some genomic catastrophe, but instead affects the expression of selected genes (Calogero *et al.*, 1999; Shen *et al.*, 1995) even when it has a global effect on chromatin structure (Barra *et al.*, 2000). The surprising conclusion from all these data, therefore, is that the nucleus uses these seemingly nonspecific DNA-binding entities both in a genome-wide way, and also to assemble unique chromatin structures at specific loci.

REGULATORY NONHISTONE COMPONENTS OF CHROMATIN

In addition to proteins that establish the structure of chromosomes, the nucleus contains a wealth of regulators that bind DNA and/or chromatin with various degrees of permanence.

Transcription Factors

All transcription factors can bind naked DNA, and also DNA within stably remodeled stretches of chromatin (e.g., DNAse I-hypersensitive sites)—this includes TBP, general transcription factors such as NF-1, Sp1, and C/EBP, and the innumerable gene- and cell-type-specific regulators coded for by mammalian genomes. In contrast, the ability to interact with a mature, unremodeled chromatin fiber, and thus become a relatively stable component of chromatin, is the realm of relatively few regulators. We review here three examples in which this capacity is thought to subserve an important aspect of the protein's biological function.

The "winged helix" transcription factor hepatocyte nuclear factor 3 (HNF3) is strikingly similar in overall structure to linker histone (Clark *et al.*, 1993). As a consequence, HNF3 has the remarkable property of binding more stably to nucleosomes than to naked DNA (Cirillo and Zaret, 1999). It asymmetrically engages the nucleosome in the linker histone binding site (Cirillo *et al.*, 1998) and thus leads to the stabilization of the histone octamer relative to the DNA in a particular position (Shim *et al.*, 1998). Such restrictive effects on nucleosome mobility are a known consequence of linker histone action on chromatin (Pennings *et al.*, 1994; Ura *et al.*, 1995). The important distinction here is that linker histones bind sequence nonspecifically, but HNF3 is a highly sequence-specific DNA binding protein! Thus, it is expected that such "mobility-restricting" action of HNF3 will occur only on a very small set of loci in the genome. Indeed, in the case of the enhancer of the mouse serum albumin gene (Zaret, 1995), HNF3 is known to engage a particular site within it, stabilize nucleosome position, and thus allow the assembly on the adjacent DNA of a nucleoprotein complex that promotes albumin transcription in a hepatocyte-specific manner (McPherson *et al.*, 1993; Shim *et al.*, 1998). The presumed rationale of HNF3's unique engagement of chromatin is that during early development, prior to the formation of the liver as a distinct organ, HNF3 invades the albumin locus, which at this point is assembled into mature, unremodeled chromatin, stabilizes nucleosomes over the enhancer in a particular position, and facilitates their subsequent remodeling by other factors (Cirillo and Zaret, 1999).

A conceptually similar functional relationship with chromatin is established by certain members of the nuclear hormone receptor superfamily—for example, the thyroid hormone receptor (TR) (Zhang and Lazar, 2000). This protein—like its many relatives, a ligand-regulated transcription factor that contains a zinc-finger protein DNA-binding domain—is a stable component of chromatin (Samuels *et al.*, 1982) and binds its target sites when presented to it on the surface of a nucleosome (Urnov and Wolffe, 2001a; Wong *et al.*, 1997). After binding to chromatin, in the absence of ligand TR drives robust transcriptional repression via multiple, functionally distinct mechanisms (Urnov *et al.*, 2000; Wong *et al.*, 1995, 1998). Thus, the rationale of TR binding to and residing on mature chromatin is to maintain repression by remaining on its target locus, and to respond rapidly to hormone by relieving this repression and up-regulating transcription. In the case of one promoter naturally responsive to TR, it is known that chromatin assembly actually potentiates binding by TR to its target response elements (Urnov and Wolffe,

2001a), highlighting the profound mutual adaptation of promoter DNA sequence, chromatin structure over it, and the structure of regulators that bind this chromatin template.

A requirement for stable maintenance of the repressed state also underlies the invasion of chromatin by the methylated DNA-binding proteins [the MBDs (Bird and Wolffe, 1999)]. Human DNA is markedly depleted of the dinucleotide CpG, and where present, CpG is very frequently symmetrically (i.e., on both strands of the DNA) methylated on position 5 of the cytosine residue, to yield 5mCpG (Robertson and Wolffe, 2000). A very large body of data implicates DNA methylation-coupled transcriptional repression to the inactivation of repetitive DNA (Bestor, 1998), monoallelic silencing of imprinted loci and dosage-compensated loci (Bartolomei and Tilghman, 1997; Panning and Jaenisch, 1998), and the aberrant silencing of gene promoters in human neoplasia (Robertson and Jones, 2000). The tiny methyl group exerts such striking effects via being bound by the MBDs that selectively bind 5mCpG rather than CpG. Once bound, the MBDs target the enzyme histone deacetylase (Jones *et al.*, 1998; Nan *et al.*, 1998) to repress transcription. One of the remarkable features of the MBD–DNA interface is its very small area (only 800Å^2, which is ~60% of what one expects from most protein–DNA complexes (Ohki *et al.*, 2001)). This explains the known ability of some MBD proteins to stably bind nucleosomes (Chandler *et al.*, 1999), because the protein needs only a small surface on the DNA major groove to bind it, and thus can attach to the half-turn of the DNA's major groove visible to it on the surface of the nucleosome (Wade and Wolffe, 2001).

From a perspective of somatic cloning, an important general point these three examples provide is that chromatin in a somatic nucleus contains not only active gene promoters, bound by TBP or related factors (Veenstra *et al.*, 2000) that can be removed during global nuclear remodeling (Kikyo *et al.*, 2000), but extensive stretches of silent chromatin stably bound by nonhistone regulators. Very little is known about how, or even if, these are affected by the reprogramming event.

Enzymes That Chemically Modify Components of Chromatin

Structural transitions in chromatin underlie every phenomenon in genome control. First evidence to this effect came from observations of massive localized insect polytene chromosome decondensation during gene induction by steroid hormones (Clever and Karlson, 1960), and from cytological and genetic data indicating that the transcriptionally inactive X chromosome in mammalian females forms a highly condensed Barr body (Lyon, 1961; Ohno *et al.*, 1959). Some 40 years later, we know these to be representative of two generalized phenomena in genome biology: (1) the synergy between ATP-dependent chromatin remodeling machines, such as SWI/SNF (Sudarsanam and Winston, 2000), and histone modifying enzymes, such as histone acetyltransferases (HATs) (Sterner and Berger, 2000), in transcriptional activation, and (2) a functional collaboration between different types of ATP-dependent chromatin remodeling machines, such as ISWI-containing complexes (Goldmark *et al.*, 2000; Vignali *et al.*, 2000), and histone deacetylases (Ng and Bird, 2000), in achieving transcriptional repression. A partial listing of complexes containing these enzymes that have been identified in mammalian cells is presented in Table 1.

Almost 40 years ago it was discovered that histones are covalently modified *in vivo* (Allfrey *et al.*, 1964). The major such modification—acetylation—occurs on the ε-amino group of certain lysine residues within the core histone tails, and is effected by histone acetyltransferases (Marmorstein and Roth, 2001; Sterner and Berger, 2000). The first HAT was cloned from the ciliate *Tetrahymena* (Brownell *et al.*, 1996) and—in testimony to the taxonomical ubiquity of chromatin modification—was found to have a yeast homolog called Gcn5p, a known transcriptional

Table 1 Complexes Formed by Chromatin Modifying and Remodeling Enzymes in Vertebrates[a]

Name	Core subunit(s) (enzymatic activity)	Other subunits	References
hGcn5/PCAF	P/CAF (HAT)	p400 (ATM), ADA2/3, Spt3, TAFs	Ogryzko et al., 1998; Vassilev et al., 1998
Tip60	Tip60 (HAT), RuvB (helicase)	p400, RuvB-like, β-actin, BAF53	Ikura et al., 2000
Sin3	HDAC1 and HDAC2	Sin3, RbAp48/46	Zhang et al., 1997
N-CoR/SMRT complex	HDAC3	N-CoR/SMRT, TBL1	Guenther et al., 2000; Li et al., 2000
BRG1 (hSWI/SNF)	BRG1 (ATPase)	β-actin, BAFs: 250, 155/170, 60, 53, 47 (INI1/SNF5)	Phelan et al., 1999; Wang et al., 1996a,b
hBrm	hBRM (ATPase)	Highly similar	Kingston and Narlikar, 1999; Sudarsanam and Winston, 2000
ACF	ISWI (ATPase)	ACF1, p175	Guschin et al., 2000a
CHRAC	ISWI (ATPase)	ACF1, p17, p15	Poot et al., 2000
Mi-2	Mi-2 (ATPase), HDAC1	MTA1, p66, RbAp48, MBD3	Wade et al., 1998, 1999
NuRD/NRD	Same	Same, plus a number of other polypeptides	Tong et al., 1998; Zhang et al., 1998
DNMT1/Rb	DNMT1 (DNA methyltransferase), HDAC1	Rb, E2F	Robertson et al., 2000

[a]See Sterner and Berger (2000) for a review on HATs, and Sudarsanam and Winston (2000) for a discussion of ATPases. The list is not comprehensive, and deliberately includes only those complexes for which a conventional biochemical purification scheme from crude extract has validated observations made by other approaches.

coactivator (Berger et al., 1992; Georgakopoulos and Thireos, 1992). This was the first data set illuminating a known correlation between locus-specific histone hyper-acetylation and transcriptional activity (Bone et al., 1994; Hebbes et al., 1988), an observation subsequently much expanded (Kuo et al., 1998; Parekh and Maniatis, 1999; Strahl and Allis, 2000). Importantly, HATs such as Gcn5 (Xu et al., 2000) and CBP/p300 (Yao et al., 1998) are broadly required for mammalian development, whereas other HATs, for exampe SRC-1, are involved in gene-specific regulatory pathways (Weiss et al., 1999; Xu et al., 1998).

A comparison of the wealth of HATs populating the nucleus yields several common themes. In vivo, these enzymes occur in large, multisubunit complexes, and the function of the additional subunits is not entirely clear. In specific cases, some of these are thought to enable one of the central phenomena in transcriptional activation: the targeting of the HAT-containing complex to a DNA-bound activator [for example, the Tra1p/TRRAP subunit of several HAT complexes in yeast and in humans mediates an interaction with acidic activation domains (Brown et al., 2001)]. Once targeted, the HATs exhibit a distinct preference for acting on particular lysine residues in specific histone tails (Sterner and Berger, 2000), and it is thought that the resulting idiosyncratic (i.e., HAT-specific) pattern of acetylation may communicate, via a "histone code" (Strahl and Allis, 2000), a message to other regulators that bind chromatin, and also that it affects chromatin structure. In addition to the core histones, HATs can act on nonhistone substrates—for example, components of the basal transcriptional machinery (Imhof et al., 1997; Roth et al., 2001). Finally, many HATs contain a conserved motif termed a "bromodomain" (Winston and Allis, 1999), which can bind the acetylated histone tail (Dhalluin et al., 1999; Jacobson et al., 2000)—i.e., the product of the reaction catalyzed by the HAT. The potential significance of this rather unusual arrangement is discussed in the next section.

The functional antagonists of the HATs are the histone deacetylases (Khochbin *et al.*, 2001; Ng and Bird, 2000). In striking parallel to the history of the discovery of the HATs (see above), when the first HDAC was purified from mammalian cell extract and subsequently cloned (Taunton *et al.*, 1996), it was found to have a homolog in budding yeast, Rpd3p, known to be involved in transcriptional control (Vidal *et al.*, 1990; Vidal and Gaber, 1991). Transcriptionally silenced loci—e.g., the inactive X chromosome in female cells (Jeppesen and Turner, 1993), other epigenetically silenced loci (Braunstein *et al.*, 1993), or genes bound by transcriptional repressors (Magdinier and Wolffe, 2001; Rundlett *et al.*, 1998)—are deacetylated *in vivo*, completing the "HAT = activation/HDAC = repression" paradigm.

As of June 2001, mammalian genomes are thought to contain 15 HDAC genes (Khochbin *et al.*, 2001) that are divided into three classes based on homology to founding members from the budding yeast genome: class I [related to Rpd3p (Taunton *et al.*, 1996; Vidal and Gaber, 1991)], class II [related to yeast Hda1p (Carmen *et al.*, 1996; Grozinger *et al.*, 1999)], and class III [related to Sir2p (Frye, 1999; Imai *et al.*, 2000)]. It is not known whether any HDAC exists as an uncomplexed monomer *in vivo*, but the predominant form of all these enzymes is in complexes with other proteins; for example, the major form of HDAC1 is a six-subunit complex called Mi-2 (Wade *et al.*, 1998; Ahringer, 2000) [in subsequent studies described as NRD/NuRD (Tong *et al.*, 1998; Zhang *et al.*, 1998)], and the major form of HDAC3 is a three-subunit nuclear receptor corepressor complex (Guenther *et al.*, 2000; Urnov and Wolffe, 2001c). In the former case, the complex unites the deacetylase with an ATP-dependent chromatin remodeling engine (the Mi-2 protein) that can slide histone octamers relative to the DNA (Guschin *et al.*, 2000b); the entire complex can create a localized domain of superhelical torsion within chromatin fibers (Havas *et al.*, 2000).

There is an intuitively obvious connection between the neutralization of positive charge of the histone tails and their ability to interact with the negatively charged DNA backbone, and there clearly exists a causal relationship between histone acetylation/deacetylation and the transcriptional activity of the DNA that resides in the locus being modified. Interestingly, the mechanism effecting this causality is unclear, and may involve both localized effects [i.e., control of nonhistone regulatory factor access to their binding sites in chromatin (Lee *et al.*, 1993; Vettese-Dadey *et al.*, 1996)] and more global effects on the level of higher order chromatin folding (Hebbes *et al.*, 1994; Tse *et al.*, 1998b). This very important issue is the current focus of intense investigation. With respect to acetylation of nonhistone targets, in specific cases it is thought to control the targeting to promoters of other regulators, whether DNA bound (Boyes *et al.*, 1998) or tethered via protein–protein interactions (Chen *et al.*, 1999b).

An exciting recent development has been a growing appreciation of the biochemistry and functional relevance of enzymes that effect nonacetylation-type modifications of chromatin (Berger, 2001; Hans and Dimitrov, 2001; Stallcup, 2001). Interestingly, and in strong support of the "histone code" model, the same chemical modification can have diametrically opposite effects: for example, methylation of histone H3 on lysine 9 creates a binding site for the heterochromatin protein HP1 (Lachner *et al.*, 2001) and thus is associated with transcriptional silencing (Nakayama *et al.*, 2001). Conversely, methylation of arginine 3 in histone H4 is driven by certain nuclear receptor coactivators (Chen *et al.*, 1999a; Strahl *et al.*, 2001) and leads to transcriptional activation (Wang *et al.*, 2001). An even more remarkable example is provided by serine 10 in histone H3: this modification occurs both during mitosis, where it is driven by kinases that promote mitotic chromosome condensation (De Souza *et al.*, 2000; Hsu *et al.*, 2000), and during gene activation in interphase (Lo *et al.*, 2000; Nowak and Corces, 2000). The mechanism whereby the same modification of the same residue can occur during, and presumably have functional consequences for, both chromatin condensation and gene activation (known to be associated with chromatin decondensation) is unclear. It is equally uncertain how the "histone code"—i.e., the targeting to idiosyncratically modified chromatin loci of other regulators—and direct effects on chromatin structure collude in mediating the functional effects of these modifications on gene behavior.

ATP-Dependent Chromatin Remodeling Engines

In the early 1970s it became apparent that the genome's basic and ubiquitous unit of structure *in vivo* is the nucleosome (Germond *et al.*, 1975; Hewish and Burgoyne, 1973; Kornberg and Thomas, 1974; Woodcock, 1973). The late 1970s saw the discovery of what appeared to be "nucleosome-free" regions of the genome (Nedospasov and Georgiev, 1980; Varshavsky *et al.*, 1978, 1979; Wu, 1980), when the probing of various target loci with nucleases revealed preferential accessibility of certain stretches ("nuclease-hypersensitive sites") within chromatin *in vivo*. At the time, available evidence correctly suggested that these represented DNA stretches active in genome control (e.g., controlling DNA replication or transcription). It was then discovered that the extent of such targeted chromatin remodeling is dynamic and positively correlates with gene expression levels (Zaret and Yamamoto, 1984). Subsequent analysis confirmed that with very few exceptions, promoters and enhancers of active genes occur within nuclease hypersensitive sites *in vivo* (Elgin, 1988; Gross and Garrard, 1988).

It was obvious that these DNA sequences were bound by nonhistone regulatory factors, but it was not clear whether their increased sensitivity to nucleases was due to simple exclusion of the core histones from a particular DNA stretch (i.e., a competition mechanism), or if some other events were taking place. The situation began to clear up when independent efforts to characterize genes mutations that make budding yeast incapable of activating inducible promoters (Neigeborn and Carlson, 1984; Stern *et al.*, 1984) unexpectedly led to the same set of loci (Peterson and Herskowitz, 1992), most prominently, a gene called *SWI2/SNF2* (the name derives from the phenotype of yeast mutant fot this locus—these strains cannot <u>swi</u>tch mating type, and are <u>s</u>ucrose <u>n</u>onfermenters). The functional relationship between this gene (and several others that when mutated endowed yeast with the same phenotype) and transcriptional activation was revealed in a pioneering set of experiments that showed how specific mutations in histone genes, or a reduction in histone content, could alleviate the effect of the *swi/snf* mutations on genes (Hirschhorn *et al.*, 1992). These experiments indicated that products of the *SWI/SNF* genes act to *alleviate* the repressive effects that chromatin has on gene activity (this is why mutating the histones obviates the need for the products of the *SWI/SNF* genes).

The exact mechanism of this action became apparent with the purification of yeast and mammalian SWI/SNF complexes (Cote *et al.*, 1994; Imbalzano *et al.*, 1994; Kwon *et al.*, 1994), which turned out to be multisubunit entities (Sudarsanam *et al.*, 1999) with a core subunit (Swi2p/Snf2p) capable of using the energy derived from ATP hydrolysis to remodel a conventional nucleosome into an entity of uncertain structure in which the DNA was significantly more accessible to binding by nonhistone factors. A causal relationship between SWI/SNF action and the formation of nuclease-hypersensitive sites was then firmly established for several loci in the yeast genome (Gregory *et al.*, 1999; Wu and Winston, 1997), and robust evidence also implicated mammalian SWI/SNF in the action of transcription factors known to remodel chromatin *in vivo* (Fryer and Archer, 1998).

The mechanism of SWI/SNF remodeling of chromatin, the structural nature of chromatin in nuclease-hypersensitive sites, and the relationship between remodeled nucleosomal templates produced by SWI/SNF *in vitro* and remodeled chromatin *in vivo* have been the subjects of intense investigation and debate (Sudarsanam and Winston, 2000; Vignali *et al.*, 2000). A very important development is the discovery (Gavin *et al.*, 2001; Havas *et al.*, 2000) that when SWI/SNF acts on chromatin fibers (its natural substrate), it twists chromatin to create a localized constrained domain of superhelical torsion. One presumes, then, that such twisting somehow translates into increased accessibility of the DNA to nonhistone factors, and also likely affects higher order chromatin folding. As described in the next section, SWI/SNF is known to be targeted to promoters by factors that can bind unremodeled chromatin, and after being targeted, it remodels histone–DNA contacts to make the DNA accessible to other regulators [this is referred to as "bimodal" gene regulation (Archer *et al.*, 1992)].

Genetic analysis of SWI/SNF action in budding yeast (Holstege *et al.*, 1998; Sudarsanam *et al.*, 1999) and in *Drosophila* (Collins and Treisman, 2000) yielded the very unexpected observation that it is required for the repression of some target genes as well as activation of a number of others. In budding yeast, there is conclusive genetic evidence that such action is direct (Sudarsanam *et al.*, 1999); a related ATP-dependent chromatin remodeling complex, RSC (Cairns *et al.*, 1996), is also required for the repression of some target genes (Moreira and Holmberg, 1999).

These observations, however difficult to account for in simple mechanistic terms (it is unclear how action by a complex that is thought to make DNA *more* accessible can lead to *repression*), receive strong support from genetic and biochemical analysis of other SWI2/SNF2-related ATP-dependent chromatin remodeling proteins in various species, and of the complexes in which these enzymes exist. Thus, the most prominent member of the CHD family of ATPases, the Mi-2 protein (the name stems from its identification as an autoantigen in dermatomyositis, an autoimmune disorder), exists in a complex with five other polypeptides, including, most prominently, the enzyme histone deacetylase 1 (HDAC1) (Wade *et al.*, 1998). Subsequent to its initial characterization in *Xenopus,* this complex was identified in mammals as well (Tong *et al.*, 1998; Zhang *et al.*, 1998) and was designated as NRD/NuRD. There is no conclusive evidence directly connecting this complex to repression or any regulatory phenomenon driven by any transcriptional or other regulator *in vivo*, but the well-established role of histone deacetylases in repression (Khochbin *et al.*, 2001; Ng and Bird, 2000) offers a strong indication that the Mi-2 (NRD/NuRD) complex may contribute to the assembly of a repressive chromatin architecture at some loci (although at present no *bona fide* Mi-2 target loci have been identified).

An different ATPase related to the SWI2/SNF2—its name, ISWI (imitation switch), reflects this relatedness—has been proved to be a transcriptional corepressor in budding yeast (Goldmark *et al.*, 2000). In this system, a combination of genetic and chromatin structure analysis showed that certain genes are inappropriately activated in the absence of ISWI function, and connected this activation to the erroneous repositioning of histone octamers over the promoters of these genes. There is an interesting evolutionary analogy to the physical association between an ATPase and HDAC1 in *Xenopus* and in mammals: in budding yeast, repression by ISWI is known to synergize with histone deacetylation driven by the yeast homolog of HDAC1, the enzyme Rpd3p (Goldmark *et al.*, 2000)—the distinction being, however, that the Mi-2 complex contains both an ATPase and an HDAC, whereas in yeast these are in separate complexes that are independently targeted by the same promoter-bound transcriptional repressor.

The biochemistry of the ISWI protein is remarkably complicated. It is found in three separate complexes in budding yeast (Tsukiyama *et al.*, 1999) and in at least three distinct complexes in metazoa (these include CHRAC, NURF, and ACF) (Vignali *et al.*, 2000). Unfortunately, evidence as to their function *in vivo* is currently lacking, and the lethality of ISWI mutations in *Drosophila* (Deuring *et al.*, 2000) aggravate this predicament. *In vitro*, the ISWI protein can effect "nucleosome mobilization"—that is, the physical repositioning in cis of a histone octamer on a piece of DNA (this process is also referred to as "nucleosome sliding"). There is *in vitro* evidence that the assembly of the ISWI protein into the various complexes modulates its properties as a nucleosome mobilization engine (Guschin *et al.*, 2000a). For example, the complex between ISWI and the protein Acf1 is evolutionarily conserved and can mediate core histone deposition and subsequent nucleosome spacing *in vitro* (Ito *et al.*, 1997, 1999).

It is quite possible that the ability of ISWI to move the histone octamer on the DNA contributes to its role as a repressor in budding yeast (Goldmark *et al.*, 2000), as well as in other organisms. The immediate relevance of this issue to the subject of somatic cloning was made conspicuous by a study that examined the biochemical requirements for somatic cell nuclei remodeling in egg extract (Kikyo and Wolffe, 2000). Classic experiments in the 1950s and 1960s (reviewed in Kikyo and Wolffe, 2000) revealed the profoundly unexpected ability of egg cytoplasm to erase the nucleoprotein history of a cell nucleus from a differentiated tissue and reestablish its totipotency. While unequivocally robust and reproducible, this experimental observation is nothing short of quasi-miraculous, because it reveals the stunning plasticity of genome behavior in chromatin. A useful experimental system exists to study the biochemical mechanisms of this process: when *Xenopus* tissue culture cell nuclei are incubated in egg extract, a sequence of events occurs that strikingly mimics remodeling in enucleated eggs. The nucleus swells massively, and after 2 hours undergoes an extensive exchange of protein components with its cytoplasmic milieu. Specifically, while core chromatin components such as histone H2B and nonhistone regulators such as MeCP2 remain resident in the nucleus, other proteins, most notably components of the basal transcription machinery such as TBP, TFIIB, and TFIIF, are eliminated from the nucleus (Kikyo *et al.*, 2000). In the specific case of TBP, a 2-hour incubation in egg extract relocalizes more than 90% of this protein into the cytoplasm! Remarkably, such sweeping changes are energy dependent (require ATP), but are not replication dependent (are insensitive to aphidicolin and araC treatment)—thus, the chromatin-wide destruction of preinitiation complexes on gene promoters is not effected by passing DNA polymerase (Kikyo *et al.*, 2000).

It was reasonable to suspect that ATP-dependent chromatin remodeling machines would contribute to nuclear reprogramming. Indeed, such a functional connection was made to ISWI, the most abundant such ATPase in *Xenopus* egg cytoplasm. When egg extract was partially fractionated, certain fractions lost the ability to remodel nuclei as gauged by continued TBP residence in the nucleus (Kikyo *et al.*, 2000). The addition of recombinant ISWI protein, or of a purified five-subunit ISWI-containing complex called ISWI-D (Guschin *et al.*, 2000a), robustly restored the remodeling capacity of the egg extract and led to TBP elimination.

The mechanisms whereby a nucleosome-mobilizing engine, or complexes thereof, can eliminate anything from chromatin *in vivo* are unknown and it is difficult to speculate on this subject in the absence of data. Several known facts about transcription factor biology can provide a foundation for future experiments: (1) TBP cannot bind to nucleosomal templates (Imbalzano *et al.*, 1994); (2) the binding of certain nonhistone regulators (the glucocorticoid and estrogen receptors, for example) to chromatin *in vivo* is highly transient and occurs via a "shuttling" or "hit-and-run" mechanism (McNally *et al.*, 2000; Shang *et al.*, 2000) (as of July, 2001, it remains unknown, however, if TBP shuttles onto and off of chromatin); (3) the departure of a nonhistone regulatory factor from its target site will rapidly lead to an ATP-dependent occlusion of this site by nucleosomes that will "slide" onto that DNA sequence (Pazin *et al.*, 1997); and (4) various complexes of the ISWI protein can effect nucleosome mobilization/sliding both in a targeted (Goldmark *et al.*, 2000) and in a nontargeted (i.e., chromatin-wide) fashion (Ito *et al.*, 1997). A model that combines these experimental observations into a highly hypothetical scenario for ISWI-driven promoter remodeling is presented in Fig. 5.

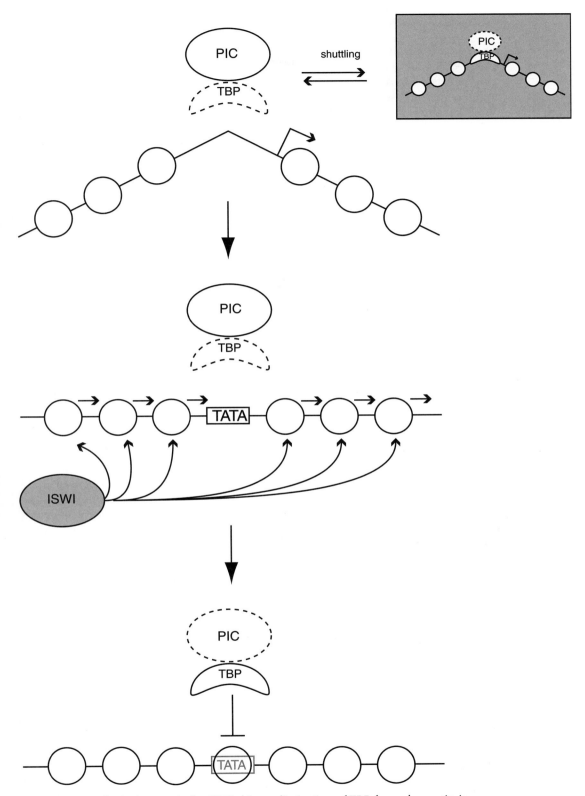

Figure 5 *Hypothetical scenario for ISWI-driven elimination of TBP from chromatin in somatic cell nuclei, as demonstrated in Kikyo et al. (2000).*

Gene Control *in Vivo*: The Substrate Is Chromatin

The second half of this section's title has been borrowed from a review by Barton and Crowe (2001) that describes advances in studying the poorly understood interplay between DNA replication and gene control. Indeed, the chromatinization of the genome imposes an extensive set of rules on gene activation and repression. While referring the reader to recent reviews for details (Gregory, 2001; Urnov and Wolffe, 2001b), we present here a few general principles.

In describing the pathways of chromatin remodeling that lead to gene activation or repression it is important to make a distinction between certain housekeeping genes and those that are more subtly developmentally regulated. The former—e.g., rDNA (Scheer and Rose, 1984; Sirri *et al.*, 2000), or heat-shock genes (Lis and Wu, 1993; Martinez-Balbas *et al.*, 1995; Rougvie and Lis, 1988; Wu, 1980)—somehow escape the totalitarian machinery of mitotic chromatin condensation and remain poised for rapid activation at all points in the cell cycle through a variety of mechanisms. Because chromatin over these genes is always stably remodeled (e.g., Martinez-Balbas *et al.*, 1995; Wu, 1980), far fewer steps are required to make a transition from silence to transcriptional activity.

In contrast, a locus destined for expression in a specific tissue—e.g., the α-globin gene locus, or the serum albumin gene—is inactive and assembled into mature chromatin during early development, until the embryo produces cells (preerythroblasts or hepatocytes, respectively) that are destined to express the cognate gene. At that point, a complicated and still poorly understood sequence of events leads to a final state of robust transcriptional activation. In the next two paragraphs, we present a highly condensed possible scenario. We note that the reverse sequence of events—the gradual silencing of transcription—is equally important in metazoan ontogeny, but much less well understood.

For some loci, it is clear that the early steps of activation involve sequence-specific nonhistone DNA-binding factors that can invade a mature chromatin infrastructure (Cirillo and Zaret, 1999; Urnov and Wolffe, 2001c). Once this invasion has occurred, the local mature heterochromatic array is thought to be perturbed (Lundgren *et al.*, 2000), and through mechanisms not currently understood, a domain of chromatin becomes unfolded (Tumbar *et al.*, 1999), hyperacetylated, and accessible to nucleases (Epner *et al.*, 1998; Hebbes *et al.*, 1988; Myers *et al.*, 2001; Reik *et al.*, 1998) before transcription begins. It is possible that this unfolding involves a functional synergy between an ATPase such as SWI/SNF and a HAT (Krebs *et al.*, 2000). Within the resulting unfolded domain (which can encompass many thousands of base pairs), gene-specific activation can now occur. Early events once again involve access to promoter/enhancer sequences by factors that can recognize their targets in the context of a nucleosomal array [e.g., the nuclear hormone receptors (Urnov and Wolffe, 2001c)], or else preemptive access by these proteins to their target sites to immature chromatin in the aftermath of replication fork progression (Felsenfeld, 1992). The factors, once bound, sequentially target an ATPase and a HAT, which facilitates binding by other regulatory factors incapable by themselves of attaching to chromatin, and a complex that mediates interaction with the basal transcriptional machinery. The order of this targeting will vary from promoter to promoter in a gene- and cell-cycle-dependent manner (Agalioti *et al.*, 2000; Cosma *et al.*, 1999; Krebs *et al.*, 1999; Shang *et al.*, 2000; Urnov and Wolffe, 2001b).

The important general points that can be obtained from the current literature are that a synergy between different types of remodeling and modification engines yields a more profound state of chromatin disruption than can be obtained by either one acting alone (Gregory *et al.*, 1998; Ito *et al.*, 2000), and that nonhistone DNA-binding proteins assemble on the DNA sequentially (Archer *et al.*, 1992; Cosma *et al.*, 1999; Yie *et al.*, 1999) in a manner coordinated with the level of chromatin disruption. A final general point is that once assembled, the transcription complex can be extraordinarily unstable and disperse on a scale of seconds (McNally *et al.*, 2000; Stenoien *et al.*, 2001), minutes (Cosma *et al.*, 1999; Shang *et al.*, 2000), or hours (Agalioti *et al.*, 2000). Mechanisms of such attenuation of response are unknown, but in some cases are thought to involve ATP-dependent disruption of protein–DNA contacts (Fletcher *et al.*, 2000), a synergy between chromatin assembly and factor binding (Urnov and Wolffe, 2001a), and covalent modification-driven regulation of factor–factor association (Chen *et al.*, 1999b).

THE NUCLEAR ENVELOPE, LAMINA, AND MATRIX: *E PLURIBUS UNUM*

However attached are scholars of transcription to a vision of their object of study as beads-on-a-string floating in space, all are aware of how simplistic this is. Not only does chromatin *in vivo* likely operate at levels of condensation much more extensive than the nucleosomal fiber (Muller *et al.*, 2001), but chromosomes do not float in space, and instead operate within a proteinaceous network inside the nucleus (the nuclear matrix), which is connected to a double lipid bilayer (the nuclear envelope) supported by an additional filamentous network (the nuclear lamina). All three entities are at present relatively poorly understood; paucity of conclusive data, however, is not a measure of the biological import of nuclear architecture in genome function, described in the present section.

THE NUCLEAR ENVELOPE AND THE LAMINA

The most distinctive cytological feature of a eukaryotic cell, the nucleus, is contained within a double lipid bilayer (Nigg, 1997), which consists of the inner nuclear membrane (INM), an outer nuclear membrane that is continuous with the endoplasmic reticulum (ER) membrane, and a perinuclear space that is continuous with the lumen of the ER. The close association with the ER severely complicates the isolation of the INM *per se*, which as a result remains very poorly characterized. Macroscopically, the nuclear envelope can be seen extensively studded with very large, eightfold symmetrical protein entities termed "nuclear pore complexes" (NPCs) (Ryan and Wente, 2000; Wente, 2000). The stunning details of NPC structure (Gorlich and Kutay, 1999) and the lovely mechanisms ensuring the directionality of nuclear import and export through the NPC driven by GTPases such as Ran (Azuma and Dasso, 2000) are outside of the scope of the present chapter. It is useful to recall, however, that several genome control pathways exploit the requirement for active import/export as an integral component: these include the sequestration of type I nuclear hormone receptors such as the glucocorticoid receptor in the cytoplasm in the absence of ligand (Hager, 2001), signal transduction pathways that control transcription factor abundance in the nucleus (Kaffman *et al.*, 1998), or similar pathways that regulate the intranuclear concentration of coregulators required for DNA-bound transcription factor function (McKinsey *et al.*, 2000).

The physicochemical features of lipid bilayers that form cell membranes necessitate a structurally robust internal framework, which is provided by the cytoskeleton; the word "skeleton" in this context was aptly chosen by analogy with the macroscopic structure familiar from vertebrate anatomy, because both cells and animals rely on it for the maintenance of shape, and for its dynamic alteration. Like the cytoplasm, the nucleus also contains a skeleton, if not several (see below). The most prominent and relatively unambiguously characterized is the "nuclear lamina" (Collas and Courvalin, 2000). This extensive proteinaceous network resides under the INM and consists, to a first approximation, of two major components

(Stuurman *et al.*, 1998), the lamin proteins and various lamin-associated polypeptides (LAPs), some of which are integral proteins of the INM.

Biochemically, the lamins—of which mammalian cells contain three major types, designated A, B1/2, and C—are all members of the intermediate filament (IF) superfamily (Fuchs and Cleveland, 1998). This widely conserved group of cytoskeletal proteins has a relatively invariant secondary structure: an individual IF protein contains an extensive α-helical domain, and two such domains can associate in parallel to form a dimer via a distinctive coil-coiled rod. Once this dimerization has been accomplished, multiple dimers aggregate "head-to-tail" to form extended wirelike structures. Four such wires then associate into a thicker entity designated as a "protofibril," and several protofibrils then coalesce into a 10-nm-thick intermediate filament. The IF network that forms under the INM is then anchored to it via a number of LAPs, most prominently the lamin B receptor (Ye and Worman, 1994), which contains both a highly basic nucleoplasmic and a transmembrane domain.

It is very likely that the nuclear lamina plays an architectural role for the nucleus: indeed, overexpression of mutated lamin B1 in cells yields very strikingly misshapen, lobulated cell nuclei (Schirmer *et al.*, 2001). A major question, however, is whether it also contributes to genome control, and available evidence suggests that it does. A striking example is offered by the genetics of Emery–Dreifuss muscular dystrophy (EDMD), a human disorder characterized by a complex pattern of muscle dysfunction, including frequently fatal cardiomyopathy (Nagano and Arahata, 2000). Remarkably, although in particular patients EDMD can be traced either to an X-linked or to an autosomal locus, both underlying genetic lesions affect the nuclear lamina: type A lamin in the latter case (Bonne *et al.*, 1999), and a LAP called "emerin" in the former (Bione *et al.*, 1994). Strikingly, genetic ablation of lamin A in the mouse does not interfere with embryonic development, and, instead, leads to muscular dystrophy caused by tissue-specific alterations in nuclear lamina integrity (Sullivan *et al.*, 1999). A possible molecular explanation for this phenotype is provided by data indicating that the lamin B receptor can bind to HP1 (Ye *et al.*, 1997; Ye and Worman, 1996), a major protein component of heterochromatin in metazoa. This finding offers a striking complement to the very well-established localization of heterochromatin domains in all eukarya to the nuclear periphery (Gasser, 2001). It is not known if heterochromatin becomes delocalized from the nuclear periphery in EDMD patients or in lamin A knockout mice, and if this localization then contributes to genome misregulation. A hypothetical model for lamina organization on the INM is presented in Fig. 6.

THE CHROMOSOME SCAFFOLD AND THE NUCLEAR MATRIX

Observations of dividing cells illuminate the seeming facility with which highly condensed mitotic chromosomes "dissolve" into the interphase nucleus, only to reappear again in a few hours. Considering the absolute length of the genome, it is very hard to envisage the chromosomes as stochastically decondensing within the extraordinarily tight spatial confines of the nucleus. The discovery in the 1970s of a "nuclear matrix" (Berezney and Coffey, 1974) and of a "chromosome scaffold" (Paulson and Laemmli, 1977)—in both cases, a nonhistone proteinaceous entity that is revealed when interphase nuclei or mitotic chromosomes, respectively, are extracted with salt or detergent and with nucleases—provided a possible solution. In the years that followed, the very existence, as well as the physiological relevance, of the matrix and the scaffold has been the subject of much debate (Cook, 2001; Wolffe, 1998), which centered primarily on whether these entities are *revealed* during, or instead are *created by*, the procedure of extraction (a debate on epistemology that goes back at least to Wordsworth's well-known stanza about man's "meddling intellect" that "murders to dissect" and thus "misshapes the beauteous

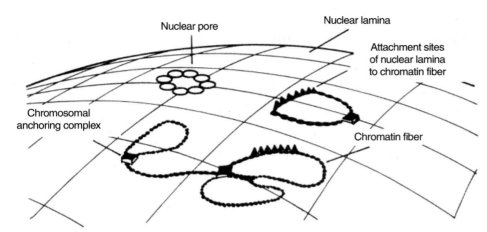

Figure 6 *A hypothetical representation of the inner nuclear membrane surface. From Wolffe (1998).*

form of things"). Although details and extensive arguments *pro* and *contra* are outside the scope of this review, several established facts from this field of research are noteworthy.

The chromosome scaffold was originally revealed as what remains in mitotic chromosomes after most of the nucleic acid and protein have been removed (Paulson and Laemmli, 1977): the entity observed largely preserved the overall shape of the chromosome. Original concerns about the draconian preparation methods have since been obviated via the introduction of milder techniques (Earnshaw and Laemmli, 1983), and the biochemical characterization of the scaffold yielded an important observation (Saitoh *et al.*, 1994): one of its major constituents (originally called ScII) was found to be related a protein already known to be required for proper mitotic transmission in budding yeast (Larionov *et al.*, 1985; Strunnikov *et al.*, 1993), a member of a family dubbed "structural maintenance of chromosomes" (SMC) proteins. The budding yeast genome contains several SMC genes (vertebrate ScII is homologous to Smc2p) (Freeman *et al.*, 2000), and homologs are contained in all taxa in which they have been studied (Hirano, 2000; Strunnikov, 1998). The primary structure of the SMCs reveals an invariant arrangement of an NTP-binding motif at the NH_2 terminus, followed by a coiled-coil region, and a "DA" box on the COOH terminus, suggesting these proteins are ATPases.

Since the original discovery of the SMCs, extensive biochemical and genetic characterization of these proteins in budding yeast, *Xenopus*, and mammalian cells have revealed two large complexes: (1) "cohesin," a complex between the ATPases Smc1p, Smc3p, and several other proteins (Hirano, 2000) that mediates sister chromatid cohesion during mitosis, and (2) "condensin," a complex between the ATPases Smc2p and Smc4p (XCAP-E and XCAP-C in *Xenopus*) and several other polypeptides (Hirano, 1999) that effects chromosome condensation during mitosis. These data, as well as chromatin immunoprecipitation experiments in budding yeast (Freeman *et al.*, 2000; Tanaka *et al.*, 1999), support the notion that a "chromosome scaffold" of some form may exist *in vivo*.

Whether cohesin and condensin play a role in the interphase nucleus remains an open question. It is tempting to speculate that the well-established role for condensins in enabling chromosome compaction during mitosis—i.e., in driving a transition from decondensed chromatin fiber in interphase to the familiar X-shaped structures seen in metaphase spreads—is also exploited by the cell during interphase to maintain particular stretches of its genome in a more condensed state. Components of the condensin complex are chromatin bound throughout the cell cycle in budding yeast (Freeman *et al.*, 2000), and genetic data indicate a requirement for condensins in enabling a functional heterochromatin–euchromatin boundary (Donze *et al.*, 1999). It is equally possible that rather than perform a condensation-related function, these proteins somehow contribute to introducing a modicum of order into the decondensed chromosome, i.e., act to maintain some sort of overall chromosome structure. It is interesting in this regard that both cohesin (Losada and Hirano, 2001) and condensin (Kimura and Hirano, 1997; Kimura *et al.*, 1999) can bind DNA *in vitro*.

It is equally noteworthy that a second major component of the "chromosome scaffold" is the enzyme topoisomerase II (Earnshaw *et al.*, 1985), and although an obvious application of its enzymatic activity would be the disentangling of DNA during condensation, a variety of data point to the organization of interphase chromosomes into small-scale "lampbrush chromosome"-like structures with large chromatin loops (Fig. 7) emanating from a central scaffold-like entity containing topoisomerase II (Cook, 2001; Gasser and Laemmli, 1986; Iarovaia *et al.*, 1996; Jackson *et al.*, 1990; Paulson and Laemmli, 1977).

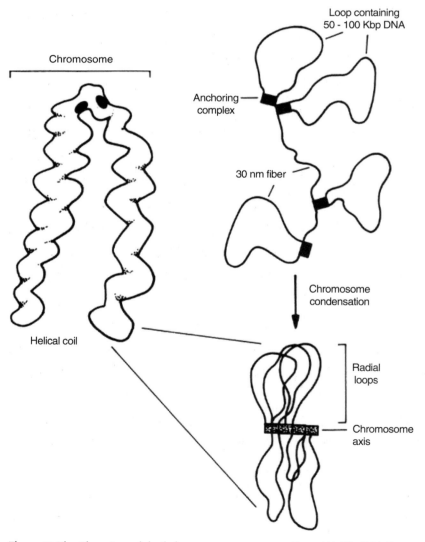

Figure 7 *The "loop" model of chromosome structure. From Wolffe (1998).*

Nuclear Neighborhoods and the Invisible Boundaries between Them

There are no known lipid bilayers within the nucleus *per se*; thus, in some sense, the inside space of the nucleus is a continuum. Functionally, however, the genome does not form a homogeneous mesh. In addition to the deservedly famous nucleolus with its well-established multiple *raisons d'être* (Pederson, 1998), cytological, genetic, and biochemical data suggest that intranuclear space contains a certain number of additional, relatively discrete compartments (Cook, 2001; Gasser, 2001). Macroscopically, the most conspicuous is the distinction between heterochromatin and euchromatin (Eissenberg and Elgin, 2000; Heitz, 1928; Jenuwein, 2001). It is clear, however, that the nucleus contains a number of additional, relatively discrete structures that are not bounded by a lipid bilayer and yet maintain a certain degree of spatial autonomy (Matera, 1999). Among others, these include Cajal bodies, thought to be storage/assembly sites for the mRNA transcription and processing machineries (Gall, 2000), and PML bodies (Maul *et al.*, 2000).

PERSPECTIVE

The progress of the past few years in the characterization of molecular machines that enable genome functionality within the nucleus is simply staggering, and illuminates the challenges that lie ahead. These include the development and application of new methods for investigating higher order structure in chromatin, and its dynamic behavior during gene control, as well as a meaningful integration of the wealth of information from various genome projects into our understanding of the biology of the nucleus.

UPDATE

Recent data demonstrate that mammalian embryos obtained by nuclear transfer from somatic and ES cell nuclei display aberrant DNA methylation patterns during early development (e.g., Dean *et al.*, 2001; reviewed in Reik *et al.*, 2002). Taken with data that the paternal progenome undergoes selective demethylation immediately postfertilization (Oswald *et al.*, 2000), these observations suggest that a causative connection may exist between chromatin remodeling over the paternal progenome postfertilization and the proper establishment and maintenance of DNA methylation patterns during early development (see Reik *et al.*, 2002).

Acknowledgments

I sincerely thank Michael Bustin and Alexander Strunnikov for very helpful clarifications and Jim Kadonaga for sharing unpublished data, and gratefully acknowledge Erica Dumais for drawing Figs. 2 and 5.

Author's Note

As the first draft of this manuscript was being completed, the unfathomable news arrived of Alan Wolffe's death in a traffic accident. A brilliant scholar and an erudite of unrivalled scope, Alan was a mentor and colleague *par excellence*. With deep sorrow and humility, this chapter is dedicated to his memory.

References

Agalioti, T., Lomvardas, S., Parekh, B., Yie, J., Maniatis, T., and Thanos, D. (2000). Ordered recruitment of chromatin modifying and general transcription factors to the IFN-beta promoter. *Cell* **103**, 667–678.

Ahringer, J. (2000). NuRD and SIN3 histone deacetylase complexes in development. *Trends Genet.* **16**, 351–356.

Allfrey, V., Faulkner, R. M., and Mirsky, A. E. (1964). Acetylation and methylation of histones and their possible role in the regulation of RNA synthesis. *Proc. Natl. Acad. Sci. U.S.A.* **51**, 786–794.

Anderson, J. D., and Widom, J. (2000). Sequence and position-dependence of the equilibrium accessibility of nucleosomal DNA target sites. *J. Mol. Biol.* **296**, 979–987.

Archer, T. K., Lefebvre, P., Wolford, R. G., and Hager, G. L. (1992). Transcription factor loading on the MMTV promoter: A bimodal mechanism for promoter activation. *Science* **255**, 1573–1576.

Azuma, Y., and Dasso, M. (2000). The role of Ran in nuclear function. *Curr. Opin. Cell Biol.* **12**, 302–307.

Baneres, J. L., Martin, A., and Parello, J. (1997). The N tails of histones H3 and H4 adopt a highly structured conformation in the nucleosome. *J. Mol. Biol.* **273**, 503–508.

Barra, J. L., Rhounim, L., Rossignol, J. L., and Faugeron, G. (2000). Histone H1 is dispensable for methylation-associated gene silencing in ascobolus immersus and essential for long life span. *Mol. Cell Biol.* **20**, 61–69.

Bartolomei, M. S., and Tilghman, S. M. (1997). Genomic imprinting in mammals. *Annu. Rev. Genet.* **31**, 493–525.

Barton, M. C., and Crowe, A. J. (2001). Chromatin alteration, transcription and replication: What's the opening line to the story? *Oncogene* **20**, 3094–3099.

Berezney, R., and Coffey, D. S. (1974). Identification of a nuclear protein matrix. *Biochem. Biophys. Res. Commun.* **60**, 1410–1417.

Berger, S. L. (2001). An embarrasment of niches: The many covalent modifications of histones in transcriptional regulation. *Oncogene* **20**.

Berger, S. L., Pina, B., Silverman, N., Marcus, G. A., Agapite, J., Regier, J. L., Triezenberg, S. J., and Guarente, L. (1992). Genetic isolation of ADA2: A potential transcriptional adaptor required for function of certain acidic activation domains. *Cell* **70**, 251–265.

Bestor, T. H. (1998). The host defence function of genomic methylation patterns. *Novartis Found. Symp.* **214**, 187–195.

Bione, S., Maestrini, E., Rivella, S., Mancini, M., Regis, S., Romeo, G., and Toniolo, D. (1994). Identification of a novel X-linked gene responsible for Emery–Dreifuss muscular dystrophy. *Nat. Genet.* **8**, 323–327.

Bird, A. P., and Wolffe, A. P. (1999). Methylation-induced repression—Belts, braces, and chromatin. *Cell* **99**, 451–454.

Bone, J. R., Lavender, J., Richman, R., Palmer, M. J., Turner, B. M., and Kuroda, M. I. (1994). Acetylated histone H4 on the male X chromosome is associated with dosage compensation in Drosophila. *Genes Dev.* **8**, 96–104.

Bonne, G., Di Barletta, M. R., Varnous, S., Becane, H. M., Hammouda, E. H., Merlini, L., Muntoni, F., Greenberg, C. R., Gary, F., Urtizberea, J. A., *et al.* (1999). Mutations in the gene encoding lamin A/C cause autosomal dominant Emery–Dreifuss muscular dystrophy. *Nat. Genet.* **21**, 285–288.

Boyes, J., Byfield, P., Nakatani, Y., and Ogryzko, V. (1998). Regulation of activity of the transcription factor GATA-1 by acetylation. *Nature* **396**, 594–598.

Braunstein, M., Rose, A. B., Holmes, S. G., Allis, C. D., and Broach, J. R. (1993). Transcriptional silencing in yeast is associated with reduced nucleosome acetylation. *Genes Dev.* **7**, 592–604.

Brown, C. E., Howe, L., Sousa, K., Alley, S. C., Carrozza, M. J., Tan, S., and Workman, J. L. (2001). Recruitment of HAT complexes by direct activator interactions with the ATM-related Tra 1 subunit. *Science* **292**, 2333–2337.

Brownell, J. E., Zhou, J., Ranalli, T., Kobayashi, R., Edmondson, D. G., Roth, S. Y., and Allis, C. D. (1996). Tetrahymena histone acetyltransferase A: A homolog to yeast Gcn5p linking histone acetylation to gene activation. *Cell* **84**, 843–851.

Bustin, M. (1999). Regulation of DNA-dependent activities by the functional motifs of the high-mobility-group chromosomal proteins. *Mol. Cell. Biol.* **19**, 5237–5246.

Bustin, M. (2001). Revised nomenclature for high mobility group (HMG) chromosomal proteins. *Trends Biochem. Sci.* **26**, 152–153.

Bustin, M., and Reeves, R. (1996). High-mobility-group chromosomal proteins: Architectural components that facilitate chromatin function. *Prog Nucleic Acid Res. Mol. Biol.* **54**, 35–100.

Cairns, B. R., Lorch, Y., Li, Y., Zhang, M., Lacomis, L., Erdjument-Bromage, H., Tempst, P., Du, J., Laurent, B., and Kornberg, R. D. (1996). RSC, an essential, abundant chromatin-remodeling complex. *Cell* **87**, 1249–1260.

Calladine, C. R., and Drew, H. R. (1992). "Understanding DNA: The Molecule and How It Works." Academic Press, San Diego.

Calogero, S., Grassi, F., Aguzzi, A., Voigtlander, T., Ferrier, P., Ferrari, S., and Bianchi, M. E. (1999). The lack of chromosomal protein Hmg1 does not disrupt cell growth but causes lethal hypoglycaemia in newborn mice. *Nat. Genet.* **22**, 276–280.

Carmen, A. A., Rundlett, S. E., and Grunstein, M. (1996). HDA1 and HDA3 are components of a yeast histone deacetylase (HDA) complex. *J. Biol. Chem.* **271**, 15837–15844.

Chandler, S. P., Guschin, D., Landsberger, N., and Wolffe, A. P. (1999). The methyl-CpG binding transcriptional repressor MeCP2 stably associates with nucleosomal DNA. *Biochemistry* 38, 7008–7018.

Chen, D., Ma, H., Hong, H., Koh, S. S., Huang, S. M., Schurter, B. T., Aswad, D. W., and Stallcup, M. R. (1999a). Regulation of transcription by a protein methyltransferase. *Science* 284, 2174–2177.

Chen, H., Lin, R. J., Xie, W., Wilpitz, D., and Evans, R. M. (1999b). Regulation of hormone-induced histone hyperacetylation and gene activation via acetylation of an acetylase. *Cell* 98, 675–686.

Cho, C., Willis, W. D., Goulding, E. H., Jung-Ha, H., Choi, Y. C., Hecht, N. B., and Eddy, E. M. (2001). Haploinsufficiency of protamine-1 or -2 causes infertility in mice. *Nat. Genet.* 28, 82–86.

Cirillo, L. A., and Zaret, K. S. (1999). An early developmental transcription factor complex that is more stable on nucleosome core particles than on free DNA. *Mol. Cell* 4, 961–969.

Cirillo, L. A., McPherson, C. E., Bossard, P., Stevens, K., Cherian, S., Shim, E. Y., Clark, K. L., Burley, S. K., and Zaret, K. S. (1998). Binding of the winged-helix transcription factor HNF3 to a linker histone site on the nucleosome. *EMBO J.* 17, 244–254.

Clark, K. L., Halay, E. D., Lai, E., and Burley, S. K. (1993). Co-crystal structure of the HNF-3/fork head DNA-recognition motif resembles histone H5. *Nature* 364, 412–420.

Clever, U., and Karlson, P. (1960). Induktion von Puff-Veraenderungen in den Speicheldruesenchromosomen von Chironomus tentans durch Ecdyson. *Exp. Cell Res.* 20, 623–626.

Collas, I., and Courvalin, J. C. (2000). Sorting nuclear membrane proteins at mitosis. *Trends Cell Biol.* 10, 5–8.

Collins, R. T., and Treisman, J. E. (2000). Osa-containing brahma chromatin remodeling complexes are required for the repression of wingless target genes. *Genes Dev.* 14, 3140–3152.

Cook, P. R. (2001). "Principles of Nuclear Structure and Function." John Wiley, New York.

Cosma, M. P., Tanaka, T., and Nasmyth, K. (1999). Ordered recruitment of transcription and chromatin remodeling factors to a cell cycle- and developmentally regulated promoter. *Cell* 97, 299–311.

Cote, J., Quinn, J., Workman, J. L., and Peterson, C. L. (1994). Stimulation of GAL4 derivative binding to nucleosomal DNA by the yeast SWI/SNF complex. *Science* 265, 53–60.

De Souza, C. P., Osmani, A. H., Wu, L. P., Spotts, J. L., and Osmani, S. A. (2000). Mitotic histone H3 phosphorylation by the NIMA kinase in *Aspergillus nidulans. Cell* 102, 293–302.

Dean, W., Santos, F., Stojkovic, M., Zakhartchenko, V., Walter, J., Wolf, E., and Reik, W. (2001). Conservation of methylation reprogramming in mammalian development: Aberrant reprogramming in cloned embryos. *Proc. Natl. Acad. Sci. U.S.A.* 98, 13734–13738.

Deuring, R., Fanti, L., Armstrong, J. A., Sarte, M., Papoulas, O., Prestel, M., Daubresse, G., Verardo, M., Moseley, S. L., Berloco, M., *et al.* (2000). The ISWI chromatin-remodeling protein is required for gene expression and the maintenance of higher order chromatin structure *in vivo. Mol. Cell* 5, 355–365.

Dhalluin, C., Carlson, J. E., Zeng, L., He, C., Aggarwal, A. K., and Zhou, M. M. (1999). Structure and ligand of a histone acetyltransferase bromodomain. *Nature* 399, 491–496.

Donze, D., Adams, C. R., Rine, J., and Kamakaka, R. T. (1999). The boundaries of the silenced HMR domain in *Saccharomyces cerevisiae. Genes Dev.* 13, 698–708.

Dou, Y., and Gorovsky, M. A. (2000). Phosphorylation of linker histone H1 regulates gene expression *in vivo* by creating a charge patch. *Mol. Cell* 6, 225–231.

Dou, Y., Mizzen, C. A., Abrams, M., Allis, C. D., and Gorovsky, M. A. (1999). Phosphorylation of linker histone H1 regulates gene expression *in vivo* by mimicking H1 removal. *Mol. Cell* 4, 641–647.

Earnshaw, W. C., and Laemmli, U. K. (1983). Architecture of metaphase chromosomes and chromosome scaffolds. *J. Cell Biol.* 96, 84–93.

Earnshaw, W. C., Halligan, B., Cooke, C. A., Heck, M. M., and Liu, L. F. (1985). Topoisomerase II is a structural component of mitotic chromosome scaffolds. *J. Cell Biol.* 100, 1706–1715.

Eissenberg, J. C., and Elgin, S. C. (2000). The HP1 protein family: Getting a grip on chromatin. *Curr. Opin. Genet. Dev.* 10, 204–210.

Elgin, S. C. (1988). The formation and function of DNase I hypersensitive sites in the process of gene activation. *J. Biol. Chem.* 263, 19259–19262.

Ellenberg, J., Siggia, E. D., Moreira, J. E., Smith, C. L., Presley, J. F., Worman, H. J., and Lippincott-Schwartz, J. (1997). Nuclear membrane dynamics and reassembly in living cells: Targeting of an inner nuclear membrane protein in interphase and mitosis. *J. Cell. Biol.* 138, 1193–1206.

Epner, E., Reik, A., Cimbora, D., Telling, A., Bender, M. A., Fiering, S., Enver, T., Martin, D. I., Kennedy, M., Keller, G., and Groudine, M. (1998). The beta-globin LCR is not necessary for an open chromatin structure or developmentally regulated transcription of the native mouse beta-globin locus. *Mol. Cell* 2, 447–455.

Felsenfeld, G. (1992). Chromatin as an essential part of the transcriptional mechanism. *Nature* 355, 219–224.

Flaus, A., and Owen-Hughes, T. (2001). Mechanisms for ATP-dependent chromatin remodelling. *Curr. Opin. Genet. Dev.* 11, 148–154.

Fletcher, T. M., Ryu, B.-W., Baumann, C. T., Warren, B. S., Fragoso, G., and Hager, G. (2000). Structure and dynamic properties of a glucocorticoid receptor-induced chromatin transition. *Mol. Cell Biol.* 20, 6466–6475.

Freeman, L., Aragon-Alcaide, L., and Strunnikov, A. (2000). The condensin complex governs chromosome condensation and mitotic transmission of rDNA. *J. Cell Biol.* **149**, 811–824.

Frye, R. A. (1999). Characterization of five human cDNAs with homology to the yeast SIR2 gene: Sir2-like proteins (sirtuins) metabolize NAD and may have protein ADP-ribosyltransferase activity. *Biochem. Biophys. Res. Commun.* **260**, 273–279.

Fryer, C. J., and Archer, T. K. (1998). Chromatin remodelling by the glucocorticoid receptor requires the BRG1 complex. *Nature* **393**, 88–91.

Fuchs, E., and Cleveland, D. W. (1998). A structural scaffolding of intermediate filaments in health and disease. *Science* **279**, 514–519.

Gall, J. G. (2000). Cajal bodies: The first 100 years. *Annu. Rev. Cell Dev. Biol.* **16**, 273–300.

Gasser, S. M. (2001). Positions of potential: nuclear organization and gene expression. *Cell* **104**, 639–642.

Gasser, S. M., and Laemmli, U. K. (1986). The organization of chromatin loops: characterization of a scaffold attachment site. *EMBO J.* **5**, 511–518.

Gavin, I., Horn, P. J., and Peterson, C. L. (2001). SWI/SNF chromatin remodeling requires changes in DNA topology. *Mol. Cell* **7**, 97–104.

Georgakopoulos, T., and Thireos, G. (1992). Two distinct yeast transcriptional activators require the function of the GCN5 protein to promote normal levels of transcription. *EMBO J.* **11**, 4145–4152.

Germond, J. E., Hirt, B., Oudet, P., Gross-Bellark, M., and Chambon, P. (1975). Folding of the DNA double helix in chromatin-like structures from simian virus 40. *Proc. Natl. Acad. Sci. U.S.A.* **72**, 1843–1847.

Goldmark, J. P., Fazzio, T. G., Estep, P. W., Church, G. M., and Tsukiyama, T. (2000). The Isw2 chromatin remodeling complex represses early meiotic genes upon recruitment by Ume6p. *Cell* **103**, 423–433.

Goodwin, G. H., Woodhead, L., and Johns, E. W. (1977). The presence of high mobility group non-histone chromatin proteins in isolated nucleosomes. *FEBS Lett.* **73**, 85–88.

Gorlich, D., and Kutay, U. (1999). Transport between the cell nucleus and the cytoplasm. *Annu. Rev. Cell. Dev. Biol.* **15**, 607–660.

Gregory, P. D. (2001). Transcription and chromatin converge: lessons from yeast genetics. *Curr. Opin. Genet. Dev.* **11**, 142–147.

Gregory, P. D., Schmid, A., Zavari, M., Lui, L., Berger, S. L., and Horz, W. (1998). Absence of Gcn5 HAT activity defines a novel state in the opening of chromatin at the PHO5 promoter in yeast. *Mol. Cell* **1**, 495–505.

Gregory, P. D., Schmid, A., Zavari, M., Munsterkotter, M., and Horz, W. (1999). Chromatin remodelling at the PHO8 promoter requires SWI-SNF and SAGA at a step subsequent to activator binding. *EMBO J.* **18**, 6407–6414.

Gross, D. S., and Garrard, W. T. (1988). Nuclease hypersensitive sites in chromatin. *Annu. Rev. Biochem.* **57**, 159–197.

Groudine, M., and Conkin, K. F. (1985). Chromatin structure and *de novo* methylation of sperm DNA: Implications for activation of the paternal genome. *Science* **228**, 1061–1068.

Grozinger, C. M., Hassig, C. A., and Schreiber, S. L. (1999). Three proteins define a class of human histone deacetylases related to yeast Hda1p. *Proc. Natl. Acad. Sci. U.S.A.* **96**, 4868–4873.

Guarente, L. (2000). Sir2 links chromatin silencing, metabolism, and aging. *Genes Dev.* **14**, 1021–1026.

Guenther, M. G., Lane, W. S., Fischle, W., Verdin, E., Lazar, M. A., and Shiekhattar, R. (2000). A core SMRT corepressor complex containing HDAC3 and TBL1, a WD40- repeat protein linked to deafness. *Genes Dev.* **14**, 1048–1057.

Guschin, D., Geiman, T. M., Kikyo, N., Tremethick, D. J., Wolffe, A. P., and Wade, P. A. (2000a). Multiple ISWI ATPase complexes from *Xenopus laevis*: Functional conservation of an ACF/CHRAC homolog. *J. Biol. Chem.* **275**, 35248–35255.

Guschin, D., Wade, P. A., Kikyo, N., and Wolffe, A. P. (2000b). ATP-Dependent histone octamer mobilization and histone deacetylation mediated by the Mi-2 chromatin remodeling complex. *Biochemistry* **39**, 5238–5245.

Hager, G. L. (2001). Understanding nuclear receptor function: From DNA to chromatin to the interphase nucleus. *Prog. Nucleic Acid Res. Mol. Biol.* **66**, 279–305.

Hans, F., and Dimitrov, S. (2001). Histone H3 phosphorylation and cell division. *Oncogene* **20**, 3021–3027.

Hansen, J. C., and Wolffe, A. P. (1994). A role for histones H2A/H2B in chromatin folding and transcriptional repression. *Proc. Natl. Acad. Sci. U.S.A.* **91**, 2339–2343.

Havas, K., Flaus, A., Phelan, M. L., Kingston, R. E., Wade, P. A., Lilley, D. M. J., and Owen-Hughes, T. (2000). Generation of superhelical torsion by ATP-dependent chromatin remodeling activities. *Cell* **103**, 1133–1142.

Hayes, J. J., and Hansen, J. C. (2001). Nucleosomes and the chromatin fiber. *Curr. Opin. Genet. Dev.* **11**, 124–129.

Hayes, J. J., Tullius, T. D., and Wolffe, A. P. (1990). The structure of DNA in a nucleosome. *Proc. Natl. Acad. Sci. U.S.A.* **87**, 7405–7409.

Hebbes, T. R., Thorne, A. W., and Crane-Robinson, C. (1988). A direct link between core histone acetylation and transcriptionally active chromatin. *EMBO J.* **7**, 1395–1402.

Hebbes, T. R., Clayton, A. L., Thorne, A. W., and Crane-Robinson, C. (1994). Core histone hyperacetylation co-maps with generalized DNase I sensitivity in the chicken β-globin chromosomal domain. *EMBO J.* **13**, 1823–1830.

Heitz, E. (1928). Das Heterochromatin der Moose. *Jahrb. Wiss. Bot.* **69**, 762.

Hellauer, K., Sirard, E., and Turcotte, B. (2001). Decreased expression of specific genes in yeast cells lacking histone h1. *J. Biol. Chem.* **276**, 13587–13592.

Hewish, D. R., and Burgoyne, L. A. (1973). Chromatin sub-structure. The digestion of chromatin DNA at regularly spaced sites by a nuclear deoxyribonuclease. *Biochem. Biophys. Res. Commun.* **52**, 504–510.

Hirano, T. (1999). SMC-mediated chromosome mechanics: A conserved scheme from bacteria to vertebrates? *Genes Dev.* **13**, 11–19.

Hirano, T. (2000). Chromosome cohesion, condensation, and separation. *Annu. Rev. Biochem.* **69**, 115–144.

Hirschhorn, J. N., Brown, S. A., Clark, C. D., and Winston, F. (1992). Evidence that SNF2/SWI2 and SNF5 activate transcription in yeast by altering chromatin structure. *Genes Dev.* **6**, 2288–2298.

Holstege, F. C., Jennings, E. G., Wyrick, J. J., Lee, T. I., Hengartner, C. J., Green, M. R., Golub, T. R., Lander, E. S., and Young, R. A. (1998). Dissecting the regulatory circuitry of a eukaryotic genome. *Cell* **95**, 717–728.

Hsu, J. Y., Sun, Z. W., Li, X., Reuben, M., Tatchell, K., Bishop, D. K., Grushcow, J. M., Brame, C. J., Caldwell, J. A., Hunt, D. F., *et al.* (2000). Mitotic phosphorylation of histone H3 is governed by Ipl1/aurora kinase and Glc7/PP1 phosphatase in budding yeast and nematodes. *Cell* **102**, 279–291.

Iarovaia, O., Hancock, R., Lagarkova, M., Miassod, R., and Razin, S. V. (1996). Mapping of genomic DNA loop organization in a 500-kilobase region of the *Drosophila* X chromosome by the topoisomerase II-mediated DNA loop excision protocol. *Mol. Cell Biol.* **16**, 302–308.

Ikura, T., Ogryzko, V. V., Grigoriev, M., Groisman, R., Wang, J., Horikoshi, M., Scully, R., Qin, J., and Nakatani, Y. (2000). Involvement of the TIP60 histone acetylase complex in DNA repair and apoptosis. *Cell* **102**, 463–473.

Imai, S., Armstrong, C. M., Kaeberlein, M., and Guarente, L. (2000). Transcriptional silencing and longevity protein Sir2 is an NAD-dependent histone deacetylase. *Nature* **403**, 795–800.

Imbalzano, A. N., Kwon, H., Green, M. R., and Kingston, R. E. (1994). Facilitated binding of TATA-binding protein to nucleosomal DNA. *Nature* **370**, 481–485.

Imhof, A., Yang, X. J., Ogryzko, V. V., Nakatani, Y., Wolffe, A. P., and Ge, H. (1997). Acetylation of general transcription factors by histone acetyltransferases. *Curr. Biol.* **7**, 689–692.

Ito, T., Bulger, M., Pazin, M. J., Kobayashi, R., and Kadonaga, J. T. (1997). ACF, an ISWI-containing and ATP-utilizing chromatin assembly and remodeling factor. *Cell* **90**, 145–155.

Ito, T., Levenstein, M. E., Fyodorov, D. V., Kutach, A. K., Kobayashi, R., and Kadonaga, J. T. (1999). ACF consists of two subunits, Acf1 and ISWI, that function cooperatively in the ATP-dependent catalysis of chromatin assembly. *Genes Dev.* **13**, 1529–1539.

Ito, T., Ikehara, T., Nakagawa, T., Kraus, W. L., and Muramatsu, M. (2000). p300-Mediated acetylation facilitates the transfer of histone H2A-H2B dimers from nucleosomes to a histone chaperone. *Genes Dev.* **14**, 1899–1907.

Iyer, V. R., Horak, C. E., Scafe, C. S., Botstein, D., Snyder, M., and Brown, P. O. (2001). Genomic binding sites of the yeast cell-cycle transcription factors SBF and MBF. *Nature* **409**, 533–538.

Jackson, D. A., Dickinson, P., and Cook, P. R. (1990). The size of chromatin loops in HeLa cells. *EMBO J.* **9**, 567–571.

Jacobson, R. H., Ladurner, A. G., King, D. S., and Tjian, R. (2000). Structure and function of a human TAFII250 double bromodomain module. *Science* **288**, 1422–1425.

Jenuwein, T. (2001). Re-SET-ting heterochromatin by histone methyltransferases. *Trends Cell Biol.* **11**, 266–273.

Jeppesen, P., and Turner, B. M. (1993). The inactive X chromosome in female mammals is distinguished by a lack of histone H4 acetylation, a cytogenetic marker for gene expression. *Cell* **74**, 281–289.

Jones, P. L., Veenstra, G. J., Wade, P. A., Vermaak, D., Kass, S. U., Landsberger, N., Strouboulis, J., and Wolffe, A. P. (1998). Methylated DNA and MeCP2 recruit histone deacetylase to repress transcription. *Nat. Genet.* **19**, 187–191.

Kaffman, A., Rank, N. M., O'Neill, E. M., Huang, L. S., and O'Shea, E. K. (1998). The receptor Msn5 exports the phosphorylated transcription factor Pho4 out of the nucleus. *Nature* **396**, 482–486.

Kaufman, P. D., and Almouzni, G. (2000). DNA replication, nucleotide excision repair, and nucleosome assembly. *In* "Chromatin Structure and Gene Expression" (S. C. R. Elgin and J. L. Workman, eds.), pp. 24–48. Oxford University Press, Oxford.

Khochbin, S., Verdel, A., Lemercier, C., and Seigneurin-Berny, D. (2001). Functional significance of histone deacetylase diversity. *Curr. Opin. Genet. Dev.* **11**, 162–166.

Kikyo, N., and Wolffe, A. P. (2000). Reprogramming nuclei: insights from cloning, nuclear transfer and heterokaryons. *J. Cell Sci.* **113**, 11–20.

Kikyo, N., Wade, P. A., Guschin, D., Ge, H., and Wolffe, A. P. (2000). Active remodeling of somatic nuclei in egg cytoplasm by the nucleosomal ATPase ISWI. *Science* **289**, 2360–2362.

Kimura, H., and Cook, P. R. (2001). Kinetics of core histones in living human cells: Little exchange of H3 and H4 and some rapid exchange of H2B. *J. Cell Biol.* **153**, 1341–1353.

Kimura, K., and Hirano, T. (1997). ATP-dependent positive supercoiling of DNA by 13S condensin: A biochemical implication for chromosome condensation. *Cell* **90**, 625–634.

Kimura, K., Rybenkov, V. V., Crisona, N. J., Hirano, T., and Cozzarelli, N. R. (1999). 13S condensin actively reconfigures DNA by introducing global positive writhe: Implications for chromosome condensation. *Cell* **98**, 239–248.

Kingston, R. E., and Narlikar, G. J. (1999). ATP-dependent remodeling and acetylation as regulators of chromatin fluidity. *Genes Dev.* **13**, 2339–2352.

Kornberg, R. D., and Thomas, J. O. (1974). Chromatin structure; oligomers of the histones. *Science* **184**, 865–868.

Kossel, A. (1884). Ueber einen peptonartigen Bestandteil des Zellkerns. *Z. Physiol. Chem.* **8**, 511.

Krebs, J. E., Kuo, M. H., Allis, C. D., and Peterson, C. L. (1999). Cell cycle-regulated histone acetylation required for expression of the yeast HO gene. *Genes Dev.* **13**, 1412–1421.

Krebs, J. E., Fry, C. J., Samuels, M. L., and Peterson, C. L. (2000). Global role for chromatin remodeling enzymes in mitotic gene expression. *Cell* **102**, 587–598.

Kuo, M. H., Zhou, J., Jambeck, P., Churchill, M. E., and Allis, C. D. (1998). Histone acetyltransferase activity of yeast Gcn5p is required for the activation of target genes *in vivo*. *Genes Dev.* **12**, 627–639.

Kwon, H., Imbalzano, A. N., Khavari, P. A., Kingston, R. E., and Green, M. R. (1994). Nucleosome disruption and enhancement of activator binding by a human SW1/SNF complex. *Nature* **370**, 477–481.

Lachner, M., O'Carroll, D., Rea, S., Mechtler, K., and Jenuwein, T. (2001). Methylation of histone H3 lysine 9 creates a binding site for HP1 proteins. *Nature* **410**, 116–120.

Lander, E. S., Linton, L. M., Birren, B., Nusbaum, C., Zody, M. C., Baldwin, J., Devon, K., Dewar, K., Doyle, M., FitzHugh, W., *et al.* (2001). Initial sequencing and analysis of the human genome. *Nature* **409**, 860–921.

Larionov, V. L., Karpova, T. S., Kouprina, N. Y., and Jouravleva, G. A. (1985). A mutant of *Saccharomyces cerevisiae* with impaired maintenance of centromeric plasmids. *Curr. Genet.* **10**, 15–20.

Lee, D. Y., Hayes, J. J., Pruss, D., and Wolffe, A. P. (1993). A positive role for histone acetylation in transcription factor access to nucleosomal DNA. *Cell* **72**, 73–84.

Lemon, B., and Tjian, R. (2000). Orchestrated response: a symphony of transcription factors for gene control. *Genes Dev.* **14**, 2551–2569.

Lever, M. A., Th'ng, J. P., Sun, X., and Hendzel, M. J. (2000). Rapid exchange of histone H1.1 on chromatin in living human cells. *Nature* **408**, 873–876.

Li, J., Wang, J., Wang, J., Nawaz, Z., Liu, J. M., Qin, J., and Wong, J. (2000). Both corepressor proteins SMRT and N-CoR exist in large protein complexes containing HDAC3. *EMBO J.* **19**, 4342–4350.

Lis, J., and Wu, C. (1993). Protein traffic on the heat shock promoter: parking, stalling, and trucking along. *Cell* **74**, 1–4.

Lo, W. S., Trievel, R. C., Rojas, J. R., Duggan, L., Hsu, J. Y., Allis, C. D., Marmorstein, R., and Berger, S. L. (2000). Phosphorylation of serine 10 in histone H3 is functionally linked *in vitro* and *in vivo* to Gcn5-mediated acetylation at lysine 14. *Mol. Cell* **5**, 917–926.

Losada, A., and Hirano, T. (2001). Intermolecular DNA interactions stimulated by the cohesin complex *in vitro*: Implications for sister chromatid cohesion. *Curr. Biol.* **11**, 268–272.

Luger, K., Mader, A. W., Richmond, R. K., Sargent, D. F., and Richmond, T. J. (1997). Crystal structure of the nucleosome core particle at 2.8 Å resolution. *Nature* **389**, 251–260.

Lundgren, M., Chow, C., Sabbattini, P., Georgiou, A., Minaee, S., and Dillon, N. (2000). Transcription factor dosage affects changes in higher order chromatin structure associated with activation of a heterochromatic gene. *Cell* **103**, 733–743.

Lyon, M. F. (1961). Gene action in the X-chromosome of the mouse. *Nature* **190**, 372–373.

Magdinier, F., and Wolffe, A. P. (2001). Selective association of the methyl-CpG binding protein MBD2 with the silent p14/p16 locus in human neoplasia. *Proc. Natl. Acad. Sci. U.S.A.* **98**, 4990–4995.

Marmorstein, R., and Roth, S. Y. (2001). Histone acetyltransferases: function, structure, and catalysis. *Curr. Opin. Genet. Dev.* **11**, 155–161.

Martinez-Balbas, M. A., Dey, A., Rabindran, S. K., Ozato, K., and Wu, C. (1995). Displacement of sequence-specific transcription factors from mitotic chromatin. *Cell* **83**, 29–38.

Matera, A. G. (1999). Nuclear bodies: multifaceted subdomains of the interchromatin space. *Trends Cell Biol.* **9**, 302–309.

Maul, G. G., Negorev, D., Bell, P., and Ishov, A. M. (2000). Review: Properties and assembly mechanisms of ND10, PML bodies, or PODs. *J. Struct. Biol.* **129**, 278–287.

McKinsey, T. A., Zhang, C. L., Lu, J., and Olson, E. N. (2000). Signal-dependent nuclear export of a histone deacetylase regulates muscle differentiation. *Nature* **408**, 106–111.

McNally, J. G., Muller, W. G., Walker, D., Wolford, R., and Hager, G. L. (2000). The glucocorticoid receptor: rapid exchange with regulatory sites in living cells. *Science* **287**, 1262–1265.

McPherson, C. E., Shim, E. Y., Friedman, D. S., and Zaret, K. S. (1993). An active tissue-specific enhancer and bound transcription factors existing in a precisely positioned nucleosomal array. *Cell* **75**, 387–398.

Merika, M., and Thanos, D. (2001). Enhanceosomes. *Curr. Opin. Genet. Dev.* **11**, 205–208.

Miescher, F. (1874). Die Spermatozoen einiger Wilbelthiere. *Verh. Naturforsch. Ges. Basel* **6**, 138.

Misteli, T., Gunjan, A., Hock, R., Bustin, M., and Brown, D. T. (2000). Dynamic binding of histone H1 to chromatin in living cells. *Nature* **408**, 877–881.

Moreira, J. M., and Holmberg, S. (1999). Transcriptional repression of the yeast CHA1 gene requires the chromatin-remodeling complex RSC. *EMBO J.* **18**, 2836–2844.

Muller, W. G., Walker, D., Hager, G. L., and McNally, J. G. (2001). Large-scale chromatin decondensation and recondensation regulated by transcription from a natural promoter. *J. Cell Biol.* **154**, 33–48.

Myers, F. A., Evans, D. R., Clayton, A. L., Thorne, A. W., and Crane-Robinson, C. (2001). Targeted and extended acetylation of histones H4 and H3 at active and inactive genes in chicken embryo erythrocytes. *J. Biol. Chem.* **276**, 20197–20205.

Nagano, A., and Arahata, K. (2000). Nuclear envelope proteins and associated diseases. *Curr. Opin. Neurol.* **13**, 533–539.

Nakagawa, T., Bulger, M., Muramatsu, M., and Ito, T. (2001). Multistep chromatin assembly on supercoiled plasmid DNA by nucleosome assembly protein-1 and ACF. *J. Biol. Chem.* **1**, 1.

Nakayama, J., Rice, J. C., Strahl, B. D., Allis, C. D., and Grewal, S. I. (2001). Role of histone H3 lysine 9 methylation in epigenetic control of heterochromatin assembly. *Science* **292**, 110–113.

Nan, X., Ng, H. H., Johnson, C. A., Laherty, C. D., Turner, B. M., Eisenman, R. N., and Bird, A. (1998). Transcriptional repression by the methyl-CpG-binding protein MeCP2 involves a histone deacetylase complex. *Nature* **393**, 386–389.

Nedospasov, S., and Georgiev, G. (1980). Non-random cleavage of SV40 DNA in the compact minichromosome and free in solution by micrococcal nuclease. *Biochem. Biophys. Res. Commun.* **29**, 532–539.

Neigeborn, L., and Carlson, M. (1984). Genes affecting the regulation of SUC2 gene expression by glucose repression in *Saccharomyces cerevisiae*. *Genetics* **108**, 845–858.

Ng, H. H., and Bird, A. (2000). Histone deacetylases: silencers for hire. *Trends Biochem Sci.* **25**, 121–126.

Nigg, E. A. (1997). Nucleocytoplasmic transport: Signals, mechanisms and regulation. *Nature* **386**, 779–787.

Nightingale, K., Dimitrov, S., Reeves, R., and Wolffe, A. P. (1996). Evidence for a shared structural role for HMG1 and linker histones B4 and H1 in organizing chromatin. *EMBO J.* **15**, 548–561.

Nowak, S. J., and Corces, V. G. (2000). Phosphorylation of histone H3 correlates with transcriptionally active loci. *Genes Dev.* **14**, 3003–3013.

Ogryzko, V. V., Kotani, T., Zhang, X., Schlitz, R. L., Howard, T., Yang, X. J., Howard, B. H., Qin, J., and Nakatani, Y. (1998). Histone-like TAFs within the PCAF histone acetylase complex. *Cell* **94**, 35–44.

Ohki, I., Shimotake, N., Fujita, N., Jee, J., Ikegami, T., Nakao, M., and Shirakawa, M. (2001). Solution structure of the methyl-cpg binding domain of human mbd1 in complex with methylated dna. *Cell* **105**, 487–497.

Ohno, S., Kaplan, W. D., and Kinosita, R. (1959). Formation of the sex chromatin by a single X-chromosome in liver cells of *Rattus norvegicus*. *Exp. Cell Res.* **18**, 415–418.

Oswald, J., Engemann, S., Lane, N., Mayer, W., Olek, A., Fundele, R., Dean, W., Reik, W., and Walter, J. (2000). Active demethylation of the paternal genome in the mouse zygote. *Curr. Biol.* **10**, 475–478.

Panning, B., and Jaenisch, R. (1998). RNA and the epigenetic regulation of X chromosome inactivation. *Cell* **93**, 305–308.

Parekh, B. S., and Maniatis, T. (1999). Virus infection leads to localized hyperacetylation of histones H3 and H4 at the IFN-beta promoter. *Mol. Cell* **3**, 125–129.

Paulson, J. R., and Laemmli, U. K. (1977). The structure of histone-depleted metaphase chromosomes. *Cell* **12**, 817–828.

Pazin, M. J., Bhargava, P., Geiduschek, E. P., and Kadonaga, J. T. (1997). Nucleosome mobility and the maintenance of nucleosome positioning. *Science* **276**, 809–812.

Pederson, T. (1998). The plurifunctional nucleolus. *Nucleic Acids Res.* **26**, 3871–3876.

Pederson, T. (2001). Protein mobility within the nucleus—What are the right moves? *Cell* **104**, 635–638.

Pennings, S., Meersseman, G., and Bradbury, E. M. (1994). Linker histones H1 and H5 prevent the mobility of positioned nucleosomes. *Proc. Natl. Acad. Sci. U.S.A.* **91**, 10275–10279.

Peterson, C. L., and Herskowitz, I. (1992). Characterization of the yeast SWI1, SWI2, and SWI3 genes, which encode a global activator of transcription. *Cell* **68**, 573–583.

Phair, R. D., and Misteli, T. (2000). High mobility of proteins in the mammalian cell nucleus. *Nature* **404**, 604–609.

Phelan, M. L., Sif, S., Narlikar, G. J., and Kingston, R. E. (1999). Reconstitution of a core chromatin remodeling complex from SWI/SNF subunits. *Mol. Cell* **3**, 247–253.

Philpott, A., and Leno, G. H. (1992). Nucleoplasmin remodels sperm chromatin in *Xenopus* egg extracts. *Cell* **69**, 759–767.

Philpott, A., Leno, G. H., and Laskey, R. A. (1991). Sperm decondensation in *Xenopus* egg cytoplasm is mediated by nucleoplasmin. *Cell* **65**, 569–578.

Poo, M., and Cone, R. A. (1974). Lateral diffusion of rhodopsin in the photoreceptor membrane. *Nature* **247**, 438–441.

Poot, R. A., Dellaire, G., Hulsmann, B. B., Grimaldi, M. A., Corona, D. F., Becker, P. B., Bickmore, W. A., and Varga-Weisz, P. D. (2000). HuCHRAC, a human ISWI chromatin remodelling complex contains hACF1 and two novel histone-fold proteins. *EMBO J.* **19**, 3377–3387.

Portugal, F. H., and Cohen, J. S. (1977). "A Century of DNA: A History of the Discovery of the Structure and Function of the Genetic Substance." MIT Press, Cambridge, MA.

Pruss, D., Bartholomew, B., Persinger, J., Hayes, J., Arents, G., Moudrianakis, E. N., and Wolffe, A. P. (1996). An asymmetric model for the nucleosome: A binding site for linker histones inside the DNA gyres. *Science* **274**, 614–617.

Ramakrishnan, V., Finch, J. T., Graziano, V., Lee, P. L., and Sweet, R. M. (1993). Crystal structure of globular domain of histone H5 and its implications for nucleosome binding. *Nature* **362**, 219–223.

Reik, A., Telling, A., Zitnik, G., Cimbora, D., Epner, E., and Groudine, M. (1998). The locus control region is necessary for gene expression in the human beta-globin locus but not the maintenance of an open chromatin structure in erythroid cells. *Mol. Cell Biol.* **18**, 5992–6000.

Reik, A., Gregory, P. D., and Urnov, F. D. (2002). Biotechnologies and therapeutics: Chromatin as a target. *Curr. Opin. Genet. Dev.* **12**, 233–242.

Ren, B., Robert, F., Wyrick, J. J., Aparicio, O., Jennings, E. G., Simon, I., Zeitlinger, J., Schreiber, J., Hannett, N., Kanin, E., *et al.* (2000). Genome-wide location and function of DNA binding proteins. *Science* **290**, 2306–2309.

Robertson, K. D., and Jones, P. A. (2000). DNA methylation: Past, present and future directions. *Carcinogenesis* **21**, 461–467.

Robertson, K. R., and Wolffe, A. P. (2000). DNA methylation in health and disease. *Nature Rev. Genet.* **1**, 11–19.

Robertson, K. D., Ait-Si-Ali, S., Yokochi, T., Wade, P. A., Jones, P. L., and Wolffe, A. P. (2000). DNMT1 forms a complex with rb, E2F1 and HDAC1 and represses transcription from E2F-responsive promoters. *Nat. Genet.* **25**, 338–342.

Roth, S. Y., Denu, J. M., and Allis, C. D. (2001). Histone acetyltransferases. *Annu. Rev. Biochem.* **70**, 81–120.

Rougvie, A. E., and Lis, J. T. (1988). The RNA polymerase II molecule at the 5′ end of the uninduced hsp70 gene of *D. melanogaster* is transcriptionally engaged. *Cell* **54**, 795–804.

Rundlett, S. E., Carmen, A. A., Suka, N., Turner, B. M., and Grunstein, M. (1998). Transcriptional repression by UME6 involves deacetylation of lysine 5 of histone H4 by RPD3. *Nature* **392**, 831–835.

Ryan, K. J., and Wente, S. R. (2000). The nuclear pore complex: A protein machine bridging the nucleus and cytoplasm. *Curr. Opin. Cell Biol.* **12**, 361–371.

Saitoh, Y., and Laemmli, U. K. (1994). Metaphase chromosome structure: Bands arise from a differential folding path of the highly AT-rich scaffold. *Cell* **76**, 609–622.

Saitoh, N., Goldberg, I. G., Wood, E. R., and Earnshaw, W. C. (1994). ScII: An abundant chromosome scaffold protein is a member of a family of putative ATPases with an unusual predicted tertiary structure. *J. Cell Biol.* **127**, 303–318.

Samuels, H. H., Perlman, A. J., Raaka, B. M., and Stanley, F. (1982). Organization of the thyroid hormone receptor in chromatin. *Recent Prog. Horm. Res.* **38**, 557–599.

Santisteban, M. S., Arents, G., Moudrianakis, E. N., and Smith, M. M. (1997). Histone octamer function *in vivo*: Mutations in the dimmer–tetramer interfaces disrupt both gene activation and repression. *EMBO J.* **16**, 2493–2506.

Scheer, U., and Rose, K. M. (1984). Localization of RNA polymerase I in interphase cells and mitotic chromosomes by light and electron microscopic immunocytochemistry. *Proc. Natl. Acad. Sci. U.S.A.* **81**, 1431–1435.

Schild, C., Claret, F. X., Wahli, W., and Wolffe, A. P. (1993). A nucleosome-dependent static loop potentiates estrogen-regulated transcription from the *Xenopus* vitellogenin B1 promoter *in vitro*. *EMBO J.* **12**, 423–433.

Schirmer, E. C., Guan, T., and Gerace, L. (2001). Involvement of the lamin rod domain in heterotypic lamin interactions important for nuclear organization. *J. Cell Biol.* **153**, 479–489.

Sewack, G. F., and Hansen, U. (1997). Nucleosome positioning and transcription-associated chromatin alterations on the human estrogen-responsive pS2 promoter. *J. Biol. Chem.* **272**, 31118–31129.

Shang, Y., Hu, X., DiRenzo, J., Lazar, M. A., and Brown, M. (2000). Cofactor dynamics and sufficiency in estrogen receptor-regulated transcription. *Cell* **103**, 843–852.

Shen, X., Yu, L., Weir, J. W., and Gorovsky, M. A. (1995). Linker histones are not essential and affect chromatin condensation *in vivo*. *Cell* **82**, 47–56.

Shim, E. Y., Woodcock, C., and Zaret, K. S. (1998). Nucleosome positioning by the winged helix transcription factor HNF3. *Genes Dev.* **12**, 5–10.

Shirakawa, H., Herrera, J. E., Bustin, M., and Postnikov, Y. (2000). Targeting of high mobility group-14/-17 proteins in chromatin is independent of DNA sequence. *J. Biol. Chem.* **275**, 37937–37944.

Sinden, R. R. (1994). "DNA Structure and Function." Academic Press, San Diego.

Sirri, V., Roussel, P., and Hernandez-Verdun, D. (2000). *In vivo* release of mitotic silencing of riboso-mal gene transcription does not give rise to precursor ribosomal RNA processing. *J. Cell Biol.* **148**, 259–270.

Stallcup, M. R. (2001). Role of protein methylation in chromatin remodeling and transcriptional regu-lation. *Oncogene* **20**, 3014–3020.

Steger, K. (1999). Transcriptional and translational regulation of gene expression in haploid spermatids. *Anat. Embryol. (Berl.)* **199**, 471–487.

Stenoien, D. L., Nye, A. C., Mancini, M. G., Patel, K., Dutertre, M., O'Malley, B. W., Smith, C. L., Belmont, A. S., and Mancini, M. A. (2001). Ligand-mediated assembly and real-time cellular dynam-ics of estrogen receptor α-coactivator complexes in living cells. *Mol. Cell Biol.* **21**, 4404–4412.

Stern, M., Jensen, R., and Herskowitz, I. (1984). Five SWI genes are required for expression of the HO gene in yeast. *J. Mol. Biol.* **178**, 853–868.

Sterner, D. E., and Berger, S. L. (2000). Acetylation of histones and transcription-related factors. *Microbiol. Mol. Biol. Rev.* **64**, 435–459.

Strahl, B. D., and Allis, C. D. (2000). The language of covalent histone modifications. *Nature* **403**, 41–45.

Strahl, B. D., Briggs, S. D., Brame, C. J., Caldwell, J. A., Koh, S. S., Ma, H., Cook, R. G., Shabanowitz, J., Hunt, D. F., Stallcup, M. R., and Allis, C. D. (2001). Methylation of histone H4 at arginine 3 occurs *in vivo* and is mediated by the nuclear receptor coactivator PMRT1. *Curr. Biol.* **11**, 996–1000.

Strunnikov, A. V. (1998). SMC proteins and chromosome structure. *Trends Cell Biol.* **8**, 454–459.

Strunnikov, A. V., Larionov, V. L., and Koshland, D. (1993). SMC1: An essential yeast gene encoding a putative head-rod-tail protein is required for nuclear division and defines a new ubiquitous protein family. *J. Cell Biol.* **123**, 1635–1648.

Stuurman, N., Heins, S., and Aebi, U. (1998). Nuclear lamins: Their structure, assembly, and interac-tions. *J. Struct. Biol.* **122**, 42–66.

Sudarsanam, P., and Winston, F. (2000). The Swi/Snf family: Nucleosome-remodeling complexes and transcriptional control. *Trends Genet.* **16**, 345–351.

Sudarsanam, P., Cao, Y., Wu, L., Laurent, B. C., and Winston, F. (1999). The nucleosome remodeling complex, Snf/Swi, is required for the maintenance of transcription *in vivo* and is partially redundant with the histone acetyltransferase, Gcn5. *EMBO J.* **18**, 3101–3106.

Sullivan, T., Escalante-Alcalde, D., Bhatt, H., Anver, M., Bhat, N., Nagashima, K., Stewart, C. L., and Burke, B. (1999). Loss of A-type lamin expression compromises nuclear envelope integrity leading to muscular dystrophy. *J. Cell Biol.* **147**, 913–920.

Tanaka, T., Cosma, M. P., Wirth, K., and Nasmyth, K. (1999). Identification of cohesin association sites at centromeres and along chromosome arms. *Cell* **98**, 847–858.

Taunton, J., Hassig, C. A., and Schreiber, S. L. (1996). A mammalian histone deacetylase related to the yeast transcriptional regulator Rpd3p. *Science* **272**, 408–411.

Tong, J. K., Hassig, C. A., Schnitzler, G. R., Kingston, R. E., and Schreiber, S. L. (1998). Chromatin deacetylation by an ATP-dependent nucleosome remodelling complex. *Nature* **395**, 917–921.

Tse, C., Fletcher, T. M., and Hansen, J. C. (1998a). Enhanced transcription factor access to arrays of histone H3/H4 tetramer·DNA complexes *in vitro*: Implications for replication and transcription. *Proc. Natl. Acad. Sci. U.S.A.* **95**, 12169–12173.

Tse, C., Sera, T., Wolffe, A. P., and Hansen, J. C. (1998b). Disruption of higher-order folding by core histone acetylation dramatically enhances transcription of nucleosomal arrays by RNA polymerase III. *Mol. Cell Biol.* **18**, 4629–4638.

Tsukiyama, T., Palmer, J., Landel, C. C., Shiloach, J., and Wu, C. (1999). Characterization of the imi-tation switch subfamily of ATP-dependent chromatin-remodeling factors in *Saccharomyces cerevisiae*. *Genes Dev.* **13**, 686–697.

Tumbar, T., Sudlow, G., and Belmont, A. S. (1999). Large-scale chromatin unfolding and remodeling induced by VP16 acidic activation domain. *J. Cell Biol.* **145**, 1341–1354.

Ura, K., Hayes, J. J., and Wolffe, A. P. (1995). A positive role for nucleosome mobility in the tran-scriptional activity of chromatin templates: restriction by linker histones. *EMBO J.* **14**, 3752–3765.

Ura, K., Nightingale, K., and Wolffe, A. P. (1996). Differential association of HMG1 and linker histones B4 and H1 with dinucleosomal DNA: Structural transitions and transcriptional repression. *EMBO J.* **15**, 4959–4969.

Urnov, F. D., and Wolffe, A. P. (2001a). An array of positioned nucleosomes potentiates thyroid hormone receptor action *in vivo*. *J. Biol. Chem.* **276**, 19753–19761.

Urnov, F. D., and Wolffe, A. P. (2001b). Chromatin remodeling and transcriptional activation: The cast (in order of appearance). *Oncogene* **24**, 2991–3006.

Urnov, F. D., and Wolffe, A. P. (2001c). A necessary good: Nuclear hormone receptors and their chromatin templates. *Mol. Endocrinol.* **15**, 1–16.

Urnov, F. D., Yee, J., Sachs, L., Collingwood, T. N., Bauer, A., Beug, H., Shi, Y. B., and Wolffe, A. P. (2000). Targeting of N-CoR and histone deacetylase 3 by the oncoprotein v-erbA yields a chromatin infrastructure-dependent transcriptional repression pathway. *EMBO J.* **19**, 4074–4090.

Varga-Weisz, P. D. (2001). ATP-dependent chromatin remodeling factors: nucleosome shufflers with many missions. *Oncogene* **20**, 3076–3085.

Varshavsky, A. J., Sundin, O. H., and Bohn, M. J. (1978). SV40 viral minichromosome: Preferential exposure of the origin of replication as probed by restriction endonucleases. *Nucleic Acids Res.* **5**, 3469–3477.

Varshavsky, A. J., Sundin, O., and Bohn, M. (1979). A stretch of "late" SV40 viral DNA about 400 bp long which includes the origin of replication is specifically exposed in SV40 minichromosomes. *Cell* **16**, 453–466.

Vassilev, A., Yamauchi, J., Kotani, T., Prives, C., Avantaggiati, M. L., Qin, J., and Nakatani, Y. (1998). The 400 kDa subunit of the PCAF histone acetylase complex belongs to the ATM superfamily. *Mol. Cell* **2**, 869–875.

Veenstra, G. J., Weeks, D. L., and Wolffe, A. P. (2000). Distinct roles for TBP and TBP-like factor in early embryonic gene transcription in *Xenopus*. *Science* **290**, 2312–2315.

Venter, J. C., Adams, M. D., Myers, E. W., Li, P. W., Mural, R. J., Sutton, G. G., Smith, H. O., Yandell, M., Evans, C. A., Holt, R. A., *et al.* (2001). The sequence of the human genome. *Science* **291**, 1304–1351.

Verreault, A. (2000). *De novo* nucleosome assembly: New pieces in an old puzzle. *Genes Dev.* **14**, 1430–1438.

Verreault, A., Kaufman, P. D., Kobayashi, R., and Stillman, B. (1998). Nucleosomal DNA regulates the core-histone-binding subunit of the human Hat1 acetyltransferase. *Curr. Biol.* **8**, 96–108.

Vettese-Dadey, M., Grant, P. A., Hebbes, T. R., Crane-Robinson, C., Allis, C. D., and Workman, J. L. (1996). Acetylation of histone H4 plays a primary role in enhancing transcription factor binding to nucleosomal DNA *in vitro*. *EMBO J.* **15**, 2508–2518.

Vidal, M., and Gaber, R. F. (1991). RPD3 encodes a second factor required to achieve maximum positive and negative transcriptional states in *Saccharomyces cerevisiae*. *Mol. Cell Biol.* **11**, 6317–6327.

Vidal, M., Buckley, A. M., Hilger, F., and Gaber, R. F. (1990). Direct selection for mutants with increased K⁺ transport in *Saccharomyces cerevisiae*. *Genetics* **125**, 313–320.

Vignali, M., Hassan, A. H., Neely, K. E., and Workman, J. L. (2000). ATP-dependent chromatin-remodeling complexes. *Mol. Cell Biol.* **20**, 1899–1910.

Wade, P. A., and Wolffe, A. P. (2001). Inherent asymmetry in recognition of a symmetrically methylated CpG dinucleotide. *Nat. Struct. Biol.* **8**, 575–577.

Wade, P. A., Jones, P. L., Vermaak, D., and Wolffe, A. P. (1998). A multiple subunit Mi-2 histone deacetylase from *Xenopus laevis* cofractionates with an associated Snf2 superfamily ATPase. *Curr. Biol.* **8**, 843–846.

Wade, P. A., Gegonne, A., Jones, P. L., Ballestar, E., Aubry, F., and Wolffe, A. P. (1999). Mi-2 complex couples DNA methylation to chromatin remodelling and histone deacetylation. *Nat. Genet.* **23**, 62–66.

Wang, W., Cote, J., Xue, Y., Zhou, S., Khavari, P. A., Biggar, S. R., Muchardt, C., Kalpana, G. V., Goff, S. P., Yaniv, M., *et al.* (1996a). Purification and biochemical heterogeneity of the mammalian SWI–SNF complex. *EMBO J.* **15**, 5370–5382.

Wang, W., Xue, Y., Zhou, S., Kuo, A., Cairns, B. R., and Crabtree, G. R. (1996b). Diversity and specialization of mammalian SWI/SNF complexes. *Genes Dev.* **10**, 2117–2130.

Wang, X., Moore, S. C., Laszckzak, M., and Ausio, J. (2000). Acetylation increases the alpha-helical content of the histone tails of the nucleosome. *J. Biol. Chem.* **275**, 35013–35020.

Wang, H., Huang, Z. Q., Xia, L., Feng, Q., Erdjument-Bromage, H., Strahl, B. D., Briggs, S. D., Allis, C. D., Wong, J., Tempst, P., and Zhang, Y. (2001). Methylation of histone H4 at arginine 3 facilitates transcriptional activation by nuclear hormone receptor. *Science* **31**, 31.

Watson, J. D., and Crick, F. H. C. (1953). A structure for deoxyribose nucleic acid. *Nature* **171**, 737.

Weiss, R. E., Xu, J., Ning, G., Pohlenz, J., O'Malley, B. W., and Refetoff, S. (1999). Mice deficient in the steroid receptor co-activator 1 (SRC-1) are resistant to thyroid hormone. *EMBO J.* **18**, 1900–1904.

Wente, S. R. (2000). Gatekeepers of the nucleus. *Science* **288**, 1374–1377.

Winston, F., and Allis, C. D. (1999). The bromodomain: A chromatin-targeting module? *Nat. Struct. Biol.* **6**, 601–604.

Wolfe, S. A., Nekludova, L., and Pabo, C. O. (2000). DNA recognition by Cys2His2 zinc finger proteins. *Annu. Rev. Biophys. Biomol. Struct.* **29**, 183–212.

Wolffe, A. P. (1998). "Chromatin Structure and Function," 3rd Ed. Academic Press, San Diego.

Wolffe, A. P. (1999). Architectural regulations and Hmg1. *Nat. Genet.* **22**, 215–217.

Wolffe, A. P., and Hansen, J. C. (2001). Nuclear visions: Functional flexibility from structural instability. *Cell* **104**, 631–634.

Wolffe, A. P., and Hayes, J. J. (1999). Chromatin disruption and modification. *Nucleic Acids Res.* **27**, 711–720.

Wong, J., Shi, Y. B., and Wolffe, A. P. (1995). A role for nucleosome assembly in both silencing and activation of the *Xenopus* TR β gene by the thyroid hormone receptor. *Genes Dev.* **9**, 2696–2711.

Wong, J., Li, Q., Levi, B. Z., Shi, Y. B., and Wolffe, A. P. (1997). Structural and functional features of a specific nucleosome containing a recognition element for the thyroid hormone receptor. *EMBO J.* **16**, 7130–7145.

Wong, J., Patterton, D., Imhof, A., Guschin, D., Shi, Y. B., and Wolffe, A. P. (1998). Distinct requirements for chromatin assembly in transcriptional repression by thyroid hormone receptor and histone deacetylase. *EMBO J.* **17**, 520–534.

Woodcock, C. L. F. (1973). Ultrastructure of inactive chromatin. *J. Cell Biol.* **59**, 368a.

Wu, C. (1980). The 5' ends of *Drosophila* heat shock genes in chromatin are hypersensitive to DNase I. *Nature* **286**, 854–860.

Wu, L., and Winston, F. (1997). Evidence that Snf-Swi controls chromatin structure over both the TATA and UAS regions of the SUC2 promoter in *Saccharomyces cerevisiae*. *Nucleic Acids Res.* **25**, 4230–4234.

Wu, J. Y., Ribar, T. J., Cummings, D. E., Burton, K. A., McKnight, G. S., and Means, A. R. (2000). Spermiogenesis and exchange of basic nuclear proteins are impaired in male germ cells lacking Camk4. *Nat. Genet.* **25**, 448–452.

Wyrick, J. J., Holstege, F. C., Jennings, E. G., Causton, H. C., Shore, D., Grunstein, M., Lander, E. S., and Young, R. A. (1999). Chromosomal landscape of nucleosome-dependent gene expression and silencing in yeast. *Nature* **402**, 418–421.

Xie, R., van Wijnen, A. J., van Der Meijden, C., Luong, M. X., Stein, J. L., and Stein, G. S. (2001). The cell cycle control element of histone h4 gene transcription is maximally responsive to interferon regulatory factor pairs irf-1/irf-3 and irf-1/irf-7. *J. Biol. Chem.* **276**, 18624–18632.

Xu, J., Qiu, Y., DeMayo, F. J., Tsai, S. Y., Tsai, M. J., and O'Malley, B. W. (1998). Partial hormone resistance in mice with disruption of the steroid receptor coactivator-1 (SRC-1) gene. *Science* **279**, 1922–1925.

Xu, W., Edmondson, D. G., Evrard, Y. A., Wakamiya, M., Behringer, R. R., and Roth, S. Y. (2000). Loss of gcn5l2 leads to increased apoptosis and mesodermal defects during mouse development. *Nat. Genet.* **26**, 229–232.

Yao, T. P., Oh, S. P., Fuchs, M., Zhou, N. D., Ch'ng, L. E., Newsome, D., Bronson, R. T., Li, E., Livingston, D. M., and Eckner, R. (1998). Gene dosage-dependent embryonic development and proliferation defects in mice lacking the transcriptional integrator p300. *Cell* **93**, 361–372.

Ye, Q., and Worman, H. J. (1994). Primary structure analysis and lamin B and DNA binding of human LBR, an integral protein of the nuclear envelope inner membrane. *J. Biol. Chem.* **269**, 11306–11311.

Ye, Q., and Worman, H. J. (1996). Interaction between an integral protein of the nuclear envelope inner membrane and human chromodomain proteins homologous to *Drosophila* HP1. *J. Biol. Chem.* **271**, 14653–14656.

Ye, Q., Callebaut, I., Pezhman, A., Courvalin, J. C., and Worman, H. J. (1997). Domain-specific interactions of human HP1-type chromodomain proteins and inner nuclear membrane protein LBR. *J. Biol. Chem.* **272**, 14983–14989.

Yie, J., Merika, M., Munshi, N., Chen, G., and Thanos, D. (1999). The role of HMG I(Y) in the assembly and function of the IFN-beta enhanceosome. *EMBO J.* **18**, 3074–3089.

Yu, Y. E., Zhang, Y., Unni, E., Shirley, C. R., Deng, J. M., Russell, L. D., Weil, M. M., Behringer, R. R., and Meistrich, M. L. (2000). Abnormal spermatogenesis and reduced fertility in transition nuclear protein 1-deficient mice. *Proc. Natl. Acad. Sci. U.S.A.* **97**, 4683–4688.

Zaret, K. S. (1995). Nucleoprotein architecture of the albumin transcriptional enhancer. *Semin. Cell Biol.* **6**, 209–218.

Zaret, K. S., and Yamamoto, K. R. (1984). Reversible and persistent changes in chromatin structure accompany activation of a glucocorticoid-dependent enhancer element. *Cell* **38**, 29–38.

Zhang, J., and Lazar, M. A. (2000). The mechanism of action of thyroid hormones. *Annu. Rev. Physiol.* **62**, 439–466.

Zhang, Y., Iratni, R., Erdjument-Bromage, H., Tempst, P., and Reinberg, D. (1997). Histone deacetylases and SAP18, a novel polypeptide, are components of a human Sin3 complex. *Cell* **89**, 357–364.

Zhang, Y., LeRoy, G., Seelig, H. P., Lane, W. S., and Reinberg, D. (1998). The dermatomyositis-specific autoantigen Mi2 is a component of a complex containing histone deacetylase and nucleosome remodeling activities. *Cell* **95**, 279–289.

Zhong, J., Peters, A. H., Lee, K., and Braun, R. E. (1999). A double-stranded RNA binding protein required for activation of repressed messages in mammalian germ cells. *Nat. Genet.* **22**, 171–174.

CHAPTER 4

NUCLEAR REPROGRAMMING: BIOLOGICAL AND TECHNOLOGICAL CONSTRAINTS

Kevin Eggan and Rudolf Jaenisch

The realization of mammalian cloning by somatic cell nuclear transfer has raised considerable interest in the mechanisms by which the mammalian oocyte returns the gene expression profile of a somatic cell to one that is appropriate and necessary for the development of a single-cell zygote. Investigations into the importance, nature, and efficiency of this process, dubbed "reprogramming," as well as its influence on the successful development of cloned animals, are in their earliest stages. In this chapter, we summarize our current understanding of the reprogramming process, as it occurs after nuclear transfer. Furthermore, we attempt to build a framework for understanding how both the technological parameters of the cloning process and the biological limitation of nuclear reprogramming might relate to the inefficiencies and phenotypes observed during the development of cloned animals.

NUCLEAR TRANSFER TECHNOLOGY AND ITS INFLUENCE ON THE DEVELOPMENT OF CLONED MAMMALS

Viable sheep (Campbell *et al.*, 1996), cows (Cibelli *et al.*, 1998), pigs (Onishi *et al.*, 2000; Polejaeva *et al.*, 2000), goats (Baguisi *et al.*, 1999), cats (Shin *et al.*, 2002), and mice (Wakayama *et al.*, 1998) have all been produced through the direct introduction of somatic nuclei into enucleated oocytes. Unfortunately, the universally low efficiency of nuclear transfer experiments has made it difficult to dissociate technical difficulties from true biological phenomena in cloning research. However, as nuclear transfer technology becomes more standardized and as research focus shifts—from simply whether cloning is possible to which parameters influence its success—a glimpse of the variables that determine a successful cloning outcome begins to emerge. In this chapter we begin by focusing on four key parameters of nuclear transfer technology that influence the development of cloned embryos: (1) the state and nature of the recipient cytoplast, (2) the cell cycle status of the donor cell, (3) the identity, differentiation state, and developmental potency of the donor cell, and, finally, (4) genetic influences on the successful development of cloned embryos. After discussing these technological constraints, we summarize some aspects of epigenetic reprogramming after nuclear transfer.

THE RECIPIENT OOCYTE AND ITS CYTOPLASM

A robust, reproducible method for nuclear transfer was first pioneered in the mouse embryo (McGrath and Solter, 1983). With this technique, a zygotic pronucleus or

Table 1 Effects of the Recipient Cytoplasm on NT Embryo Development

Recipient cytoplast	Species	Efficiency of *in vitro* development to morulae/blastocyst	Developmental potency after embryo transfer	References
Zygote	Mouse	0	None	McGrath and Solter, 1984; Wakayama *et al.*, 2000b; Tani *et al.*, 2001
	Cow	0	None	
MII oocyte then serial transfer to zygote	Mouse	29%	Adult	Ono *et al.*, 2001
	Pig	N/A	Adult	Polejaeva *et al.*, 2000
Oocyte MII	Mouse	70%	Adult	Wakayama *et al.*, 2000b
	Cow	65%	Adult	Wells *et al.*, 1999
	Sheep	20–40%	Adult	Wilmut *et al.*, 1997
	Pig	30%	Adult	Onishi *et al.*, 2000
	Goat	N/A	Adult	Baguisi *et al.*, 1999
Activated oocyte, (anaphase or telophase)	Mouse	6%	None	Wakayama *et al.*, 2000b
	Sheep	12%	Adult	Campbell *et al.*, 1996
	Goat	N/A	Adult	Baguisi *et al.*, 1999

embryonic nucleus was transferred as a karyoplast via Sendai virus-mediated membrane fusion into another enucleated zygote. Zygotes reconstituted in this way, with the pronucleus of another zygote, were competent to develop to the blastocyst stage with high efficiency and on to birth after embryo transfer (McGrath and Solter, 1983). However, when the same nuclear transfer methodology was used to introduce nuclei—from either cleavage stage blastomeres or the inner cell mass (ICM)—into zygotes, the cloned embryo rarely completed more than a few cell divisions (Table 1) (McGrath and Solter, 1984).

When conceptually similar methods, utilizing electrically mediated membrane fusion, were used to introduce either embryonic or adult ovine cells into unactivated, metaphase II (MII) oocytes, the preimplantation embryos developed to the blastocyst stage with a reasonable efficiency and cloned sheep were produced (Wells *et al.*, 1997; Wilmut *et al.*, 1997). Nuclear transfer of both embryonic and adult cells into MII oocytes (Table 1) has also been used to produce cloned cattle (Sims and First, 1994; Wells *et al.*, 1999; Kato *et al.*, 1998). The positive outcome of these efforts relative to earlier attempts in the mouse seems to lie in the nature of the recipient cytoplasm, suggesting that some property of the MII oocyte is critical for cloning success.

The most thorough analysis of the nature of the recipient cytoplasm and the development of cloned embryos examined the effects of cumulus cell nuclear transfer (NT) into murine MII oocytes, activated telophase oocytes, and zygotes (Wakayama *et al.*, 2000b). These experiments, utilizing a direct injection method for nuclear transfer (Wakayama *et al.*, 1998), demonstrated that the efficiency of *in vitro* embryonic development dropped off markedly as the oocyte cytoplasm transitioned from meiosis to G1 of the first cell cycle (Tables 1 and 2). Nuclear transfer into MII oocytes followed by either immediate or delayed activation resulted in efficient preimplantaion development and full-term development. Preactivated oocytes, however, developed *in vitro* with a very low efficiency after NT, and cloned mice could not be produced. As observed previously (McGrath and Solter, 1984), nuclear transfer into zygotes resulted in very early cleavage failure of the cloned embryo. When the chromosomes of the embryos generated by NT into enucleated zygotes were analyzed, all contained gross karyotypic abnormalities. In contrast, 70% of embryos produced by NT into enucleated MII oocytes possessed an

Table 2 *Influence of Cell Cycle Stage on NT Embryo Development*

Cell type	Cell cycle stage	Reconstructed embryos developing to blastocyst stage	References
Cumulus	G1, G0	70%	Wakayama *et al.*, 1998
Sertoli	G1, G0	65%	Ogura *et al.*, 2000
Fibroblast	G1, G0	58%	Wakayama and Yanagimachi, 1999
ES cells	(Log) high serum	10–20%	Wakayama *et al.*, 1999; Rideout *et al.*, 2000; Eggan *et al.*, 2001
	Small (G1?), low serum	55%	Wakayama *et al.*, 1999
	Large cell–nocodozol (G2–M), high serum	43%	Wakayama *et al.*, 1999

intact diploid complement of chromosomes. These experiments suggest the metaphase environment is somehow critical for preparing the somatic chromosomes for preimplantation development after NT.

In the MII oocyte, metaphase/maturation promoting factor (MPF) levels are high. After fertilization or artificial oocyte activation, MPF levels decrease, allowing the embryo to complete meiosis and commence development (for review, see Fulka *et al.*, 1996). High MPF levels in the oocyte cytoplasm lead to somatic cell nuclear envelope breakdown and premature chromosome condensation (Wakayama *et al.*, 1998). These events do not occur after NT into enucleated zygotes (Wakayama *et al.*, 2000b). It therefore seems likely that either the physical condensation of the chromosomes, induced cell cycle synchronization of the donor nucleus and oocyte cytoplasm, or some other consequence of high MPF levels must be responsible for protecting the somatic DNA from damage after NT. Whether cytoplasm rich in MPF has intrinsic properties necessary for developmental reprogramming, other than those bestowing protection from DNA damage, has yet to be determined. Interestingly, cloned goats (Baguisi *et al.*, 1999 144) and sheep (Campbell *et al.*, 1996) have been generated by introducing somatic nuclei into activated oocytes (Table 1), indicating that exposing the donor nucleus to cytoplasm rich in MPF may not be strictly required for the successful development of cloned animals. Further investigation will be required to determine whether this discrepancy reflects interesting species-specific differences in nuclear remodeling after NT, or simply variation in MPF half-life following oocyte activation.

CELL CYCLE STATUS OF THE DONOR CELL

Comparative analysis of cloning experiments with a variety of murine cells has begun to demonstrate a clear correlation between the efficiency of NT embryo development to the blastocyst stage and the proportion of donor cells in the G1 phase of the cell cycle (Table 2). When Sertoli (Ogura *et al.*, 2000), cumulus (Wakayama *et al.*, 1998), and serum-starved fibroblast cells (Wakayama and Yanagimachi, 1999), all primarily in a G1 (2N) state, were used for NT a majority of activated embryos developed to the blastocyst stage. In contrast, when rapidly cycling embryonic stem (ES) cells were used as nuclear donors (Wakayama *et al.*, 1999; Rideout *et al.*, 2000; Eggan *et al.*, 2001), only a small percentage of NT embryos completed cleavage development. Because most ES cells in a given population are in S phase (K. Eggan and A. Jaenisch, unpublished observation), this may be the cause of their poor *in vitro* development after nuclear transfer. Consistent with this interpretation, methods that force a higher proportion of ES cells into the G1 phase of the cell cycle, such as partial serum withdrawal, also increase their *in vitro* developmental potential after NT (Wakayama *et al.*, 1999) (Table 2).

Presumably, the importance of the donor nucleus cell cycle state is directly linked to its compatibility with the oocyte cytoplasm. As mentioned previously, high MPF levels in the oocyte cause donor nuclear envelope breakdown and premature chromosome condensation (Fulka *et al.*, 1996). Chromatin structure in the S phase of the cell cycle may be incompatible with this condensation, leading to DNA damage and zygotic arrest. In contrast, Chromatin in G1 and G2/M phases, it seems, is compatible with this condensation.

In sheep (Wilmut *et al.*, 1997) and cattle (Cibelli *et al.*, 1998), nuclear transfer procedures often utilize simultaneous fusion and activation. In the mouse (Wakayama *et al.*, 1998), delayed activation combined with chemicals preventing polar body extrusion ensure that the entire chromosomal content of the G1 (2N) donor cell remains in the oocyte. Further evidence that the numeral chromosome content of the donor cell is critical, rather than a particular cell cycle state per se, also comes from experiments using ES cells as nuclear donors (Wakayama *et al.*, 1999). In these studies, ES cell nuclei were arrested in mitosis by nocadozol treatment and transferred into MII oocytes. Several hours after NT, a functional metaphase spindle was observed in a subset of the reconstructed embryos. However, complete retention of all DNA from a G2/M phase cell would create a tetraploid (4N) embryo, a chromosome content incompatible with full-term development (Kaufman and Webb, 1990). To prevent a tetraploid state, oocyte activation was performed in the absence of cytoskeleton depolymerizing agents. After activation, part of the presumed 4N donor cell chromosomal complement was shown to be extruded into the polar body. Similarly activated embryos developed to the blastocyst stage and to term with a high efficiency.

In conclusion, a G0 cell cycle state is clearly not necessary for the successful development of cloned embryos, as once suggested (Wilmut *et al.*, 1997). However, the somatic nucleus must be in a conformation compatible with nuclear envelope breakdown and chromosome condensation. In addition, careful consideration of the donor cell cycle state, combined with methods for ensuring correct numeral chromosome content after activation, such as inclusion or exclusion of drugs preventing polar body extrusion, are critical.

THE INHERENT DEVELOPMENTAL POTENTIAL OF THE DONOR CELL

Comparing the efficiency of nuclear transfer experiments using both somatic and embryonic cells lends support to the hypothesis that intrinsic developmental potency of the donor cell plays a role in the cloning outcome (Table 3). When mouse cumulus (Wakayama *et al.*, 1998) and tail-tip cells (Wakayama and Yanagimachi, 1999) were used as nuclear donors, only 1–3 and 0.5%, respectively, of embryos transferred to surrogate mothers developed to term. In contrast, 5–25% of blastocysts generated by nuclear transfer with ES cell (Wakayama *et al.*, 1999; Rideout *et al.*, 2000; Eggan *et al.*, 2001) or blastomere (Tsunoda and Kato, 1997) nuclei survived until birth.

This increased developmental efficiency suggests that the ES cell and blastomere genome must be either more amenable to, or require less, epigenetic reprogramming than that of somatic cells. For instance, it would be expected that an ES cell nucleus would already express genes critical for early development, such as *Oct4* (Nichols *et al.*, 1998). In cumulus cell or tail-tip nuclei, these genes are not expressed and must be reactivated after NT. Furthermore, cumulus cells and tail-tip fibroblasts have chromatin configurations and DNA methylation levels appropriate for expression of genes necessary for maintenance of their differentiated state. Presumably the chromatin and methylation configurations that drive expression of these genes must

Table 3 *Donor Cell Developmental Potency and NT Embryo Development*

Species	Cell type	Morulae/blastocysts developing to term	References
Mouse	Cumulus	2–3%	Wakayama *et al.*, 1998
	Sertoli	3.5%	Ogura *et al.*, 2000
	Fibroblast	1%	Wakayama and Yanagimachi, 1999
	ES cells	5–21%	Wakayama *et al.*, 1999; Rideout *et al.*, 2000; Eggan *et al.*, 2001
	Blastomere	25%	Tsunoda and Kato, 1997
Sheep	Mammary	3.5%	Wilmut *et al.*, 1997
	Fetal fibroblast	7.5%	Wilmut *et al.*, 1997 McCreath *et al.*, 2000
	Embryo cells	4.5%	Wilmut *et al.*, 1997
	ICM ES-like	4%	Wells *et al.*, 1997
Cow	Cumulus	10%	Wells *et al.*, 1999
	Oviductal	15%	Kato *et al.*, 1998
	Fetal fibroblast	14%	Cibelli *et al.*, 1998
	ICM ES-like	12%	Sims and First, 1994
Pig	Granulosa (serial NT)	7%	Polejaeva *et al.*, 2000
	Fetal fibroblast (single NT)	1%	Onishi *et al.*, 2000; Betthauser *et al.*, 2000
Goat	Fetal fibroblast	3.5%	Baguisi *et al.*, 1999

be reset to an embryonic state in order to ensure that these genes are not inappropriately expressed either during cleavage or later in development. It may be that maternal stores of embryonic proteins and RNAs, present in the oocyte, support development of NT embryos derived from differentiated cells to the morula–blastocyst stage. Inability to reprogram gene expression appropriately, followed by depletion of these stores, could lead to the frequent developmental failure observed shortly thereafter (Wakayama and Yanagimachi, 2001).

In farm animals such as cattle and sheep, there is no apparent difference in cloning efficiency when using embryonic (Wells *et al.*, 1997; Sims and First, 1994) rather than somatic cells (Wells *et al.*, 1999; Wilmut *et al.*, 1997) as nuclear donors. However, this observation could reflect that the embryonic cells used for these studies, derived from both cattle and sheep embryos, had decreased intrinsic developmental potential relative to murine ES cells. Therefore, the inconsistency between experiments in mouse and cattle/sheep may again result from differences in experimental design rather than reflecting species-specific differences in the reprogramming process.

It may be that not all nuclei have the potency to direct embryogenesis after NT and that surviving clones are derived from rare somatic stem cells, with increased developmental potential, present at a low frequency in the donor cell population. Thus the great inefficiency of cloning might merely reflect the rare nature of those donor cells with the developmental capacity to direct development after NT. However, experiments using differentiated, mature B and T lymphocytes as donors have now demonstrated that even highly differentiated cells can give rise to ES cell lines and to cloned mice after nuclear transfer, albeit at an extremely low efficiency (Hochedlinger and Jaenisch, 2002). At the other extreme, it may be that all somatic nuclei are competent to direct development of cloned offspring after NT. If this is the case, inherent inefficiencies and errors in the reprogramming process itself must cause erratic and/or stochastic expression of genes, critical for embryogenesis, resulting in developmental failure. Consistent with erratic reprogramming, it has been shown that some bovine NT blastocysts failed to express appropriate levels of fibroblast growth factor and receptor (FGF-4, FGF-2r) and interleukin 6 (IL-6) (Daniels *et al.*, 2000). Furthermore, some bovine NT blastocysts were shown to have DNA methylation levels, at certain repetitive elements, more similar to the donor cell population than control blastocysts (Kang *et al.*, 2001). Recently, it has been suggested that tissue-specific stem cells in the adult may have greater plasticity then previously believed and that it is possible to isolate these cells for directed experimentation (Clarke *et al.*, 2000). Although the nature of these cells is still controversial, comparing the efficiency of cloning using adult stem cells and more differentiated cells from the same tissue may shed light on which aspects of cloning are limited by the developmental potential of the donor cell.

GENETIC INFLUENCES ON THE CLONING PROCESS

The outbred nature of livestock precludes genetic analysis of cloning in cattle, goats, pigs, or sheep. In contrast, the mouse is an ideal model for the genetic dissection of nuclear reprogramming. When ES cells from various strains were used as nuclear donors, inbred 129, C57/B6, and F1 ES cell lines all gave rise to newborn clones (Wakayama *et al.*, 1999; Rideout *et al.*, 2000; Eggan *et al.*, 2001). Interestingly, however, clones from four independent inbred ES cell lines died shortly after birth due to respiratory failure (Wakayama *et al.*, 1999; Rideout *et al.*, 2000; Eggan *et al.*, 2001), whereas most clones derived from five different F1 ES cell lines survived to adulthood (Table 4) (Rideout *et al.*, 2000; Eggan *et al.*, 2001).

Neonatal lethality has also been reported in mice entirely derived from inbred ES cells injected into tetraploid blastocysts (ES cell–tetraploid) (Nagy *et al.*, 1990, 1993). Like inbred clones, ES cell–tetraploid pups derived from inbred ES cell lines died shortly after delivery with signs of respiratory distress (Nagy *et al.*, 1990; Eggan *et al.*, 2001). In contrast, most ES cell–tetraploid neonates derived from six F1 ES cell lines developed into fertile adults (Eggan *et al.*, 2001). These results suggest that the death of inbred ES cell clones is not a direct result of the nuclear transfer procedure but instead is due to the limited developmental potential of inbred ES cells. It is possible that the decreased developmental potential and respiratory failure observed in inbred ES NT embryos could be due to delayed developmental timing relative to their F1 counterparts. However, experiments demonstrating that NT embryos derived from both F1 (Wakayama *et al.*, 1998) and inbred 129 cumulus

Table 4 Genetic Influences on NT Embryo Development

Cell type	Genetic background	Survival to term	Survival from term to adult	References
Cumulus	Various F1	1–2%	85%	Wakayama *et al.*, 1998; Wakayama and Yanagimachi, 2001
	Inbred C57	0	N/A	Wakayama and Yanagimachi, 2001
	Inbred 129	1.5%	95%	Wakayama and Yanagimachi, 2001
ES cells	Various F1	8–17%	50–85%	Wakayama *et al.*, 1999; Rideout *et al.*, 2000; Eggan *et al.*, 2001
	Inbred C57	17%	0	Eggan *et al.*, 2001
	Inbred 129	1–18%	0	Eggan *et al.*, 2001
	Inbred 129	1.6%	20% ($n = 1$)	Wakayama *et al.*, 1999
ES cells into 4N blasts	Various F1	18%	85%	Eggan *et al.*, 2001
	Inbred C57	4%	0	Eggan *et al.*, 2001
	Inbred 129	10%	0	Eggan *et al.*, 2001

cells (Wakayama and Yanagimachi, 2001) survive from birth to adulthood at a high frequency seem to argue against this conclusion. It is therefore reasonable to speculate that inbred ES cells may suffer some ill effect due to long-term cell culture, which ultimately leads to the demise of ES cell-derived offspring. Experiments suggesting that prolonged *in vitro* passage of ES cells can further aggravate these phenotypes, in ES cell-derived offspring, lend further support to this hypothesis (Nagy *et al.*, 1993). Interestingly, an F1 genetic background seems to be protective against these detrimental effects, because offspring from all F1 genotypes tested survived to adulthood, even after prolonged *in vitro* culture (Eggan *et al.*, 2001). Cloned animals of other species, including sheep, also often display signs of respiratory distress at birth, especially following extended donor cell *in vitro* culture (McCreath *et al.*, 2000; Denning *et al.*, 2001). It is therefore possible that these unknown effects of long-term donor cell *in vitro* culture, observed in mice cloned from ES cells, may also be relevant to the survival of other cloned animals.

EPIGENETIC REPROGRAMMING AFTER NUCLEAR TRANSFER

The most interesting issue in cloning by nuclear transfer is the problem of epigenetic reprogramming (Gurdon and Colman, 1999). For clones to complete development, genes normally expressed during embryogenesis but silent in the somatic donor cell must be reactivated. To date, the efficiency of deriving live cloned animals has been low and independent of the source of the cell type used as nuclear donor, with one notable exception. Nuclei isolated from ES cells and embryonic blastomeres generated viable cloned animals with a significantly higher efficiency compared to any somatic donor cell type (Table 3). As stated above, this observation is consistent with the notion that the genome of pluripotent embryonic cells is easier to reprogram or requires less reprogramming than that of a somatic cell. If this is the case, what is the nature of the reprogramming that must occur for a somatic cell clone to survive?

GERM LINE TRANSMISSION RELIEVES EPIGENETIC DAMAGE ACCUMULATED DURING THE CLONING PROCESS

Remarkably, it has proved possible to clone mice from cumulus cell nuclei for up to six generations (Wakayama *et al.*, 2000a). However, with each round of reproductive cloning, the generation of offspring became increasingly more difficult (Wakayama *et al.*, 2000a). Although these adult mice displayed few overt developmental defects they may have accumulated genetic or epigenetic damage that prevented their further propagation by nuclear transfer. The observation that the sexual offspring of cloned animals are normal and fertile suggests that the abnormalities observed in clones are epigenetic rather than genetic in nature (Wakayama *et al.*, 1998). In a similar experiment, bovine embryos were created by nuclear transfer, cultured to the blastocyst stage, disaggregated, and subjected to repeated rounds of cloning (Peura *et al.*, 2001). The results showed that repeated cloning led to a gradual decrease in the production of blastocysts (Peura *et al.*, 2001). Together, these serial cloning experiments suggest that bypassing the normal removal and reestablishment of epigenetic information, normally occurring during gametogenesis, can aggravate the developmental inefficiency observed when cloning animals.

Even more insidious than evidence that incomplete epigenetic reprogramming leads to developmental failure (Daniels *et al.*, 2000; Kang *et al.*, 2001) are signs that considerable epigenetic damage may be tolerated by mammalian development and only manifested in cloned animals late in life or after multiple rounds of cloning (Jaenisch and Wilmut, 2001; Humpherys *et al.*, 2001; Wakayama *et al.*, 2000a). Thus epigenetic abnormalities are present in clones and may cause the many abnormal phenotypes characteristic of cloned animals (Rideout *et al.*, 2001). Such epigenetic abnormalities may be the cause of obesity and shortened life span reported for some cloned mice (Tamashiro *et al.*, 2002; Ogonuki *et al.*, 2002).

In normal development, the reprogramming of the genome occurs during gametogenesis, a complex process that assures that the genome of the two gametes, when combined at fertilization, can faithfully activate early embryonic genes (Hilscher, 1999; Barton *et al.*, 1984; Kafri *et al.*, 1992). In cloning, reprogramming has to occur in the short interval between transfer of the donor nucleus into the egg and the onset of cellular differentiation, a context dramatically different from normal fertilization. The challenge now lies in identifying which epigenetic errors, arising either as a result of faulty reprogramming after NT or incurred by the donor cell nucleus during *in vitro* cultivation and aging, are responsible for cloning phenotypes. For the following discussion it is useful to distinguish epigenetic information coded before formation of the zygote from that which is established in the developing embryo after fertilization. For example, parent-specific changes in DNA methylation and chromatin structure that lead to monoallelic expression of imprinted genes are programmed during gametogenesis and maintained after fertilization (Tilghman, 1999; Bartolomei and Tilghman, 1997). In contrast, X-chromosome inactivation (Lyon, 1999a,b) and telomere length adjustment (DePinho, 2000) are processes occurring after fertilization. We first discuss advances in our understanding of how reprogramming of zygotically controlled epigenetic processes, such as X inactivation and telomere length adjustment, occur after nuclear transfer. In addition, we comment on the identity of candidate genes and processes, such as maintenance of imprinted gene expression, the regulation of which is controlled by information established during gametogenesis and the improper or incomplete reprogramming of which may lead to the phenotypes observed in cloned animals.

Reprogramming X-Chromosome Inactivation

Dosage compensation in mammals is achieved by extinguishing gene expression from one X chromosome in female somatic cells, a process known as X inactivation (Lyon, 1999a,b). The inactive X chromosome (X_i) differs from both the active chromosomes X (X_a) and the autosomes in its heterochromatic and transcriptionally silent state. The characteristics of a heterochromatic state exemplified by the X_i include increased concentration of macrohistone H2a (Csankovszki *et al.*, 1999), histone H4 hypoacetylation (Jeppesen and Turner, 1993), and increased DNA methylation at promoter sequences (Norris *et al.*, 1991).

Dosage compensation in female embryos by X inactivation is a developmentally regulated process, commencing during preimplantation cleavage with both X chromosomes transcriptionally active, occurring in a stereotypical series of molecular events (Panning *et al.*, 1997). First, an unidentified parental epigenetic mark, or imprint, causes preferential inactivation of the paternal X in the extraembryonic trophectoderm (TE), a tissue that contributes to the placenta and is necessary for implantation of the embryo (Marahrens *et al.*, 1997; Takagi and Sasaki, 1975; Lyon, 1999a). After TE allocation and extraembryonic X inactivation, the paternal imprint is subsequently lost in those cells giving rise to the embryonic (epiblast) lineage where X-chromosome choice is random (Epstein *et al.*, 1978; Kratzer and Gartler, 1978). Once chromosomal choice and transcriptional silencing have occurred, the heterochromatic state imposed on the X_i is stable, with the X_i state being faithfully passed on to all of the mitotic daughters of the respective chromosome chosen for inactivation (Lyon, 1999b).

The mere existence of cloned female animals indicated that dosage compensation must occur following nuclear transfer of a female donor nucleus. However, it was unclear whether X inactivation would be random, as in normal animals, or nonrandom, as in the TE lineage (Gurdon and Colman, 1999). The issue was whether the heterochromatic state of the X_i in the donor nucleus would determine the state of this chromosome in all cells of the clone, resulting in nonrandom inactivation. Alternatively, the epigenetic mark imposed on the X chromosomes during embryonic development of the donor animal could have been erased after nuclear transfer and randomly reestablished on either of the two chromosomes at the time of inactivation.

Using genetically marked X chromosomes it was demonstrated that X inactivation occurs randomly in the epiblast lineage of cloned mice (Eggan *et al.*, 2000). These data indicate that the epigenetic marks that distinguish X_a and X_i in somatic cells are removed and reestablished on either X during the cloning process, resulting in random X inactivation in the cloned animal. In contrast, epigenetic marks on X_a and X_i in the somatic donor cell are not removed in the TE lineage of the clone and predispose the X_a of the donor cell to be active and the donor X_i to be inactive. Thus, TE X inactivation occurs in cloned animals largely as it does in normal development. The gametic imprint-like mark, present in the somatic cell, is removed only after allocation of the TE lineage, leading to imprinted extraembryonic X inactivation (Lyon, 1999a).

In contrast to somatic cells, both X chromosomes are active in female ES cells, suggesting the lack of an imprint-like epigenetic mark on either chromosome (Panning *et al.*, 1997; Wutz and Jaenisch, 2000). When ES cells were used as nuclear donors, random X inactivation was observed in the epiblast as well as the TE of the cloned embryo. This observation is consistent with the existence of a mechanism that assures random X inactivation in the TE lineage in the absence of an epigenetic mark on the donor chromosomes (Eggan *et al.*, 2000).

These observations indicate that the different epigenetic information present on the somatic donor X chromosomes is not immediately removed following nuclear transfer, but is interpreted in the TE lineage resulting in nonrandom inactivation. After TE allocation, the epigenetic marks distinguishing the two donor X chromosomes are removed and randomly reestablished in the epiblast lineage, allowing for random inactivation in the somatic cells of the cloned female. In summary, the process of X inactivation is faithfully recapitulated in cloned female embryos and largely resembles that in normal females (Eggan *et al.*, 2000).

Telomere Length Adjustment in Cloned Embryos

A characteristic of most somatic cells dividing *in vivo* or *in vitro* is the progressive shortening of their telomeres (DePinho, 2000). Once telomeres shorten below a critical length, apoptotic pathways are activated and the cell loses its ability to divide (DePinho, 2000). It is known that telomerase, the enzyme that maintains telomere length, is expressed during gametogenesis and in embryonic cells consistent with maximal telomere length in the early embryo (DePinho, 2000). Importantly, it has been shown that expression of telomerase in aging cells can restore telomere length and rescue a cell from mitotic senescence (Counter *et al.*, 1998).

The observation that Dolly and other cloned sheep had shortened telomeres (Shiels *et al.*, 1999) raised the question of whether the gradual erosion of telomere length that occurs during both the *in vivo* and *in vitro* aging of the donor cell nucleus might lead to premature aging of cloned animals. However, in contrast to the shortened telomeres observed in cloned sheep, cloned cattle have telomere lengths either similar to or longer than their age-matched controls. When near-senescent embryonic bovine cells were used as nuclear donor, telomere length and cellular proliferative life span were completely restored or even enhanced by the cloning process (Lanza *et al.*, 2000). Further studies investigating telomere length in ear biopsies taken directly from a large numbers of cloned and control offspring concluded that telomere length in cloned neonatal cattle was nearly identical to that of newborn controls (Tian *et al.*, 2000; Kato *et al.*, 2000). Notably, it was determined that telomere length was similar in fibroblasts derived both from clones that died shortly after delivery and from clones that survived to adulthood, suggesting that neither telomere length maintenance nor premature cellular aging plays a role in the neonatal mortality of cloned cows (Tian *et al.*, 2000). It has also been demonstrated that telomerase activity is reactivated in cloned preimplantation bovine embryos with kinetics slightly delayed yet similar to those in normal fertilized controls (Betts *et al.*, 2001).

Species-specific differences in the mechanisms of either telomere length maintenance or reconstruction between cows and sheep are possible explanations for the differences observed. However, the observation that telomere length in cloned mice is also normal (Wakayama *et al.*, 2000a) suggests that exposing the telomeres to telomerase early in embryogenesis (Betts *et al.*, 2001), similar to telomerase overexpression in senescent cells (Counter *et al.*, 1998), is sufficient to restore normal telomere length and extend cellular life span.

Loss of Normal Imprinted Gene Expression as a Potential Cause of the Phenotypes Observed in Cloned Animals

Unlike X inactivation and telomere length maintenance, epigenetic information controlling imprinted gene expression is primarily established during gametogenesis, with allele-specific marks being maintained only postzygotically (Bartolomei and Tilghman, 1997; Latham *et al.*, 1994; Tremblay *et al.*, 1995). It therefore seems likely that any alterations in epigenetic information, controlling imprinted gene expression, incurred by the donor cell nucleus would be present in any resulting cloned embryo. It has been realized that *in vitro* maintenance of both mammalian embryos and cell lines can lead to changes in DNA methylation that disrupt normal regulation of imprinted genes (Dean *et al.*, 1998; Doherty *et al.*, 2000). Because the most common phenotypes observed in cloned animals are neonatal death and fetal growth abnormalities, this loss of epigenetic information is of particular relevance in the preparation of donor cells used for nuclear transfer (Eggan *et al.*, 2001; Wakayama and Yanagimachi, 1999; McCreath *et al.*, 2000; Hill *et al.*, 1999). Similar phenotypes have been observed in both human patients and in mice as a consequence of both naturally occurring and targeted mutations disrupting imprinted gene expression (Tilghman, 1999; Jaenisch, 1997). These apparent similarities suggest that aberrant imprinting might be one of the causal molecular mechanisms for the abnormal phenotypes observed in cloned animals.

Evidence now suggests that abnormal imprinted gene expression might indeed be the source of some phenotypes observed in cloned animals. Analysis of imprinted gene expression in neonatal mice derived by nuclear transfer of ES cells revealed improper expression of several imprinted genes (Humpherys *et al.*, 2001). Most strikingly, *H19* expression was extinguished and *Igf2* levels were increased in both the placentas and the organs of many clones. Other clones expressed intermediate or normal levels of these imprinted genes. In all cases the methylation status at the *H19* differentially methylated region correlated well with that gene's expression. Stochastic losses in expression of other imprinted genes, such as *Meg1/Grb10* and *Peg1/Mest* were also observed. In contrast, some imprinted genes seemed to retain their normal expression (*Snrpn* and *Peg3*) and methylation levels [*Igf2r* differentially methylated region of intron 2 (DMR2)]. Strikingly, all of the epigenetic abnormalities found within the cloned animals could also be found in the donor ES cell populations from which they were derived. Furthermore, when imprinted gene expression was analyzed in mice generated from the same donor cells by tetraploid embryo complementation, similar abnormalities in gene expression were observed. As a whole, these results suggest that for the subset of imprinted genes analyzed, epigenetic abnormalities were present in the ES cells prior to nuclear transfer and were not likely a direct result of the cloning procedure *per se* (Humpherys *et al.*, 2001).

Loss of information encoding imprinted gene expression *in vivo* seems rare but appears to be aggravated by *in vitro* culture (Dean *et al.*, 1998; Doherty *et al.*, 2000).

It might be expected that animals cloned from noncultured cumulus cells possess imprinted gene expression patterns more like those of normal offspring, whereas clones derived from cultured cells, such as tail-tip or embryonic fibroblasts, may show disruptions in imprinted gene expression similar those observed in ES cell NT animals.

However, genome-wide expression analysis on placentas from mice cloned from both cumulus and ES cell donor cells has revealed abnormal expression of both imprinted and nonimprinted genes (Humpherys, unpublished observations). Strikingly, there were disruptions in gene expression both common and unique to animals cloned from these cultured and primary cells. These results indicate that both the *in vitro* and *in vivo* history of the donor cell may affect both development as well as postnatal phenotypes that are unique to animals cloned from a particular cell type (Ogonuki *et al.*, 2002; Tamashiro *et al.*, 2002).

Thus far no clear correlation can be made between disruption in expression of any one imprinted gene with the phenotypes observed in clones (Humpherys *et al.*, 2001). However, the apparently stochastic nature of epigenetic insults incurred during ES cell culture, and cellular aging combined with the reciprocal functions of many imprinted genes, may complicate this analysis (Humpherys *et al.*, 2001; Dean *et al.*, 1998). Interestingly, and consistent with this hypothesis, mounting evidence in both sheep and mice suggests that prolonged *in vitro* culture of donor cell populations prior to nuclear transfer may increase the likelihood of neonatal complications in cloned offspring (Eggan *et al.*, 2001; McCreath *et al.*, 2000; Denning *et al.*, 2001). It will therefore be important to determine whether long-term culture of somatic cells can cause disruptions in imprinted gene expression similar to those seen in embryonic stem cells and if phenotypes in clones can be linked directly to any of these disruptions.

What Molecular Mechanisms Control Reprogramming?

It is now clear that even nuclei from highly differentiated cells, such as lymphocytes, can be reprogrammed after nuclear transfer to a pluripotent state compatible with embryonic development (Hochedlinger and Jaenisch, 2002). Furthermore, although several epigenetic characteristics of a differentiated cell are returned to an embryonic ground state by nuclear transfer (Lanza *et al.*, 2000; Eggan *et al.*, 2000), the mechanisms by which this reprogramming is accomplished remain largely unknown. In fact, it remains unclear as to what extent reprogramming is a passive process, involving dilution of factors required for maintenance of a differentiated state into the oocyte cytoplasm, or an active process requiring specific factors within the oocyte that actively remodel the differentiated chromatin to reestablish pluripotency.

Some combination of these two mechanisms seems likely. For instance, restoration of telomere length must be an active process, recruiting telomerase and other factors required for telomere polymerization and extension. However, silencing of tissue-specific genes and X chromosome reactivation after nuclear transfer may be accomplished primarily by dilution of tissue-specific transcription and heterochromatin factors, which are not found in the oocyte. Meanwhile, other epigenetic information established during gametogenesis, such as allele-specific DNA methylation encoding imprinted gene expression may be completely refractory to reprogramming. If reprogramming indeed requires enzymatic work by molecular machines within the mammalian oocyte, the identification of these enzymes, as well as the chromatin components or genes that they target in the donor nucleus, are obviously of considerable interest.

Because of cloning's inherent inefficiencies, nuclear transfer experiments using lymphocytes and other terminally differentiated cells as nuclear donors must be carefully designed and carried out if we are to understand the mechanisms by which epigenetic reprogramming is accomplished. In addition, as suggested by recent NT studies with the mouse, increased standardization of nuclear transfer technology from lab to lab, and if possible from organism to organism, will substantially improve our ability to synthesize experimental data into a coherent hypothesis regarding why cloning by nuclear transfer succeeds or fails.

REFERENCES

Baguisi, A., Behboodi, E., Melican, D. T., Pollock, J. S., Destrempes, M. M., Cammuso, C., Williams, J. L., Nims, S. D., Porter, C. A., Midura, P., Palacios, M. J., Ayres, S. L., Denniston, R. S., Hayes, M. L., Ziomek, C. A., Meade, H. M., Godke, R. A., Gavin, W. G., Overstrom, E. W., and Echelard, Y. (1999). Production of goats by somatic cell nuclear transfer. *Nat. Biotechnol.* **17**, 456–461.

Bartolomei, M. S., and Tilghman, S. M. (1997). Genomic imprinting in mammals. *Annu. Rev. Genet.* **31**, 493–525.

Barton, S. C., Surani, M. A., and Norris, M. L. (1984). Role of paternal and maternal genomes in mouse development. *Nature* **311**, 374–376.

Betthauser, J., Forsberg, E., Augenstein, M., Childs, L., Eilertsen, K., Enos, J., Forsythe, T., Golueke, P., Jurgella, G., Koppang, R., Lesmeister, T., Mallon, K., Mell, G., Misica, P., Pace, M., Pfister-Genskow, M., Strelchenko, N., Voelker, G., Watt, S., Thompson, S., and Bishop, M. (2000). Production of cloned pigs from *in vitro* systems. *Nat. Biotechnol.* **18**, 1055–1059.

Betts, D., Bordignon, V., Hill, J., Winger, Q., Westhusin, M., Smith, L., and King, W. (2001). Reprogramming of telomerase activity and rebuilding of telomere length in cloned cattle. *Proc. Natl. Acad. Sci. U.S.A.* **98**, 1077–1082.

Campbell, K. H., McWhir, J., Ritchie, W. A., and Wilmut, I. (1996). Sheep cloned by nuclear transfer from a cultured cell line. *Nature* **380**, 64–66.

Cibelli, J. B., Stice, S. L., Golueke, P. J., Kane, J. J., Jerry, J., Blackwell, C., Ponce de Leon, F. A., and Robl, J. M. (1998). Cloned transgenic calves produced from nonquiescent fetal fibroblasts. *Science* **280**, 1256–1258.

Clarke, D. L., Johansson, C. B., Wilbertz, J., Veress, B., Nilsson, E., Karlström, H., Lendahl, U., and Frisén, J. (2000). Generalized potential of adult neural stem cells. *Science* **288**, 1660–1663.

Counter, C. M., Hahn, W. C., Wei, W., Caddle, S. D., Beijersbergen, R. L., Lansdorp, P. M., Sedivy, J. M., and Weinberg, R. A. (1998). Dissociation among *in vitro* telomerase activity, telomere maintenance, and cellular immortalization. *Proc. Natl. Acad. Sci. U.S.A.* **95**, 14723–14728.

Csankovszki, G., Panning, B., Bates, B., Pehrson, J. R., and Jaenisch, R. (1999). Conditional deletion of Xist disrupts histone macroH2A localization but not maintenance of X inactivation. *Nat. Genet.* **22**, 323–324.

Daniels, R., Hall, V., and Trounson, A. O. (2000). Analysis of gene transcription in bovine nuclear transfer embryos reconstructed with granulosa cell nuclei. *Biol. Reprod.* **63**, 1034–1040.

Dean, W., Bowden, L., Aitchison, A., Klose, J., Moore, T., Meneses, J. J., Reik, W., and Feil, R. (1998). Altered imprinted gene methylation and expression in completely ES cell-derived mouse fetuses: Association with aberrant phenotypes. *Development* **125**, 2273–2282.

Denning, C., Burl, S., Ainslie, A., Bracken, J., Dinnyes, A., Fletcher, J., King, T., Ritchie, M., Ritchie, W. A., Rollo, M., de Sousa, P. A., Travers, A., Wilmet, T., and Clark, A. J. (2001). Deletion of the alpha (1,3) galactosyltransferase (GGTA1) gene and the prion protein (Prp) gene in sheep. *Nat. Biotech.* **19**, 559–562.

DePinho, R. A. (2000). The age of cancer. *Nature* **408**, 248–254.

Doherty, A. S., Mann, M. R., Tremblay, K. D., Bartolomei, M. S., and Schultz, R. M. (2000). Differential effects of culture on imprinted H19 expression in the preimplantation mouse embryo. *Biol. Reprod.* **62**, 1526–1535.

Eggan, K., Akutsu, H., Hochedlinger, K., Rideout, W., 3rd, Yanagimachi, R., and Jaenisch, R. (2000). X-Chromosome inactivation in cloned mouse embryos. *Science* **290**, 1578–1581.

Eggan, K., Akutsu, H., Loring, J., Jackson-Grusby, L., Klemm, M., Rideout, W. M., 3rd, Yanagimachi, R., and Jaenisch, R. (2001). Hybrid vigor, fetal overgrowth, and viability of mice derived by nuclear cloning and tetraploid embryo complementation. *Proc. Natl. Acad. Sci. U.S.A.* **98**, 6209–6214.

Epstein, C. J., Smith, S., Travis, B., and Tucker, G. (1978). Both X chromosomes function before visible X-chromosome inactivation in female mouse embryos. *Nature* **274**, 500–503.

Fulka, J., Jr., First, N. L., and Moor, R. M. (1996). Nuclear transplantation in mammals: Remodelling of transplanted nuclei under the influence of maturation promoting factor. *BioEssays* **18**, 835–840.

Gurdon, J. B., and Colman, A. (1999). The future of cloning. *Nature* **402**, 743–746.

Hill, J. R., Roussel, A. J., Cibelli, J. B., Edwards, J. F., Hooper, N. L., Miller, M. W., Thompson, J. A., Looney, C. R., Westhusin, M. E., Robl, J. M., and Stice, S. L. (1999). Clinical and pathologic features of cloned transgenic calves and fetuses (13 case studies). *Theriogenology* **51**, 1451–1465.

Hilscher, W. (1999). Some remarks on the female and male Keimbahn in the light of evolution and history. *J. Exp. Zool.* **285**, 197–214.

Hochedlinger, K., and Jaenisch, R. (2002). Monoclonal mice generated by nuclear transfer from mature B and T donor cells. *Nature* **415**, 1035–1038.

Humpherys, D., Eggan, K., Akutsu, H., Hochedlinger, K., Rideout, W. M., 3rd, Biniszkiewicz, D., Yanagimachi, R., and Jaenisch, R. (2001). Epigenetic instability in ES cells and cloned mice. *Science* **293**, 95–97.

Jaenisch, R. (1997). DNA methylation and imprinting: Why bother? *Trends Genet.* **13**, 323–329.

Jaenisch, R., and Wilmut, I. (2001). Don't clone humans! *Science* **291**, 2552.

Jeppesen, P., and Turner, B. M. (1993). The inactive X chromosome in female mammals is distinguished by a lack of histone H4 acetylation, a cytogenetic marker for gene expression. *Cell* **74**, 281–289.

Kafri, T., Ariel, M., Brandeis, M., Shemer, R., Urven, L., McCarrey, J., Cedar, H., and Razin, A. (1992). Developmental pattern of gene-specific DNA methylation in the mouse embryo and germ line. *Genes Dev.* **6**, 705–714.

Kang, Y. K., Koo, D. B., Park, J. S., Choi, Y. H., Chung, A. S., Lee, K. K., and Han, Y. M. (2001). Aberrant methylation of donor genome in cloned bovine embryos. *Nat. Genet.* **28**, 173–177.

Kato, Y., Tani, T., Sotomaru, Y., Kurokawa, K., Kato, J., Doguchi, H., Yasue, H., and Tsunoda, Y. (1998). Eight calves cloned from somatic cells of a single adult. *Science* **282**, 2095–2098.

Kato, Y., Tani, T., and Tsunoda, Y. (2000). Cloning of calves from various somatic cell types of male and female adult, newborn and fetal cows. *J. Reprod. Fertil.* **120**, 231–237.

Kaufman, M. H., and Webb, S. (1990). Postimplantation development of tetraploid mouse embryos produced by electrofusion. *Development* **110**, 1121–1132.

Kratzer, P. G., and Gartler, S. M. (1978). HGPRT activity changes in preimplantation mouse embryos. *Nature* **274**, 503–504.

Lanza, R. P., Cibelli, J. B., Blackwell, C., Cristofalo, V. J., Francis, M. K., Baerlocher, G. M., Mak, J., Schertzer, M., Chavez, E. A., Sawyer, N., Lansdorp, P. M., and West, M. D. (2000). Extension of cell life-span and telomere length in animals cloned from senescent somatic cells. *Science* **288**, 665–669.

Latham, K. E., Doherty, A. S., Scott, C. D., and Schultz, R. M. (1994). Igf2r and Igf2 gene expression in androgenetic, gynogenetic, and parthenogenetic preimplantation mouse embryos: Absence of regulation by genomic imprinting. *Genes Dev.* **8**, 290–299.

Lyon, M. F. (1999a). Imprinting and X-chromosome inactivation. *Cell Differ.* **25**, 73–90.

Lyon, M. F. (1999b). X-chromosome inactivation. *Curr. Biol.* **9**, R235–R237.

Marahrens, Y., Panning, B., Dausman, J., Strauss, W., and Jaenisch, R. (1997). Xist-deficient mice are defective in dosage compensation but not spermatogenesis. *Genes. Dev.* **11**, 156–166.

McCreath, K. J., Howcroft, J., Campbell, K. H., Colman, A., Schnieke, A. E., and Kind, A. J. (2000). Production of gene-targeted sheep by nuclear transfer from cultured somatic cells. *Nature* **405**, 1066–1069.

McGrath, J., and Solter, D. (1983). Nuclear transplantation in the mouse embryo by microsurgery and cell fusion. *Science* **220**, 1300–1302.

McGrath, J., and Solter, D. (1984). Inability of mouse blastomere nuclei transferred to enucleated zygotes to support development *in vitro*. *Science* **226**, 1317–1319.

Nagy, A., Gocza, E., Diaz, E. M., Prideaux, V. R., Ivanyi, E., Markkula, M., and Rossant, J. (1990). Embryonic stem cells alone are able to support fetal development in the mouse. *Development* **110**, 815–821.

Nagy, A., Rossant, J., Nagy, R., Abramow-Newerly, W., and Roder, J. C. (1993). Derivation of completely cell culture-derived mice from early-passage embryonic stem cells. *Proc. Natl. Acad. Sci. U.S.A.* **90**, 8424–8428.

Nichols, J., Zevnik, B., Anastassiadis, K., Niwa, H., Klewe-Nebenius, D., Chambers, I., Scholer, H., and Smith, A. (1998). Formation of pluripotent stem cells in the mammalian embryo depends on the POU transcription factor Oct4. *Cell* **95**, 379–391.

Norris, D. P., Brockdorff, N., and Rastan, S. (1991). Methylation status of CpG-rich islands on active and inactive mouse X chromosomes. *Mamm. Genome* **1**, 78–83.

Ogonuki, N., Inoue, K., Yamamoto, Y., Noguchi, Y., Tanemura, K., Suzuki, O., Nakayama, H., Doi, K., Ohtomo, Y., Satoh, M., Nishida, A., and Ogura, A. (2002). Early death of mice cloned from somatic cells. *Nat. Genet.* **30**, 253–254.

Ogura, A., Inoue, K., Takano, K., Wakayama, T., and Yanagimachi, R. (2000). Birth of mice after nuclear transfer by electrofusion using tail tip cells. *Mol. Reprod. Dev.* **57**, 55–59.

Onishi, A., Iwamoto, M., Akita, T., Mikawa, S., Takeda, K., Awata, T., Hanada, H., and Perry, A. C. (2000). Pig cloning by microinjection of fetal fibroblast nuclei. *Science* **289**, 1188–1190.

Ono, Y., Shimozawa, N., Ito, M., and Kono, T. (2001). Cloned mice from fetal fibroblast cells arrested at metaphase by a serial nuclear transfer. *Biol. Reprod.* **64**, 44–50.

Panning, B., Dausman, J., and Jaenisch, R. (1997). X chromosome inactivation is mediated by Xist RNA stabilization. *Cell* **90**, 907–916.

Peura, T. T., Lane, M. W., Lewis, I. M., and Trounson, A. O. (2001). Development of bovine embryo-derived clones after increasing rounds of nuclear recycling. *Mol. Reprod. Dev.* **58**, 384–389.

Polejaeva, I. A., Chen, S. H., Vaught, T. D., Page, R. L., Mullins, J., Ball, S., Dai, Y., Boone, J., Walker, S., Ayares, D. L., Colman, A., and Campbell, K. H. (2000). Cloned pigs produced by nuclear transfer from adult somatic cells. *Nature* **407**, 86–90.

Rideout, W. M., 3rd, Wakayama, T., Wutz, A., Eggan, K., Jackson-Grusby, L., Dausman, J., Yanagimachi, R., and Jaenisch, R. (2000). Generation of mice from wild-type and targeted ES cells by nuclear cloning. *Nat. Genet.* **24**, 109–110.

Rideout, W. M., 3rd, Eggan, K., and Jaenisch, R. (2001). Nuclear cloning and epigenetic reprogramming of the genome. *Science* **293**, 1093–1098.

Shiels, P. G., Kind, A. J., Campbell, K. H., Waddington, D., Wilmut, I., Colman, A., and Schnieke, A. E. (1999). Analysis of telomere lengths in cloned sheep. *Nature* **399**, 316–317.

Shin, T., Kraemer, D., Pryor, J., Liu, L., Rugila, J., Howe, L., Buck, S., Monphy, K., Lyons, L., and Westhusin, M. (2002). A cat cloned by nuclear transplantation. *Nature* **415**, 859–860.

Sims, M., and First, N. L. (1994). Production of calves by transfer of nuclei from cultured inner cell mass cells. *Proc. Natl. Acad. Sci. U.S.A.* **91**, 6143–6147.

Takagi, N., and Sasaki, M. (1975). Preferential inactivation of the paternally derived X chromosome in the extraembryonic membranes of the mouse. *Nature* **256**, 640–642.

Tamashiro, K. L., Wakayama, T., Akutso, H., Yamazaki, Y., Lachey, J. L., Wortman, M. D., Seeley, R. J., D'Alessio, D. A., Woods, S. C., Yanagimachi, R., and Sakai, R. R. (2002). Cloned mice have an obese phenotype not transmitted to their offspring. *Nat. Med.* **8**, 262–267.

Tani, T., Kato, Y., and Tsunoda, Y. (2001). Direct exposure of chromosomes to nonactivated ovum cytoplasm is effective for bovine somatic cell nucleus reprogramming. *Biol. Reprod.* **64**, 324–330.

Tian, X. C., Xu, J., and Yang, X. (2000). Normal telomere lengths found in cloned cattle. *Nat. Genet.* **26**, 272–273.

Tilghman, S. M. (1999). The sins of the fathers and mothers: Genomic imprinting in mammalian development. *Cell* **96**, 185–193.

Tremblay, K. D., Saam, J. R., Ingram, R. S., Tilghman, S. M., and Bartolomei, M. S. (1995). A paternal-specific methylation imprint marks the alleles of the mouse H19 gene. *Nat. Genet.* **9**, 407–413.

Tsunoda, Y., and Kato, Y. (1997). Full-term development after transfer of nuclei from 4-cell and compacted morula stage embryos to enucleated oocytes in the mouse. *J. Exp. Zool.* **278**, 250–254.

Wakayama, T., and Yanagimachi, R. (1999). Cloning of male mice from adult tail-tip cells. *Nat. Genet.* **22**, 127–128.

Wakayama, T., and Yanagimachi, R. (2001). Mouse cloning with nucleus donor cells of different age and type. *Mol. Reprod. Dev.* **58**, 376–583.

Wakayama, T., Perry, A. C., Zuccotti, M., Johnson, K. R., and Yanagimachi, R. (1998). Full-term development of mice from enucleated oocytes injected with cumulus cell nuclei. *Nature* **394**, 369–374.

Wakayama, T., Rodriguez, I., Perry, A. C., Yanagimachi, R., and Mombaerts, P. (1999). Mice cloned from embryonic stem cells. *Proc. Natl. Acad. Sci. U.S.A.* **96**, 14984–14989.

Wakayama, T., Shinkai, Y., Tamashiro, K. L., Niida, H., Blanchard, D. C., Blanchard, R. J., Ogura, A., Tanemura, K., Tachibana, M., Perry, A. C., Colgan, D. F., Mombaerts, P., and Yanagimachi, R. (2000a). Cloning of mice to six generations. *Nature* **407**, 318–319.

Wakayama, T., Tateno, H., Mombaerts, P., and Yanagimachi, R. (2000b). Nuclear transfer into mouse zygotes. *Nat. Genet.* **24**, 108–109.

Wells, D. N., Misica, P. M., Day, T. A., and Tervit, H. R. (1997). Production of cloned lambs from an established embryonic cell line: A comparison between *in vivo-* and *in vitro*-matured cytoplasts. *Biol. Reprod.* **57**, 385–393.

Wells, D. N., Misica, P. M., and Tervit, H. R. (1999). Production of cloned calves following nuclear transfer with cultured adult mural granulosa cells. *Biol. Reprod.* **60**, 996–1005.

Wilmut, I., Schnieke, A. E., McWhir, J., Kind, A. J., and Campbell, K. H. (1997). Viable offspring derived from fetal and adult mammalian cells. *Nature* **385**, 810–813.

Wutz, A., and Jaenisch, R. (2000). A shift from reversible to irreversible X inactivation is triggered during ES cell differentiation. *Mol. Cell* **5**, 695–705.

PLASTICITY OF SOMATIC NUCLEUS BY EPIGENETIC REPROGRAMMING VIA CELL HYBRIDIZATION

Takashi Tada, Masako Tada, and M. Azim Surani

INTRODUCTION

More than 30 years ago, Henry Harris (1965) pioneered the use of cell fusion techniques to generate heterokaryons between HeLa cells and rat lymphocytes, hen erythrocytes, and rabbit macrophages to examine the behavior of differentiated nuclei. Later, this technique of cell fusion was widely used for many diverse purposes. Examples in the life sciences include gene mapping with radiation hybrid panels produced by interspecific hybrid cells (McCarthy, 1996), as well as cell differentiation manifested by the extinguishment of cell type-specific transcription (Boshart *et al.*, 1993), production of monoclonal antibody (Hiatt, 1991), and the identification of tumor suppressor genes (Anderson and Stanbridge, 1993). More significantly, cell fusion experiments between differentiated cell types were used to study genomic plasticity (Blau *et al.*, 1985; Baron and Maniatis, 1986; Blau and Baltimore, 1991). Indeed, other aspects of genomic reprogramming were also studied by this approach, including X-chromosome inactivation and reactivation (Takagi, 1993) and switching of parental-origin-specific marks of imprinted genes (Tada *et al.*, 1997).

Somatic hybrid cells demonstrated that the expression of cell type-specific genes, but not housekeeping genes, is selectively extinguished, whereas activation of other cell type-specific genes is detected. The phenomenon of transcriptional downregulation is well known and is called *extinction*. For example, extinction of liver-specific transcripts was observed in hepatoma × fibroblast hybrids, pituitary gland-specific transcript in pituitary × fibroblast hybrids, and immunoglobuline genes in B cell × fibroblast hybrids; examples have been summarized by Boshart *et al.* (1993). The extinction of transcription is at least in part controlled by trans-acting regulatory factors. As factors involved in the extinction, three tissue-specific extinguisher (*Tse*) loci have been identified in mice and humans (Killary and Fournier, 1984; Petit *et al.*, 1986; Chin and Fournier, 1989). The *Tse* loci may function as dominant-negative regulators of tissue-specific genes. However, the precise suppression mechanism concerned with the extinction phenomenon still remains unknown.

X-Chromosome inactivation is essential to equalize the X-linked gene dosage between XX females and XY males. The X-chromosome inactivation occurs during early embryogenesis of mammalian females (Lyon, 1961). In early preimplantation embryos, two X chromosomes are active and then X-chromosome inactivation takes place when cells undergo differentiation (Monk and Harper, 1979). The inactive status of an X chromosome in a female cell is faithfully inherited by daughter cells in the somatic lineage, except for the germ cell lineage. An inactive X chromosome is reactivated in female primordial germ cells soon after they enter into the developing gonads in the mouse (Tam *et al.*, 1994). This X-chromosome reactivation is one aspect of the overall genomic reprogramming events that is linked to toti- or pluripotential competence in cells. Reactivation of an inactive X chromosome is also observed in preimplantation embryos, following transplantation of a somatic nucleus into enucleated oocytes to generate a clone (Eggan *et al.*, 2000). More importantly, reactivation of an inactive X chromosome is also observed in hybrid cells between mouse embryonal carcinoma (EC) cells and thymocytes (Takagi *et al.*, 1983). The phenotype of the hybrid cells resembles the parental EC cell and not the somatic cells. To address whether EC hybrid cells contribute to developing embryos when introduced into host blastocysts, near-tetraploid hybrid cells between multipotential mouse EC cells and thymocytes were used for making chimeras. However, no contribution of the hybrid cells to normal embryogenesis was detected, though teratocarcinomas were successfully derived following their subcutaneous injection into adult recipients (Martin *et al.*, 1984). Thus, nuclear reprogramming of thymocytes was not proved in the near-tetraploid EC hybrid cells. Subsequently we carried out experiments to address the ability of pluripotential stem cells to reprogram somatic nuclei in hybrid cells. These studies indicate reprogramming of a somatic nucleus to pluripotency (Tada *et al.*, 1997; Tada *et al.*, 2001).

NUCLEAR REPROGRAMMING OF SOMATIC CELLS BY HYBRIDIZATION WITH EMBRYONIC STEM CELLS

The embryonic stem (ES) cell is a pluripotential stem cell derived from the inner cell mass cells of blastocysts. The phenotype of an ES cell is characterized by a large nucleus, small amount of cytoplasm, a robust capacity for self-renewal, and differentiation capability into a variety of cell types, similar to EC cells. Unlike EC cells, ES cells generally retain a normal set of chromosomes and effectively participate in normal embryonic development, as shown by contribution to various types of tissues, including germ cells in chimeras produced by reintroduction of ES cells into a recipient blastocyst. To study the capacity of ES cells to induce reprogramming of a somatic nucleus, we made hybrid cells between ES cells and thymocytes by a technique of electric fusion (Tada *et al.*, 2001) (Fig. 1). In these ES hybrid cells, expression of the thymocyte-specific transcript of *Thy-1.2* was extinguished, just as reported previously in a similar experiment using EC × thymocyte hybrid cells (Martin *et al.*, 1984).

Mammalian adult tissues contain somatic stem cells, which are capable of trans-differentiating to pluripotential stem cells (Fuchs and Segre, 2000). Thymocytes consist of a variety of cell types, including T cell precursors, B cell precursors, and lymphoid cells. Prior to expression of antigen-specific receptors, the T cell receptor and immunoglobulin genes must be assembled from V, J, and in some cases, D gene segments by the process of V(D)J recombination (Fowlkes and Pardoll, 1989). Thus, the DNA rearrangement is one of the significant signs of thymocyte differentiation to lymphoid cells. To examine if ES cells were hybridized with somatic stem cells in thymocytes, the lymphoid-cell-specific DNA rearrangement was polymerase chain reaction (PCR)-amplified with primer sets specific to the V(D)J recombination region in the ES hybrid cell lines. There was no sign of rearrangement of the immunoglobulin locus in the DNA from ES cells. However, in more than 50% of the ES hybrid

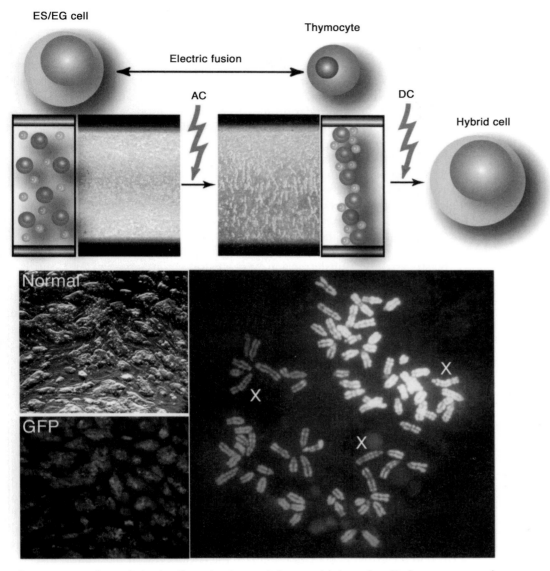

Figure 1 ES and gEG hybrid cell production and the reactivation of an X chromosome and Oct4–GFP reporter gene derived from somatic cells via cell fusion. ES and gEG cells are hybridized with thymocytes by electric fusion. In all ES hybrid cell colonies, the Oct4–GFP gene, which is repressed in thymocytes, is reactivated. An X chromosome, which is inactivated in the female somatic cell, is reactivated in the ES hybrid cell as visualized by synchronous replication of all three X chromosomes.

cell lines examined, DNA rearrangement in at least one segment of the V, J, and D regions was detected in addition to intact DNA from ES cells. Therefore, the ES cells had been hybridized with thymocytes differentiated to lymphoid cells.

We also examined if the inactive X chromosome present in the female somatic nucleus can be reactivated in ES–somatic hybrid cells. Two X chromosomes are active in a female ES cell, but one of them is inactive in a somatic cell. Facultative heterochromatin formation of an X chromosome is facilitated by epigenetic modification, including delay of DNA replication to late S phase (Takagi *et al.*, 1982), DNA hypermethylation (Jones, 1999), and histone hypoacetylation (Turner, 2000). X-Chromosome replication timing is detected by acridine orange staining following continuous incorporation of bromodeoxyuridine (BrdU) through the second half of S phase (Takagi *et al.*, 1982). The active X chromosome and the autosomes are seen as banded red and green elements, whereas the inactive X chromosome is uniformly dull red in female somatic cells due to delayed replication. In hybrid cell lines between male ES cells and female thymocytes, an inactive X chromosome present in thymocytes was reactivated, as exhibited by three synchronously replicating X chromosomes (active X chromosomes) in metaphase (Fig. 1), suggesting that histone hypoacetylation and DNA methylation through an entire X chromosome are dramatically reprogrammed by trans-acting factors derived from ES cells.

The reactivation of X chromosome in ES hybrid cells may be a reflection of an extensive reprogramming of the somatic nucleus, which may result in the restoration of pluripotency to the somatic nucleus. To address this question, ES cells were hybridized with thymocytes collected from 129/Sv-Rosa26 transgenic mice carrying the *neo/lacZ* reporter gene, the expression of which is ubiquitously detected in all tissues (Friedrich and Soriano, 1991). The hybrid cells carrying the full chromosomal components of the tetraploid cell were introduced into a blastocoel cavity of mouse blastocysts and transferred into uteri of foster mothers (Fig. 2). Contribution of the ES hybrid derivatives was detected as blue cells by X-gal staining of E7.5 chimeric embryos. Differentiation potential of the ES hybrid cells was represented by contribution of the donor cells to ectodermal, mesodermal, and endodermal tissues (Fig. 2). However, tetraploid (ES–somatic) hybrid cells can make only a limited contribution in the presence of normal host diploid cells in chimeras. ES tetraploid hybrid cells may have a tendency to contribute more readily to the extraembryonic tissues than to the fetus (Nagy *et al.*, 1990).

We went on to examine whether nuclear reprogramming of somatic cells occurred soon after cell hybridization with ES cells. To visualize reprogramming of somatic nuclei, thymocytes collected from the *Oct4–GFP* (green fluorescent protein) transgenic mice were applied for making ES–somatic hybrid cells. The expression of the *Oct4* gene is uniquely confined to germ cells, preimplantation embryos, and the epiblast of early postimplantation embryos (Yeom *et al.*, 1996), and it is required for the maintenance of these pluripotential cells (Niwa *et al.*, 2000). Temporal and spatial expression patterns of GFP in the *Oct4–GFP* transgenic mice correspond to those of the endogenous *Oct4* (Yoshimizu *et al.*, 1999). Thus, the activity of GFP provides an ideal marker for the identification of totipotential and pluripotential cells. When we fused ES cells with the thymocytes carrying *Oct4–GFP*, which is transcribed in the growing oocytes but not in the thymocytes of the adult transgenic mice, successive observation of the hybrid cell culture revealed reactivation of GFP, which is repressed in thymocytes. Expression was first detected 48 hours after cell hybridization (Fig. 1). Within the next 24 hours, several additional GFP-positive colonies were observed. Furthermore, DNA demethylation was associated with the reactivation of somatic *Oct4–GFP*, suggesting that transcriptional reactivation of *Oct4* is achieved through change in chromatin structure.

The combined evidence from our experiments shows that epigenetic reprogramming of the somatic nucleus occurs following fusion with ES cells. This includes reactivation of the X chromosome and *Oct4* gene, as well as restoration of pluripotency. Interestingly, the epigenetic modifications associated with imprinted genes were not, however, affected in the ES–thymocyte hybrid cells. Parental-origin-specific expression and repression of imprinted genes are also regulated by epigenetic modification, including allele-specific methylation (Brannan and Bartolomei, 1999). DNA methylation associated with imprint of the imprinted genes, including insulin-like growth factor 2 receptor (*Igf2r*), was not changed before and after cell hybridization (Tada *et al.*, 2001). Maternal methylation of the intronic region of *Igf2r*, which is paternally repressed, is well characterized as the primary imprint, which is essential for establishing its imprinting status (Stoger *et al.*, 1993). The

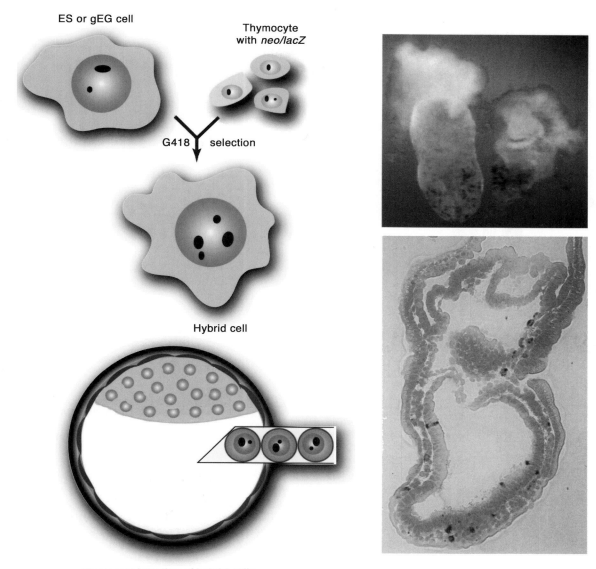

Figure 2 Contribution of ES and gEG hybrid cells in chimeric embryos. ES and gEG cells are hybridized with thymocytes carrying the neo/lacZ reporter gene. Hybrid cells are reintroduced into the blastocoel cavity of normal blastocysts. Contribution of these cells is visualized as blue cells by X-gal staining in E7.5 chimeric embryos. Hybrid cell derivatives contribute to the three primary embryonic germ layers as seen in a semithin section.

primary methylation imprint is maintained through preimplantation development. Reprogramming of the imprints requires passage through germ cell development (Tucker *et al.*, 1996). Maternal methylation of *Igf2r* is maintained in host ES cells, similar to the inner cell mass cells of mouse blastocysts (Fig. 3). Thus, the maintenance of epigenetic status of imprinted genes in somatic nuclei in ES–somatic hybrids is consistent with the observation that these changes can occur only in the germ cell lineage; the property to induce such changes is absent from ES cells. Nevertheless, ES cells have the ability to induce extensive reprogramming of somatic nuclei in hybrid cells, as demonstrated in our studies. Thus, the preimplantation embryo-type reprogramming machinery is active in ES cells.

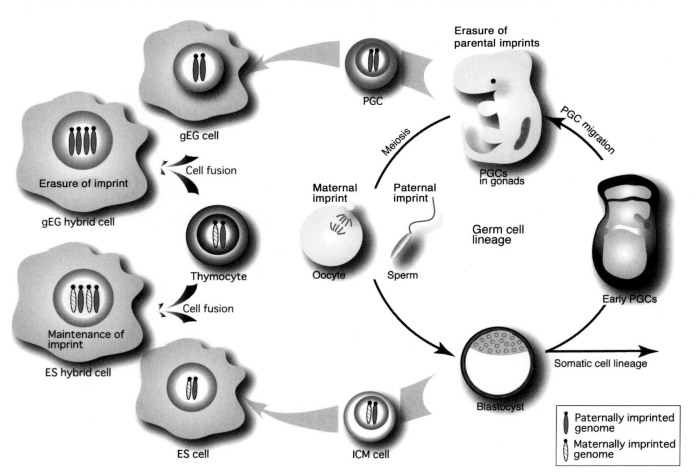

Figure 3 *Reprogramming of genomic imprinting in the mouse life cycle and in a hybrid cell. Parental imprints from sperm and egg are maintained in ES cells derived from the inner cell mass (ICM) cells of blastocysts. In contrast to ES cells, parental imprints are erased in gEG cells derived from primordial germ cells (PGCs) after they enter the gonads in E11.5–12.5 embryos. The distinction between parental chromosomes as a result of imprints is lost at this time. Fusion of a somatic cell with an ES or gEG cell confers pluripotency to the somatic nuclei. However, the fate of parental imprints in hybrid cells with ES and gEG differs, because parental imprints are erased from somatic nuclei only after fusion with gEG but not in ES–somatic hybrid cells. This is consistent with the differences in the properties of the precursor cells for ES and gEG cells, which are epiblast and primordial germ cells, respectively.*

NUCLEAR REPROGRAMMING OF SOMATIC CELLS BY HYBRIDIZATION WITH EMBRYONIC GERM CELLS

As indicated previously, the germ cell lineage may possess additional potential for inducing reprogramming of somatic nuclei, particularly of imprinted genes. The embryonic germ (EG) cell is a pluripotential stem cell derived from a primordial germ cell (PGC) (Matsui *et al.*, 1992; Resnick *et al.*, 1992). The phenotype of EG cells closely resembles that of ES cells. Mouse EG cells, which are similar to ES cells, are capable of effectively contributing to a variety of tissues, including germ cells, when reintroduced into a recipient blastocyst. Interestingly, EG cells can be established from PGCs at various developmental stages. Mouse EG cells derived from migrating PGCs of E8.5–9.5 embryos give rise to no detectable developmental defects in chimeras (Labosky *et al.*, 1994). By contrast, EG cells derived from PGCs after their entry into gonads in E11.5–12.5 embryos (gEG) (Fig. 3) exert detrimental effects on embryonic development characterized by overgrowth and skeletal malformation in chimeras (Tada *et al.*, 1998). These effects are at least in part caused by biallelic expression or repression of some imprinted genes due to erasure of parental imprint. In fact, allele-specific DNA methylation of many imprinted genes is lost and both alleles are modified to hypomethylation. This biallelic hypomethylation state is irreversible even after the gEG cells undergo differentiation into cells in the somatic cell lineage in chimeras. The phenotypic characteristics of gEG cells are attributed to their epigenotype but not their genotype, as shown by their switching to a normal epigenotype in the next generation.

Because the gEG cells may have additional potential for inducing epigenetic modifications, we examined gEG–thymocyte hybrid cells. As with ES cells, we first observed extinction of thymocyte-specific transcript, *Thy-1.2* and reactivation of an X chromosome, which is silent in the female somatic cell, by shift of X-chromosome replication timing from late to early in S phase. Pluripotential competence was similarly confirmed by chimera formation with gEG hybrid cells carrying the *lacZ/neo* transgene (Fig. 2). The X-gal-positive gEG–somatic hybrid cell derivatives contributed to the three primary germ layers—ectoderm, mesoderm, and endoderm—of E7.5–10.5 chimeric embryos produced by the technique of blastocyst injection. These data indicate that thymocyte nuclei were reprogrammed and acquired pluripotential competence (Tada *et al.*, 1997).

The most striking effect in hybrids with gEG cells, which was not observed with ES cells, was the reprogramming of thymocyte imprinting status in the gEG hybrid cells (Tada *et al.*, 1997). Contrary to the maintenance of the allele-specific gametic methylation associated with the imprinted gene of *Igf2r* in ES hybrid cells, there was disruption of allele-specific DNA methylation of three paternally expressed imprinted genes, *Igf2*, *Peg3*, and *Peg1/Mest*, and three maternally expressed imprinted genes, *H19*, *Igf2r*, and *p57kip2*, of thymocyte nuclei. Both parental alleles became hypomethylated in gEG hybrid cells (Fig. 3). To investigate whether the disruption of allele-specific DNA methylation gives rise to reactivation of a repressed allele of imprinted gene, thymocytes from the mutant mice in which exons 3 to 8 of *Peg1/Mest* were replaced with a IRES/β-*geo* cassette (Lefebvre *et al.*, 1998) were used to make gEG hybrid cells. The mutated allele, which is maternally repressed and methylated in somatic cells (Lefebvre *et al.*, 1997), was demethylated in the gEG hybrid cells. Reactivation of the maternally mutated allele was visualized as X-gal-positive cells in chimeric embryos produced by blastocyst injection of the gEG hybrid cells (Tada *et al.*, 1997). Demethylation was not just confined to imprinted genes, but also affected other nonimprinted genes such as *Pgk-1*, *Pgk-2*, β-*globin*, and *Aprt*. Furthermore, the minor satellite DNA sequences, *MR150*, which are scattered around the chromosome arms and centromeres (Pietras *et al.*, 1983), were also extensively demethylated. Thus, genome-wide demethylation of thymocyte nuclei takes place by the activity of *trans*-acting factors from the gEG cells.

The genomic modifications observed in the gEG hybrid cells with thymocyte nuclei are consistent with the changes that occur in the germ cell lineage. Erasure of bulk methylation is detected prior to meiosis in germ cells (Kafri *et al.*, 1992; Brandeis *et al.*, 1993). The germ-cell-specific demethylation event is accompanied by reactivation of inactive X chromosome in female germ cells (Tam *et al.*, 1994) and erasure of parental imprints (Szabo and Mann, 1995; Tada *et al.*, 1998; Obata *et al.*, 1998). Thus, a germ cell type of reprogramming takes place in somatic nuclei by hybridization with gEG cells.

These studies reveal similarities and differences between the pluripotential ES and gEG cells in their capacity to reprogram somatic nuclei in hybrid cells. Both of them can apparently restore pluripotential competence to somatic nuclei. However, the capability to reprogram parental imprints is found only in gEG cells, not in ES cells, suggesting that the regulatory mechanism controlling DNA methylation status differs between them. To investigate which mechanism is dominant, hybrid cells between ES and gEG cells were produced. In ES × gEG hybrid cells, allele-specific methylation of *Igf2r* derived from ES cells disappeared and all alleles became hypomethylated. Thus, the demethylation activity present in gEG cells is dominant to the methylation imprint maintenance seen in ES cells (Tada *et al.*, 2001).

CLONING BY SOMATIC CELL TRANSPLANTATION INTO OOCYTES

The ability to reprogram a somatic nucleus to totipotency was first demonstrated in cloned frogs generated by nuclear transfer from the intestinal endoderm of feeding tadpoles into activated enucleated eggs (Gurdon, 1962). Similar nuclear reprogramming capacity of the oocyte has been shown in mammals, resulting in the production of cloned sheep, cows, mice, and pigs (Wilmut *et al.*, 1997; Kato *et al.*, 1998; Wakayama *et al.*, 1998; Onishi *et al.*, 2000; Polejaeva *et al.*, 2000). Thus, nuclear reprogramming of somatic nuclei in oocytes is a common phenomenon among animals. This plasticity of the somatic nucleus is widely found using various types of somatic cells. Similarly, such nuclear reprogramming of somatic cells is seen when they are fused with ES and gEG cells. However, the molecular mechanisms involved in these processes are poorly understood. The role of factors present in oocytes may be crucial for the modifications of the parental genomes in the zygote and for early development. How they act to confer totipotency to a somatic nucleus remains to be elucidated. Nuclear reprogramming factors may not be for conferring toti- or pluripotential competence upon somatic nuclei, but they may function as important molecules in early postfertilization events.

CONCLUSION

Pluripotential stem (ES and gEG) cells are capable of reprogramming somatic nuclei by cell fusion, similar to the nuclear reprogramming represented by transplantation of somatic nuclei into enucleated oocytes in various animal species. This plasticity of the somatic nucleus is induced by *trans*-acting factors derived from ES and gEG cells. Nuclear reprogramming of somatic cells in hybrid cells is demonstrated by (1) reactivation of the inactive X chromosome present in a female somatic cell, (2) contribution of ES and gEG hybrid cells to the primary germ layers of chimeric embryos, and (3) reactivation of the somatic-cell-derived *Oct4–GFP* marker gene, which is specifically expressed in pluripotent cells. Apart from the ability of both ES and gEG cells to confer pluripotency to somatic nuclei, the gEG cells have an additional capacity not found in ES cells, which is the ability to erase DNA methylation associated with imprinted genes. Nuclear reprogramming of somatic nuclei in gEG hybrid cells corresponds to the germ-cell-type reprogramming; nuclear reprogramming of somatic nuclei in ES hybrid cells corresponds to the preimplantation embryo type (Fig. 3). The latter process is perhaps similar to what occurs during the process of somatic cell cloning via nuclear transplantation.

REFERENCES

Anderson, M. J., and Stanbridge, E. J. (1993). Tumor suppressor genes studied by cell hybridization and chromosome transfer. *FASEB J.* 7, 826–833.

Baron, M. H., and Maniatis, T. (1986). Rapid reprogramming of globin gene expression in transient heterokaryons. *Cell* 46, 591–602.

Blau, H. M., and Baltimore, D. (1991). Differentiation requires continuous regulation. *J. Cell Biol.* 112, 781–783.

Blau, H. M., Pavlath, E. C., Hardeman, C.-P., Chiu, L., Silberstein, S. G., Webster, S. C., and Miller, C. (1985). Plasticity of the differentiated state. *Science* 230, 758–766.

Boshart, M., Nitsch, D., and Schutz, G. (1993). Extinction of gene expression in somatic cell hybrids— A reflection of important regulatory mechanisms? *Trends Genet.* 9, 240–245.

Brandeis, M., Kafri, T., Ariel, M., Chaillet, J. R., McCarrey, J., Razin, A., and Cedar, H. (1993). The ontogeny of allele-specific methylation associated with imprinted genes in the mouse. *EMBO J.* **12**, 3669–3677.

Brannan, C. I., and Bartolomei, M. S. (1999). Mechanisms of genomic imprinting. *Curr. Opin. Genet. Dev.* **9**, 164–170.

Chin, A. C., and Fournier, R. E. (1989). Tse-2: A trans-dominant extinguisher of albumin gene expression in hepatoma hybrid cells. *Mol. Cell. Biol.* **9**, 3736–3743.

Eggan, K., Akutsu, H., Hochedlinger, K., Rideout, W., 3rd, Yanagimachi, R., and Jaenisch, R. (2000). X-Chromosome inactivation in cloned mouse embryos. *Science* **290**, 1578–1581.

Fowlkes, B. J., and Pardoll, D. M. (1989). Molecular and cellular events of T cell development. *Adv. Immunol.* **44**, 207–264.

Friedrich, G., and Soriano, P. (1991). Promoter traps in embryonic stem cells: A genetic screen to identify and mutate developmental genes in mice. *Genes Dev.* **5**, 1513–1523.

Fuchs, E., and Segre, J. A. (2000). Stem cells: A new lease on life. *Cell* **100**, 143–155.

Gurdon, J. (1962). The developmental capacity of nuclei taken from intestinal epithelial cells of feeding tadpoles. *J. Embryol. Exp. Morphol.* **10**, 622–640.

Harris, H. (1965). Behaviour of differentiated nuclei in heterokaryons of animal cells from different species. *Nature* **206**, 583–588.

Hiatt, A. C. (1991). Monoclonal antibodies, hybridoma technology and heterologous production systems. *Curr. Opin. Immunol.* **3**, 229–232.

Jones, P. A. (1999). The DNA methylation paradox. *Trends Genet.* **15**, 34–37.

Kafri, T., Ariel, M., Brandeis, M., Shemer, R., Urvan, L., McCarrey, J., Cedar, H., and Razin, A. (1992). Developmental pattern of gene-specific DNA methylation in the mouse embryo and germ line. *Genes Dev.* **6**, 705–714.

Kato, Y., Tani, T., Sotomaru, Y., Kurokawa, K., Kato, J., Doguchi, H., Yasue, H., and Tsunoda, Y. (1998). Eight calves cloned from somatic cells of a single adult. *Science* **282**, 2095–2098.

Killary, A. M., and Fournier, R. E. (1984). A genetic analysis of extinction: Trans-dominant loci regulate expression of liver-specific traits in hepatoma hybrid cells. *Cell* **38**, 523–534.

Labosky, P. A., Barlow, D. P., and Hogan, B. L. M. (1994). Mouse embryonic germ (EG) cell lines: Transmission through the germline and differences in the methylation imprint of insulin-like growth factor 2 receptor (Igf2r) gene compared with embryonic stem (ES) cell lines. *Development* **120**, 3197–3204.

Lefebvre, L., Viville, S., Barton, S. C., Ishino, F., and Surani, M. A. (1997). Genomic structure and parent-of-origin-specific methylation of Peg1. *Hum. Mol. Genet.* **6**, 1907–1915.

Lefebvre, L., Viville, S., Barton, S. C., Ishino, F., Keverne, E. B., and Surani, M. A. (1998). Abnormal maternal behaviour and growth retardation associated with loss of the imprinted gene Mest. *Nat. Genet.* **20**, 163–169.

Lyon, M. F. (1961). Gene action in the X-chromosome of the mouse (*Mus musculus* L.). *Nature* **22**, 372–373.

Martin, G. M., Ogburn, C. E., Au, K., and Disteche, C. M. (1984). Altered differentiation, indefinite growth potential, diminished tumorigenicity, and suppressed chimerization potential of hybrids between mouse teratocarcinoma cells and thymocytes. *J. Exp. Pathol.* **1**, 103–133.

Matsui, Y., Zsebo, K., and Hogan, B. L. (1992). Derivation of pluripotential embryonic stem cells from murine primordial germ cells in culture. *Cell* **70**, 841–847.

McCarthy, L. C. (1996). Whole genome radiation hybrid mapping. *Trends Genet.* **12**, 491–493.

Monk, M., and Harper, M. I. (1979). Sequential X chromosome inactivation coupled with cellular differentiation in early mouse embryos. *Nature* **281**, 311–313.

Nagy, A., Gocza, E., Diaz, E. M., Prideaux, V. R., Ivanyi, E., Markkula, M., and Rossant, J. (1990). Embryonic stem cells alone are able to support fetal development in the mouse. *Development* **110**, 815–821.

Niwa, H., Miyazaki, J., and Smith, A. G. (2000). Quantitative expression of Oct-3/4 defines differentiation, dedifferentiation or self-renewal of ES cells. *Nat. Genet.* **24**, 372–376.

Obata, Y., Kaneko-Ishino, T., Koide, T., Takai, Y., Ueda, T., Domeki, I., Shiroishi, T., Ishino, F., and Kono, T. (1998). Disruption of primary imprinting during oocyte growth leads to the modified expression of imprinted genes during embryogenesis. *Development* **125**, 1553–1560.

Onishi, A., Iwamoto, M., Akita, T., Mikawa, S., Takeda, K., Awata, T., Hanada, H., and Perry, A. C. (2000). Pig cloning by microinjection of fetal fibroblast nuclei. *Science* **289**, 1188–1190.

Petit, C., Levilliers, J., Ott, M. O., and Weiss, M. C. (1986). Tissue-specific expression of the rat albumin gene: Genetic control of its extinction in microcell hybrids. *Proc. Natl. Acad. Sci. U.S.A.* **83**, 2561–2565.

Pietras, D. F., Bennett, K. L., Siracusa, L. D., Woodworth-Gutai, M., Chapman, V. M., Gross, K. W., Kane-Haas, C., and Hastie, N. D. (1983). Construction of a small Mus musculus repetitive DNA library: Identification of a new satellite sequence in *Mus musculus*. *Nucleic Acids Res.* **11**, 6965–6983.

Polejaeva, I. A., Chen, S. H., Vaught, T. D., Page, R. L., Mullins, J., Ball, S., Dai, Y., Boone, J., Walker, S., Ayares, D. L., Colman, A., and Campbell, K. H. (2000). Cloned pigs produced by nuclear transfer from adult somatic cells. *Nature* **407**, 86–90.

Resnick, J. L., Bixler, L. S., Cheng, L., and Donovan, P. (1992). Long-term proliferation of mouse primordial germ cells in culture. *Nature* **359,** 550–551.

Stoger, R., Kubicka, P., Liu, C.-G., Kafri, T., Razin, A., Cedar, H., and Barlow, D. P. (1993). Maternal-specific methylation of the imprinted mouse Igf2r locus identifies the expressed locus as carrying the imprinting signal. *Cell* **73,** 61–71.

Szabo, P. E., and Mann, J. R. (1995). Biallelic expression of imprinted genes in the mouse germ line: implications for erasure, establishment, and mechanisms of genomic imprinting. *Genes Dev.* **9,** 1857–1868.

Tada, M., Tada, T., Lefebvre, L., Barton, S. C., and Surani, M. A. (1997). Embryonic germ cells induce epigenetic reprogramming of somatic nucleus in hybrid cells. *EMBO J.* **16,** 6510–6520.

Tada, T., Tada, M., Hilton, K., Barton, S. C., Sado, T., Takagi, N., and Surani, M. A. (1998). Epigenotype switching of imprintable loci in embryonic germ cells. *Dev. Gene Evol.* **207,** 551–561.

Tada, M., Takahama, Y., Abe, K., Nakatsuji, N., and Tada, T. (2001). Nuclear reprogramming of somatic cells by *in vitro* hybridization with ES cells. *Curr. Biol.* **11,** 1553–1558.

Takagi, N. (1993). Variable X chromosome inactivation patterns in near-tetraploid murine EC × somatic cell hybrid cells differentiated in vitro. *Genetica* **88,** 107–117.

Takagi, N., Sugawara, O., and Sasaki, M. (1982). Regional and temporal changes in the pattern of X-chromosome replication during the early post-implantation development of the female mouse. *Chromosoma* **85,** 275–286.

Takagi, N., Yoshida, M. A., Sugawara, O., and Sasaki, M. (1983). Reversal of X-inactivation in female mouse somatic cells hybridized with murine teratocarcinoma stem cells *in vitro*. *Cell* **34,** 1053–1062.

Tam, P. L., Zhou, S. X., and Tan, S.-S. (1994). X-Chromosome activity of the mouse primordial germ cells revealed by the expression of an X-linked lacZ transgene. *Development* **120,** 2925–2932.

Tucker, K. L., Beard, C., Dausman, J., Jackson-Grusby, L., Laird, P. W., Lei, H., Li, E., and Jaenisch, R. (1996). Germ-line passage is required for establishment of methylation and expression patterns of imprinted but not of nonimprinted genes. *Gene Dev.* **10,** 1008–1020.

Turner, B. M. (2000). Histone acetylation and an epigenetic code. *BioEssays* **22,** 836–845.

Wakayama, T., Perry, A. F. C., Zuccotti, M., Johnson, K. R., and Yanagimachi, R. (1998). Full-term development of mice from enucleated oocytes injected with cumulus cell nuclei. *Nature* **394,** 369–374.

Wilmut, I., Schnieke, A. E., McWhir, J., Kind, A. J., and Campbell, K. H. S. (1997). Viable offspring derived from fetal and adult mammalian cells. *Nature* **385,** 810–813.

Yeom, Y. I., Fuhrmann, G., Ovitt, C. E., Brehm, A., Ohbo, K., Gross, M., Hubner, K., and Scholer, H. R. (1996). Germline regulatory element of Oct-4 specific for the totipotent cycle of embryonal cells. *Development* **122,** 881–894.

Yoshimizu, T., Sugiyama, N., De Felice, M., Yeom, Y. I., Ohbo, K., Masuko, K., Obinata, M., Abe, K., Scholer, H. R., and Matsui, Y. (1999). Germline-specific expression of the Oct-4/green fluorescent protein (GFP) transgene in mice. *Dev. Growth Differ.* **41,** 675–684.

CHAPTER 6

DETERMINANTS OF PLURIPOTENCY IN MAMMALS

Michele Boiani and Hans R. Schöler[1]

INTRODUCTION

Until recently, the concept of pluripotency was inextricably linked to the classic concepts of cell lineage and differentiation. It was unclear, however, whether stem cells should be defined by their host tissue or their degree of potency. In light of recent reports on adult stem cells, showing that cells from one tissue can give rise to cells of unrelated tissues, one may ask whether distinguishing between different levels of potency in mammals still makes sense (van der Kooy and Weiss, 2000). If a blood stem cell can colonize the brain and give rise to authentic brain cells, and if brain stem cells can colonize various tissues and organs and produce cells according to the new environment, then the fate of these cells seems conditional. If this is true, the search for cellular determinants of pluripotency could stop short of its goal; the stem cell would be like a blank slate, a *tabula rasa*, that is subjected to empirical determination according to the local environment. Predictability of cell differentiation along lineages and pathways would then be lost, and the conventional germ layer terminology could be abandoned. Such changes would entail a revolution in biology, challenging concepts that have been held as dogmas for a long time.

Such a radical revision of our classic concepts needs further support. For this reason, cell potency in mammals is reviewed here in light of recent discoveries but on the well-established grounds of classic embryology. First, definitions of pluripotency and totipotency are compared. Because both terms are often used synonymously, their differences are deciphered by approaching each term from various viewpoints, which will set the stage for the rest of this review. Second, an explanation is offered as to why pluripotency is crucial for the generation of a multicellular organism while allowing the genetic information to flow from one generation to the next. Also discussed are the fields and levels of stem cell potency, and embryoid bodies and chimeras are described as experimental systems for its measure. This discussion is particularly relevant because stem cell technology has recently resulted in exciting biomedical applications, such as therapeutic cloning, thereby moving to the front line of scientific and public interest. Finally, it is shown how nuclear transplantation has been used as a tool to dissect the molecular basis of pluripotency, at the nuclear level, and as a renewable source of one's own stem cells. Also provided is a list of novel (as well as established) markers, to measure the potency of these stem cells and their derivatives. Although, in the future, stem cells may be defined

---

[1]To whom correspondence should be addressed.

in terms of the repertoire of genes that is or not expressed, the current genetic characterization is insufficient and must be complemented by a functional definition of stem cells. The discussion is concluded by revisiting cloning from the standpoint of pluripotency (McLaren, 2000).

DEFINING PLURIPOTENCY BY MEANS OF TOTIPOTENCY

Potency in biology can be defined in terms of a developmental, cellular, or nuclear standpoint. Here potency is addressed from the cellular perspective and the other two viewpoints are referred to when appropriate. Defining cell potential is difficult because there is often no clearcut distinction between different potency levels, as, for example, between totipotency and pluripotency. This obscurity has sometimes been blurred by the indiscriminate use of inappropriate definitions across various research fields. It is rather difficult to prove that a cell is totipotent, i.e., it can form everything, because this requires testing in all possible environments. However, it is possible to prove that a cell is at least pluripotent, i.e., it can form almost everything. To provide a basis for understanding pluripotency, some of the common meanings of totipotency are outlined.

THE BASIC DEFINITION

Potency can be defined as the range of developmental capabilities of a cell that is in a permissive or supportive environment. A unipotent cell can just make one type of cell, whereas an oligopotent cell, such as a hematopoietic stem cell, can make several cell types. A multipotent cell can be considered as having the potential to generate multiple cell types, such as the derivatives of one germ layer. Finally, a cell is considered to be totipotent if it displays the full range of developmental capabilities when it is placed in a permissive environment.

A popular variant of this definition that applies to mammals refers to the capacity of a cell to produce all the cells of the organism, including the extraembryonic tissues. In this scenario, the uterus is the environment in which the embryonal stem cells develop into an organism. Although this definition is the only one that reflects what actually happens in nature, it provides a quite narrow viewpoint of totipotency. Strictly speaking, the zygote and the earliest blastomeres are the only cells in mammals that are capable of accomplishing this, such that once separated, each can give rise to the whole organism, including the extraembryonic tissues. One may ask, however, what are the functions of the earliest extraembryonic lineages in the mammalian embryo? The functions of the first extraembryonic lineage, the trophectoderm, are to implant into the uterine epithelium, mediate nutrition by the maternal tissue, and provide proliferative signals (chorionic gonadotropin in primates, interferon tau in other mammals). Embryonic viability would be compromised if either the trophectoderm or the maternal uterus were missing.

The nutritive-type cells of the earliest extraembryonic lineage are different from the later somatic cells that are required to induce specific lineages in the embryo, such as the germ cell or hematopoietic lineages. One may thus ask whether there is any other difference between the maternal and the trophectodermal supporting tissues besides their origin. If there is none, one should focus on what can generate an embryo proper, and consequently consider a cell as totipotent if it can do so. In this view, embryonic stem (ES) cells would be totipotent, because mice can be generated from ES cells when they are placed in the appropriate environment (Nagy *et al.*, 1990, 1993). In this particular case, the appropriate environment consists of two components, a trophectodermal "container" that is built from tetraploid blastomeres, and the maternal uterus. An embryo would die for the same reason—namely, the absence of a supportive environment—regardless of which one of these two components was missing. A discussion of the embryo having to be self-supportive by forming its own trophoblast is futile in this respect. This experiment even allows us to consider a single ES cell as totipotent because this one cell can generate a clone of ES cells that is used for aggregation, which in turn will generate a mouse. If this initial ES cell carried a mutation, every cell in the mouse would also carry it. So far, ES cells have proved to be a more pliable and powerful system than the other embryo-derived cells, the embryonic carcinoma (EC) and embryonic germ (EG) cells (see later).

THE CONTRIBUTIVE DEFINITION

According to a more flexible progeny test, a cell is totipotent if it is capable of contributing to all cell lineages of the organism. Contribution means that the candidate cell does not necessarily have to give rise to all cells of the new organism. As with the basic definition, an important issue concerns whether to include or exclude the trophectoderm lineage. To test this definition, the candidate cell is examined in the presence of other cells. For example, injection of cells, such as ES cells, into the cavity of a blastocyst, which is then carried through gestation in a female host, has been shown to result in cell contribution to the embryo proper and its annexes (yolk sac, amnion, allantois), but not to the extraembryonic annexes, such as the chorion and the fetal part of the placenta. According to classic concepts in developmental biology, a cell would be expected to contribute to embryo development according to its cell lineage. As long as prelineage cells, such as the early embryonic, the inner cell mass (ICM), or ES cells, were tested, the possibility of a differentiated cell contributing to a cell lineage other than its own could not be entertained. However, it has recently been discerned that adult stem cells (ASCs) have the ability to cross cell lineage boundaries into ectodermal, mesodermal, and endodermal lineages by an event known as "transdetermination" or "transdifferentiation" (see later).

The contributive definition of totipotency depends on the cellular environment, which can either enhance or depress the survival of test cells, leading to a selection process that may not necessarily be meaningful in terms of intrinsic cell potency. Lessons regarding cell potency have been learned from uniparental mouse embryos (reviewed in Pedersen, 1994). These rarely develop beyond day 10 of gestation (Kaufman *et al.*, 1977). However, donor cells from parthenogenetic (homozygous female pronuclei), gynogenetic (heterozygous female pronuclei), or androgenetic (heterozygous male pronuclei) embryos can be rescued by complementation (aggregation) with normal biparental embryos (Anderegg and Markert, 1986). The uniparental cells are capable of contributing to most cell lineages, including the germ line, of the resulting chimera. A female chimera of this type has successfully bred from her parthenogenetic component (Stevens *et al.*, 1977; Stevens, 1978), proving that fertilization is not a necessary condition for germ cell formation and that the oocyte is potentially totipotent in that, besides forming somatic cells, it can even form germ cells (McLaren, 1981). Yet, no uniparental embryo could fully develop on its own (Mann *et al.*, 1990). Nuclear adequacy of parthenogenetic mouse embryos to support development has also been reported (Hoppe and Illmensee, 1982).

The spatial analysis of cell distribution in chimeras showed preferential contribution of the gynogenetic cells to the embryo proper and of the androgenetic cells to the extraembryonic tissues. The uniparental cells that colonize the host embryo are titrated out during development and almost vanish by adulthood. Because these cells are taken over by the others, they do not get the opportunity to realize their potential. However, there seems to be no specific point during development at which uniparental embryos fail and no specific developmental process that they cannot undergo. Rather, failure seems to occur continuously. The possibility that restricted development results from poor placental development was disproved by transplantation of parthenogenetic ICMs into fertilized blastocysts deprived of their own ICM

(McLaren, 1981). Another possibility is that restricted development of uniparental embryos is caused by an imbalance of imprinted gene expression that regulates cell differentiation and proliferation in the developing embryo and fetus. Defining gene expression has become increasingly difficult since epigenetic regulation has been extended to include the level of gene transcription in addition to the binary on–off state. This issue is explored in detail when we present *Oct4*, a nonimprinted gene; the level of expression of this gene controls the fate of embryonic cells (see later).

THE "GENETIC FLOW" DEFINITION

In a minimalist view, the cell can be thought of as nothing but a vector for genetic information. It is the genetic program encoded in the nucleus that determines the potency of a cell. Therefore, we can define totipotency as the contribution of genetic information from one cell to all the cells of the organism. This understanding is reminiscent of the requirements of the basic definition. However, it does not require the donor cell to give rise to (all) cells of the organism in one generation. In fact, once the flow of genetic information becomes part of the germ line, transmission to all cell lineages is accomplished by the next generation, provided that the soma is supportive to fertility. In this view, a cell is totipotent if it can give rise to germ cells. For example, the genetic information of ES cells that enter the germ line will flow along this lineage in the next generation. Chimeras of normal with uniparental embryos have produced functional gametes derived from the latter (reviewed in Pedersen, 1994). This indicates that germ line contribution is characteristic of, but not specific to, totipotency, as defined by "genetic flow." However, it is important to test if the resulting gametes can generate a healthy organism.

THE TOTIPOTENCY DILEMMA

We have attempted to outline carefully the different views of potency. The major difference between these views pertains to how a cell's "environment" is considered. One can call a mammalian ICM cell totipotent if the trophectoderm is considered only to be an additional "environment" that is required for the development of the embryo. In contrast, if embryo proper is seen as just one of numerous integral components, the cell would be considered pluripotent. In the latter view, a mammalian pluripotent cell may give rise to the same lineages as does a cell in a nonmammalian species. In nonmammals, the cell would be considered totipotent, because there is no trophectoderm in nonmammalian species. We respect both views and could thus continue by using the word hybrid toti/pluripotency to emphasize this. We decided to continue with the term pluripotency, realizing that once one would be able to generate an embryo from only ES cells, such a distinction would become obsolete. We use the term totipotency when it is agreeable with both perspectives.

WHY PLURIPOTENCY?

ORIGIN OF PLURIPOTENCY

As life on Earth evolved from unicellularity to pluricellularity, like in *Volvox*, the need for pluripotency became established (Kirk, 1994). Because of the large number of cells making up the organism, they were not all required to carry out the full range of functions. On the other hand, it would be impossible to generate an organism if all body cells were unipotent in the same way. Pluripotency was a solution to this as well as a convenient adaptation to replace body cells lost by senescence or injury. As organisms were generated through processes more complex than cleavage or budding, culminating in sexual reproduction, pluripotency became essential to generate a complete body. The germ line developed as a distinct cell lineage, enabling the genetic flow from one generation to the next.

Because body cells in mammals carry the same amount of genetic information as the zygote, the potential for pluripotency theoretically resides in all of them (there are exceptions, such as lymphocytes). Destruction of up to 85% of the cells within a mouse embryo, prior to organogenesis, does not necessarily prevent formation of a normal mouse (Snow and Tam, 1979). This plasticity is thought to reside in the capacity of pluripotent embryonic cells to reprogram their development in response to environmental cues. At subsequent stages, the observation of more fixed cell phenotypes indicates that the actual potential of most body cells has been epigenetically, but not irreversibly, restricted and streamed into one path. The epigenetic unipotency of mammalian cells is resistant to change, as shown by the extreme difficulty of obtaining cloned mammals by somatic cell nucleus transplantation (SCNT) (Campbell, 1999; Wakayama and Yanagimachi, 1999). In some cases, however, cells exhibit a natural ability to adopt new developmental fates. This ensures homeostasis and eliminates the biological demand of maintaining all body cells capable of regenerating. These cells, known as stem cells, can generate both a replacement stem cell and a differentiating cell. To realize their potential, stem cells must be exposed to the appropriate microenvironmental cues (reviewed in Fuchs and Segre, 2000).

CONTEMPORARY PLURIPOTENCY

Mammalian stem cells have recently moved into the mainstream of scientific and public interest due to at least three reasons: (1) embryonic stem or ES-like cell lines have been derived from various species, including humans, from fertilized (Reubinoff *et al.*, 2000; Thomson *et al.*, 1998) or cloned (Kawase *et al.*, 2000; Munsie *et al.*, 2000; Wakayama *et al.*, 2001) embryos; (2) adult stem cells can transdifferentiate into developmentally unrelated cell types of all three germ layers (Bouwens, 1998); and (3) stem cells that are out of biological control in the adult may be one source for tumorigenesis (Sell, 1993).

COMMON THEMES

Stem cells, whether they are in the embryo or in the adult, share the key feature of pluripotency. We defined the potency of a cell as its potential to follow a number of different paths if it is in the appropriate environment. Until recently, the differentiation potential into ectoderm, mesoderm, and endoderm had been attributed only to ES cells. However, the finding that ASCs can transdifferentiate when moved to a new microenvironment indicates that they can either redefine their fate or are already pluripotent but are prevented from expressing their potential (reviewed in Fuchs and Segre, 2000). Gaining insight into the biological basis of cell potency will help to determine whether ASCs acquire or express the new potential directly according to the microenvironment, or pass through an intermediate embryonic stem cell-like state first. A fascinating question is whether ES and germ cells can directly differentiate into ASCs or vice versa. The identification of intrinsic and specific markers will provide the necessary tools to address these questions and will help explore how cell fate commitments are made in the embryo and in the adult.

THE FIELDS OF PLURIPOTENCY

THE NATIVE EMBRYO

The study of pluripotency was originally approached from the perspective of the embryo as a convenient milieu that allows presumptive pluripotent cells to express their potency after transplantation. This followed earlier studies on embryonic cell potency performed by monitoring the fate of single blastomeres. Obvious quantitative limitations were innate of such an approach. Blastomeres were either labeled and tracked during embryo development (clonal analysis) or dissociated and cultured individually (Tarkowski, 1959). In one of its most significant advances, this approach addressed the process of X-chromosome inactivation. With the current technology, this has been done in clone mouse embryos by the X-linked green fluorescent protein (GFP) (Eggan *et al.*, 2000). As the female embryo develops into a blastocyst, both X chromosomes are active in the undifferentiated ICM cells, but only one is active in the differentiated trophectoderm (TE) cells. Models of X-chromosome inactivation and the origin of the germ line have been proposed (McMahon *et al.*, 1983).

CELL LINES

Since the 1970s, new experimental tools and systems have become available for the study of pluripotency. The isolation and stable propagation of EC, ES, and EG cells were paramount achievements that heightened research and applications in biology and medicine. By culturing these cells *in vitro*, it became possible to modify their genome and to perpetuate these modifications through the germ line via chimeric embryos. Cell-specific antigens (including the latest transgene products) (Uhm *et al.*, 2000) help track the contribution of donor cells to host embryo lineages, and provide a measure of cell potency. Contribution to the germ line is the most desirable feature of chimeras (see section later). Apart from the integration of stem cells into recipient embryos, the use of endogenous cell markers (e.g., glucose phosphate isomerase) (Chapman *et al.*, 1972), labeling (e.g., peroxidase) (Balakier and Pedersen, 1982), or infection (e.g., provirus) (Soriano and Jaenisch, 1986) of individual blastomeres is a way to follow the progeny of selected embryonic cells during development.

EMBRYOID BODIES

The formation of embryoid bodies (EBs) from aggregated ES cells, in culture, has been viewed as a unique model system to study pluripotency (Doetschman *et al.*, 1985). ES cells are used alone or in combination with other cell types. A nice example of the latter is the coculture of ES cells with primitive endodermal-like cells, allowing the molecular dissection of the signals that are required for apoptosis (Coucouvanis and Martin, 1995, 1999). The most striking feature of EBs is their ability to differentiate into three-dimensional structures with a germ layer arrangement similar to that of postimplantation embryos. EBs and ES cells are a pliable and powerful experimental system, as shown by their ability to generate muscle, neuronal, and hematopoietic cells. But EBs differ from conventional ES cell cultures, which give rise only to colonies or layers of cells and thus present less differentiation options. So far, germ cells have not been derived from EBs. During normal development, the germ cell lineage arises and is set aside subsequent to the stage when ES cells are derived from the embryo and aggregated to form EBs, raising the possibility of deriving germ cells from EBs. Why has this not been possible thus far? Favorable conditions for development seem to be contained within EBs, as evident by the generation of endodermal, mesodermal, and ectodermal germ layer derivatives, awaiting appropriate triggers for commitment. However, so far, knowledge of how EBs differentiate is rather limited. Because of their embryo-like complexity, EBs are considered to be an important tool with which to develop biomedical applicable systems for tissue and organ generation in transplantation medicine.

ADULT STEM CELLS

ASCs are localized in specific niches of the body and ensure the homeostasis of adult tissues (Firulli and Olson, 1997; Orkin and Zon, 1997; Orkin, 1998; Edlund and Jessell, 1999). ASCs may represent the best alternative to therapeutic cloning or umbilical cord storage as a source of individual-specific stem cells, being useful for a number of medical conditions and diseases in which the availability of patient-compatible cells is required (Temple, 1999). Although they display broad fate plasticity, it has not yet been possible to isolate ASCs pure enough to be used for clinical applications. Enriched ASC populations have been obtained and tested for development and colonization, respectively, by injection into blastocysts and tissues. In several cases, changes of cell fate and transdifferentiation have been described. Testing hematopoietic stem cells (HSC) for repopulation in lethally irradiated hosts showed that advancing from the aorta–gonad–mesonephric region to fetal liver to adult bone marrow results in better repopulation (van der Kooy and Weiss, 2000). Further on, cells found in the bone marrow and dubbed multipotent adult progenitor cells (MAPCs) have been reported to differentiate into everything that ES cells can differentiate into, but without forming tumors once introduced *in vivo* (*New Scientist*, press release, Jan. 23, 2002). This suggests quite a counterintuitive scenario—namely, the appearance or maturation of stem cells in the adult—and hints of the concept that stem cells may not be the "legacy" of the embryo to the adult, but that they may also appear later in development.

THE POTENCY OF A SINGLE CELL

Potency is assessed by testing whether a single cell can contribute to all cell lineages of the body. It is possible that potency is apparent only when groups of cells are transplanted or aggregated with others. Limited information is available on the developmental competence and differentiation capability of individual ES cells microinjected into mouse morulae and blastocysts (Illmensee and Mintz, 1976; Saburi *et al.*, 1997). Nuclear transplantation into the oocyte seems to provide a clearcut experimental strategy that eliminates the possibility that the differentially committed pluripotential cell types present in the culture account for the variety of differentiated derivatives produced.

CLONING TO DISSECT THE MOLECULAR BASIS OF CELL POTENCY

Stem cells, chimeras, and EBs are multicellular approaches to study cell potency. Their potency is verified through differentiation into multiple cell lineages in culture and by their "genetic flow" through germ line contribution in chimeric offspring. Why cloning, then? The strictest verification of pluripotency requires a single cell to give rise to all fully differentiated somatic tissues in one generation. Cloning by nuclear transplantation fulfills the single cell requirement, but because it tests the single cell nucleus rather than the whole donor cell, it has been questioned as a suitable method to address cellular potency. Cloning may actually be the best way to study how potency is reestablished. Cloning by nuclear transfer allows the study of the molecular basis of cellular potency, because this is restored following transplantation of the nucleus of a unipotent cell into an oocyte. Preimplantation embryo development after somatic cell nuclear transfer demonstrates the ability of the transplanted nucleus to adopt a zygotelike developmental program. Postimplantation development accounts for the accomplishment of pluripotency. Because the starting material is a single nucleus introduced into a single cell, this process is initially cell autonomous and the analysis of genes involved in resetting cell potency is facilitated. As we will outline later, to ensure pluripotency, one of the genes that must be expressed in the mouse embryos as well as in clones, is *Oct4*. Mouse embryos cannot develop without Oct4 being expressed (Nichols *et al.*, 1998), and ES cells have been shown to differentiate into different lineages, or fail to do so, depending on Oct4 levels (Niwa *et al.*, 2000). Characterization of clone-derived ES cells is underway (Wakayama *et al.*, 2001) to test their functional normality, as measured by the ability to participate in chimeras and give rise to somatic and germ cells. The *Oct4* gene provides one possible parameter to explore these processes.

LEVELS OF CELL POTENCY

LANDMARKS OF CELL POTENCY

Mammals are formed from a single totipotent founder cell, the zygote. During development, the potency of cells becomes progressively restricted. In the mouse, a single blastomere from a four-cell or eight-cell embryo cannot generate a normal mouse on its own, whereas this is possible in the rabbit and sheep (reviewed in Pedersen, 1994). It has been proposed that the mouse embryo may be unable to recover from splitting by the blastocyst stage, and have a total cell number too small to support embryogenesis. As the embryo develops into morula and blastocyst, two cellular compartments with distinct differentiation commitments can be distinguished: the inner and the outer blastomeres at the morula stage and the inner cell mass and trophectoderm at the blastocyst stage. A minimal cell number must be attained for the ICM to be developmentally competent (McLaren, 1981; Papaioannou and Ebert, 1995). An archetypal stem cell population is found in the ICM of the blastocyst, from which all cell lineages found in the adult body, including the germ cell lineage, are derived. At or shortly after birth, most tissues are not to be remodeled or diversified any further, except for adult stem and germ cells. In fully developed mammals, in the context of their normal position within the organism, most adult cells are unipotent. ASCs can respond to specific cues from the environment, causing them to begin proliferation and differentiation. The germ line retains the information to direct the formation of gametes, which are the precursors of the zygote, which in turn will generate a new organism. A few decades ago it was generally believed that the spectrum of fates of mammalian cells could only become more restricted, except for the germ cells that link one generation to the next. The possibility of changing a cell's fate, or reverting the differentiative program of a cell, was inconceivable, as shown by schemes in Fig. 1.

POTENCY AT THE CELLULAR LEVEL

Prior to the advent of stem cell and cloning techniques, the classic perspective on cell potency arose from work on dissociated embryos. Studies on isolated blastomeres showed that two-cell blastomeres of mouse and rabbit embryos are totipotent. In contrast, one or two cleavages later, most individual blastomeres from four-cell and eight-cell embryos are impaired at development (Tarkowski and Wroblewska, 1967), forming undersized embryos consisting of trophoblast vesicles lacking an ICM. This observation led to the so-called inside–outside hypothesis (Graham, 1971). This postulate states that differentiation of cells into ICM and TE is determined by the inner or outer position of blastomeres at the morula stage. It has been suggested that the seal provided by the cell junctions between the outer cells makes differentiation possible by creating a chemically distinct interior environment. However, chimeras produced from aggregated 8-cell and 16-cell embryos gave rise to normal fertile offspring (Tarkowski, 1998), indicating that commitment is reversible and cell–cell interactions control the fate of blastomeres. The development of techniques to introduce cells into blastocysts (Gardner, 1968) allowed the testing the potency of cells other than blastomeres. With the exception of a few single-cell reports (Illmensee and Mintz, 1976; Saburi *et al.*, 1997), studies that involved the simultaneous transfer of multiple cells led to cell–cell interactions that hindered the understanding of the cell-autonomous bases of potency. It was not until nuclear transfer studies that we could start to dissect potency at a deeper level.

STEM CELL POTENCY

Despite the widespread and increasing use of stem cells from the mouse, and repeated attempts to obtain them from other mammalian species, knowledge about

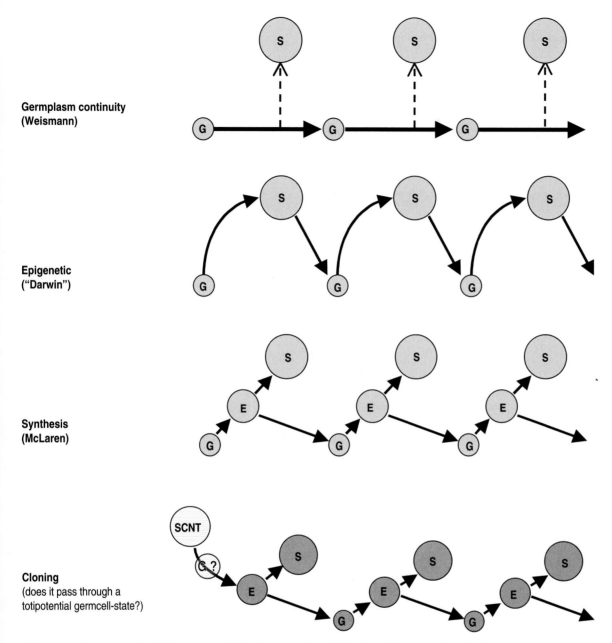

**Germplasm continuity
(Weismann)**

**Epigenetic
("Darwin")**

**Synthesis
(McLaren)**

Cloning
(does it pass through a
totipotential germcell-state?)

Figure 1 *The origin of cell lineages and the first question of cloning. How can the development of a nucleus-transplanted oocyte be reconciled with the classic schemes of development? S, Soma; E, epiblast; G, germ line; SCNT, somatic cell nuclear transfer.*

their origin and biology is limited. Stemlike cells from nonmouse species have shown heterogeneous biochemical and genetic features (see Tables 1 and 2 and references therein). The only definite traits of stem cells pertain to their function: the ability to self-renew, proliferate, and give rise to specialized cells. Stem cells can be pluripotent or oligopotent, as shown by differentiation of stem cells in culture, formation of EBs, and chimeras. A particular kind of chimera was produced by Nagy and colleagues (Nagy *et al.*, 1990), in which ES cells were introduced into a tetraploid blastocyst. Based on the successful generation of all-ES cell-derived newborns, the authors concluded that ES cells alone could support full development and are, therefore, totipotent. The newborns failed to survive after birth despite any obvious

Table 1 How to Make Stem Cells: An Essential Overview

Source	Type produced	Specificity	Potency
The inner cell mass (ICM)	Embryonic stem (ES) cells	Heterologous	All cell lineages
Hematopoietic and other "organ-specific" stem cells	Adult stem cells (ASC)	Auto/heterologous	Multiple cell types of one lineage; transdifferentiation from one lineage to another
Therapeutic cloning	ES or ES-like cells	Autologous	All cell lineages
Embryoid bodies	Stem cells later than ES stage	Auto/heterologous	Multiple cell lineages (including germ line?)

abnormality. In a follow-up study, Nagy and colleagues (Nagy *et al.*, 1993) obtained adult fertile mice. This technique has now been adapted to the use of normal (non-tetraploid) recipient blastocysts (Amano *et al.*, 2000) and employed as a control in the cloning procedure (Eggan *et al.*, 2001). In the mouse, ES cells do not normally differentiate into the TE lineage, *in vitro* or *in vivo*, and it is accepted that once ICM and TE have split, their fates cannot be reconciled (Beddington and Robertson, 1989).

POTENCY AT THE NUCLEAR LEVEL

The importance of nuclear transfer studies was evident from the work on amphibian embryos by Briggs and King (1950s) and Gurdon (1960s), and on mammals by Illmensee and Hoppe (1970s–1980s) and McGrath and Solter (1980s). Until recently, however, only nuclei taken from very early embryos were effective (although this procedure is basically a nucleus–cytoplasm reshuffling because of the similarity between the ooplasm and the early embryonic cytoplasm). Clone development was still often abnormal, for reasons that are not fully understood. Despite recurrent failures, the achievement of cloning highlights few, but crucial, points (Table 3). In the mouse, although the two-, four-, and eight-cell blastomeres are considered broadly pluripotent, if not totipotent, in that they can contribute to many different tissues in the adult, the nuclei behave rather differently. An early but not a late two-cell mouse nucleus can support development after transplantation into an enucleated zygote. However, eight-cell stage nuclei support full development once transferred into enucleated two-cell recipients (McLaren, 1981). Is this proof for different nuclear potencies? The interaction between the nucleus and cytoplasm is probably responsible for the degree of cellular potency. Because the nucleus is a cellular component, which is modified during this interaction, it does not have any potency *a priori*. The ability to support development is rather accomplished by nuclear–cytoplasmic interaction in a process referred to as (epigenetic) "reprogramming." Recent nuclear transfer experiments have proved that even the nuclei of terminally differentiated cells of adult mammals are capable of supporting full development when reprogrammed in "enucleated" oocytes. By this method, live clones of five different species (sheep, mouse, cattle, goat, and pig) were generated (reviewed in Solter, 2000). At the beginning of 2002, a report on cats adds to the list of the mammalian species cloned to date (Shin *et al.*, 2002). This contrasts the immense difficulties experienced by previous studies in obtaining embryos from embryonic nuclei. Wakayama and Yanagimachi (Wakayama and Yanagimachi, 2001) reviewed the cloning results obtained with mouse cells to test the hypothesis that cloning success is so low because random mutations accumulate in cells during the life of an animal (noteworthy is the shortening of chromosome telomeres also involved in the aging process). These authors' data could not support a correlation between cloning performance and the age and type of nucleus donor. So, what makes the cloning success rate differ between adult, young, fetal, and embryonic nucleus donor cells? One possible explanation that we discuss in this chapter is related to cellular and nuclear potency. A differential epigenetic state of the nucleus, which makes the gene expression profile different from cell to cell, is one possible way of thinking of the cellular bases of pluripotency. Because the efficiency of cloning by nuclear transfer significantly increases when ES cells are the donors, it has been argued that previous success in cloning from adult somatic cells was actually due to pluripotent stem cells, which were incidentally picked up from the cell population (Liu, 2001). If stem cells require an appropriate microenvironment in order to fulfill their potency, the better results obtained with ES cells may hint that the oocyte cytoplasm is an appropriate microenvironment for the ES cell nuclei but not for other cells. If this is the case, the search for ways to improve the efficiency of cloning could stop short of its goal—unless the oocyte recipient could be changed to match the incoming nucleus or *vice versa*. This would imply that the nucleus is predisposed or permissive for cloning rather than being reprogrammed.

THE CELLULAR BASES OF POTENCY

GENERAL CONCEPTS

What directs cell potency into cell fate is still an open question in embryology. In particular, the issue of whether the developmental decision of embryonic cells to differentiate into germ line or somatic lineages is preimposed or is regulated along the way has yet to be resolved. In the 1800s, two possible scenarios of germ cell and soma relationship were drawn (Fig. 1). The continuity of the germinal cytoplasm, or germ plasm, through the generations reflects the idea of Weismann (Weissman, 1885). It postulates that a stock of totipotential cells are set aside from the prospective soma very early in development and are fated as germ cells. The other cells excluded from the germ plasm lose their totipotency and give rise to the body structures that support the germ line—the body is the "slave" of its immortal germ plasm. This implies that the germ line can modify the soma (but not vice versa). Contrasting the Weismannist view is the scheme based on the epigenetic, or Darwinist, view (McLaren, 1981). In this scheme, the germ line is discontinuous and germ cells transform into the soma. Later in development, some of the body's cells regain totipotency epigenetically and transform into germ cells. Thus, the soma can potentially influence the germ line.

THE SIMPLEST SITUATION

Volvox offers an example of a simple cell fate specification in development. In this green alga, which is related to *Chlamydomonas*, it is the difference in cell size rather than any difference in cytoplasmic quality that is important for cell fate (Kirk, 1994). At the sixth cell divisions, if a *Volvox* cell is larger than $8\,\mu m$ in diameter, it becomes a potentially immortal germ cell. If it is smaller, it is doomed to somatic fate and certain death.

DETERMINANTS IN ANIMAL EMBRYOS

In many animals, the spatial patterning of the early embryo is largely determined by determinants within the egg. In such species, the germinal cytoplasm, or germ plasm, can be identified cytologically. Mammals might be an exception to this concept. It has been argued that body axes formation begins too late in mammalian development to be specified by material in the egg or in the zygote. In fact, mammals lack obvious, localized determinants in the cytoplasm of their oocytes (but germ cells are often characterized by electron-dense bodies known as "nuage") (McLaren, 1981) and the germ line is formed relatively late in development. Only limited spatial heterogeneity can be detected in the oocyte. This is exemplified by the cortical struc-

Table 2 Pluripotent Cells in Various Species[a]

Species	Type	Source	Classical assays					
			Evaluation by morphology	Differentiation into cell lineages	Chimerism of eye/coat color	Germ line in chimeras	Response to LIF	Passages in vitro
Mouse[b]	ES	ICM 3.5 dpc	+	In vitro, in vivo, EB	+	+	+	?
	EC	Tumor	+	In vivo	+	+	+	?
	EG	PGC 8.5 dpc	+	Chimera	+	+	?	?
Rat[c]	ES-like	Blastocyst 5 dpc	+	In vitro, in vivo, EB	+	−	−/+	>40
Rabbit[d]	ES-like	Day 5 embryo	+	In vitro, in vivo, EB	+	−	?	23
	GC/EG	Fetus 18–22 dpc	+	In vivo	+/−	?	?	?
Hamster[e]	ES-like	Blastocyst	+	In vitro, EB	−	?	?	>25
Mink[f]	ES-like	Morula–blastocysts	+	In vitro, in vivo (mouse host), EB/tumor formation (mouse host)	−	−	?	?
Pig[g]	ES-like	Blastocyst	+	In vitro, in vivo, EB	+	−	?	?
	GC/EG	Fetus 25 dpc	+	In vitro, in vivo, EB	+	−	?	>29
Cow[h]	ES-like	ICM 8–9 dpc	+	In vitro, in vivo	+	?	?	?
	GC/EG	Fetus 35–175 dpc	?	ICM incorporation in vitro	?	?	?	?
Sheep[i]	ES-like	Blastocyst (ICM)	+	?	Blood	?	?	?
Monkey[j] (rhesus, marmoset)	ES	Blastocyst	+	In vitro, in vivo (mouse host), EB	−	?	−	3 months
Human[k]	ES	Blastocyst (ICM)	+	In vitro, in vivo (mouse host), EB	?	?	−	~300
	EC	Testicular tumor	+	In vitro, in vivo (mouse host)	?	?	−	?
	EG	PGC	+	EB	?	?	+/−	>20
Chicken[l]	ES-like	Blastoderm	+	In vitro, in vivo (chicken host)	+	+	+	?
Fish[m]	EG	?	?	?	?	?	?	?
Zebra fish	ES-like	Blastula	+	In vivo, in vitro	?	?	?	?
Medaka fish	ES-like	Blastula	+	In vivo, in vitro	+	?	?	20–60

Antigenic profile (undifferentiated)												
AP	SSEA-1	SSEA-3	SSEA-4	TRA-1-80	TRA-1-81	GCTM-2	EMA-1	EMA-6	ECMA-7	Oct4	High telomerase	Nuclear transfer (stage attained)
+	+	−	−	−	−	−	+	+	+	+	+	+ (adult)
+	+	−	−	−	−	−	+	+	−	+	?	+ (morula)
+	+	−	−	?	?	?	+	+	?	+	?	+ (midgestation)
+	+	?	?	?	?	?	?	?	+	+	?	−
+	?	?	?	?	?	?	?	?	?	?	?	+ (blastocysts not transferred)
+	+	?	?	?	?	?	?	?	?	?	?	+ (blastocysts not transferred)
?	?	?	?	?	?	?	?	?	?	?	?	−
?	?	?	?	?	?	?	?	?	?	?	?	−
?	?	?	?	?	?	?	?	?	?	?	?	−
+	?	?	?	?	?	?	?	?	?	?	?	−
−/+	−	?	?	?	?	?	?	?	−	?	?	+ (calves)
+	?	?	?	?	?	?	?	?	?	?	?	+ (blastocyst, limited postimpl. +1 calf)
?	?	?	?	?	?	?	?	?	?	?	?	−
+	−	+	+	+	+	?	?	?	?	?	?	+ (cleavage only)
+	−	+	+	+	+	+	?	?	?	+	+	−
+	−	+	+	+	+	+	?	?	?	+	?	−
+	+	+	+	+	+	?	?	?	?	?	?	−
+	+	+	?	?	?	?	+	+	?	?	+	−
+	+	?	?	?	?	?	+	+	?	?	?	−
+	?	?	?	?	?	?	?	?	?	?	?	?
+	?	?	?	?	?	?	?	?	?	?	?	?

Table 3 The Heritage of Cloning—Five Years after Dolly (1996–2001)

Issue	Pros	Cons
Technical	Proof of principle	The vast majority of clones do not develop
Biological	Somatic cell differentiation or specialization does not involve irreversible genetic change—the differentiative program of a cell can be reversed (plasticity)	Loss of diversity after skipping meiotic recombination
Medical	Possibility to derive individual-specific stem cells for therapy	The production-scale and time may not match the needed applications
Ethical	Hope for sufferers of certain diseases	Use of potential embryo as a source of cells

ture of the cytoskeleton and the absence of microvilli in the region of the polar body extrusion (Nicosia *et al.*, 1977), or the presence of sperm antigens on the cell membrane up to the eight-cell stage close to the original sperm point of entry (Gabel *et al.*, 1979). However, because parthenogenetic and intracytoplasmic sperm-injected (ICSI) embryos can develop to midgestation and to term, respectively, doubts are cast on the relevance of these heterogeneities.

INVERTEBRATES

A very different situation occurs in invertebrates, such as worms and insects. The oocyte of *Caenorhabditis elegans* contains cytoplasmic P granules that become localized to germ cell progenitors shortly after fertilization. Although the granules still remain somewhat mysterious, a gene, *mes-1*, has been implicated in the control

[a]Symbols and abbreviations: +, positive; −, negative; −/+ or +/−, contrasting reports, the first more likely; ?, unknown, not tested or not available. ES, embryonic stem; EC, embryonic carcinoma; EG, embryonic germ; GC, germ cell; ICM, inner cell mass; EB, embryoid body; dpc, days postconception.
[b]Selected references: Brinster, *Cancer Res.* **36**, 3412–3414, 1976; Donovan, *Curr. Top. Dev. Biol.* **29**, 189–225, 1994; Evans and Kaufman, *Nature* **292**(5819), 154–156, 1981; Illmensee and Mintz, *Proc. Natl. Acad. Sci. U.S.A.* **73**, 549–553, 1976; Kato *et al.*, *Development* **126**, 1823–1832, 1999; Martin, *Proc. Natl. Acad. Sci. U.S.A.* **78**(12), 7634–7638, 1981; Matsui, Zsebo, and Hogan, *Cell* **70**, 841–847, 1992; Mintz and Illmensee, *Proc. Natl. Acad. Sci. U.S.A.* **72**, 3585–3589, 1975; Modlinski *et al.*, *Development* **108**, 337–348, 1990; Stewart, Gadi, and Bhatt, *Dev. Biol.* **161**, 626–628, 1994; Vassilieva *et al.*, *Exp. Cell Res.* **258**, 361–373, 2000.
[c]Selected references: Brenin *et al.*, *Transplant. Proc.* **29**, 1761–1765, 1997; Iannaccone *et al.*, *Dev. Biol.* **85**, 124–125, 1997; Iannaccone *et al.*, *Dev. Biol.* **163**, 288–292, 1994; Ouhibi *et al.*, *Mol. Reprod. Dev.* **40**, 311–324, 1995; Stranzinger, *Int. J. Exp. Pathol.* **77**, 263–267, 1996; Vassilieva *et al.*, *Exp. Cell Res.* **258**, 361–373, 2000.
[d]Selected references: Du *et al.*, *J. Reprod. Fertil.* **104**, 219–223, 1995; Graves and Moreadith, *Mol. Reprod. Dev.* **36**, 424–433, 1993; Moens *et al.*, *Mol. Reprod. Dev.* **43**, 38–46, 1996; Moens *et al.*, *Differentiation* **60**, 339–345, 1996; Moens *et al.*, *Zygote* **5**, 47–60, 1997; Schoonjams *et al.*, *Mol. Reprod. Dev.* **45**, 439–443, 1996.
[e]Selected references: Doetschman *et al.*, *Dev. Biol.* **127**, 224–227, 1988.
[f]Selected references: Sukoyan *et al.*, *Mol. Reprod. Dev.* **36**, 148–158, 1993.
[g]Selected references: Shim *et al.*, *Biol. Reprod.* **57**, 1089–1095, 1997; Wheeler, *Reprod. Fertil. Dev.* **6**, 563–568, 1994.
[h]Selected references: Cherny *et al.*, *Reprod. Fertil. Dev.* **6**, 569–575, 1994; Delhaise *et al.*, *Reprod. Fertil. Dev.* **7**, 1217–1219, 1995; First *et al.*, *Reprod. Fert. Dev.* **6**, 553–562, 1994; Iwasaki *et al.*, *Biol Reprod.* **62**, 470–475, 2000; Sims and First, *Proc. Natl. Acad. Sci. U.S.A.* **91**, 6143–6147, 1994; Strelchenko, *Theriogenology* **45**, 131–140, 1996; Van Stekelenburg *et al.*, *Mol. Reprod. Dev.* **40**, 444–454, 1995; Zakhrtachenko *et al.*, *Mol. Reprod. Dev.* **52**, 421–426, 1999.
[i]Selected references: Handyside *et al.*, *Roux's Arch. Dev. Biol.* **198**, 185–190, 1987; Notarianni *et al.*, *J. Reprod. Fertil.* Suppl. **43**, 255–260, 1991.
[j]Selected references: Thomson *et al.*, *Proc. Natl. Acad. Sci. U.S.A.* **92**, 7844–7848, 1995; Thomson *et al.*, *Biol. Reprod.* **55**, 254–259, 1996; Thomson and Marshall, *Curr. Top. Dev. Biol.* **38**, 133–165, 1998; Wolf *et al.*, *Biol. Reprod.* **60**, 199–204, 1999.
[k]Selected references: Sekiya *et al.*, *Differentiation* **29**, 259–267, 1985; Shamblott *et al.*, *Proc. Natl. Acad. Sci. U.S.A.* **95**, 13726–13731, 1998; Thomson *et al.*, *Science* **282**, 1145–1147, 1998.
[l]Selected references: Pain *et al.*, *Development* **122**, 2339–23489, 1996; Pain *et al.*, *Cells Tissues Organs* **165**, 212–219, 1999.
[m]Selected references: Hong *et al.*, *Proc. Natl. Acad. Sci. U.S.A.* **95**, 3679–3684, 1998; Sun *et al.*, *Mol. Mar. Biol. Biotechnol.* **4**, 193–199, 1995.

of their asymmetry (Strome *et al.*, 1994). Mislocalization of the germ-line-specific protein Pie-1 has been described to cause a change in the overall gene expression profile of the early somatic blastomeres, driving them into a germ cell fate (Mello *et al.*, 1996; Seydoux *et al.*, 1996). The somatic cell precursors excluded from the germ plasm may undergo chromatin diminution, as seen in *Ascaris lumbricoides*. This is not a haphazard ripping apart of the genome, but an orchestrated rearrangement whereby certain regions of DNA are degraded and new telomeres are constructed (Muller *et al.*, 1991). *Drosophila* oocytes contain germ determinants in the posterior cytoplasm, or pole plasm, which is marked with polar granules (Lehmann and Ephrussi, 1994).

VERTEBRATES AND MAMMALS

Among vertebrates, only the anuran amphibians—the frog and the toads—contain germ plasm in their eggs, which is marked by dark-staining granules. Curiously, in contrast to anurans, there is no evidence of a germ plasm in urodele amphibians (salamander) (Pesce *et al.*, 1998). This applies to birds as well. Experimental findings, so far, have failed to support the existence of gradients of heritable determinants in mammalian oocytes. This suggests that the main decision of cell potency, germ versus soma, is accomplished in mammals in a regulative manner rather than by determinants preimposed during oogenesis. The only report showing that a classic, deterministic mechanism may act in mammals was the observation that a gradient of the leptin/STAT3 proteins is formed in the mouse oocyte and passes through the early cleavage stages (Antczak and Van Blerkom, 1997). Leptin and STAT3 occur in polarized domains in mouse and human oocytes, prior to fertilization. These domains become differently distributed between blastomeres in the cleavage stages and between the ICM and TE in the blastocyst. Although absence of STAT3 leads to early embryonic lethality in the mouse (Takeda *et al.*, 1997), the consequences of disrupting its polarization are not known.

Recent observations have suggested that the asymmetry of the mouse embryo could be laid down long before the morphological appearance of the anteroposterior (AP) axis at gastrulation. The sperm point of entry into the mouse oocyte seems to lie along the equator that will separate the ICM and the TE (Piotrowska and Zernicka-Goetz, 2001). It has been reported that the first blastomere of the two-cell stage to divide contributes disproportionately more cells to the ICM (Graham and Deussen, 1978; Kelly *et al.*, 1978). If this early allocation were not due to chance but to cytoplasmic localization of ICM/TE determinants in the egg, it would be difficult to reconcile it with the regulative model.

It has been shown that the position of the second polar body that is formed after fertilization and that remains associated with the embryo up to the blastocyst stage correlates with the prospective AP axis of the mouse embryo (Gardner, 1997, 2001). An asymmetrical distribution or proliferation of cells forming the nascent visceral endoderm (VE), between the region of the inner cell mass close to the polar body and the region away from it, is established as early as day 5.5 of embryonic development (Weber *et al.*, 1999). Interestingly, VE cells arising from the ICM close to the polar body cover the embryonic ectoderm and migrate from the prospective posterior pole toward the anterior. VE cells originating from distal ICM cells mainly cover the extraembryonic ectoderm (Weber *et al.*, 1999). In contrast, VE cells covering the epiblast and migrating toward the anterior have been found to express a number of genes that are supposed to play a role in the establishment of the AP axis prior to the beginning of gastrulation (Beddington and Robertson, 1999). Therefore, the finding that the VE is formed according to an AP mode prior to 5.5 days postcoitum (dpc), suggests that the overall AP structure of the embryo could be established during preimplantation development.

The leptin/STAT3 polarization, the sperm point of entry, and the polar body extrusion site, correlating with embryonic axes, hint at the existence of classical determinants in mammalian eggs (Ciemerych *et al.*, 2000). But the paradigmatic situation in mammals remains that positional information is not prelocalized. Rather, it arises during development by virtue of the reciprocal positions of the cells in the embryo.

REGULATIVE DEVELOPMENT IN MAMMALS

A recurrent motif in development is that cells are sheltered in a protective enclave that prevents their exposure to signals from the environment. Positional information leads embryonic cells to have their potency either retained or restricted. For example, the inner cells of the morula will become ICM, while the outer cells become trophectoderm (TE), in the blastocyst. The manner in which this allocation occurs during cleavage is apparently left to chance. As cells divide, some of them end up on the inside rather than on the outside of the morula. In fact, chimeras obtained by aggregation of eight-cell stage mouse embryos demonstrate that cell position is not fixed and cell fate is not predetermined.

The signal(s) that trigger the inner and outer blastomeres to different functions are not known. Restriction of *Oct4* expression and, thereby, totipotency during early mouse embryogenesis appears to be regulated by positional effects likely due to the establishment of gene expression asymmetry in the preimplantation embryo. Only cells that are located inside the blastocyst seem to be able to maintain *Oct4* expression and, hence, pluripotency, whereas cells that are excluded from this position differentiate into TE. In addition, cells of the ICM that differentiate into primitive endoderm transiently up-regulate *Oct4* and activate a gene expression cascade distinct from that of the outer cells. A unique cellular identity of the blastomeres is maintained until compaction of the morula. At this stage, an inner compartment of cells is segregated from the outer layer of cells that will form the TE. It has been found that a differential gene expression pattern in the outer cells of the compacted morula is established prior to the differentiation of these cells to TE (Handyside and Johnson, 1978). One possibility is that this gradient may be due to the different polarity and cell contacts established between outer cells of the compacted morula in relation to the inner cells. Therefore, genes such as *Oct4*, which are initially equally expressed in all the cells of the embryo, may be repressed as a consequence of specific cell-to-cell adhesion, thereby allowing the up-regulation of genes involved in the formation of the TE.

Failure of TE formation in E-cadherin-deficient mice has indicated that E-cadherin is required for early mouse development, setting the stage as to how inner/outer blastomere differentiation may be initiated (Pesce *et al.*, 1998). During compaction, E-cadherin accumulates specifically at the junctional complexes that form between the blastomeres at this stage. On reciprocal interaction of E-cadherin with β-catenin and transcription factors, such as LEF-1, β-catenin is translocated to the nucleus, where it alters gene expression. Because the outer cells of the compacted morula keep one part of their surface free of intercellular contacts, external cells may have a different response to β-catenin nuclear translocation. This may in turn generate a pattern of differential gene expression. The extent of cell contacts may influence the level at which *Oct4* is expressed, as adhesive interactions, such as those between cadherins, activate signal transduction pathways that regulate gene expression (Pesce and Schöler, 2000). Two scenarios can be envisaged. In a "two-signals" situation, the independent activation of TE-specific genes in outer cells and ICM-specific genes in inner cells may occur. A "single-signal" situation is also possible, whereby all totipotent cells differentiate unless pluripotency is preserved in the ICM. This latter scenario resembles the Weismannist scheme, except that totipotency instead of the germ plasm would account for continuity. *Oct4* is the best candidate gene as a gatekeeper of totipotency. A "totipotent cycle" has been proposed (Pesce *et al.*, 1999a), in which cells losing *Oct4* expression during embryonic development are fated to somatic lineages, whereas cells maintaining *Oct4* expression retain totipotency and competence to segregate as germ cells. Evidence supporting this model came from Niwa and colleagues (Niwa *et al.*, 2000), who showed how ES cells respond to different amounts of Oct4.

CELL POTENCY IN SOMATIC CELLS

While we understand more and more about the potency of the germ line, knowledge of what determines the potency and fate of somatic cells remains scanty. Studies on ASCs indicate that moving a cell from one environment to another may induce significant changes in gene expression that ultimately lead to transdifferentiation (reviewed in Fuchs and Segre, 2000). What is the link between environmental cues and phenotype modification? ASCs may sense the contact with the intercellular matrix (integrin) or soluble factors released by local cells (Cirulli *et al.*, 2000; Giet *et al.*, 2001). The screening of genes expressed before and after the change of interest can potentially reveal the master genes involved in pluripotency. Messenger RNA differential display or microarray technology (Kelly and Rizzino, 2000) analysis can be performed to point out differences. ES cells may help to establish a correlation between the expression levels of certain genes, such as *Oct4*, *GCNF*, and *PEM* (see later), and cell fate.

PLURIPOTENTIAL STEM CELLS AND CLONING

STEM CELLS AT WORK

The isolation of pluripotent stem cells has been a paramount achievement in biology and medicine, because these cells have widespread implications for developmental biology, drug discovery and testing, and transplantation medicine. Stem cells can help us unravel the complex events that occur during mammalian development, thereby identifying the factors involved in the cellular decision-making process that leads to cell specialization. Human ES cells, in particular, could offer insight into developmental events that cannot be studied directly in humans *in utero* or fully understood using animal models. Cancer disease and certain birth defects are due to abnormal cell division and specialization, which result from an altered decision-making process during embryogenesis and/or adulthood (Sachs, 1986). Stem cell research can also dramatically change the way we develop drugs and test their safety profile. Pluripotent human stem cells could be used to test drugs in more cell types, and, thus, partly reduce testing in whole animals and humans. Identification of gene targets and tests for drug toxicity or teratogenicity could become more feasible. Drug testing on cultured human embryonic stem cells could also reduce the risk of drug-related birth defects.

A SPECIAL SOURCE OF STEM CELLS: NONREPRODUCTIVE CLONING

The ultimate goal of research on pluripotent stem cells is the generation of cells and tissue that could be used for so-called cell therapies and organ replacements, substituting human body cell populations destroyed by disease (Lanza *et al.*, 1999). To date, stem cells have been derived from human embryos and fetal tissue, yet, there is the risk of rejection following transplantation. Drugs are currently used to suppress cell-mediated rejection mechanisms following human allografts. The most ambitious application of stem cells is the generation of transplantable organs. The ability to culture effectively multipotent somatic stem cells and direct them along different lineages is the first step required to unfold these possibilities. Although prospects are good, it is important not to raise people's hopes falsely. Before these cells/organs can be used for transplantation, the problem of immune rejection must be overcome. Future research must focus on the modification of stem cells to minimize histoincompatibility, on the isolation of an individual's own ASCs, or on the generation of individual-specific stem cells by therapeutic cloning. Cells obtained from cloned blastocysts would be almost genetically identical to the patient's cells, and would not stimulate an immune response. The critical shortage of human organs available for transplantation has greatly accelerated research in therapeutic cloning and stem cell technology. If successful, a multitude of therapeutic uses could be anticipated, providing a renewable source of replacement cells and tissue for diseases, conditions, and disabilities, such as Parkinson's and Alzheimer's diseases, spinal cord injury, stroke, burns, heart disease, diabetes, osteoarthritis, and rheumatoid arthritis. After scientists learn how the egg reprograms a cell to make a stem cell line, they may even be able to create stem cells directly from patients without creating an embryo.

STEM CELLS AND CLONING: CONCURRENT BUT DISTINCT PROCEDURES

Although the above-mentioned applications show how stem cell technology and cloning are coupled to reach a common goal (Fig. 2), we should stress that their original aims and biological processes are quite different. The aim of stem cell technology is to identify environmental conditions that will lead stem cells to differentiate into one specific cell type. The aim of replacing the "nucleus" of an unfertilized egg with the nucleus of another cell—what most scientists refer to as cloning—is to dissect cell potency at the nuclear level, and to find experimental conditions that will allow a specialized cell nucleus to substitute for the embryonic genome and drive the formation of all cell lineages. The intrinsic simplicity of cloning by SCNT renders this method suitable to investigate the genes involved in regulation of cell potency.

ETHICAL ASPECTS: INTERFERING WITH NATURE?

Scientists believe that stem cell therapy offers enormous potential to treat and cure degenerative disorders and illnesses by replacing diseased cells or tissue. But because it involves research on human embryonic cells and cloning, it has infuriated antiabortion rights groups and sparked an ethical debate about how far scientists should interfere with nature. Therapeutic cloning incites many of the same emotions, as well as fears that it will eventually lead to human reproductive cloning, which most scientists vehemently deny (Jaenisch and Wilmut, 2001).

METHODS OF ADDRESSING CELL POTENCY

CELLULAR MODELS OF PLURIPOTENCY

Because of the limited number of available stem cells from natural sources, *in vitro*/artificial model systems have been devised to analyze pluripotency in development.

Pluripotent cell lines have been derived from the ICM of embryos at the blastocyst stage (ES cells) or isolated from aborted fetuses. SCNT is another method used to produce pluripotent stem cells. For this, an oocyte's chromosomes (often ill-defined as the "nucleus") are removed. The remaining material contains nutrients, energy-producing systems (mitochondria) and, most importantly, maternal transcripts that support the beginning of development. Then, either the somatic cell is placed next to the "enucleated" oocyte and the two cells are fused, or the nucleus of the somatic cell is isolated and injected into the "enucleated" oocyte. The result-

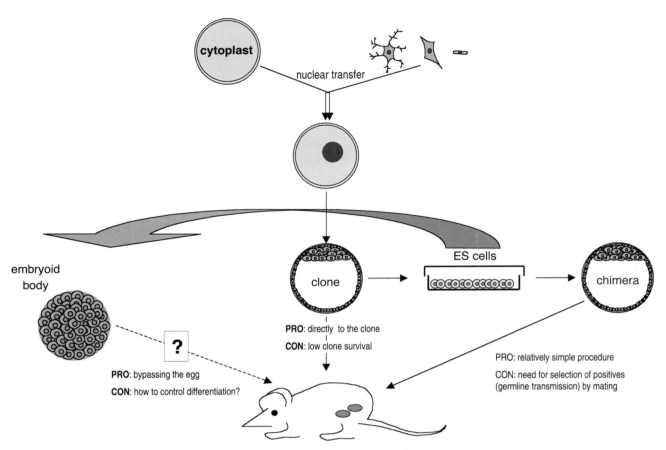

Figure 2 *A bridge between cloning and stem cell technology. Stem cells can be either a start or an end point for cloning. When used as nucleus donors, they perform better than somatic cells, possibly due to their pluripotency, and make the cloning process more efficient. When derived, they can be obtained from natural embryos and used to target the germ line (chimeras), or from cloned embryos and used for individual-specific cell therapy (therapeutic cloning). Stem cells can be aggregated into embryoid bodies that allow derivation into somatic cell lineages.*

ing reconstructed oocyte, and its immediate descendants, are regarded to have the full potential to develop into an entire animal, and hence are considered totipotent. Given the actual state of cloning, it is relatively easy to obtain preimplantation-stage embryos from somatic cell nuclei, as opposed to the low viability of postimplantation-stage embryos (reviewed in Solter, 2000). Cells from the ICM of blastocyst clones could, in theory, be used to develop individual-specific pluripotent stem cell lines. Alternatively, stem cells could be obtained from the adult, but the identification, selection, and stable propagation of ASCs in culture pose major problems.

PLURIPOTENT EMBRYONIC CELL LINES: EC, ES, AND EG CELLS

Classic work on mouse teratocarcinomas in the 1970s by Martin and Evans (Martin and Evans, 1974) provided the concept of a pluripotent stem cell (embryonal carcinoma) that can give rise to multiple types of tissues found in tumors (O'Shea, 1999). The teratocarcinomas used as the cell source either originated spontaneously from male or female germ cells or were induced experimentally by grafting early postimplantation conceptuses to ectopic sites in syngeneic hosts. EC cells are pluripotent, as shown by the variety of tissues into which they can differentiate.

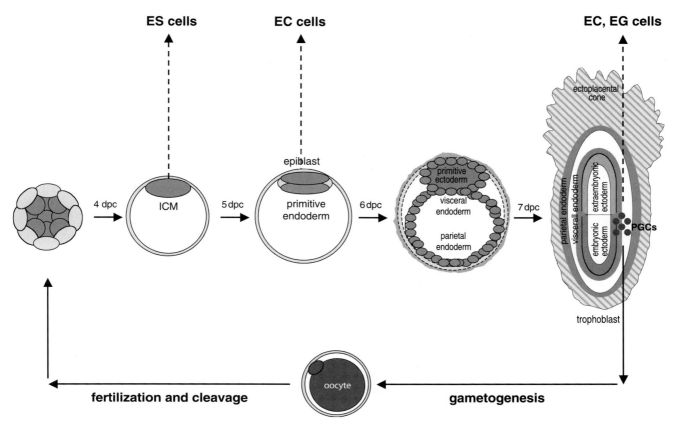

Figure 3 *Embryonic and derived pluripotent stem cells in the mouse. The earliest embryonic stem (ES) cell population can be derived from the embryo as early as the blastocyst stage. ES cells are totipotent. Following this stage, restriction of pluripotency allows derivation of pluripotent rather than totipotent stem cells. These are the embryonic carcinoma (EC) and the embryonic germ (EG) cells. ICM, Inner cell mass.*

There is a consensus that EC cells more closely resemble epiblast than ICM cells (Fig. 3) and are closer to the path that leads to germ cells, as corroborated by the presence of two active X chromosomes in female EC cells (Martin *et al.*, 1978). EC cells can be maintained in culture or propagated *in vivo* (ascites) and retain the ability to participate in normal development. Once they become integrated in an embryo, they support formation of viable and fertile mosaic animals. Tumor formation in the adults is sporadic. This suggests that a normal embryonic environment is sufficient for the restoration of orderly gene expression in EC cells—with some limitation. In fact, EC cells preferably differentiate into egg cylinder ectoderm (F9) or fibroblast, muscle, and neuronal lineages (P19), and they have limited ability to colonize the germ line (Stewart and Mintz, 1981).

Subsequent to embryonic days 3.5–4.5, when mouse ES cells can be derived (Fig. 3), the ICM splits into epiblast and hypoblast, which have lost the ability to participate in development, as shown by blastocyst injection of isolated cells. Four fundamental properties render mouse ES cells extremely valuable: (1) They are derived from the pluripotent cell population of the ICM and (2) they are genetically normal and undergo karyotypically normal mitotic divisions, *in vitro*, resulting in stable diploidy, as demonstrated by generation of normal mice. (3) They replicate in the primitive embryonic state to an extent far beyond the usual number of culture passages of cultured mammalian cells (HeLa, 3T3, 293, fibroblasts, etc.). Unlike other cell lines, which have to be mutagenized or transformed by viruses to achieve indefinite replication, ES cells are naturally regulated to replicate without the usual

"Hayflick limit." However, long-term *in vitro* maintenance of mouse stem cells shows that their proliferative capacity is limited. (4) ES cells spontaneously differentiate into multiple cell lines representative of all three embryonic cell layers and can potentially generate any cell type in the body, including the germ line, when transplanted into a host blastocyst. However, the general finding is that with increasing time in culture, ES lines tend to lose the ability to form gametes.

A single cell should meet all four criteria, but, so far, only populations of mouse ES cells have come close to meeting all the requirements for pluripotency. Human ES cells, unable to be tested for the master proof of perpetuation (germ line transmission through chimeras), seem capable of longer term proliferation *in vitro*, possibly reflecting inherent species differences in senescence.

Although ES cell line derivation is based on a clonal procedure, cell–cell heterogeneity may occur with time (passages) and account for the varying competence to colonize the germ line. Stochastic processes (mutations) may also occur during culture *in vitro*. Moreover, when deriving ES cells, it is standard practice to pool all undifferentiated colonies that are obtained from an individual conceptus, based on the assumption that a cell line derived from a single embryo is by definition a clonal population in terms of karyotype and genotype. The possibility that ES cell progenitors are already heterogeneous in developmental potential or in other respects has not been entertained much.

Mouse ES cell line establishment is more readily obtained, and germ line contribution is most efficient, with two specific strains of mice (129 and B6). Once ES cells are established, their maintenance in an undifferentiated, proliferative, and pluripotent state depends on a subtle balance of signals provided by feeder cells (fibroblasts). In particular, the STAT3 signal cascade must be activated and the transcription factor Oct4 expressed. The STAT3 cascade is initiated by the leukemia inhibitory factor (LIF), also known as differentiation inhibiting agent (DIA) (Smith *et al.*, 1992), which is secreted by feeder cells and supplemented to the media. If one of these two conditions—STAT3 cascade and *Oct4* expression—is not met, ES cells undergo differentiation, forming predominantly endoderm and mesoderm and, to a lesser extent, hematopoietic cells (Mummery *et al.*, 1990). These differentiation paths are facilitated by the presence of undefined factors in serum. In contrast, when ES cells are grown in serum-free medium, they rapidly lose their ES cell phenotype without forming mesoderm and develop, together with other cell types, into neuroectoderm. ES cells form cell aggregates under the conditions described above (EB precursors), thus it is difficult to assess whether neuroectoderm differentiation is cell autonomous or involves cell–cell interactions. It is possible that neuroectoderm development is supported by factors synthesized by ES cells or from newly formed endoderm developing around the EB. In species other than the mouse, ES-like cells have been derived from rat (Vassilieva *et al.*, 2000), chicken, monkey, and human embryos (see Table 2 and references therein), but, so far, none has been shown to colonize the germ line.

Early germ cells isolated from fetuses can be placed in culture, where they will give rise to embryonic germ (EG) cells. In the mouse, the process of EG cell derivation is most efficient when germ cells are collected from the posterior region of fetuses on days 8–8.5 (Fig. 3). In the human species, germ cells are recovered from aborted fetuses. Derived EG cells resemble immature germ cells or precursors of germ cells (primordial germ cells, PGCs) in that their epigenotype is hypomethylated (Surani, 1999). Mouse EG cells can contribute to chimeras and to the germ line when they are introduced into host blastocysts. Although it has been proved that mouse EG cells from 8.5-dpc PGCs can contribute to chimeras (Labosky *et al.*, 1994), it is not clear whether primordial germ cells or fetal germ cells have a potency to produce chimeras without previous exposure to *in vitro* culture. Applications of human EG cells have been limited to EB technology due to ethical constraints that prevent extensive manipulation of human embryos.

THE EXISTENCE OF ADULT STEM CELLS AND THEIR COMPARTMENTS

Although cell potency becomes progressively restricted during mammalian development, there is now evidence that some cells may retain an unexpectedly broad differentiation repertoire. Even after an animal is fully grown, many tissues and organs have the capacity to replenish those cells that are lost either by pycnosis (injury) or apoptosis (natural death). This is accomplished through ASCs. In contrast to ES cells, ASCs in mammals normally have more limited differentiative options. Typically, they are localized in specific compartments or niches and are committed to differentiate into the tissues in which they reside. Examples include the epidermal (skin), follicular (hair), intestinal (crypts), neural (brain and nerve), and hematopoietic (bone marrow and blood) niches (Fig. 4). However, additional data indicate that ASCs from one tissue or organ can also repopulate heterologous cell systems. Eglitis and Mezey, in 1997, first challenged the notion of the niche-restricted fate of ASCs (Eglitis and Mezey, 1997). They stunned the scientific community by showing what they called the "transformation of blood into brain": cells of the bone marrow differentiating into microglia and macroglia in the brain of adult recipient mice. This finding spurred further reports on cell fate transition. Putative neural stem cells were expanded *in vitro* and injected intravenously into irradiated mice. From 5 to 12 months later, donor-derived cells were found in the hematopoietic tissues (Bjornson *et al.*, 1999). When skeletal muscle cells of adult mice were transplanted into lethally irradiated recipients, a 56% contribution of muscle cell progeny to peripheral blood was found after 6 to 12 weeks (Jackson *et al.*, 1999). When putative neural stem cells from the adult mouse brain were introduced into mouse and chick embryos, chimeric embryos were formed in which the donor cells had contributed to most germ layers (Clarke *et al.*, 2000). This plasticity of adult stem cells, flexible enough to trespass the biological species border, has been referred to as transdetermination (Wei *et al.*, 2000). Although these observations are fascinating, the basis for transdetermination is unclear. Transdetermination raises a provocative question regarding the embryonic germ layers: if blood stem cells can contribute to brain after intravenous injection, and neural stem cells can contribute to most tissues and organs in chimeric embryos, what is the significance of the classic distinction between ectoderm, mesoderm, and endoderm? Another question is whether ASCs are homogeneous, i.e., are they all equally pluripotent or do they exhibit different preferences for differentiation that can be recognized in subpopulations? Candidate ASCs should be tested individually from pure populations. To date, ASCs have not been isolated from all body tissues and are often present in minute quantities. For human applications, stem cells would first have to be isolated from the patient and then grown in sufficient amounts, but for some acute disorders time may be limiting. In addition, the number of ASCs decreases with age.

HOW STEM CELLS ARE ESTABLISHED, SELECTED, AND TESTED *IN VITRO* AND *IN VIVO*

Stem cells typically depend on feeder cells to provide the appropriate cues for survival and growth. However, other chaperone cells can be present in culture and may easily overgrow stem cells. For this reason, stem cells are modified by the introduction of a selectable marker into their genome. Differential expression of the selectable marker enables preferential survival and/or division of the desired stem cells compared to nonstem cells.

Expression of the selectable marker in the cell line may be achieved by inserting it into an expression construct followed by stable or transient transformation. The genetic marker may be introduced into the cells by a variety of ways, includ-

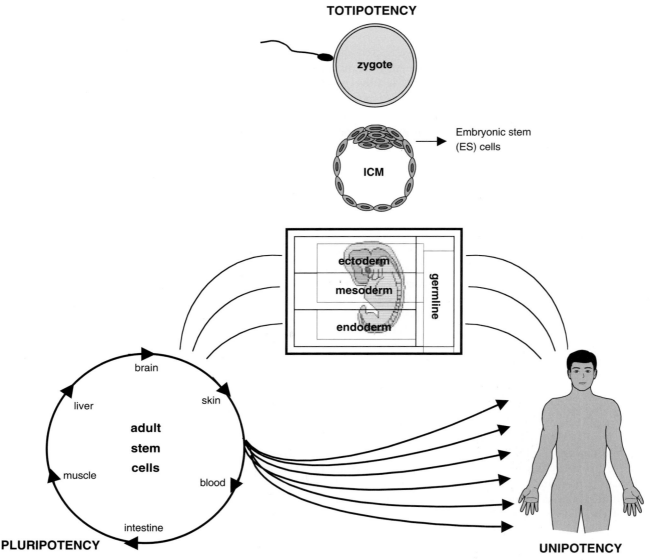

Figure 4 Cell potencies in mammals. Pure totipotency is a feature of the earliest embryonic stages only. Subsequently, only pluri- and unipotent cells are found in the embryo, fetus, or adult. The degree of pluripotency is variable. Adult stem cells seem to retain special flexibility in their fate and are able to trans-differentiate along pathways other than their own. ICM, Inner cell mass.

ing injection, transfection, lipofection, electroporation, or infection with a viral vector. Stable transformation can be improved by inserting the selectable marker into an endogenous cellular gene (homologous recombination). Markers are either designed to protect the desired cells from the effect of an inhibiting factor present in the culture medium (antibiotic resistance), to selectively permit the growth of stem cells (growth factor), or to allow purification/depletion of expressing cells (surface antigen used for immune cell sorting).

For the generation of pluripotential ES cells, the expression constructs preferably comprise a DNA sequence encoding a selectable marker linked to or targeting a genetic element, which is associated with a stage of embryonic development prior to differentiation. Most preferably, the genetic control elements are derived from a

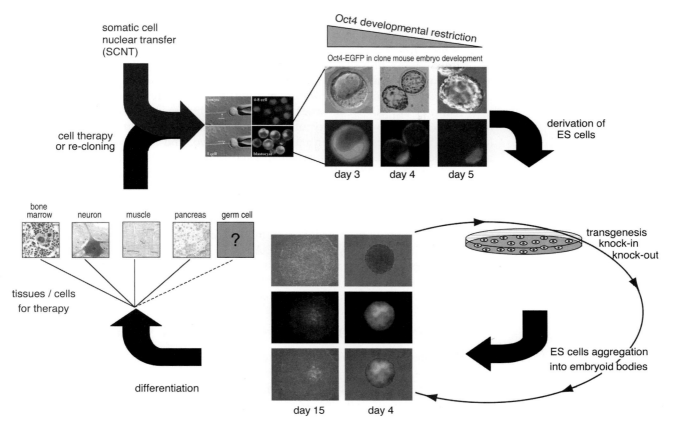

Figure 5 *Oct4: a marker for embryo development and differentiation. Oct4 is a powerful marker to study cell potency during development. Here it is shown in conjunction with a green fluorescent protein (EGFP) reporter. On nucleus transplantation, the nucleus of a somatic cell is reprogrammed within the ooplasm and the previously silent somatic Oct4 is reactivated. Expression follows the time pattern expected as for normal development. ES cells can be derived from the inner cell mass of the clone blastocyst and aggregated to form embryoid bodies. These can be induced to differentiate into different cell lineages to be used for cell therapy.*

gene specifically active in the ICM of the blastocyst, primitive ectoderm, and primordial germ cells of the early postimplantation embryo.

One example of a gene that displays a suitably restricted stem cell expression pattern, and therefore may provide suitable stem cell-specific regulatory elements for the expression of a selectable marker, is *Oct4*. Selectable marker genes under the control of the *Oct4* promoter may be useful for the isolation of ES cell lineages. Furthermore, reports describing low levels of Oct4 expression in some adult tissues (Takeda *et al.*, 1992) may extend the utility of these expression constructs beyond ES cells to include other stem cells that are essential for tissue homeostasis and repair. These cells might be the ASCs. One approach used to generate the desired spatial and temporal restriction of marker gene expression in transgenic animals is to make a construct using the selectable marker in conjunction with the appropriate promoter parts of the gene of interest. One example (Fig. 5) is the Oct4–EGFP (enhanced green fluorescent protein) construct (Yoshimizu *et al.*, 1999). Another approach uses the endogenous *Oct4* gene locus, and, therefore, the associated *Oct4* gene regulatory elements, to link a selectable marker gene expression as closely as possible to the endogenous *Oct4* gene expression profile. ES cells can then be transfected with the construct or targeted for homologous recombination. Transgenic or knocked-in ES cells can be used for the production of chimeras or EBs, or as nucleus donors for cloning.

Unlike mouse ES cells, which can be cultured using DIA/LIF in the absence of a feeder layer of mitotically inactivated fibroblasts, human ES cells are not supported by supplementation of culture medium with DIA/LIF in the absence of feeder cells (Amit *et al.*, 2000). In order to test for pluripotency, ES cells are most commonly injected into the cavity of day-3.5 blastocysts to form chimeric embryos (Hogan, 1994). These are then reimplanted, *in vivo*, and allowed to develop. Pups are selected for coat color chimerism and germ line transmission. Alternatively, EBs are produced from ES cells by aggregation in the absence of LIF. Dissociated ES cells are seeded in 10- to 30-μl drops (1 drop of medium at a density of 100 to 500 cells/drop). Arrays of drops are plated on the lid of a 10-cm tissue culture dish and inverted over the base of the dish, which contains water, phosphate-buffered saline (PBS), or medium in order to maintain humidity. The hanging drops are cultured at 37°C in a 7% CO_2 atmosphere. After 40 hours in the hanging drops culture, the aggregates are transferred into suspension culture in the absence of selection. After several days, EBs form, each containing a variety of morphologically differentiated cells (Karin Hübner, personal communication).

POSTIMPLANTATION DEVELOPMENT MODELS OF STEM CELLS

The pluripotency of EC, ES, and EG cells can be demonstrated by differentiation, *in vivo*. The simplest way of testing for pluripotency of stem cells is to inject them into host severe combined immune-deficient (SCID) or athymic mice and check for production of various cell types (Amit *et al.*, 2000). Usually, approximately 10^6 cells are injected into the rear leg muscle or the peritoneal cavity of the animal, and teratoma formation is assayed after 3–4 months. The variety of cell types formed is evidence of pluripotency of the original injected cell population.

When cells are transplanted into the perivitelline space of a precompaction morula or into the cavity of a blastocyst, they become integrated, forming a chimera. This may develop, once reimplanted *in vivo*, into offspring in which some phenotypic traits are derived from the transplanted cells. A functional germ line contribution may sometimes not be possible, given that embryos of certain genotypes (e.g., *W/W*) form mice with all somatic tissues but contain a germ cell deficiency. The number of transplanted cells seems to have an impact on the success rate of germ line transmission. Transmission is achieved only when approximately 10–15 cells are injected per blastocyst (Gardner and Brook, 1997). Why it is necessary to transplant so many cells in order to secure functional colonization of the germ line is not clear, particularly because respectable levels of somatic chimerism can be achieved with substantially fewer cells. One possibility is that, during the culture of ES cells, genomic changes occur that are too subtle to discern karyotypically, and although not obviously impairing the ability to contribute to somatic chimerism, may prevent gametogenesis. Particular gene expression levels and patterns may comprise the critical determinants involved in this dichotomy, as shown here by Oct4.

PGCs from day-8.5 pc mouse fetuses have thus far been unable to give rise to any embryonic cell lineage when injected into blastocysts, but they transform into EG cells on culture *in vitro* (Labosky *et al.*, 1994). EG cells look very similar to ES cells and can be used in the same way to generate chimeras. But they differ from EC and ES cells in that their epigenotype (Surani, 1999) does not mirror the somatic situation. In fact, PGCs have the original gametic imprint modified. Its erasure occurs during migration of PGCs to the gonadal ridge, and it is then reset according to the sex of the embryo. EG cells are able to progress *in vivo* or *in vitro*, forming teratocarcinomas when injected subcutaneously into nude mice and EBs when cultured. To what extent does the imprinting of the donor nucleus matter for developmental pluripotency? The simple observation that gametes have to complement each other's imprints in order to establish zygotic totipotency argues that the potentiality of stem cells may be affected by modification(s) of the original imprints during the time-long derivation, selection, and propagation process. In fertilized mouse embryos, *in vitro* culture for preimplantation development seems to alter the expression of imprinted genes (Doherty *et al.*, 2000; Khosla *et al.*, 2001). EG cells are even more sensitive to this point, because they are derived and then cultured from PGCs that are in the process of resetting the original imprint (8.5 to 12.5 dpc). EG cells that were derived from PGCs collected during this time frame were tested for chimerism, and indicated that germ line transmission was achieved no matter what was the derivation stage. This suggests that both 8.5- and 12.5-dpc embryo-derived EG cells are totipotent, although their imprint status is different (Labosky *et al.*, 1994). Allele methylation analysis of the *Igf2r* gene has shown that demethylation and erasure of the somatic imprint has already begun by day 8.5 pc. Thus, the imprint status apparently does not correlate with cell's ability to transmit through the germ line. We can speculate that this depends on the way chimeric embryos are obtained—that is, from blastocyst injection of a subpopulation of 10 to 20 stem cells rather than from a single cell with a unique imprint status. Cloning by SCNT would be the ideal procedure to test the relevance of imprinting for totipotency at the single-cell level.

To date, the ability of somatic cells to contribute to chimeras has not been demonstrated. In fact, chimerism probably does not involve reprogramming of nuclei as cloning does, but source cells must have the potential to become pluripotent when transplanted. We can speculate that unipotent somatic cells are either irreversibly specialized or the growth rate of the host embryo does not allow sufficient time for adaptive changes to occur. Indirect evidence from cloning experiments suggests that somatic cells may exhibit some degree of (latent) totipotency. After nucleus transplantation from TE cells into mouse oocytes and production of TE-cloned blastocysts, portions of the ICM were isolated from the TE-clones and injected into the cavity of fertilized blastocysts, resulting in the production of chimeric embryos (Sotomaru *et al.*, 1999). The original TE cells had to be transformed into ICM cells by cloning, implying that reprogramming rather than original totipotency was the key factor for success. Nuclear transfer technology has been applied to the production of chimeric mice, by transferring the embryonic nucleus into one two-cell stage blastomere (Kato and Tsunoda, 1995).

One can induce stem cell differentiation to an advanced degree, entirely *in vitro*. In principle, it should be possible to culture ES cells in the presence of instructive molecules, such as growth and differentiation factors and get them to form a desired cell type. Although scientists have appreciated this potential for about 15 years, only recently have realistic systems been developed. On release from a differentiation-inhibiting environment (feeder cells or exogenous LIF), mouse ES cells are cultured on low cell-adhesion substrates (suspension culture dishes, methyl cellulose, hanging drops) and aggregate into clumps (Fig. 5). Under the appropriate conditions, these EBs undergo a differentiation program that is reminiscent of normal embryogenesis, with the ordered appearance of primitive endoderm, primitive ectoderm, and their derivatives (Desbaillets *et al.*, 2000). Fluid-filled yolk-sac-like cysts containing α-fetoprotein can form within EBs, followed by the occasional appearance of blood islands. Differentiation can be induced by agents such as dimethyl sulfoxide (DMSO), dibutyryl cyclic AMP (db-cAMP), forskolin, or retinoic acid (RA) and is enhanced in high-serum media. The initial number of cells aggregated is also critical in controlling the differentiation pathway. With the proper combination of growth and differentiation factors, mouse ES and EG cell cultures can generate cells of the hematopoietic lineage, cardiomyocytes, skeletal muscle, neurons, and vascular endothelial cells (Desbaillets *et al.*, 2000; O'Shea, 1999). To date, the most extensively studied differentiation pathway in EBs has been that of hematopoiesis.

However, triggering these events is chaotic, implying the importance of appropriate cellular organization and cell–cell interactions in orchestrating development. Current efforts are aimed at defining the conditions that support the proper and controllable differentiation of EBs. But, with more than 2000 growth factors known so far (reviewed in Fuchs and Segre, 2000), the chance of succeeding by just exploring new formulas for culture media may be farfetched. For this reason, some scientists think that the best way to prove totipotency in the EB system is to push the EBs into germ line differentiation rather than attempt to obtain all other cell lineages.

MARKERS OF PLURIPOTENCY

There are a wide variety of antigenic and molecular markers that, by virtue of being present in EC, ES and epiblast cells, are believed to be specific to pluripotent cells (Table 2). The assessment of such markers typically consists of a non-viable procedure. The availability of either viable or non-viable markers to assess the potentiality of clone embryo cells is limited. This renders a better understanding of cloning failure more difficult.

SURFACE ANTIGENIC MARKERS ENABLE AN IMMEDIATE BUT NOT VIABLE CHARACTERIZATION

Mouse EC, ES, and EG cells share specific cell surface glycoproteins, such as SSEA-1 (Solter and Knowles, 1978). Because pluripotent stem cell lines have also been reported in chickens, minks, hamsters, pigs, common marmosets, rhesus monkeys, and, most recently, humans (Pera *et al.*, 2000; Reubinoff *et al.*, 2000), defining additional markers has become essential. The presence of SSEA-1 is widely accepted as a diagnostic feature of pluripotent cells, although, the antigen was originally characterized using F9 EC cells, which are clearly restricted in their developmental potential. SSEA-1 was also found in primitive endoderm and epiblast cells (Knowles *et al.*, 1978). Although candidate pluripotent stem cells generally meet the basic criteria (indefinite propagation, karyotypic normality), a battery of markers used to characterize these cells has shown that more particular similarities may not cross the species boundaries. In fact, human EC and ES cells were found to be negative for SSEA-1, and a group of antibodies directed against glycoproteins and glycolipids (SSEA-3, SSEA-4, TRA-1-60, and TRA-1-81) has identified human, but not mouse, EC, ES, and EG cells (Pera *et al.*, 2000). Thus, for practical purposes, the most useful property of stem cells turns out to be their ability to differentiate *in vitro* into ectodermal, endodermal, and mesodermal derivatives. In this respect, EB formation is a very convenient assay, and it would be especially so for the human species, where contribution to the germ line cannot be tested by chimerism. Immunohistochemical analysis of ES cells and EBs after differentiation shows more conserved results between species. Neurofilaments, muscle actin, and the α-fetoprotein are "universal" markers for ectoderm, mesoderm, and endoderm derivatives, respectively (Itskovitz-Eldor *et al.*, 2000). If a true pluripotential cell marker were available, it would be more useful if its presence or absence could be detected in living cells without compromising their viability. Such a strategy would entail selection for pluripotent cells by coupling the control sequence of a gene, the expression of which was found to be restricted to these cells, to the coding region of a suitable reporter gene. This has been done with the *Oct4* gene.

CERTAIN ENZYMATIC ACTIVITIES ARE CHARACTERISTIC OF BUT NOT SPECIFIC TO STEM CELLS

Embryonal stem cells have been defined by the presence of high levels of alkaline phosphatase (AP) (Hass *et al.*, 1979), which is characteristic of but not specific to them (in fact, it is also expressed by PGCs). In more recent studies, the activity of the ribonucleoproteic enzyme telomerase has been pointed out as a feature shared by highly replicative cells, among which are the stem cells. The shortening of telomeres at each cell division reduces the replicative potential and signals cell senescence ("mitotic clock") (Allsopp *et al.*, 1992), unless telomerase counteracts the shortening. Although most adult cells have no detectable telomerase activity, in some cell types telomerase is active and telomeres are stabilized during divisions with no net shortening. This can prevent aging of the cell, allowing indefinite divisions, by maintaining telomere length without disruption of the genome. Cells with high replicative capacity, such as germ cells, embryonic stem cells, and tumor cells, show high telomerase activity (Kim *et al.*, 1994; Wright *et al.*, 1996; Thomson *et al.*, 1998). Normal somatic cells gradually lose telomeric DNA and become senescent after 50 population doublings (PDs). In human somatic cells, terminal restriction fragments (TRFs) shorten, on average, from 10–15 to 5–7 kb. In contrast, ES cells preserve their TRFs unshortened. These results demonstrate the proliferative capacity of ES cells. Human ES cells have been clonally derived from single precursor cells and passaged for 8 months (almost 300 PDs) without losing their pluripotency. Extending their life-span from 50 to 300 PDs suggests that these cells are able replicate without the usual "Hayflick limit" and are virtually immortal (Amit *et al.*, 2000). Nevertheless, a farther limit may derive from the accumulation of genetic mutations over a period of time. Telomere length has also been implied as a factor affecting cloning efficiency. Results conflict as to the presence of shortened telomeres in cloned farm animals (Shiels *et al.*, 1999; Lanza *et al.*, 2000; Tian *et al.*, 2000).

GENETIC MARKERS

During development, cell fate is defined by transcription factors that act as molecular switches to activate or repress specific gene expression programs (Niwa *et al.*, 2000). Relatively few embryonic genes have been proved to be crucial for development.

Oct4 is one such gene that has received an inordinate amount of attention by virtue of the fact that it represents the earliest null phenotype (at the blastocyst stage) and is required for embryonic cell differentiation. Emphasizing the importance of *Oct4* in development is the finding that embryos lacking it die soon after implantation. *Oct4* encodes a developmentally regulated transcription factor and has been postulated to maintain the Oct4-expressing cells in a totipotential state— Oct4 is a "gatekeeper" of totipotency (Pesce *et al.*, 1999a). Oct4-deficient mice have compromised blastocysts as a consequence of aberrant ICM formation (Nichols *et al.*, 1998). *In vitro*, an alteration in the level of Oct4 expression causes mouse ES cells to acquire a trophectodermal or an endodermal phenotype (see later).

OCT4, A MODERN TOOL FOR ANALYSIS OF PLURIPOTENCY

OCT4 IN DEVELOPMENT

Oct4 is the first transcription factor to be expressed during mammalian embryogenesis and, as such, represents a potential controlling determinant of early mouse lineage decisions. Its expression profile follows a strict developmentally regulated pattern, possibly involving positional information, and is suggestive of a critical role in the determination of embryonic cell fate. This maternally inherited transcript is expressed at low levels in all blastomeres until the four-cell stage, at which time it undergoes zygotic activation, resulting in high Oct4 protein levels in the nuclei of all blastomeres until morula compaction (Palmieri *et al.*, 1994; Yeom *et al.*, 1991). After cavitation, Oct4 is involved in the first separation of embryonic and somatic lineages that occurs in the mouse embryo; its expression is maintained in ICM cells of the blastocyst (3.5–4.5 dpc), which will give rise to the embryo proper and is down-regulated in the outer cells of the morula on differentiation into trophecto-derm, which is required for implantation and placental development (Okamoto *et al.*, 1990; Rosner *et al.*, 1990; Schöler *et al.*, 1990a) (Fig. 5). Following implanta-tion, at the second lineage split, the ICM differentiates into the primitive ectoderm (epiblast) and primitive endoderm (hypoblast), accompanied by restricted *Oct4* expression in the epiblast (5.5–6.5 dpc), although it is transiently increased in the forming hypoblast before it differentiates into parietal/visceral endoderms, which do not express *Oct4* (Palmieri *et al.*, 1994; Schöler, 1991). These results resound the same notion, namely, that Oct4 is involved in the maintenance of an undiffer-entiated, pluripotent embryonic cell state during the first and second lineage deter-minations in the early embryo. Not surprisingly, *Oct4* expression must be tightly observed for development to proceed according to its specified plan. Mouse embryos lacking Oct4 experience early lethality due to lack of a proper ICM compartment formation with all the cells of the morula, including the inner cells, thus differenti-ating into the trophectodermal lineage and thereby annihilating any totipotential stem cells (Nichols *et al.*, 1998). The presence of Oct4 in the ICM is therefore a fundamental necessity in preventing ICM somatic differentiation. Finally, during gastrulation, *Oct4* is down-regulated in an anterior to posterior manner in the somatic lineages and expression becomes restricted in primordial germ cells (7.5–8.5 dpc), precursors of the gametes in the extraembryonic mesoderm (Yeom *et al.*, 1996), lending credence to the postulate that Oct4 is also a likely factor for the germ line lineage (in germ cell totipotency), by preventing their differentiation to a somatic cell phenotype during gastrulation. *Oct4* is also expressed in embryonal stem, carcinoma, and germ cell lines, of broad differentiative potential, that are derived from the ICM, epiblast, and PGCs, respectively, and is down-regulated fol-lowing RA-induced cell differentiation (Minucci *et al.*, 1996), implying that it is a crucial component of cell potency. The basic rule of thumb regarding *Oct4* expres-sion is that it correlates with the pluripotent phenotype of embryonic and germ cells (and cell lines derived thereof), whereas cells that are driven into embryonic and extraembryonic somatic differentiation lose *Oct4* expression. The protein is postu-lated to exhibit its function by differential regulation of genes involved in the main-tenance of a pluripotent cell state versus those necessary for the differentiation of somatic cell lineages (Pesce *et al.*, 1998, 1999a; Pesce and Schöler, 2000). In light of such findings, it is no wonder that Oct4, which so far has only been detected in mammals (Sooden-Karamath and Gibbins, 2001), represents the only known pluripotential cell marker in the mouse.

STRUCTURE OF THE *Oct4* GENE

Oct4, also named Oct3, Pou5f1, and NF-A3, belongs to the class V of the Pit/Oct/Unc (POU) transcription factors (TFs), all members of which are expressed during early embryonic development (Ryan and Rosenfeld, 1997). Similar to Oct4, all POU TFs, with the exception of Oct1, are differentially regulated such that they are expressed in a lineage-specific fashion in order to fulfill numerous cell type-specific roles. Oct4 was initially defined as a DNA binding protein in extracts from EC and ES cells (Lenardo *et al.*, 1989; Schöler *et al.*, 1989a,b). A closer examina-tion of the Oct4 gene becomes imperative to understanding its DNA binding prop-erties and hence its role in embryogenesis (Fig. 6).

The mouse Oct4 gene is located on chromosome 17, near the t-complex, an interesting finding considering that a number of t-complex alleles display early lethality that is associated with Oct4-expressing cells (Schöler *et al.*, 1990a). The gene is organized into five exons (Yeom *et al.*, 1991; Okazawa *et al.*, 1991) and encodes an mRNA transcript of 1.5 kb and a protein of 352 amino acids (Okamoto *et al.*, 1990; Rosner *et al.*, 1990; Schöler *et al.*, 1990b). The protein contains a region

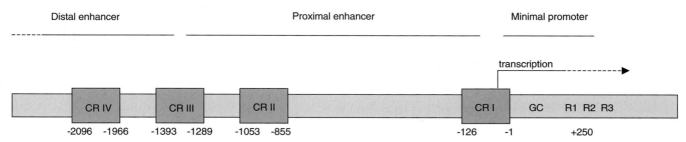

Figure 6 *Structure of the Oct4 gene in the mouse. The conserved regions (CR) in the promoter are boxed. The GC box and the hormone-responsive elements (R) are indicated in the minimal promoter (see Nordhoff et al., 2001).*

of approximately 150 amino acids comprising the DNA binding domain, known as the POU domain, which is conserved among all POU TFs. The POU domain is further subdivided into two subdomains: the POU-specific (POU$_S$) domain, a distinguishing feature of POU TFs composed of four α-helices (Herr *et al.*, 1988; Herr and Cleary, 1995), and the POU homeodomain (POU$_H$), a more general region that bears wide resemblance to selector homeotic proteins in *Drosophila* (McGinnis *et al.*, 1984; Scott and Weiner, 1984) and is composed of three α-helices. The POU$_S$ and POU$_H$ of Oct4 are connected by a linker region of 17 amino acids, the length and sequence of which are variable for different POU TFs. Oct4 and other POU TFs recognize and bind to the 8-bp DNA sequence ATGCAAAT, termed the octamer motif. This motif and variations thereof are found in many ubiquitously expressed (e.g., small nuclear RNA and histone H2B) and cell-specifically expressed (e.g., immunoglobulin) genes (Landolfi *et al.*, 1986; Fletcher *et al.*, 1987; Murphy *et al.*, 1992). Both POU$_S$ and POU$_H$ subdomains can bind to DNA independently of each other (Botfield *et al.*, 1992); however, simultaneous binding results in high-affinity sequence-specific binding (Ingraham *et al.*, 1990; Klemm *et al.*, 1994). Extrapolating from the crystal structure of the Oct1 protein bound to the octamer sequence, the POU$_S$ domain binds to the first four bases of the octamer motif (ATGC), whereas the POU$_H$ binds to the last four bases of the octamer motif (AAAT). Binding of Oct4 to the octamer motif, in the regulatory regions of key developmental genes, is likely to be the first of a series of steps resulting in regulation of gene expression that ensures totipotentiality of embryonic cells.

The genetic elements mediating differential expression of *Oct4*, in the different cell lineages, have been characterized. The Oct4 promoter lacks a TATA box, and transcription can be initiated from multiple sites (Okazawa *et al.*, 1991). The minimal promoter encompasses the first 250 bp within a major transcription initiation site and contains a cluster of overlapping binding sites: a GC box binding Sp1 and Sp3 TFs, and a hormone-responsive element (HRE) binding numerous members of the steroid–thyroid receptor family as well as the nuclear orphan receptor family (COUP-TF1, ARP-1, EAR-2). With the use of transgenic mice containing constructs of various *Oct4* regulatory sequences and *LacZ* reporter, we have identified two distinct *Oct4* enhancers located upstream of the minimal promoter. The distal enhancer (DE) (~2 kb) and the proximal enhancer (PE) (~1.2 kb), so-named due to their position relative to promoter (Fig. 6), exhibit a reciprocal pattern of activity during embryogenesis that leads to correct cell type-specific *Oct4* expression (Okazawa *et al.*, 1991; Yeom *et al.*, 1996). More specifically, the DE drives *Oct4* expression in the morula, ICM, PGCs, and ES cells, whereas the PE activates *Oct4* in the epiblast and EC cells (Yeom *et al.*, 1996). A comparative analysis of the mouse *Oct4* gene has specified the precise regulatory motifs that may mediate these interactions (Nordhoff *et al.*, 2001). The most predominant of these, the CCC(A/T)CCC motif, is present in both orientations all throughout the mouse *Oct4* upstream regions, including in the 2A site of the DE (5'-ACCCTGC**CCCT**C**CCCCC**A-3') and in the 1A site of the PE (5'-CACAGGAATG**GGGGGAGGG**GTG-3') (Nordhoff *et al.*, 2001). These sites display a differential transcription factor occupancy pattern in undifferentiated versus differentiated ES and EC cells (Minucci *et al.*, 1996) and, as such, represent crucial *Oct4* regulatory sites during development.

Oct4 IN OTHER SPECIES

Orthologs to the mouse *Oct4* gene have been identified in cells of other mammalian species, including bovines (van Eijk *et al.*, 1999) and humans (Takeda *et al.*, 1992). The common organization of conserved regulatory elements, high degree of sequence homology, gene organization, and chromosomal location all reverberate the notion that *Oct4* bears an evolutionarily conserved function during embryogenesis in these species (Nordhoff *et al.*, 2001; Pesce and Schöler, 2000). For instance, the upstream promoter regions of the human, bovine, and mouse *Oct4* orthologs exhibit four conserved regions (range 105–199 bp) of high sequence homology, in particular, the Sp1/Sp3 binding sites and the HRE of the CR1 region, suggesting that the *Oct4* gene might be under the control of similar regulative mechanisms (Nordhoff *et al.*, 2001). Similarly, the murine *Oct4* gene is highly conserved in humans and cows, as evidenced by 87 and 81.7% overall protein sequence identity, genomic organization into five exons, and chromosomal mapping to the major histocompatibility complex (Takeda *et al.*, 1992; Abdel-Rahman *et al.*, 1995; van Eijk *et al.*, 1999). In agreement with these findings, comparable *Oct4* expression in embryonic stem and germ cells in mice and humans strongly indicates that Oct4 fulfills a similar role in the maintenance of a totipotent cell phenotype, in both species, during embryonic development (Hansis *et al.*, 2000, 2001). However, the predicted function of Oct4 as the gatekeeper of pluripotency may discriminately apply to certain mammalian species. *Oct4* expression in the bovine and the porcine systems is not limited to pluripotent cells of the early embryo, but appears to have a much broader expression pattern than in the mouse (Kirchhof *et al.*, 2000; van Eijk *et al.*, 1999). However, bovine and porcine blastocysts, used in the latter study, are peculiar, and their implantation is delayed compared with that of the mouse. Candidate *Oct4*-related genes are postulated exist in monotreme and marsupial species, with a factor of 90% sequence similarity to the POU domain of murine *Oct4* isolated from the urodele amphibian *Axolotl* (Pesce *et al.*, 1998), and *Oct4*-expressing rat embryonic stemlike cell lines have been identified (Vassilieva *et al.*, 2000). The marsupial homolog of the mammalian *Oct4* gene has also been identified and found to share 74, 78, and 79% protein identity with murine, human, and bovine *Oct4*, as well as to exhibit a comparable expression pattern (Frankenberg *et al.*, 2001). The *Oct4*-related sequences have not been found in other vertebrates, such as chicken, frog, and fish (Yeom *et al.*, 1991), with the later confirmation that there is indeed no *Oct4* in chicken, which may resound differences in PGC development between these latter species and those of other vertebrates, such as the mouse (Soodeen-Karamath and Gibbins, 2001). The function of Oct4 in the mouse and in some other mammalian species thus appears to be integral to the avoidance of somatic cell differentiation processes, commitment to the PGC lineage, and completion of the totipotent cycle.

PRECISE OCT4 LEVELS DEFINE CELL FATE

Oct4-deficient embryos have been generated to assess the phenotypic effects of the loss of *Oct4* (Nichols *et al.*, 1998). Briefly, an *Oct4* null allele was first created in ES cells and heterozygous mice were then obtained by blastocyst injection of the *Oct4* +/− ES cells. Although these mice were normal and fertile, their offspring showed a skewed frequency of the wild-type allele, with about one-third of the offspring being +/+ and two-thirds of the offspring being +/−, but none being −/−. *In vitro* experiments revealed that the –/– embryos actually formed but could not develop much further. Although cavitation occurs and a putative ICM is observed in the *Oct4* null embryos, development does not proceed beyond implantation due to the formation of a nonpluripotent ICM, and trophectodermal function is sufficient to induce the decidual reaction by the uterine endometrium. This observation led researchers to ask whether the level of Oct4 was controlling the fate of embryonic cells. Niwa and co-workers addressed this issue and extended our findings (Niwa *et al.*, 2000). They reported the use of a conditional expression and repression system that allowed the determination of the cellular phenotype of mouse ES cells containing variable Oct4 levels. ES cells, in which one allele of the endogenous *Oct4* gene had been inactivated (Mountford *et al.*, 1994), were transfected with constructs for a tetracycline (Tc)-regulated transactivator (tTA) (Gossen and Bujard, 1992) and a tTA-responsive *Oct4* transgene, with varying amounts of Tc in order to modulate *Oct4* expression levels precisely. Results indicate that addition of Tc led to an almost 50% reduction in endogenous Oct4 levels, accompanied by differentiation of ES cells into clones of trophectoderm, as evidenced by induction of Hand1 and Cdx2 mRNAs, two factors involved in trophoblast differentiation (Niwa *et al.*, 2000). This result is consistent with *Oct4* down-regulation in outer cells of the morula on differentiation into TE, during the first lineage split in the embryo (Okamoto *et al.*, 1990; Rosner *et al.*, 1990; Schöler *et al.*, 1990a), and differentiation of otherwise ICM-destined cells to the trophoblast lineage, in Oct4$^{-/-}$ embryos (Nichols *et al.*, 1998). On the other hand, in the absence of Tc, Oct4 protein levels increase by almost 50% over the endogenous levels in ES cells, with the concomitant differentiation of cells to form clones of primitive endoderm and mesoderm, as evidenced by activation of markers *Gata4* and *brachyury* mRNA, respectively (Niwa *et al.*, 2000). This is in agreement with the transiently increased Oct4 protein levels in the forming hypoblast, observed at the second lineage split when the ICM differentiates into the primitive ectoderm (epiblast) and primitive endoderm (hypoblast) (Palmieri *et al.*, 1994; Schöler, 1991). Less pronounced changes in Oct4 levels were shown to maintain ES cell self-renewal and developmental potency (Niwa *et al.*, 2000). Thus, Oct4 up-regulation and down-regulation by 50% relative to that of undifferentiated mouse ES cells can act as a trigger toward ES cell differentiation to somatic lineages, whereas less subtle fluctuations maintain ES cell identity (Niwa *et al.*, 2000). Additional evidence also suggests that the fate of early human blastomeres may also depend on the precise Oct4 levels, as noted by the progressive restriction of blastomeres that highly express Oct4 during early cleavage stages (Hansis *et al.*, 2001). These observations thus suggest that Oct4 transcriptional regulatory effects are not only defined by a binary (on/off) switch but may also entail intermediate degrees of transcriptional regulation.

Oct4 REGULATION

Our analysis of transgenic mice carrying the LacZ reporter gene, linked to various *Oct4* upstream sequences, has led to the elucidation of three major upstream cis-regulatory regions determining *Oct4* expression: the proximal promoter, the proximal enhancer, and the distal enhancer (Fig. 6) (Yeom *et al.*, 1996). *Oct4* expression, in the developing mouse embryo, is the culmination of the activities of these promoter and enhancer regions that are executed in a precise temporal and spatial fashion. It has been postulated that Sp1 (or Sp3), which binds to the GC box of *Oct4* and mediates basal promoter activity (Pesce *et al.*, 1999b), may be involved in initiating transcription from TATA-less promoters (Pugh and Tjian, 1991), such as that of *Oct4*. This is in agreement with the abolition of *Oct4* promoter expression observed in the ES and EC cells bearing a mutated GC box (Minucci *et al.*, 1996). However, the stem-cell-specific *Oct4* expression pattern appears to be exemplified by the stage-specific activity of the *Oct4* enhancers. Thus, DE activity is observed in undifferentiated cells (ICM) of the preimplantation embryo and later on in development in PGCs, whereas PE activity is required for *Oct4* expression in the primitive ectoderm or epiblast. The differential *in vivo* activity of the *Oct4* enhancers is reverberated in ES, EG, and EC stem cells that resemble cells of the ICM, PGCs, and the epiblast, respectively. The molecular switch that directs activity of the DE to the PE at the time of implantation and back to the DE after gastrulation is unknown. The noncommitment of proximal epiblast cells to the germ cell lineage until their destination to the extraembryonic mesoderm, at the base of the allantois, suggests that reprogramming of *Oct4* expression is mediated by local signals (Pesce *et al.*, 1999a). It has, however, been confirmed that the activity of the DE is restricted to a totipotent and pluripotent cell phenotype. The next question becomes whether the activity of these enhancer elements is mediated via the differential recruitment and binding of transcription factors. *In vivo* footprinting experiments and electrophoretic mobility shift assays (EMSAs) confirm that these enhancers (in particular, the 1A site of the PE and the 2A site of the DE) bind to transcription factors in undifferentiated ES and EC cells in an identical fashion, and are both released from such binding on RA-induced differentiation of either cell type (Minucci *et al.*, 1996). These results strongly suggest that other upstream elements, in conjunction with these enhancers, are involved in mediating their stem-cell-specific activities.

Comparative analysis of the entire upstream regulatory regions of the murine, human, and bovine *Oct4* genes reveals the conservation of numerous sequence motifs, including CCC(A/T)CCC, the E-box consensus (5'-CANNTG-3'), and ATTA homeobox motif that may bind candidate factors, such as Sp1 and β enolase repressor factor 1 (BERF-1), Mash-2, and Pem transcription factors, respectively (Nordhoff *et al.*, 2001). A further question pertains as to what upstream elements and binding factors are involved in *Oct4* down-regulation, following stem cell differentiation. The answer promises to be complex and may be mediated directly or indirectly, and may vary depending on the cell type and time following differentiation. For example, a transiently induced factor (TRIF) binds to the proximal promoter about 12 hours after RA induction and is responsible for the initial *Oct4* down-regulation during EC cell differentiation (Fuhrmann *et al.*, 1999). After this time, other factors, including members of the nuclear hormone receptor and orphan receptor families, which can negatively regulate *Oct4* expression, may take over. It is noteworthy that the 1A site of the PE of P19 EC cells, which closely resemble epiblast cells, has deemed necessary for RA-mediated *Oct4* repression (for discussion see Ovitt and Schöler, 1998). Finally, preliminary evidence indicates that the germ cell nuclear factor (GCNF), an orphan nuclear repressor, represses *Oct4* gene activity by specifically binding within the proximal promoter, and, as such, presents a prime candidate for the confinement of the mammalian germ line and restriction of embryonic stem cell potency (Fuhrmann *et al.*, 2001).

Oct4 AND METHYLATION

The differential expression pattern of *Oct4* during embryonic development is not subject to the epigenetic marking mechanism of genomic imprinting; however, it may potentially be affected by methylation. It has been proposed that DNA methylation is important in somatic lineages but is not involved in embryonic lineages, including the germ line (Jaenisch, 1997). A wave of *de novo* methylation has been reported to occur in somatic cells of the embryo (Jaenisch, 1997), which lose *Oct4* expression during gastrulation. Similarly, the loss of *Oct4* gene activity in stem cell fibroblast hybrid cells is accompanied by a rapid methylation of CpG islands in the *Oct4* promoter and PE regions (Ben-Shushan *et al.*, 1993). This is in agreement with increased methylation and changes in chromatin structure of the *Oct4* upstream regions, which are associated with *Oct4* down-regulation, following embryonal carcinoma cell differentiation *in vitro* (Ben-Shushan *et al.*, 1993). Methylation of CpG sequences may inhibit gene transcription—in this case, that of *Oct4*—either directly by interfering with the binding of specific transcription factors, such as cyclic AMP-responsive element-binding protein (CREB), or indirectly by the binding of proteins, such as methyl-CpG-binding protein 1 (MeCP1) to methylated DNA, in turn preventing binding of specific transcription factors (for discussion see Ben-Shushan *et al.*, 1993). MeCP complexes have also been shown to comprise histone deacetylases, which induce chromatin condensation and thereby repress gene expression (Leonhardt and Cardoso, 2000). In contrast, the genome of PGCs is maintained in an undermethylated state (Monk *et al.*, 1987) and is subject to a potent undermethylation activity (Surani, 1998). It is therefore conceivable that methylation of *Oct4* gene regulatory sequences is a regulatory cue mediating *Oct4* shutdown during gastrulation, and that segregation of PGCs in an extraembryonic tissue, such as the extraembryonic mesoderm, may prevent methylation of their genome and concomitant *Oct4* down-regulation and germ cell differentiation. This suggests that maintenance of *Oct4* expression in PGCs, and thereby of mammalian germ line totipotency, may be a consequence of germ cells escaping from the general epigenetic reprogramming of the chromatin that occurs in epiblastic cells at the time of gastrulation and/or of their intrinsic demethylation activity.

ONE OR MORE PATHWAYS FOR PLURIPOTENCY?

The crucial role of Oct4 in the maintenance of embryonic cell potency has been firmly established. Recall that embryonic lethality of Oct4-deficient embryos attests to the absolute necessity of Oct4 in maintaining the totipotency of embryonic cells of the preimplantation embryo during the development of the first somatic lineage. *Oct4* expression in the epiblast, during the formation of the second somatic lineage, and restriction of *Oct4* expression to PGCs and germ cells postgastrulation raise the question as to whether Oct4 is an essential factor in the potency of subsequent embryonic stages. Because of the death of embryos lacking Oct4, a conditional homologous recombination approach is required to assess the precise role of Oct4 in the determination of the germ cell fate. It is possible that local factors and/or pathways, alone or in concert with Oct4, contribute to the potency at specific stages of development or, for that matter, in different mammalian species.

It is noteworthy that the propagation of murine ES cells, which resemble ICM cells, in an undifferentiated and pluripotent state sensitive to Oct4 levels is dependent on LIF, a factor belonging to the interleukin (IL)-6 cytokine family. Human ES cells, however, although expressing high levels of Oct4, do not rely on an exogenous supply of LIF to sustain their proliferation as do undifferentiated stem cells. Curiously, LIF-deficient mouse embryos develop quite normally (Stewart *et al.*, 1992). The effect of LIF is mediated through a cell surface receptor complex composed of low-affinity LIF receptor (LIFR) and GP130, a common receptor subunit of the IL-6 cytokine family (Niwa *et al.*, 1998). As LIF binds to its receptor, LIFR heterodimerizes with GP130, thereby initiating a signal cascade. Both monomeric LIFR and GP130 lack phosphorylation activity; however, their heterodimer activates a protein tyrosine kinase, Janus kinase (Jak). On tyrosine phosphorylation of LIFR and GP130 by Jak, molecules of the STAT family are recruited onto the LIFR/GP130 heterodimer and phosphorylated by Jak. Phosphorylated STAT molecules dimerize and are then translocated to the cell nucleus where they bind to DNA and induce transcription of specific genes. In particular, self-renewal and prevention of differentiation of ES cells have been shown to depend on STAT3. A dominant interfering mutation of STAT3 (STAT3F) was produced and ES cells were episomally supertransfected with STAT3F to express this molecule constitutively. ES cells overexpressing STAT3F, in the presence of LIF, stopped self-renewing and underwent differentiation (Niwa *et al.*, 1998). STAT3-deficient mouse embryos undergo a rapid degeneration between days 6.5 and 7.5 of development, resulting in early embryonic lethality (Takeda *et al.*, 1997). These results are quite interesting, considering the finding that a gradient of leptin/STAT3 proteins may be formed within the oocyte or in early cleavage stage blastomeres (Antczak and Van Blerkom, 1997), which may establish a differential gene expression pattern in the inner cells destined to a totipotent phenotype versus the outer cells destined to somatic lineages.

GENETIC INTERACTIONS OF OCT4

As a key regulator of embryonic cell totipotency, the transcription factor Oct4 is involved in the setting of appropriate gene expression pattern during embryogenesis. Oct4 is postulated to bind, either alone or in a cooperative interaction with other factors, to the regulatory regions of different target genes and, thereby, activate the transcription of genes required in maintaining an undifferentiated totipotent state, while preventing the transcription of genes that are activated during stem cell differentiation (Pesce *et al.*, 1998, 1999a; Pesce and Schöler, 2000). Several Oct4 target genes have been identified, including *FGF-4*, *Rex-1*, *OPN*, *hCG*, and *Utf-1* (see below).

Oct4 activates gene transcription from octamer motifs located proximally or distally to the transcription start sites. However, it can activate transcription in a distance-independent manner only in undifferentiated ES and embryonal carcinoma (EC) cells, which express coactivators bridging the remotely bound Oct4 to the transcriptional machinery (Brehm *et al.*, 1999; Schöler *et al.*, 1991). In differentiated cells, the adenoviral protein E1A, which functions as an Oct4 coactivator, mimics this activity (Schöler *et al.*, 1991). Oct4 contains two potent transactivation domains, the N-terminus domain and C-terminus domain. The N-terminus domain is active in various cultured cell types, whereas the C-terminus domain is cell-type-specific and is modulated via phosphorylation (Brehm *et al.*, 1997).

Fibroblast growth factor-4 (FGF-4) is expressed in the ICM of the blastocyst subsequent to Oct4 expression, where it plays a role in the proliferation/survival of the ICM of the preimplantation embryo and in the establishment of primitive endoderm. This temporal expression pattern, as well as the absence of or reduced amounts of FGF-4 in Oct4-deficient embryos (Nichols *et al.*, 1998), suggest that Oct4 acts as a positive regulator of *FGF-4* transcription. Confirmatory evidence for this comes from studies of EC cells showing that Oct4, and, in cooperation, the transcription factor Sox-2, bind to the octamer and high-mobility group (HMG) box found in the enhancer of the *FGF-4* gene, respectively, to synergistically activate *FGF-4* expression (Ambrosetti *et al.*, 1997; Yuan *et al.*, 1995).

The undifferentiated embryonic cell transcription factor-1 (Utf-1) is coexpressed with Oct4 in undifferentiated ES and EC cell lines, and is rapidly down-regulated following RA-induced cell differentiation. In mouse embryos, *Utf-1* is expressed in the ICM and confined to embryonic and extraembryonic ectoderm, where it is postulated to function as a tissue-specific transcriptional coactivator during early embryogenesis (Okuda *et al.*, 1998). As in the case of *FGF-4* gene, Oct4 and Sox-2 bind to the 3′ untranslated region (UTR) of the *Utf-1* gene and thereby synergistically activate its expression (Nishimoto *et al.*, 1999).

Rex-1 is a zinc-finger transcription factor specifically expressed in undifferentiated stem cells (Hosler *et al.*, 1993). Its expression pattern parallels that of *Oct4*, in that they are both elevated in EC cells and in the ICM, and both diminish during differentiation of the former and development of the latter, suggesting that Oct4 also positively regulates *Rex-1* expression. It has been demonstrated that Oct4 cooperates with a novel stem-cell-specific activity named Rox-1 to activate *Rex-1* gene transcription (Ben-Shushan *et al.*, 1998; Hosler *et al.*, 1993), and further proposed that the cellular levels of Oct4 determine the type of regulation. At low levels, Oct4 activates the Rex-1 promoter, whereas at high levels, Oct4 represses the Rex-1 promoter (Ben-Shushan *et al.*, 1998). The cellular environment appears to play a role in the transcriptional outcome, in that the presence of cofactors may selectively enhance or prevent Oct4 recruitment to the *Rex-1* promoter. Another possibility is that at high levels, Oct4 may form dimers on the *Rex*-1 promoter, which may hold a direct repressive activity or indirectly prevent access of Rox-1 to its binding site.

Osteopontin (OPN) is an extracellular matrix phosphoprotein postulated to play a role in the migration of the primitive endoderm, and is coexpressed with *Oct4* in the preimplantation embryo and during differentiation of embryonal cell lines (Botquin *et al.*, 1998). Enrichment of F9 EC cell cross-linked chromatin fragments containing the *OPN* enhancer, following immunoprecipitation with Oct4 antibodies, strongly suggests that Oct4 regulates *OPN* expression during preimplantation development (Botquin *et al.*, 1998). The OPN enhancer contains a palindromic Oct4 binding sequence (PORE) with the ATGCAAAT octamer motif and an inverted half-site CAAAT, to which the Oct4 protein was shown to bind as a dimer, in a distinct configuration that uses only the homeodomain sites. Oct4 dimer formation on the PORE element, oligomerized in front of a minimal promoter in F9 and P19 EC cells, activates reporter gene expression and strongly implies that the Oct4 dimer similarly mediates *OPN* expression in the ICM and hypoblast (Botquin *et al.*, 1998). However, unlike in the *FGF-4* enhancer, Sox-2 binds to its cognate site adjacent to the PORE in the *OPN* enhancer, thereby, repressing Oct4-mediated *OPN* gene transactivation (Botquin *et al.*, 1998). Preliminary evidence demonstrates that Sox-2 probably interferes with Oct4 dimer formation (Botquin *et al.*, 1998). Because *OPN* and *Oct4* expression overlap only in preimplantation embryos and *Oct4* expression is not necessary for *OPN* expression in all cell types tested, it is likely that Oct4 is responsible for regulating *OPN* expression only during preimplantation development.

In addition to its activation properties, Oct4 may also inhibit the transcription of genes that are activated during stem cell differentiation. Oct4 represses the transcription of α and β human chorionic gonadotropin (*hCG*) genes of choriocarcinomal cells, which are required for implantation and maintenance of pregnancy in primates (Liu and Roberts, 1996; Liu *et al.*, 1997). It is likely that Oct4 downregulation during TE differentiation at the surface on the compacted morula relieves the repression of the *hCG* genes, thereby enabling the TE to produce the chorionic gonadotropin, which prevents regression of the corpus luteum during early pregnancy (Liu and Roberts, 1996; Liu *et al.*, 1997). These results are in accordance the findings of Oct4-deficient morulae, which differentiate into trophoblastic cells (Nichols *et al.*, 1998).

The expression of the Pem transcription factor shows a strong negative correlation with that of *Oct4*. Murine *Pem* is an X-linked homeobox gene expressed in the preimplantation embryo and in a lineage-restricted fashion following implantation. Its forced expression in ES cells blocks their differentiation *in vitro* and *in vivo*, a phenomenon that appears to be cell autonomous, as evidenced by failure of Pem[+/+] ES cells to differentiate when cocultured with normal ES cells (Fan *et al.*, 1999). Although no direct interaction between Oct4 and Pem has been noted thus far, it is possible that Pem is negatively regulated by Oct4 and itself plays a role in regulating the transition between undifferentiated and differentiated cells of the early mouse embryo (Fan *et al.*, 1999).

Finally, the abolishment of Oct4 in mouse ES cells results in a marked change in cell morphology such that cells resemble trophoectodermal stem cells (Niwa *et al.*, 2000). In accordance with this phenotype, TE cells express numerous genetic markers (Cdx2, Esrrb, Mash-2, Tpbp, Pl1) (Niwa *et al.*, 2000) that represent potential negatively regulated candidate target genes of Oct4.

LESSONS FROM CLONING

Studies on cloning by nuclear transplantation have shown that not only early embryonic nuclei but also fetal and adult somatic cell nuclei can be induced to develop fully and become live young (sheep, cattle, mice, goats, pigs, and cats). So far, germ cell nucleus transplantation has not been successful. The production of clones by nuclear transplantation has provoked new thoughts about the potency of (some) adult nucleus donor cells. Previous information about animal development provided no basis for the expectation that adult nuclei (other than gamete nuclei) could support clonal development to adulthood. Additional insight has been gained on the biological issues of differentiation and development. Because adult nuclei can be induced to recapitulate development by transfer to oocyte cytoplasm, factors responsible for their stable differentiation in the original tissue environment are apparently subject to alteration. The nucleus may be more or less permissive once transplanted into the oocyte microenvironment. This is reminiscent of the modern concept of stem cell fate, i.e., that stem cells have a conditional fate depending on the cues from the microenvironment. Nuclear transfer provides new awareness regarding normal development of fertilized embryos: sperm and oocytes are not unique in possessing nuclear totipotency, but are unique in their capacity to activate eggs or to evoke nuclear potential for embryogenesis, respectively. However, the low efficiency of somatic nuclei at recapitulating development (1–2% development to term by somatic cell nuclei versus 50–75% for fertilized oocytes), and the related high rates of embryonic waste, fetal and neonatal deaths, and birth defects, hinder broader application of the current procedure to biotechnology, not to mention large-scale engineering in farm species or the madness of thinking about cloning humans (Jaenisch and Wilmut, 2001).

It is well-established that a significant proportion of cloned offspring produced by nuclear transplantation exhibit abnormal phenotypes. Abnormal placentation, overgrowth, and lengthened gestation are just a few examples. Survivors to birth exhibit pulmonary hypertension leading to insufficient pulmonary perfusion and respiratory distress syndrome, enlarged/dilated right ventricles, and patent ductus arteriosus. Due to the short time of generation, these phenotypes have been described best in the mouse (Eggan *et al.*, 2001; Ogonuki *et al.*, 2002). Besides the developmental abnormalities, observed prior to or at birth, it is clear that embryos produced by nuclear transplantation also result in decreased pregnancy rates following embryo transfer, and fetal losses (abortions/resorptions) are significantly higher. The cause of abnormal embryo/fetal development in clones remains unknown. Increased birth weights have previously been attributed to *in vitro* culture conditions (Sinclair *et al.*, 2000). This suggests that developmental abnormalities observed in animals produced by nuclear transplantation may be caused by the culture environment and not by the nuclear transfer procedure. Clone embryos reimplanted *in vivo*, immediately or shortly after microsurgery (two-cell and eight-cell stages), exhibit the same anomalies. This is based on the assumption that the genital tract is not hostile to clone embryo development and the clone is a real embryo. Although differences in gene expression profiles most probably exist between embryos produced *in vivo*, *in vitro*, or by nuclear transplantation, developmental aberrations are much more common in animals produced by nuclear transplantation compared to those produced using standard procedures for *in vitro* oocyte maturation, *in vitro* fertilization, and *in vitro* embryo culture. Although the culture environment may contribute to clone defects, it is unlikely to be the major factor involved. Defects would also be difficult to explain by the notion that successful cloning reflects the incidental selection of stem cells, which are already toti- or pluripotent and therefore presumably adequate for cloning. Conversely, defects would be more likely if the nuclei underwent a stochastic alteration of their epigenetic marks. It might be possible to estimate the core number of essential genes by the rate of survival of clone embryos.

Because primordial germ cells fail to support clone embryo development after nuclear transplantation (Kato *et al.*, 1999), but their derivatives (EG cells) contribute to fertile chimeras regardless of their stage and imprinting state (Labosky *et al.*, 1994), imprinting imbalance has been addressed as a main cause for cloning failure. However, from the standpoint of reprogramming, the initial state of the nucleus— say, the array of methylated DNA loci—should not make any difference. If a nucleus can be reprogrammed from a terminally differentiated state, even from a lymphocyte (Hochedlinger and Jaenisch, 2002), everything else should be easier to accomplish. Support for this view comes from parthenogenetic diploid embryos. Their nuclei undergo X-chromosome inactivation as if they were normal biparental embryos, indicating that parthenotes can reprogram a major part of their genome. Why, then, do nuclei from germ cells (at a not much earlier stage) fail to undergo reprogramming after transplantation into ooplasm? A different degree of permissiveness for reprogramming might exist at different stages of differentiation and/or in different cell types, or the oocyte could have a limited capacity to reset the epigenetic marks, i.e., it could reset general methylation but not the imprints. Interestingly, all these embryos (uniparental embryos and germ and somatic cell clones) can reach the blastocyst stage. Major developmental differences occur later, supposedly as a consequence of imprinting imbalance. For this reason, it may be argued that the making of a blastocyst is a developmental default and not very meaningful. Our unpublished data argue that anomalies in gene expression may occur from the very beginning of clone embryo development and are related to a nonimprinted gene, *Oct4*. In conclusion, insights on pluripotency, obtained through stem cell and cloning technology, do not support the Weismannist view of continuity. Recent advances in the field of embryoid bodies outline something that resembles the Darwinist view—that everything can transform into everything else under the appropriate environmental or experimental conditions. However, EB technology is far from being able to convert stem cells, somatic cells, and germ cells into each other— at least with current knowledge and techniques. But cloning has shown that in order to make germ cells, the soma does not necessarily have to be derived from germ cells. The relationship between soma and germ might involve more than a dotted line.

ACKNOWLEDGMENTS

We wish to thank Karin Hübner, Katharina Lins, Stefan Schlatt, and Alexey Tomilin, and we are especially grateful to Maurizio Pesce and Areti Malapetsa for critically reading the manuscript. Photographic material prepared by Michele Boiani and Karin Hübner.

REFERENCES

Abdel-Rahman, B., Fiddler, M., Rappolee, D., and Pergament, E. (1995). Expression of transcription regulating genes in human preimplantation embryos. *Hum. Reprod.* **10**, 2787–2792.

Allsopp, R. C., Vaziri, H., Patterson, C., Goldstein, S., Younglai, E. V., Futcher, A. B., Greider, C. W., and Harley, C. B. (1992). Telomere length predicts replicative capacity of human fibroblasts. *Proc. Natl. Acad. Sci. U.S.A.* **89**, 10114–10118.

Amano, T., Nakamura, K., Tani, T., Kato, Y., and Tsunoda, Y. (2000). Production of mice derived entirely from embryonic stem cells after injecting the cells into heat treated blastocysts. *Theriogenology* **53**, 1449–1458.

Ambrosetti, D. C., Basilico, C., and Dailey, L. (1997). Synergistic activation of the fibroblast growth factor 4 enhancer by Sox2 and Oct-3 depends on protein–protein interactions facilitated by a specific spatial arrangement of factor binding sites. *Mol. Cell. Biol.* **17**, 6321–6329.

Amit, M., Carpenter, M. K., Inokuma, M. S., Chiu, C. P., Harris, C. P., Waknitz, M. A., Itskovitz-Eldor, J., and Thomson, J. A. (2000). Clonally derived human embryonic stem cell lines maintain pluripotency and proliferative potential for prolonged periods of culture. *Dev. Biol.* **227**, 271–278.

Anderegg, C., and Markert, C. L. (1986). Successful rescue of microsurgically produced homozygous uniparental mouse embryos via production of aggregation chimeras. *Proc. Natl. Acad. Sci. U.S.A.* **83**, 6509–6513.

Antczak, M., and Van Blerkom, J. (1997). Oocyte influences on early development: The regulatory proteins leptin and STAT3 are polarized in mouse and human oocytes and differentially distributed within the cells of the preimplantation stage embryo. *Mol. Hum. Reprod.* **3**, 1067–1086.

Balakier, H., and Pedersen, R. A. (1982). Allocation of cells to inner cell mass and trophectoderm lineages in preimplantation mouse embryos. *Dev. Biol.* **90**, 352–362.

Beddington, R. S., and Robertson, E. J. (1989). An assessment of the developmental potential of embryonic stem cells in the midgestation mouse embryo. *Development* **105**, 733–737.

Beddington, R. S., and Robertson, E. J. (1999). Axis development and early asymmetry in mammals. *Cell* **96**, 195–209.

Ben-Shushan, E., Pikarsky, E., Klar, A., and Bergman, Y. (1993). Extinction of Oct-3/4 gene expression in embryonal carcinoma × fibroblast somatic cell hybrids is accompanied by changes in the methylation status, chromatin structure, and transcriptional activity of the Oct-3/4 upstream region. *Mol. Cell. Biol.* **13**, 891–901.

Ben-Shushan, E., Thompson, J. R., Gudas, L. J., and Bergman, Y. (1998). Rex-1, a gene encoding a transcription factor expressed in the early embryo, is regulated via Oct-3/4 and Oct-6 binding to an octamer site and a novel protein, Rox-1, binding to an adjacent site. *Mol. Cell. Biol.* **18**, 1866–1878.

Bjornson, C. R., Rietze, R. L., Reynolds, B. A., Magli, M. C., and Vescovi, A. L. (1999). Turning brain into blood: A hematopoietic fate adopted by adult neural stem cells *in vivo*. *Science* **283**, 534–537.

Botfield, M. C., Jancso, A., and Weiss, M. A. (1992). Biochemical characterization of the Oct-2 POU domain with implications for bipartite DNA recognition. *Biochemistry* **31**, 5841–5848.

Botquin, V., Hess, H., Fuhrmann, G., Anastassiadis, C., Gross, M. K., Vriend, G., and Schöler, H. R. (1998). New POU dimer configuration mediates antagonistic control of an osteopontin preimplantation enhancer by Oct-4 and Sox-2. *Genes Dev.* **12**, 2073–2090.

Bouwens, L. (1998). Transdifferentiation versus stem cell hypothesis for the regeneration of islet beta-cells in the pancreas. *Microsc. Res. Tech.* **43**, 332–336.

Brehm, A., Ohbo, K., and Schöler, H. R. (1997). The carboxy-terminal transactivation domain of Oct-4 acquires cell specificity through the POU domain. *Mol. Cell. Biol.* **17**, 154–162.

Brehm, A., Ohbo, K., Zwerschke, W., Botquin, V., Jansen-Durr, P., and Schöler, H. R. (1999). Synergism with germ line transcription factor Oct-4: Viral oncoproteins share the ability to mimic a stem cell-specific activity. *Mol. Cell. Biol.* **19**, 2635–2643.

Campbell, K. H. (1999). Nuclear transfer in farm animal species. *Semin. Cell Dev. Biol.* **10**, 245–252.

Chapman, V. M., Ansell, J. D., and McLaren, A. (1972). Trophoblast giant cell differentiation in the mouse: expression of glucose phosphate isomerase (GPI-1) electrophoretic variants in transferred and chimeric embryos. *Dev. Biol.* **29**, 48–54.

Ciemerych, M. A., Mesnard, D., and Zernicka-Goetz, M. (2000). Animal and vegetal poles of the mouse egg predict the polarity of the embryonic axis, yet are nonessential for development. *Development* **127**, 3467–3474.

Cirulli, V., Beattie, G. M., Klier, G., Ellisman, M., Ricordi, C., Quaranta, V., Frasier, F., Ishii, J. K., Hayek, A., and Salomon, D. R. (2000). Expression and function of alpha(v)beta(3) and alpha(v)beta(5) integrins in the developing pancreas: Roles in the adhesion and migration of putative endocrine progenitor cells. *J. Cell Biol.* **150**, 1445–1460.

Clarke, D. L., Johansson, C. B., Wilbertz, J., Veress, B., Nilsson, E., Karlstrom, H., Lendahl, U., and Frisen, J. (2000). Generalized potential of adult neural stem cells. *Science* **288**, 1660–1663.

Coucouvanis, E., and Martin, G. R. (1995). Signals for death and survival: A two-step mechanism for cavitation in the vertebrate embryo. *Cell* **83**, 279–287.

Coucouvanis, E., and Martin, G. R. (1999). BMP signaling plays a role in visceral endoderm differentiation and cavitation in the early mouse embryo. *Development* **126**, 535–546.

Desbaillets, I., Ziegler, U., Groscurth, P., and Gassmann, M. (2000). Embryoid bodies: An *in vitro* model of mouse embryogenesis. *Exp. Physiol.* **85**, 645–651.

Doetschman, T. C., Eistetter, H., Katz, M., Schmidt, W., and Kemler, R. (1985). The in vitro development of blastocyst-derived embryonic stem cell lines: Formation of visceral yolk sac, blood islands and myocardium. *J. Embryol. Exp. Morphol.* **87**, 27–45.

Doherty, A. S., Mann, M. R., Tremblay, K. D., Bartolomei, M. S., and Schultz, R. M. (2000). Differential effects of culture on imprinted H19 expression in the preimplantation mouse embryo. *Biol. Reprod.* **62**, 1526–1535.

Edlund, T., and Jessell, T. M. (1999). Progression from extrinsic to intrinsic signaling in cell fate specification: A view from the nervous system. *Cell* **96**, 211–224.

Eggan, K., Akutsu, H., Hochedlinger, K., Rideout, W., Yanagimachi, R., and Jaenisch, R. (2000). X-Chromosome inactivation in cloned mouse embryos. *Science* **290**, 1578–1581.

Eggan, K., Akutsu, H., Loring, J., Jackson-Grusby, L., Klemm, M., Rideout, W. M. 3rd, Yanagimachi, R., and Jaenisch, R. (2001). Hybrid vigor, fetal overgrowth, and viability of mice derived by nuclear cloning and tetraploid embryo complementation. *Proc. Natl. Acad. Sci. U.S.A.* **98**, 6209–6214.

Eglitis, M. A., and Mezey, E. (1997). Hematopoietic cells differentiate into both microglia and macroglia in the brains of adult mice. *Proc. Natl. Acad. Sci. U.S.A.* **94**, 4080–4085.

Fan, Y., Melhem, M. F., and Chaillet, J. R. (1999). Forced expression of the homeobox-containing gene Pem blocks differentiation of embryonic stem cells. *Dev. Biol.* **210**, 481–496.

Firulli, A. B., and Olson, E. N. (1997). Modular regulation of muscle gene transcription: A mechanism for muscle cell diversity. *Trends Genet.* **13**, 364–369.

Fletcher, C., Heintz, N., and Roeder, R. G. (1987). Purification and characterization of OTF-1, a transcription factor regulating cell cycle expression of a human histone H2b gene. *Cell* **51**, 773–781.

Frankenberg, S., Tisdall, D., and Selwood, L. (2001). Identification of a homologue of POU5F1 (OCT3/4) in a marsupial, the brushtail possum. *Mol. Reprod. Dev.* **58**, 255–261.

Fuchs, E., and Segre, J. A. (2000). Stem cells: A new lease on life. *Cell* **100**, 143–155.

Fuhrmann, G., Sylvester, I., and Schöler, H. R. (1999). Repression of Oct-4 during embryonic cell differentiation correlates with the appearance of TRIF, a transiently induced DNA-binding factor. *Cell. Mol. Biol. (Noisy-le-grand)* **45**, 717–724.

Fuhrmann, G., Chung, A. C., Jackson, K. J., Hummelke, G., Baniahmad, A., Sutter, J., Sylvester, I., Scholer, H. R., and Cooney, A. J. (2001). Mouse germline restriction of Oct4 expression by germ cell nuclear factor. *Dev. Cell* **1**(3), 377–387.

Gabel, C. A., Eddy, E. M., and Shapiro, B. M. (1979). After fertilization, sperm surface components remain as a patch in sea urchin and mouse embryos. *Cell* **18**, 207–215.

Gardner, R. L. (1968). Mouse chimeras obtained by the injection of cells into the blastocyst. *Nature* **220**, 596–597.

Gardner, R. L. (1997). The early blastocyst is bilaterally symmetrical and its axis of symmetry is aligned with the animal–vegetal axis of the zygote in the mouse. *Development* **124**, 289–301.

Gardner, R. L. (2001). Specification of embryonic axes begins before cleavage in normal mouse development. *Development* **128**, 839–847.

Gardner, R. L., and Brook, F. A. (1997). Reflections on the biology of embryonic stem (ES) cells. *Int. J. Dev. Biol.* **41**, 235–243.

Giet, O., Huygen, S., Beguin, Y., and Gothot, A. (2001). Cell cycle activation of hematopoietic progenitor cells increases very late antigen-5-mediated adhesion to fibronectin. *Exp. Hematol.* **29**, 515–524.

Gossen, M., and Bujard, H. (1992). Tight control of gene expression in mammalian cells by tetracycline-responsive promoters. *Proc. Natl. Acad. Sci. U.S.A.* **89**, 5547–5551.

Graham, C. F. (1971). The design of the mouse blastocyst. *In* "Control Mechanisms of Growth and Differentiation" (D. Davis and M. Balls, Eds.). Cambridge University Press, Cambridge.

Graham, C. F., and Deussen, Z. A. (1978). Features of cell lineage in preimplantation mouse development. *J. Embryol. Exp. Morphol.* **48**, 53–72.

Handyside, A. H., and Johnson, M. H. (1978). Temporal and spatial patterns of the synthesis of tissue-specific polypeptides in the preimplantation mouse embryo. *J. Embryol. Exp. Morphol.* **44**, 191–199.

Hansis, C., Grifo, J. A., and Krey, L. C. (2000). Oct-4 expression in inner cell mass and trophectoderm of human blastocysts. *Mol. Hum. Reprod.* **6**, 999–1004.

Hansis, C., Tang, Y. X., Grifo, J. A., and Krey, L. C. (2001). Analysis of Oct-4 expression and ploidy in individual human blastomeres. *Mol. Hum. Reprod.* **7**, 155–161.

Hass, P. E., Wada, H. G., Herman, M. M., and Sussman, H. H. (1979). Alkaline phosphatase of mouse teratoma stem cells: Immunochemical and structural evidence for its identity as a somatic gene product. *Proc. Natl. Acad. Sci. U.S.A.* **76**, 1164–1168.

Herr, W., and Cleary, M. A. (1995). The POU domain: Versatility in transcriptional regulation by a flexible two-in-one DNA-binding domain. *Genes Dev.* **9**, 1679–1693.

Herr, W., Sturm, R. A., Clerc, R. G., Corcoran, L. M., Baltimore, D., Sharp, P. A., Ingraham, H. A., Rosenfeld, M. G., Finney, M., Ruvkun, G., *et al.* (1988). The POU domain: A large conserved region in the mammalian pit-1, oct-1, oct-2, and *Caenorhabditis elegans* unc-86 gene products. *Genes Dev.* **2**, 1513–1516.

Hochedlinger, K., and Jaenisch, R. (2002). Monoclonal mice generated by nuclear transfer from mature B and T donor cells. *Nature* **415**, 1035–1038.

Hogan, B., Beddington, R., Costantini, F., and Lacy, E. (1994). "Manipulating the Mouse Embryo: A Laboratory Manual." CSH Press, Cold Spring Harbor.

Hoppe, P. C., and Illmensee, K. (1982). Full-term development after transplantation of parthenogenetic embryonic nuclei into fertilized mouse eggs. *Proc. Natl. Acad. Sci. U.S.A.* **79**, 1912–1916.

Hosler, B. A., Rogers, M. B., Kozak, C. A., and Gudas, L. J. (1993). An octamer motif contributes to the expression of the retinoic acid-regulated zinc finger gene Rex-1 (Zfp-42) in F9 teratocarcinoma cells. *Mol. Cell. Biol.* **13**, 2919–2928.

Illmensee, K., and Mintz, B. (1976). Totipotency and normal differentiation of single teratocarcinoma cells cloned by injection into blastocysts. *Proc. Natl. Acad. Sci. U.S.A.* **73**, 549–553.

Ingraham, H. A., Flynn, S. E., Voss, J. W., Albert, V. R., Kapiloff, M. S., Wilson, L., and Rosenfeld, M. G. (1990). The POU-specific domain of Pit-1 is essential for sequence-specific, high affinity DNA binding and DNA-dependent Pit-1-Pit-1 interactions. *Cell* **61**, 1021–1033.

Itskovitz-Eldor, J., Schuldiner, M., Karsenti, D., Eden, A., Yanuka, O., Amit, M., Soreq, H., and Benvenisty, N. (2000). Differentiation of human embryonic stem cells into embryoid bodies comprising the three embryonic germ layers. *Mol. Med.* **6**, 88–95.

Jackson, K. A., Mi, T., and Goodell, M. A. (1999). Hematopoietic potential of stem cells isolated from murine skeletal muscle. *Proc. Natl. Acad. Sci. U.S.A.* **96**, 14482–14486.

Jaenisch, R. (1997). DNA methylation and imprinting: Why bother? *Trends Genet.* **13**, 323–339.

Jaenisch, R., and Wilmut, I. (2001). Developmental biology. Don't clone humans! *Science* **291**, 2552.

Kato, Y., and Tsunoda, Y. (1995). Germ cell nuclei of male fetal mice can support development of chimeras to midgestation following serial transplantation. *Development* **121**, 779–783.

Kato, Y., Rideout, W. M. 3rd, Hilton, K., Barton, S. C., Tsunoda, Y., and Surani, M. A. (1999). Developmental potential of mouse primordial germ cells. *Development* **126**, 1823–1832.

Kaufman, M. H., Barton, S. C., and Surani, M. A. (1977). Normal postimplantation development of mouse parthenogenetic embryos to the forelimb bud stage. *Nature* **265**, 53–55.

Kawase, E., Yamazaki, Y., Yagi, T., Yanagimachi, R., and Pedersen, R. A. (2000). Mouse embryonic stem (ES) cell lines established from neuronal cell-derived cloned blastocysts. *Genesis* **28**, 156–163.

Kelly, D. L., and Rizzino, A. (2000). DNA microarray analyses of genes regulated during the differentiation of embryonic stem cells. *Mol. Reprod. Dev.* **56**, 113–123.

Kelly, S. J., Mulnard, J. G., and Graham, C. F. (1978). Cell division and cell allocation in early mouse development. *J. Embryol. Exp. Morphol.* **48**, 37–51.

Khosla S., Dean W., Brown, D., Reik, W., and Feil R. (2001) Culture of preimplantation mouse embryos affects fetal development and the expression of imprinted genes. *Biol. Reprod.* **64**, 918–926

Kim, N. W., Piatyszek, M. A., Prowse, K. R., Harley, C. B., West, M. D., Ho, P. L., Coviello, G. M., Wright, W. E., Weinrich, S. L., and Shay, J. W. (1994). Specific association of human telomerase activity with immortal cells and cancer. *Science* **266**, 2011–2015.

Kirchhof, N., Carnwath, J. W., Lemme, E., Anastassiadis, K., Schöler, H., and Niemann, H. (2000). Expression pattern of oct-4 in preimplantation embryos of different species. *Biol. Reprod.* **63**, 1698–1705.

Kirk, D. (1994). Germ cell specification in *Volvox carteri*. *In* "Germline Development" (J. Marsh and J. Goode, Eds.), pp. 2–15. John Wiley and Sons, England.

Klemm, J. D., Rould, M. A., Aurora, R., Herr, W., and Pabo, C. O. (1994). Crystal structure of the Oct-1 POU domain bound to an octamer site: DNA recognition with tethered DNA-binding modules. *Cell* **77**, 21–32.

Knowles, B. B., Aden, D. P., and Solter, D. (1978). Monoclonal antibody detecting a stage-specific embryonic antigen (SSEA-1) on preimplantation mouse embryos and teratocarcinoma cells. *Curr. Top. Microbiol. Immunol.* **81**, 51–53.

Labosky, P., Barlow, D. P., and Hogan, B. L. M. (1994). Embryonic germ cell lines and their derivation from mouse primordial germ cells. *In* "Germline Development" (J. Marsh and J. Goode, Eds.). John Wiley and Sons, England.

Landolfi, N. F., Capra, J. D., and Tucker, P. W. (1986). Interaction of cell-type-specific nuclear proteins with immunoglobulin VH promoter region sequences. *Nature* **323**, 548–551.

Lanza, R. P., Cibelli, J. B., and West, M. D. (1999). Human therapeutic cloning. *Nat. Med.* **5**, 975–977.

Lanza, R. P., Cibelli, J. B., Blackwell, C., Cristofalo, V. J., Francis, M. K., Baerlocher, G. M., Mak, J., Schertzer, M., Chavez, E. A., Sawyer, N., *et al.* (2000). Extension of cell life-span and telomere length in animals cloned from senescent somatic cells. *Science* **288**, 665–669.

Lehmann, R., and Ephrussi, A. (1994). Germ plasm formation and germ cell determination in Drosophila. *In* "Germline Development" (J. Marsh and J. Goode, Eds.). John Wiley and Sons, England.

Lenardo, M. J., Staudt, L., Robbins, P., Kuang, A., Mulligan, R. C., and Baltimore, D. (1989). Repression of the IgH enhancer in teratocarcinoma cells associated with a novel octamer factor. *Science* **243**, 544–546.

Leonhardt, H., and Cardoso, M. C. (2000). DNA methylation, nuclear structure, gene expression and cancer. *J. Cell Biochem.* (Suppl.) **35**, 78–83.

Liu, L. (2001). Cloning efficiency and differentiation. *Nat. Biotechnol.* **19**, 406.

Liu, L., and Roberts, R. M. (1996). Silencing of the gene for the beta subunit of human chorionic gonadotropin by the embryonic transcription factor Oct-3/4. *J. Biol. Chem.* **271**, 16683–16689.

Liu, L., Leaman, D., Villalta, M., and Roberts, R. M. (1997). Silencing of the gene for the alpha-subunit of human chorionic gonadotropin by the embryonic transcription factor Oct-3/4. *Mol. Endocrinol.* **11**, 1651–1658.

Mann, J. R., Gadi, I., Harbison, M. L., Abbondanzo, S. J., and Stewart, C. L. (1990). Androgenetic mouse embryonic stem cells are pluripotent and cause skeletal defects in chimeras: Implications for genetic imprinting. *Cell* **62**, 251–260.

Martin, G. R., and Evans, M. J. (1974). The morphology and growth of a pluripotent teratocarcinoma cell line and its derivatives in tissue culture. *Cell* **2**, 163–172.

Martin, G. R., Epstein, C. J., Travis, B., Tucker, G., Yatziv, S., Martin, D. W., Jr., Clift, S., and Cohen, S. (1978). X-Chromosome inactivation during differentiation of female teratocarcinoma stem cells *in vitro*. *Nature* **271**, 329–333.

McGinnis, W., Levine, M. S., Hafen, E., Kuroiwa, A., and Gehring, W. J. (1984). A conserved DNA sequence in homoeotic genes of the *Drosophila* Antennapedia and bithorax complexes. *Nature* **308**, 428–433.

McLaren, A. (1981). Germ cells and soma: A new look at an old problem. Mrs. Hepsa Ely Silliman Memorial Lectures. Yale University Press, New Haven and London.

McLaren, A. (2000). Cloning: pathways to a pluripotent future. *Science* **288**, 1775–1780.

McMahon, A., Fosten, M., and Monk, M. (1983). X-Chromosome inactivation mosaicism in the three germ layers and the germ line of the mouse embryo. *J. Embryol. Exp. Morphol.* **74**, 207–220.

Mello, C. C., Schubert, C., Draper, B., Zhang, W., Lobel, R., and Priess, J. R. (1996). The PIE-1 protein and germline specification in *C. elegans* embryos. *Nature* **382**, 710–712.

Minucci, S., Botquin, V., Yeom, Y. I., Dey, A., Sylvester, I., Zand, D. J., Ohbo, K., Ozato, K., and Schöler, H. R. (1996). Retinoic acid-mediated down-regulation of Oct3/4 coincides with the loss of promoter occupancy *in vivo*. *EMBO J.* **15**, 888–899.

Monk, M., Boubelik, M., and Lehnert, S. (1987). Temporal and regional changes in DNA methylation in the embryonic, extraembryonic and germ cell lineages during mouse embryo development. *Development* **99**, 371–382.

Mountford, P., Zevnik, B., Duwel, A., Nichols, J., Li, M., Dani, C., Robertson, M., Chambers, I., and Smith, A. (1994). Dicistronic targeting constructs: Reporters and modifiers of mammalian gene expression. *Proc. Natl. Acad. Sci. U.S.A.* **91**, 4303–4307.

Muller, F., Wicky, C., Spicher, A., and Tobler, H. (1991). New telomere formation after developmentally regulated chromosomal breakage during the process of chromatin diminution in *Ascaris lumbricoides*. *Cell* **67**, 815–822.

Mummery, C. L., Feyen, A., Freund, E., and Shen, S. (1990). Characteristics of embryonic stem cell differentiation: A comparison with two embryonal carcinoma cell lines. *Cell Differ. Dev.* **30**, 195–206.

Munsie, M. J., Michalska, A. E., O'Brien, C. M., Trounson, A. O., Pera, M. F., and Mountford, P. S. (2000). Isolation of pluripotent embryonic stem cells from reprogrammed adult mouse somatic cell nuclei. *Curr. Biol.* **10**, 989–992.

Murphy, S., Yoon, J. B., Gerster, T., and Roeder, R. G. (1992). Oct-1 and Oct-2 potentiate functional interactions of a transcription factor with the proximal sequence element of small nuclear RNA genes. *Mol. Cell. Biol.* **12**, 3247–3261.

Nagy, A., Gocza, E., Diaz, E. M., Prideaux, V. R., Ivanyi, E., Markkula, M., and Rossant, J. (1990). Embryonic stem cells alone are able to support fetal development in the mouse. *Development* **110**, 815–821.

Nagy, A., Rossant, J., Nagy, R., Abramow-Newerly, W., and Roder, J. C. (1993). Derivation of completely cell culture-derived mice from early-passage embryonic stem cells. *Proc. Natl. Acad. Sci. U.S.A.* **90**, 8424–8428.

Nichols, J., Zevnik, B., Anastassiadis, K., Niwa, H., Klewe-Nebenius, D., Chambers, I., Schöler, H., and Smith, A. (1998). Formation of pluripotent stem cells in the mammalian embryo depends on the POU transcription factor Oct4. *Cell* **95**, 379–391.

Nicosia, S. V., Wolf, D. P., and Inoue, M. (1977). Cortical granule distribution and cell surface characteristics in mouse eggs. *Dev. Biol.* **57**, 56–74.

Nishimoto, M., Fukushima, A., Okuda, A., and Muramatsu, M. (1999). The gene for the embryonic stem cell coactivator UTF1 carries a regulatory element which selectively interacts with a complex composed of Oct-3/4 and Sox-2. *Mol. Cell. Biol.* **19**, 5453–5465.

Niwa, H., Burdon, T., Chambers, I., and Smith, A. (1998). Self-renewal of pluripotent embryonic stem cells is mediated via activation of STAT3. *Genes Dev.* **12**, 2048–2060.

Niwa, H., Miyazaki, J., and Smith, A. G. (2000). Quantitative expression of Oct-3/4 defines differentiation, dedifferentiation or self-renewal of ES cells. *Nat. Genet.* **24**, 372–376.

Nordhoff, V., Hubner, K., Bauer, A., Orlova, I., Malapetsa, A., and Schöler, H. R. (2001). Comparative analysis of human, bovine, and murine Oct-4 upstream promoter sequences. *Mamm. Genome* **12**, 309–317.

Ogonuki, N., Inoue, K., Yamamoto, Y., Noguchi, Y., Tanemura, K., Suzuki, O., Nakayama, H., Doi, Kunio, Ohtomo, Y., Satoh, M., Nishida, A., and Ogura, A. (2002). Early death of mice cloned from somatic cells. *Nature Genetics* **30**, 253–254.

Okamoto, K., Okazawa, H., Okuda, A., Sakai, M., Muramatsu, M., and Hamada, H. (1990). A novel octamer binding transcription factor is differentially expressed in mouse embryonic cells. *Cell* **60**, 461–472.

Okazawa, H., Okamoto, K., Ishino, F., Ishino-Kaneko, T., Takeda, S., Toyoda, Y., Muramatsu, M., and Hamada, H. (1991). The oct3 gene, a gene for an embryonic transcription factor, is controlled by a retinoic acid repressible enhancer. *EMBO J.* **10**, 2997–3005.

Okuda, A., Fukushima, A., Nishimoto, M., Orimo, A., Yamagishi, T., Nabeshima, Y., Kuro-o, M., Nabeshima, Y., Boon, K., Keaveney, M., *et al.* (1998). UTF1, a novel transcriptional coactivator expressed in pluripotent embryonic stem cells and extraembryonic cells. *EMBO J.* **17**, 2019–2032.

Orkin, S. H. (1998). Embryonic stem cells and transgenic mice in the study of hematopoiesis. *Int. J. Dev. Biol.* **42**, 927–934.

Orkin, S. H., and Zon, L. I. (1997). Genetics of erythropoiesis: Induced mutations in mice and zebrafish. *Annu. Rev. Genet.* **31**, 33–60.

O'Shea, K. (1999). Embryonic stem cells as models of development. *Anat. Rec. (New Anat.)* **257**, 32–41.

Ovitt, C. E., and Schöler, H. R. (1998). The molecular biology of Oct-4 in the early mouse embryo. *Mol. Hum. Reprod.* **4**, 1021–1031.

Palmieri, S. L., Peter, W., Hess, H., and Schöler, H. R. (1994). Oct-4 transcription factor is differentially expressed in the mouse embryo during establishment of the first two extraembryonic cell lineages involved in implantation. *Dev. Biol.* **166**, 259–267.

Papaioannou, V. E., and Ebert, K. M. (1995). Mouse half embryos: Viability and allocation of cells in the blastocyst. *Dev. Dyn.* **203**, 393–398.

Pedersen, R. (1994). Mammalian embryogenesis. *In* "The Physiology of Reproduction" (E. Knobil, J. D. Neill, G. S. Greenwald, C. L. Markert, and D. W. Pfaff, Eds.), pp. 319–390. Raven Press, New York.

Pera, M. F., Reubinoff, B., and Trounson, A. (2000). Human embryonic stem cells. *J. Cell Sci.* **113**, 5–10.

Pesce, M., and Schöler, H. R. (2000). Oct-4: Control of totipotency and germline determination. *Mol. Reprod. Dev.* **55**, 452–457.

Pesce, M., Gross, M. K., and Schöler, H. R. (1998). In line with our ancestors: Oct-4 and the mammalian germ. *BioEssays* **20**, 722–732.

Pesce, M., Anastassiadis, K., and Schöler, H. R. (1999a). Oct-4: Lessons of totipotency from embryonic stem cells. *Cells Tissues Organs* **165**, 144–152.

Pesce, M., Marin Gomez, M., Philipsen, S., and Schöler, H. R. (1999b). Binding of Sp1 and Sp3 transcription factors to the Oct-4 gene promoter. *Cell. Mol. Biol. (Noisy-le-grand)* **45**, 709–716.

Piotrowska, K., and Zernicka-Goetz, M. (2001). Role for sperm in spatial patterning of the early mouse embryo. *Nature* **409**, 517–521.

Pugh, B. F., and Tjian, R. (1991). Transcription from a TATA-less promoter requires a multisubunit TFIID complex. *Genes Dev.* **5**, 1935–1945.

Reubinoff, B. E., Pera, M. F., Fong, C. Y., Trounson, A., and Bongso, A. (2000). Embryonic stem cell lines from human blastocysts: somatic differentiation *in vitro*. *Nat. Biotechnol.* **18**, 399–404.

Rosner, M. H., Vigano, M. A., Ozato, K., Timmons, P. M., Poirier, F., Rigby, P. W., and Staudt, L. M. (1990). A POU-domain transcription factor in early stem cells and germ cells of the mammalian embryo. *Nature* **345**, 686–692.

Ryan, A. K., and Rosenfeld, M. G. (1997). POU domain family values: Flexibility, partnerships, and developmental codes. *Genes Dev.* **11**, 1207–1225.

Saburi, S., Azuma, S., Sato, E., Toyoda, Y., and Tachi, C. (1997). Developmental fate of single embryonic stem cells microinjected into 8-cell-stage mouse embryos. *Differentiation* **62**, 1–11.

Sachs, L. (1986). Cell differentiation and malignancy. *Cell Biophys.* **9**, 225–242.

Schöler, H. R. (1991). Octamania: The POU factors in murine development. *Trends Genet.* **7**, 323–329.

Schöler, H. R., Balling, R., Hatzopoulos, A. K., Suzuki, N., and Gruss, P. (1989a). Octamer binding proteins confer transcriptional activity in early mouse embryogenesis. *EMBO J.* **8**, 2551–2557.

Schöler, H. R., Hatzopoulos, A. K., Balling, R., Suzuki, N., and Gruss, P. (1989b). A family of octamer-specific proteins present during mouse embryogenesis: Evidence for germline-specific expression of an Oct factor. *EMBO J.* **8**, 2543–2550.

Schöler, H. R., Dressler, G. R., Balling, R., Rohdewohld, H., and Gruss, P. (1990a). Oct-4: A germline-specific transcription factor mapping to the mouse t-complex. *EMBO J.* **9**, 2185–2195.

Schöler, H. R., Ruppert, S., Suzuki, N., Chowdhury, K., and Gruss, P. (1990b). New type of POU domain in germ line-specific protein Oct-4. *Nature* **344**, 435–439.

Schöler, H. R., Ciesiolka, T., and Gruss, P. (1991). A nexus between Oct-4 and E1A: Implications for gene regulation in embryonic stem cells. *Cell* **66**, 291–304.

Scott, M. P., and Weiner, A. J. (1984). Structural relationships among genes that control development: Sequence homology between the Antennapedia, Ultrabithorax, and fushi tarazu loci of Drosophila. *Proc. Natl. Acad. Sci. U.S.A.* **81**, 4115–4119.

Sell, S. (1993). Cellular origin of cancer: Dedifferentiation or stem cell maturation arrest? *Environ. Health Perspect.* **101**(Suppl. 5), 15–26.

Seydoux, G., Mello, C. C., Pettitt, J., Wood, W. B., Priess, J. R., and Fire, A. (1996). Repression of gene expression in the embryonic germ lineage of *C. elegans*. *Nature* **382**, 713–716.

Shiels, P. G., Kind, A. J., Campbell, K. H., Waddington, D., Wilmut, I., Colman, A., and Schnieke, A. E. (1999). Analysis of telomere lengths in cloned sheep. *Nature* **399**, 316–317.

Shin, T., Kraemer, D., Pryor, J., Liu, L., Rugila, J., Howe, L., Buck, S., Murphy, K., Lyons, L., and Westhusin, M. (2002). A cat cloned by nuclear transplantation. *Nature* **415**, 859.

Sinclair, K. D., Young, L. E., Wilmut, I., and McEvoy, T. G. (2000). *In-utero* overgrowth in ruminants following embryo culture: Lessons from mice and a warning to men. *Hum. Reprod.* **15**, 68–86.

Smith, A. G., Nichols, J., Robertson, M., and Rathjen, P. D. (1992). Differentiation inhibiting activity (DIA/LIF) and mouse development. *Dev. Biol.* **151**, 339–351.

Snow, M. H., and Tam, P. P. (1979). Is compensatory growth a complicating factor in mouse teratology? *Nature* 279, 555–557.

Solter, D. (2000). Mammalian cloning: advances and limitations. *Nat. Rev.* 1, 199–207.

Solter, D., and Knowles, B. B. (1978). Monoclonal antibody defining a stage-specific mouse embryonic antigen (SSEA-1). *Proc. Natl. Acad. Sci. U.S.A.* 75, 5565–5569.

Soodeen-Karamath, S., and Gibbins, A. M. (2001). Apparent absence of oct 3/4 from the chicken genome. *Mol. Reprod. Dev.* 58, 137–148.

Soriano, P., and Jaenisch, R. (1986). Retroviruses as probes for mammalian development: Allocation of cells to the somatic and germ cell lineages. *Cell* 46, 19–29.

Sotomaru, Y., Kato, Y., and Tsunoda, Y. (1999). Induction of pluripotency by injection of mouse trophectoderm cell nuclei into blastocysts following transplantation into enucleated oocytes. *Theriogenology* 52, 213–220.

Stevens, L. C. (1978). Totipotent cells of parthenogenetic origin in a chimaeric mouse. *Nature* 276, 266–267.

Stevens, L. C., Varnum, D. S., and Eicher, E. M. (1977). Viable chimaeras produced from normal and parthenogenetic mouse embryos. *Nature* 269, 515–517.

Stewart, T. A., and Mintz, B. (1981). Successive generations of mice produced from an established culture line of euploid teratocarcinoma cells. *Proc. Natl. Acad. Sci. U.S.A.* 78, 6314–6318.

Stewart, C. L., Kaspar, P., Brunet, L. J., Bhatt, H., Gadi, I., Kontgen, F., and Abbondanzo, S. J. (1992). Blastocyst implantation depends on maternal expression of leukaemia inhibitory factor. *Nature* 359, 76–79.

Strome, S., Garvin, C., Paulsen, J., Capowski, E., Martin, P., and Beanan, M. (1994). Specification and development of the germline in *Caenorhabditis elegans*. In "Germline Development" (J. Marsh and J. Goode, Eds.), pp. 31–45. John Wiley and Sons, England.

Surani, M. A. (1998). Imprinting and the initiation of gene silencing in the germ line. *Cell* 93, 309–312.

Surani, M. A. (1999). Reprogramming a somatic nucleus by trans-modification activity in germ cells. *Semin. Cell. Dev. Biol.* 10, 273–277.

Takeda, J., Seino, S., and Bell, G. I. (1992). Human Oct3 gene family: cDNA sequences, alternative splicing, gene organization, chromosomal location, and expression at low levels in adult tissues. *Nucleic Acids Res.* 20, 4613–4620.

Takeda, K., Noguchi, K., Shi, W., Tanaka, T., Matsumoto, M., Yoshida, N., Kishimoto, T., and Akira, S. (1997). Targeted disruption of the mouse Stat3 gene leads to early embryonic lethality. *Proc. Natl. Acad. Sci. U.S.A.* 94, 3801–3804.

Tarkowski, A. K. (1959). Experiments on the development of isolated blastomeres of mouse eggs. *Nature* 184, 1286–1287.

Tarkowski, A. K. (1998). Mouse chimaeras revisited: Recollections and reflections. *Int. J. Dev. Biol.* 42, 903–908.

Tarkowski, A. K., and Wroblewska, J. (1967). Development of blastomeres of mouse eggs isolated at the 4- and 8-cell stage. *J. Embryol. Exp. Morphol.* 18, 155–180.

Temple, S. (1999). CNS development: The obscure origins of adult stem cells. *Curr. Biol.* 9, R397–R399.

Thomson, J. A., Itskovitz-Eldor, J., Shapiro, S. S., Waknitz, M. A., Swiergiel, J. J., Marshall, V. S., and Jones, J. M. (1998). Embryonic stem cell lines derived from human blastocysts. *Science* 282, 1145–1147.

Tian, X. C., Xu, J., and Yang, X. (2000). Normal telomere lengths found in cloned cattle. *Nat. Genet.* 26, 272–273.

Uhm, S. J., Kim, N. H., Kim, T., Chung, H. M., Chung, K. H., Lee, H. T., and Chung, K. S. (2000). Expression of enhanced green fluorescent protein (EGFP) and neomycin resistant [Neo(R)] genes in porcine embryos following nuclear transfer with porcine fetal fibroblasts transfected by retrovirus vector. *Mol. Reprod. Dev.* 57, 331–337.

van der Kooy, D., and Weiss, S. (2000). Why stem cells? *Science* 287, 1439–1441.

van Eijk, M. J., van Rooijen, M. A., Modina, S., Scesi, L., Folkers, G., van Tol, H. T., Bevers, M. M., Fisher, S. R., Lewin, H. A., Rakacolli, D., *et al.* (1999). Molecular cloning, genetic mapping, and developmental expression of bovine POU5F1. *Biol. Reprod.* 60, 1093–1103.

Vassilieva, S., Guan, K., Pich, U., and Wobus, A. M. (2000). Establishment of SSEA-1- and Oct-4-expressing rat embryonic stem-like cell lines and effects of cytokines of the IL-6 family on clonal growth. *Exp. Cell Res.* 258, 361–373.

Wakayama, T., and Yanagimachi, R. (1999). Cloning the laboratory mouse. *Semin. Cell Dev. Biol.* 10, 253–258.

Wakayama, T., and Yanagimachi, R. (2001). Mouse cloning with nucleus donor cells of different age and type. *Mol. Reprod Dev.* 58, 376–383.

Wakayama, T., Tabar, V., Rodriguez, I., Perry, A. C., Studer, L., and Mombaerts, P. (2001). Differentiation of embryonic stem cell lines generated from adult somatic cells by nuclear transfer. *Science* 292, 740–743.

Weber, R. J., Pedersen, R. A., Wianny, F., Evans, M. J., and Zernicka-Goetz, M. (1999). Polarity of the mouse embryo is anticipated before implantation. *Development* 126, 5591–5598.

Wei, G., Schubiger, G., Harder, F., and Muller, A. M. (2000). Stem cell plasticity in mammals and transdetermination in *Drosophila*: Common themes? *Stem Cells* 18, 409–414.

Weissman, A. (1885). "Die Continuitaet des Keimplasmas als Grundlage Einer Theorie der Vererbung." Fisher-Verlag, Jena.

Wright, W. E., Piatyszek, M. A., Rainey, W. E., Byrd, W., and Shay, J. W. (1996). Telomerase activity in human germline and embryonic tissues and cells. *Dev. Genet.* 18, 173–179.

Yeom, Y. I., Ha, H. S., Balling, R., Schöler, H. R., and Artzt, K. (1991). Structure, expression and chromosomal location of the Oct-4 gene. *Mech. Dev.* 35, 171–179.

Yeom, Y. I., Fuhrmann, G., Ovitt, C. E., Brehm, A., Ohbo, K., Gross, M., Hubner, K., and Schöler, H. R. (1996). Germline regulatory element of Oct-4 specific for the totipotent cycle of embryonal cells. *Development* 122, 881–894.

Yoshimizu, T., Sugiyama, N., De Felice, M., Yeom, Y. I., Ohbo, K., Masuko, K., Obinata, M., Abe, K., Schöler, H. R., and Matsui, Y. (1999). Germline-specific expression of the Oct-4/green fluorescent protein (GFP) transgene in mice. *Dev. Growth Differ.* 41, 675–684.

Yuan, H., Corbi, N., Basilico, C., and Dailey, L. (1995). Developmental-specific activity of the FGF-4 enhancer requires the synergistic action of Sox2 and Oct-3. *Genes Dev.* 9, 2635–2645.

PART II
METHODS

CHAPTER 7

MICROMANIPULATION TECHNIQUES FOR CLONING

Raymond L. Page

INTRODUCTION

Because many of the detailed theoretical considerations pertaining to cloning are covered in other chapters of this volume, this chapter focuses primarily on the mechanics and practical details associated with nuclear transfer experiments. However, some of the differences between species as they pertain to technical considerations are discussed. Although structured mostly as an introduction and training manual for the novice, it is hoped that experienced investigators will find some useful information as well. The goals of most nuclear transfer experiments vary widely, but in most cases the micromanipulation procedures remain constant and are merely a tool used to achieve these goals. Separate sections are devoted to the major topics of toolmaking, equipment, microscopy, manipulation media, enucleation, cell transfer, and fusion, with an additional section covering piezoelectric nuclear transfer. Consideration for species is given for the most commonly used laboratory and livestock models for experimental embryology. In addition, there is a section on technical improvements such as chemical enucleation and nuclear transfer without micromanipulators, to present alternative methods aimed at simplification of some of the technically demanding procedures.

MAKING MANIPULATION TOOLS

As with many technical procedures, about 75% of the success of cloning experiments is due to having good manipulation tools. Although there are several commercial sources of micromanipulation tools, most skilled and experienced people still prefer to make their own. The two largest suppliers for premade micromanipulation tools are Cook Veterinary Products USA (Bloomington, IN) and Humagen Fertility Diagnostics Inc. (Charlottesville, VA). Both manufacturers offer custom services, so it is possible to obtain unique tools from commercial sources. However, once mastered, custom toolmaking does not take too much time and allows the investigator to tailor the tools on the spot for various experimental situations. Also, individual preferences about the way tools are shaped and polished make it difficult and nearly impossible for commercial suppliers to make micromanipulation tools to suit everyone. This section describes the equipment and methods used to make all of the glassware needed for all types of embryo manipulation for nuclear transfer.

GLASSWARE

Micromanipulation pipettes are fashioned from thin-walled glass capillaries, typically 1.0 mm OD × 0.75 mm ID × 15 cm long. Some investigators use different types of glass for holding and enucleation pipettes; I find this to be unnecessary and use the above type of glass for all manipulation tools. This simplifies the procedures and eliminates the need to use different pipette-puller settings for different tools. The capillaries used for nuclear transfer tools are different from those for microinjection in that they do not contain an internal filament, which allows for backloading of material that is to be injected. Several manufacturers make glass capillaries, which vary in cost and quality. It is worthwhile to invest in good quality glass because any additional cost of material will be easily justified in less time spent making tools with substandard capillaries. Suppliers of capillary glass are Sutter Instruments (Novato, CA) and World Precision Instruments (Sarasota, FL).

EQUIPMENT

Equipment needed to make micromanipulation tools consists of a pipette puller, microforge, pipette grinder, diamond-tip pencil, and gas burner. Selection of the best equipment for making manipulation tools is a difficult process, because personal references play a large role. Many investigators tend to stay with familiar brands or brands that were used by the laboratories where they received their training. Unfortunately, many of the suppliers do not offer in-house demonstrations, so purchases may be made before really knowing much about the performance of an instrument. No single supplier of any equipment for micromanipulation offers a complete line of instruments with all of the preferred features. Therefore, the preferred setup may consist of bits and pieces from nearly all manufacturers.

The pipette puller is used to provide finely drawn pipettes from capillary glass; the pipettes can then be used to make precision tools for micromanipulation. Both vertical and horizontal pullers are available. Most vertical pullers rely partially on gravity to make the pull, in addition to a solenoid-driven mechanical device to provide precise user-adjustable force to shape the pipette. Horizontal pullers that pull the pipette from both directions, such as the Sutter Instruments P-series (Novato, CA), allow for two identical pipettes to be made from the same piece of glass, which saves time and glass. The Sutter P-97 model has a chamber that contains the heating element; the chamber is flushed with dry air prior to each pull, thus giving a greater level of consistency between pulls. Other popular suppliers of pipette pullers are David Kopf Instruments (Tujunga, CA), Narishige USA (East Meadow, NY), and Campden Instruments (Leicester, UK). Different pullers allow the user to control various aspects of the pull and therefore allow various types of tips to be made.

The microforge is used to finish and polish the tips of holding and nuclear transfer pipettes. The most important features of this tool are flexibility in moving the pipette or heating element in all three axes, quality and magnification of the optics, and the ability to attach multiple sizes of wire to the heating element. The axis of movement is important so that the pipette tip can be easily placed into close proximity to the heating wire; fine control over positioning without drift is important as well. In order to bring the pipette tip and heating wire rapidly into focus at the same time, an optical headpiece with a magnification range is desired. Zoom optical heads from stereomicroscopes are available with some microforge models, but rarely do these provide a high enough magnification for precision work, especially for measuring the size of a pipette tip. When it is important to make pipette tips of reliably uniform and known size, particularly smaller ones used for piezoelectric injection (8–10 μm in diameter), the Narishige MF-900 with 33× objective and 15× eyepieces is a good choice. It is also useful to have a micrometer scale attached to one of the eyepieces for making these measurements. Other sources of microforges are Energy Beam Sciences (DeFonbrune type, Agawam, MA) and World Precision Instruments (Sarasota, FL).

The pipette grinder for making beveled tips optimally should offer a variable speed, nonbumpy grinding surface with a way of dispensing water over the surface. The Sutter Instruments Model BV-10 has a magnetic drive that controls the grinding surface, which floats on a thin film of oil. As opposed to belt- or gear-driven systems, this allows the wheel to rotate with minimal vertical movement across the surface, which would be induced if it were directly connected to the drive system. A water dispenser is important because it minimizes the amount of ground glass fragments that stick to the end of pipettes and cause problems during fire polishing. Some grinders have the ability to attach a stereomicroscope head such that the grinding operation can be visualized. For nuclear transfer pipettes this may be unnecessary because the pipette is broken to the correct size on the microforge prior to grinding, and the bevel can be easily controlled by grinding for a specific amount of time. Pipette bevellers are also available from Narishige and World Precision Instruments.

For making embryo-moving pipettes and for putting bends into micromanipulation pipettes, a gas source with a fine flame is desired. A standard laboratory Bunsen burner can be used and the flame may be modified by attaching a blunted 18- or 20-gauge hypodermic needle to the gas port. If an in-house natural gas supply is not available, KISAG Instruments (Switzerland; available from Pfingst & Company, South Plainfield, NJ) makes a convenient portable butane gas burner. Diamond-tip pencils for making precise cuts in glass capillaries are available from most bulk laboratory suppliers, such as Fisher, VWR, and Thomas Scientific.

HOLDING PIPETTES

Holding pipettes are used to position the oocyte for enucleation and cell transfer. The optimal size depends on personal preferences and the species of oocyte. A good holder for mouse oocytes will give problems holding in place the much larger bovine eggs. If the outside diameter of the pipette is too large, optimal focus of the oocyte cannot be made without the pipette scraping the bottom of the manipulation chamber. If the outside diameter is too small, there will be difficulty holding the oocyte firmly in position during enucleation or cell transfer. A rule of thumb for sizing holding pipettes is for the outside diameter and inside diameter to be about 90 and 20% of the oocyte diameter, respectively. This size of inside diameter will enable the oocyte to be held firmly without too much of it being sucked into the holder, thus causing deformation and damage.

To make holding pipettes, first pull the capillary on the pipette puller, placing the filament exactly in the middle of the pipette. A glass bead should be first placed on the microforge heating filament by melting a discarded capillary with high heat (about 80% of maximum). The size of bead is determined experimentally but is usually about three to four times the width of the tip at the position to be broken. The function of the bead is to aid in breaking the tip at the desired width. Position the heating filament on the microforge such that it is in focus, then bring it slightly toward the back. Place the pipette in the microforge and lower it in front of the filament and in focus until the filament is located at the required breaking point. Now bring the heating filament forward until it just touches the pipette (Fig. 1A). With the heat set to about 40 to 50% of maximum, turn on the heat; release the heat as soon as the pipette melts to the glass bead (begins to bend slightly). As the glass cools, the pipette will stick to the glass bead, and as the filament relaxes from its expansion, the pipette should break evenly (Fig. 1B). If it does not break, moving the filament slightly should break it. Alternatively, a diamond-tip pencil can be used to score the glass at the desired outside diameter. This technique is quick, but will

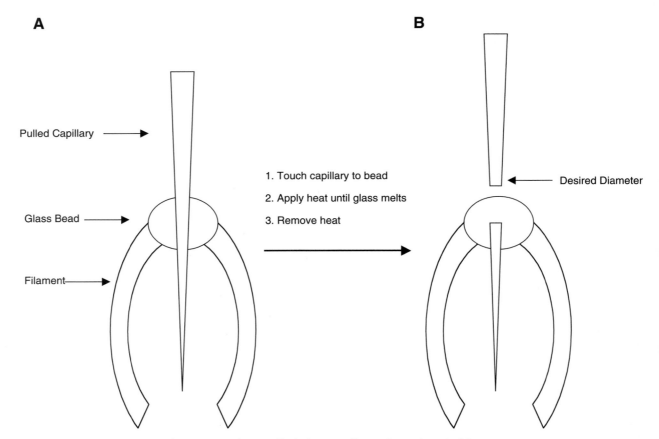

Figure 1 *Breaking pulled glass capillaries for making holding pipettes.*

take some practice and trial and error at first. Once broken, remove any attached capillary glass from the bead and position the pipette tip over the heating element, such that when the heat is activated the bead will not touch the pipette. Activate the heat at about 80% of maximum and carefully move the filament closer to the pipette tip. As it gets closer, the glass will melt and the inside diameter will shrink (polishing the tip). Use the micrometer scale in the eyepiece to judge when to stop polishing.

TOOLS FOR ENUCLEATION AND CELL TRANSFER

Pipettes for enucleation and cell transfer are essentially the same except that the enucleation pipette is usually smaller, depending on the size of cell to be transferred. Some people use the same tools for both procedures, but preferably the cell transfer pipette should be just slightly larger than the cell. This provides a good way to measure the cells that are transferred, because within a population there are cells of many sizes.

The first steps in tool preparation are exactly the same as those for holding pipettes, up to the point where the tip is broken. Here the tip is broken to yield a tip size slightly larger than the ultimate desired size, to account for size reduction during polishing. To make the bevel, attach the pipette to the grinder at the desired angle (45–55°; preferably 50°). Use a permanent marker to indicate on the capillary the direction that will register the location of the long point of the bevel. With the wheel wet and set to about 80% of maximum speed, lower the pipette to the surface until it just touches and a slight inflexion in the pipette tip is noticed. Keep the wheel wet during the grinding operation and continue until the bevel is cut com-

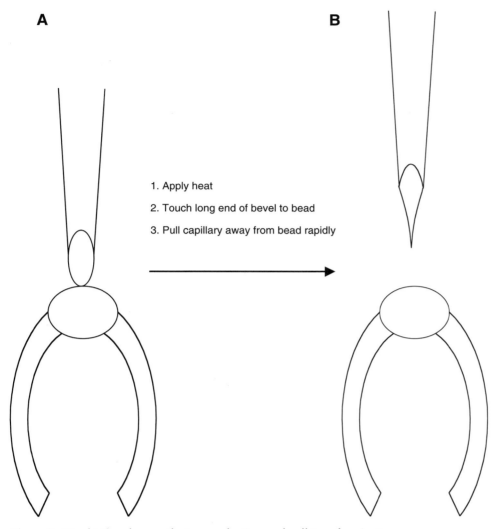

A

B

1. Apply heat

2. Touch long end of bevel to bead

3. Pull capillary away from bead rapidly

Figure 2 *Attaching a sharp spike to enucleation and cell transfer pipettes.*

pletely across the pipette tip. Trial and error may be required to determine how much time this takes, but it is usually about 30–45 seconds. Stop the water flow to the wheel before the pipette is removed such that, as the wheel dries, any water sucked into the tip is removed. Remove the tip from the wheel just as the tip is clear of water. This will ensure that any ground bits of glass are removed from the tip as well. The next and final step is to attach a spike to the end of the beveled tip to facilitate penetrating the zona pellucida. Mount the capillary onto the microforge as before, except this time position the tip just over the glass bead, with both in focus. With the heat set at about 45% of maximum, use the micromanipulator on the microforge to barely touch the filament to the glass (Fig. 2A). As soon as the glass visibly melts, turn off the heat and move the filament straight away from the pipette. If this is all done with a continuous motion, a sharp spike will have been placed on the tip and any rough edges will be polished away (Fig. 2B). The pipette is now ready for use.

TOOLS FOR CELL TRANSFER BY PIEZOELECTRIC INJECTION

The procedure for making piezo tips is much the same as described previously, except they are much smaller and blunt at the tip and are bent such that the tip is parallel to the micromanipulation chamber. This is where good-quality high magnification on the microforge is essential.

For breaking the tip, a small heating wire (150 µm diameter) on the microforge will give better performance. It is not essential to fire polish the tip after breaking, but if there are sharp edges left it may lead to poorer oocyte survival after cell injection. Fire polishing should be done in the same manner as that for holding pipettes, except that much lower heat is used and only for long enough to take off the sharp edges. If the tip is polished too much, the tip may not function properly with the piezo device. The location of the dogleg is about 7 mm from the tip at an angle of about 20–30°. This is done by placing the pipette in the microforge perpendicular to the filament, with the location of the desired bend directly over the filament. The heat is turned on and then off when the desired bend is reached. Care must be taken as to not allow the pipette to touch the filament.

MICROSCOPY AND EQUIPMENT FOR MICROMANIPULATION

There are four major manufacturers of inverted microscopes for micromanipulation: Olympus (Melville, NY), Nikon (Melville, NY), Carl Zeiss (Thornwood, NY), and Leica (Depew, NY). These are all good-quality microscopes and will work for micromanipulation, yet each brand has unique features that may be desirable depending on personal preference. Models currently available all have infinity-corrected optics, which allows fluorescence attachments without having to add extra blank magnification. Perhaps more important than the microscope is the table on which it sits. It is essential that the workstation be isolated from room vibration produced from high-velocity air conditioning systems and other equipment such as refrigerators and freezers. Even a seemingly low amount of vibration will be transmitted to the micromanipulation pipette and will interfere with procedures. Any noticeable tip vibration at a resting state will make piezoelectric micromanipulation nearly impossible. This is a difficult problem, because many investigators do not have the choice of where and in what building their lab is situated. Vibration-damping tables are available from a variety of manufacturers, tend to be quite expensive, and do not always solve the problem. Placing tennis balls under the table legs often can provide the same results as a $10,000 vibration-damping table.

The microscope for nuclear transfer is minimally equipped with a low- and high-power set of objectives with specialized optical attachments for contrast enhancement of bright-field images (discussed below). The inverted microscope is preferred because the image is not inverted as it is for upright compound microscopes, so the image plane is oriented identically to the position of the specimen and micromanipulation tools. In addition, long working-distance condensers and objectives are available, which enables plenty of room for the micromanipulation tools to be placed over the specimen. Epifluorescence using a mercury arc lamp is used to facilitate visualization of the chromosomes for species such as bovine, sheep, and goats, which have oocytes with opaque cytoplasm. A standard ultraviolet light filter cube can be used. However, a narrow-band excitation filter (350 ± 20 nm) combined with a long-pass (450 nm) barrier filter will provide adequate fluorescence while enabling simultaneous visualization of the specimen using Nomarski Differential Interference Contrast or Hoffman Modulation Contrast optical systems. This type of illumination can be enhanced further by placing a 0.5 neutral density filter in the transillumination light path. This way, the transillumination need not be turned off for visualization of the chromosomes by epiillumination.

At a minimum, the microscope should have a low-power (2× or 4×) and a high-power (20× or 40×) objective lens. The low-power objective is useful for visualizing the whole manipulation field. Therefore, it makes setup and loading/unloading oocytes and cells easier. The high-power objective is used for the manipulation pro-cedures. The choice of 20× or 40× depends on personal preference. A magnification somewhere in between may be preferred. For this, the Olympus IX70 has a 1.5× magnification slider in the light path, which permits observation at 300× total mag-nification when a 20× objective is used with 10× eyepieces. An important feature to keep in mind is that these long working-distance objectives come with either adjustable correction collars or detachable lens caps to correct for the thickness of glass or plastic used, which is usually about 1 mm in thickness. Depending on whether a plastic dish or glass slide is used, either Hoffman Modulation Contrast or Differential Interference Contrast optics are used.

CONTRAST OPTICAL SYSTEMS

In order to visualize live cells without artificial contrast enhancement (staining), either Hoffman Modulation Contrast (HMC) or Nomarski Differential Interference Contrast (DIC) optical systems are employed. Each system offers unique advantages, but unfortunately at the expense of undesirable characteristics. Ultimately, the choice lies with user preferences and a decision as to what degree of flexibility is required to image different types of samples. HMC optics gives otherwise clear specimens a three-dimensional appearance by detecting optical gradients and then converting them into variations of light intensity. In essence, a shadow effect is created from differences in thickness and refractive index within the specimen by passing the light through a series of specialized filters and polarizers. In HMC optics, the specimen is not located between two polarizers, so plastic dishes, which react to polarized light, can be used without diminishing the optical quality.

Nomarski DIC optics creates an apparent three-dimmensional image by split-ting polarized beams using a modified Wollaston prism (Nomarski prism), and when passed through the specimen the parallel beams are altered in length according to differences in refraction and gradients within the specimen. The two beams of dif-ferent length are then brought back together by a second Nomarski prism and resolved to the same axis by a second polarizer. This causes different sides of spec-imen detail to appear different in intensity or color. Because the specimen is located between the polarizers, glass must be used for the micromanipulation chamber because plastic reacts to polarized light and severely diminishes the image quality. Despite this inconvenience, the optical quality with the Nomarski DIC seems to be far superior to that of the HMC, especially for assessing subcellular detail. For example, chromosomes in metaphase-stage mouse oocytes are readily visible using the DIC optics. Furthermore, membrane integrity may be assessed easily and the vacuoles inside somatic cells are seen readily with DIC. This type of detail is not easily visible using the HMC optical system. However, another drawback is that Nomarski DIC optics cost about twice as much as HMC optics.

STEREOMICROSCOPE

For general preparation and observation of oocytes and embryos, a good-quality stereomicroscope is needed. The magnification is generally continuously adjustable from about 2× to 90×. The lower magnifications are used for general sorting and moving eggs in and out of culture dishes and drops. The higher magnification ranges are used for observation of oocyte and embryo quality. A transillumination base is preferred to older models with topical illumination. A good-quality stereomicro-scope can provide enough resolution at higher magnification to permit observation of the polar body to confirm nuclear maturation. The stereomicroscope is also used to align the oocyte–cell couplets with the electrodes for fusion, as well as for con-firming that fusion has taken place after the procedure. The stereomicroscope in the author's lab consists of an Olympus SZX-12 head with 1× objective mounted onto

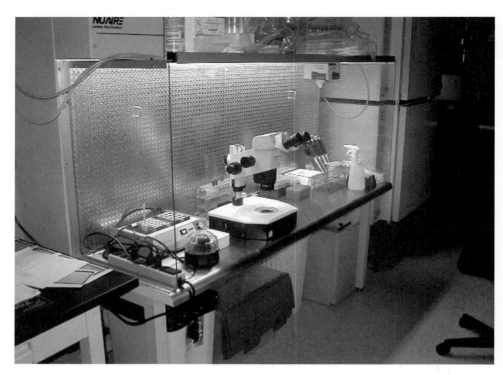

Figure 3 *Stereomicroscope and laminar flow hood.*

a Diagnostic Instruments model TLB D4.1 (Sterling Heights, MI) transmitted light base. The microscope is located inside a laminar flow tissue culture hood to help avoid contamination while working in open dishes (Fig. 3).

MICROMANIPULATORS

The four major suppliers of micromanipulators are Eppendorf (Brinkman Instruments, Westbury, NY), Narishige, Burleigh (Victor, NY), and Leica, and the choice depends entirely on personal preference and budget. They are each adequate for the task and can be mounted on any inverted microscope, but some of the specific differences make some slightly better suited for certain operations. For example, the Eppendorf and Burleigh systems are more rigid at the point of pipette attachment, thus piezoelectric energy transmission is slightly better. This is probably more important if a fluid other than mercury is used in the pipette. The Eppendorf and Burleigh systems are electronically driven, with the Burleigh having a special piezoelectric drive system. Although providing exceptionally smooth movement, this system is slightly limited by its 300-μm maximum range of motion. The Narishige system is hydraulically driven, which allows for flexibility in positioning the joysticks relative to the manipulators. The modular design of the Narishige system also offers the largest degree of flexibility in mounting and positioning the pipettes. The Leica manipulators are mechanical and are mounted on large, heavy blocks to minimize vibration. Eppendorf, Leica, and Narishige systems all offer an adjustable range of motion.

MICROINJECTORS

To permit aspiration and expulsion of fluid in the holding and nuclear transfer pipettes, a syringe system is attached to fluid-filled lines that are then attached to the butt end of the pipette. Paraffin oil is most commonly used in the syringe and

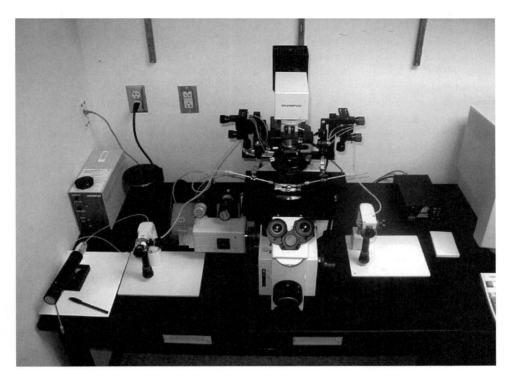

Figure 4 *Micromanipulation setup.*

lines. Most systems, particularly for the enucleation or cell transfer side, are devoid of air, which, due to compressibility, interferes with fine control over fluid movement. This is not as critical for the holding pipette. The most widely known commercial sources of syringe-based thread-drive microinjectors are Eppendorf, Narishige, and Sutter Instruments. These systems are supplied with pipette holders to enable mounting to most micromanipulators. These are particularly useful for appropriate positioning of the spike on enucleation and cell transfer pipettes after they are positioned in the micromanipulation chamber. The complete setup for micromanipulation is shown in Fig. 4.

MICROMANIPULATION PROCEDURES

MANIPULATION MEDIA

The choice of culture medium with which to perform the micromanipulation procedures depends primarily on the oocyte species and experimental parameters. However, in general, manipulation media are typically hydroxyethylpiperazinesulfonic acid (HEPES)-buffered versions of bicarbonate-buffered embryo culture media, such that pH balance outside the incubator is achieved. Many investigators opt for standard media developed in the industry, such as M2 (Hogan *et al.*, 1994), originally developed for mouse embryos, or TL-HEPES (Parish *et al.*, 1986), originally developed for bovine embryos. These relatively simply media require few components and are easy to make. Due to widespread applicability and use of KSOM (a medium composed of six stock solutions and additional components) (Lawitts and Biggers, 1991) and recent modifications (Biggers *et al.*, 2000), commercial sources of KSOM and a HEPES-buffered counterpart, flushing holding medium (FHM), for micromanipulation are available (Cell and Molecular Technologies, Specialty Media Division, Phillipsburg, NJ). Also, in recent years embryo culture media specialized for optimum development for embryos from individual species have been developed—for bovines, synthetic oviductal fluid (SOF) (Gardner *et al.*, 1994); for mice, Chatot/Ziomek/Bavister (CZB), medium (Chatot *et al.*, 1989); and for pigs, NCSU-23 (Petters and Wells, 1993). Concomitantly, some investigators have chosen to customize the holding or manipulation medium accordingly. In general, this philosophy is largely intuitive and designed to keep the environment surrounding the oocytes as consistent as possible, whether inside or outside of the incubator. Examples include HEPES-buffered CZB for mice (Wakayama *et al.*, 1998), HEPES-buffered SOF for bovines (Wells *et al.*, 1999), and phosphate-buffered NCSU-23 for pigs (Polejaeva *et al.*, 2000). These modifications might be important for improved survival, particularly when the oocyte membrane has been compromised, as in nuclear transfer procedures, because the oocyte is not exposed to a dramatically different ionic environment when transferred from manipulation medium to the incubator.

PREPARATION OF THE MICROMANIPULATION CHAMBER

The micromanipulation chamber consists of a several drops of manipulation medium covered with paraffin oil to prevent evaporation and infection. The drop system further allows for more convenient management of the groups of oocytes compared to an open dish filled with medium. A multidrop system permits media with different composition, such as fluorescent dye, microfilament inhibitors, etc., to be present in the same chamber, which alleviates the need to change chambers for different operations. The material used to construct the chamber depends on the choice of Nomarski DIC (glass) or Hoffman Modulation Contrast (plastic) optical systems. For Nomarski DIC, a standard glass slide with a raised silicone ring to hold the oil can be used (Fig. 5). In this case, it is necessary to siliconize the slide (e.g., using Sigmacoat) to prevent undesired spreading of the microdrops. For Hoffman Modulation Contrast, the lid of a standard 100-mm petri dish is commonly used. The lid is used because the edge is shallower, facilitating placement of the micromanipulation tools without too much of an angle. It is preferred that one of the drops contain manipulation medium supplemented with 10% polyvinylpyrrolidone (PVP) to prewash the enucleation and cell transfer pipettes. This is a technique borrowed from piezoelectric nuclear transfer (Wakayama *et al.*, 1998), wherein this solution is used to either immobilize sperm or lyse the cells. The PVP seems to coat the glass with a thin film that dramatically reduces fouling of the pipette tip with cell debris after multiple procedures.

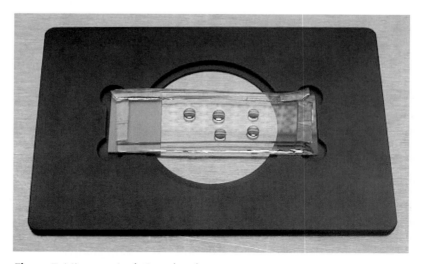

Figure 5 *Micromanipulation chamber.*

ENUCLEATION

If genetic material is to be transferred from one cell to another, the DNA from the recipient cell must first be removed. In the case of oocytes, which are enclosed by the zona pellucida, this is usually done when the oocyte is arrested at metaphase II. At this stage, the chromosomes are condensed in the form of the metaphase plate (meiotic spindle) and in most cases the spindle is located at the edge of the oocyte in close proximity to the extruded polar body (Fig. 6A). In most livestock species the metaphase chromosomes are not visible by light microscopy due to the presence of cytoplasmic lipid (Fig. 6B). Therefore, the intercalating dye Hoechst 33342 (bis-benzimide) is used to confirm that the chromosomes were removed from the oocyte (Westhusen *et al.*, 1992). It is not recommended that epiillumination of the oocyte in the presence of DNA-binding fluorescent dye be used to locate the chromosomes prior to enucleation, because this may damage the oocyte (Bradshaw *et al.*, 1995), perhaps reducing developmental competence. Figure 6B is used only to illustrate the position desired for enucleation. However, in many cases the metaphase chromosomes may not be adjacent to the polar body, which necessitates ultraviolet light (UV) illumination to locate the chromosomes. If this is the case, then a narrow-band UV excitation filter should be used to eliminate exposing the oocyte to high-energy UV, even for a brief period of time. Figure 6A shows the preferred alignment of the oocyte, polar body, and enucleation pipette for removal of the metaphase chromosomes. For optimum performance, the inner zona surface, polar body (thus metaphase plate), and pipette tip are all in the same focal plane. To achieve this position, the oocyte is held lightly by the holding pipette and the enucleation pipette is used to rotate the egg accordingly. Once proper alignment is achieved, the holding pipette is tightened and the enucleation pipette is inserted into the zona pellucida, with the opening placed near the oocyte membrane adjacent to where the chromosomes should be (Fig. 6C). Using a microinjector, cytoplasm is gently aspirated into the enucleation pipette until the polar body follows a section of cytoplasm (Fig. 6D). It is important to note that the oocyte membrane is not actually compromised. Cytochalasin B at 5–10 µg/ml is used in the micromanipulation drop to destabilize the actin cytoskeleton to prevent lysis and damage during this procedure. The enucleation pipette is gently withdrawn, which results in the section of ooplasm containing the chromosomes pinching off from the remainder of the oocyte (Fig. 6E). The enucleated oocyte (Fig. 6F) is moved out of the field of view and epiillumination is used to confirm that the chromosomes and polar body are removed (Fig. 6G). Enucleation should be done in small batches of oocytes at a time (10–20) to minimize the time they are exposed to the manipulation environment (artificial light, suboptimal temperature, etc.). Optimally, enucleated oocytes are placed into the incubator in culture medium until all of the oocytes for the session are finished. This allows the membrane and cytoskeleton to recover from the enucleation procedure.

CELL TRANSFER

Cells to be used are harvested according to the cell type (e.g., trypsinization for adherent cells). Centrifugation at 800 *g* for 5 minutes is used to wash out the culture medium and concentrate the cells. The cell pellet is suspended in manipulation medium and cells are loaded into a fresh manipulation drop containing enucleated oocytes. Several cells at a time may be loaded into the cell transfer tool to speed up the process. A larger field of view will facilitate selecting which cells to transfer based on size and morphological characteristics; 200× total magnification may be used for this procedure. It is useful to start loading cells such that the oil/medium interface inside the cell transfer pipette is visible. As more cells are loaded and the interface moves further from the tip, fine control over the movement of cells within

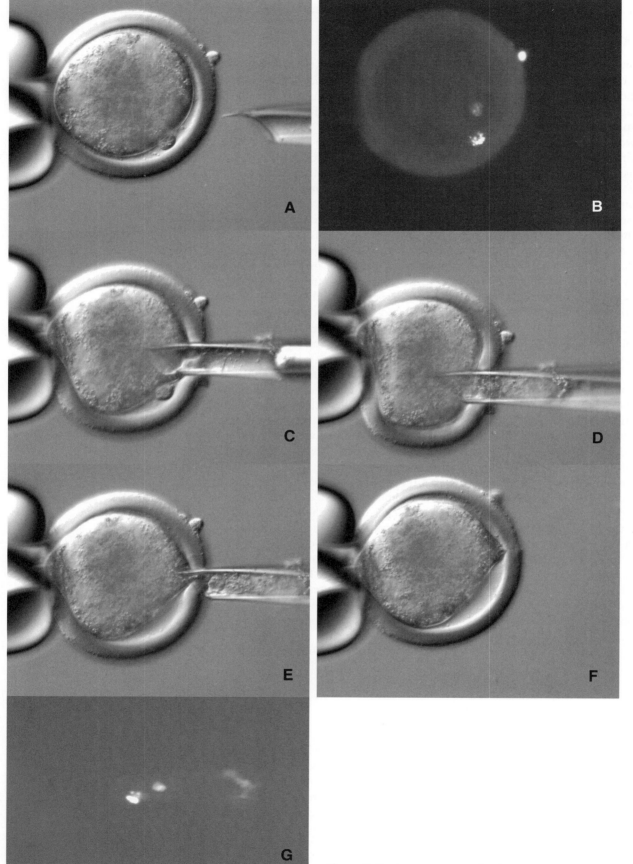

Figure 6 Bovine oocyte enucleation.

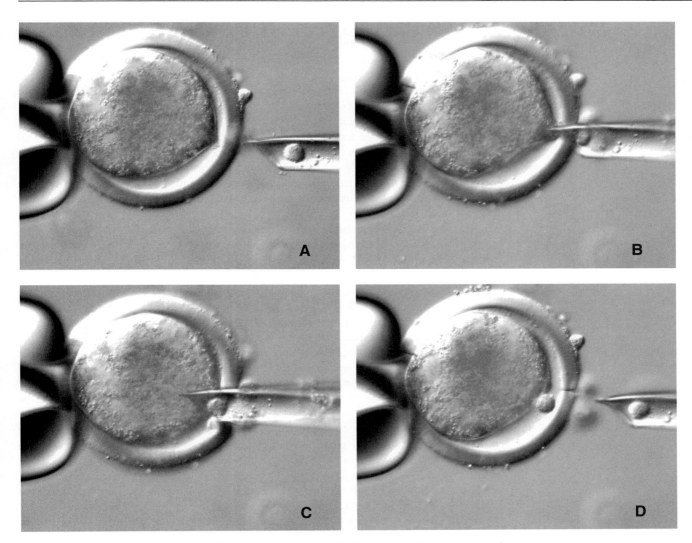

Figure 7 *Transfer of cultured adult fibroblast into the perivitelline space of an enucleated bovine oocyte.*

the pipette will be diminished. This is due to the oil/medium interface being located in a section of the cell transfer pipette with a larger diameter.

Cells are transferred to the perivitelline space by positioning the oocyte such that the slit in the zona pellucida made during enucleation is proximal to the tip of the cell transfer pipette. Before the zona is penetrated, a cell is positioned close to the tip of the pipette (Fig. 7B) to minimize the amount of culture medium blown into the perivitelline space. Entry through the same hole serves two major functions. It will prevent the oocyte from being squeezed and it will prevent embryonic material from hatching through two locations later in development. Once inside the zona (Fig. 7C), the cell is transferred, using the microinjector to push it out of the pipette such that good contact between the cell and oocyte membrane is maintained (Fig. 7D). Batches of about 20 to 40 oocytes at a time are done, then fusion should be started while good contact is still present.

COUPLET FUSION

Fusion may be accomplished by electrical stimulation, polyethylene glycol (PEG), or Sendai virus, with electrical pulse being by far the most popular method. For electrofusion, the cell–oocyte couplets are placed in a largely nonionic, slightly hypotonic medium between two parallel electrodes and a high-voltage direct-current pulse delivered such that membrane breakdown is achieved (reviewed in Zimmermann and Vienken, 1982). The pulse strength and duration must be determined experimentally, because conditions will vary slightly among different laboratories. The electrical pulse results in inversion of membrane phospholipids (breakdown), thus creating holes in the membrane; on healing, the membranes are fused. Therefore, too high a voltage for too long will result in lysis. However, if the voltage or duration is too low, poor fusion rates will be result. A rule of thumb for bovine nuclear transfer is to adjust the fusion parameters such that about 10% of the oocytes will lyse. On average, fusion parameters are about 1.25–1.5 kv/cm for 10–50 µsec. However, some investigators prefer delivering several shorter pulses compared to one longer one.

Another variable is the fusion chamber to be used, which will affect the raw settings due to the dimensions and positions of the electrodes. Suppliers of electrofusion equipment, such as BTX Instruments (San Diego, CA) and Eppendorf (Brinkman Instruments), provide several chambers to chose from and provide methods to calculate the electric field strength for each chamber. The temperature of the fusion medium will also affect results, so this should be kept constant either by using a heated stage for the stereomicroscope or by maintaining a constant base temperature. The latter may be more difficult because the intensity of the light for transillumination will affect the base temperature and may be variable for different users. In addition, small volumes of fusion medium and higher temperatures will result in rapid evaporation, leading to increased osmolarity, which leads to couplet detachment. Therefore, if a small chamber is used, the fusion medium should be replaced often.

A common basis for fusion medium is $0.3 M$ mannitol, $0.1 mM$ $MgSO_4$, and $0.05 mM$ $CaCl_2$ (Iwasaki *et al.*, 1989), with slight modifications. The osmolarity may be lowered by adjustment of the mannitol concentration to $0.25–0.28 M$. If an activation stimulus at the time of fusion is not desired, then the calcium is deleted. The magnesium is important because it provides a divalent cation source to help maintain membrane contact between the cell and oocyte. To prevent the oocytes from adhering to the glass, $0.1 mg/ml$ of a macromolecule such as bovine serum albumin, polyvinyl alcohol, or PVP may be added. Better results may be achieved using a macromolecule with a molecular weight high enough as to not cross the zona pellucida, which can interfere with membrane contact between the couplets. Most investigators do not buffer the fusion medium, but this has been done using $0.5 mM$ HEPES (Wells *et al.*, 1999), with good results.

Typically, the fusion medium is of higher density than manipulation medium, so when oocytes are added they float until equilibrated. It is preferable to use a four-well plate to prepare couplets for fusion. Well 1 contains manipulation medium; well 2, a 1:1 mixture of manipulation medium; well 3, fusion medium; and well 4, manipulation medium. Couplets are transferred from the manipulation microscope to well 1, then to wells 2 and 3, and are allowed to settle to the bottom in each well. It is preferable to transfer 5 at a time in overall batches of 20 to the fusion chamber from well 3. Once fused, they are placed into well 4. Once the batch of 20 is complete, all oocytes are placed into the incubator in culture medium, and a new batch is started. It is also useful to have a setup with two stereomicroscopes side by side for the fusion operation. One microscope is used to hold and prepare the couplets and the other has the fusion chamber attached. This allows for rapid transfer into and out of the fusion chamber, thus minimizing the time oocytes spend in fusion medium. For maximum field strength at the site of membrane contact, the couplets are aligned in the chamber such that the area of membrane contact is parallel to the electrodes. For this reason, generally couplets are fused in small batches of 10 or less, because it is difficult to align more than this many quickly in a field of few given by a magnification large enough to visualize the cells. Oocyte activation and embryo culture are the next steps in the process. There are many choices for the method of artificial activation and many theories as to the optimal timing between fusion and activation as well as for embryo culture. Other chapters in this volume deal with the nuances of species, and theories on the cell cycle as it pertains to the timing of fusion and activation.

PIEZOELECTRIC ASSISTED NUCLEAR TRANSFER

The use of a piezoelectric inertial impact device for mouse nuclear transfer was pioneered by Wakayama (Wakayama *et al.*, 1998) using modifications of a technique originally developed for intracytoplasmic sperm injection (Kimura and Yanagimachi, 1995). Although this technique was developed originally for the mouse, and it has been supposed that this is the only way to clone mice, in principle it should be applicable to any species. Direct nuclear injection offers advantages of minimal cytoplasmic transfer from the donor somatic cells and eliminates the need for fusion. However, it is a technically more demanding procedure requiring additional skill and practice to master. Piezoelectric material is a specialized ceramic that has unique electromechanical properties in that exposure to an electric field induces directional distortion of the material. When such material is attached to a micropipette holder, very powerful and precise directional movement of the pipette tip is achieved. Application of defined pulses in repeated succession allows a flat pipette tip to literally cut through the zona pellucida. A lower amplitude single pulse is used to break gently the cytoplasmic membrane of the oocyte, permitting transfer of a somatic cell nucleus. The two major manufacturer of piezoelectric devices for micromanipulation are Primetech (Ibaraki, Japan) and Burleigh Instruments (Victor, NY). Devices from each manufacturer mount to the pipette differently and operate on slightly different principles, mostly pertaining to the directionality, but each functions adequately for the intended purpose. Both devices are equipped with foot pedal activation and switching between two independently adjustable channels to alternate between two different sets of parameters. Typically, one channel is set for zona drilling and the other is set for membrane penetration.

The procedures for piezoelectric nuclear transfer are modified because the cell nucleus with some surrounding cytoplasm is actually injected into the ooplasm, in contrast to fusion of the whole cell with the oocyte. The pipette is filled with a dense fluid such as mercury or fluorinert (perfluoro compound FC-77) rather than oil to enhance energy transfer. Oocytes to be enucleated are incubated in a medium containing 5 μg/ml cytochalasin B for 15 minutes. The oocyte is positioned such that the metaphase plate, inner zona surface, and enucleation pipette are all in the same focal plane (Fig. 8A). Optimally, the oocyte is also pinned to the bottom of the injection chamber. It is further desired that the region of penetration have a large amount of perivitelline space to reduce the risk of oocyte damage during zona drilling. All at the same time, a slight deflection is made in the zona, slight suction pressure applied to the enucleation pipette, and the piezo device is activated. As the zona is cut, the deflection will disappear and a core will be pulled into the pipette (Fig. 8B). The moment the core is released, the piezo device is switched off. The enucleation pipette can then be easily inserted through the zona hole and pressed against the oolemma adjacent to the metaphase chromosomes (Fig. 8C). Using gentle suction, a small amount of cytoplasm surrounding the metaphase chromosomes is aspirated into the enucleation pipette and the pipette is slowly withdrawn, pinching off a karyoplast containing the chromosomes (Fig. 8D). Oocytes are typically enucleated in batches of about 20 at a time.

For cell transfer, cells are placed into a drop of medium with 10% PVP and stirred continuously until mixed very well. It takes about 30 minutes for the oocytes to settle to the bottom of the drop in this highly viscous solution. Cells are loaded

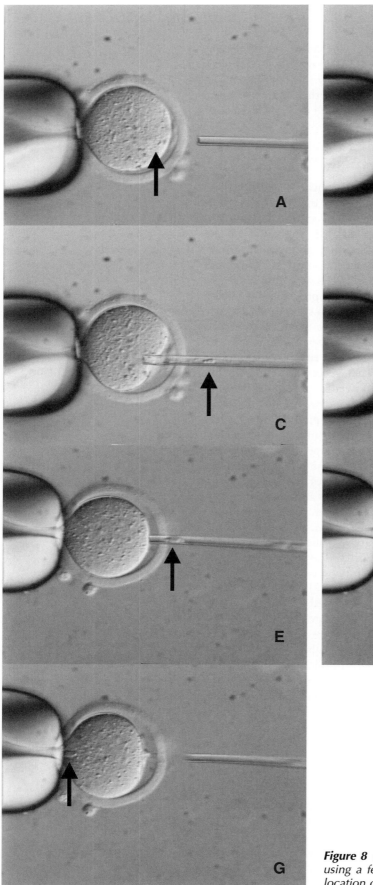
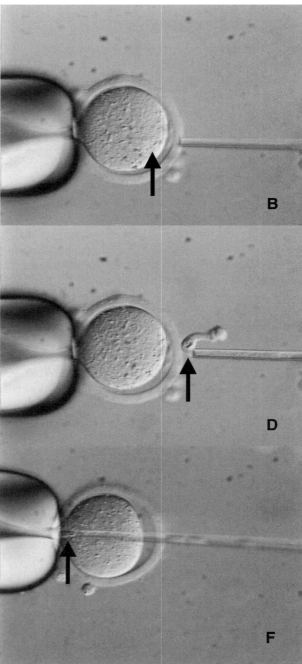

Figure 8 *Piezoelectric nuclear transfer in a mouse oocyte using a fetal fibroblast as nuclear donor. Arrow indicates location of chromosomes (A–D) or donor nucleus (E–G).*

into the cell transfer tool and the membrane is broken by shear forces generated by repeated aspiration in and out of the pipette tip. The cell transfer pipette tip is therefore smaller than the diameter of the cells, but larger than the nucleus, to prevent nuclear damage during aspiration. Five to 10 cells are loaded at one time. It is difficult to load many more than this because the energy generated by the piezo device is diminished as the mercury or fluorinert interface moves further from the tip.

The pipette is then moved to a drop of fresh medium containing enucleated oocytes. It is useful to first pass through a wash drop to remove PVP solution from the outside of the pipette. Optimally, the oocyte is positioned to permit entry though the hole in the zona created during enucleation (Fig. 8E). This is not always easy and a new hole can be made if necessary. In one continuous motion, the cell is positioned at the very tip of the pipette and the pipette is moved into the oocyte such that the membrane is invaginated nearly to the other side. Once in place, gentle suction is used to apply tension to the membrane and the piezo is activated with a single pulse (Fig. 8F). If broken, the membrane will be seen to relax along the pipette. As soon as this occurs the cell is deposited into the cytoplasm and the pipette is withdrawn slowly with a continuous motion (Fig. 9G). Care must be taken not to transfer any extra fluid into the oocyte, because this will impair membrane healing and lead to lysis. The oocyte is immediately released from the holder and nudged out of the view field using the transfer pipette. This immediate release seems to help the oocyte heal by releasing all tension from the membrane. Details on subsequent oocyte activation and culture are given in Wakayama *et al.* (1998).

A variation on this procedure involves transfer of the cell prior to enucleation (Munsie *et al.*, 2000). This method resulted in reduced lysis purportedly due to the membrane being broken prior to being weakened by the enucleation process. Because there were two sets of chromosomes present in the oocyte, the morphological difference between the endogenous and exogenous chromosomes was used to ensure that the endogenous chromosomes were removed on enucleation. This work led to the derivation of embryonic stem cell lines from embryos produced by nuclear transfer, which were the basis of a mouse model for therapeutic cloning.

TECHNICAL IMPROVEMENTS

Nuclear transfer is a technically demanding process, particularly when large numbers of oocytes are to be processed to achieve either large-scale production of clones or when large data sets comparing experimental parameters are desired. Therefore, there have been attempts to eliminate micromanipulation for some of the most cumbersome procedures, such as enucleation and cell transfer. Alternatives have included gradient centrifugal enucleation (Tatham *et al.*, 1995), chemical enucleation (Fulka and Moor, 1993), functional enucleation (Wagoner *et al.*, 1996), telophase enucleation (Bordingnon and Smith, 1998), and nuclear transfer without micromanipulators (Peura *et al.*, 1998; Vajita *et al.*, 2001).

Centrifugal enucleation takes advantage of the fact that the genetic material is heavier than the cytoplasm. If zona-free oocytes are centrifuged in an appropriate gradient in the presence of a cytoskeletal inhibitor, a karyoplast containing the metaphase plate can be separated from the cytoplasm. The major advantage is the virtually unlimited number of oocytes that can be processed at one time. Chemical enucleation is another means to solve the same problem of enucleation, but may have biological consequences as well. It is accomplished by using a chemical to prevent disjunction of the chromosomes at the second meiosis following activation. This way, all of the genetic material is expelled form the oocytes in the second polar body. It is thought that this may be additionally advantageous in that spindle-associated factors important for development may be left in the cytoplast. If the spindle is removed prior to activation by micromanipulation, most of these factors may be removed during enucleation. However, this theory has yet to be supported by large statistical data sets. Telophase enucleation potentially offers similar biological advantages because the oocytes are enucleated following activation. Presumably, spindle-associated material would be released into the cytoplasm and not removed with the chromosomes. In addition, because the chromosomes would be closely associated with the second polar body, they may be removed without the use of chemical dyes such as Hoechst 33342. Functional enucleation involves denaturing the DNA using UV light and Hoechst 33342, then centrifugation to remove the denatured DNA from the cytoplast. However, the chromosomes were located in the perivitelline space in only about one-half of the oocytes, and both centrifugation and exposure to UV light/Hoechst 33342 caused reduced developmental competence. Another means to avoid the use of fluorescent dyes employs the Pol-Scope (Oldenberg, 1996); this has been applied to visualization and enucleation of mammalian oocytes with very good success (Liu *et al.*, 2000). However, it remains to be demonstrated whether this equipment will enable visualization of the meiotic spindle in more opaque oocytes from species such as pigs and cows.

Nuclear transfer without manipulators is a recent achievement that involves using bisection of zona-free oocytes for enucleation using a standard stereomicroscope. Contact between the cell and the oocyte is facilitated using phytohemagglutinin; fusion can then be done at large scale. This was first done to evaluate the effect on development of adding additional cytoplasm to the oocyte to compensate for the loss during enucleation using blastomere nuclear transfer as a model (Peura *et al.*, 1998). This technique was later applied to somatic cell nuclear transfer largely to improve the oocyte throughput (Vajta *et al.*, 2001). These technical improvements have not yet achieved widespread use, but if shown to be more effective than traditional micromanipulation techniques, they hold promise to dramatically streamline the cloning process for both production and experimentation.

REFERENCES

Biggers, J. D., McGinnis, L. K., and Raffin, M. (2000). Amino acids and preimplantation development of the mouse in protein-free potassium simplex optimized medium. *Biol. Reprod.* **63**, 281–293.

Bordingnon, V., and Smith, L. C. (1998). Telophase enucleation: An improved method to prepare recipient cytoplasts for use in bovine nuclear transfer. *Mol. Reprod. Dev.* **49**, 29–36.

Bradshaw, J., Jung, T., Fulka, J., Jr., and Moor, R. M. (1995). UV irradiation of chromosomal DNA and its effect upon MPF and meiosis in mammalian oocytes. *Mol. Reprod. Dev.* **41**, 503–512.

Chatot, C. L., Ziomek, C. A., Bavister, B. D., Lewis, J. L., and Torres, I. (1989). An improved culture medium supports development of random-bred 1-cell mouse embryos in vitro. *J. Reprod. Fertil.* **86**, 679–688.

Fulka, J., Jr., and Moor, R. M. (1993). Noninvasive chemical enucleation of mouse oocytes. *Mol. Reprod. Dev.* **34**, 427–430.

Gardner, D. K., Lane, M., Spitzer, A., and Batt, P. (1994). Enhanced rates of cleavage and development for sheep zygotes cultured to the blastocyst stage *in vitro* in the absence of serum and somatic cells: Amino acids, vitamins and culturing embryos in groups stimulate development. *Biol. Reprod.* **50**, 390–400.

Hogan, B., Beddington, R., Costantini, F., and Lacy, E. (1994). " Manipulating the Mouse Embryo: A Laboratory Manual," 2nd Ed. Cold Spring Harbor Laboratory Press, Cold Spring Harbor.

Iwasaki, S., Kono, T., Fakatsu, H., and Nakahara, T. (1989). Production of bovine tetraploid embryos by electrofusion and their developmental capability *in vitro. Gamete Res.* **24**, 261–267.

Kimura, Y., and Yanagimachi, R. (1995). Intracytoplasmic sperm injection in the mouse. *Biol. Reprod.* **52**, 709–720.

Lawitts, J. A., and Biggers, J. D. (1991). Optimization of mouse embryo culture media using simplex methods. *J. Reprod. Fertil.* **91**, 543–546.

Liu, L., Oldenberg, R., Trimarchi, J. R., and Keefe, D. L. (2000). A reliable, noninvasive technique for spindle imaging and enucleation of mammalian oocytes. *Nat. Biotechnol.* **18**, 223–225.

Munsie, M. J., Michalska, A. E., O'Brien, C. M., Trounson, A. O., Pera, M. F., and Mountford, P. S. (2000). Isolation of pluripotent embryonic stem cells from reprogrammed adult mouse somatic cell nuclei. *Curr. Biol.* **10**, 989–992.

Oldenberg, R. (1996). A new view on polarization microscopy. *Nature* **381**, 811–812.

Petters, R. M., and Wells, K. D. (1993). Culture of pig embryos. *J. Reprod. Fertil. (Suppl.)* **48**, 61–73.

Peura, T. T., Lewis, I. M., and Trounson, A. O. (1998). The effect of recipient oocyte volume on nuclear transfer in cattle. *Mol. Reprod. Dev.* **50**, 185–191.

Polejaeva, I. A., Chen, S. H., Vaught, T. D., Page, R. L., Mullins, J., Ball, S., Gai, Y., Boone, J., Walker, S., Ayares, D. L., Colman, A., and Campbell, K. H. (2000). Cloned pigs produced by nuclear transfer from adult somatic cells. *Nature* **407**, 86–90.

Tatham, B. G., Dowsing, A. T., and Trounson, A. O. (1995). Enucleation by centrifugation of *in vitro*-matured bovine oocytes for nuclear transfer. *Biol. Reprod.* **53**, 1088–1094.

Vajta, G., Lewis, I. M., Hyttel, P., Thouas, G. A., and Trounson, A. O. (2001). Somatic cell cloning without micromanipulators. *Cloning* **3**, 89–95.

Wagoner, E. J., Rosenkrans, C. F., Jr., Gliedt, D. W., Pierson, J. N., and Munyon, A. L. (1996). Functional enucleation of bovine oocytes: Effects of centrifugation and ultraviolet light. *Theriogenology* **46**, 279–284.

Wakayama, T., Perry, A. C. F., Zuccotti, M., Johnson, K. R., and Yanagimachi, R. (1998). Full-term development of mice from enucleated oocyte injected with cumulus cell nuclei. *Nature* **394**, 369–374.

Wells, D. N., Misica, P. M., and Tervit, H. R. (1999). Production of cloned calves following nuclear transfer with cultured adult mural granulose cell. *Biol. Reprod.* **60**, 996–1005.

Westhusen, M. E., Levanduski, M. J., Scarborough, R., Looney, C. R., and Bondioloi, K. R. (1992). Viable embryos and normal calves after nuclear transfer into Hoechst stained demi-oocytes of cows. *J. Reprod. Fertil.* **95**, 475–480.

Zimmermann, U., and Vienken, J. (1982). Electrical field-induced cell-to-cell fusion. *J. Membr. Biol.* **67**, 165–182.

CHAPTER 8

MICROINSEMINATION AND NUCLEAR TRANSFER WITH MALE GERM CELLS

Atsuo Ogura, Narumi Ogonuki, and Kimiko Inoue

Different types of male germ cells, from primordial germ cells in fetal gonads to spermatozoa in adult testes, can participate in the formation of early embryos after microinsemination or nuclear transfer. Microinsemination with mature spermatozoa has already been successful in several mammalian species, including humans. Microinsemination with round spermatids leads to normal births in the mouse, rabbit, and human. Furthermore, in the mouse, secondary and primary spermatocytes also support full-term development after incorporation into immature or mature oocytes. Nuclear transfer using mouse primordial germ cells from male fetal gonads during days 12.5 to 16.5 of pregnancy produces cloned fetuses that arrest their development on day 10, probably due to erasure of the parental memory of imprinted genes. It is apparent that the data obtained from these micromanipulation and nuclear transfer techniques using male germ cells will certainly further understanding of the mechanisms of germ cell differentiation and germ line continuity. However, the microinsemination and nuclear transfer techniques still require a high degree of skill, especially in the mouse. More general use of these techniques and significant advances in germ cell biology will require technical improvements.

INTRODUCTION

As male germ cells develop from primordial germ cells in fetuses to mature spermatozoa in the adult testis, they undergo marked cellular, biochemical, and genetic changes. Genetic and epigenetic changes play the most integral part in the acquisition of the properties of male gametes, including haploidization of the chromosomes and epigenetic modifications, such as genomic imprinting. Thus, studies of the mechanisms of these genetic changes are expected to provide the basis for a better understanding of the differentiation of germ cells and germ line continuity in mammals. *In vitro* studies with isolated male germ cells have actively sought to identify functional proteins and genes that modify their genetic and epigenetic status during gametogenesis. The DNA methylation patterns of imprinted genes have also been analyzed [primordial germ cells (Ueda *et al.*, 2000); spermatozoa (Arial *et al.*, 1995)]. In addition to these direct approaches involving isolated germ cells, analysis of embryos constructed with the same cells allows more systematic examination of their biological properties. The embryo-constructing techniques of microinsemination and nuclear transfer (cloning) are extremely effective for analyzing the fertilizing ability of germ cells, and for studying the role of genome imprinting in embryo development. These techniques have also become practical for con-

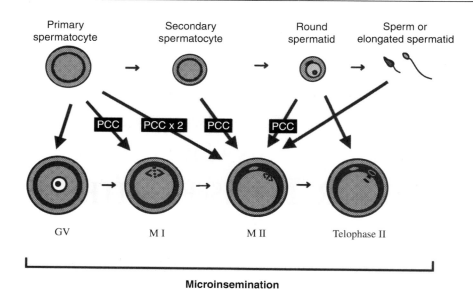

Figure 1 *Combinations of oocytes and male germ cells used to construct diploid embryos. Although the embryos thus constructed develop into fetuses around day 10, only those obtained by microinsemination develop to term offspring, because the imprinting status of their genome critically affects fetal development (see Table 1). PGC, Primordial germ cell; PCC, premature chromosome condensation; GV, germinal vesicle; MII, metaphase II.*

serving genetic resources, generating transgenic animals, and treating male-factor infertility.

In mammals, only fully mature male gametes (spermatozoa) acquire fertilizing ability in the epididymis and female genital tract. However, modern microinsemination techniques have made it possible to use immature male germ cells as male gametes. In mice, elongated spermatids, round spermatids (Ogura *et al.*, 1994; Kimura and Yanagimachi, 1995a), secondary spermatocytes (Kimura and Yanagimachi, 1995b), and late primary spermatocytes (pachytene and diplotene stages) (Ogura *et al.*, 1998; Kimura *et al.*, 1998) have all produced diploid embryos that have developed into healthy offspring. In other words, male germ cells have the ability to "fertilize" oocytes before they reach meiosis I. Therefore, the term "fertilizing ability" can be defined more broadly as "the ability to form a haploid set of paternal chromosomes within an oocyte." Theoretically, early spermatocytes (in the leptotene and zygotene stages) cannot enter meiosis I, because they have not yet completed synapsis (Handel, 1998). Spermatogonia and male primordial germ cells are at the earliest stages of male gamete development. Like somatic cells, they have diploid sets of chromosomes, so they can form diploid embryos by cloning (nuclear transfer into enucleated oocytes). In the mouse, embryos constructed from male primordial germ cells and gonocytes develop into blastocysts or fetuses (Kato *et al.*, 1999; Lee *et al.*, 2002). Collectively, diploid embryos can be produced by either microinsemination or nuclear transfer, using male germ cells ranging from primordial germ cells to mature spermatozoa (Fig. 1). Both microinsemination and nuclear transfer involve the use of unfertilized oocytes, but the techniques differ, both in the stages of recipient oocytes required to produce diploid embryos and in the epigenetic conditions required to support normal embryo development (Table 1). This chapter is divided into two sections: microinsemination techniques using spermatozoa, spermatids, and spermatocytes, and nuclear transfer techniques using male primordial germ cells.

Table 1 Male Germ Cells Used for Microinsemination and Nuclear Transfer: Requirements for Subsequent Development

Technique	Male germ cells	Requirements for development into a diploid embryo	Requirements for development into an offspring
Microinsemination	Spermatozoon, spermatid, spermatocyte	Formation of a haploid set of chromosomes in a recipient oocyte	Completion of paternal genomic imprinting
Nuclear transfer	Spermatogonium, primordial germ cell	Synchronization of the cell cycle with a recipient oocyte	Reprogramming of the genome and presence of parent-specific imprinting memory

MICROINSEMINATION

Microinsemination generally refers to intracytoplasmic sperm injection (ICSI), in which a sperm cell is injected directly into the cytoplasm of an oocyte, although it also includes techniques such as partial zona dissection and subzonal insemination, which aid the passage of spermatozoa through the zona pellucida (Iritani, 1991).

MICROINSEMINATION WITH SPERMATOZOA

Mammalian ICSI was first carried out successfully in golden hamsters by Uehara and Yanagimachi in 1976. Pronuclear oocytes were readily formed after sperm head injection, clearly demonstrating that oocyte activation and pronucleus formation are not dependent on sperm-specific properties, such as sperm motility, acrosome reaction, or sperm–oocyte membrane fusion. With the ICSI technique, the timing of fertilization can be controlled precisely, and heterologous fertilization can be successfully performed. Thus, ICSI can be modified to study the biochemical basis of fertilization and sperm function. These early experiments used mature hamster and rat oocytes, because of the high rate of survival of these oocytes after sperm injection. Since about 1990, ICSI has been successfully used to produce offspring in rabbits (Hosoi *et al.*, 1988), bovines (Goto *et al.*, 1990), and humans (Palermo *et al.*, 1992), confirming that fertilized oocytes produced by ICSI are biologically normal. Among the mammalian species so far examined, humans have yielded the highest number of ICSI births, with more than 10,000 since the introduction of the technique in 1992. Reliable ICSI techniques with a high degree of reproducibility were established for use in mice by Kimura and Yanagimachi (1995c). Mouse oocytes are very sensitive to puncturing: they burst instantly when a hole greater than 1 μm in diameter is made at the time of nucleus injection. This problem can be overcome by using a device called a piezo micromanipulator. Mice have unusual centrosome kinetics during fertilization: in mice, the microtubule-organizing centers of zygotes originate from oocytes, whereas in other animals, the microtubule-organizing centers originate predominantly from the centrioles of sperm (Navara *et al.*, 1995). Thus, a mouse embryo can be produced by removing the spermatozoan tail (which includes the centrioles) and injecting only the head (Kuretake *et al.*, 1996). This has now become the prevailing method of performing mouse ICSI, because the survival rate of mouse oocytes injected with only heads is higher than that of those injected with whole sperm, and subsequent embryo development is not compromised by the lack of sperm tail components. Injection of the sperm head alone should also be sufficient for embryogenesis in the rat and golden hamster, because the centrioles of zygotes also originate from oocytes in these animals. Very recently, normal pups were born after sperm heads were injected into rat and hamster oocytes, but the efficiency seems to be very low (Hirabayashi *et al.*, 2002). Intracytoplasmic sperm injection has been successfully performed in other animals, including sheep (Gomez *et al.*, 1998), cats (Pope *et al.*, 1998), rhesus monkeys (Hewitson *et al.*, 1999), pigs (Martin, 2000), and rabbits (Deng and Yang, 2001). Because rhesus monkeys are most closely related to humans, the aberrant behavior of monkey sperm and oocyte components seen in ICSI experiments raised some concerns regarding the safety of clinical application of ICSI to humans. These abnormalities included the physical effects of ICSI stimulation on oocyte chromosomes (Hewitson *et al.*, 1999), uneven sperm nucleus decondensation (Hewitson *et al.*, 1999), and contamination with foreign DNA (Chan *et al.*, 2000). Rabbit ICSI was developed to provide an animal model for the study of treatment for infertility at advanced maternal age by combining *in vitro* fertilization with germinal vesicle transfer (Li *et al.*, 2001).

MICROINSEMINATION WITH SPERMATIDS

The use of sperm injection to produce offspring in several mammalian species has demonstrated that fertilization does not require oocyte–sperm membrane fusion and that sperm chromosomes can be directly transferred to the cytoplasm of oocytes. This ICSI success was followed by trial microinsemination with round spermatids, the most immature haploid male germ cells, which were subsequently successfully transferred into the cytoplasm of mature oocytes by electrofusion and by intracytoplasmic injection. Normal fertilized oocytes were first obtained in injection and electrofusion experiments with hamsters (Ogura and Yanagimachi, 1993; Ogura *et al.*, 1993). Normal offspring were first obtained with mice (Ogura *et al.*, 1994) followed by rabbits (Sofikitis *et al.*, 1994) and humans (Tesarik *et al.*, 1995). This showed that the genomic and chromosomal changes required for male gamete function are completed by the round spermatid stage, and normal expression of imprinted genes in mouse embryos constructed with round spermatids has been confirmed (Shamanski *et al.*, 1999). In most mammals, microinsemination was first performed with spermatozoa, then by microinsemination with round spermatids. In mice, however, offspring were first obtained from round spermatids, because no intracytoplasmic injection method had been established for mouse oocytes by 1994, and electrofusion was therefore employed (Ogura *et al.*, 1994). In 1995, immediately after a report of successful mouse ICSI with a piezo-driven micromanipulator (Kimura and Yanagimachi, 1995c), the same technique was employed for round spermatid injection. The introduction of the piezo-driven micromanipulator method increased the fertilization rate to about 90%, and also led to improved birth rates after embryo transfer (Kimura and Yanagimachi, 1995a). Microinsemination with round spermatids has since been performed in bovines (Goto *et al.*, 1996), pigs (Lee *et al.*, 1998), and cynomolgus monkeys (Ogura *et al.*, 2000a), and the resulting embryos have developed into blastocysts. In cynomolgus monkeys, ongoing pregnancy was achieved after transfer of these spermatid-derived embryos into the uterus (N. Ogonuki *et al.*, unpublished observations).

Because the biological and biochemical characteristics of round spermatids differ from those of mature spermatozoa, they require different microinsemination techniques (for review, see Ogura and Yanagimachi, 1999). In certain species, the most important difference between spermatozoa and round spermatids is their oocyte-activating capacity. Mouse round spermatids are unable to activate oocytes, so the oocytes need to be artificially activated about 30 minutes before microinsemination (Ogura *et al.*, 1994; Kimura and Yanagimachi, 1995a) or after spermatid chromosome condensation (Ogura *et al.*, 1999). In contrast, human and monkey round spermatids can activate oocytes, and then synchronize oocyte and spermatid chromosomes (Tesarik *et al.*, 1995; Ogonuki *et al.*, 2001). Dissociation in timing has been found between the appearance of oocyte-activating capacity and that of Ca^{2+} oscillation-inducing capacity during spermatogenesis (Yazawa *et al.*,

2001; Ogonuki *et al.*, 2001). The elongated spermatids of most animal species are believed to activate oocytes, but microinsemination with elongated spermatids is technically more difficult to perform than microinsemination with round spermatids, because far fewer elongated spermatids are found in normal seminiferous epithelium. In humans, microinsemination with elongated spermatids has been reported for the treatment of nonobstructive azoospermia, because human ICSI, in principle, favors the male germ cells that are closer to mature spermatozoa. Rates of embryo development and pregnancy following microinsemination are higher for elongated spermatids than for round spermatids (Fishel *et al.*, 1997; Vanderzwalmen *et al.*, 1997). Some clinicians experienced extremely poor outcome after injection with round spermatids from azoospermic patients (Ghazzawi *et al.*, 1999; Levran *et al.*, 2000).

MICROINSEMINATION WITH SPERMATOCYTES

Kimura and Yanagimachi first performed microinsemination with secondary spermatocytes that were just about to undergo meiosis II (Kimura and Yanagimachi, 1995b). Like metaphase II (MII) oocytes, secondary spermatocytes have a $2n$ haploid set of chromosomes. Thus, these two cells can easily synchronize to form a diploid embryo. The injection alone does not activate mouse oocytes, so the levels of maturation-promoting factors (MPFs) remain high after injection of the secondary spermatocyte. As a result, maternal and paternal chromosomes completely synchronize, and after artificial oocyte activation, maternal and paternal chromosomes release polar bodies to form a diploid embryo. The fact that a high percentage of these embryos develop to term offspring indicates that secondary spermatocytes can function as substitute gametes, although a reliable method for differentiating secondary spermatocytes from other round testicular cells needs to be developed.

Unlike secondary spermatocytes, primary spermatocytes are easily identified in testicular cell suspension, but to construct diploid embryos with them requires relatively complicated techniques. To safely advance the meiotic division of primary spermatocytes (which have a $4n$ diploid set of chromosomes) within oocytes, the spermatocytes used for microinsemination must be ready for homologous chromosome segregation at meiosis I (i.e., in the late pachytene or diplotene stage). These spermatocytes can be distinguished from others by their large size (18–20 μm diameter). Primary spermatocytes are in the G2 phase of the cell cycle. Therefore, to synchronize the cell cycles of primary spermatocytes and oocytes, the oocytes used should be in the G2 or M phase; specifically, they should be in the germinal vesicle (GV), MI, or MII phase. We have obtained normal offspring from MI oocytes prepared by subjecting immature oocytes to cytochalasin to arrest the cell cycle at MI (Ogura *et al.*, 1998). Arresting these oocytes at MI allows maternal and paternal chromosomes to be completely synchronized at MI (Fig. 2). Offspring cannot be obtained by this method without cytochalasin, perhaps because spermatocytes in prophase I cannot synchronize with oocytes in prometaphase I (Ogura *et al.*, 1997). Kimura and colleagues obtained offspring by serial transfer to MII oocytes (Kimura *et al.*, 1998). The rate of healthy offspring development is low in their experiments and in ours, and this may be partially attributable to premature segregation of sister spermatocyte chromatids during meiosis I within oocytes (because sister chromatids normally segregate during meiosis II) (Fig. 3). The low efficiency may also result from incomplete genomic imprinting at this stage, but most genes have completed genomic imprinting specific for male germ cells by this point, as demonstrated by analysis of the allele-specific expression of imprinted genes in spermatocyte-derived fetuses (A. Ogura, unpublished observations). Nonetheless, such microinsemination experiments with primary spermatocytes may

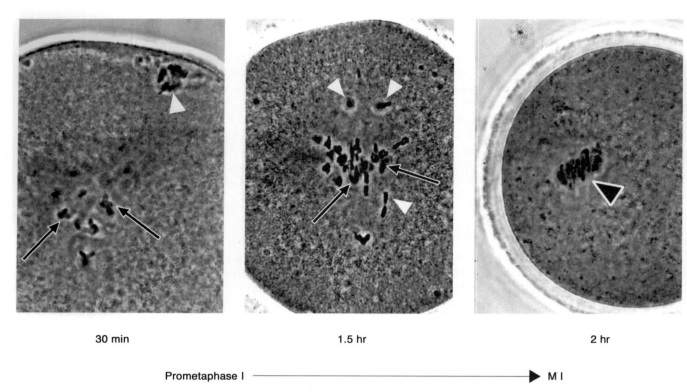

30 min   1.5 hr   2 hr

Prometaphase I ————————————————————————▶ M I

Figure 2 Microinsemination with primary spermatocytes. A spermatocyte nucleus is intro-duced into a prometaphase I oocyte, which is then treated with cytochalasin to arrest matu-ration at metaphase I (MI). The condensed chromosomes from the oocyte and spermatocyte form a single chromosome mass at MI (black arrowhead at 2 hr), and thus their stages are syn-chronized completely. The MI chromosome mass is an 8n tetraploid, which will be reduced to a 4n diploid at MII after extrusion of the first polar body, and to a 2n diploid after extru-sion of the second polar body, to form a normal zygote. White arrowheads, spermatocyte chromosomes; arrows, oocyte chromosomes. [From Ogura et al. (1998). Development of normal mice from metaphase I oocytes fertilized with primary spermatocytes. Proc. Natl. Acad. Sci. U.S.A. 95, 5611–5615. Copyright 1998 National Academy of Sciences, U.S.A.]

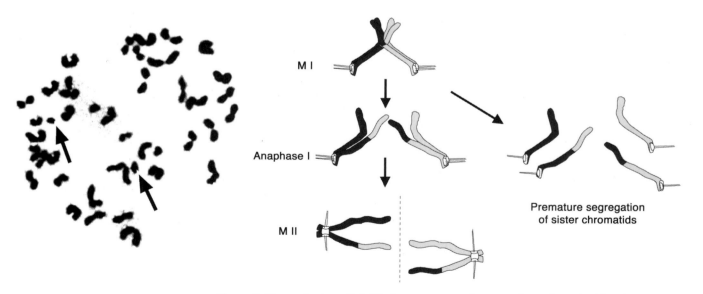

Figure 3 The metaphase II (MII) chromosomes of oocytes inseminated with primary sperma-tocytes at MI (left) and the presumptive behavior of the spermatocyte chromosomes during meiosis I within oocytes (right). This premature segregation of spermatocyte sister chromatids is probably the main cause of the frequent postimplantation death of fetuses derived from microinsemination with primary spermatocytes.

be very useful in determining the levels of genomic and chromosomal maturation of male gametes.

PRACTICAL APPLICATIONS OF MICROINSEMINATION TECHNIQUES FOR HUMANS AND ANIMALS

Due to its high fertilization rates, microinsemination is widely used in human clinics, not only with spermatozoa, but also with spermatids at various maturational stages. Although this technique gives consistently successful results, concerns have been raised about its potential biological and technical risks. Biological risks include the use of fertilization-incompetent spermatozoa and immature male germ cells, and technical risks include direct physical contact with gametes. These issues have been reviewed by others in detail (Schatten *et al.*, 1998; Yanagida *et al.*, 2000).

Microinsemination techniques have also been applied to biological research and to the animal industry. The generation of transgenic mice by ICSI with DNA-treated spermatozoa is an example. Because DNA easily adheres to spermatozoa, exogenous DNA is introduced into oocytes via a conventional *in vitro* fertilization (IVF) procedure (Lavitrano *et al.*, 1989). However, the reproducibility of this technique is very limited. The ICSI-mediated transgenesis procedure developed in mice by Perry and colleagues (1999) is more reliable, and may be applicable to livestock. The EGFP transgene thus introduced into the founder mice was transmitted to following generations at a very high rate. Another novel technique that has been developed for mouse transgenesis involves *in vivo* transfection of DNA into spermatogenic cells (Yamazaki *et al.*, 1998). When exogenous DNA [the enhanced green fluorescent protein (EGFP) gene] is successfully transfected into spermatogonial stem cells, they continue to produce transgenic spermatozoa. Ideally, transfected germ cells would carry markers, such as fluorescent proteins, so that those cells could be selected for microinsemination, to produce transgenic mice most efficiently. The *mit-YFP* gene has been used as a reporter gene to identify spermatozoa or spermatids carrying a transgene (Huang *et al.*, 2000). Those experiments found that the EGFP gene could not be used as a reporter gene, because excessive expression of EGFP proteins seemed to be detrimental to immature sperm cells.

Cryopreservation of embryos and gametes is an effective technique for the preservation of valuable animal species and experimental animal lines. Cryopreservation of spermatozoa is more efficient in terms of both cost and space than is cryopreservation of embryos, and this technique may be useful for banking transgenic or knockout mouse lines that are not needed immediately. However, IVF with thawed cryopreserved spermatozoa is less efficient, owing to sperm motility problems. The fertilization rate for the cryopreserved spermatozoa of C57BL/6 mice is particularly low after thawing, so a partial zona dissection method was developed that is easy to perform (Nakagata *et al.*, 1997). However, in ICSI, the motility of spermatozoa is not essential, so a wider range of sperm cell conditions is acceptable. In fact, offspring have been obtained from spermatozoa that were freeze-dried and then stored at room temperature or 4°C (Wakayama and Yanagimachi, 1998). Spermatids can also be stored frozen; healthy offspring have been obtained from mouse spermatids that were cryopreserved and then thawed (Ogura *et al.*, 1996a). This technique has been used to store the genes of mice with oligospermia caused by mutations, systemic disease (Ogura *et al.*, 1996b), or aging (Tanemura *et al.*, 1997). Live human births from cryopreserved spermatids have also been reported (Antinori *et al.*, 1997). Offspring have even been obtained by ICSI techniques from oocytes that were not capable of being naturally inseminated, as is the case with CD9-knockout mice (Miyado *et al.*, 2000).

NUCLEAR TRANSFER WITH MALE PRIMORDIAL
GERM CELLS

As mentioned above, spermatocytes before the pachytene stage cannot be used to construct embryos, owing to the inability of their chromosomes to segregate correctly (Handel, 1998). Therefore, the next step in research on the properties of male germ cells involves analysis with spermatogonia (Type A, or gonocytes) and primordial germ cells (PGCs), which have not entered the prophase of meiosis I. Because the chromosomal constitution of these early spermatogonia and PGCs is the same as that of somatic cells, diploid embryos can be constructed by the nuclear transfer technique using these cells. Although the embryos thus constructed are $2n$ diploid as fertilized oocytes, they do not necessarily develop normally to term, because whether the embryonic genome supports full-term development depends on its parent-specific imprinting memory (see Table 1). The DNA methylation patterns of the differentially methylated regions (DMRs) on imprinted genes are thought to indicate imprinting status (Reik and Walter, 2001). Advances in techniques for analyzing DNA methylation patterns have demonstrated that demethylation of DMRs in PGCs clearly occurs shortly before or after they enter the gonads, and that the gametic patterns are reestablished before meiosis I according to the sex of the gonads (Reik and Walter, 2001). However, defining the term "genomic imprinting" as differential expression from parental alleles may be a more precise way to interpret the imprinting status when examining gene expression patterns in fetuses, in which most imprinted genes are actively expressed. The parent-specific memory, probably marked on the DMRs of the genome of donor cells, is not affected by the reprogramming process during nuclear transfer (Inoue *et al.*, 2002). Therefore, the expression patterns of imprinted genes in PGC-derived fetuses probably reflect the imprinting status of the PGCs accurately. Fetuses have reportedly been obtained from male primordial germ cells during days 14.5–16.5 of development, by nuclear transfer and by a combination of nuclear transfer and chimera techniques (Kato and Tsunoda, 1995; Kato *et al.*, 1999), but with both these methods the development of fetuses arrested at day 9.5 of gestation. Analyses of the DNA methylation and *in situ* expression patterns of these fetuses showed that *H19* was expressed from both alleles, and that *Igf2*, *Igf2r*, and *Mash2*, the genes essential for fetal and placental development, were not expressed from either allele (Kato *et al.*, 1999). This indicates that the parent-specific imprinting memory is erased by day 14.5 of gestation. This is consistent with the finding that developmental arrest of PGC-derived fetuses occurs at day 9.5, when some imprinted genes start to be transcribed (see Fig. 4). To further examine when and how the imprinting memory is erased from the PGC genome, we analyzed fetuses derived from PGCs during days 11.5–13.5. We employed the so-called Honolulu technique developed by Wakayama and colleagues (see Chapter 16, this volume; Wakayama *et al.*, 1998) for nuclear transfer with PGCs. When we compared this method to nuclear transfer with somatic cells reported in our previous studies (Ogura *et al.*, 2000b,c), we found that the rates of embryonic and fetal development after nuclear transfer with PGCs were remarkably higher than those after nuclear transfer with somatic cells (A. Ogura, unpublished observation). The PGCs that we used had the genotype of (B6 × JF1)F$_1$, a hybrid between laboratory and wild strains of mice, so the active alleles of the imprinted genes could be identified by their polymorphism. Our experiments demonstrated that the parent-specific memory in the PGC genome is erased by day 12.5, and the development of fetuses reconstructed with day 11.5 PGCs was greatly improved compared with that of fetuses from day 12.5 PGCs (Lee *et al.*, 2002). This clearly indicates that fetal development is highly dependent on the expression of imprinted genes from the correct parental alleles. A bovine embryo produced from the male germ cells of a 52-day-old fetus reportedly developed into a normal calf, although it died of respiratory failure within 24 hours after birth

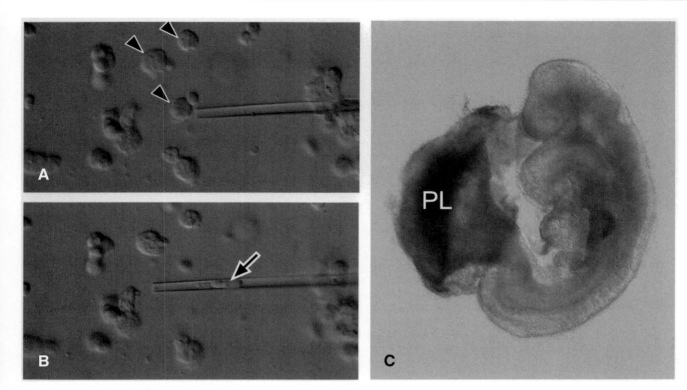

Figure 4 *(A and B) Male primordial germ cells (PGCs) isolated from a day 14.5 fetus (arrowheads) and (C) a day 9.5 fetus cloned from a day 14.5 male PGC. Because PGCs are soft and their nucleus is easily isolated in injection pipettes (arrow), they are very suitable for cloning by the microinjection method. Fetuses cloned from male PGCs at days 12.5–16.5 arrest development before day 10, although they show a heartbeat and placental formation. PL, Placenta.*

(Zakhartchenko *et al.*, 1999). Whether the pattern of its imprinted gene expression was specific for male germ cells is unclear.

REFERENCES

Antinori, S., Versaci, C., Dani, G., Antinori, M., and Selman, H. A. (1997). Successful fertilization and pregnancy after injection of frozen-thawed round spermatids into human oocytes. *Hum. Reprod.* **12**, 554–556.

Ariel, M., Robinson, E., McCarrey, J. R., and Cedar, H. (1995). Gamete-specific methylation correlates with imprinting of the murine *Xist* gene. *Nat. Genet.* **9**, 312–315.

Chan, A. W. S., Luetjens, C. M., Dominko, T., Ramalho-Santos, J., Simerly, C. R., Hewitson, L., and Schatten, G. (2000). Foreign DNA transmission by ICSI; injection of spermatozoa bound with exogenous DNA results in embryonic GFP expression and live Rhesus monkey births. *Mol. Hum. Reprod.* **6**, 26–33.

Deng, M., and Yang, X. (2001). Full term development of rabbit oocytes fertilized by intracytoplasmic sperm injection. *Mol. Reprod. Dev.* **59**, 38–43.

Fishel, S., Green, S., Hunter, A., Lisi, F., Rinaldi, L., Lisi, R., and McDermott, H. (1997). Human fertilization with round and elongated spermatids. *Hum. Reprod.* **12**, 336.

Ghazzawi, I. M., Alhasani, S., Taher, M., and Souso, S. (1999). Reproductive capacity of round spermatids compared with mature spermatozoa in a population of azoospermic men. *Hum. Reprod.* **14**, 736–740.

Gomez, M. C., Catt, J. W., Evans, G., and Maxwell, W. M. C. (1998). Cleavage, development and competence of sheep embryos fertilized by intracytoplasmic sperm injection and *in vitro* fertilization. *Theriogenology* **49**, 1143–1154.

Goto, K., Kinoshita, A., Takuma, Y., and Ogawa, K. (1990). Fertilisation of bovine oocytes by the injection of immobilized, killed spermatozoa. *Vet. Rec.* **127**, 517–520.

Goto, K., Kinoshita, A., Nakanishi, Y., and Ogawa, K. (1996). Blastocyst formation following intracytoplasmic injection of *in-vitro* derived spermatids into bovine oocytes. *Hum. Reprod.* **11**, 824–829.

Handel, M. A. (1998). Monitoring meiosis in gametogenesis. *Theriogenology* **49**, 423–430.

Hewitson, L., Dominko, T., Takahashi, D., Martinovich, C., Ramalho-Santos, J., Sutovsky, P., Fanton, J., Jacob, D., Monteith, D., Neuringer, M., Battaglia, D., Simerly, C., and Schatten, G. (1999). Unique checkpoints during the first cell cycle of fertilization after intracytoplasmic sperm injection in rhesus monkeys. *Nat. Med.* **5**, 431–433.

Hirabayashi, M., Kato, M., Aoto, T., Sekimoto, A., Ueda, M., Miyoshi, I., Kasai, N., and Hochi, S. (2002). Offspring derived from intracytoplasmic injection of transgenic rat sperm. *Transgen. Res.* **11**, 221–228.

Hosoi, Y., Miyake, M., Utsumi, K., and Iritani, A. (1988). Development of rabbit oocytes after microinjection of spermatozoa. *Proc. 11th Int. Congr. Anim. Reprod.* **3** (Abstr.), 331.

Huang, Z., Sakurai, T., Chuma, S., Saito, T., and Nakatsuji, N. (2000). *In vivo* transfection of testicular germ cells and transgenesis by using the mitochondrially localized jellyfish fluorescent protein gene. *FEBS Lett.* **487**, 248–251.

Inoue, K., Kohda, T., Lee, J., Ogonuki, N., Mochida, K., Noguchi, Y., Tanemura, K., Kaneko-Ishino, T., Ishino, F., and Ogura, A. (2002). Faithful expression of imprinted genes in cloned mice. *Science* **295**, 297.

Iritani, A. (1991). Micromanipulation of gametes for *in vitro* assisted fertilization. *Mol. Reprod. Dev.* **28**, 199–207.

Kato, Y., and Tsunoda, Y. (1995). Germ cell nuclei of male fetal mice can support development of chimeras to midgestation following serial transplantation. *Development* **121**, 779–783.

Kato, Y., Rideout III, W. M., Hilton, K., Barton, S. C., Tsunoda, Y., and Surani, M. A. (1999). Developmental potential of mouse primordial germ cells. *Development* **126**, 1823–1832.

Kimura, Y., and Yanagimachi, R. (1995a). Mouse oocytes injected with testicular spermatozoa or round spermatids can develop into normal offspring. *Development* **121**, 2397–2405.

Kimura, Y., and Yanagimachi, R. (1995b). Development of normal mice from oocytes injected with secondary spermatocyte nuclei. *Biol. Reprod.* **53**, 855–862.

Kimura, Y., and Yanagimachi, R. (1995c). Intracytoplasmic sperm injection in the mouse. *Biol. Reprod.* **52**, 709–720.

Kimura, Y., Tateno, H., Handel, M. A., and Yanagimachi, R. (1998). Factors affecting meiotic and developmental competence of primary spermatocyte nuclei injected into mouse oocytes. *Biol. Reprod.* **59**, 871–877.

Kuretake, S., Kimura, Y., Hoshi, K., and Yanagimachi, R. (1996). Fertilization and development of mouse oocytes injected with isolated sperm heads. *Biol. Reprod.* **55**, 789–795.

Lavitrano, M., Camaioni, A., Fazio, V. M., Dolci, S., Farace, M. G., and Spadafora, C. (1989). Sperm cells as vectors for introducing foreign DNA into eggs: Genetic transformation of mice. *Cell* **57**, 717–723.

Lee, J., Inoue, K., Ono, R., Ogonuk, R., Kohda, K., Kaneko-Ishino, T., Ogura, A., and Ishino, F. (2002). Erasing genomic imprinting memory in mouse clone embryos produced from day 11.5 primordiral germ cells. *Development* **129**, 1807–1817.

Lee, J. W., Kim, N.-H., Lee, H. T., and Chung, K. S. (1998). Microtubule and chromatin organization during the first cell-cycle following intracytoplasmic injection of round spermatid into porcine oocytes. *Mol. Reprod. Dev.* **50**, 221–228.

Levran, D., Nahum, H., Farhi, J., and Weissman, A. (2000). Poor outcome with round spermatid injection in azoospermic patients with maturation arrest. *Fertil. Steril.* **74**, 443–449.

Li, G. P., Chen, D. Y., Lian, L., Sun, Q. Y., Wang, M. K., Liu, J. L., Li, J. S., and Han, Z. M. (2001). Viable rabbits derived from reconstructed oocytes by germinal vesicle transfer after intracytoplasmic sperm injection (ICSI). *Mol. Reprod. Dev.* **58**, 180–185.

Martin, M. J. (2000). Development of *in vivo*-matured porcine oocytes following intracytoplasmic sperm injection. *Biol. Reprod.* **63**, 109–112.

Miyado, K., Yamada, G., Yamada, S., Hasuwa, H., Nakamura, Y., Ryu, F., Suzuki, K., Kosai, K., Inoue, K., Ogura, A., Okabe, M., and Mekada, E. (2000). Requirement of CD9 on the egg plasma membrane for fertilization. *Science* **287**, 321–324.

Nakagata, N., Okamoto, M., Ueda, O., and Suzuki, H. (1997). Positive effect of partial zona-pellucida dissection on the *in vitro* fertilizing capacity of cryopreserved C57BL/6J transgenic mouse spermatozoa of low motility. *Biol. Reprod.* **57**, 1050–1055.

Navara, C. S., Wu, G.-J., Simerly, C., and Schatten, G. (1995). Mammalian model systems for exploring cytoskeletal dynamics during fertilization. *Curr. Top. Dev. Biol.* **31**, 321–342.

Ogonuki, N., Sankai, T., Yagami, K., Shikano, T., Oda, S., Miyazaki, S., and Ogura, A. (2001). Activity of a sperm-borne oocyte-activating factor in spermatozoa and spermatogenic cells from cynomolgus monkeys and its localization after oocyte activation. *Biol. Reprod.* **65**, 351–357.

Ogura, A., and Yanagimachi, R. (1993). Round spermatid nuclei injected into hamster oocytes form pronuclei and participate in syngamy. *Biol. Reprod.* **48**, 219–225.

Ogura, A., and Yanagimachi, R. (1999). Microinsemination using spermatogenic cells in mammals. *In* "Male Sterility for Motility Disorders" (S. Hamamah, R. Mieusset, F. Olivennes, and R. Frydman, eds.), pp. 189–202. Springer-Verlag, New York.

Ogura, A., Yanagimachi, R., and Usui, N. (1993). Behavior of hamster and mouse round spermatid nuclei incorporated into mature oocytes by electrofusion. *Zygote* **1**, 1–8.

Ogura, A., Matsuda, J., and Yanagimachi, R. (1994). Birth of normal young following fertilization of mouse oocytes with round spermatids by electrofusion. *Proc. Natl. Acad. Sci. U.S.A.* **91**, 7460–7462.

Ogura, A., Matsuda, J., Asano, T., Suzuki, O., and Yanagimachi, R. (1996a). Mouse oocytes injected with cryopreserved round spermatids can develop into normal offspring. *J. Assist. Reprod. Genet.* **13**, 431–434.

Ogura, A., Yamamoto, Y., Suzuki, O., Takano, K., Wakayama, T., Mochida, K., and Kimura, H. (1996b). *In vitro* fertilization and microinsemination with round spermatids for propagation of nephrotic genes in mice. *Theriogenology* **45**, 1141–1149.

Ogura, A., Wakayama, T., Suzuki, O., Shin, T.-Y., Matsuda, J., and Kobayashi, Y. (1997). Chromosomes of mouse primary spermatocytes undergo meiotic divisions after incorporation into homologous immature oocytes. *Zygote* **5**, 177–182.

Ogura, A., Suzuki, O., Tanemura, K., Mochida, K., Kobayashi, Y., and Matsuda, J. (1998). Development of normal mice from metaphase I oocytes fertilized with primary spermatocytes. *Proc. Natl. Acad. Sci. U.S.A.* **95**, 5611–5615.

Ogura, A., Inoue, K., and Matsuda, J. (1999). Spermatid nuclei can support full term development after premature chromosome condensation within mature oocytes. *Hum. Reprod.* **14**, 1294–1298.

Ogura, A., Inoue, K., Ogonuki, N., Suzuki, O., Mochida, K., Matsuda, J., and Sankai, T. (2000a). Recent advances in the microinsemination of laboratory animals. *Int. J. Androl.* **23**, 60–62.

Ogura, A., Inoue, K., Ogonuki, N., Noguchi, A., Takano, K., Nagano, R., Suzuki, O., Lee, J., Ishino, F., and Matsuda, J. (2000b). Production of male clone mice from fresh, cultured, and cryopreserved immature Sertoli cells. *Biol. Reprod.* **62**, 1579–1584.

Ogura, A., Inoue, K., Takano, K., Wakayama, T., and Yanagimachi, R. (2000c). Birth of mice after nuclear transfer by electrofusion using tail tip cells. *Mol. Reprod. Dev.* **57**, 55–59.

Palermo, G., Joris, H., Debroey, P., and Van Steirteghem, A. C. (1992). Pregnancies after intracytoplasmic injection of single spermatozoon into an oocyte. *Lancet* **340**, 17–18.

Perry, A. C. F., Wakayama, T., Kishikawa, H., Kasai, T., Okabe, M., Toyoda, Y., and Yanagimachi, R. (1999). Mammalian transgenesis by intracytoplasmic sperm injection. *Science* **284**, 1180–1183.

Pope, C. E., Johnson, C. A., McRae, M. A., Keller, G. L., and Dresser, B. L. (1998). Development of embryos produced by intracytoplasmic sperm injection of cat oocytes. *Anim. Reprod. Sci.* **53**, 221–236.

Reik, W., and Walter, J. (2001). Genomic imprinting: Parental influence on the genome. *Nat. Rev. Genet.* **2**, 21–32.

Schatten, G., Hewitson, L., Simerly, C., Sutovsky, P., and Huszar, G. (1998). Cell and molecular biological challenges of ICSI: ART before science? *J. Law Med. Ethics* **26**, 29–37.

Shamanski, F. L., Kimura, Y., Lavoir, M.-C., Pedersen, R. A., and Yanagimachi, R. (1999). Status of genomic imprinting in mouse spermatids. *Hum. Reprod.* **14**, 1050–1056.

Sofikitis, N. V., Miyagawa, I., Agapitos, E., Pasyianos, P., Toda, T., Hellstrom, W. J. G., and Kawamura, H. (1994). Reproductive capacity of the nucleus of the male gamete after completion of meiosis. *J. Assist. Reprod. Genet.* **11**, 335–341.

Tanemura, K., Wakayama, T., Kuramoto, K., Hayashi, Y., Sato, E., and Ogura, A. (1997). Birth of normal young by microinsemination with frozen-thawed round spermatids collected from aged azoospermic mice. *Lab. Anim. Sci.* **47**, 203–204.

Tesarik, J., Mendoza, C., and Testart, J. (1995). Viable embryos from injection of round spermatids into oocytes. *New Engl. J. Med.* **333**, 525.

Ueda, T., Abe, K., Miura, A., Yuzuriha, M., Zubair, M., Noguchi, M., Niwa, K., Kawase, Y., Kono, T., Matsuda, Y., Fujimoto, H., Shibata, H., Hayashizaki, Y., and Sasaki, H. (2000). The paternal methylation imprint of the mouse *H19* locus is acquired in the gonocyte stage during foetal testis development. *Genes Cells* **5**, 649–659.

Uehara, T., and Yanagimachi, R. (1976). Microsurgical injection of spermatozoa into hamster eggs with subsequent transformation of sperm nuclei into male pronuclei. *Biol. Reprod.* **15**, 467–470.

Vanderzwalmen, P., Zech, H., Birkenfeld, A., Yemini, M., Bertin, G., Lejeune, B., Nijs, M., Segal, L., Stecher, A., Vandamme, B., vanRoosendaal, E., and Schoysman, R. (1997). Intracytoplasmic injection of spermatids retrieved from testicular tissue: Influence of testicular pathology, type of selected spermatids and oocyte activation. *Hum. Reprod.* **12**, 1203–1213.

Wakayama, T., and Yanagimachi, R. (1998). Development of normal mice from oocytes injected with freeze-dried spermatozoa. *Nat. Biotechnol.* **16**, 639–641.

Wakayama, T., Perry, A. C. F., Zuccotti, M., Johnson, K. R., and Yanagimachi, R. (1998). Full-term development of mice from enucleated oocytes injected with cumulus cell nuclei. *Nature* **394**, 369–374.

Yamazaki, Y., Fujimoto, H., Ando, H., Ohyama, T., Hirota, Y., and Noce, T. (1998). *In vivo* gene transfer to mouse spermatogenic cells by deoxyribonucleic acid injection into seminiferous tubules and subsequent electroporation. *Biol. Repord.* **59**, 1439–1444.

Yanagida, K., Katayose, H., Yazawa, H., Hayashi, S., Kimura, Y., and Sato, A. (2000). Limits to the clinical application of ICSI. *J. Mammal. Ova Res.* **17**, 77–83.

Yazawa, H., Yanagida, K., Katayose, H., Hayashi, S., and Sato, A. (2001). Comparison of oocyte activation and oscillation-inducing abilities of round/elongated spermatids of mouse, hamster, rat, and human assessed by mouse oocyte activation assay. *Hum. Reprod.* **15**, 2582–2590.

Zakhartchenko, V., Durcova-Hills, G., Schernthaner, W., Stojkovic, M., Reichenbach, H.-D., Mueller, S., Steinborn, R., Mueller, M., Wenigerkind, H., Prelle, K., Wolf, E., and Brem, G. (1999). Potential of fetal germ cells for nuclear transfer in cattle. *Mol. Reprod. Dev.* **52**, 421–426.

CHAPTER 9

DEVELOPMENT OF VIABLE MAMMALIAN EMBRYOS *IN VITRO*: EVOLUTION OF SEQUENTIAL MEDIA

David K. Gardner and Michelle Lane

INTRODUCTION

The field of modern embryo culture spans several decades (Biggers *et al.*, 1965; Whitten and Biggers, 1968; Whittingham, 1971; Biggers, 1987), although it was in the final decade of the twentieth century that a resurgence of interest in embryo physiology and metabolism led to significant improvements in this field. Two discrete approaches led to the formulations of different and yet relatively effective types of culture media. The first approach relied heavily on lessons learned from both embryo and maternal physiology and led to the formulation of sequential media. The second approach used a computer model to generate culture media formulations using a process of simplex optimization. The former approach was used extensively in our laboratory, whereas the laboratory of John Biggers (Lawitts and Biggers, 1993) championed the latter. The history of simplex optimization (SOM) began initially with the formulation of SOM, then led to KSOM and finally to KSOMaa; this has been reviewed elsewhere (Lawitts and Biggers, 1993; Biggers *et al.*, 2000).

The term "viable embryo" is used deliberately in the title of this Chapter and throughout, because, as we shall demonstrate, although mammalian embryos can be cultured throughout the preimplantation period in a number of different conditions, the resultant viability of the blastocysts can be extremely different. Viability is here defined as the ability of the blastocyst to implant and develop successfully into a fetus. The emphasis of this treatise is on sequential media; data are presented to support the move to more than one medium formulation in order to culture the mammalian preimplantation embryo throughout the preimplantation period to the viable blastocyst stage.

DYNAMICS OF EMBRYO AND MATERNAL PHYSIOLOGY

During the preimplantation period of mammalian embryo development, the conceptus undergoes significant changes in its physiology, metabolism, and genetic control. These changes are so dramatic that the starting point of development, the zygote, and the final stage, the blastocyst, can be likened to two very different

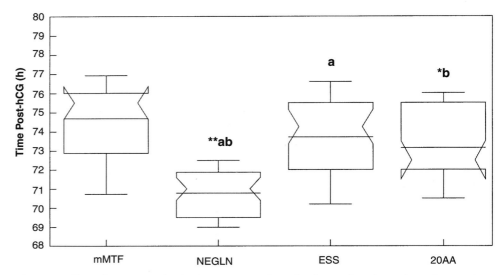

Figure 1 *Effect of amino acids on cleavage rates from the four-cell to the eight-cell stage (from Lane and Gardner, 1997b, J. Assist. Reprod. Genet., with permission). There were at least 30 embryos per treatment group. Notches represent the interquartile range, therefore including 50% of the data; whiskers represent 5 and 95% quartiles; line across the box represents the mean. NEGLN–mMTF supplemented with nonessential amino acids and glutamine; ESS–mMTF supplemented with essential amino acids without glutamine; 20AA–mMTF supplemented with all 20 amino acids. *, Significantly different from mMTF (P < 0.05); **, significantly different from mMTF (P < 0.01). a, Like pairs significantly different (P < 0.01); b, like pairs significantly different (P < 0.05). Culture with nonessential amino acids stimulated cleavage rates to the eight-cell stage. Addition of essential amino acids to the nonessential amino acids and glutamine group negated this benefit, plausibly due to competition for specific transporters.*

somatic cell types. The zygote, like the oocyte from which it was derived, has a low metabolic activity, exhibiting low levels of oxygen consumption and low QO_2, therefore being likened to relatively quiescent adult tissue such as brain (Leese, 1991). Rather than using glucose as its primary energy source, the zygote and cleavage stages utilize the carboxylic acids pyruvate and lactate oxidatively (Biggers *et al.*, 1967), with relatively low levels of glycolysis. At this stage the embryo is under the control of the maternal genome, the embryonic genome being sequentially activated from the two- to the eight-cell stage, depending on the species. Significantly, prior to compaction and the generation of the first transporting epithelium, the individual cells of the embryos are only loosely associated and readily disaggregate if the zona is removed. At this time the individual cells exhibit a physiology typical of unicellular organisms and are consequently more susceptible to their environment (Gardner, 1998a,b; Lane, 2001). Prior to compaction, the embryo in culture benefits from the presence of specific amino acids—for example, alanine, aspartate, asparagine, glycine, glutamate, glutamine, proline, and serine (Gardner, 1994; Gardner and Lane, 1993a; Lane and Gardner, 1997a,b; Steeves and Gardner, 1999). These amino acids have been shown to stimulate cleavage rates and compaction in the mouse (Figs. 1 and 2) and cow, and are known to fill several niches in the embryo's physiology, serving as osmolytes and buffers of pH_i (Table 1).

A major change in the embryo occurs at the time of compaction due to the formation of an epithelium, with the cells of the embryo beginning to take on a more somatic-cell physiology. With the establishment of an epithelium the embryo is no longer as dependent on specific amino acids to serve as osmolytes (Hammer *et al.*, 2000; Lane, 2001) and buffers of pH_i (Edwards *et al.*, 1998a), but rather can better regulate its internal environment (Figs. 3 and 4). With the establishment of basolaterally positioned ATPases (Benos and Biggers, 1981) the embryo actively creates

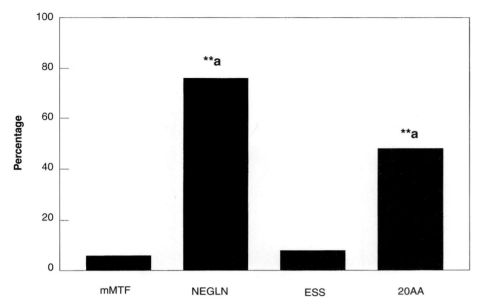

Figure 2 *Effect of amino acids in the medium on the compaction of eight-cell embryos after 78 hours post-hCG (57 hours of culture) (from Lane and Gardner, 1997b, J. Assist. Reprod. Genet., with permission). Values are mean; n = at least 30 embryos per treatment. NEGLN—mMTF were supplemented with nonessential amino acids and glutamine; ESS–mMTF were supplemented with essential amino acids without glutamine; 20AA–mMTF were supplemented with all 20 amino acids.**, Significantly different from mMTF and ESS (P < 0.01); the letter "a" denotes like pairs that were significantly different (P < 0.05).*

a blastocoel, the composition of which is regulated by the epithelium. Significantly, the blastocyst stage has a high oxidative capacity and QO_2 (similar to that of active skeletal muscle) and readily uses glucose as its primary energy source, though is able to adapt and use alternative nutrients should the need arise (Gardner and Leese, 1988); i.e., the embryo adapts to the environment in which it is placed. This trait is referred to as embryo plasticity. The blastocyst of all mammalian species studied exhibits a capacity for aerobic glycolysis, defined as the conversion of glucose to lactate even in the presence of sufficient oxygen for its complete oxidation (Wales, 1969; Wales *et al.*, 1987; Gardner and Leese, 1990), although the levels of aerobic glycolysis do vary between species (Hardy *et al.*, 1989; Gardner and Leese, 1990; Rieger *et al.*, 1992; Gardner *et al.*, 1993) and the medium used for metabolic analysis (Lane and Gardner, 1998). At this point the embryo has two distinct cell types, the inner cell mass and trophectoderm, which exhibit different carbohydrate (Hewitson and Leese, 1993) and amino acid (Lane and Gardner, 1997a) requirements. Table 2 highlights the differences between pre- and postcompaction embryos.

Table 1 Functions of Amino Acids during Preimplantation Mammalian Embryo Development

Function	Reference
Biosynthetic precursors	Crosby *et al.*, 1988
Sources of energy	Rieger *et al.*, 1992
Regulators of energy metabolism	Gardner and Lane, 1993b
Osmolytes	Van Winkle *et al.*, 1990
Buffers of pH_i	Edwards *et al.*, 1998a
Antioxidants	Liu and Foote, 1995
Chelators	Lindenbaum, 1973

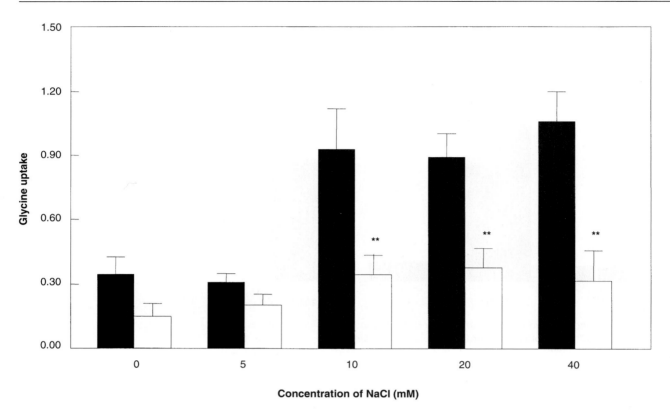

Figure 3 *Effect of compaction on glycine uptake by mouse embryos when exposed to increasing salt concentrations in the medium. Glycine uptake by mouse embryos at the eight-cell (solid bars) and compacted eight-cell stages (open bars). **, Significantly different from eight-cell embryos (P < 0.01). As the concentration of sodium chloride increases up to 10 mM there is an increase in the uptake of the amino acid glycine in order to counter the effect of high salt concentrations. Data from Lane (2001).*

As well as changes in embryo physiology and metabolism, the environment in the female reproductive tract is also under a state of flux, with differences occurring within the same region of the tract due to endocrine changes (Nichol *et al.*, 1992; Gardner *et al.*, 1996), and differences between regions of the tract, specifically the oviduct and uterus (Fischer and Bavister, 1993; Gardner *et al.*, 1996) (Table 3). The mammalian embryo is therefore exposed to gradients of nutrients (both carbohydrates and amino acids), oxygen, and pH as it develops *in vivo*.

Table 2 Differences in Embryo Physiology Pre- and Postcompaction

Precompaction	Postcompaction
Low biosynthetic activity	High biosynthetic activity
Low QO_2	High QO_2
Pyruvate preferred nutrient	Glucose preferred nutrient
Nonessential amino acids	Nonessential + essential amino acids
Maternal genome	Embryonic genome
Individual cells	Transporting epithelium
One cell type	Two distinct cell types (ICM and trophectoderm)

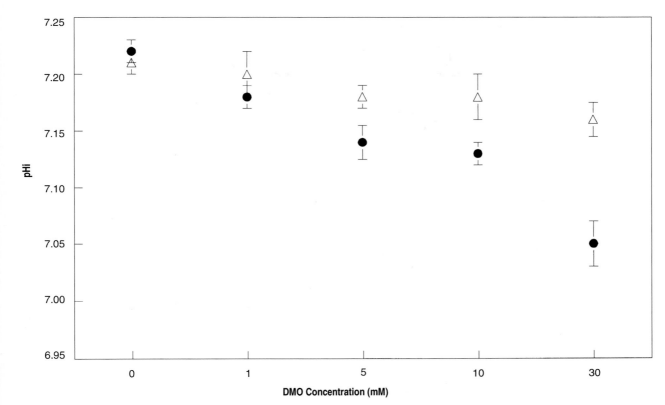

Figure 4 *Effect of compaction on intracellular pH (pH$_i$) of mouse embryos incubated at increasing DMO (a nonmetabolizable acid) concentrations, showing the pH$_i$ of eight-cell embryos (●) and the pH$_i$ of compacted embryos (△). Data were taken from different experiments from Edwards et al. (1998a) and therefore no statistics are present. However, it is evident that the decrease in pH$_i$ was greater in eight-cell embryos than in those that had undergone compaction. See Edwards et al. (1998a) for further experiments that demonstrate that experimental decompaction results in a decrease in the embryo's ability to regulate pH$_i$.*

Table 3 Differences in Oviduct and Uterine Environments

Component	Oviduct	Uterus
Glucose concentration	0.5 mM[a]	3.15 mM[a]
Pyruvate concentration	0.32 mM[a]	0.10 mM[a]
Lactate concentration	10.5 mM[a]	5.2 mM[a]
Oxygen concentration	8%[b]	1.5%[b]
Carbon dioxide concentration	12%[c]	10%[d]
pH	7.5[c]	7.1[e]
Glycine concentration	2.77 mM[f]	19.33 mM[g]
Alanine concentration	0.5 mM[f]	1.24 mM[g]
Serine concentration	0.32 mM[f]	0.80 mM[g]

[a]Gardner *et al.* (1996).
[b]Fischer and Bavister (1993).
[c]Maas *et al.* (1977).
[d]Garris (1984).
[e]Dale *et al.* (1998).
[f]Iritani *et al.* (1971).
[g]Miller and Schultz (1987).

METABOLISM OF THE EMBRYO

There have been numerous treatises on the metabolism of the mammalian embryo (Biggers and Stern, 1973; Brinster, 1973; Rieger, 1984, 1992; Biggers *et al.*, 1989; Leese, 1991, 1995; Barnett and Bavister, 1996; Gardner, 1998a,b, 1999; Gardner, *et al.*, 2000a). However, historically the majority of experiments on embryo metabolism have been performed in simple culture media (Whittingham, 1971; Brinster, 1973), which lack amino acids. It has subsequently been demonstrated that amino acids are both metabolized and regulate the utilization of carbohydrates (Gardner and Lane, 1993b). Therefore data from early studies using simple culture conditions need to be interpreted with this in mind. In the following section the current understanding of the utilization of carbohydrates and amino acids by the mammalian preimplantation embryo is reviewed.

GLUCOSE

There is a misconception that glucose is toxic to the mammalian embryo in culture. This stems from work on the hamster, which showed that in the presence of phosphate, 5.5 mM glucose induced developmental arrest in culture (Schini and Bavister, 1988; Seshagiri and Bavister, 1989). However, such developmental arrest can be attributed to phosphate alone (Seshagiri and Bavister, 1989; Biggers and McGinnis, 2001). Significantly, glucose is present in both oviduct and uterine fluids (Gardner *et al.*, 1996), and the oocyte and embryo have a specific carrier for this hexose (Gardner and Leese, 1988; Hogan *et al.*, 1991; Aghayan *et al.*, 1992; Dan-Goor *et al.*, 1997). Furthermore, glucose is used by the embryo throughout the preimplantation period for the synthesis of triacylglycerols and phospholipids and to provide precursors for complex sugars of mucopolysaccharides and glycoproteins. Glucose metabolized by the pentose phosphate pathway (PPP) generates ribose moieties required for nucleic acid synthesis and the NADPH required for the biosynthesis of lipids and other complex molecules (Hume and Weidemann, 1979; Reitzer *et al.*, 1980; Morgan and Faik, 1981). The production of nucleic acids is an important biosynthetic role of glucose. NADPH is also required for the reduction of intracellular glutathione, an important antioxidant for the embryo (Rieger, 1992).

In the rodent and human it appears that during invasive implantation there is little vasculature in the vicinity of the implantation site for several hours (Rogers *et al.*, 1982a,b). Therefore, during this period, glycolysis (anaerobic) will be the only available means of energy production for the blastocyst. It is conceivable that during this time the blastocyst uses glycogen, its endogenous glucose store. Certainly from the eight-cell stage to the blastocyst there is a loss of viability if the mouse embryo is cultured without glucose (Gardner and Lane, 1996). This suggests that the embryo is forced to use its endogenous glycogen store to generate free glucose for subsequent metabolism, thereby compromising the embryo at implantation. A similar result was subsequently reported for the hamster, when one-cell embryos were cultured to the blastocyst stage in the presence or absence of 0.5 mM glucose. Hamster embryos cultured for the entire preimplantation period in the absence of glucose had a significantly reduced viability compared to those embryos exposed to 0.5 mM glucose (Ludwig *et al.*, 2001). This experiment shows for the first time that rather than being detrimental to the hamster embryo, glucose is actually beneficial.

It has been shown that the concentration of glucose in the culture medium affects its rate of consumption by the embryo (Vella *et al.*, 1997). Therefore, increasing the concentration of glucose in the medium can result in increased glucose uptake and utilization. The genes for the glucose transporter are transcribed in the human oocyte and embryo (Hogan *et al.*, 1991; Aghayan *et al.*, 1992; Dan-Goor *et al.*, 1997), and kinetic studies have indicated the presence of the glucose transporter in the mouse embryo throughout development (Gardner and Leese, 1988).

The maximal activities of several key enzymes required for glucose metabolism have been determined in the mouse and human embryo (Brinster, 1968, 1971; Chi *et al.*, 1988; Martin *et al.*, 1993; Gardner *et al.*, 2000a). In all cases the activities of the three rate-limiting enzymes of glucose metabolism (hexokinase, phosphofructokinase, and pyruvate kinase) have been determined to be higher than the amount of glucose utilized during the preimplantation period. In all probability glucose utilization during the preimplantation period is not regulated by transport across the plasma membrane or the absence of sufficient enzyme activity. Rather substrate availability (concentration) and the specific regulation of enzyme activity appear to control glucose utilization by the preimplantation embryo (Biggers *et al.*, 1989). In a series of experiments, Barbehenn *et al.* (1974, 1978), attempted to locate the control points in glycolysis in individual mouse embryos by first starving the embryo for 60 minutes and then refeeding the embryo with either glucose alone or glucose with pyruvate. The levels of metabolic intermediates within an embryo were then quantitated using the technique of enzyme cycling (described by Gardner and Leese, 1999). Using this approach it was determined that at least two enzymes between the two-cell and morula stages were potentially rate limiting in the glycolytic pathway—hexokinase and 6-phophofructokinase. Both of these enzymes appear to be present at sufficient levels for glucose metabolism, thus intracellular control of such enzymes should be considered (Gardner, 1998a; Gardner *et al.*, 2000a).

PYRUVATE AND LACTATE

Pyruvate readily enters the embryo by means of a facilitated carrier (Gardner and Leese, 1988; Butcher *et al.*, 1998) and is the preferred nutrient of the cleavage-stage embryo of several species (Leese and Barton, 1984; Hardy *et al.*, 1989; Gardner *et al.*, 1993; Thompson *et al.*, 1996). Although lactate is readily taken up, and can be metabolized to some degree, it cannot support the first cleavage division in the mouse (Biggers *et al.*, 1967; Cross and Brinster, 1973). Inside the embryo pyruvate and lactate are interconverted by the enzyme lactate dehydrogenase (LDH) through the following reaction:

$$\text{Pyruvate} + \text{NADH} + \text{H}^+ \rightleftharpoons \text{Lactate} + \text{NAD}^+$$

A primary function of pyruvate conversion to lactate in cells is to regenerate NAD^+ for subsequent use in glycolysis when under anaerobic conditions, and therefore this conversion is of greatest significance at the blastocyst stage. This process is required because the cytoplasmic and mitochondrial pools of NADH are not shared. In order to transfer reducing power between these two distinct cellular compartments, a specific shuttle is required. In mammalian cells this is the malate:aspartate shuttle. Studies have revealed that this shuttle has little or no activity at the one-cell stage in the mouse embryo, but that there is significant activity from the two-cell stage onward. Furthermore, a reduction of activity of this shuttle at the blastocyst stage is associated with aberrant levels of lactate production by blastocysts developed *in vitro* (Gardner *et al.*, 2000a), and therefore the malate:aspartate shuttle is involved in the regulation of embryo metabolism.

Lane and Gardner (2000a) demonstrated that the mouse zygote and blastocyst differ in their ability to metabolize pyruvate and lactate, and that such differences could be accounted for only by a change in the intracellular NAD^+:NADH ratio, which in turn can be affected by the ratio of pyruvate:lactate. This example shows that by changing the ratio of certain medium components one can inadvertently change the ratio of intracellular regulators. Gardner and Sakkas (1993) have previously shown that changing the concentration of lactate in the culture medium can have a significant effect on mouse embryo viability and that this effect is stage specific. Such studies bring into question the potential pitfalls of using one of these carboxylic acids in the absence of the other.

As well as being used as an energy source throughout development, pyruvate is also a powerful antioxidant, being able to reduce intracellular levels of hydrogen peroxide in the embryo (Kouridakis and Gardner, 1995; O'Fallon and Wright, 1995). Its presence in embryo culture medium therefore confers a significant degree of protection as well as serving as a vital energy source. Furthermore, being weak acids, both pyruvate and lactate can reduce the pH_i of the embryo when they are present in culture media at high concentrations, i.e., >1 mM (Gibb *et al.*, 1997; Edwards *et al.*, 1998b). This is particularly pertinent for lactate, which can be present in culture media at over 20 mM. Lactate routinely comes in the form of D and L isomers, both of which can decrease pH_i (Edwards *et al.*, 1998b). It is therefore important to use only the biologically active form, the L isomer, in order to reduce effects on pH_i.

AMINO ACIDS

The majority of mammalian embryos can develop in culture in the absence of amino acids and give rise to offspring. This stems from work initially performed during the 1960s on mouse embryos derived from F_1 hybrid strains (Whitten and Biggers, 1968). Embryos from such strains are relatively insensitive to culture conditions. As a result of this the significance of amino acids during preimplantation embryo development was not considered for many years. Up to the early 1990s amino acids were conspicuously missing from the media designed to support mammalian preimplantation embryos in culture; this was in spite of the fact that amino acids are abundant in the fluids of the female reproductive tract (Casslen, 1987; Miller and Schultz, 1987; Moses *et al.*, 1997). Specific amino acids such as glycine and taurine are present in millimolar amounts. Furthermore, the oocyte and embryo maintain an endogenous pool of amino acids (Schultz *et al.*, 1981) and possess specific transport systems to take up amino acids from their surroundings (Van Winkle, 1988). The first indication that amino acids had a role in embryonic development came from Gwatkin (1966), who showed that amino acids were required for the attachment and outgrowth of mouse blastocysts. Gwatkin and Haidri (1973) went on to show that glutamine, isoleucine, methionine, and phenylalanine promoted nuclear maturation of the hamster oocyte. Subsequently Juetten and Bavister (1983) initiated research on the effects of this group of amino acids on hamster embryo development. Further studies on the rat (Zhang and Armstrong, 1990), the mouse (Mehta and Kiessling, 1990; Gardner and Sakkas, 1993; Gardner and Lane, 1993a; Lane and Gardner, 1994), and sheep (Gardner *et al.*, 1994) determined that amino acids were not only beneficial during the culture of various stages of development, but also significantly increased the resultant viability of embryos.

Subsequent studies have revealed a biphasic requirement for amino acids during the preimplantation period (Lane and Gardner, 1997a; Steeves and Gardner, 1999). The zygote and cleavage-stage embryo benefit from the inclusion of Eagle's nonessential amino acids and glutamine. Significantly, this group of amino acids bears a striking homology to those present at high levels in the female reproductive tract. Although the terms "nonessential" and "essential" amino acids as defined by Eagle (1959) have nothing to do with the requirements of the mammalian embryo, they serve as convenient groups in which to place amino acids. Eagle's nonessential amino acids could perhaps be best classified as facilitators of blastomere function.

By contrast, after the eight-cell stage the mammalian embryo benefits from the presence of a more complex array of amino acids (Steeves and Gardner, 1999), with Eagle's essential amino acids being found to stimulate the development of the inner cell mass (ICM) (Lane and Gardner, 1997a; Lane *et al.*, 2001). Significantly, equivalent rates of implantation to *in vivo*-developed blastocysts were obtained when mouse zygotes were cultured with nonessential amino acids up to the eight-cell stage followed by culture to the blastocyst stage in the presence of 20 amino acids (Lane and Gardner, 1997a).

DECREASING INTRACELLULAR STRESS

The significance of embryo plasticity is that it enables the mammalian embryo to develop *in vitro* in a wide variety of culture conditions, which has been fortuitous for those working in embryology. The variety of culture media used in mammalian embryology is quite staggering (Gardner and Lane, 1993c, 1999; Pool *et al.*, 1998). However, adaptation by an embryo to less than optimal culture conditions is not without cost, the cost paid by the embryo being a reduction in viability. One gross manifestation of embryo adaptation that follows from altered patterns of nutrient consumption is reduced energy production (Menke and McLaren, 1970; Gardner and Leese, 1990). Even a transient exposure (6 hours) of an *in vivo*-developed mouse blastocyst to inappropriate conditions leads to a reduction in viability (Lane and Gardner, 1998). It is evident that those culture conditions that do not place a stress on the developing embryo, and do not cause it to adapt to less than optimum conditions, are required for the development of viable embryos. For culture media to sustain the viability of an embryo it is essential that metabolic, homeostatic, and osmotic stress are minimized (Gardner, 1998a; Gardner *et al.*, 2000a; Lane, 2001; Lane and Gardner, 2000b, 2001a). This can be achieved by the use of carbohydrate gradients and an increasingly complex array of amino acids as development proceeds, reflecting the changes in physiology (Tables 1–3).

EVOLUTION OF SEQUENTIAL EMBRYO CULTURE MEDIA

Although different culture conditions have been used in sequence to support mammalian embryos (Bavister, 1999), we have applied the concept of sequential media that are designed specifically with the changing needs of the embryo in mind (Gardner and Leese, 1990). The design of such media focused on the dynamics of embryo physiology and metabolism, and the reduction of intracellular stress. Furthermore, it took into account data obtained on the environment within the female reproductive tract. Two media, G1 and G2, were therefore formulated for the pre- and postcompacted embryo (Gardner, 1994; Barnes *et al.*, 1995). These media have subsequently been modified (Gardner *et al.*, 1998) (Series II) and their formulations are shown in Table 4.

In spite of species differences in embryo physiology and metabolism, because the sequential media G1 and G2 were designed to minimize intracellular stress, thereby facilitating normal cellular function, these media have been able to support the development of viable blastocysts from a wide variety of mammalian species, including humans, mice, and cows (Tables 5–7) (Figs. 5 and 6).

Significantly, not all culture conditions support equivalent levels of blastocyst viability, even though the percentage of pronucleate embryos reaching this stage may be similar. This apparent paradox is exemplified in Fig. 7, in which it is evident that the formation of a blastocyst *in vitro* does not equate to the formation of a viable blastocyst. Unfortunately the vast majority of research on culture media formulations did not culminate in embryo transfer experiments, leaving open the question of whether the resultant embryos were viable. From the data presented it is evident that those conditions that support optimal blastocyst differentiation in culture actually compromise the development of the zygote (see Figs. 1, 2, and 7). Furthermore, those conditions that are favorable for the zygote, such as nonessential amino acids and EDTA, do not support the development of viable blastocysts (Figs. 7 and 8).

Experiments have been performed to compare the efficacy of sequential media with a leading single medium formulation, KSOMaa (Biggers *et al.*, 2000), on mouse

Table 4 Composition of Sequential Culture Media G1 and G2 (Version II)

Medium	Component	Concentration (mM)
G1 (cleavage stage development)	Sodium chloride	90.08
	Potassium chloride	5.5
	Sodium phosphate	0.25
	Magnesium sulfate	1.0
	Sodium bicarbonate	25.0
	Calcium chloride	1.8
	Glucose	0.5
	Lactate	10.5
	Pyruvate	0.32
	Alanyl-glutamine	0.5
	Alanine	0.1
	Aspartate	0.1
	Asparagine	0.1
	Glutamate	0.1
	Glycine	0.1
	Proline	0.1
	Serine	0.1
	Taurine	0.1
	EDTA	0.01
G2 (blastocyst development)	Sodium chloride	90.08
	Potassium chloride	5.5
	Sodium phosphate	0.25
	Magnesium sulfate	1.0
	Sodium bicarbonate	25.0
	Calcium chloride	1.8
	Glucose	3.15
	Lactate	5.87
	Pyruvate	0.10
	Alanyl-glutamine	1.0
	Alanine	0.1
	Aspartate	0.1
	Asparagine	0.1
	Glutamate	0.1
	Glycine	0.1
	Proline	0.1
	Serine	0.1
	Arginine	0.6
	Cystine	0.1
	Histidine	0.2
	Isoleucine	0.4
	Leucine	0.4
	Lysine	0.4
	Methionine	0.1
	Phenylalanine	0.2
	Threonine	0.4
	Tryptophan	0.5
	Tyrosine	0.2
	Valine	0.4
	Choline chloride	0.0072
	Folic acid	0.0023
	Inositol	0.01
	Nicotinamide	0.0082
	Pantothenate	0.0042
	Pyridoxal	0.0049
	Riboflavin	0.00027
	Thiamine	0.00296

Table 5 Efficacy of the Sequential Media G1 and G2 in Supporting Human Blastocyst Development and Implantation Using Donated Oocytes[a]

Parameter	Measurement
Number of patients	211
Number of pronulceate embryos	15.2 ± 0.4 (mean ± SEM)
Blastocyst development on day 5	51.7%
Blastocyst development on day 6	8.3%
Total blastocyst development	60.0%
Implantation rate (fetal sac)	62.1%
Implantation rate (fetal heart)	60.8%
Clinical pregnancy rate	79.6%

[a]Data from Gardner *et al.* (2002). Significantly, the implantation rate of human blastocysts developed in sequential media is equivalent to that reported by Buster *et al.* (1985), who were able to obtain human blastocysts developed *in vivo* and flushed from the uteri of donors. When transferred to recipient patients an implantation rate of 60% was attained. Such data imply that present culture conditions are reaching optimum levels.

embryos obtained from CF1 females mated to CF1 males (Table 6, Fig. 9). This is in contrast to previously published studies on KSOMaa in which CF1 females were mated with hybrid males (B6D2F1/CrlBR) (Biggers *et al.*, 2000). This is an important issue because the genotypes of both the female and the male have a significant impact on embryo development (Shire and Whitten, 1980a,b). When CF1 females are mated with hybrid males, subsequent embryo development is superior to that obtained when the CF1 females are mated with males of the same strain (Table 6) (Lynette Scott, personal communication). To ensure that the two media were treated in a similar way, embryos in medium KSOMaa were transferred to fresh medium after 48 hours of culture at the time when embryos cultured in medium G1 were moved to medium G2. This ensured that there was no extra buildup of ammonium in the KSOMaa medium drops. It can be seen from Fig. 9 and Table 6 that all parameters measured were significantly lower for mouse embryos cultured in KSOMaa compared to those in sequential media. Furthermore, there were clear differences in the morphology of the resultant blastocysts (Fig. 10).

Table 6 Efficacy of Sequential Media G1 and G2 vs. KSOMaa in Supporting CF1 × CF1 Mouse Zygotes in Culture and Their Subsequent Viability[a]

Culture system	Number of 2PN	≥ Eight cells on day 3 (%)	Compaction at 72 hours post-hCG (%)	Blastocyst at 117 hours post-hCG (%)	Hatching at 117 hours post-hCG (% of total)	Implantation (%)	Fetal development (%)
Sequential media	226	78.2	29.9	67.7	38.9	75	55.6
KSOMaa	220	42.1[b]**	7.3[b]**	45.5[b]**,c	16.8[b]	50*	36.1

[a]All media were supplemented with 5 mg/ml HSA. Number of embryos transferred, 36 per treatment; implantation and fetal development were determined on day 15 of pregnancy, day 1 being the day of copulation plug.

[b]Significantly different from sequential media: *, $P < 0.05$; **, $P < 0.01$.

[c]Blastocyst development in KSOMaa (45.5%) is lower than that reported by Biggers *et al.* (2000), who obtained 82% blastocyst development when CF1 females were mated to hybrid males. In our laboratory, when CF1 females are mated to hybrid males, embryo development is increased by around 30% over embryos derived from a CF1 × CF1 mating. Using this "correction factor" the data obtained in KSOMaa, as given here, are very similar to previously reported data (Biggers *et al.*, 2000).

Table 7 Efficacy of the Sequential Media G1 and G2 in Supporting Bovine Embryo Development and Viability

Culture system[a]	Number of oocytes	Blastocyst/ oocyte (%)	Blastocyst/ cleaved (%)	Total cell number[b]	ICM number	Trophectoderm number	ICM (%)
Coculture with B2	1388	22.3	33.8	169 ± 9a	63 ± 5a	106 ± 6a	37
G1/G2	1557	16.8	26.1	207 ± 13b	81 ± 7b	126 ± 9b	39
KSOMaa	1497	16.1	25.6	169 ± 10a	61 ± 7a	98 ± 6a	37

[a]The medium for all treatments was renewed after 72 hours of culture.

[b]Cell numbers were obtained from expanded blastocysts. Different letters (a or b) following the cell numbers within a column represent a significant difference ($P < 0.05$). Embryo viability following culture in the sequential media G1/G2 was determined by the transfer of 234 blastocysts to recipients; 51.3% (120/234) of the recipients became pregnant. When grade 1 blastocysts were transferred the pregnancy rate was 56.4% (88/156). These data compare favorably with data obtained from blastocysts developed *in vivo*, flushed from the uterus, and transferred to recipient females (67% pregnancy for all grades of blastocysts and 76% for grade 1) (Hasler, 1998). When blastocysts were cultured in G1/G2 and subsequently frozen (in $1.5\,M$ ethylene glycol), thawed, and transferred to Heifer recipients, the pregnancy rate was 47.5% (19/40). Viability data are courtesy of Dr. John Hasler.

A further experiment was performed to compare the rates of mouse embryo development *in vivo* and in sequential media. Zygotes from naturally ovulated CF1 females mated to CF1 males were collected from half the females and placed into culture, while embryos from the remaining half were collected on the morning of implantation (day 4 post mating). After 72 hours of culture, embryos from both groups were photographed and cell number and allocation were determined. In Fig. 11 the morphology of blastocysts developed *in vivo* and *in vitro* can be compared and are seen to be similar. The cell number of expanded blastocysts developed *in vivo* was 49.8 ± 1.4, compared to 50.8 ± 2.0 for *in vitro*-developed embryos. In these two groups of blastocysts the percentage of blastomeres in the ICM was equivalent (29.0 ± 1.2 for *in vivo*, 35.4 ± 1.6 for *in vitro*). Such data are most

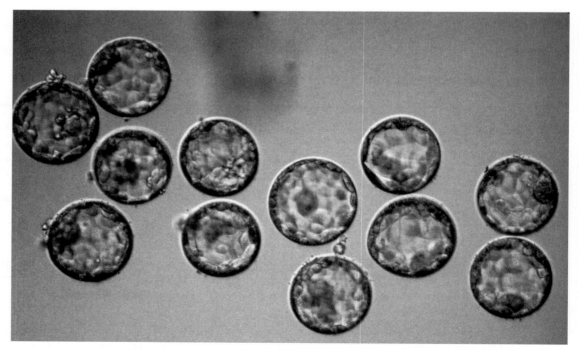

Figure 5 *Photomicrograph of human blastocysts cultured in sequential media G1 and G2. Pronucleate embryos were cultured for 48 hours in medium G1 followed by culture for 48 hours in medium G2. Both media were used at 6% CO_2, 5% O_2, and 89% N_2.*

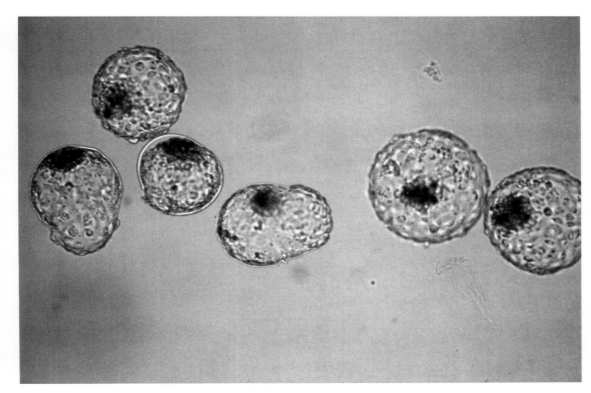

Figure 6 Photomicrograph of bovine blastocysts cultured in sequential media G1 and G2. Putative zygotes were cultured for 72 hours in medium G1 followed by culture for 72 hours in medium G2. Both media were used at 6% CO_2, 5% O_2, and 89% N_2.

encouraging as they indicate that CF1 mouse embryos cultured in sequential media develop at a rate similar to that of embryos *in vivo*.

FACTORS OTHER THAN MEDIUM FORMULATION THAT IMPACT EMBRYO DEVELOPMENT AND VIABILITY

MACROMOLECULES

Historically, embryo culture media have been supplemented with protein in the form of either serum albumin or serum. Under stringent culture conditions, including the presence of amino acids, embryos can be cultured to the blastocyst stage in the absence of protein (Bavister, 1995; Gardner *et al.*, 1999). However, the inclusion of protein does facilitate gamete and embryo manipulation *in vitro* by acting as a surfactant, while also conferring benefit to the embryo by the chelation of potential toxins (Flood and Shirley, 1991). Significantly, albumin is the most abundant protein in the female reproductive tract (Leese, 1988) and has been shown to maintain embryo physiology and metabolism *in vitro* compared to embryos cultured in the presence of a synthetic macromolecule, polyvinyl alcohol (PVA) (Eckert *et al.*, 1998; Thompson *et al.*, 1998). Unfortunately a problem with serum albumin, serum, or any biological product is the risk of disease transmission and contamination. This alone is reason to consider alternatives. Furthermore, there is considerable variation in the composition of serum albumin from batch to batch (Batt *et al.*, 1991; Gray *et al.*, 1992; McKiernan and Bavister, 1992), making standardization of procedures difficult. Should albumin be used in an embryo culture system, it is important to ensure that it is fatty acid free.

As well as the reasons listed above, the inclusion of serum in embryo culture systems, whether it be in media or for use in coculture, can no longer be considered

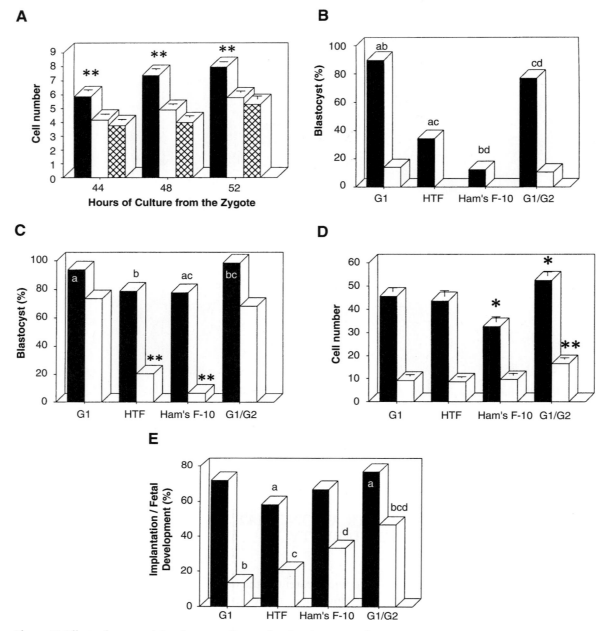

Figure 7 *Effect of sequential culture media on the development of F₁ (C57BL/6 × CBA/Ca) mouse zygotes in vitro. Zygotes were collected at 20 hours post-hCG. All media were supplemented with BSA (2 mg/ml). All embryos were transferred to fresh medium after 48 hours of culture, with the exception of embryos in medium G1; these embryos were transferred to either medium G1 or G2. To compensate for this, twice the number of embryos were originally cultured in medium G1, although only a designated 50% of these embryos were used in the statistical analysis of the data set covering 44 to 52 hours. (A) Embryo cell number after 44, 48, and 52 hours of culture. Values are mean ± SEM; n = 200 embryos/medium. Media: G1 (solid bar); HTF (open bar); Ham's F-10 (hatched bar). **, Significantly different from other media (P < 0.01). (B) Embryo development after 72 hours of culture; n = 150 embryos/medium; G1/G2, embryos cultured for 48 hours in medium G1 and then transferred to medium G2; blastocyst, solid bar; hatching blastocysts (as a percentage of total blastocysts), open bar. Letters a, c, and d, like pairs are significantly different (P < 0.05); b (P < 0.01). (C) Embryo development after 92 hours of culture; n = 150 embryos/medium; G1/G2, embryos cultured for 48 hours in medium G1 and then transferred to medium G2; blastocyst, solid bar; hatching blastocysts (as a percentage of total blastocysts), open bar. Letters a, c, and d, like pairs*

acceptable. It is important to emphasize that mammalian embryos are not exposed to serum *in vivo*. Oviduct and uterine fluids are not simple serum transudates. Rather, serum is a pathological fluid and by default contains abundant growth factors, such as platelet-derived growth factor and transforming growth factor-α, released during platelet aggregation, as well as a host of other growth factors. In spite of this, the addition of serum to culture medium does add a certain degree of protection to the embryo via the minimization of transient pH shifts and chelation of potential toxins. It is owing to this ability to confer a degree of robustness to the culture medium that its use has persisted. However, data on the development of sheep and cattle blastocysts in the presence of serum have raised serious issues regarding the use of serum for embryo culture (Gardner, 1994; Gardner and Lane, 1999). Serum can adversely affect the development of embryos at several levels:

1. Precocious blastocoel formation (Thompson *et al.*, 1995; Walker *et al.*, 1992).
2. Sequestration of lipid (Dorland *et al.*, 1994; Thompson *et al.*, 1995).
3. Abnormal mitochondrial ultrastructure (Dorland *et al.*, 1994; Thompson *et al.*, 1995).
4. Perturbations in metabolism (Gardner *et al.*, 1994).
5. Association with abnormally large offspring in sheep (Thompson *et al.*, 1995).

Therefore, in attempts to define embryo culture media, Bavister advocated the use of PVA (Bavister, 1981) to replace serum or serum albumin. This approach has worked for the *in vitro* development of embryos from several mammalian species. However, the use of such synthetic macromolecules cannot be said to be physiological, and as described above PVA is not able to maintain the physiology and metabolism of the embryo. Furthermore, bovine embryos cultured in the presence of PVA did not survive cryopreservation as well as those cultured in the presence of albumin (Eckert *et al.*, 1998).

Recombinant human albumin has become available and has been shown to be as effective as blood-derived albumin in supporting embryo development (Gardner and Lane, 2000; Hooper *et al.*, 2000). Significantly, embryos cultured in the presence of recombinant albumin exhibit an increased tolerance to cryopreservation

*are significantly different (P < 0.05); **, significantly different from medium G1 and G1/G2 (P < 0.01). (D) Cell allocation in the blastocyst after 92 hours of culture; n = 150 embryos/medium; G1/G2, embryos cultured for 48 hours in medium G1 and then transferred to medium G2; trophectoderm, solid bars; inner cell mass, open bars; *, significantly different from other media (P < 0.05); **, (P < 0.01). (E) Viability of cultured blastocysts; n = at least 60 blastocysts transferred per treatment; G1/G2, embryos cultured for 48 hours in medium G1 and then transferred to medium G2; implantation, solid bar; fetal development per implantation, open bar. Letters a and d, like pairs are significantly different (P < 0.05); b and c (P < 0.01). When mouse embryos were cultured in medium G1 for the entire preimplantation period up to the blastocyst stage, although the embryos formed healthy-looking blastocysts, most implantations were lost, i.e., they did not have a sufficient inner cell mass to form a viable fetus. The lack of adequate inner cell mass development stems from both the lack of sufficient glucose and the presence of EDTA (both affect glycolysis) and the omission of essential amino acids. In contrast, those mouse embryos that were switched to medium G2 after 48 hours of culture formed blastocysts at the same rate and with morphologies equivalent to those in medium G1 for the entire culture period. However, for those blastocysts developed in medium G1 and switched to medium G2, very few implantations were lost due to the development of a large inner cell mass, thereby maintaining a very high pregnancy rate. [From Gardner, D. K., and Lane, M. (1998). Culture of viable human blastocysts in defined sequential serum-free media. Hum. Reprod. **13(Suppl. 3)**, 148–159. © European Society of Human Reproduction and Embryology. Reproduced by permission of Oxford University Press/Human Reproduction.]*

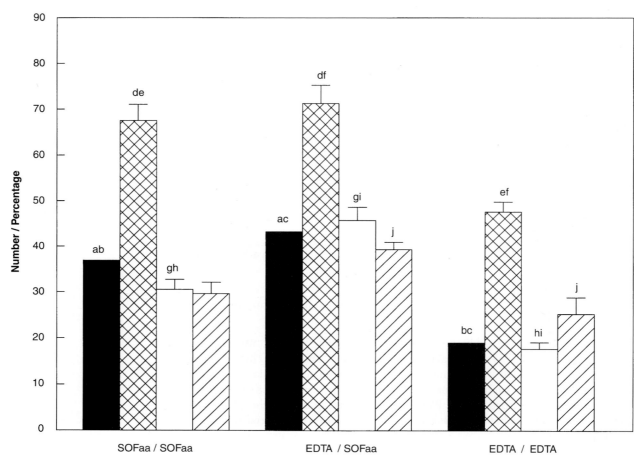

Figure 8 *Effect of EDTA on bovine blastocyst development and differentiation. Percent blastocyst development after 144 hours of culture, solid bars; trophectoderm cell number, hatched bars; inner cell mass cell number, open bars; percent inner cell mass cells of total cells, striped bars. Like pairs of letters are significantly different (a, d, h, and j, P < 0.05; b, c, e, f, g, and i, P < 0.01). For SOFaa/SOFaa, embryos were cultured for 72 hours in medium SOF supplemented with amino acids but lacking EDTA (SOFaa) (Gardner et al., 1994) for 72 hours, and then transferred to fresh SOFaa for a further 72 hours. For EDTA/SOFaa, embryos were cultured for 72 hours in SOFaa and 100 μM EDTA, and then transferred to fresh SOFaa (no EDTA) for a further 72 hours. For EDTA/EDTA, embryos were cultured for 72 hours in SOFaa and 100 μM EDTA, and then transferred to fresh SOFaa and 100 μM EDTA for a further 72 hours. Data from Gardner et al. (2000b). EDTA has been shown to stimulate the cleavage-stage embryo of both the mouse (Abramczuk et al., 1977; Mehta and Kiessling, 1990; Gardner and Lane, 1996) and the cow (Gardner et al., 2000b). It has been determined that EDTA is an inhibitor of glycolytic kinases and helps prevent aberrant levels of glycolysis (Lane and Gardner, 2001b). However, the inclusion of EDTA in the medium for embryos from the eight-cell stage onward (medium G2) is not advisable, because the embryo becomes increasingly glycolytic in nature, especially the inner cell mass (Hewitson and Leese, 1993). Therefore, continued exposure to EDTA negatively impacts blastocyst development and impairs inner cell mass formation (Gardner et al., 2000b).*

(Gardner *et al.*, 2001). Another macromolecule present in the female reproductive tract is hyaluronan, which in the mouse increases at the time of implantation (Zorn *et al.*, 1995). Hyaluronan is a high-molecular-mass polysaccharide and can be obtained endotoxin and prion free from a yeast fermentation procedure. It has been demonstrated that not only can hyaluronan replace albumin in a mouse and bovine embryo culture system, but that its use for embryo transfer results in a significant increase in embryo implantation (Gardner *et al.*, 1999). Furthermore, similar to

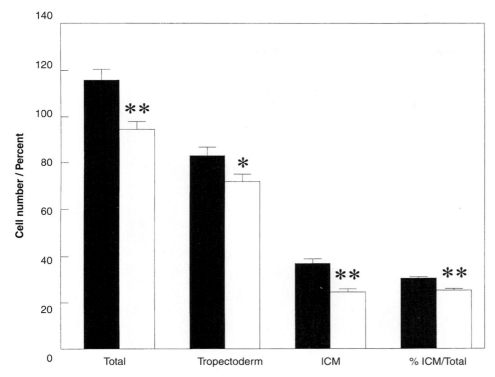

Figure 9 *Blastocyst cell number and differentiation of mouse (CF1 × CF1) embryos cultured in sequential media (solid bars) or KSOMaa (open bars). Significantly different from G1/G2: *, P < 0.05; **, P < 0.01.*

results with recombinant albumin, the presence of hyaluronan in the culture medium increases the cryosurvivability of blastocysts (Gardner *et al.*, 2001; Stojkovic *et al.*, 2001). It has been found that the recombinant albumin and hyaluronan confer a synergistic benefit to the embryo (Gardner *et al.*, 1999; Hooper *et al.*, 2000).

AMMONIUM

It has been demonstrated that upon incubation at 37°C, amino acids spontaneously deaminate to release ammonium into the culture medium. Furthermore, the embryos actively deaminate amino acids when they are metabolized, leading to a further buildup of ammonium. The significance of this is that it has been shown that ammonium not only retards embryo development in culture (Gardner and Lane, 1993a), but can also induce fetal retardation and neural tube defects in mice (Lane and Gardner, 1994). Interestingly, there appears to be a link between the concentration of ammonium in serum and the induction of fetal oversize in sheep (McEvoy *et al.*, 1997; Sinclair *et al.*, 1998). Therefore it is imperative to renew the culture medium used at least every 48 to 72 hours in order to circumvent the toxicity of ammonium. The main culprit with regard to deamination and ammonium release is glutamine. However, this amino acid can be replaced with the dipeptide alanylglutamine, which is stable at 37°C, and its inclusion significantly reduces ammonium release into the culture medium (Lane *et al.*, 2001).

INCUBATION VOLUMES

It is essential to consider the culture system as a whole and not simply focus on the culture media, because all aspects of the system (gas phase, embryo incubation volume and group size, macromolecule supplementation, etc.) interact (Gardner and Lane, 1999, 2001). The culture of mammalian embryos in reduced volumes of

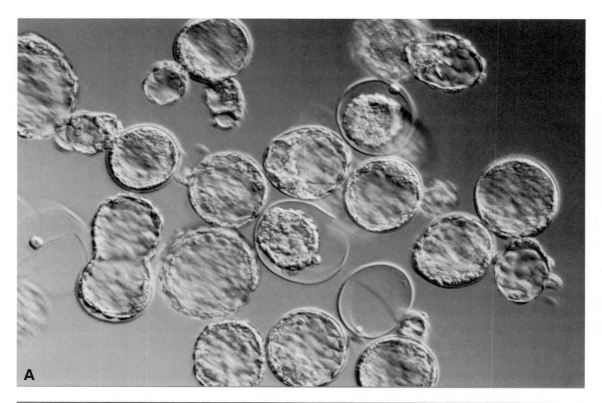

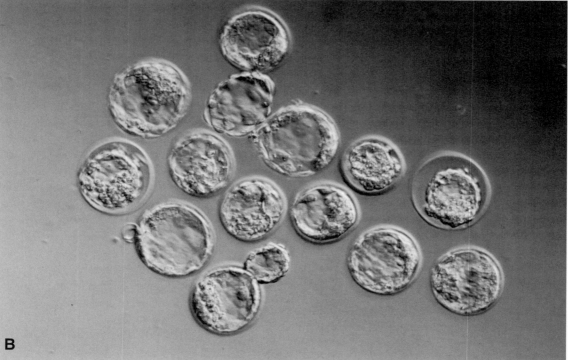

Figure 10 Photomicrographs of CF1 × CF1 mouse embryos cultured from the pronucleate stage for 96 hours in (A) sequential media G1 and G2 or (B) KSOMaa. Note the increased expansion and hatching of blastocysts cultured in sequential media.

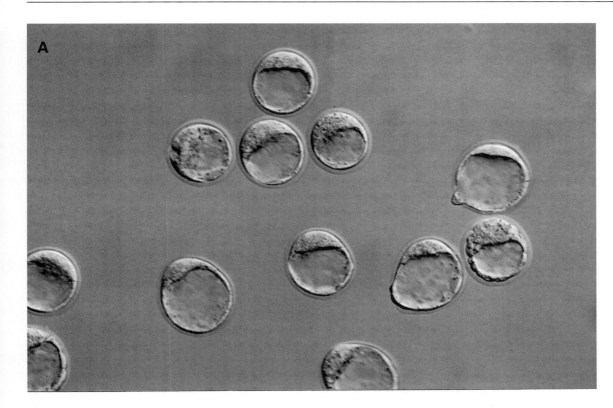

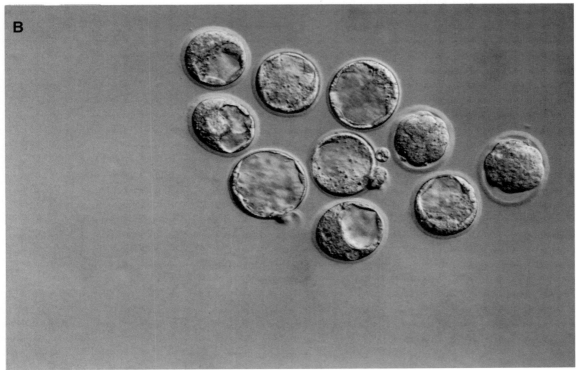

Figure 11 Photomicrographs of CF1 × CF1 blastocysts. Both groups of embryos came from mice mated at the same time. (A) Embryos developed in vivo. (B) Embryos developed in media G1 and G2. The photographs were taken after 72 h of culture (based on in vitro group time), i.e., the morning of day 4 post mating. Note that the in vivo developed blastocysts are fully expanded and are beginning to hatch from the zona pellucida.

medium and/or in groups significantly increases blastocyst development (Wiley *et al.*, 1986; Paria and Dey, 1990; Lane and Gardner, 1992; Salahuddin *et al.*, 1995) and blastocyst cell number (Lane and Gardner, 1992). Furthermore, culturing embryos in reduced volumes increases subsequent viability after transfer (Lane and Gardner, 1992). It has been proposed that the benefit of growing embryos in small volumes and/or in groups is due to the production of specific embryo-derived autocrine/paracrine factors that stimulate development. The culture of embryos in large volumes will result in a dilution of such a factor so that it becomes ineffectual (Gardner, 1994). This phenomenon is not confined to the mouse, in which several embryos reside in the female tract at one time, but has also been reported for the sheep and cow, which, like humans, are monovular (Gardner *et al.*, 1994; Ahern and Gardner, 1998). It has been shown in both the mouse and the cow that increasing the embryo:incubation volume ratio specifically stimulates the development of the ICM. This explains the increased viability of embryos cultured in reduced volumes in groups (Fig. 12) (Ahern and Gardner, 1998).

GAS PHASE

The concentration of oxygen in the lumen of the rabbit oviduct is reported to be 2–6%, (Mastroianni and Jones, 1965; Ross and Graves, 1974) and 8% in the oviduct of hamster, rabbit, and rhesus monkey (Fischer and Bavister, 1993). Furthermore, the oxygen concentration in the uterus is significantly lower than in the oviduct, ranging from 5% in the hamster and rabbit to 1.5% in the rhesus monkey (Maas *et al.*, 1976; Fischer and Bavister, 1993). Studies on different mammalian species have demonstrated that culture in a reduced oxygen concentration, especially embryos of ruminants, results in enhanced embryo development *in vitro*. Several studies have shown that a reduced oxygen concentration (between 5 and 8%) enhances development to the blastocyst stage in mice (Quinn and Harlow, 1978; Umaoka *et al.*, 1992; Gardner and Lane, 1996), rabbits (Li and Foote, 1993), sheep (Thompson *et al.*, 1990), goats (Batt *et al.*, 1991), and cows (Thompson *et al.*, 1990).

The concentration of CO_2 employed in the culture system has a direct impact on medium pH. Although most media work over a wide range of pH (7.2–7.4), it is preferable to ensure that pH does not go over 7.4, considering that intracellular pH is actually 7.2 (Lane and Gardner, 2001a). Therefore it is advisable to use a CO_2 concentration of between 6 and 7% (Lawitts and Biggers, 1993; Gardner and Lane, 2001). We routinely culture embryos at 6% CO_2/5% O_2/89% N_2 in groups in microliter drops of medium under oil. For mice we culture embryos in groups of 10 in 20-μl drops and replace the medium after 48 hours. For ruminant embryos we culture embryos in groups of 4 in 50-μl drops and replace the medium after 72 hours. For human embryos we culture embryos in groups of 4 in 50-μl drops and replace the medium after 48 hours. The macromolecules of choice are serum albumin (preferably recombinant albumin) (Gardner and Lane, 2000; Hooper *et al.*, 2000) and hyaluronan (Gardner *et al.*, 1999).

INHERENT PROBLEMS OF COCULTURE

Coculture systems were initially developed in an attempt to make culture systems more physiological (Gandolfi and Moor, 1987). However, not only does coculture not accommodate the changing nutrient requirements of the developing embryo, but one is faced with trying to culture two totally different cell types (somatic and embryonic) in the same culture medium. This is not a feasible proposition, because it is not possible to supply the nutrient requirements of two different cell types. Therefore some compromises in cell viability are inevitable. Furthermore, not only do coculture systems typically rely on the use of tissue culture media rather than

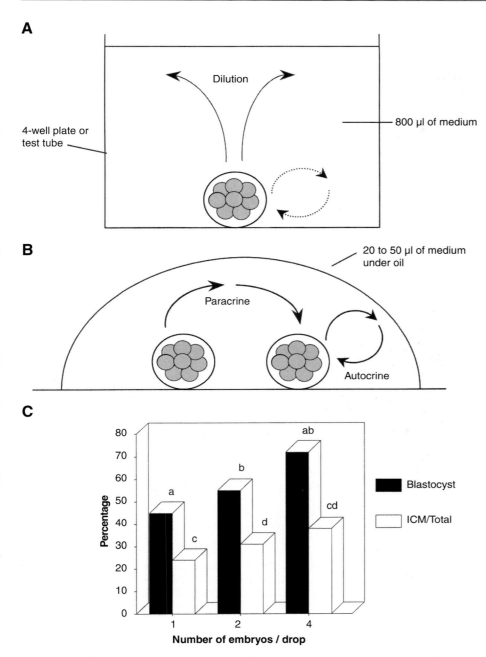

Figure 12 *Effect of incubation volume and embryo grouping on embryo development and differentiation [reproduced from Gardner (1998c) with permission of Martin Dunitz Press]. (A) A single embryo cultured in a four-well plate or test tube, any factor produced by the embryo will become ineffectual due to dilution. (B) Culture of embryos in reduced volumes and/or in groups increases the effective concentration of embryo-derived factors, facilitating their action in either a paracrine or an autocrine manner. (C) Effect of embryo grouping on bovine blastocyst development and differentiation. Bovine embryos were cultured either individually or in groups of two or four in 50-μl drops of medium (Ahern and Gardner, 1998). Like pairs of letters are significantly different (P < 0.05). (Reproduced from Gardner, 1998c.)*

embryo culture media, they usually employ serum as the macromolecule (Gardner, 1998b). The perils of including serum in an embryo culture system were discussed previously.

The growing body of evidence is showing that coculture is not required to develop viable blastocysts *in vitro* (Mullaart *et al.*, 2001), irrespective of the species being studied (Gardner, 1998b). Certainly embryo development in culture, and subsequent viability after transfer, are as good if not better following culture in sequential media as compared to coculture (Table 7).

CONCLUSIONS

In this chapter we have presented evidence that in order to optimize mammalian embryo development in culture, sequential media are required, each designed to meet the changing requirements of the developing embryo (Gardner and Leese, 1990). Conditions that are good for the early stage are not optimal for blastocyst differentiation. In contrast, those conditions that do support good blastocyst differentiation and maintain embryo viability are not optimal for the development of the zygote (Gardner and Lane, 1998; Lane and Gardner, 1997a).

Supplying the embryo with gradients of carbohydrates and amino acids not only provides the changing nutrients required but also reduces intracellular stress (Gardner, 1998a; Gardner *et al.*, 2000a; Lane, 2001; Lane and Gardner, 2000a). The ability to minimize intracellular stress is a significant factor in being able to maintain embryo viability in culture. This premise is supported by the fact that the sequential culture media G1 and G2 can support the development of viable blastocysts of the primate, rodent, and ruminant, in spite of species differences in embryo physiology.

Using sequential media, blastocyst developmental rates *in vitro* and subsequent implantation rates are very close to those observed *in vivo* (Table 5). Indeed, the ability of sequential media to support the development of viable inner cell masses was fundamental in the establishment of the first human embryonic stem cell lines (Thomson *et al.*, 1998).

REFERENCES

Abramczuk, J., Solter, D., and Koprowski, H. (1977). The beneficial effect EDTA on development of mouse one-cell embryos in chemically defined medium. *Dev. Biol.* **61**, 378–383.

Aghayan, M., Rao, L. V., Smith, R. M., Jarett, L., Charron, M. J., Thorens, B., and Heyner, S. (1992). Developmental expression and cellular localization of glucose transporter molecules during mouse preimplantation development. *Development* **115**, 305–312.

Ahern, T. J., and Gardner, D. K. (1998). Culturing bovine embryos in groups stimulates blastocyst development and cell allocation to the inner cell mass. *Theriogenology* **49**, 194.

Barbehenn, E. K., Wales, R. G., and Lowry, O. H. (1974). The explanation for the blockade of glycolysis in early mouse embryos. *Proc. Natl. Acad. Sci. U.S.A.* **71**, 1056–1060.

Barbehenn, E. K., Wales, R. G., and Lowry, O. H. (1978). Measurement of metabolites in single preimplantation embryos; a new means to study metabolic control in early embryos. *J. Embryol. Exp. Morphol.* **43**, 29–46.

Barnes, F. L., Crombie, A., Gardner, D. K., Kausche, A., Lacham-Kaplan, O., Suikkari, A. M., Tiglias, J., Wood, C., and Trounson, A. O. (1995). Blastocyst development and birth after *in-vitro* maturation of human primary oocytes, intracytoplasmic sperm injection and assisted hatching. *Hum. Reprod.* **10**, 3243–3247.

Barnett, D. K., and Bavister, B. D. (1996). What is the relationship between the metabolism of preimplantation embryos and their developmental competence? *Mol. Reprod. Dev.* **43**, 105–133.

Batt, P. A., Gardner, D. K., and Cameron, A. W. (1991). Oxygen concentration and protein source affect the development of preimplantation goat embryos *in vitro*. *Reprod. Fertil. Dev.* **3**, 601–607.

Bavister, B. D. (1981). Substitution of a synthetic polymer for protein in a mammalian gamete culture system. *J. Exp. Zool.* **217**, 45–51.

Bavister, B. D. (1995). Culture of preimplantation embryos: Facts and artifacts. *Hum. Reprod. Update* **1**, 91.

Bavister, B. D. (1999). Stage-specific culture media and reactions of embryos to them. *In* "Towards Reproductive Certainty. Fertility and Genetics beyond 1999" (R. Jansen and D. Mortimer, eds.), pp. 367–377. Parthenon, New York.

Benos, D., and Biggers, J. D. (1981). Blastocyst fluid formation. *In* "Fertilization and Embryonic Development *In vitro*" (L. J. Mastroianni and J. D. Biggers, eds.), pp. 283–297. Plenum Press, New York.

Biggers, J. D. (1987). Pioneering mammalian embryo culture. *In* "The Mammalian Preimplantation Embryo Regulation of Growth and Differentiation *in vitro*" (B. D. Bavister, ed.), pp. 1–22. Plenum Press, New York.

Biggers, J. D., and McGinnis, L. K. (2001). Evidence that glucose is not always an inhibitor of mouse preimplantation development *in vitro*. *Hum. Reprod.* **16**, 153–163.

Biggers, J. D., and Stern, S. (1973). Metabolism of the preimplantation mammalian embryo. *Adv. Reprod. Physiol.* **6**, 1–59.

Biggers, J. D., Moore, B. D., and Whittingham, D. G. (1965). Development of mouse embryos *in vivo* after cultivation from two-cell ova to blastocysts *in vitro*. *Nature* **206**, 734–735.

Biggers, J. D., Whittingham, D. G., and Donahue, R. P. (1967). The pattern of energy metabolism in the mouse oocyte and zygote. *Proc. Natl. Acad. Sci. U.S.A.* **58**, 560–567.

Biggers, J. D., Gardner, D. K., and Leese, H. J. (1989). Control of carbohydrate metabolism in preimplantation mammalian embryos. *In* "Growth Factors in Mammalian Development" (I. Y. Rosenblum and S. Heyner, eds.), pp. 19–32. CRC Press, Boca Raton.

Biggers, J. D., McGinnis, L. K., and Raffin, M. (2000). Amino acids and preimplantation development of the mouse in protein-free potassium simplex optimized medium. *Biol. Reprod.* **63**, 281–293.

Brinster, R. L. (1968). Hexokinase activity in the preimplantation mouse embryo. *Enzymologia* **34**, 304–308.

Brinster, R. L. (1971). Phosphofructokinase activity in the preimplantation mouse embryo. *Wilhelm Roux Arch. Entwicklungsmech. Org.* **166**, 300–302.

Brinster, R. L. (1973). Nutrition and metabolism of the ovum, zygote, and blastocyst. *In* "Handbook of Physiology," Vol. 2 (S. R. Geiger, ed.), pp. 165–184. Waverly Press, Maryland.

Buster, J. E., Bustillo, M., Rodi, I. A., Cohen, S. W., Hamilton, M., Simon, J. A., Thorneycroft, I. H., and Marshall, J. R. (1985). Biologic and morphologic development of donated human ova recovered by nonsurgical uterine lavage. *Am. J. Obstet. Gynecol.* **153**, 211–217.

Butcher, L., Coates, A., Martin, K. L., Rutherford, A. J., and Leese, H. J. (1998). Metabolism of pyruvate by the early human embryo. *Biol. Reprod.* **58**, 1054–1056.

Casslen, B. G. (1987). Free amino acids in human uterine fluid. Possible role of high taurine concentration. *J. Reprod. Med.* **32**, 181–184.

Chi, M. M., Manchester, J. K., Yang, V. C., Curato, A. D., Strickler, R. C., and Lowry, O. H. (1988). Contrast in levels of metabolic enzymes in human and mouse ova. *Biol. Reprod.* **39**, 295–307.

Crosby, I. M., Gandolfi, F., and Moor, R. M. (1988). Control of protein synthesis during early cleavage of sheep embryos. *J. Reprod. Fertil.* **82**, 769–775.

Cross, P. C., and Brinster, R. L. (1973). The sensitivity of one-cell mouse embryos to pyruvate and lactate. *Exp. Cell Res.* **77**, 57–62.

Dale, B., Menezo, Y., Cohen, J., DiMatteo, L., and Wilding, M. (1998). Intracellular pH regulation in the human oocyte. *Hum. Reprod.* **13**, 964–970.

Dan-Goor, M., Sasson, S., Davarashvili, A., and Almagor, M. (1997). Expression of glucose transporter and glucose uptake in human oocytes and preimplantation embryos. *Hum. Reprod.* **12**, 2508–2510.

Dorland, M., Gardner, D. K., and Trounson, A. (1994). Serum in synthetic oviduct fluid causes mitochondrial degeneration in ovine embryos. *J. Reprod. Fertil.* (Abstr. Ser.) **13**, 70.

Eagle, H. (1959). Amino acid metabolism in mammalian cell cultures. *Science* **130**, 432–437.

Eckert, J., Pugh, P. A., Thompson, J. G., Niemann, H., and Tervit, H. R. (1998). Exogenous protein affects developmental competence and metabolic activity of bovine pre-implantation embryos *in vitro*. *Reprod. Fertil. Dev.* **10**, 327–332.

Edwards, L. J., Williams, D. A., and Gardner, D. K. (1998a). Intracellular pH of the mouse preimplantation embryo: Amino acids act as buffers of intracellular pH. *Hum. Reprod.* **13**, 3441–3448.

Edwards, L. J., Williams, D. A., and Gardner, D. K. (1998b). Intracellular pH of the preimplantation mouse embryo: Effects of extracellular pH and weak acids. *Mol. Reprod. Dev.* **50**, 434–442.

Fischer, B., and Bavister, B. D. (1993). Oxygen tension in the oviduct and uterus of rhesus monkeys, hamsters and rabbits. *J. Reprod. Fertil.* **99**, 673–679.

Flood, L. P., and Shirley, B. (1991). Reduction of embryotoxicity by protein in embryo culture media. *Mol. Reprod. Dev.* **30**, 226–231.

Gandolfi, F., and Moor, R. M. (1987). Stimulation of early embryonic development in the sheep by coculture with oviduct epithelial cells. *J. Reprod. Fertil.* **81**, 23–28.

Gardner, D. K. (1994). Mammalian embryo culture in the absence of serum or somatic cell support. *Cell Biol. Int.* **18**, 1163–1179.

Gardner, D. K. (1998a). Changes in requirements and utilization of nutrients during mammalian preimplantation embryo development and their significance in embryo culture. *Theriogenology* **49**, 83–102.

Gardner, D. K. (1998b). Embryo development and culture techniques. *In* "Animal Breeding: Technology for the 21st Century" (J. R. Clark, ed.), pp. 13–46. Harwood Academic Publishers, London.

Gardner, D. K. (1998c). Improving embryo culture and enhancing pregnancy rate. *In* "Female Infertility Therapy: Current Practice" (Z. Shoham, C. Howles, and H. Jacobs, eds.), pp. 283–299. Martin Dunitz, London.

Gardner, D. K. (1999). Development of serum-free culture systems for the ruminant embryo and subsequent assessment of embryo viability. *J. Reprod. Fertil.* (Suppl.) **54**, 461–475.

Gardner, D. K., and Lane, M. (1993a). Amino acids and ammonium regulate mouse embryo development in culture. *Biol. Reprod.* **48**, 377–385.

Gardner, D. K., and Lane, M. (1993b). The 2-cell block in CF1 mouse embryos is associated with an increase in glycolysis and a decrease in tricarboxylic acid (TCA) cycle activity: Alleviation of the 2-cell block is assocated with the restoration of *in vivo* metabolic pathway activities. *Biol. Reprod.* **48**(Suppl. 1), 152.

Gardner, D. K., and Lane, M. (1993c). Embryo culture systems. *In* "Handbook of In Vitro Fertilization" (D. K. Gardner and A. O. Trounson, eds.), pp. 85–114. CRC Press, Boca Raton.

Gardner, D. K., and Lane, M. (1996). Alleviation of the "2-cell block" and development to the blastocyst of CF1 mouse embryos: Role of amino acids, EDTA and physical parameters. *Hum. Reprod.* **11**, 2703–2712.

Gardner, D. K., and Lane, M. (1998). Culture of viable human blastocysts in defined sequential serum-free media. *Hum. Reprod.* **13**(Suppl. 3), 148–159.

Gardner, D. K., and Lane, M. (1999). Embryo culture systems. *In* "Handbook of In Vitro Fertilization," 2nd Ed. (A. O. Trounson and D. K. Gardner, eds.), pp. 205–264. CRC Press, Boca Raton.

Gardner, D. K., and Lane, M. (2000). Recombinant human serum albumin and hyaluronan can replace blood-derived albumin in embryo culture media. *Fertil. Steril.* **74**(Suppl. 3), 0–086.

Gardner, D. K., and Lane, M. (2001). Embryo culture. *In* "Textbook of Assisted Reproductive Techniques" (D. K., Gardner, A., Weissman, C. Howles, *et al.* eds.), pp. 203–222. Martin Dunitz, London.

Gardner, D. K., and Leese, H. J. (1988). The role of glucose and pyruvate transport in regulating nutrient utilization by preimplantation mouse embryos. *Development* **104**, 423–429.

Gardner, D. K., and Leese, H. J. (1990). Concentrations of nutrients in mouse oviduct fluid and their effects on embryo development and metabolism *in vitro*. *J. Reprod. Fertil.* **88**, 361–368.

Gardner, D. K., and Leese, H. J. (1999). Assessment of embryo metabolism and viability. *In* "Handbook of *In Vitro* Fertilization," 2nd Ed. (A. O. Trounson and D. K. Gardner, eds.), pp. 347–372. CRC Press, Boca Raton.

Gardner, D. K., and Sakkas, D. (1993). Mouse embryo cleavage, metabolism and viability: Role of medium composition. *Hum. Reprod.* **8**, 288–295.

Gardner, D. K., Lane, M., and Batt, P. (1993). Uptake and metabolism of pyruvate and glucose by individual sheep preattachment embryos developed *in vivo*. *Mol. Reprod. Dev.* **36**, 313–319.

Gardner, D. K., Lane, M., Spitzer, A., and Batt, P. A. (1994). Enhanced rates of cleavage and development for sheep zygotes cultured to the blastocyst stage *in vitro* in the absence of serum and somatic cells: Amino acids, vitamins, and culturing embryos in groups stimulate development. *Biol. Reprod.* **50**, 390–400.

Gardner, D. K., Lane, M., Calderon, I., and Leeton, J. (1996). Environment of the preimplantation human embryo *in vivo*: Metabolite analysis of oviduct and uterine fluids and metabolism of cumulus cells. *Fertil. Steril.* **65**, 349–353.

Gardner, D. K., Schoolcraft, W. B., Wagley, L., Schlenker, T., Stevens, J., and Hesla, J. (1998). A prospective randomized trial of blastocyst culture and transfer in in-vitro fertilization. *Hum. Reprod.* **13**, 3434–3440.

Gardner, D. K., Rodriegez-Martinez, H., and Lane, M. (1999). Fetal development after transfer is increased by replacing protein with the glycosaminoglycan hyaluronan for mouse embryo culture and transfer. *Hum. Reprod.* **14**, 2575–2580.

Gardner, D. K., Pool, T. B., and Lane, M. (2000a). Embryo nutrition and energy metabolism and its relationship to embryo growth, differentiation, and viability. *Semin. Reprod. Med.* **18**, 205–218.

Gardner, D. K., Lane, M. W., and Lane, M. (2000b). EDTA stimulates cleavage stage bovine embryo development in culture but inhibits blastocyst development and differentiation. *Mol. Reprod. Dev.* **57**, 256–261.

Gardner, D. K., Maybach, J. M., and Lane, M. (2001). Hyaluronan and rHSA increase blastocyst cryosurvival. *In* "Proceedings of the 17th World Congress on Fertility and Sterility," p. 226. Melbourne, Australia.

Gardner, D. K., Lane, M., and Schoolcraft, W. B. (2002). Physiology and culture of the human blastocyst. *J. Reprod. Immunol.* **55**, 85–100.

Garris, D. R. (1984). Uterine blood flow, pH, and pCO_2 during nidation in the guinea pig: Ovarian regulation. *Endocrinology* **114**, 1219–1224.

Gibb, C. A., Poronnik, P., Day, M. L., and Cook, D. I. (1997). Control of cytosolic pH in two-cell mouse embryos: Roles of H(+)-lactate cotransport and Na^+/H^+ exchange. *Am. J. Physiol.* **273**, C404–C419.

Gray, C. W., Morgan, P. M., and Kane, M. T. (1992). Purification of an embryotrophic factor from commercial bovine serum albumin and its identification as citrate. *J. Reprod. Fertil.* **94**, 471–480.

Gwatkin, R. B. (1966). Amino acid requirements for attachment and outgrowth of the mouse blastocyst *in vitro. Cell Comp. Physiol.* **68**, 335–344.

Gwatkin, R. B., and Haidri, A. A. (1973). Requirements for the maturation of hamster oocytes *in vitro. Exp. Cell Res.* **76**, 1–7.

Hammer, M. A., Kolajova, M., Leveille, M., Claman, P., and Baltz, J. M. (2000). Glycine transport by single human and mouse embryos. *Hum. Reprod.* **15**, 419–426.

Hardy, K., Hooper, M. A., Handyside, A. H., Rutherford, A. J., Winston, R. M., and Leese, H. J. (1989). Non-invasive measurement of glucose and pyruvate uptake by individual human oocytes and preimplantation embryos. *Hum. Reprod.* **4**, 188–191.

Hasler, J. F. (1998). The current status of oocyte recovery, *in vitro* embryo production, and embryo transfer in domestic animals, with an emphasis on the bovine. *J. Anim. Sci.* **76**(Suppl. 3), 52–74.

Hewitson, L. C., and Leese, H. J. (1993). Energy metabolism of the trophectoderm and inner cell mass of the mouse blastocyst. *J. Exp. Zool.* **267**, 337–343.

Hogan, A., Heyner, S., Charron, M. J., Copeland, N. G., Gilbert, D. J., Jenkins, N. A., Thorens, B., and Schultz, G. A. (1991). Glucose transporter gene expression in early mouse embryos. *Development* **113**, 363–372.

Hooper, K, Lane, M., and Gardner, D. K. (2000). Toward defined physiological embryo culture media: replacement of BSA with recombinant albumin. *Biol. Reprod.* **62**(Suppl. 1), 249.

Hume, D. A., and Weidemann, M. J. (1979). Role and regulation of glucose metabolism in proliferating cells. *J. Natl. Cancer Inst.* **62**, 3–8.

Iritani, A., Nishikawa, Y., Gomes, W. R., and VanDemark, N. L. (1971). Secretion rates and chemical composition of oviduct and uterine fluids in rabbits. *J. Anim. Sci.* **33**, 829–835.

Juetten, J., and Bavister, B. D. (1983). The effects of amino acids, cumulus cells, and bovine serum albumin on *in vitro* fertilization and first cleavage of hamster eggs. *J. Exp. Zool.* **227**, 487–490.

Kouridakis, K., and Gardner, D. K. (1995). Pyruvate in embryo culture media acts as antioxidant. *Proc. Fert. Soc. Aus.* **14**, 29.

Lane, M. (2001). Mechanisms for managing cellular and homeostatic stress *in vitro. Theriogenology* **55**, 225–236.

Lane, M., and Gardner, D. K. (1992). Effect of incubation volume and embryo density on the development and viability of mouse embryos *in vitro. Hum. Reprod.* **7**, 558–562.

Lane, M., and Gardner, D. K. (1994). Increase in postimplantation development of cultured mouse embryos by amino acids and induction of fetal retardation and exencephaly by ammonium ions. *J. Reprod. Fertil.* **102**, 305–312.

Lane, M., and Gardner, D. K. (1997a). Differential regulation of mouse embryo development and viability by amino acids. *J. Reprod. Fertil.* **109**, 153–164.

Lane, M., and Gardner, D. K. (1997b). Nonessential amino acids and glutamine decrease the time of the first three cleavage divisions and increase compaction of mouse zygotes *in vitro. J. Assist. Reprod. Genet.* **14**, 398–403.

Lane, M., and Gardner, D. K. (1998). Amino acids and vitamins prevent culture-induced metabolic perturbations and associated loss of viability of mouse blastocysts. *Hum. Reprod.* **13**, 991–997.

Lane, M., and Gardner, D. K. (2000a). Lactate regulates pyruvate uptake and metabolism in the preimplantation mouse embryo. *Biol. Reprod.* **62**, 16–22.

Lane, M., and Gardner, D. K. (2000b). Regulation of ionic homeostasis by mammalian embryos. *Semin. Reprod. Med.* **18**, 195–204.

Lane, M., and Gardner, D. K. (2001a). Blastomere homeostasis. *In* "ART and the Human Blastocyst" (D. K. Gardner and M. Lane, eds.), Springer-Verlag, New York, pp 69–90.

Lane, M., and Gardner, D. K. (2001b). Inhibiting 3-phosphoglycerate kinase by EDTA stimulates the development of the cleavage stage mouse embryo. *Mol. Reprod. Dev.* **60**, 233–240.

Lane, M., Hooper, K., and Gardner, D. K. (2001). Effect of essential amino acids on mouse embryo viability and ammonium production. *J. Assist. Reprod. Genet.* **18**, 519–525.

Lawitts, J. A., and Biggers, J. D. (1993). Culture of preimplantation embryos. *Methods Enzymol.* **225**, 153–164.

Leese, H. J. (1988). The formation and function of oviduct fluid. *J. Reprod. Fertil.* **82**, 843–856.

Leese, H. J. (1991). Metabolism of the preimplantation mammalian embryo. *Oxf. Rev. Reprod. Biol.* **13**, 35–72.

Leese, H. J. (1995). Metabolic control during preimplantation mammalian development. *Hum. Reprod. Update.* **1**, 63–72.

Leese, H. J., and Barton, A. M. (1984). Pyruvate and glucose uptake by mouse ova and preimplantation embryos. *J. Reprod. Fertil.* **72**, 9–13.

Li, J., and Foote, R. H. (1993). Culture of rabbit zygotes into blastocysts in protein-free medium with one to twenty per cent oxygen. *J. Reprod. Fertil.* **98**, 163–167.

Lindenbaum, A. (1973). A survey of naturally occurring chelating ligands. *Adv. Exp. Med. Biol.* **40**, 67–77.

Liu, Z., and Foote, R. H. (1995). Development of bovine embryos in KSOM with added superoxide dismutase and taurine and with five and twenty percent O_2. *Biol. Reprod.* **53**, 786–790.

Ludwig, T. E., Lane, M., and Bavister, B. D. (2001). Differential effect of hexoses on hamster embryo development in culture. *Biol. Reprod.* **64**, 1366–1374.

Maas, D. H., Storey, B. T., and Mastroianni, L. J. (1976). Oxygen tension in the oviduct of the rhesus monkey (*Macaca mulatta*). *Fertil. Steril.* **27**, 1312–1317.

Maas, D. H., Storey, B. T., and Mastroianni, L. J. (1977). Hydrogen ion and carbon dioxide content of the oviductal fluid of the rhesus monkey (*Macaca mulatta*). *Fertil. Steril.* **28**, 981–985.

Martin, K. L., Hardy, K., Winston, R. M., and Leese, H. J. (1993). Activity of enzymes of energy metabolism in single human preimplantation embryos. *J. Reprod. Fertil.* **99**, 259–266.

Mastroianni, L. J., and Jones, R. (1965). Oxygen tension in the rabbit fallopian tube. *J. Reprod. Fertil.* **9**, 99.

McEvoy, T. G., Robinson, J. J., Aitken, R. P., Findlay, P. A., and Robertson, I. S. (1997). Dietary excesses of urea influence the viability and metabolism of preimplantation sheep embryos and may affect fetal growth among survivors. *Anim. Reprod. Sci.* **47**, 71–90.

McKiernan, S. H., and Bavister, B. D. (1992). Different lots of bovine serum albumin inhibit or stimulate *in vitro* development of hamster embryos. *In Vitro Cell Dev. Biol.* **28A**, 154–156.

Mehta, T. S., and Kiessling, A. A. (1990). Development potential of mouse embryos conceived *in vitro* and cultured in ethylenediaminetetraacetic acid with or without amino acids or serum. *Biol. Reprod.* **43**, 600–606.

Menke, T. M., and McLaren, A. (1970). Mouse blastocysts grown *in vivo* and *in vitro*: Carbon dioxide production and trophoblast outgrowth. *J. Reprod. Fertil.* **23**, 117–127.

Miller, J. G., and Schultz, G. A. (1987). Amino acid content of preimplantation rabbit embryos and fluids of the reproductive tract. *Biol. Reprod.* **36**, 125–129.

Morgan, M. J., and Faik, P. (1981). Carbohydrate metabolism in cultured animal cells. *Biosci. Rep.* **1**, 669–686.

Moses, D. F., Matkovic, M., Cabrera-Fisher, E., and Martinez, A. G. (1997). Amino acid contents of sheep oviductal and uterine fluids. *Theriogenology* **47**, 336.

Mullaart, E., Merton, S. M., de Ruigh, L., Hendriksen, P. J. M., and van Wagtendonk, A. M. (2001). Pregnancy rates and calf characteristics following transfer of embryos produced from oocytes collected by OPU after FSH pre-stimulation. *Theriogenology* **55**, 434.

Nichol, R., Hunter, R. H., Gardner, D. K., Leese, H. J., and Cooke, G. M. (1992). Concentrations of energy substrates in oviductal fluid and blood plasma of pigs during the peri-ovulatory period. *J. Reprod. Fertil.* **96**, 699–707.

O'Fallon, J. V., and Wright, R. W. J. (1995). Pyruvate revisited: A non-metabolic role for pyruvate in preimplantation embryo development. *Theriogenology* **43**, 288.

Paria, B. C., and Dey, S. K. (1990). Preimplantation embryo development *in vitro*: Cooperative interactions among embryos and role of growth factors. *Proc. Natl. Acad. Sci. U.S.A.* **87**, 4756–4760.

Pool, T. B., Atiee S. H., and Martin, J. E. (1998). Oocyte and embryo culture: Basic concepts and recent advances. *Infertility* **9**, 181–203.

Quinn, P., and Harlow, G. M. (1978). The effect of oxygen on the development of preimplantation mouse embryos *in vitro*. *J. Exp. Zool.* **206**, 73–80.

Reitzer, L. J., Wice, B. M., and Kennell, D. (1980). The pentose cycle: Control and essential function in HeLa cell nucleic acid synthesis. *J. Biol. Chem.* **255**, 5616–5626.

Rieger, D. (1984). The measurement of metabolic activity as an approach to evaluating viability and diagnosing sex in early embryos. *Theriogenology* **21**, 138–149.

Rieger, D. (1992). Relationship between energy metabolism and development of the early embryo. *Theriogenology* **37**, 75–93.

Rieger, D., Loskutoff, N. M., and Betteridge, K. J. (1992). Developmentally related changes in the metabolism of glucose and glutamine by cattle embryos produced and co-cultured *in vitro*. *J. Reprod. Fertil.* **95**, 585–595.

Rogers, P. W., Murphy, C. R., and Gannon, B. J. (1982a). Absence of capillaries in the endometrium surrounding the implanting rat blastocyst. *Micron* **13**, 373–374.

Rogers, P. W., Murphy, C. R., and Gannon, B. J. (1982b). Changes in the spatial organization of the uterine vasculature during implantation in the rat. *J. Reprod. Fertil.* **65**, 211–214.

Ross, R. N., and Graves, C. N. (1974). O_2 levels in the female rabbit reproductive tract. *J. Anim. Sci.* **39**, 994.

Salahuddin, S., Ookutsu, S., Goto, K., Nakanishi, Y., and Nagata, Y. (1995). Effects of embryo density and co-culture of unfertilized oocytes on embryonic development of *in-vitro* fertilized mouse embryos. *Hum. Reprod.* **10**, 2382–2385.

Schini, S. A., and Bavister, B. D. (1988). Two-cell block to development of cultured hamster embryos is caused by phosphate and glucose. *Biol. Reprod.* **39**, 1183–1192.

Schultz, G. A., Kaye, P. L., McKay, D. J., and Johnson, M. H. (1981). Endogenous amino acid pool sizes in mouse eggs and preimplantation embryos. *J. Reprod. Fertil.* **61**, 387–393.

Seshagiri, P. B., and Bavister, B. D. (1989). Phosphate is required for inhibition by glucose of development of hamster 8-cell embryos *in vitro*. *Biol. Reprod.* **40**, 607–614.

Shire, J. G., and Whitten, W. K. (1980a). Genetic variation in the timing of first cleavage in mice: Effect of maternal genotype. *Biol. Reprod.* **23**, 369–376.

Shire, J. G., and Whitten, W. K. (1980b). Genetic variation in the timing of first cleavage in mice: Effect of paternal genotype. *Biol. Reprod.* **23**, 363–368.

Sinclair, K. D., McEvoy, T. G., and Carolan, C. (1998). Conceptus growth and development following *in vitro* culture of ovine embryos in media supplemented with bovine sera. *Theriogenology* **49**, 218.

Steeves, T. E., and Gardner, D. K. (1999). Temporal and differential effects of amino acids on bovine embryo development in culture. *Biol. Reprod.* **61**, 731–740.

Stojkovic, M., Peinl, S., Stojkovic, P., Zakhartchenko, V., Thompson, J. G., and Wolf, E. (2001). High concentration of hyaluronic acid in culture medium increases the survival rate of frozen/thawed *in vitro* produced bovine blastocysts. *Theriogenology* **55**, 317.

Thompson, J. G., Simpson, A. C., Pugh, P. A., Donnelly, P. E., and Tervit, H. R. (1990). Effect of oxygen concentration on *in-vitro* development of preimplantation sheep and cattle embryos. *J. Reprod. Fertil.* **89**, 573–578.

Thompson, J. G., Gardner, D. K., Pugh, P. A., McMillan, W. H., and Tervit, H. R. (1995). Lamb birth weight is affected by culture system utilized during *in vitro* pre-elongation development of ovine embryos. *Biol. Reprod.* **53**, 1385–1391.

Thompson, J. G., Partridge, R. J., Houghton, F. D., Cox, C. I., and Leese, H. J. (1996). Oxygen uptake and carbohydrate metabolism by *in vitro* derived bovine embryos. *J. Reprod. Fertil.* **106**, 299–306.

Thompson, J. G., Sherman, A. N., Allen, N. W., McGowan, L. T., and Tervit, H. R. (1998). Total protein content and protein synthesis within pre-elongation stage bovine embryos. *Mol. Reprod. Dev.* **50**, 139–145.

Thomson, J. A., Itskovitz-Eldor, J., Shapiro, S. S., Waknitz, M. A., Swiergiel, J. J., Marshall, V. S., and Jones, J. M. (1998). Embryonic stem cell lines derived from human blastocysts. *Science* **282**, 1145–1147.

Umaoka, Y., Noda, Y., Narimoto, K., and Mori, T. (1992). Effects of oxygen toxicity on early development of mouse embryos. *Mol. Reprod. Dev.* **31**, 28–33.

Van Winkle, L. J. (1988). Amino acid transport in developing animal oocytes and early conceptuses. *Biochim. Biophys. Acta* **947**, 173–208.

Van Winkle, L. J., Haghighat, N., and Campione, A. L. (1990). Glycine protects preimplantation mouse conceptuses from a detrimental effect on development of the inorganic ions in oviductal fluid. *J. Exp. Zool.* **253**, 215–219.

Vella, P., Lane, M., and Gardner, D. K. (1997). Induction of glycolysis in the day-3 mouse embryo by glucose. *Biol. Reprod.* **57**(Suppl. 1), 26.

Wales, R. G. (1969). Accumulation of carboxylic acids from glucose by the pre-implantation mouse embryo. *Aust. J. Biol. Sci.* **22**, 701–707.

Wales, R. G., Whittingham, D. G., Hardy, K., and Craft, I. L. (1987). Metabolism of glucose by human embryos. *J. Reprod. Fertil.* **79**, 289–297.

Walker, S. K., Heard, T. M., and Seamark, R. F. (1992). *In vitro* culture of sheep embryos without co-culture: Success and perspectives. *Theriogenology* **37**, 111–126.

Whitten, W. K., and Biggers, J. D. (1968). Complete development *in vitro* of the pre-implantation stages of the mouse in a simple chemically defined medium. *J. Reprod. Fertil.* **17**, 399–401.

Whittingham, D. G. (1971). Culture of mouse ova. *J. Reprod. Fertil.* (Suppl.) **14**, 7–21.

Wiley, L. M., Yamami, S., and Van Muyden, D. (1986). Effect of potassium concentration, type of protein supplement, and embryo density on mouse preimplantation development *in vitro*. *Fertil. Steril.* **45**, 111–119.

Zhang, X., and Armstrong, D. T. (1990). Presence of amino acids and insulin in a chemically defined medium improves development of 8-cell rat embryos *in vitro* and subsequent implantation *in vivo*. *Biol. Reprod.* **42**, 662–668.

Zorn, T. M., Pinhal, M. A., Nader, H. B., Carvalho, J. J., Abrahamsohn, P. A., and Dietrich, C. P. (1995). Biosynthesis of glycosaminoglycans in the endometrium during the initial stages of pregnancy of the mouse. *Cell. Mol. Biol.* **41**, 97–106.

CHAPTER 10

GENETIC AND PHENOTYPIC SIMILARITY AMONG MEMBERS OF MAMMALIAN CLONAL SETS

George E. Seidel, Jr.

"Your dog [pronounced *dawg* by most Texans] is dead." This is the succinct answer given by Mark Westhusin, in a presentation at the 2001 annual meeting of the International Embryo Transfer Society, to the question of "How similar would a clone be to my pet dog who just died?" "You're history" might be a similarly appropriate response to people who wish to clone *themselves* to achieve immortality: there are myriad scientifically defensible applications of cloning, but recreating an animal, except in a genetic sense, is not one of them. Perfect genetic identity is actually unachievable, although one can get close for some practical purposes. An example is cloning a valuable bull; most valuable bulls are kept for their sperm production, and a clone will produce a population of sperm identical to that of the original animal, except for mutations and the mitochondrial genome. Phenotype is of little consequence as long as fertile sperm are produced.

Phenotypic identity is most elusive. Phenotype obviously is determined by genotype, environment, epigenetic effects (often with a chance component), chance itself, and interactions of all these factors (Seidel, 1983). It has become increasingly clear over the past few decades that chance epigenetic effects can have substantial influence for some traits (Reik *et al.*, 1993; Finch and Kirwood, 2000). These ideas are the basis for this chapter.

DEFINITION OF CLONING

The word "clone" is derived from the Greek word for twig. This etymology is easily understood, because our species has known for millennia that many species of plants can be propagated from somatic tissues such as twigs. Clonal reproduction is routine for dozens of domesticated plants, such as potatoes, and is common in nature as well, e.g., groves of aspen trees. A variation that is fairly common in nearly all mammalian species is identical multiplets; the incidence of identical twins and triplets, etc., is low in most species, but in others, such as armadillos, this occurs with every pregnancy. For practical purposes, identical multiplets are the "gold standard" of the maximum degree achievable of phenotypic identity of mammals. None of the methods of cloning will achieve greater identity, and cloning by nuclear transplantation/cell fusion results in considerably less identity.

But are identical twins or triplets clones? Perhaps terms such as "clone" or "cloning" should be used less frequently, and instead more specific descriptions should be used, embracing concepts such as making genetically identical sets by blastomere separation, splitting postcompaction embryos, or nuclear transplantation. Some scientists define nuclear transplantation/cell fusion as "true cloning," suggesting that other procedures should not come under the umbrella term "cloning," even though some of these other procedures result in animal sets that are more identical, compared to sets produced by nuclear transplantation/cell fusion.

CYTOPLASMIC GENETICS

The clearest example of cytoplasmic inheritance in animal cells is the mitochondrial genome. The approximately 16,000-base-pair circular mitochondrial genome has genes for ribosomal RNAs, transfer RNAs, and approximately a dozen mitochondrial proteins (Cummins, 1998). Although the vast majority of mitochondrial proteins are specified by chromosomal genes, the mitochondrial genome has considerable variability, and mutates more readily than chromosomal genes, in part due to poor proofreading during DNA synthesis (Cummins, 1998). The resulting mutations are the source of considerable (cytoplasmic) genetic disease, as well as phenotypic variation in normal mitochondrial function. Although there are rare exceptions (Cummins, 1998), for the most part, mitochondrial genomes are inherited via the maternal ooplasm. The relatively few mitochondria introduced by the sperm usually degenerate, and in any case become so dilute that they usually would not end up in the relatively few cells of the blastocyst that differentiate into the resulting animal.

The maternal lineages of mitochondria in the recipient oocyte and in the donor nucleus often will be different. Simply by the process of dilution, the animal resulting from cloning procedures using cell fusion usually ends up with mitochondrial genes of the oocyte, not the donor nucleus. Of course, mitochondria will be heteroplasmic initially; sometimes this situation persists, and rarely the donor mitochondria out-compete the recipient oocyte mitochondria, but usually the mitochondria of the recipient oocyte prevail (Evans *et al.*, 1999). It is also possible that donor mitochondria do not survive well in ooplasm because of the very specially differentiated state of mitochondria in oocytes (Cummins, 1998). Experiments to sort this out are progressing, and it is obvious that answers will have a statistical quality rather than a simple outcome. Note that in the future, it is likely that heteroplasmy of mitochondrial genomes can be eliminated by selective elimination of donor or recipient mitochondria by chemical or other means.

Cytoplasmic inheritance also occurs for centrosomes, usually via the sperm in mammals (Stearns, 2001). To my knowledge, no cytoplasmic nucleic acid sequence information is involved, but the semiconservative nature of centriole duplication (Stearns, 2001) does have implications for cloning, particularly if procedures become more sophisticated. For example, centrosomal material might be provided from sperm parts rather than from donor cell cytoplasm.

Cytoplasmic inheritance of viruses occurs in some situations. In these cases, there is a nucleic acid sequence specifying a cytoplasmic component.

EPIGENETIC EFFECTS

A simple definition of "epigenetic" is elusive, although many epigenetic qualities are captured by the definition: different genetic and/or phenotypic outcomes from the same DNA sequence. During embryonic and neonatal development, there clearly are dozens and perhaps hundreds of epigenetic events that influence phenotype differently in genetically identical animals. Several examples of epigenetic phenomena are given in Table 1. There is also evidence of epigenetic instability when embryonic stem cells are used as donor nuclei (Humpherys *et al.*, 2001), and possibly when there are several rounds of nuclear recycling (Peura *et al.*, 2001).

One of the most obvious differences between genetically identical animals is coat color patterns in strains that are not one solid color. Although the broad pattern

Table 1 Examples of Epigenetic Effects

Migration patterns of melanoblasts and primordial germ cells during fetal development
Random inactivation of X chromosomes in female mammals
Numbers of beta cells in the endocrine pancreas, or oocytes in ovaries
Gametic imprinting
Telomere length
Differential methylation of cytosines in DNA of different tissues

of coloration is similar among genetic identicals, the specific pattern can be quite different. This is mostly a consequence of how melanoblasts, originating from the neural crest, invade hair follicles; there obviously is not a genetic instruction for each follicle (Seidel, 1985). Similarly, the numbers of primordial germ cells that invade each gonad have a statistical quality (Tam and Snow, 1981), as also would be expected for many other aspects of embryonic development. There are many such examples, some of which easily distinguish one identical twin from another. One example is iris pigmentation patterns (Daugman, 2001). An extreme example is monozygous human twins with different hair color, thought to be due to differential melanoblast migration (Gringras, 1999), illustrating the imprecision of the term "identical twins."

During organogenesis in female mammals, the maternal or paternal X chromosome is inactivated in individual cells (Gardner *et al.*, 1985); once this decision is made, progeny of cells maintain the initial inactivation, either the maternal or paternal X chromosome (Eggan *et al.*, 2000). By chance, a given tissue or organ from genetically identical females may have more cells with paternal or maternal X chromosomes inactivated. This is perhaps best illustrated with tricolor female cats, in which red or black hair color is an elegant marker for which parental X chromosome is active in melanocytes of any given area of cat skin. Identical twin tricolor cats would have very different patterns of red and black due to superimposition of two epigenetic effects: X-chromosome inactivation patterns and melanoblast migration. Similar, less easily visualized epigenetic phenomena occur frequently during embryonic development (Reik *et al.*, 1993; Finch and Kirwood, 2000).

Just as there are genetic × environmental interactions, there are epigenetic × environmental interactions, nicely illustrated by the integrating work of Barker (e.g., Barker, 2000). The main concept is that maternal nutrition during gestation can have profound effects on phenotype for chronic diseases decades later. Maternal undernutrition, for example, leads to increased incidence of diabetes, heart disease, and depression. This has an epigenetic basis via the numbers of cells that are allocated to different tissues, or how they differentiate within tissues. Doubtless these events are mediated via modifications of histones (Berger, 2001), such as methylation and acetylation, or differential methylation of cytosines in the DNA of these tissues. Although maternal undernutrition or overnutrition has consequences, one would not expect identical sequellae in each member of a set of clones.

Gametic imprinting, by which expression of genes depends on sex of parental origin rather than DNA sequence, is one of the more fascinating examples of epigenetic inheritance (Surani *et al.*, 1986; Moore, 2001). Theoretically, this should not lead to epigenetic or phenotypic differences among clones. However, there are some indications that imprinted genes become reprogrammed inappropriately during nuclear transplantation to clone mammals (Moore, 2001; Wrenzycki *et al.*, 2001), possibly explaining some of the huge differences in the sizes of placentas and fetuses among genetically identical offspring.

Telomere length is another epigenetic trait that might be quite relevant in donor chromosomes. There is little debate that telomere length decreases with each mitotic cell division in many somatic tissues in a number of species, and thus decreases with aging. However, there is conflicting information on the degree of restoration of telomere length with cloning procedures. Results range from slight shortening (Shiels *et al.*, 1999) to no effect (Kubota *et al.*, 2000) to significant lengthening (Lanza *et al.*, 2000). These differences in findings could be due to cloning procedures, sources of donor nuclei, species, or chance with individual clonal sets. Some donors have been quite aged (Kubota *et al.*, 2000), but there have been no observations of premature aging reported to date. Perhaps none will occur; knockout mice for telomerase had to go through six generations to observe a phenotype (Blasco *et al.*, 1997) and mice have been recloned through six generations with no buildup problems (Wakayama *et al.*, 2000). The situation could be quite different, however, with longer lived species with shorter telomeres than are found in mice.

UTERINE EFFECTS

All procedures for cloning mammals require embryo transfer, and thus exposure to the vagaries of different uterine environments. In species in which inbred lines are available, sets of recipients can be made essentially genetically identical. F_1 crosses often are used. They have the special advantages of hybrid vigor and being genetically identical sets without being inbred. However, these theoretical advantages are moot for most nonresearch applications because inbred lines are available for only a few species. The relatively expensive option now is available, however, of cloning animals to be recipients so that uteri of recipient sets are genetically identical. However, even these would likely be of different ages.

Litter-bearing species present numerous challenges for achieving identical environments within the same uterus, or between uteri. Litter size affects birth weight and gestation length, and thus stage of maturity at birth. Because of the unpredictability of pregnancy rates per embryo, cloning procedures exacerbate variability in litter size. When transferring embryos with low probabilities of survival, carrier embryos of a different genotype, but high probability of survival, are often used (sometimes provided by just mating the recipient). Although carrier embryos make litter size more uniform, they compete for implantation sites and affect the local uterine environment in various ways—for example, in placenta size.

When litters are large, there often are one or two runts of the litter. These frequently are from implantations crowded into the tips of uterine horns, where vasculature did not develop as well as in the rest of the uterus. The runt phenotype generally remains with the animal for life.

Normally, there is a spectrum of masculinity and femininity in litter-bearing animals that is due in part to the gender of fetuses implanted on either side of another fetus (vom Saal, 1989). An oversimplified view, which illustrates the point, is that an embryo in the miduterine horn could have two females, two males, or one of each sex for immediate neighbors. Apparently the hormones and other factors that neighbors secrete make a permanent difference in degree of gender phenotype. Nongender phenotypic effects, such as growth characteristics, likely also occur, but are largely unexplored.

Obviously, these kinds of effects are minimized with clonal sets of embryos in the absence of carrier embryos. However, they do become an issue if one is trying to copy a particular animal. One could imagine a female pig that was gestated next to two males, and therefore grew faster and was somewhat more aggressive in feeding. Litters of clones of this pig would not have those different gender-related neighbor-induced characteristics. These kinds of issues are especially relevant when attempting to copy extreme phenotypes, because they are not entirely genetically based.

Uterine variation also is present in monotocous species, and is exacerbated by twin gestations, whether fraternal or identical. Note that because of low pregnancy rates per embryo, twinning with embryo transfer is a frequent practice in cattle, and twin-clone gestations are not so unusual. Such gestations are shorter than normal and produce smaller calves than do singleton gestations. Birth weight also is affected by breed, parity, age, and size of recipients. Vertical transmission of pathogens can also affect offspring.

NEONATAL ENVIRONMENT

A plethora of differences in neonatal environments can influence phenotype, ranging from nutrition to climate, from prevalence of disease-causing organisms to competition from siblings. All mammals require milk to thrive neonatally, and the amount and composition of suckled milk will vary greatly among recipients. Passive transfer of immunity via immunoglobulins in colostrum and/or across the placenta varies enormously. Maternal behavior also affects neonatal health, well-being, and socialization.

LARGE-OFFSPRING SYNDROME

Large-offspring syndrome is one of the most spectacular examples of phenotypic variation within genetically identical clutches of cloned animals. This is illustrated for birth weight of calves cloned (Table 2) using blastomeres of morulae as donor cells (Garry et al., 1996). Ironically, there often is more variation in birth weights within genetically identical clonal sets than among full sibs (Green et al., 1996; Gärtner et al., 1998). The syndrome was first described in the context of cloning by nuclear transfer by Willadsen et al. (1991) and was documented thoroughly by Wilson et al. (1995). It is poorly understood and frequently misunderstood. Large-offspring syndrome has been documented primarily in sheep and cattle (Sinclair et

Table 2 Characteristics of Clonal Sets of Calves[a]

Set	Birth wt. (kg)	Gestation length (days)	Set	Birth wt. (kg)	Gestation length (days)
Set 1	30.9	271	Set 5	42.3[b]	273
	30.9	279		47.3[b]	288
	33.6	283			
	42.7	289			
Set 2	41.8	275	Set 6	37.3	291
	43.6[b]	293		40.0	296
	55.5[b]	289		45.5	297
	47.3[b]	288		44.1	299
				46.8	310
Set 3	44.5	283	Set 7	42.7	294
	53.6	293		40.0	295
	65.5	293			
Set 4	57.7	291	Set 8	32.7	273
	62.7	298		41.8	273
	54.1[b]	293		30.0[b]	282
				45.5	283
				46.4	289

[a]From Garry et al. (1996).
[b]Indicates calf died neonatally.

al., 2000) and may not occur in all species, although it clearly can occur in mice (Eggan *et al.*, 2001). There may be multiple etiologies, but the primary lesion in most cases likely is an epigenetic disturbance of placental function (Hill *et al.*, 2000b; De Sousa *et al.*, 2001). Thus, fetuses may become abnormal solely due to the placental environment, much as occurs with macrosomic babies born to diabetic mothers.

What is terribly misleading about the syndrome is that the high incidence of neonatal death and morbidity of animals derived via cloning by nuclear transplantation often is ascribed to birthing difficulty due to large size. Far more serious are epigenetic disturbances of metabolism (e.g., Garry *et al.*, 1996), again most likely of placental origin and not correlated with size of offspring. The metabolic disturbances are exacerbated by an abnormal parturition process that is prolonged, among other problems. Also, the metabolic problems result in the ultimate in phenotypic variation, life or death, whether it be embryonic, fetal, neonatal, or postnatal death. In addition, the incidence of congenital abnormalities is increased (Hill *et al.*, 1999). That birthing difficulty due to size is not the fundamental problem is clearly illustrated by Garry *et al.* (1996), who derived cloned calves by elective cesarean section when parturition was imminent. Note that this is a common practice for cloned calves (e.g., Wells *et al.*, 1999) as well as for calves produced commercially by *in vitro* oocyte maturation, fertilization, and culture of embryos. Oddly, elective cesarean section a few days preterm actually decreases phenotypic variation among cloned calves.

Large-offspring syndrome and related problems occur in many situations (Khosla *et al.*, 2001) unrelated to cloning, albeit usually at a lower incidence. However, problems can be huge, just with *in vitro* maturation, fertilization, and culture of embryos. For example, in the study of Behboodi *et al.* (1995), seven of eight calves died neonatally, and Walker *et al.* (1992) had 20% losses of lambs compared to 3% for controls, just from culturing *in vivo*-produced embryos. *In vivo* vs. *in vitro*-matured and -fertilized oocytes also can contribute to large-offspring syndrome (Behboodi *et al.*, 2001). With cloning, *in vitro* oocyte maturation and *in vitro* culture and embryo transfer are superimposed on the cloning procedures. Under some circumstances, effects may be multiplicative.

It is important to note that these epigenetic effects will decrease as more is learned. Removing serum from culture media can help (e.g., Thompson *et al.*, 1995) as can avoiding coculture systems. The problem will never disappear, because large-offspring syndrome occurs at a low incidence (<1% in most breeds) with naturally mating cattle, and can be exacerbated in numerous ways, such as hormone treatments and asynchronous embryo transfer.

MUTATIONS

Because billions of base pairs per cell are replicated to produce hundreds of billions to trillions of cells in an animal, there will be some mistakes. Other mutations are environmentally induced—for example, from gamma rays or from intracellular peroxidation events. The majority of consequential mutations are corrected or eliminated by cell death. Nevertheless, mutations are bound to accumulate, particularly in aged, somatic cells. They even accumulate in the germ line. One experiment with frozen semen collected from the same bull at young and old ages (Hill *et al.*, 2000a) suggested that the quality of embryos produced with this semen declined with age of the bull at the time the semen was frozen. Perhaps the most striking example of a mutation affecting phenotype of genetically identical individuals is human monozygotic twins of opposite sex (Wachtel *et al.*, 2001).

There are two separate issues with mutations: (1) different genotypes among clonal donor cells and (2) mutations in organisms postcloning. Both will result in differing genotypes and, often, phenotypes among clonal sets. An intriguing question concerns what can be done to minimize genetic differences among clones due to mutations. This is obviously related to questions of aging (Seidel, 2000). Very likely, there will be fewer mutations in cells of some tissues than in others. Germ line cells may have more mechanisms to minimize mutations compared to somatic cells (Seidel, 2000). Cells that are constantly dividing, such as hematopoietic cells, probably accumulate more mutations than do those that rarely divide, such as Sertoli cells. Granulosa and cumulus cells in ovarian follicles arise from follicle cells that have been quiescent since fetal life, and have been dividing for only a few months prior to ovulation. They may have dual advantages of fewer mutations and being more readily reprogrammed compared to other cells. A final, obvious tactic is to use younger rather than older donor cells; with planning, young cells from animal cloning candidates can be cryopreserved for cloning at a later date.

CULTURAL INHERITANCE

Although largely anecdotal, there also is solid evidence that some learned behavioral traits are passed on from generation to generation in various animal species, generally through the maternal line (Avital and Jablonka, 2000). This cultural transmission is easily broken with embryo transfer to surrogate recipients, thus imposing a different maternal learning paradigm. An example is cattle grazing extensively in the semiarid foothills and mountains of western North America. Cows have a choice of dozens of microenvironments over thousands of acres, and cows seem to teach their calves where and what to graze, or even browse brush in some cases. This is passed on from year to year maternally (essentially all males are sold for fattening). One practical consequence is simply finding the cattle at round-up time each fall; searches on horseback for hundreds of cattle occur over hundreds of square kilometers, yet certain cows and their progeny and grand progeny can be found, year after year, hidden in the same areas. To the extent that such phenomena occur, the sorts of phenotypes described obviously will not be the same within sets of clones unless the surrogate recipients come from the same maternal pod.

HOW SIMILAR?

All sorts of mechanisms have been described that can cause genetic, epigenetic, and phenotypic differences within clonal sets. However, just how much variability will there be for practical purposes? This obviously will depend on the environment, the species, and the trait or traits of interest. We already have huge amounts of information to answer such questions from inbred strains of animals and their F_1 crosses, and from identical multiplets as well as from clonal sets to a limited degree (Green et al., 1996; Gärtner et al., 1998). Genetically identical rodents living in standardized, noncrowded environments are very similar in traits such as body weight at given younger ages. However, as they age, large differences occur in various traits, particularly longevity (Finch and Kirwood, 2000). An example of practical behavioral variability is provided by the multiple cages of male rodents kept for purposes such as breeding superovulated females. Some males are better performers than others, even though they are of the same strain and age, managed as identically as practicable.

The extreme of natural variability among clonal sets is human identical twins reared apart. Although a discussion of this literature is beyond the scope of this chapter, there is considerable phenotypic variability between such individuals, as well as among identical twins reared together (Bouchard et al., 1990).

Between the two extremes just described is where most cloning applications fall, and where genetic and phenotypic differences and similarities of clones are most relevant. The major immediate application of cloning by nuclear transfer is research, such as studies on reprogramming DNA. This special case is discussed in detail elsewhere in this volume. The nonresearch applications, other than to replace defective human organs and tissues, and production of transgenic animals to produce pharmacological products, seem to fall into three broad categories: replacing cherished companion animals, copying valuable athletic animals, and producing agricultural animals. Businesses already exist to address these opportunities. The first cloned kitten, announced by a group in Texas, has been produced to promote a commercial endeavor.

Although cloning a pet or a valuable racing camel will not result in the same animal, such animals will certainly be very similar to the donor in many phenotypic respects. Most pet owners likely will be quite satisfied with the product, even though there likely would be clearly measurable phenotypic differences between the donor and the clone. If the animal is of a spotted strain, the spots will not be identical, but otherwise will probably be sufficiently similar to satisfy the customer in most cases. The racing camel owner is, however, likely to be disappointed if the objective is racing ability. There likely will be a considerable regression toward the mean for traits such as athletic ability. Such traits, of course, have huge environmental components, which can be controlled to a considerable extent. However, there also are likely to be substantial epigenetic effects that are impossible to control. If a prized animal were being copied for breeding purposes because of being a proved sire, then success is probable because a genetic trait is the objective. Cloning provides the opportunity for several classes of sterile animals to reproduce or be reproduced. An intact copy of a castrated male can be made; this might be especially relevant for gelded horses. A second example is mules, hinneys, and the like.

Perhaps animal agriculture is considered by most people as the obvious, most legitimate large-scale application of cloning technology. Use of this technology for improving meat and milk production has been considered carefully by Smith (1989) and Van Vleck (1999), among others. The numerous studies on identical twin cattle also provide information about the maximal degree of identity that is possible in agricultural production traits. Basically there are two fundamental questions for quantitative agricultural traits: (1) Accuracy—to what extent is an animal identified as truly phenotypically superior and, in turn, genetically superior? (2) Heritability— to what extent is the superior phenotype due to genetics on the average and, therefore, how superior will the average clone be?

At least for bulls, when they have thousands of offspring, it is already possible to know genetic superiority for certain traits with up to 99% accuracy. It will also be possible eventually to determine the true superiority of any individual by making large clonal sets. However, in the absence of such tests, the true genetic value of an individual superior animal will only be estimated imprecisely for most traits (Van Vleck, 1999). This means that selecting an animal with an outstanding phenotype for a trait, such as milk production, and cloning it will often lead to disappointment, simply because this animal was not truly superior genetically. As mentioned earlier, clones of animals with extreme phenotypes, for example 3 or more standard deviations from the mean, likely will not live up to expectations. On the average, such clonal sets will be superior, but not as superior as expected (Van Vleck, 1999), and some will not even be superior to the mean. Thus, a testing step costing considerable time (around one generation) will be needed in most cases to determine true phenotypic superiority of a clonal set. To the extent that there is available genetic information on relatives of the superior animal in question, one can bolster confidence that this animal is truly superior. An extreme example is the bull with recorded information on thousands of offspring. Often techniques such as marker-assisted selection and genotyping for specific alleles also can be used to improve chances that the animal being cloned is good starting material.

The second important aspect of cloning from an agricultural production standpoint is similarity among members of clonal sets, which is governed by the relationship between genotype and phenotype, or heritability (abbreviated H^2 for the sum of all modes of inheritance). For some traits H^2 is virtually 100% (e.g., coat color) and for others H^2 is close to 0% (e.g., fertility). For most important agricultural traits, such as growth rates, milk production, meat or milk composition, or docility, H^2 usually is on the order of 20–50%. This has two consequences. First, the clone mates of a truly phenotypically superior animal (e.g., a cow producing 14,000 kg of milk/year in a herd with an average milk production of 10,000 kg) will average only 11,000 kg of milk/year if H^2 is 25%. Second, there will be huge variation from clone to clone. Note that increasing average milk production from 10,000 to 11,000 kg in one step is a huge gain relative to other breeding techniques. However, it will not be the 4000 kg gain that many people expect. The situation is better than that just described to the extent that environments are standardized because of genotype × environment interactions. Thus, clone mates on the same farm or among farms where the environment is exceedingly similar will be somewhat more similar to the original animal and to each other than an H^2 of 25% predicts. The bottom line, however, is that genetic principles still operate with cloning, that cloning does not copy phenotypes, and that, for many traits, most phenotypic variation is not due to genetic variation.

SUMMARY AND PERSPECTIVE

Members of clonal sets of mammals produced by nuclear transplantation will be genetically identical except for mutations and mitochondrial genes (if the same maternal line is not used for donor nucleus and recipient cytoplast). Numerous chance epigenetic phenomena, some of which are mediated via differential methylation of cytosines, will affect members of the same clonal set differently. Superimposed on these effects are uterine effects and other environmental vagaries postbirth, even under relatively controlled conditions. For some species, situations, and traits, there will only be small phenotypic differences among members of cloned sets. For other species, situations, and traits, phenotypic variation will be fairly large.

For many applications of cloning by nuclear transplantation/cell fusion, the phenotypic variation among clones will be of minimal consequence. For other applications, the phenotypic variation within clonal sets may negate most of the potential benefits of the procedure. Cloning by nuclear transplantation is an exceedingly valuable tool; however, it is a tool, not a panacea, and it never can recreate the same animal.

REFERENCES

Avital, E., and Jablonka, E. (2000). Animal traditions. "Behavioural Inheritance in Evolution." Cambridge University Press, New York.

Barker, D. J. B. (2000). *In utero* programming of cardiovascular disease. *Theriogenology* 53, 555–574.

Behboodi, E., Anderson, G. B., BonDurant, R. H., Cargill, S. L., Kreuscher, B. R., Medrano, J. F., and Murray, J. D. (1995). Birth of large calves that developed from *in vitro* derived bovine embryos. *Theriogenology* 44, 227–232.

Behboodi, E., Groen, W., Destrempes, M. M., Williams, J. L., *et al.* (2001). Transgenic production from *in vivo*-derived embryos: Effect on calf birth weight and sex ratio. *Mol. Reprod. Dev.* 60, 27–37.

Berger, S. L. (2001). The histone modificate cirus. *Science* 292, 64–65.

Blasco, M. A., Lee, H.-W., Hande, M. P., Samper, E., Lansdorp, P. M., DePinko, R. A., and Greider, C. W. (1997). Telomere shortening and tumor formation by mouse cells lacking telomerase RNA. *Cell* 91, 25–34.

Bouchard, T. M., Lykken, D. T., McGui, M., Segal, N. L., and Tellegen, A. (1990). Sources of human psychological differences: The Minnesota study of twins reared apart. *Science* 250, 223–228.

Cummins, J. (1998). Mitochondrial DNA in mammalian reproduction. *Rev. Reprod.* 3, 172–182.

Daugman, J. (2001). Iris recognition. *Am. Sci.* 89, 326–333.

De Sousa, P. A., King, T., Harkness, L., Young, L. E., Walker, S. K., and Wilmut, I. (2001). Evaluation of gestational deficiencies in cloned sheep fetuses and placentae. *Biol. Reprod.* 65, 23–30.

Eggan, K., Akutsu, H., Hochedlinger, K., Rideout, W. III, Yanagimachi, R., and Jaenisch, R. (2000). X-Chromosome inactivation in cloned mouse embryos. *Science* 290, 1578–1581.

Eggan, K., Akutsu, H., Loring, J., Jackson-Grusby, L., Klemm, M., Rideout, W. III, Yanagimachi, R., and Jaenisch, R. (2001). Hybrid vigor, fetal overgrowth, and viability of mice derived by nuclear cloning and tetraploid embryo complementation. *Proc. Natl. Acad. Sci. U.S.A.* **98**, 6209–6214.

Evans, M. J., Gurer, C., Loike, J. D., Wilmut, I., Schnieke, A. E., and Schon, E. A. (1999). Mitochondrial DNA genotypes in nuclear transfer-derived cloned sheep. *Nature Genet.* **23**, 90–93.

Finch, C. E., and Kirwood, B. L. (2000). *"Chance, Development, and Aging."* Oxford University Press, New York.

Gardner, R. L., Lyon, M. F., Evans, E. P., and Burtenshaw, M. D. (1985). Clonal analysis of X-chromosome inactivation and the origin of the germ line in the mouse embryo. *J. Embryol. Exp. Morphol.* **88**, 349–363.

Garry, F. B., Adams, R., McCann, J. P., and Odde, K. G. (1996). Postnatal characteristics of calves produced by nuclear transfer cloning. *Theriogenology* **45**, 141–157.

Gärtner, K., Bondioli, K., Hall, K., and Rapp, K. (1998). High variability of body size within nucleus-transfer-clones of calves: Artifacts or a biological feature. *Reprod. Domest. Anim.* **33**, 67–75.

Green, R. D., Diles, J. J. B., Hughes, L. J., Shepard, H. H., and Matthews, G. L. (1996). Variation in birth and weight traits of identical calves produced through nuclear transplantation. *Prof. Anim. Sci.* **12**, 238–243.

Gringras, P. (1999). Identical differences. *Lancet* **353**, 562.

Hill, J. R., Roussel, A. J., Cibelli, J. B., Edwards, J. F., Hooper, N. L., Miller, M. W., Thompson, J. A., Looney, C. R., Westhusin, M. E., Robl, J. M., and Stice, S. L. (1999). Clinical and pathologic features of cloned transgenic calves and fetuses (13 case studies). *Theriogenology* **51**, 1451–1465.

Hill, B. R., Lefort, M., L'arrivee, R., *et al.* (2000a). A retrospective study of the effects of a single sire on embryo transfer results. *Theriogenology* (Abstr.) **53**, 310.

Hill, J. R., Burghardt, R. C., Jones, K., Long, C. R., Looney, C. R., Shin, T., Spencer, T. E., Thompson, J. A., Winger, Q. A., and Westhusin, M. E. (2000b). Evidence for placental abnormality as the major cause of mortality in first-trimester somatic cell cloned bovine fetuses. *Biol. Reprod.* **63**, 1787–1794.

Humpherys, D., Eggan, K., Akutsu, H., Hochedlinger, K., Rideout, W. M. III, Biniszkiewicz, D., Yanagimachi, R., and Jaenisch, R. (2001). Epigenetic instability in ES cells and cloned mice. *Science* **293**, 95–97.

Khosla, S., Dean, W., Brown, D., Reik, W., and Fell, R. (2001). Culture of preimplantation mouse embryos affects fetal development and the expression of imprinted genes. *Biol. Reprod.* **64**, 918–926.

Kubota, C., Yamakuchi, H., Todoroki, J., Mizoshita, K., Tabara, N., Barber, M., and Yang, X. (2000). Six cloned calves produced from adult fibroblast cells after long-term culture. *Proc. Natl. Acad. Sci. U.S.A.* **97**, 990–995.

Lanza, R. P., Cibelli, J. B., Blackwell, C., Cristofolo, V. J., Francis, M. K., Boerlocher, G. M., Mak, J., Schutzer, M., Chavez, E. A., Sawyer, N., Lansdorp, P. M., and West, M. D. (2000). Extension of cell life-span and telomere length in animals cloned from senescent somatic cells. *Science* **288**, 665–669.

Moore, T. (2001). Genetic conflict, genomic imprinting, and the establishment of the epigenotype in relation to growth. *Reproduction* **122**, 185–193.

Peura, T. T., Lane, M. W., Lewis, I. M., and Trounson, A. O. (2001). Development of bovine embryo-derived clones after increasing rounds of nuclear recycling. *Mol. Reprod. Dev.* **58**, 384–389.

Reik, W., Römer, I. Barton, S. C., Surani, M. A., Howlett, S. K., and Klose, J. (1993). Adult phenotype in the mouse can be affected by epigenetic events in the early embryo. *Development* **119**, 933–942.

Roemer, I., Reik, W., Dean, W., and Klose, J. (1997). Epigenetic inheritance in the mouse. *Curr. Biol.* **7**, 277–280.

Seidel, G. E., Jr. (1983). Production of genetically identical sets of mammals: Cloning? *J. Exp. Zool.* **228**, 347–354.

Seidel, G. E., Jr. (1985). Are identical twins produced from micromanipulation always identical? "Proc. Annu. Conf. Artif. Insem. Embryo Transfer in Beef Cattle," pp. 50–53. Natl. Assoc. Anim. Breeders, Columbia, MO.

Seidel, G. E., Jr. (2000). Reproductive biotechnologies and the "big" biological questions. *Theriogenology* **53**, 187–194.

Shiels, P. G., Kind, A. J., Campbell, K. H. S., Waddington, D., Wilmut, I., Colman, A., and Schnieke, A. E. (1999). Analysis of telomere lengths in cloned sheep. *Nature* **398**, 316–317.

Sinclair, K. D., Young, L. E., Wilmut, I., and McEvoy, T. G. (2000). In-utero overgrowth in ruminants following embryo culture: Lessons from mice and a warning to men. *Hum. Reprod.* **15** (Suppl. 5), 68–86.

Smith, C. (1989). Cloning and genetic improvement of beef cattle. *Anim. Prod.* **49**, 49–62.

Stearns, T. (2001). Centrosome duplication: A centriolar pas de deux. *Cell* **105**, 417–420.

Surani, M. A. H., Barton, S. C., and Norris, M. L. (1986). Nuclear transplantation in the mouse: Heritable differences between parental genomes after activation of the embryonic genome. *Cell* **45**, 127–136.

Tam, P. P. L., and Snow, M. H. L. (1981). Proliferation and migration of primordial germ cells during compensatory growth in mouse embryos. *J. Embryol. Exp. Morphol.* **64**, 133–147.

Thompson, J. G., Gardner, D. K., Pugh, P. A., McMillan, W. H., and Tervit, H. R. (1995). Lamb birth weight is affected by culture system utilized during *in vitro* pre-elongation development of ovine embryos. *Biol. Reprod.* **53**, 1385–1391.

Van Vleck, L. D. (1999). Implications of cloning for breed improvement strategies: Are traditional methods of animal improvement obsolete? *J. Anim. Sci.* **77** (Suppl. 2), 111–121.

vom Saal, F. S. (1989). Sexual differentiation in litter-bearing mammals: Influence of sex of adjacent fetuses *in utero. J. Anim. Sci.* **67**, 1824–1840.

Wachtel, S. S., Somkuti, S. G., and Schufield, J. S. (2001). Monozygotic twins of opposite sex. *Cytogenet. Cell Genet.* **91**, 293–295.

Wakayama, T., Shinkai, Y., Tamashiro, K. L. K., Niida, H., Blanchard, D. C., Blanchard, R. J., Ogura, A. Tanemura, K., Tachibana, M., Perry, A. C. F., Colgan, D. F., Mombaerts, P., and Yanagimachi, R. (2000). Cloning of mice to six generations. *Nature* **407**, 318–319.

Walker, S. K, Heard, T. M., and Seamark, R. F. (1992). *In vitro* culture of sheep embryos without co-culture: Successes and perspectives. *Theriogenology* **37**, 111–126.

Wells, D. N., Misica, P. M., and Tervit, H. R. (1999). Production of cloned calves following nuclear transfer with cultured adult mural granulosa cells. *Biol. Reprod.* **60**, 996–1005.

Willadsen, S. M., Janzen, R. E., McAlister, R. J., Shea, B. F., Hamilton, G., and McDermand, D. (1991). The viability of late morulae and blastocysts produced by nuclear transplantation in cattle. *Theriogenology* **35**, 161–170.

Wilson, J. M., Williams, J. D., Bondioli, K. R., Looney, C. R., Westhusin, M. E., and McCalla, D. F. (1995). Comparison of birth weight and growth characteristics of bovine calves produced by nuclear transfer (cloning), embryo transfer and natural mating. *Anim. Reprod. Sci.* **38**, 73–83.

Wrenzycki, C., Wells, D., Herrmann, D., Miller, A., Oliver, J., Tervit, R., and Niemann, H. (2001). Nuclear transfer protocol affects messenger RNA expression patterns in cloned bovine blastocysts. *Biol. Reprod.* **65**, 309–317.

CHAPTER 11

GENETIC MODIFICATION AND CLONING IN MAMMALS

Patrick W. Dunne and Jorge A. Piedrahita

GENERAL INTRODUCTION

Creating transgenic animals by pronuclear injection and by transformation of somatic cells followed by cloning is now becoming routine. The challenge now and into the foreseeable future will be to create transgenic animals with targeted alterations of specific genes as quickly and reliably as is now possible with mice. This chapter first reviews the conventional techniques for making transgenic livestock, and their associated limitations. Next, the mechanisms that determine whether a targeting vector integrates randomly in the genome (nonhomologous end joining) or precisely at the targeted gene (homologous recombination) are examined and possible means of manipulating each pathway to achieve enhanced rates of gene targeting are discussed. Finally, the problem of senescence of somatic cells in culture is reviewed and some possible solutions are suggested.

HISTORICAL PERSPECTIVE

The initial introduction of genes into live animals relied on the technique of pronuclear injection, and to a lesser extent retroviral vectors. The technique of pronuclear injection, or the introduction of new genes into the pronucleus of fertilized embryos by microinjection, was the first technique to be broadly utilized for the generation of transgenic mice. In 1980 and 1981, six papers detailed studies on DNA microinjection into fertilized mouse embryos (Gordon et al., 1980; Wagner et al., 1981a,b; Costantini and Lacy, 1981; Brinster et al., 1981). Since then, the generation of transgenic mice, and other mammals to a much lesser extent, has become routine.

Pronuclear injections, the use of viral vectors, and more recently electroporation/lipofection of DNA into somatic cells all rely on a random insertion approach. Ideally, the transgene expression should mimic endogenous patterns of expression with respect to tissue, developmental and temporal specificity, and levels of expression, but that is rarely the case. This is due to the effects of the neighboring chromatin on the behavior of the transgene. Insertion of the transgene in heterochromatic regions, with their closed chromatin configuration, can affect the accessibility of transcription factor binding to the regulatory regions and affect the proper expression pattern, and expression levels, of the transgene. As a result, transgenic animals made by random insertion approaches can be radically different from each other, even when the same transgene is used. This requires the establishment of different transgenic founder lines that can be analyzed separately.

In addition, the problem of random insertion is also associated with insertional inactivation, the process whereby the transgene inserts within a critical gene, disrupting it. This effect can prove deleterious or even lethal to the developing embryo

(Schnieke *et al.*, 1983). Fortunately the frequency of insertional inactivation leading to an observable phenotype is quite low. In the technique of pronuclear injection the efficiency of transgene integration can be quite variable, especially in domestic species. Ebert and Schindler (1993) reported a low efficiency for the production of transgenic farm animals (porcine, ovine, bovine, caprine), with a range of 0.0–4.0%. This compares with a transgenic efficiency of 10–40% in mice (Palmiter and Brinster, 1986). Thus, the production of transgenic animals by pronuclear injection is hindered by its low efficiency and its variable transgene regulation.

Fortunately, developments in the area of cloning with somatic cells provide a more efficient alternative to pronuclear injection. Using cloning as the basis for generation of transgenic animals, somatic cells are genetically modified and selected *in vitro*, and only those containing the transgene are used for the nuclear transfer procedure. Thus, all animals born from the procedure are transgenic (100% efficiency). Moreover, with advances in manipulation of somatic cells, and the development of sensitive and rapid molecular detection techniques, it is possible to determine rapidly the approximate number of copies inserted into each cell line and whether one or more integration sites are present. As a result, issues of copy number and number of integrations can be resolved prior to cloning the transgenic animals. Cloning also affords an additional benefit in that animals generated in this manner are 100% derived from the transgenic cell. In contrast, animals generated by pronuclear injection or retroviral vectors tend to be mosaics of both transgenic and nontransgenic cells, thus requiring the breeding and segregation of the different genotypes (Bradl *et al.*, 1991). The increase in efficiency of the nuclear transfer procedure has been demonstrated in sheep by Schnieke *et al.* (1997). Unfortunately, issues associated with random insertion, such as the positional effect, are not resolved by the cloning procedure.

In order to reduce the positional effect, several cis-acting elements can be included in transgenic constructs to promote a position-independent expression pattern. These elements include locus control regions (LCRs), insulators, and matrix attachment regions (MARs) (Reitman *et al.*, 1993; McKnight *et al.*, 1992; Krnacik *et al.*, 1995).

A more powerful approach to resolving the issues of random insertion is the utilization of techniques that allow the placement of the transgene in specific regions of the chromosome. These techniques include gene targeting by homologous recombination, and site-directed recombination mediated by enzymes such as Cre and Flp (Ringrose *et al.*, 1998). The system with the most flexibility and applicability is gene targeting by homologous recombination, because it allows insertions, deletions, and introduction of point mutations in any chromosomal region. Unfortunately, due to its high degree of specificity, the frequency of accurate homologous recombination in embryonic stem (ES) cells is on the order of one accurate event per million cells treated. As a result, in order to introduce precise modifications into the germ line, cell lines are required that can be easily manipulated *in vitro* yet retain the ability to generate an animal once the modification has been completed. In mice, the carrier cell lines of choice has been the embryonic stem cells. These cell lines are derived from the inner cell mass of the preimplantation blastocyst and, under the proper conditions, can be cultured for an indefinite period of time in an undifferentiated state. Yet, even after prolonged culture they retain the ability to differentiate into derivatives of all three embryonic germ layers when injected into a host blastocyst. It is the ability to culture indefinitely without senescence or loss of differentiative capacity that has made the ES cell such a valuable tool for the modification of the mouse germ line.

Although the ES cell technology has been very successful for the generation of gene-targeted mice, attempts at developing the ES technology in other mammalian species have been disappointing. Doetschman *et al.* (1988) showed that ES cells could be isolated from hamster embryos using feeders composed of murine primary embryonic fibroblasts. However, the isolated ES cells were not capable of participating in chimera formation (J. Piedrahita, unpublished observations). Our group and others (Evans *et al.*, 1990; Gerfen and Wheeler, 1995; Notarianni *et al.*, 1990; Piedrahita *et al.*, 1990; Strojek *et al.*, 1990), using an STO cell line as feeder layers, have reported the isolation of porcine embryo-derived cell lines with ES-like morphology and a limited ability to differentiate *in vitro* and *in vivo*. However, to date there have been no reports of germ line transmission of the ES genotype in swine, or any species other than mice.

Alternate sources of carrier cells are cell lines derived from primordial germ cells (Labosky *et al.*, 1994; Stewart *et al.*, 1994). These cells, referred to as embryonic germ cells (EG) or primordial germ cell (PGC)-derived cells, are indistinguishable from ES cells in terms of markers of the undifferentiated state as well as their ability to colonize the germ line following injection into a host blastocyst. Shim *et al.* (1997) reported the ability of PGC-derived cell lines to contribute to the formation of a porcine chimera, demonstrating the pluripotential characteristics of these cell lines. We extended this observation by demonstrating the ability of genetically transformed PGCs to contribute to chimera formation (Piedrahita *et al.*, 1998). These results demonstrated the ability of EG cells to participate in chimera formation, but we have not been able to demonstrate germ line transmission of the EG component. At this point, therefore, neither ES nor EG cell lines are appropriate alternatives for generation of targeted animals in domestic species. Presently their value lies in their pluripotential abilities and their potential use for tissue engineering and repair.

Fortunately, breakthroughs in the area of cloning using somatic cells as donor nuclei have provided alternate cell lines for genetic manipulation (Campbell *et al.*, 1996; Wells *et al.*, 1997; Wilmut *et al.*, 1997). Both fetal and adult fibroblasts as well as other cell types can be maintained in culture for a limited period of time, manipulated genetically, and still retain the ability to generate live offspring. In our experience, however, the targeting efficiency of somatic cells is considerably lower than that of ES cells. Notwithstanding this inefficiency, there have been reports of successful targeting in domestic animal somatic cells (McCreath *et al.*, 2000; Denning *et al.*, 2001).

In summary, recent advances in the isolation and culture of PGC-derived cells from domestic species, combined with advances in somatic cell cloning, provide the carrier cell lines required to undertake the process of homologous recombination in the domestic species.

CONVENTIONAL ENRICHMENT PROTOCOLS TO ENHANCE HOMOLOGOUS RECOMBINATION

The process of homologous recombination involves the replacement of an endogenous DNA fragment with the fragment of interest. The primary concern in dealing with homologous recombination is the low efficiency at which foreign DNA is correctly inserted into the host chromosome. In ES cells this frequency is in the range of 10^{-5} to 10^{-6}. In order to increase the ease with which targeted cells are identified, selectable markers such as neomycin resistance and thymidine kinase genes are utilized. In addition, several enrichment approaches can be utilized:

   1. Positive–negative selection (PNS) process (Mansour *et al.*, 1988). This utilizes a positive marker for identification of random and targeted cell lines and a negative selection marker to select against random events.
   2. Promoterless selectable markers (Charron *et al.*, 1990). Promoterless vectors are effective in that they are only activated if they "trap" the promoter of

the gene one is attempting to target. To be effective it requires expression of the target gene in the cell lines being modified.

3. Polyadenylation signalless marker (Joyner et al., 1989). Based on the ability of the polyA to stabilize the messenger RNA. A successful targeting event uses the polyA from the target gene.

4. Length of homology. Although it has been possible to induce homologous recombination with as little as 1.2 kb of homology (Doetschman et al., 1987), others have demonstrated that targeting frequencies can be augmented by increasing the length of uninterrupted homology between the incoming targeting plasmid and the endogenous locus (Hasty et al., 1991; Deng and Capecchi, 1992). In general, homologies in the range of 4–8 kb have been found to yield homologous recombinants at a high frequency.

5. Isogenic DNA. The use of isogenic DNA can be extremely important, depending on the locus of interest. In general, targeting DNA that is isogenic to the locus being modified undergoes the correct event with a higher efficiency than does nonhomologous DNA (Deursen and Wieringa, 1992; Riele et al., 1992). The reduced targeting frequency with nonisogenic DNA has been attributed to the presence of scattered base mismatches in the regions involved in the homologous recombination event between the incoming plasmid and the endogenous locus. Although this is not a problem in inbred mouse strains, it is a considerable problem in outbred domestic species. Fortunately, this problem can be overcome by synthesizing isogenic DNA in domestic animals by long-range polymerase chain reaction (PCR). Long-range PCR has been used previously for obtaining isogenic DNA for the generation of targeting constructs (Randolph et al., 1996). The technology is based on the combination of DNA polymerases and can produce amplicons 10–12 kb in length with few if any mismatches. Targeting with long-range PCR-derived constructs has been accomplished in mice with no differences in targeting rates observed between long-range-generated constructs and conventionally (libraries) derived constructs (Barnes, 1994).

MODULATION OF HOMOLOGOUS RECOMBINATION AND NONHOMOLOGOUS END JOINING DNA REPAIR PATHWAYS

As an alternative to improving the means of selecting for targeted events, and as a way of overcoming the low frequency of homologous targeted events, we have developed a method to increase homologous recombination. In addition to the problems associated with the use of nonisogenic DNA for gene targeting, there are reports indicating that the absolute targeting frequency is higher in ES cells than in other cell types. Adair et al. (1989) reported targeting of a nonexpressed gene in Chinese hamster somatic cells using a conventional, transcriptionally active targeting construct. The frequency of recombination in the latter case, however, was very low, with only 1 of 4000 transgenic clones analyzed being a targeted event. Also, there is an increase in the homologous recombination rate in transformed versus non-transformed somatic cells (Thygaryajan et al., 1996a). Our own results reinforce the difficulties associated with targeting somatic cells. Results of targeting in murine ES cells by our group are in accordance with frequencies reported by other groups and range between 1 and 60 targeted events per million cells electroporated. In contrast, the frequency of targeting in porcine and bovine somatic cells is 100–1000× lower in both expressed and nonexpressed genes (P. Dunne, B. Mir, G. Zaunbrecher, and J. Piedrahita, unpublished observations). However, a successful targeting of fetal fibroblasts cells, followed by cloning, has been reported at the expressed sheep COL1A1 locus as well as the prion protein and 1,3-galactosyl transferase loci (McCreath et al., 2000; Denning et al., 2001), indicating that even though difficult, the process is doable.

GENE TARGETING: EXPLOITING MECHANISMS OF DNA DOUBLE-STRAND BREAK REPAIR IN MAMMALIAN CELLS

A DNA double-strand break (DSB) that remains unrepaired can lead to extensive chromosome instability in cells and/or the induction of cell death. DSBs can be produced by exogenous agents such as ionizing radiation (for example, gamma rays), or as a part of normal cellular metabolism. For example, if during DNA replication, a nick in either the leading or lagging strand were encountered by the replication machinery, a double-strand break would occur (Kowalczykowski, 2000). Mammalian cells have evolved two major pathways to repair DSBs: homologous recombination (HR) and nonhomologous end joining (NHEJ). Homologous recombination repairs the lesion by using a sister chromatid or a homologous chromosome as a template to effect an error-free repair. It acts in late S phase or G2 when a sister chromatid template is available or at a stalled replication fork (Takata *et al.*, 1998; Bierne and Michel, 1994).

Nonhomologous end joining (or illegitimate recombination) requires only short regions of microhomology (several bases) around the site of the DSB, but typically results in insertions or deletions at the repaired break (Roth and Wilson, 1986). NHEJ is active during G1–S phase of the cell cycle (Takata *et al.*, 1998). In contrast to yeast, genetic screens of mammalian cells sensitive to formation of DSBs by ionizing radiation yield only mutants in the NHEJ pathway, not mutants in the HR repair pathway (Jackson and Jeggo, 1995). Moreover, when mammalian cells are modified to measure the relative contributions of HR and NHEJ to DSB repair, the predominant repair mechanism is NHEJ (Godwin *et al.*, 1994; Sargent *et al.*, 1997). Together these results suggest that mammalian cells greatly prefer to repair DSBs by NHEJ. This preference for repairing DSBs by NHEJ can drastically affect the ratio of targeted versus random integrands (Hanson and Sedivy, 1995) (Fig. 1).

MODELS OF HOMOLOGOUS RECOMBINATION REPAIR

Two models have been proposed to account for HR DSB repair: the DSB repair model of Szostak *et al.* (1983) and the synthesis-dependent strand-annealing (SDSA) model of Nassif *et al.* (1994). The Szostak model (Fig. 2) predicts that resolution of the two Holliday junctions formed following invasion of a 3′ end from the broken strand can yield either cross-over or noncross-over products with equal frequency. Because each 3′ end of the broken duplex acts as a priming site for repair synthesis, both the donor duplex (unbroken dsDNA) and the recipient duplex (broken dsDNA) will contain a newly synthesized strand. Two difficulties have been observed for this model. Heteroduplex DNAs derived from mismatches between the broken recipient dsDNA and its donor template are rarely observed in the donor, but are frequently seen in the recipient, suggesting that information flow is usually unidirectional from donor to recipient (Paques *et al.*, 1998; Richardson *et al.*, 1998). Second, the great majority of DSB repairs in mammalian cells are mediated by gene conversion without cross-over of flanking markers (Sonada *et al.*, 1998; Johnson and Jasin, 2000; Richardson *et al.*, 1998). How mammalian cells suppress reciprocal exchange in homologous recombination repair is unknown, but it has been suggested that such a mechanism would inhibit undesirable alterations such as loss of heterozygosity associated with carcinogenesis in mammalian cells (Moynahan and Jasin, 1997). Several variants of the SDSA model have been proposed that explain these observations without invoking resolution of Holliday junctions (Nassif *et al.*, 1994; Paques *et al.*, 1998; Richardson *et al.*, 1998; Holmes and Haber, 1999). In these models recombination repair is coupled with DNA replication. Strand invasion and new DNA synthesis can be initiated from either 3′ end of the DSB (Fig. 3). Leading-strand synthesis occurs within the displaced D-loop, with newly repli-

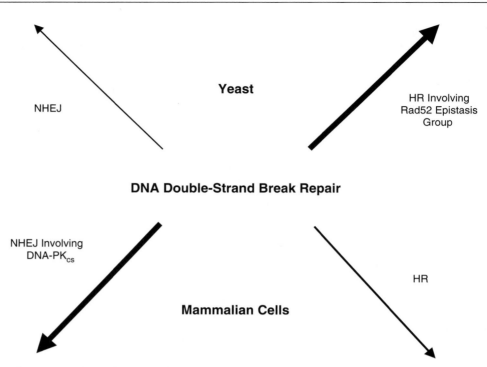

Figure 1 *Nonhomologous end joining. Yeast vs. mammalian double-strand bread repair. DSB repair is conserved between yeast and mammalian cells but the predominant pathways utilized by each organism differ. Yeast cells repair the majority of its DSBs by homologous recombination, whereas mammalian cells preferentially use a DNA-PK$_{cs}$-dependent non-homologous end-joining mechanism. After Jeggo (1997).*

cated strand is then displaced from the template, annealing to the complementary broken strand. The 3′ end of the opposite end of the DSB can then either anneal to the newly synthesized strand, with strand extension by lagging-strand synthesis using the newly synthesized strand as template (gene conversion without cross-over), or in rare cases anneal with the D-loop to form a pair of Holliday junctions (reciprocal exchange). In this model almost all of the newly synthesized DNA would be contained in the recipient (unidirectional gene conversion). An implication of the SDSA model is that a targeting construct that is linearized becomes a recipient of genetic information from the target locus, rather than the other way round.

BIOCHEMISTRY OF HR REPAIR

The proteins that participate in HR repair in mammalian cells are predicted to be the same for both models of homologous recombination outlined above, although recombination is tied explicitly to DNA replication in the strand-annealing model. The HR proteins were originally identified in mutant screens in yeast based on their sensitivity to X rays and were placed in the same pathway by epistasis analysis, usually referred to as the Rad52 epistasis group (Game, 1993)

Rad51

The homologous recombination pathway integrates the targeting DNA at the homologous site in the genome by first scanning for the targeted gene through the homology-seeking and strand-exchange protein Rad51 (Baumann *et al.*, 1996). Once the targeted gene is aligned with the genetically altered gene in the targeting vector, Rad51 initiates the strand-exchange process while other associated proteins of the HR pathway complete the exchange process. Rad51 is essential for HR repair in mammals, because a knockout is lethal (Lim and Hasty, 1996). The gene is expressed at elevated levels in proliferative tissues such as spleen, thymus, ovary, and testis, suggesting an association between HR repair and DNA replication (Shinohara *et al.*, 1993; Yamamoto *et al.*, 1996).

Rad52

Like Rad51, BRCA1, and BRCA2, Rad52 shows elevated levels of expression in proliferative tissue such spleen and testes, but unlike Rad51, BRCA1, and BRCA2, a mouse Rad52 knockout had only marginal effects on DNA repair and homologous recombination (Rijkers *et al.*, 1998), although targeted integration frequencies were reduced in Rad52 homozygous knockout cell lines (Yamaguchi-Iwai *et al.*, 1998). Rad51 and Rad52 have been shown to interact (Shinohara *et al.*, 1992; Milne and Weaver, 1993) suggesting that they cooperate in HR. Moreover, Rad52 binding to Rad51 stimulates Rad51 strand exchange *in vitro* (Shinohara and Ogawa, 1998) and Rad52 has been shown to bind to DSBs *in vitro*, protecting the ends from degradation and perhaps targeting Rad51 to DSBs (Van Dyck *et al.*, 1999).

Rad54

Mouse ES cells in which both copies of the Rad54 gene were disrupted showed reduced levels of homologous recombination as indicated by gene-targeting frequency, whereas knockout mice were viable (Essers *et al.*, 1997). Sister chromatid exchange was also significantly reduced in Rad54 and Rad51 double-knockout mammalian cell lines (Sonoda *et al.*, 1999) Targeted integration in mammalian cell lines was also greatly reduced in a Rad54 double knockout (Bezzubova *et al.*, 1997). *In vitro*, Rad54 appears to interact with the Rad51 filament prior to homology search, promoting strand exchange and joint molecule formation (Solinger *et al.*, 2001).

Rad50, Mre11, AND p95 COMPLEX

Mre11 together with Rad50 and p95 associate to form a complex with exonuclease activity that is essential for processing dsDNA ends in homologous recombination and in NHEJ (Dolganov *et al.*, 1996; Carney *et al.*, 1998). Disruption of the Rad50 gene is lethal in ES cells, and knockout mice show a deficiency in proliferation, and display embryonic lethality at E6.5 when rapid proliferation occurs, suggesting that sister chromatid-based HR repair of spontaneously arising DSBs is inoperative in Rad50 knockouts (Luo *et al.*, 1999). Goedecke *et al.* (1999) showed that the truncated Mre11 could directly interact with Ku70 in somatic cells. Truncated Mre11 should sequester much of cellular Ku70 in an inactive Mre11trunc/Ku70 dimer, preventing formation of functional Ku70/Ku80 heterodimers.

BRCA1 AND BRCA2

These proteins are known to interact with Rad51 (Scully *et al.*, 1997) as a positive regulator, perhaps by acting as a bridge between Rad51 and the Mre11/Rad50/Nbs1 complex (Zhong *et al.*, 1999). Like Rad51, both BRCA1 and BRCA2 are highly expressed in proliferating cells (Rajan *et al.*, 1997) and in response to DNA damage colocalize with Rad51 at replication forks during S phase (Scully *et al.*, 1997), consistent with the SDSA model of HR coupled with replication. BRCA1 is also

A

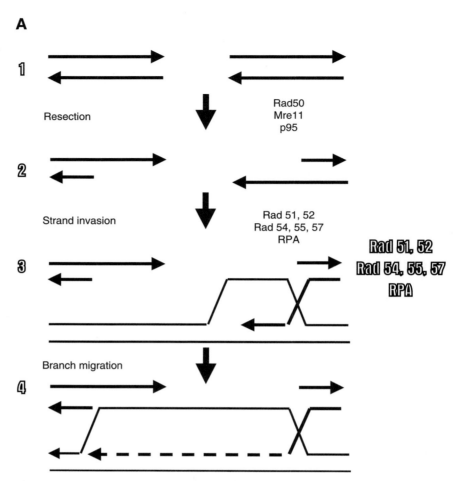

Figure 2 *Double-strand break repair model of Szostak. (A) Szostak et al. (1983) proposed that the two ends of the double-strand break (1) are resected by a 5′ → 3′ exonuclease activity [now identified as the Rad50/Mre11 complex (2)], which then permits the 3′ ends to invade the intact or donor homologous template to form a holiday junction (3) that undergoes extension by a DNA polymerase (4). The strand displaced by the D-loop can pair with sequences from the other end of the double-strand break (5), forming a primer for DNA repair synthesis. (B) The two Holliday junctions formed (5) can be resolved independently, yielding without a cross-over of flanking markers (6) or with an accompanying cross-over (7) with equal probability.*

expressed in sperm cells undergoing meiosis (Zabludoff *et al.*, 1996). More direct evidence for the involvement of BRCA1 in homologous recombination comes from knockout experiments indicating that BRCA minus ES cells have elevated levels of NHEJ and reduced levels of homologous recombination (Snouwaert *et al.*, 1999). Expression of BRCA1 and BRCA2 is not, however, involved with NHEJ (Wang *et al.*, 2001).

OVEREXPRESSION OF NORMAL AND DOMINANT-NEGATIVE FORMS OF HR PROTEINS ENHANCE HOMOLOGOUS RECOMBINATION

More than 25 individual proteins have so far been identified that participate in HR in *Escherichia coli* cells (the best-studied organism), and many of these proteins have functional homologs in mammalian cells (Kowalczykowski and Zarling, 1995).

B

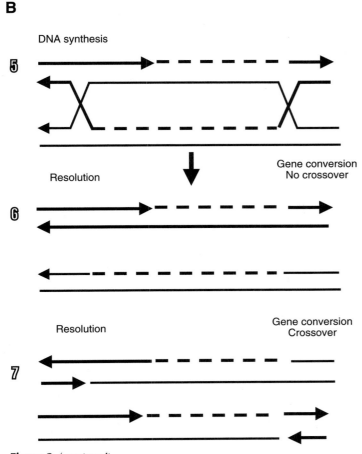

Figure 2 *(continued)*

Several of these proteins, such as the *E. coli* strand-exchange protein RecA, have been shown to enhance HR when overexpressed in eukaryotic cells. RecA protein plays a central role in homologous recombination in *E. coli* and the extensive characterization of its structure and function has served as a model for the homology-search and strand-exchange reactions of its eukaryotic counterpart, Rad51 (West, 1992). RecA protein can pair any two homologous DNA molecules and initiate strand exchange leading to the production of heteroduplex DNA recombination intermediates. Overexpression of RecA in both plant protoplasts and a mammalian cell line stimulated HR 10-fold compared to nontransfected cells (Reiss *et al.*, 1996; Shcherbakova *et al.*, 2000). Moreover, the effect in mammalian cells directly measured the gene-targeting rate, rather than the less relevant rate of intrachromosomal recombination. Rad51 plays a critical role in the mammalian homologous recombination pathway, mediating the homology-search and strand-exchange reaction (see above). It acts in concert with Rad52, Rad50, and several other proteins to complete error-free recombination repair (Fig. 4). Rad52 binds directly to double-stranded breaks *in vitro*, protecting ends from endonuclease digestion and promoting end-to-end interactions (Van Dyck *et al.*, 1999). Van Dyck *et al.* (1999) propose that Rad52 competes with Ku, the end-binding protein of the NHEJ pathway, by directing DSBs into the homologous recombination pathway rather than the alternative NHEJ pathway. This hypothesis suggests that overexpression of Rad52 will augment HR whereas inhibition of Ku will suppress NHEJ.

RuvC is a Holliday junction-specific endonuclease (Takahagi *et al.*, 1991) that promotes crossed-strand resolution by introducing a pair of nicks in strands of the

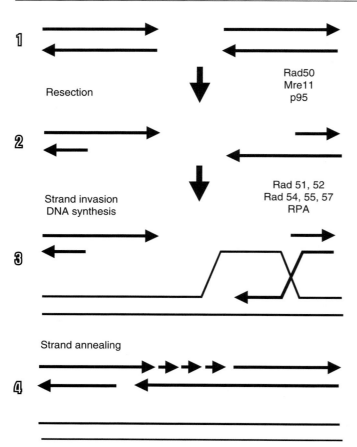

Figure 3 Synthesis-dependent strand annealing model. In this model, the proteins that participate in the resection (2) and the strand invasion (3) steps are the same as in the Szostak model. In contrast to that model, this model predicts that the invasion by one of the resected 3′ ends is transient, undergoing leading-strand DNA synthesis followed by annealing to the complementary strand on the opposite side of the double-strand break (4). Lagging-strand synthesis using the newly synthesized strand as template completes the repair of the broken recipient strand. In this model, there is no Holliday junction and the donor (unbroken) duplex is completely intact following the DNA repair.

same polarity. The nicked duplexes are sealed by DNA ligase. RuvC binds to the junction as a dimer (Ariyoshi *et al.*, 1994) and apparently without preference for DNA sequence (Bennett *et al.*, 1993). Shalev *et al.* (1999) have demonstrated that overexpression of RuvC can stimulate homologous recombination in eukaryotic cells when localized to the cell nucleus. Experiments have been carried out overexpressing Rad51 and Rad52 in mammalian cells with similar dramatic effects on HR (Vispe *et al.*, 1998; Johnson *et al.*, 1996).

An alternative to stimulating individual members involved in HR is to activate the entire HR pathway simultaneously by inactivating a suppressor of this pathway. The feasibility of this approach is suggested by the observation that expression of several mutant forms of the tumor suppressor protein p53 significantly elevates HR (Mekeel *et al.*, 1997). The gene for p53 is a classic tumor suppressor gene that is mutated in over 50% of human tumors (Harris, 1993; Friend, 1994). It functions as a checkpoint gene that can initiate cell cycle arrest in G1/S in response to DNA damage, or in some cell types stimulate the apoptosis pathway. As a transcription factor, p53 mediates its effect as a transactivator or repressor of transcription for many genes (Levine, 1997). Because p53 appears to bind DNA as a tetramer (Sturzbecher *et al.*, 1992), this may explain how some of the commonest point mutations in human tumors cause tumorigenesis through a dominant-negative genetic mechanism by forming inactive p53 heterooligomers (Hollstein *et al.*, 1994). Moreover, several of these single point mutations have been shown to act as dominant-negative mutations that can derepress the homologous recombination pathway in cell culture by as much as 100-fold over wild-type p53 (Mekeel *et al.*, 1997). Moreover, Buchop *et al.* (1997) report direct interaction between normal p53 and the central HR recombinase Rad51.

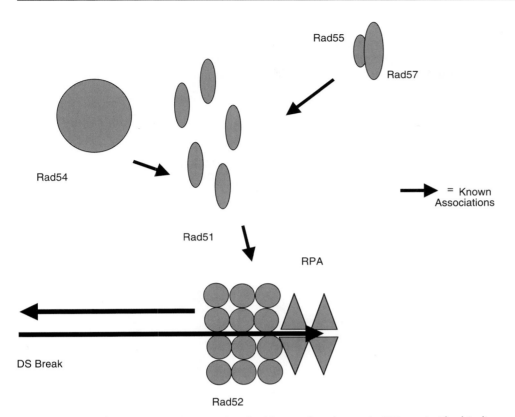

Figure 4   Recombination proteins associated with strand exchange in HR repair. The binding of replication protein A (RPA), a single-strand DNA-binding protein, to the resected single-strand 3' end of the double-strand break protects the end from degradation and removes secondary structure. The strand-exchange protein Rad51 is then recruited to the DNA by the end-binding protein Rad52 with the aid of Rad54 and the Rad55/57 heterodimer. Resection of the double-strand break to create a single-strand 3' overhang is generated by the Rad50/Mre11/p95 complex in mammalian cells (not illustrated).

RANDOM INSERTION OF TARGETING CONSTRUCTS MEDIATED BY NONHOMOLOGOUS END JOINING

More than 10 proteins have been identified that play a role in the NHEJ pathway in eukaryotes including the Mre11 complex (Chu, 1997; Critchlow and Jackson, 1998) (Fig. 5). The best characterized are the three proteins that cooperate to initiate the process: Ku70, Ku80, and DNA-PK$_{cs}$. Ku70 and Ku80 form a tightly associated heterodimer of ~70- and ~80-kDa subunits that bind to DNA ends (Jin and Weaver, 1997). On binding, the heterodimer recruits the ~470-kDa DNA-dependent protein kinase catalytic subunit and induces its kinase activity (Yoo and Dynan, 1998). In addition, it appears that the final 12 amino acids of Ku80 are essential for its specific interaction with DNA-PK$_{cs}$ (Gell and Jackson, 1999). X-Ray-sensitive rodent mutants allowed identification of Ku70 (XRCC6), Ku80 (XRCC5), and DNA-PK$_{cs}$ (XRCC7) as part of a complex defective in DSBs and other processes mediated by NHEJ, such as V(D)J recombination (Featherstone and Jackson, 1999; Zdzienicka, 1999). Other components of NHEJ include the Mre11/Rad50/p95 complex that also processes the ends of DSBs in HR (Tsukamoto et al., 1996). In mammalian cells, in addition, ligation of the DSBs requires DNA ligase IV and XRCC4 (Grawunder et al., 1998) and the recruitment of the XRCC4/ligase IV complex to DNA ends by the Ku heterodimer (Nick-McElhinny et al., 2000).

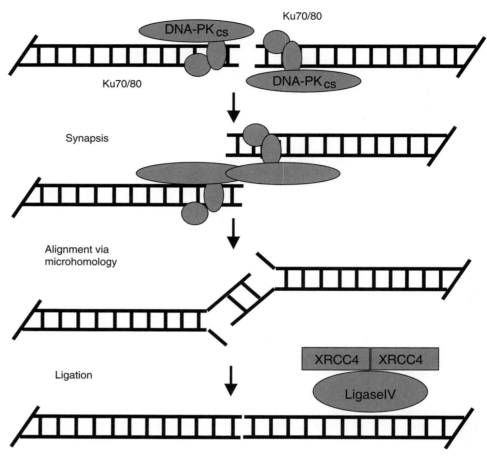

Figure 5 *The nonhomologous end-joining (NHEJ) model. The DNA double-strand break is recognized by the Ku70/80 heterodimer, which then recruits the DNA-dependent protein kinase catalytic subunit to form an active complex, permitting the synapsis of the two ends of the DSB. Following synapsis, transphosphorylation of Ku and DNA-PK$_{cs}$ on the opposite end releases DNA-PK$_{cs}$ and promotes the helicase activity of Ku. Activated Ku unwinds the DNA to allow microhomology base pairing. The XRCC4/ligase IV complex is then recruited to the DNA ends by Ku to ligate the paired ends together. HR, Homologous repair.*

SUPPRESSING NONHOMOLOGOUS END JOINING PATHWAY BY INHIBITING KEY NHEJ PROTEINS

To date no one has utilized gene disruption, or other inactivating strategies, of any of the members of the DNA–PK complex to suppress NHEJ for the purpose of promoting gene targeting. The hypothesis is straightforward: transient inhibition of any member of the DNA–PK complex should suppress random integration of targeting DNA by inhibiting the enabling NHEJ pathway.

It is clear, therefore, that with our current state of knowledge of the molecular mechanism of DNA repair, and previous observations by other groups, it will be feasible to enhance the rate of homologous recombination in somatic cells. We have now tested this approach and results to date are extremely encouraging.

REPLICATIVE SENESCENCE

Although ES and/or EG cells have properties such as cellular pluripotentiality, as defined by the ability to generate germ line chimeras, and the ability to survive in culture for prolonged periods of time, they have been very difficult to isolate and maintain in species such as cattle and swine. Breakthroughs in the area of cloning using somatic cells as donor nuclei have provided alternative cell lines for genetic manipulation (Campbell *et al.*, 1996; Wells *et al.*, 1997; Wilmut *et al.*, 1997). Both fetal and adult fibroblasts as well as other cell types can be maintained in culture for a limited period of time, manipulated genetically, and still retain the ability to generate live offspring. The efficiency of the procedure and the normal development of the fetuses remain to be improved, but it is realistic to consider somatic cells as appropriate carriers for genetic manipulation via homologous recombination. One of the problems of using somatic cells is their inability to overcome senescence and retain their normal chromosomal constitution. Thus, from a technical standpoint, it is necessary to complete all the required steps within 20–30 population doublings. Such a limitation may permit single inactivation events but will not allow any genetic manipulation, such as cre-mediated events, which require two rounds of selection.

Although the problem of senescence of somatic cells is not trivial, approaches to overcome this process are available. Bodnar *et al.* (1998) reported the immortalization of human diploid fibroblasts by transfection with the catalytic telomerase reverse transcriptase (TERT) component. Not only was senescence overcome using this system, but even after 280 passages, no phenotypic or chromosomal abnormalities associated with cellular transformation could be detected (Jiang *et al.*, 1999; Morales *et al.*, 1999). This approach would be worthwhile to try in the domestic animal system. Introduction of a TERT expression plasmid that can be regulated so as to maintain the cell growing in culture until completion of all genetic modifications, followed by silencing of the TERT and nuclear transfer, may provide a system as flexible as that of ES and EG cells. This temporary activation of telomerase has been successful in extending the life span of human diploid fibroblasts (Steinert *et al.*, 2000). Xiang *et al.* (2000) demonstrated the ability of the hTERT to interact with the rabbit telomerase template RNA. This interaction was demonstrated both by the telomere repeat amplification protocol (TRAP) assay and by coimmunoprecipitation of hTERT with the rabbit telomerase RNA. This suggests that the human TERT is capable of interacting with telomeres from other mammalian species.

Another approach that has been partially successful in extending the life span of cells in culture is the protection from oxidative damage by the use of antioxidants. In our previous experiments with PGCs we were able to demonstrate that antioxidants could protect against apoptosis in culture (Lee *et al.*, 2000). Subsequently we have demonstrated that antioxidants protect cells when they are in a quiescent state prior to being utilized in nuclear transfer (Lee and Piedrahita, 2001). Although both of these situations appear to be acting through inhibition of apoptosis, there seems to be a more complex action of antioxidants. An example of the effect of antioxidants on replicative senescence was presented by Atamma *et al.* (2000), who reported the extension of life span in human lung fibroblasts by addition of several antioxidant molecules, including α-phenyl *N*-tert-butylnitrone and *N*-(tert-butyl) hydroxylamine. Similarly, culture of human diploid fibroblasts (HDFs) in reduced oxygen levels increased the number of doubling by 30% over control cells (Chen *et al.*, 1995; Packer and Fuerh, 1977). In contrast, exposure of early-passage HDFs to hydrogen peroxide for as little as 2 hours induced a cellular state indistinguishable from senescence (Chen and Ames, 1994; Chen *et al.*, 1998). In combination, all these data demonstrate the importance of preventing oxidative damage as a way to increase the life span of cells in culture.

We feel the increase in life-span extension is critical for practical and scientific reasons. At present the life span of the cells is sufficient to allow for single gene inactivation or knockout experiments, but it does not allow for modifications requiring two rounds of selection. As a result, techniques such as double knockouts (disruption of both alleles in a single cell) and cre-loxP-based manipulations (such as deletion of the selectable cassette) are not possible. This inability to use more complex genetic manipulation techniques will impair the ability to utilize transgenic technologies in domestic species.

RELATIONSHIP BETWEEN SENESCENCE AND DNA REPAIR

It is becoming evident that the process of DNA repair is intimately associated with the process of telomere length maintenance. Both direct and indirect evidence indicates that several of the repair proteins involved in DNA repair are also involved in protection of the telomere (Hsu *et al.*, 2000; Lundblad, 2000; McAinsh *et al.*, 1999). In particular, the Ku protein is known to be actively involved in both NHEJ and telomere maintenance (Hsu *et al.*, 2000). Lundblad (2000) put forward the concept of the telomeres acting as sinks for proteins involved in DNA damage. Thus the proteins protect the telomere in a healthy cell, but DNA damage would shift the proteins from the telomere into the NHEJ DNA repair pathway. Such an arrangement is supported by evidence indicating that treatments or manipulations known to increase the rate of DNA damage, such as high oxidation levels (hydrogen peroxide), drastically reduce telomere lengths (Chen and Ames, 1994; Chen *et al.*, 1998). Conversely, compounds known to inhibit DNA damage increase life span by helping maintain telomere length (Packer and Fuehr, 1977; Atamma *et al.*, 2000). These results strongly support the interrelationship between DNA damage and senescence.

What has not been explored to date is the relationship between telomerase expression and the rate of homologous recombination. We have been puzzled by the absolute concordance between high levels of homologous recombination and the expression of telomerase reverse transcriptase component. For instance, ES cells, known to target at high frequency, are TERT positive. A comparison of targeting in ES cells and telomerase-positive somatic cells (Hatada *et al.*, 2000) indicated that there were no differences in the rate of homologous recombination between the two. Similarly, during liver regeneration there is a short and transitory increase in the rate of homologous recombination (Thyagarajan *et al.*, 1996b). This increase mirrors a short and transitory increase in TERT expression (Yamaguchi *et al.*, 1998). Also, one of the few adult cells expressing high levels of TERT are testicular cells undergoing meiosis (a process involving homologous recombination). In these cells, the expression of TERT is down-regulated during the postmeiotic phase (Yamamoto *et al.*, 1999). Finally, Fanconi's anemia cells are known to have enhanced rates of homologous recombination (Thyagarajan and Campbell, 1997). Recent biochemical analysis indicates that these cells have a fivefold greater level of expression of TERT (Leteurtre *et al.*, 1999). Thus, in every instance described above there is a concordance between the expression of TERT and the rate of homologous recombination. Knowing that proteins involved in telomere elongation and maintenance are also used in DNA repair (Lundblad, 2000), it is possible that proteins interacting with TERT shift the DNA repair mechanism in the cell, from one of NHEJ to one of homologous recombination. Conversely, cells not expressing TERT would have a very active NHEJ pathway and a reduced DNA recombination pathway. Although this concept is based on indirect evidence, and is thus speculative in nature, it is easy to test in our system by measuring the rate of homologous versus NHEJ recombination in control fibroblasts versus fibroblasts transformed with the TERT gene. If successful, this would solve both the problem of senescence and the low targeting rates of somatic cells such as the ones we are using.

SUMMARY

It is evident from recent advances in cloning of somatic cells that the goal of precise and efficient modification of the domestic animal genome is within reach. Yet several obstacles still must be overcome before there can be widespread applicability of the technology. Among these is the inefficiency of the cloning procedure per se, i.e., the high incidence of embryonic, fetal, and even postnatal losses. There are significant species differences in this respect, but the fact remains that in all domestic species a high degree of inefficiency still exists. The low rate of homologous recombination in somatic cells is another obstacle to success. From our experience and that of others the targeting efficiency in somatic cells in considerably lower than in ES cells. Approaches to overcome this inefficiency are being developed by us and others, but it is still too early to tell what effect, if any, this manipulation will have on the cloning efficiency of the targeted cell lines. Finally, for the full applicability of the technology, cells capable of overcoming senescence are needed. This will allow for more complex and precise genetic manipulation than is possible at present, and will facilitate the screening, isolation, and expansion of targeted cell lines. Isolation of unique cell lines that do not undergo senesce, can target at high efficiency, and can be cloned at high efficiency should remain a priority. In the meantime, artificial "immortalization" of cell lines by the use of TERT should be an adequate alternative.

Beyond the technical limitations of gene targeting in domestic species, it is important to utilize the information being generated by gene mapping, linkage analysis/quantitative trait loci analysis, and gene expression profiling to start identifying genes that can have a significant beneficial effect in agricultural and medical phenotypes. Moreover, procedures for bringing desired modification to the homozygous state as rapidly as possible must also be explored.

In summary, although advances in the field of genetic modification of domestic animals continue at a fast pace, it is also clear that there are tremendous opportunities for improvement of the technology, both from a basic biology perspective (i.e., understanding nuclear reprogramming) and with regard to practical applications of domestic animal transgenics.

REFERENCES

Adair, G. M., Nairn, R. S., Wilson, J. H., Seidman, M. M., Brotherman, K. A., MacKinnon, C., and Scheerer, J. B. (1989). Targeted homologous recombination at the endogenous adenine phosphoribosyltransferase locus in Chinese hamster cells. *Proc. Natl. Acad. Sci. U.S.A.* **86**, 4574–4578.

Ariyoshi, M., Vassylyev, D. G., Iwasaki, H., Nakamura, H., Shinagawa, H., and Morikawa, K. (1994). Atomic structure of the RuvC resolvase: A holliday junction-specific endonuclease from *E. coli. Cell* **78**, 1063–1072.

Atamna, H., Paler-Martinez, A., and Ames, B. N. (2000). *N-t*-Butyl hydroxylamine, a hydrolysis product of α-phenyl-*N-t*-butyl nitrone, is more potent in delaying senescence in human lung fibroblasts. *J. Biol. Chem.* **275**, 6741–6748.

Barnes, W. M. (1994). PCR amplification of up to 35-kb DNA with high fidelity and high yield from bacteriophage templates. *Proc. Natl. Acad. Sci. U.S.A.* **91**, 2216–2220.

Baumann, P., Benson, F. E., and West, S. C. (1996). Human RAD51 protein promotes ATP-dependent homologous pairing and strand transfer reactions *in vitro. Cell* **87**, 757–766.

Bennett, R. J., Dunderdale, H. J., and West, S. C. (1993). Resolution of Holliday junctions by RuvC resolvase: Cleavage specificity and DNA distortion. *Cell* **74**, 1021–1023.

Bezzubova, O., Silbergleit, A., Yamaguchi-Iwai, Y., Takeda, S., and Buerstedde, J.-M. (1997). Reduced x-ray resistance and homologous recombination frequencies in a Rad54–/– mutant of the chicken DT40 cell line. *Cell* **89**, 185–193.

Bierne, H., and Michel, B. (1994). When replication forks stop. *Mol. Microbiol.* **13**, 17–23.

Bodnar, A. G., Ouellette, M., Frolkis, M., Holt, S., Chiu, C. P., Morin, G. B., Harley, C. B., Shay, J. W., Lichtsteiner, S., and Wright, W. E. (1998). Extension of life-span by introduction of telomerase into normal human cells. *Science* **279**, 349–352.

Bradl, M., Larue, L., and Mintz, B. (1991). Clonal coat color variation due to a transforming gene expressed in melanocytes of transgenic mice. *Proc. Natl. Acad. Sci. U.S.A.* **88**(15), 6447–6451.

Brinster, R. L., Chen, H. Y., Trumbauer, M., Senear, A. W., Warren, R., and Palmiter, R. D. (1981). Somatic expression of herpes thymidine kinase in mice following injection of a fusion gene into eggs. *Cell* **27**, 223–231.

Buchop, S., Gibson, M. K., Wang, X. W., Wagner, P., Sturzbecher, H.-W., and Harris, C. C. (1997). Interaction of p53 with the human Rad51 protein. *Nucleic Acids Res.* **25**, 3868–3874.

Campbell, K. H. S., McWhir, J., Ritchie, W. A., and Wilmut, A. (1996). Sheep cloned by nuclear transfer from a cultured cell line. *Nature* **380**, 64.

Carney, J. P., Maser, R. S., Olivares, H., Davis, E. M., Le Beau, M., Yates, J. R., Hays, L., Morgan, W. F., and Petrini, J. H. J. (1998). The hMre11/Rad50 protein complex and Nijmegen Breakage Syndrome: Linkage of double-strand break repair to the cellular DNA damage response. *Cell* **93**, 477–486.

Charron, J., Malynn, B. A., Robertson, E. J., Goff, S. P., and Alt, F. W. (1990). High-frequency disruption of the N-*myc* gene in embryonic stem and pre-B cell lines by homologous recombination. *Mol. Cell. Biol.* **10**, 1799–1804.

Chen, Q., and Ames, B. (1994). Senescence-like growth arrest induced by hydrogen peroxide in human diploid fibroblast F65 cells. *Proc. Natl. Acad. Sci. U.S.A.* **91**, 1994, 4130–4134.

Chen, Q., Fischer, A., Reagan, J. D., Yan, L. J., and Ames, B. N. (1995). Oxidative DNA damage and senescence of human diploid fibroblast cells. *Proc. Natl. Acad. Sci. U.S.A.* **92**, 4337–4341.

Chen, Q. M., Bartholomew, J. C., Campisi, J., Acosta, M., Retgan, J. D., and Ames, B. N. (1998). Molecular analysis of H202-induced senescent-like growth arrest in normal human fibroblasts: p53 and Rb control G1 arrest but not cell replication. *Biochem. J.* **332**, 43–50.

Chu, G. (1997). Double strand break repair. *J. Biol. Chem.* **272**, 24097–24100.

Costantini, F., and Lacy, E. (1981). Introduction of a rabbit beta-globin gene into the mouse germ line. *Nature* **294**, 92–94.

Critchlow, S. E., and Jackson, S. P. (1998). DNA end-joining: From yeast to man. *Trends Biochem. Sci.* **23**, 394–398.

Deng, C., and Capecchi, M. R. (1992). Reexamination of gene targeting frequency as a function of the extent of homology between the targeting vector and the target locus. *Mol. Cell. Biol.* **12**, 3365–3371.

Denning, C., Burs, S., Ainslie, A., Bracken, J., Dinnyes, A., Fletcher, J., King, T., Ritchie M., Ritchie, W. A., Rollo, M., de Sousa, P., Travers, A., Wilmut, I., and Clark, A. J. (2001). Deletion of the alpha(1,3) galactosyl transferase (GGTA1) gene and the prion protein (PrP gene in sheep). *Nat. Biotechnol.* **19**(6), 559–562.

Deursen, J. V., and Wieringa, B. (1992). Targeting of the creatine kinase M gene in embryonic stem cells using isogenic and nonisogenic vectors. *Nucleic Acids Res.* **20**, 3815–3820.

Doetschman, T., Gregg, R. G., Maeda, N., Hooper, M. L., Melton, D., Thompson, W., and Smithies, O. (1987). Targeted correction of a mutant HPRT gene in mouse embryonic stem cells. *Nature* **330**, 576–578.

Doetschman T., Williams, P., and Maeda, N. (1988). Establishment of hamster blastocyst-derived embryonic stem ES cells. *Dev. Biol.* **127**, 224–227.

Dolganov, G. M., Maser, R. S., Novikov, A., Tosto, L., Chong, S., Bressan, D. A., and Petrini, J. H. J. (1996). Human Rad50 is physically associated with hMre11: Identification of a conserved multiprotein compleximplicated in recombinational DNA repair. *Mol. Cell. Biol.* **16**, 4832–4841.

Ebert, K. M., and Schindler, J. E. S. (1993). Transgenic farm animals: Progress report. *Theriogenology* **39**, 121–135.

Essers, J., Hendriks, R. W., Swagemakers, S. M. A., Troelstra, C., de Wit, J., Bootsma, D., Hoeijmakers, J. H. J., and Kanaar, R. (1997). Disruption of mouse Rad54 reduces ionizing radiation resistance and homologous recombination. *Cell* **89**, 195–204.

Evans, M. J., Notarianni, E., Laurie, S., and Moor, R. M. (1990). Derivation and preliminary characterization of pluripotent cell lines from porcine and bovine blastocysts. *Theriogenology* **33**, 125–128.

Featherstone, C., and Jackson, S. P. (1999). Ku, a DNA repair protein with multiple cellular functions? *Mutat. Res.* **434**, 3–15.

Friend, S. (1994). p53: A glimpse at the puppet behind the shadow play. *Science* **265**, 334–335.

Game, J. C. (1993). DNA double-strand breaks and the Rad50–Rad57 genes in *Saccharomyces*. *Semin. Cancer Biol.* **4**, 73–83.

Gell, D., and Jackson, S. P. (1999). Mapping of protein-protein interactions within the DNA-dependent protein kinase complex. *Nucleic Acids Res.* **27**, 3494–3502.

Gerfen, R. W., and Wheeler, M. B. (1995). Isolation of embryonic cell-lines from porcine blastocysts. *Animal Biotechnol.* **6**(1), 1–14.

Godwin, A. R., Bollag, R. J., Christie, D.-M., and Liskay, R. M. (1994). Spontaneous and restriction enzyme-induced chromosomal recombination in mammalian cells. *Proc. Natl. Acad. Sci. U.S.A.* **91**, 12554–12558.

Goedecke W, Eijpe, M., Offenberg, H. H., van Aalderen, M., and Heyting, C. (1999). Mre11 and Ku70 interact in somatic cells, but are differentially expressed in early meiosis. *Nat. Genet.* **23**(2), 194–198.

Gordon, J. W., Scangos, G. A., Plotkin, D. J., Barbosa, J. A., and Ruddle, F. H. (1980). Genetic transformation of mouse embryos by microinjection of purified DNA. *Proc. Natl. Acad. Sci. U.S.A.* **77**, 7380–7384.

Grawunder, U., Zimmer, D., Kulesza, P., and Lieber, M. (1998). Requirement for an interaction of XRCC4 with DNA Ligase IV for wild-type V(D)J recombination and DNA double-strand break repair *in vivo*. *J. Biol. Chem.* **273**, 24708–24714.

Hanson, K. D., and Sedivy, J. M. (1995). Analysis of biological selections for high-efficiency gene targeting. *Mol. Cell. Biol.* **15**, 45–51.

Harris, C. C. (1993). p53: At the crossroads of molecular carcinogenesis and risk assessment. *Science* **262**, 1980–1981.

Hasty, P., Rivera-Perez, J., and Bradley, A. (1991). The length of homology required for gene targeting in embryonic stem cells, *Mol. Cell. Biol.* **11**, 4509–4517.

Hatada, S., Nikkuni, K., Bentley, S. A., Kirby, S., and Smithies, O. (2000). Gene correction in hematopoietic progenitor cells by homologous recombination. *Proc. Natl. Acad. Sci. U.S.A.* **97**(25), 13807–13811.

Hollstein, M., Rice, K., Greenblatt, M. S., Soussi, T., Fuchs, R., Sorlie, T., Hovig E., Smith-Sorensen, B., Montesano, R., and Harris, C. C. (1994). Database of p53 gene somatic mutations in human tumors and cell lines. *Nucleic Acids Res.* **22**, 3551–3555.

Holmes, A. M., and Haber, J. E. (1999). Double-strand break repair in yeast requires both leading and lagging strand DNA polymerases. *Cell* **96**, 415–424.

Hsu, H. L., Gilley, D., Galande, S. A., Hande, M. P., Allen, B., Kim, S. H., Li, G. C., Campisi, J., Kowhi-Shigematsu, T., and Chen, D. J. (2000). Ku acts in a unique way at the mammalian telomere to prevent end joining. *Genes Dev.* **14**(22), 2807–2812.

Jackson, S. P., and Jeggo, P. A. (1995). DNA double-strand break repair and V(D)J recombination: Involvement of DNA-PK. *Trends Biochem. Sci.* **20**, 412–415.

Jeggo, P. A. (1997). DNA-PK: At the crossroads of biochemistry and genetics. *Mutat. Res.* **384**, 1–14.

Jiang, X., Jimenez, G., Chang, E., Frolkis, M., Kusler, B., Sage, M., Beeche, M., Bodnar, A. G., Wahl, G. M., Tlsty, T. D., and Chiu, C. (1999). Telomerase expression in human somatic cells does not induce changes associated with a transformed phenotype. *Nature Genet.* **21**, 111–114.

Jin, S., and Weaver, D. T. (1997). Double-strand break repair by Ku70 requires heterodimerization with Ku80 and DNA binding functions. *EMBO J.* **16**, 6874–6885.

Johnson, R. D., and Jasin, M. (2000). Sister chromatid gene conversion is a prominent double-strand break repair pathway in mammalian cells. *EMBO J.* **19**, 3398–3407.

Johnson, B., Thyagarajan, L., Krueger, B., Hirsch, C., and Campbell, C. (1996). Elevated levels of recombinational DNA repair in human somatic cells expressing the *Saccharomyces cerevisiae* RAD52 gene. *Mutat. Res.* **363**, 179–189.

Joyner, A. L., Skarnes, W. C., and Rossant, J. (1989). Production of a mutation in mouse *En-2* gene by homologous recombination in embryonic stem cells. *Nature* **338**, 153–156.

Kowalczykowski, S. C. (2000). Initiation of genetic recombination and replication-dependent replication. *Trends Biochem. Sci.* **25**, 156–164.

Kowalczykowski, S. C., and Zarling, D. A. (1995). Homologous recombination proteins and their potential applications in gene targeting technology. *In* "Gene Targeting" (M. A. Vega, ed.), pp. 167–210. CRC Press, Boca Raton, Florida.

Krnacik, M. J., Li, S., Liao, J., and Rosen, J. M. (1995). Position-independent expression of whey acidic protein transgenes. *J. Biol. Chem.* **270**, 11119–11129.

Labosky, P. A., Barlow, D. P., and Hogan, B. L. M. (1994). Mouse embryonic germ (EG) cell lines: Transmission through the germline and differences in the methylation imprint of insulin-like growth factor 2 receptor (*Igf2r*) gene compared with embryonic stem (ES) cell lines. *Development* **120**, 3197–3204.

Lee, C. K., and Piedrahita, J. A. (2001). Inhibition of apoptosis in serum starved porcine embryonic fibroblasts. *Sixth Intl. Conf. On Pig Repro.*, pp. 146.

Lee, C. K., Weaks, R. L., Johnson, G. A., Bazer, F. W., and Piedrahita, J. A. (2000). Effects of protease inhibitors and antioxidants on *in vitro* survival of porcine primordial germ cells. *Biol. Reprod.* **63**(3), 887–897.

Leteurtre, F., Li, X., Guardioloa, P., Le Roux, G., Sergere, J. C., Richard, P., Carosella, E. D., and Gluckman, E. (1999). Accelerated telomere shortening and telomerase activation in Fanconi's anaemia. *Br. J. Haematol.* **105**(4), 883–93.

Levine, A. J. (1997). P53, the cellular gatekeeper for growth and division. *Cell* **88**(3), 323–331.

Lim, D.-S., and Hasty, P. (1996). A mutation in mouse rad51 results in an early embryonic lethal that is suppressed by a mutation in p53. *Mol. Cell. Biol.* **16**, 7133–7143.

Lundblad, V. (2000). A tale of ends. *Nature* **403**, 149–151.

Luo, G., Yao, M. S., Bender, C. F., Mills, M., Bladl, A. R., Bradley, A., and Petrini, J. H. J. (1999). Disruption of mRad50 causes embryonic stem cell lethality, abnormal embryonic development and sensitivity to ionizing radiation. *Proc. Natl. Acad. Sci. U.S.A.* **96**, 7376–7381.

Mansour, S. L., Thomas, K. R., and Capecchi, M. R. (1988). Disruption of the proto-oncogene *int-2* in mouse embryo-derived stem cells: A general strategy for targeting mutations to non-selectable genes. *Nature* **336**, 348–352.

McAinsch, A. D., Scott-Drew, S., Murray, J. A. H., and Jackson, S. P. (1999). DNA damage triggers disruption of telomeric silencing and mec1p-dependent relocation of Sir3p. *Curr. Biol.* **9**, 963–966.

McCreath, K. J., Howcroft, J., Campbell, K. H. S., Colman, A., Schnieke, A. E., and Kind, A. J. (2000). Production of gene-targeted sheep by nuclear transfer from cultured somatic cells. *Nature* **405**, 1066–1069.

McKnight, R. A., Shamay, A., Sankaran, L., Wall, R. J., and Hennighausen, L. (1992). Matrix-attachment regions can impart position-independent regulation of a tissue-specific gene in transgenic mice. *Proc. Natl. Acad. Sci. U.S.A.* **89**, 6943–6947.

Mekeel, K. L., Tang, W., Kachnic, L. A., Lueo, C.-M., DeFrank, J. S., and Powell, S. N. (1997). Inactivation of p53 results in high rates of homologous recombination. *Oncogene* **14**, 1847–1857.

Milne, G. T., and Weaver, D. T. (1993. Dominant-negative alleles of Rad52 reveal a DNA repair recombination complex including Rad51 and Rad52. *Genes Dev.* **7**, 1755–1765.

Morales, C. P., Holt, S. E., Ouellette, M., Kaur, K. J., Yan, Y., Wilson, K. S., White, M. A., Wright, W. E., and Shay, J. W. (1999). Absence of cancer-associated changed in human fibroblasts immortalized with telomerase. *Nat. Genet.* **21**, 115–118.

Moynahan, M. E., and Jasin, M. (1997). Loss of heterozygosity induced by a chromosomal double-strand break. *Proc. Natl. Acad. Sci. U.S.A* **94**, 8988–8993.

Nassif, N., Penney, J., Pal, S., Engels, W. R., and Gloor, G. B. (1994). Efficient copying of nonhomologous sequences from ectopic sites via P-element gap repair. *Mol. Cell. Biol.* **14**, 1613–1625.

Nick-McElhinny, S., Snowden, C., McCarville, J., and Ramsden, D. (2000). Ku recruits the XRCC4–Ligase IV complex to DNA ends. *Mol. Cell. Biol.* **20**, 2996–3003.

Notarianni, E., Laurie, S., Moor, R. M., and Evans, M. J. (1990). Maintenance and differentiation in culture of pluripotential embryonic cell lines from pig blastocysts. *J. Reprod. Fertil. (Suppl.)* **41**, 51–56.

Packer, L., and Fuehr, K. (1977). Low oxygen concentration extends the lifespan of cultured human diploid cells. *Nature* **267**, 423–425.

Palmiter, R. D., and Brinster, R. L. (1986). Germ-line transformation of mice. *Annu. Rev. Genet.* **20**, 465–499.

Paques, F., Leung, W.-Y., and Haber, J. E. (1998). Expansions and contractions in a tandem repeat induced by double-strand break repair. *Mol. Cell. Biol.* **18**, 2045–2054.

Piedrahita, J. A., Anderson, G. B., and BonDurant, R. H. (1990). On the isolation of embryonic stem cells: Comparative behavior of murine, porcine, and ovine embryos. *Theriogenology* **34**, 879–900.

Piedrahita, J. A., Moore, K., Oetama, B., Lee, C.-K., Scales, N., Ramsoondar, J., Bazer, F., and Ott, T. (1998). Generation of transgenic porcine chimeras using primordial germ cell (PGC)-derived colonies. *Biol. Reprod.* **58**, 1321–1329.

Rajan, J., Marquis, S., Gardner, H., and Chodosh, L. (1997). Developmental Expression of Brca2 colocalizes with Brca1 and is associated with proliferation and differentiation in multiple tissues. *Dev. Biol.* **184**, 385–401.

Randolph, D. A., Verbsky, J. W., Yang, L., Fang, Y., Hakem, R., and Fields, L. E. (1996). PCR-based gene targeting of the inducible nitric oxide synthase (NOS2) locus in murine ES cells, a new and more cost-effective approach. *Transgen. Res.* **5**, 413–420.

Reiss, B., Klemm, M., Kosak, H., and Schell, J. (1996). RecA protein stimulates homologous recombination in plants. *Proc. Natl. Acad. Sci. U.S.A.* **93**, 3094–3098.

Reitman, M., Lee, E., Westphal, H., and Felsenfeld, G. (1993). An enhancer/locus control region is not sufficient to open chromatin. *Mol. Cell. Biol.* **13**, 3990–3998.

Richardson, C., Moynahan, M. E., and Jasin, M. (1998). Double-strand break repair by interchromosomal recombination: Suppression of chromosomal translocations. *Genes Dev.* **12**, 3831–3842.

Riele, H. T., Maandag, E. R., and Berns, A. (1992). Highly efficient gene targeting in embryonic stem cells through homologous recombination with isogenic DNA constructs. *Proc. Natl. Acad. Sci. U.S.A.* **89**, 5128–5132.

Rijkers, T., van den Ouweland, J., Morolli, B., Rolink, A. G., Baarends, W. M., van Sloun, P. P. H., Lohman, P. H. M., and Pastink, A. (1998). Targeted inactivation of mouse RAD52 reduces homologous recombination but not resistance to ionizing radiation. *Mol. Cell. Biol.* **18**, 6423–6429.

Ringrose, L., Lounnas, V., Ehrlich, L., Buchholz, F., Wade, R., and Stewart, A. F. (1998). Comparative kinetic analysis of FLP and cre recombinases: Mathematical models for DNA binding and recombination. *J. Mol. Biol.* **284**(2), 363–384.

Roth, D. B., and Wilson, J. H. (1986). Nonhomologous recombination in mammalian cells: Role for short sequence homologies in the joining reaction. *Mol. Cell. Biol.* **6**, 4295–4304.

Sargent, R. G., Brenneman, M. A., and Wilson, J. A. (1997). Repair of site-specific double-strand breaks in a mammalian chromosome by homologous and illegitimate recombination. *Mol. Cell. Biol.* **17**, 267–277.

Schnieke, A., Harbers, K., and Jaenisch, R. (1983). Embryonic lethal mutation in mice induced by retrovirus insertion into the alpha 1(I) collagen gene. *Nature* **304**, 315–320.

Schnieke, A. E., Kind, A. J., Ritchie, W. A., Mycock, K., Scott, A. R., Ritchie, M., Wilmut, I., Colman, A., and Campbell, K. H. (1997). Human factor IX transgenic sheep produced by transfer of nuclei from transfected fetal fibroblasts. *Science* **278**, 2130–2133.

Scully, R., Chen, J., Ochs, R., Keegan, K., Hoekstra, M., Feunteun, J., and Livingston, D. (1997). Dynamic changes of BRCA1 subnuclear location and phosphorylation state are initiated by DNA damage. *Cell* **90**, 425–435.

Shalev, G., Sitrit, Y., Avivi-Ragolski, N., Lichtenstein, C., and Levy, A. A. (1999). Stimulation of homologous recombination in plants by expression of the bacterial resolvase RuvC. *Proc. Natl. Acad. Sci. U.S.A.* **96**, 7398–7402.

Shcherbakova, O. G., Lanzov, V. A., Ogawa, H., and Filatov, M. V. (2000). Overexpression of bacterial RecA protein stimulates homologous recombination in somatic mammalian cells. *Mutat. Res.* **459**, 65–71.

Shim, H., Gutierrez-adan, A., Chen, L. R., Bondurant, R. H., Behboodi, E., and Anderson, G. B. (1997). Isolation of pluripotent stem cells from cultured porcine primordial germ cells. *Biol. Reprod.* **57**, 1089–1095.

Shinohara, A., and Ogawa, T. (1998). Stimulation by Rad52 of yeast Rad51-mediated recombination. *Nature* **391**, 404–407.

Shinohara, A., Ogawa, H., and Ogawa, T. (1992). Rad51 protein involved in repair and recombination in *Saccharomyces cerevisiae* is a RecA-like protein. *Cell* **69**, 457–470.

Shinohara, A., Ogawa, H., Matsuda, Y., Ushio, N., Ikeo, K., and Ogawa, T. (1993). Cloning of hguman, mouse and fission yeast recombination genes homologous to Rad51 and RecA. *Nat. Genet.* **4**, 239–243.

Snouwaert, J. N., Gowen, L. C., Latour, A. M., Mohn, A. R., Xiao, A., DiBiase, L., and Koller, B. H. (1999). BRCA1 deficient embryonic stem cells display a decreased homologous recombination frequency and an increased frequency of non-homologous recombination that is corrected by expression of a brca1 transgene. *Oncogene* **18**(55), 7900–7907.

Solinger, J. A., Lutz, G., Sugiyama, T., Kowalczykowski, S. C., and Heyer, W.-D. (2001). Rad54 protein stimulates heteroduplex DNA formation in the synaptic phase of DNA strand exchange via specific interactions with the presynaptic Rad51 nucleoprotein filament. *J. Mol. Biol.* **307**, 1207–1221.

Sonoda, E., Sasaki, M. S., Buerstedde, J.-M., Bezzubova, O., Shinohara, A., Ogawa, H., Takata, M., Yamaguchi-Iwai, Y., and Takeda, S. (1998). Rad51-deficient vertebrate cells accumulate chromosomal breaks prior to cell death. *EMBO J.* **17**, 598–608.

Sonoda, E., Sasaki, M. S., Morrison, C., Yamaguchi-Iwai, Y., Takata, M., and Takeda, S. (1999). Sister chromatid exchanges are mediated by homologous recombination in vertebrate cells. *Mol. Cell. Biol.* **19**, 5166–5169.

Steinert, S., Shay, J. W., and Wright, W. E. (2000). Transient expression of human telomerase extends the life span of normal human fibroblasts. *Biochem. Biophys. Res. Commun.* **273**, 1095–1098.

Stewart, C. L., Gadi, I., and Blatt, H. (1994). Stem cells from primordial germ cells can reenter the germline. *Dev. Biol.* **161**, 626–628.

Strojek, R. M., Reed, M. A., Hoover, J. L., and Wagner, T. A. (1990). A method for cultivating morphologically undifferentiated embryonic stem cells from porcine blastocysts. *Theriogenology* **33**, 901.

Sturzbecher, H. W., Brain, R., Addison, C., Rudge, K., Remm, M., Grimaldi, M., Keenan, E., and Jenkins, J. R. (1992). A C-terminal alpha-helix plus basic region motif is the major structural determinant of p53 tetramerization. *Oncogene* **7**, 1513–1523.

Szostak, J. W., Orr-Weaver, T. L., and Rothstein, R. J. (1983). The double-strand-break repair model for recombination. *Cell* **33**, 25–35.

Takahagi, M., Iwasaki, H., Nakata, A., and Shinagawa, H. (1991). Molecular analysis of the *Escherichia coli ruvC* gene, which encodes a Holliday junction-specific endonuclease. *J. Bacteriol.* **173**, 5747–5753.

Takata, M., Sasaki, M. S., Sonada, E., Morrison, C., Hashimoto, M., Utsumi, H., Yamaguchi-Iwai, Y., Shinohara, A., and Takeda, S. (1998). Homologous recombination and non-homologous end joining pathways of DNA double-strand break repair have overlapping roles in the maintenance of chromosomal integrity in vertebrate cells. *EMBO J.* **17**, 5497–5508.

Thyagarajan, B., and Campbell, C. (1997). Elevated homologous recombination activity in Fanconia anemia fibroblasts. *J. Biol. Chem.* **272**, 23328–23333.

Thygarajan, B., McCormick-Graham, M., Romero, D. P., and Campbell, C. (1996a). Characterization of homologous DNA recombination activity in normal and immortal mammalian cells. *Nucleic Acids Res.* **24**, 4084–4091.

Thyagarajan, B., Cruise, J. L., and Campbell, C. (1996b). Elevated levels of homologous DNA recombination activity in the regenerating rat liver. *Somat. Cell Mol. Genet.* **22**(1), 31–39.

Tsukamoto, Y., Kato, J., and Ikeda, H. (1996). Effects of mutations of RAD50, RAD51, and related genes on illegitimate recombination in *Saccharomyces cerevisiae*. *Genetics* **142**, 383–391.

Van Dyck, E., Stasiak, A. Z., Stasiak, A., and West, S. C. (1999). Binding of double-strand breaks in DNA by human RAD52 protein. *Nature* **398**, 728–731.

Vispe, S., Cazaux, C., Lesca, C., and Defais, M. (1998). Overexpression of Rad51 protein stimulates homologous recombination and increases resistance of mammalian cells to ionizing radiation. *Nucleic Acids Res.* **26**, 2859–2864.

Wagner, E. F., Stewart, T. A., and Mintz, B. (1981a). The human beta-globin gene and a functional viral thymidine kinase gene in developing mice. *Proc. Natl. Acad. Sci. U.S.A.* **78**, 5016–5020.

Wagner, T. E., Hoppe, P. C., Jollick, J. D., Scholl, D. R., Hodinka, R. L., and Gault, J. B. (1981b). Microinjection of a rabbit beta-globin gene into zygotes and its subsequent expression in adult mice and their offspring. *Proc. Natl. Acad. Sci. U.S.A.* **78**, 6376–6380.

Wang, H., Zhao-Chong, Z., Tu-Anh, B., DiBiase, S., Qin, W., Xia, F., Powell, S., and Iliakis, G. (2001). Nonhomologous end-joining of ionizing radiation-induced DNA double-stranded breaks in human tumor cells deficient in BRCA1 or BRCA2. *Cancer Res.* **61**, 270–277.

Wells, D. N., Misica, P. M., Day, T. A., and Tervit, H. R. (1997). Production of cloned lambs from an established embryonic cell line—A comparison between *in vivo*- and *in vitro*-matured cytoplasts. *Biol. Reprod.* **57**, 385–393.

West, S. C. (1992). Enzymes and molecular mechanisms of genetic recombination. *Annu. Rev. Biochem.* **61**, 603–640.

Wilmut, I., Schuleke, A. E., McWhir, J., Kind, A. J., and Campbell, K. H. S. (1997). Viable offspring derived from fetal and adult mammalian cells. *Nature* **385**, 810–813.

Xiang, H., Wang, J., Mao, Y.-W., and Li, D. W. (2000). hTERT can function with rabbit telomerase RNA regulation of gene expression and attenuation of apoptosis. *Biochem. Biophys. Res. Commun.* **278**, 503–510.

Yamaguchi, Y., Nozawa, K., Savoysky, E., Hayakawa, N., Nimura, Y., and Yoshida, S. (1998). Change in telomerase activity of rat organs during growth and aging. *Exp. Cell Res.* **242**(1), 120–127.

Yamaguchi-Iwai, Y., Sonada, E., Buerstedde, J.-M., Bezzubova, O., Morrison, C., Takata, M., Shinohara, A., and Takeda, S. (1998). Homologous recombination but not DNA repair is reduced in vertebrate cells deficient in Rad52. *Mol. Cell. Biol.* **18**, 6430–6435.

Yamamoto, A., Taki, T., Yagi, H., Habu, T., Yoshida, K., Yoshimura, Y., Yamamoto, K., Matsushiro, A., Nishimune, Y., and Morita, T. (1996). Cell cycle-dependent expression of the mouse Rad51 gene in proliferating cells. *Mol. Gen. Genet.* **251**, 1–12.

Yamamoto, Y., Sofikitis, N., Ono, K., Kaki, T., Isoyama, T., Suzuki, N., and Miyagawa, I. (1999). Post-meiotic modifications of spermatogenic cells are accompanied by inhibition of telomerase activity. *Urol. Res.* **27**, 336–345.

Yoo, S., and Dynan, W. S. (1998). Characterization of the RNA binding properties of Ku protein. *Biochemistry* **37**, 1336–1343.

Zabludoff, S. D., Wright, W. W., Harshman, K., and Wold, B. J. (1996). BRCA1 mRNA is expressed highly during meiosis and spermatogenesis but not during mitosis of male germ cells. *Oncogene* **13**, 649–653.

Zdzienicka, M. Z. (1999). Mammalian x-ray-sensitive mutants which are defective in non-homologous (illegitimate) DNA double-strand break repair. *Biochimie* **81**, 107–116.

Zhong, Q., Chen, C.-H., Li, S., Chen, Y., Wang, C.-C., Xiao, J., Chen, P.-L., Sharp, Z. D., and Lee, W.-H. (1999). Association of BRCA1 with the hRad50-hMre11-p95 complex and the DNA damage response. *Science* **285**, 747–750.

CHAPTER 12

PREGNANCY AND NEONATAL CARE OF CLONED ANIMALS

Jonathan R. Hill and Pascale Chavatte-Palmer

The number of somatic cell cloned animals born worldwide since 1997 has rapidly increased into the hundreds. Since the first sheep were cloned from adult and fetal cells (Wilmut *et al.*, 1997), the technique has been duplicated worldwide and in several species. The overwhelming majority of clones alive today are healthy and apparently normal (Fig. 1), but the efficiency of cloning is still very poor due to important embryonic and fetal losses (most laboratories report about 2% efficiency after fusion). Improvement of the efficiency of cloning may come through the improvement of cloning techniques and through the careful clinical management of recipient animals and newborn clones. This chapter describes clinical abnormalities identified in clones and provides clues on how to manage these precious individuals.

OCCURRENCE OF LARGE-OFFSPRING SYNDROME

IN VITRO EMBRYO PRODUCTION

Problems linked to advanced reproductive techniques were described initially in relation to *in vitro* embryo production by *in vitro* fertilization (IVF). Gestational abnormalities such as an increase in embryo and fetal losses, abortions, gestation length, birth weight, placental anomalies, and reduced neonatal survival have been described (Hasler *et al.*, 1995; Behboodi *et al.*, 1995; Farin *et al.*, 1995; Walker *et al.*, 1996; Schmidt *et al.*, 1996; Kruip *et al.*, 1997). These problems, however, do not seem to have occurred in the same proportion, reflecting differences in the laboratories. The culture medium and in particular the use of coculture and the addition of serum have been pointed out as the probable cause for the syndrome now currently known as the large-offspring syndrome (LOS) (Thompson *et al.*, 1995; Walker *et al.*, 1996; Young *et al.*, 1998; Sinclair *et al.*, 1999). In a large-scale analysis of IVF calves by Holland Genetics, pregnancies generated by coculture conditions displayed a higher incidence of congenital abnormalities such as hydrallantois and limb deformities (3.7 vs. 0.6% for artificial insemiration calves) (van Wagtendonk-de Leeuw *et al.*, 2000). Perinatal mortality was also higher in calves derived from cocultures compared to those obtained by multiple ovulation and embryo transfer (MOET) (7.5 vs. 4.6%). In ovine pregnancies derived from coculture systems, the incidence of hydrallantois increased to 23% (Sinclair *et al.*, 1999). These figures provide a basis on which to evaluate the incidence of abnormalities that occur in cloned animals.

Figure 1 A clone of 9 genetically identical calves at INRA. The three older animals are derived from an embryonic cell line and the others are derived from skin cells from the ear of one of the embryonic clones.

NUCLEAR TRANSFER

When livestock were first cloned from embryonic cells, it became apparent that an increased standard of care was required to initiate and maintain pregnancies and then to produce viable offspring (Barnes, 2000). Embryo cloning using nuclear transfer produced thousands of calves in the late 1980s and early 1990s (Seidel, 1995). Animals cloned from embryonic cells appear to have an increased incidence of problems compared to those derived from *in vitro* fertilization (Barnes *et al.*, 1993; Willadsen *et al.*, 1991; Wilson *et al.*, 1995; Kruip *et al.*, 1997; Keefer *et al.*, 1994; Garry *et al.*, 1996). The use of alternative donor cell types for nuclear transfer started in the mid-1990s. Although these experiments produced first-trimester pregnancies, live births proved elusive (Stice *et al.*, 1996). A high rate of gestational and perinatal loss was observed in the first somatic cell cloning programs in sheep, cattle, and mice (Wilmut *et al.*, 1997; Cibelli *et al.*, 1998; Wakayama *et al.*, 1998). Once again, problems similar to those seen in the *in vitro*-produced offspring occurred at a higher rate in cloning experiments that used somatic, germ, or stem cells as donor (Campbell *et al.*, 1996; Wilmut *et al.*, 1997; Schnieke *et al.*, 1997; Wells *et al.*, 1997, 1999; Cibelli *et al.*, 1998; Kato *et al.*, 1998; Kubota *et al.*, 2000; Onishi *et al.*, 2000; Polejaeva *et al.*, 2000; Wakayama *et al.*, 1998, 1999; Baguisi *et al.*, 1999; Vignon *et al.*, 1998), although subsequent data show that pregnancies obtained by nuclear transfer from adult somatic cells are more susceptible to losses than are those obtained by embryo cloning (Heyman *et al.*, 2002).

Table 1 Outcomes of Somatic Cell Cloning Programs by Species[a]

Species	Clones born	Clones alive >1 week
Ovine	13	10 (77%)
Bovine	128	89 (69%)
Murine	105	96 (90%)
Caprine	9	8 (89%)
Porcine	14	14 (100%)
Feline	1	1 (100%)
Rabbit	6	4 (67%)

[a]Pregnancies and neonatal survival results were derived from published nuclear transfer programs in five species that produced at least one surviving clone. The following species were studied: ovine (Wilmut *et al.*, 1997; Schnieke *et al.*, 1997), bovine (Cibelli *et al.*, 1998; Vignon *et al.*, 1998; Wells *et al.*, 1998, 1999; Hill *et al.*, 2000b; Kato *et al.*, 1998; Zakhartchenko *et al.*, 1999; Kubota *et al.*, 2000; Shiga *et al.*, 1999; Pace *et al.*, 2001), murine (Ogura *et al.*, 2000; Wakayama *et al.*, 1998, 1999, 2001), caprine (Baguisi *et al.*, 1999; Keefer *et al.*, 2001), porcine (Betthauser *et al.*, 2000; Onishi *et al.*, 2000; Polejaeva *et al.*, 2000), feline (Shin *et al.*, 2002), and rabbit (Chesné *et al.*, 2002).

In additional cloning experiments early gestational losses have remained consistently high. However, viability at birth has been close to normal in mice, goats, and pigs, and some bovine cloning programs have reported excellent neonatal viability (Table 1) (Wakayama *et al.*, 2001; Keefer *et al.*, 2001; Polejaeva *et al.*, 2000; Wells *et al.*, 1999). Although fetal losses during the rest of pregnancy and at birth are not as dramatic, they still remain much higher than expected for either *in vivo*- or *in vitro*-produced embryos. It is encouraging to see, however, that pig and goat cloning programs have not shown an increased rate of abnormalities at birth (Keefer *et al.*, 2001; Polejaeva *et al.*, 2000; Betthauser *et al.*, 2000). Differing techniques may also play part in neonatal viability; some experiments have shown 100% cloned calf survival with minimal problems at birth (Wells *et al.*, 1999).

It is currently assumed that the increased losses described in cloning experiments are due to LOS, indicating that the same problems affect *in vitro*-produced and cloned fetuses and calves, clones being more severely affected. It is not certain, however, that the cloning procedure does not add supplementary abnormalities and that losses in cloned animals can be attributed only to LOS.

GESTATIONAL MONITORING

PREGNANCY DIAGNOSIS AND FIRST TRIMESTER DEVELOPMENT

Losses

Following transfer of cloned bovine embryos, day 30 pregnancy rates per recipient can approach 50%, whether one or several embryos have been transferred into each recipient (Fig. 2) (Heyman *et al.*, 2000, 2001). After this initial pregnancy time period, embryo losses greater than 50% are common for nuclear transfer pregnancies in sheep, cattle, and goats and especially for clones produced from somatic cells (Wilmut *et al.*, 1997; Wells *et al.*, 1997, 1999; Cibelli *et al.*, 1998; Kubota *et al.*, 2000; Wakayama *et al.*, 1998; Hill *et al.*, 2000b; Heyman *et al.*, 2000). In contrast

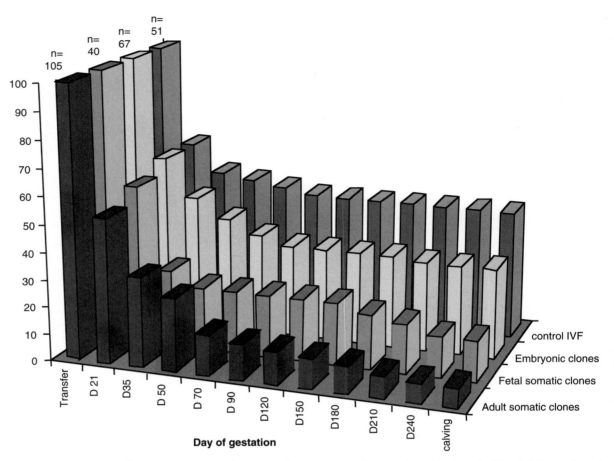

Figure 2 *In vivo development of pregnancies after transfer of in vitro-fertilized (IVF) or cloned embryos (from embryonic, fetal, or adult somatic cells) at INRA. After Heyman et al. (2002).*

only 2–4% of naturally conceived early first-trimester (day 30) bovine pregnancies and about 10% of *in vitro*-produced embryos have failed by day 60. Survival of cloned embryos to term is approximately one-half to one-fourth of that of *in vitro*-fertilized embryos, with most of these losses occurring in the first trimester (Alexander *et al.*, 1995; Forar *et al.*, 1996; Hasler *et al.*, 1995; Wells *et al.*, 1999; Heyman *et al.*, 2000).

Pathology

In naturally conceived animals, early fetal losses may be due to abnormalities of the embryo or its placenta, alterations in maternal uterine environment, or feto-maternal interactions (Wilmut *et al.*, 1986). Possible mechanisms include chromosomal anomalies, hormonal changes, environmental influences, asynchronous embryo transfer, and immunological rejection. In cloned pregnancies, Stice *et al.* (1996) observed a lack of placentome development in day 35–50 bovine fetuses cloned from embryonic cells and suggested that this caused a high rate of first-trimester death. Subsequently, it appears that in somatic-cell-cloned animals placental abnormalities occur at a high incidence in early and late-term cloned fetuses. First-trimester somatic-cell-cloned fetuses display a wide variety of placental morphologies, with a variable number and development of cotyledons, allantoic epithelium, and vascularization (Hill *et al.*, 2000b; DeSousa *et al.*, 2001; Wells *et al.*, 1999) (Fig. 3). Similar placental abnormalities occur in first-trimester IVF fetuses (Peterson and McMillan, 1998; Thompson and Peterson, 2000). In addition to

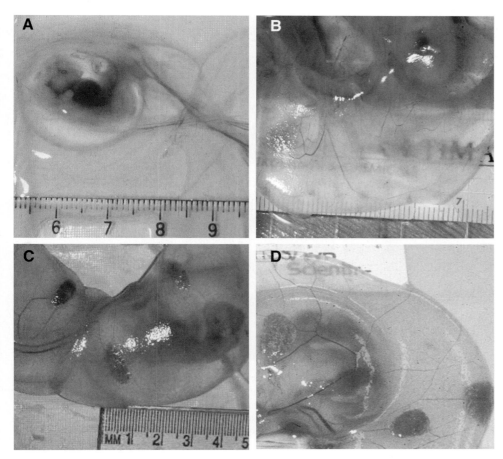

Figure 3 *First-trimester cloned bovine and ovine placentas may exhibit reduced vascularization and placentome numbers. (A) A hypoplastic type of chorioallantois from a day 40 bovine nuclear transfer pregnancy; (B) incomplete development of a day 45 placenta with a reduced number of small cotyledons; (C) apparently normal development of a day 50 nuclear transfer placenta; (D) a placenta from a day 53 in vitro-produced fetus. From Hill et al. (2000b) (Biology of Reproduction by Hill, Burghardt, Jones, Long, et al. Copyright 2000 by Soc. for the Study of Reproduction. Reproduced with permission of Soc. for the Study of Reproduction).*

underdeveloped placentas, overdevelopment has been reported in cloned mice, with placentas twice the normal weight (Wakayama and Yanagimachi, 1999, 2001). Even in nuclear transfer fetuses that survive beyond day 50, the number of placentomes may be reduced from normal by as much as 80% (Hill *et al.*, 1999), which suggests that the completeness of placental development varies widely in cloned animals. Fetuses recovered shortly after death are grossly normal and thus it appears that lack of normal placentation, rather than fetal abnormalities, is the likely cause of embryonic death. This most likely is due to inadequate placental development, which results in starvation due to inadequate maternal–fetal contact and poor transfer of nutrients.

Detection of the Losses

Ultrasonography in farm animals has been useful to demonstrate that a high rate of fetal death in cloned pregnancies occurs from the first to third month of gestation (Hill *et al.*, 2000b; DeSousa *et al.*, 2000; Wells *et al.*, 1999; Heyman *et al.*, 2001) (Fig. 2), although in some cattle and goat nuclear transfer programs these losses have not been apparent (Kato *et al.*, 1998; Keefer *et al.*, 2001). The fetuses that are destined to die become progressively more undersized for their age,

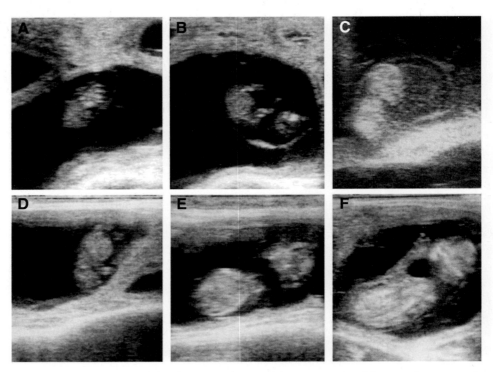

Figure 4 *A comparison of development of nonviable (A–C) and viable (D–F) bovine nuclear transfer fetuses using transrectal ultrasonography at days 45 (A, D), 55 (B, E), and 65 (C, F). The nonviable fetus was alive at day 45 (heartbeat detected) and dead at day 55 (no heartbeat); at day 65 increased echogenicity of the amniotic fluid was apparent as the fetus degenerated. The crown–rump lengths, in centimeters, for the nonviable and viable fetuses at each stage were, respectively, 3 vs. 2.8, 3.7 vs. 5.7, and 3.7 vs. 7.5 cm.*

although it is difficult by ultrasound monitoring to guess which fetuses are destined to die (Fig. 4). A reasonable rate for rectal scanning is every 15 days, beginning at day 35.

In the first trimester, placental development can also be evaluated by maternal levels of pregnancy proteins such as pregnancy-specific protein B (PSPb) (Sasser *et al.*, 1986), pregnancy-associated glycoprotein (bPAG) (Zoli *et al.*, 1992), and pregnancy serum protein 60, a protein of 60 kDa (PSP60) (Mialon *et al.*, 1993). Except for some biochemical differences, these proteins are similar. PSP60, secreted by the binucleate cells of the placenta, is a specific marker of pregnancy in cattle and is easily assayed from a blood sample of the mother. Decreasing concentrations over a period of 15 days (two samples at a 2-week interval) are indicative of fetal death, whereas the presence of higher concentrations at day 50 was correlated with subsequent fetal death (Heyman *et al.*, 2002).

SECOND- AND THIRD-TRIMESTER DEVELOPMENT

During the second and third trimesters there are sporadic losses of cloned fetuses, often accompanied by the development of major placental abnormalities such as hydrallantois. In third-trimester cloned bovine pregnancies, placental abnormalities such as edema and hydrallantois may occur at a rate similar to that seen in cocultured IVF-generated fetuses (Hill *et al.*, 1999; Sinclair *et al.*, 1999). A decreased number of enlarged placentomes may be observed indirectly using transabdominal and transrectal ultrasonography or directly following birth (Figs. 5–7). Euthanasia of pathological cases showed that placentome enlargement sometimes seems to affect only part of the placenta, consisting of areas with normal and areas with ede-

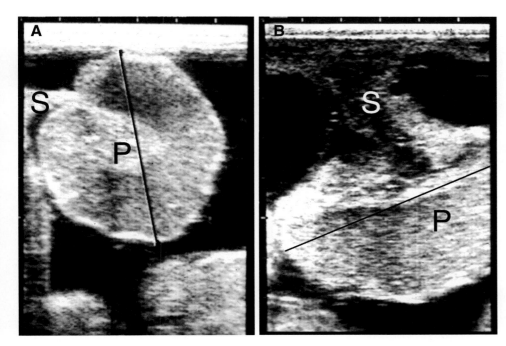

Figure 5 *Transrectal ultrasonographs of bovine placentomes from normal (A) and abnormal (B) cloned pregnancies. (B) The placentome (P) was hypertrophied (8 vs. 5 cm in diameter), with a thickened stalk (S), and on clinical examination there was excessive allantoic fluid (hydrallantois). The placentome is composed of maternal (caruncle) and fetal (cotyledon) tissue. The caruncular stalk attaches the placentome to the uterus and in large placentomes, such as in B, the blood vessels are also enlarged. The normal pregnancy calf was vigorous at birth, but the calf from the abnormal pregnancy required intensive care to correct neonatal hypoxia.*

matous placentomes. The lesions probably spread as gestation continues. The reduction in placentome numbers may not harm fetal viability if the total surface area for nutrient exchange remains constant by hypertrophy of the remaining placentomes (Bazer *et al.*, 1979; Hill *et al.*, 2001). Such an increase in placentome size has been achieved in sheep following surgical carunculectomies of the uterus prior to getting the females pregnant, but it was not sufficient for the fetuses to grow at a normal rate (Robinson *et al.*, 1979). It also appears that placental function is compromised in some cloned pregnancies because late-gestation cloned fetuses have been found to be hypoxic (Garry *et al.*, 1998). Histological observations of edematous placentomes from clones show that although edema is present, there is little or no inflammatory process and fetal and maternal tissue appear normal (Fig. 8). As a result of incomplete placental development, a delicate balance exists between the capacity of the placenta to supply nutrients and the demands of a rapidly growing fetus. Together with hydrallantois, fetal lesions are present, including omphalocele, ascites, cardiac enlargement, liver steatosis (Fig. 9), and asynchronous growth of organs.

IDENTIFICATION OF THE HIGH-RISK PREGNANCY—PRENATAL MONITORING

High-risk pregnancies can be identified by external examination and clinical observations in conjunction with ultrasonographic imaging (e.g., abnormal edematous placentation, increased allantoic fluid) and clinical tests (e.g., urinary ketones, elevated PSP60 concentrations). The presence of placental edema or excessive allantoic fluid often indicates a poor outlook for the fetus. Monitoring of body weight,

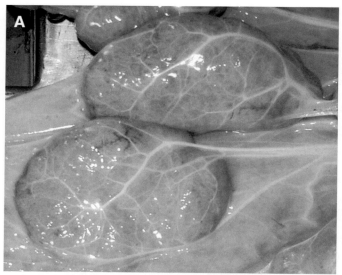

Figure 6 *Control (A) and cloned (B) bovine placentomes. In these photographs the view is that seen by the fetus with the chorioallantois (cotyledon) overlaying the maternal caruncle to form the placentome. There is extensive chorioallantoic edema in B. This placenta is from a 7-month cloned pregnancy that developed severe hydrallantois and placental edema that resulted in fetal death. Histopathology showed incomplete penetration of fetal villi into the maternal crypts, which suggests placental function was severely impaired. Photo B courtesy of Dr. D. H. Schlafer, Cornell University.*

feed intake, abdominal circumference, heart/respiratory rate, and body temperature provides valuable data to assess the health of the recipient animal. Some cows carrying cloned fetuses have a fever of unexplained origin during the last few weeks of gestation. The presence of ketonuria is not necessarily a bad prognostic sign, but it does warn of potential for trouble. Significant reduction in feed intake and large changes in body weight (rapid gain indicates hydroallantois) are important signs.

Transabdominal ultrasonography can be used to monitor fetal viability and as an aid in detecting increased allantoic fluid volumes. Fetal heart rate, however, was not statistically significant between clones and controls in a large-scale study, but it could not be measured in very severe cases, in which the amount of liquid was so great that it became impossible to locate the fetus (Heyman *et al.*, 2001). In addition, although aortic diameter has been reported to be related to fetal size in the horse (Pipers and Adams-Brendemuehl, 1994; Reef *et al.*, 1996), it was not very useful in detecting very large cloned fetuses because normal-sized clones may have an enlarged heart, thus making interpretation of the measurements difficult (Heyman *et al.*, 2002). Finally, the appearance rather than the size of the placentomes as seen on the screen (Fig. 5) was indicative of the pathology. Although at

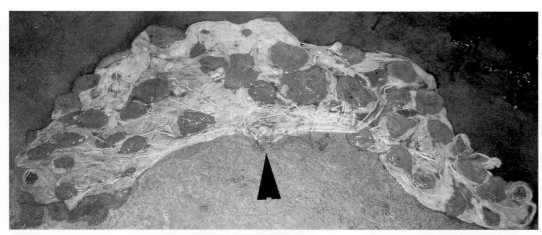

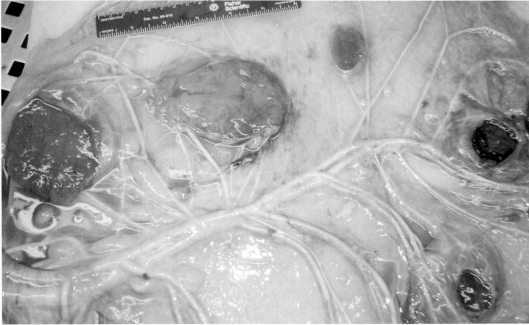

Figure 7 *A mildly abnormal placenta from a viable cloned calf; the placenta was passed within 3 hours of birth. The top photograph shows a slightly reduced number of placentomes (48) spread throughout both the gravid (left) and nongravid horns. A normal-sized umbilicus is present (arrow). The closeup photograph shows normal chorionic vasculature, with an artery and vein supplying each cotyledon, lack of edema, and variation in size of the fetal cotyledons (3–8 cm in diameter).*

present there is no useful therapy for improving the prognosis for severe cases of hydrallantois (Peek, 1997), less severe cases may be nursed through gestation by paying careful attention to the feed intake and metabolic status of the dam. Ruminants in particular are very prone to metabolic upsets (ketosis and fatty liver) due to the increased uterine volume and reduced rumen capacity that accompany hydrallantois. A differential diagnosis for enlarged abdomen is always multiple fetuses when multiple embryos have been transferred. Rapidly growing cloned fetuses may easily cause metabolic problems in ruminants and regular monitoring of urinary ketones will allow early intervention. The time of onset of the pathology (with regard to gestational age) is an important prognostic information: in the author's

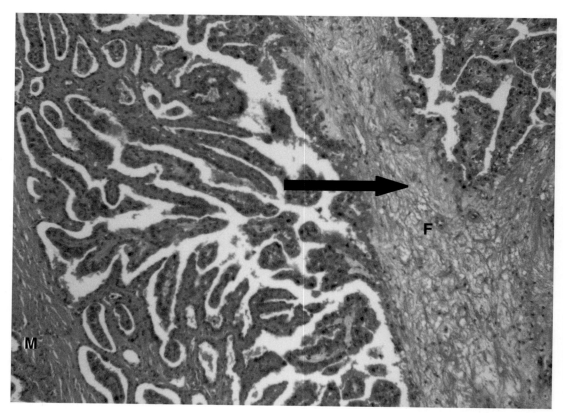

Figure 8 Histological structure of one abnormal (245 days of gestation) placentome from a clone. Both fetal and maternal cellular structures appear normal and inflammatory processes are not present. Slight interstitial edema, however, is indicated by the arrow in the fetal mesenchyme (F), but is not present in the maternal tissue (M). Tissue processing by Farah Kesri (INRA) and photographs by Nathalie Cordonnier (ENVA, France).

Figure 9 Abnormal enlarged liver from one cloned fetus suffering from hydroallantois (left); histological photograph (hematoxylin/eosin/saffron stain) of the liver with marked lesions of steatosis (confirmed with Sudan Black) (right). Histological preparation by Farah Kesri (INRA); photographs and analysis by Nathalie Cordonnier (ENVA).

lab (INRA), it is considered that the fetus may come to term if the pathology is apparent in the last few weeks prior to term. In severe cases, however, although the cow may be delivered with a live fetus close to term, the survival of the offspring is often compromised.

Poorly viable third-trimester fetuses may have significant circulatory abnormalities (hypoxia, liver congestion, placental edema). This is likely to be secondary to inadequate placental development and thus the placental circulatory system requires more careful evaluation. Because abnormal placental circulation would place undue circulatory and metabolic stresses on the fetus, it is important that prospective studies on cloned pregnancies should be performed to evaluate placental blood flow, pressure, fetal blood chemistry, endocrinology, and nutrient metabolism so that treatments may be formulated to improve neonatal viability. These studies are necessarily invasive with attendant risk of pregnancy termination and should be conducted in the way that they have been performed on *in vitro*-produced and embryo-cloned fetuses (Sangild *et al.*, 2000; Garry *et al.*, 1998).

PERINATAL MANAGEMENT

SELECTING THE DAY OF BIRTH

Many cloned pregnancies progress beyond their due date (Heyman *et al.*, 2001). This may not be a problem in itself and may suggest that clones require more time *in utero* for full maturation. However, prolonged gestations are usually associated with increased birth weight and increased risks of dystocias. Rapidly increasing fetal demands, possibly associated with inadequate nutrient supply from a suboptimal placenta, argue against letting the gestation go beyond term. Thus, induction is commonly performed around the anticipated due date (Hill *et al.*, 1999; Chavatte-Palmer *et al.*, 2002) or even before the due date (Lewis *et al.*, 2000). Induction also helps to ensure that birth occurs when there is maximum skilled assistance available. Parturition induction has been commonly employed in livestock animal production with minimal risk of morbidity/mortality (Roberts, 1986). Induction protocols utilizing either a single injection of dexamethasone or prostaglandin $F_{2\alpha}$ 24 hours prior to cesarean section have been shown to induce final pulmonary maturation in bovine fetuses prior to birth (Zaremba *et al.*, 1997). This promotes production of lung surfactant, which is critical to permit lung inflation. However, it has become apparent that when routine induction protocols are used for cloned pregnancies poor lung maturation may still occur. It is unclear if this is the result of the induction protocol or from abnormal physiology of the cloned fetus or placenta. For example, we are uncertain of the functional state of the pituitary–adrenal axis in abnormal clones, although adrenal function seems normal in neonatal cloned calves (Chavatte-Palmer *et al.*, 2002).

In cloned pregnancies it is possible that multiple injections of dexamethasone over a 48-hour period in conjunction with prostaglandin may be more efficient at inducing fetal maturation, but to our knowledge this has not yet been tested. In human infants, the analysis of phospholipids in the amniotic fluid is an indication of pulmonary maturity and readiness for birth (Eigenmann *et al.*, 1984), but so far this has not been used in the bovine species. One indirect, although less sensitive, method of predicting fetal readiness for birth is to monitor steroid hormones in the final 1–2 weeks. These values can be compared against well-established normal values, in which there is a slow decrease in progesterone and elevation in estrogens in the 2 weeks prior to birth (Henricks *et al.*, 1972; Smith *et al.*, 1973; Schams *et al.*, 1972). Two to three days prior to birth, progesterone decreases abruptly and estrogens increase rapidly (Fig. 10). By using this data, induction can be delayed until maternal progesterone has fallen below 3–4 ng/ml and estradiol has risen above 200 pg/ml. This provides evidence that parturition is expected within the next few days.

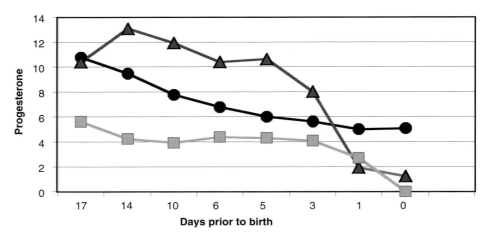

Figure 10 *Plasma progesterone in the last 2 weeks of gestation in three cloned bovine pregnancies showing an abnormal progesterone profile (—●—) and two normal profiles (—▲—, —■—). (—●—) This profile resulted in the birth of poorly viable offspring, whereas offspring from profiles —▲— and —■— were both viable. The cow from profile —●— showed many classic signs of stage 1 labor (flaccid vulva, cervix soft and 8 cm dilated, tail elevation, moderate straining), which prompted us to perform a cesarean section, yet we later found that her progesterone levels had not decreased below 5 ng/ml. The cow from profile —▲— was induced with dexamethasone 60 hours prior to birth. The cow from profile —■— was not induced and entered labor naturally.*

Our current recommendations on induced parturition in cloned bovine pregnancies follow the protocols developed in sheep as a model for inducing pulmonary maturity in premature human infants (Liggins and Howie, 1972). At the author's lab (INRA), steroid concentrations are not monitored but cows are induced on the day of term with dexamethasone, and the cows are delivered by elective cesarean section 30 hours postinjection. The expected gestational length is calculated based on the mean gestational time of control cows on the farm. If a pathology is detected, induction may begin within 7 days of the expected due date. If it is possible to attend calving at any time with competent staff, intramuscular injections of dexamethasone (25 mg) may be given every 24 hours until parturition occurs, usually 48–72 hours after the first dexamethasone injection. It is possible that multiple doses of corticosteroids are more beneficial than a single dose. Prostaglandin (e.g., 25 mg of Lutalyse) may be given with the second dexamethasone injection. Betamethasone is an alternative to dexamethasone (Ikegami *et al.*, 1997). Finally, it may also be beneficial to administer a small dose (5 mg) of dexamethasone 1 week prior to the anticipated day of birth.

MANAGEMENT OF A HIGH-RISK CLONED NEONATE

A large number of pathologies have been described in cloned animals (Table 2). Cloned neonates are at risk of being incompletely prepared for the critical transition to breathing room air. The most controlled method of delivery for the neonate is usually by cesarean section. This is the most appropriate method unless the neonate is known to be of normal size and has a very high likelihood of passing quickly through the pelvic canal. The cesarean should ideally be performed at the beginning of second-stage labor, when the cervix is fully dilated. Waiting until second-stage labor gives the fetus maximum opportunity to be ready for birth.

One important procedure at birth that is not routinely performed in animals is to clamp the umbilicus prior to removal of calf from the uterus. The umbilicus is often enlarged in clones and when broken the two umbilical arteries do not constrict. If the cord is not clamped, blood loss may be life threatening. Surgical removal

Table 2 Body Systems in Which Abnormalities Have Occurred in Cloned Neonates

System	Abnormality
Respiratory	Surfactant deficiency, meconium aspiration, pneumonia
Cardiovascular	Pulmonary hypertension, enlarged umbilicus, septicemia
Hematopoietic	Immunodeficiency, anemia
Metabolic	Hypoglycemia, diabetes, obesity, idiopathic hyperthermia
Gastrointestinal	Gastritis/enteritis
Musculoskeletal	Contracted tendons, oversized, joint infection
Reproductive	Placentation: hydroallantois and/or edema, reduced number of placentomes, enlarged placentomes; overweight placenta
Endocrine	Delayed or absent signs of parturition, low postnatal milk production, elevated leptin
Urinary	Hydronephrosis (lambs)

of the umbilical stump immediately follows delivery to avoid hemorrhagic and infectious incidents that may result from the omphalocele.

Cloned calves are in general less vigorous at birth, and without assistance this lack of viability can easily result in a reduced chance of survival (Kato *et al.*, 1998; Hill *et al.*, 1999; Kubota *et al.*, 2000; Kato *et al.*, 2000). Unlike the birth of naturally conceived calves, management of cloned neonates should be proactive. The high probability of reduced neonatal viability means that it is better to overtreat the neonates than to try to administer aid after they have become hypoxic and acidotic in the first few hours after birth.

Cloned calves produced from different laboratories and from different cell lines appear to show variation in neonatal viability. It is very difficult to pinpoint the reasons for this, because there are many differences between laboratories in methods of producing the cloned animals. There are obvious and subtle differences throughout the nuclear transfer procedure, including oocyte source, nuclear transfer technique, activation protocols, embryo culture conditions, and choice of donor cell genotype or phenotype.

In general, the majority of cloned calves and lambs require intensive care at birth, and so we broadly recommend that they receive treatments normally reserved for high-risk neonates (Martens, 1982; Whitsett *et al.*, 2000). This may be modified according to clinical experience and prior knowledge of the expected survival rate of the cloned offspring. In all cases, the choice of therapies needs to be made quickly. Many cloned calves and lambs need extra care for the first 1–5 days of life or they will die. At birth, 25% of cloned calves require specialist level (i.e., referral veterinary hospital) intensive care for several days, 50% require intensive care for 1–24 hours, and only 25% behave like normal calves. Many are hypoxic, acidotic, lack vigor, and have a weak suckle reflex. To prevent these already compromised neonates from spiraling into respiratory and cardiovascular collapse, we routinely provide ventilation support immediately after birth. Oxygen is administered initially by face mask, which is then continued using either an endotracheal tube or by nasal catheter (Fig. 11). In most cases, oxygen therapy should be continued for at least 1 hour. Arterial blood gases and pH should be ascertained if possible within 15 minutes after birth and should ideally be monitored at least hourly until satisfactory O_2, CO_2, HCO_3, and pH levels have been achieved. A conservative approach is to continue intensive care unless the calf has made very vigorous efforts to rise, has successfully stood on its own, has a very vigorous suckle reflex, and has an arterial blood $O_2 > 60 \, mmHg$, $CO_2 < 50 \, mmHg$, and $pH > 7.3$. The safest option is to maintain every cloned calf on O_2 for at least the first 24 hours.

This preventative therapy is aimed at preventing or reducing the effects of hypoxic pulmonary hypertension. Persistent pulmonary hypertension of the new-

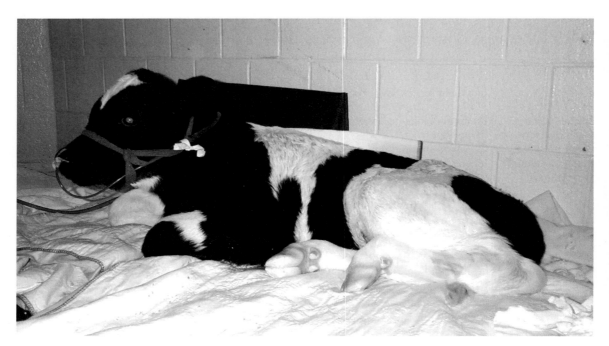

Figure 11 *This 3-hour-old cloned calf was born by assisted vaginal delivery. Prior to taking this photograph, this calf was treated with oxygen, 80 ppm nitric oxide (via a nasal cannula), commercial bovine surfactant (Infasurf, via endotracheal tube), aminophylline, plasma transfusion, and antibiotics. Her arterial blood gases at 20 minutes after birth showed adequate ventilation, with a slightly elevated CO_2. Ventilation improved markedly over the next 3 hours (20 minutes: $PO_2 = 76$ mmHg, $PCO_2 = 56$ mmHg, pH = 7.36, $HCO_3 = 32$ mEq/liter; 3 hours: $PO_2 = 208$ mmHg, $PCO_2 = 48$ mmHg, pH = 7.42, $HCO_3 = 31$ mEq/liter). At 9 days of age she developed a rotoviral diarrhea that was treated symptomatically and since then has remained healthy.*

born is characterized by sustained elevations of pulmonary vascular resistance after birth, leading to right-to-left shunting of blood across the ductus arteriosus or foramen ovale and resulting in severe hypoxemia (Levin *et al.*, 1976). This is the equivalent of persistent fetal circulation and we have documented both reverse flow through the ductus arteriosus and foramen ovale using echocardiography (Hill *et al.*, 1999, 2000a). Therapies aimed at improving pulmonary function include pulmonary surfactant, oxygen, positive-pressure ventilation, pulmonary arterial vasodilators, and bronchodilators. They should be used when one suspects pulmonary hypertension from the clinical examination.

Pulmonary surfactant registered for human use may be prohibitively expensive when used at the recommended dose of 100 mg/kg (e.g., Infasurf, ONY Inc.; Amherst, NY). To reduce this cost, the dose may be reduced according to clinical response (e.g., to 15 mg/kg); a lower price may be negotiated directly with the company. Natural surfactant may be isolated following alveolar lavage and multiple centrifugation steps. For pulmonary arterial vasodilation, maximal efficacy of nitric oxide is achieved when 80 ppm is administered for the first 30 minutes after birth and when surfactant is instilled within the first hour (Zayek *et al.*, 1993). The nitric oxide is most conveniently administered by using separate tanks of oxygen (100%) and nitric oxide (800 ppm), then running the flow rates at 9 and 1 liters/minute, respectively so that 80 ppm nitric oxide is delivered to the calf. Pulmonary hypertension has also been treated with varying degrees of success using tolazoline administered as a 1 mg/kg intravenous bolus then as an intravenous infusion at 0.1–0.2 mg/kg/hour (Bressack and Bland, 1981; Curtis *et al.*, 1996).

Systemic blood pressures may drop significantly with this treatment. Nebulization with bronchodilators (e.g., albuterol, aminophyllin + acetylcysteine) may be very beneficial. If meconium aspiration is suspected, nebulization with bronchodilators together with corticosteroid treatment is indicated. The above therapies are not indicated in all cases and their use is dependent on clinical evaluation, experience, and facilities.

Additional supportive therapy includes maintenance of body temperature (which may rapidly drop or increase), precautionary antibiotics (e.g., penicillin or aminoglycosides, although this is not approved for animals that will enter the food chain), colostrum feeding by bottle or nasogastric tube, intravenous fluids, glucose, and possibly serum transfusion to provide an immediate source of antibodies. Colostrum should preferably be fed by bottle, although suckling strength is often inadequate. Colostrum is a potential source of contamination with common bacteria and viruses such as *Mycobacterium johnei*, *Clostridium perfringens*, and bovine leukosis virus, and it is therefore ideal to use colostrum from a health-tested donor, collected in an aseptic manner. As an added precaution, the colostrum may be sterilized using radiation or by very careful pasteurization. However, inadequately prepared colostrum can denature the immunoglobulins and render it worthless. The donor should be fully vaccinated, including vaccination against *C. perfringens*. A higher than normal incidence of necrotizing enterocolitis (often caused by *C. perfringens*) in cloned calves has occurred; this may be due to altered gastrointestinal function caused by hypoxia, hypothermia, or prematurity.

Some cloned neonates may require specialized care that is available only from a large veterinary hospital. These facilities use constant (at least hourly) monitoring of vital signs (temperature, pulse, respiration, demeanor), together with clinical details (auscultation, palpation, etc), laboratory tests (metabolic profiles, hematology), blood gas measurements (arterial samples, pulse oximetry), and radiology (chest and abdominal X rays) and echocardiography. The information gained from these tests enables rapid response to treat abnormalities that occur in the pulmonary (correct inadequate ventilation or perfusion), cardiovascular, and gastrointestinal systems (e.g., abomasal ulceration, congenital abnormalities, infection), and metabolic or electrolyte disturbances (e.g., blood glucose).

POSTNATAL MONITORING

In general, cloned animals that progress through the early neonatal period have a high rate of survival. Depending on management conditions, these animals will then be challenged by a wide variety of environmental pathogens that may cause septicemia and gastrointestinal or respiratory disease. In the postneonatal period, cloned animals have succumbed to diseases likely related to the cloning procedure, such as blood cell aplasia or cardiac disease (Renard *et al.*, 1999; Hill *et al.*, 1999). In addition, there appears to be a higher than expected incidence of gastrointestinal disease. Sudden death with no premonitory signs, supposedly due to cardiac failure, has been observed in calves 2 to 3 months old. Finally, paradoxical hyperthermia (up to 41°C), with no response to medical treatment but mechanical cooling with cold showers, is often observed. It is therefore prudent to evaluate methodically each body system for abnormalities, at least for the first month after birth.

At present there is only a small body of published data on objective characteristics of cloned animals. The first somatic cell clone, Dolly, has produced and reared lambs each year she was bred. Cloned mice have been found to be overweight, which suggests that very detailed evaluations of cloned animals may reveal subtle abnormalities (Tamashiro *et al.*, 2000). In farm animals, production-related data such as growth rates, milk production, and fertility appear to be within normal ranges (Pace *et al.*, 2001; Enright *et al.*, 2001).

SUMMARY

The urgency to produce cloned livestock efficiently has increased with commercialization of the technique. The first cloning programs in sheep, cattle, and mice showed a high rate of gestational and perinatal loss (Wilmut *et al.*, 1997; Cibelli *et al.*, 1998; Wakayama *et al.*, 1998). Cloning experiments still show higher than normal early gestational losses, although viability at birth has moved closer to normal in mice, goats, and pigs (Wakayama and Yanagimachi, 2001; Keefer *et al.*, 2001; Polejaeva *et al.*, 2000), and bovine cloning programs have reported excellent neonatal viability (Wells *et al.*, 1999). Progress is thus being made in reducing the problems that beset the embryo cloning industry in the 1990s.

Current observations strongly suggest that placental abnormalities are the cause of lowered cloned neonatal viability. Cloned calves and lambs at birth may show signs of a stressful uterine environment (meconium staining, hypoxia). Placental reserve capacity is therefore likely to be limited due to inadequate development. The combination of a usually larger than normal fetus and suboptimal placentation means cloned fetuses may share many of the attributes of fetuses suffering from placental insufficiency. As we define the physiology of these prenatal placental problems more carefully, outcomes should improve. Defining the origins of these abnormalities may also lead to prevention, through alterations in the cloning technique. The variation in neonatal viability between experiments and between species is intriguing. Whether these differences are due to technique, to breed, or to placental type remains to be determined.

ACKNOWLEDGMENTS

Much of the clinical information presented here was derived from interactions with the internal medicine specialists and pathologists at Texas A&M and Cornell Universities and work at INRA in Jouy en Josas, France, and the research farm of Bressonvilliers, with the help of the pathology department of Alfort Veterinary school. In particular, we would like to thank Allen Roussel, John Edwards, Tom Divers, and Don Schlafer in the United States and Christophe Richard, Patrice Laigre, Yvan Heyman, Farah Kesri, and Nathalie Cordonnier in France.

REFERENCES

Alexander, B. M., Johnson, M. S., Guardia, R. O., Graaf, W. L. V. D., Senger, P. L., and Sasser, R. G. (1995). Embryonic loss from 30 to 60 days post breeding and the effect of palpation per rectum on pregnancy. *Theriogenology* **43**, 551–556.

Baguisi, A., Behboodi, E., Melican, D. T., Pollock, J. S., Destrempes, M. M., Cammuso, C., Williams, J. L., Nims, S. D., Porter, C. A., Midura, P., Palacios, M. J., Ayres, S. L., Denniston, R. S., Hayes, M. L., Ziomek, C. A., Meade, H. M., Godke, R. A., Gavin, W. G., Overstrom, E. W., and Echelard, Y. (1999). Production of goats by somatic cell nuclear transfer. *Nat. Biotechnol.* **17**, 456–461.

Barnes, F. L. (2000). The effects of the early uterine environment on the subsequent development of embryo and fetus. *Theriogenology* **53**, 649–658.

Barnes, F. L., Collas, P., Powell, R., King, W. A., Westhusin, M., and Shepherd, D. (1993). Influence of recipient oocyte cell cycle stage on DNA synthesis, nuclear envelope breakdown, chromosome constitution, and development in nuclear transplant bovine embryos. *Mol. Reprod. Dev.* **36**, 33–41.

Bazer, F. W., Roberts, R. M., Basha, S. M., Zavy, M. T., Caton, D., and Barron, D. H. (1979). Method for obtaining ovine uterine secretions from unilaterally pregnant ewes. *J. Anim. Sci.* **49**, 1522–1527.

Behboodi, E., Anderson, G. B., Bondurant, R. H., Cargill, S. L., Kreuscher, B. R., Medrano, J. F., and Murray, J. D. (1995). Birth of large calves that developed from *in vitro*-derived bovine embryos. *Theriogenology* **44**, 227–232.

Behboodi, E., Ayres, S. L., Reggio, B. C., O'Coin, M. D., Gavin, W. G., Denniston, R. S., Landry, A. M., Meade, H. M., and Echelard, Y. (2002). Pregnancy profile and health status of goats derived by somatic cell nuclear transfer. *Theriogenology* **57**, 395.

Betthauser, J., Forsberg, E., Augenstein, M., Childs, L., Eilertsen, K., Enos, J., Forsythe, T., Golueke, P., Jurgella, G., Koppang, R., Lesmeister, T., Mallon, K., Mell, G., Misica, P., Pace, M., Pfister-Genskow, M., Strelchenko, N., Voelker, G., Watt, S., Thompson, S., and Bishop, M. (2000). Production of cloned pigs from *in vitro* systems. *Nat. Biotechnol.* **18**, 1055–1059.

Bondioli, K., Ramsoondar, J., Williams, B., Costa, C., and Fodor, W. (2001). Cloned pigs generated from cultured skin fibroblasts derived from a H-transferase transgenic boar. *Mol. Reprod. Dev.* 60(2), 189–195.

Bressack, M. A., and Bland, R. D. (1981). Intravenous infusion of tolazoline reduces pulmonary vascular resistance and net fluid filtration in the lungs of awake, hypoxic newborn lambs. *Am. Rev. Respir. Dis.* 123, 217–221.

Campbell, K. H., McWhir, J., Ritchie, W. A., and Wilmut, I. (1996). Sheep cloned by nuclear transfer from a cultured cell line. *Nature* 380, 64–66.

Chavatte-Palmer, P., Heyman, Y., Richard, C., Laigre, P., Monget, P., Vignon, X., LeBourhis, D., Wiebe, J., Hill, J., Kann, G., Chilliard, Y., Ponter, A., and Renard, J. P. (2002). Clinical, hormonal and hematological characteristics of fetuses and calves born after nuclear transfer of a somatic cell. *Biol. Reprod.* 66(6), in press.

Chesné, P., Adenot, P. G., Viglietta, C., Baratte, M., Boulanger, L., and Renard, J.-P. (2002). Cloned rabbits produced by nuclear transfer from somatic cells. *Nature Biotech* 20, 366–369.

Cibelli, J. B., Stice, S. L., Golueke, P. J., Kane, J. J., Jerry, J., Blackwell, C., Ponce de Leon, F. A., and Robl, J. M. (1998). Cloned transgenic calves produced from nonquiescent fetal fibroblasts. *Science* 280, 1256–1258.

Curtis, J., Palacino, J. J., and O'Neill, J. T. (1996). Production of pulmonary vasodilation by tolazoline, independent of nitric oxide production in neonatal lambs. *J. Pediatr.* 128, 118–124.

Denning, C., Burl, S., Ainslie, A., Bracken, J., Dinnyes, A., Fletcher, J., King, T., Ritchie, M., Ritchie, W. A., Rollo, M., De Sousa, P., Travers, A., Wilmut, I., and Clark, A. J. (2001). Deletion of the alpha(1,3)galactosyl transferase (GGTA1) gene and the prion protein (PrP) gene in sheep. *Nat. Biotechnol.* 19, 559–562.

DeSousa, P. A., Walker, S., King, T. J., Young, L. E., Harkness, L., Ritchie, W. A., Travers, A., Ferrier, P., and Wilmut, I. (2001). Evaluation of gestational deficiencies in cloned sheep fetuses and placentae. *Biol. Reprod.* 65, 23–30.

Eigenmann, U. J. E., Schoon, H. A., Jahn, D., and Grunert, E. (1984). Neonatal distress syndrome in the calf. *Vet. Rec.* 114, 141–144.

Enright, B. P., Taneja, M., Schreiber, D., Riesen, J., Tian, X., and Yang, X. (2001). Puberty and early follicular dynamics in adult somatic cell derived cloned heifers. *Theriogenology* 55, 267.

Farin, P. W., and Farin, C. E. (1995). Transfer of bovine embryos produced *in vivo* or *in vitro*: Survival and fetal development. *Biol. Reprod.* 52, 676–682.

Forar, A. L., Gay, J. M., Hancock, D. D., and Gay, C. C. (1996). Fetal loss frequency in ten holstein dairy herds. *Theriogenology* 45, 1505–1513.

French, A. J., Hall, V. J., Korfiatis, N. A., Ruddock, N. T., Vajta, G., Lewis, J. M., and Trounson, A. O. (2002). Viability of cloned bovine embryos following OPS vitrification. *Theriogenology* 57, 413.

Garry, F. B., Adams, R., McCann, J. P., and Odde, K. G. (1996). Postnatal characteristics of calves produced by nuclear transfer cloning. *Theriogenology* 45, 141–152.

Garry, F. B., Adams, R., Holland, M. D., Hay, W. W., McCann, J. P., Wagner, A., and Seidel, J. G. E. (1998). Arterial oxygen, metabolite and energy regulatory hormone concentrations in cloned bovine fetuses. *Theriogenology* 49, 321.

Gibbons, J., Arat, S., Rzucidlo, J., Miyoshi, K., Waltenburg, R., Respess, D., Venable, A., and Stice, S. (2002). Enhanced survivability of cloned calves derived from roscovitine-treated adult somatic cells. *Biol. Reprod.* 66 (in press).

Hasler, J. F., Henderson, W. B., Hurtgen, P. J., Jin, Z. Q., McCauley, A. D., Mower, S. A., Neely, B., Shuey, L. S., Stokes, J. E., and Trimmer, S. A. (1995). Production, freezing and transfer of bovine IVF embryos and subsequent calving results. *Theriogenology* 43, 141–152.

Henricks, D. M., Dickey, J. F., Hill, J. R., and Johnston, W. E. (1972). Plasma estrogen and progesterone levels after mating, and during late pregnancy and postpartum in cows. *Endocrinology* 90, 1336–1342.

Heyman, Y., Chavatte-Palmer, P., Lebourhis, D., Deniau, F., Laigre, P., Vignon, X., and Renard, J. P. (2000). Evolution of pregnancies after transfer of cloned bovine blastocysts derived from fetal or adult somatic cells. *Proc. Assoc. Eur. Transf. Embryon* Lyon, France, p. 166.

Heyman, Y., Chavatte-Palmer, P., LeBourhis, D., Camous, S., Vignon, X., and Renard, J. P. (2002). Frequency and occurrence of late gestation losses from cattle cloned embryos. *Biol. Reprod.* 66, 6–13.

Hill, J. R., Roussel, A. J., Cibelli, J. B., Edwards, J. F., Hooper, R. N., Miller, M. W., Thompson, J. A., Looney, C. R., Westhusin, M. E., Robl, J. M., and Stice, S. L. (1999). Clinical and pathologic features of cloned transgenic calves and fetuses (13 case studies). *Theriogenology* 51, 1451–1465.

Hill, J. R., Winger, Q. A., Long, C. R., Looney, C. R., Thompson, J. A., and Westhusin, M. E. (2000a). Development rates of male bovine nuclear transfer embryos derived from adult and fetal cells. *Biol. Reprod.* 62, 1135–1140.

Hill, J. R., Burghardt, R. C., Jones, K., Long, C. R., Looney, C. R., Shin, T., Spencer, T. E., Thompson, J. A., Winger, Q. A., and Westhusin, M. E. (2000b). Evidence for placental abnormality as the major cause of mortality in first-trimester somatic cell cloned bovine fetuses. *Biol. Reprod.* 63, 1787–1794.

Hill, J. R., Edwards, J. F., Sawyer, N., Blackwell, C., and Cibelli, J. B. (2001). Placental anomalies in a viable cloned calf. *Cloning* 3, 81–86.

Ikegami, M., Jobe, A. H., Newnham, J., Polk, D. H., Willet, K. E., and Sly, P. (1997). Repetitive prenatal glucocorticoids improve lung function and decrease growth in preterm lambs. *Am. J. Respir. Crit. Care Med.* **156**, 178–184.

Kato, Y., Tetsuya, T., Sotomaru, Y., Kurokawa, K., Kato, J., Doguchi, H., Yasue, H., and Tsunoda, Y. (1998). Eight calves cloned from somatic cells of a single adult. *Science* **282**, 2095–2098.

Kato, Y., Tani, T., and Tsunoda, Y. (2000). Cloning of calves from various somatic cell types of male and female adult, newborn and fetal cows. *J. Reprod. Fertil.* **120**, 231–237.

Keefer, C. L., Stice, S. L., and Matthews, D. L. (1994). Bovine inner cell mass cells as donor nuclei in the production of nuclear transfer embryos and calves. *Biol. Reprod.* **50**, 935–939.

Keefer, C. L., Baldassarre, H., Keyston, R., Wang, B., Bhatia, B., Bilodeau, A. S., Zhou, J. F., Leduc, M., Downey, B. R., Lazaris, A., and Karatzas, C. N. (2001). Generation of dwarf goat (*Capra hircus*) clones following nuclear transfer with transfected and nontransfected fetal fibroblasts and *in vitro*-matured oocytes. *Biol. Reprod.* **64**, 849–856.

Keefer, C. L., Keyston, R., Lazaris, A., Bhatia, B., Begin, I., Bilodeau, A. S., Zhou, F. J., Kafidi, N., Wang, B., Baldassarre, H., and Karatzas, C. N. (2002). Production of cloned goats after nuclear transfer using adult somatic cells. *Biol. Reprod.* **66**, 199–203.

Kishi, M., Itagaki, Y., Takakura, R., Imamura, M., Sudo, T., Yoshinari, M., Tanimoto, M., Yasue, H., and Kashima, N. (2000). Nuclear transfer in cattle using colostrum-derived mammary gland epithelial cells and ear-derived fibroblast cells. *Theriogenology* **54**, 675–684.

Kruip, T. A. M., Daas, J. H. G. D., and Den Daas, J. H. G. (1997). *In vitro* produced and cloned embryos: Effects on pregnancy, parturition and offspring. *Theriogenology* **47**, 43–52.

Kubota, C., Yamakuchi, H., Todoroki, J., Mizoshita, K., Tabara, N., Barber, M., and Yang, X. (2000). Six cloned calves produced from adult fibroblast cells after long-term culture. *Proc. Natl. Acad. Sci. U.S.A.* **97**, 990–995.

Lanza, R. P., Cibelli, J. B., Faber, D., Sweeney, R. W., Henderson, B., Nevala, W., West, M. D., and Wettstein, P. J. (2001). Cloned cattle can be healthy and normal. *Science* **294**, 1893–1894.

Levin, D. L., Heymann, M. A., Kitterman, J. A., Gregory, G. A., Phibbs, R. H., and Rudolph, A. M. (1976). Persistent pulmonary hypertension of the newborn. *J. Pediatr.* **89**, 626–633.

Lewis, I. M., Peura, T. T., Owens, J. L., Ryan, M. F., Diamente, M. G., Pushett, D. A., Lane, M. W., Jenkin, G., Coleman, P. J., and Trounson, A. O. (2000). Outcomes from novel simplified nuclear transfer techniques in cattle. *Theriogenology* **53**, 233.

Liggins, G. C., and Howie, R. N. (1972). A controlled trial of antepartum glucocorticoid treatment for the prevention of the respiratory distress syndrome in premature infants. *Pediatrics* **50**, 515–525.

Martens, R. J. (1982). Neonatal respiratory distress: A review with particular emphasis on the horse. *Compend. Contin. Edu.* **4**, S23–S33.

McCreath, K. J., Howcroft, J., Campbell, K. H., Colman, A., Schnieke, A. E., and Kind, A. J. (2000). Production of gene-targeted sheep by nuclear transfer from cultured somatic cells. *Nature* **405**, 1066–1069.

Mialon, M. M., Camous, S., Renand, G., Martal, J., and Menissier, F. (1993). Peripheral concentration of a 60-kDa pregnancy specific protein during gestation and after calving in relationship to embryonic mortality in cattle. *Reprod. Nutr. Dev.* **33**, 269–282.

Ogura, A., Inoue, K., Ogonuki, N., Noguchi, A., Takano, K., Nagano, R., Suzuki, O., Lee, J., Ishino, F., and Matsuda, J. (2000). Production of male cloned mice from fresh, cultured, and cryopreserved immature Sertoli cells. *Biol. Reprod.* **62**, 1579–1584.

Onishi, A., Iwamoto, M., Akita, T., Mikawa, S., Takeda, K., Awata, T., Hanada, H., and Perry, A. C. (2000). Pig cloning by microinjection of fetal fibroblast nuclei. *Science* **289**, 1188–1190.

Pace, M. M., Mell, G., Forsberg, E., Betthauser, J., Strelchenko, N., Golueke, P., Childs, L., Jurgella, G., Koppang, R., and Bishop, M. (2001). Cloning using somatic cells: An analysis of 75 calves. *Theriogenology* **55**, 281.

Peek, S. F. (1997). Dropsical conditions affecting pregnancy. *In* "Current Therapy in Large Animal Theriogenology," 1st Ed. (R.S. Youngquist, ed.), pp. 400–403. W. B. Saunders, Philadelphia.

Peura, T. T., Barritt, Sr., Kleeman, D. O., and Walker, S. K. (2002). Nutrition of the oocyte donor affects the *in vivo* viability of somatic cell clones in sheep. *Theriogenology* **57**, 444.

Peterson, A. J., and McMillan, W. H. (1998). Allantoic aplasia—A consequence of *in vitro* production of bovine embryos and the major cause of late gestation embryo loss. *Proc. 29th Conf. Aust. Soc. Reprod. Biol.* **29**, p. 4.

Pipers, F. S., and Adams-Brendemuehl, C. S. (1994). Techniques and applications of transabdominal ultrasonography in the pregnant mare. *J. Am. Vet. Med. Assoc.* **185**(7), 766–771.

Polejaeva, I. A., Chen, S. H., Vaught, T. D., Page, R. L., Mullins, J., Ball, S., Dai, Y., Boone, J., Walker, S., Ayares, D. L., Colman, A., and Campbell, K. H. (2000). Cloned pigs produced by nuclear transfer from adult somatic cells. *Nature* **407**, 86–90.

Reef, V. B., Vaala, W. E., Worth, L. T., Sertich, P. L., and Spencer, P. A. (1996). Ultrasonographic assessment of fetal well-being during late gestation: Development of an equine biophysical profile. *Equine Vet. J.* **28**(3), 200–208.

Reggio, B. C., James, A. N., Green, H. L., Gavin, W. G., Behboodi, E., Echelard, Y., and Godke, R. A. (2001). Cloned transgenic offspring resulting from somatic cell nuclear transfer in the goat: Oocytes derived from both follicle-stimulating hormone-stimulated and nonstimulated abattoir-derived ovaries. *Biol. Reprod.* **65**, 1528–1533.

Renard, J. P., Chastant, S., Chesné, P., Richard, C., Marchal, J., Cordonnier, N., Chavatte, P., and Vignon, X. (1999). Lymphoid hypoplasia and somatic cloning. *Lancet* **353**, 1489–1491.

Roberts, S. J. (1986). Parturition. *In* "Veterinary Obstetrics and Genital Diseases," 3rd Ed. (S. J. Roberts, ed.), pp. 245–276. S. J. Roberts, Vermont.

Robinson, J. S., Kingston, E. J., Jones, C. T., and Thorburn, G. D. (1979). Studies on experimental growth retardation in sheep. The effect of removal of endometrial caruncles on fetal size and metabolism. *J. Dev. Physiol.* **1**, 379–398.

Sangild, P. T., Schmidt, M., Jacobsen, H., Fowden, A. L., Forhead, A., Avery, B., and Greve, T. (2000). Blood chemistry, nutrient metabolism, and organ weights in fetal and newborn calves derived from *in vitro*-produced bovine embryos. *Biol. Reprod.* **62**, 1495–1504.

Sasser, R. G., Ruder, C. A., Ivani, K. A., Butler, J. E., and Hamilton, W. C. (1986). Detection of pregnancy by radioimmunoassay of a novel pregnancy-specific protein in serum of cows and a profile of serum concentrations during gestation. *Biol. Reprod.* **35**, 936–942.

Schams, D., Hoffmann, B., Fischer, S., Marz, E., and Karg, H. (1972). Simultaneous determination of LH and progesterone in peripheral bovine blood during pregnancy, normal and corticoid-induced parturition and the post-partum period. *J. Reprod. Fertil.* **29**, 37–48.

Schmidt, M., Greve, T., Avery, B., Beckers, J. F., Sulon, J., and Hansen, H. B. (1996). Pregnancies, calves and calf viability after transfer of *in vitro* produced bovine embryos. *Theriogenology* **46**, 527–539.

Schnieke, A. E., Kind, A. J., Ritchie, W. A., Mycock, K., Scott, A. R., Ritchie, M., Wilmut, I., Colman, A., and Campbell, K. H. (1997). Human factor IX transgenic sheep produced by transfer of nuclei from transfected fetal fibroblasts. *Science* **278**, 2130–2133.

Seidel, G. E. (1995). Sexing, bisection and cloning: Perspectives and applications to animal breeding. (1995). *Proc. Symp. Reprod. Anim. Breeding*, pp. 147–154. Milan, Italy.

Shiga, K., Fujita, T., Hirose, K., Sasae, Y., and Nagai, T. (1999). Production of calves by transfer of nuclei from cultured somatic cells obtained from Japanese black bulls. *Theriogenology* **52**, 527–535.

Shin, T., Kraemer, D., Pryor, J., Liu, L., Rugila, J., Howe, L., Buck, S., Murphy, K., Lyons, L., and Westhusin, M. (2002). A cat cloned by nuclear transplantation. *Nature* **415**, 859.

Sinclair, K. D., McEvoy, T. G., Maxfield, E. K., Maltin, C. A., Young, L. E., Wilmut, I., Broadbent, P. J., and Robinson, J. J. (1999). Aberrant fetal growth and development after *in vitro* culture of sheep zygotes. *J. Reprod. Fertil.* **116**, 177–186.

Smith, V. G., Edgerton, L. A., Hafs, H. D., and Convey, E. M. (1973). Bovine serum estrogens, progestins and glucocorticoids during late pregnancy parturition and early lactation. *J. Anim. Sci.* **36**, 391–396.

Stice, S. L., Strelchenko, N. S., Keefer, C. L., and Matthews, L. (1996). Pluripotent bovine embryonic cell lines direct embryonic development following nuclear transfer. *Biol. Reprod.* **54**, 100–110.

Tamashiro, K. L., Wakayama, T., Blanchard, R. J., Blanchard, D. C., and Yanagimachi, R. (2000). Postnatal growth and behavioral development of mice cloned from adult cumulus cells. *Biol. Reprod.* **63**, 328–334.

Thompson, J. G., and Peterson, A. J. (2000). Bovine embryo culture *in vitro*: New developments and post-transfer consequences. *Hum. Reprod.* **15**, 59–67.

Thompson, J. G., Gardner, D. K., Pugh, P. A., McMillan, W. H., and Tervit, H. R. (1995). Lamb birth weight is affected by culture system utilized during *in vitro* pre-elongation development of ovine embryos. *Biol. Reprod.* **53**, 1385–1391.

van Wagtendonk-de Leeuw, A. M., Mullaart, E., de Roos, A. P., Merton, J. S., den Daas, J. H., Kemp, B., and de Ruigh, L. (2000). Effects of different reproduction techniques: AI, MOET or IVP, on health and welfare of bovine offspring. *Theriogenology* **53**, 575–597.

Vignon, X., Chesne, P., Le Bourhis, D., Flechon, J. E., Heyman, Y., and Renard, J. P. (1998). Developmental potential of bovine embryos reconstructed from enucleated matured oocytes fused with cultured somatic cells. *C.R. Acad. Sci. III* **321**, 735–745.

Wakayama, T., and Yanagimachi, R. (1999). Cloning of male mice from adult tail-tip cells. *Nat. Genet.* **22**, 127–128.

Wakayama, T., and Yanagimachi, R. (2001). Mouse cloning with nucleus donor cells of different age and type. *Mol. Reprod. Dev.* **58**, 376–383.

Wakayama, T., Perry, A. C., Zuccotti, M., Johnson, K. R., and Yanagimachi, R. (1998). Full-term development of mice from enucleated oocytes injected with cumulus cell nuclei. *Nature* **394**, 369–374.

Walker, S. K., Hartwich, K. M., and Seamark, R. F. (1996). The production of unusually large offspring following embryo manipulation—Concepts and challenges. *Theriogenology* **45**, 111–120.

Wells, D. N., Misica, P. M., Day, T. A., and Tervit, H. R. (1997). Production of cloned lambs from an established embryonic cell line: A comparison between *in vivo*- and *in vitro*-matured cytoplasts. *Biol. Reprod.* **57**, 385–393.

Wells, D. N., Pavla, P. M., Tervit, H. R., and Vivanco, W. H. (1998). Adult somatic cell nuclear transfer is used to preserve the last surviving cow of the Enderby Island cattle breed. *Reprod. Fertil. Dev.* **10**, 369–378.

Wells, D. N., Misica, P. M., and Tervit, H. R. (1999). Production of cloned calves following nuclear transfer with cultured adult mural granulosa cells. *Biol. Reprod.* **60**, 996–1005.

Whitsett, J. A., Pryhuber, G. S., Rice, W. R., Warner, B. B., and Wert, S. E. (2000). Acute respiratory disorders. *In* "Neonatology. Pathophysiology and Management of the Newborn," 5th Ed. (G. B. Avery, M. A. Fletcher, and M. G. MacDonald, eds.), pp. 485–501. Lippincott Willams & Wilkins, Philadelphia.

Willadsen, S. M., Janzen, R. E., McAllister, R. J., Shea, B. F., Hamilton, G., and McDermand, D. (1991). The viabilty of late morulae and blastocysts produced by nuclear transplantation in cattle. *Theriogenology* **35**, 161–170.

Wilmut, I., Sales, D. I., and Ashworth, C. J. (1986). Maternal and embryonic factors associated with prenatal loss in mammals. *J. Reprod. Fertil.* **76**, 851–864.

Wilmut, I., Schnieke, A. E., McWhir, J., Kind, K. L., and Campbell, K. H. S. (1997). Viable offspring derived from fetal and adult mammalian cells. *Nature* **385**, 810–813.

Wilson, J. M., Williams, J. D., Bondioli, K. R., Looney, C. R., Westhusin, M. E., and McCalla, D. F. (1995). Comparison of birth weight and growth characteristics of bovine calves produced by nuclear transfer (cloning), embryo transfer and natural mating. *Anim. Reprod. Sci.* **38**, 73–83.

Young, L. E., Sinclair, K. D., and Wilmut, I. (1998). Large offspring syndrome in cattle and sheep. *Rev. Reprod.* **3**, 155–163.

Zakhartchenko, V., Alberio, R., Stojkovic, M., Prelle, K., Schernthaner, W., Stojkovic, P., Wenigerkind, H., Wanke, R., Duchler, M., Steinborn, R., Mueller, M., Brem, G., and Wolf, E. (1999). Adult cloning in cattle: Potential of nuclei from a permanent cell line and from primary cultures. *Mol. Reprod. Dev.* **54**, 264–272.

Zakhartchenko, V., Mueller, S., Alberio, R., Schernthaner, W., Stojkovic, M., Wenigerkind, H., Wanke, R., Lassnig, C., Mueller, M., Wolf, E., and Brem, G. (2001). Nuclear transfer in cattle with nontransfected and transfected fetal or cloned transgenic fetal and postnatal fibroblasts. *Mol. Reprod. Dev.* **60**, 362–369.

Zaremba, W., Grunert, E., and Aurich, J. E. (1997). Prophylaxis of respiratory distress syndrome in premature calves by administration of dexamethasone or a prostaglandin f-2-alpha analogue to their dams before parturition. *Am. J. Vet. Res.* **58**, 404–407.

Zayek, M., Cleveland, D., and Morin, F. C. (1993). Treatment of persistent pulmonary hypertension in the newborn lamb by inhaled nitric oxide. *J. Pediatr.* **122**, 743–750.

Zoli, A. P., Guilbault, L. A., Delahaut, P., Benitez-Ortiz, W., and Beckers, J. F. (1992). Radioimmunoassay of a bovine pregnancy associated glycoprotein in serum: Its application for pregnancy diagnosis. *Biol. Reprod.* **46**, 83–92.

Zou, X., Chen, Y., Wang, Y., Luo, J., Zhang, Q., Yang, Y., Ju, H., Shen, Y., Lao, W., Xu, S., and Du, M. (2001). Production of cloned goats from enucleated oocytes injected with cumulus cell nuclei or fused with cumulus cells. *Cloning* **3**, 31–37.

CHAPTER 13

Donor Cell Type and Cloning Efficiency in Mammals

Y. Tsunoda and Y. Kato

DEVELOPMENTAL POTENTIAL OF GERM LINE NUCLEI AT VARIOUS CELL CYCLE STAGES

The first reliable successful nuclear transfer in mammals was reported by McGrath and Solter (1983), who obtained mice after the exchange of pronuclei between two types of zygotes. The methods included enucleation and discarding of pronuclei with a small volume of cytoplasm (pronuclei karyoplast) from recipient zygotes, removal of pronuclei karyoplasts from donor zygotes, and fusion of donor pronuclei karyoplasts with enucleated zygote cytoplasm using inactivated Sendai virus (strain HVJ, the hemagglutinating virus of Japan). Several laboratories confirmed that a high proportion of pronuclei-exchanged zygotes developed into live offspring. (DiBerardino, 1997). Enucleated zygotes receiving nuclei from two-, four-, and eight-cell mouse embryos, however, did not develop *in vitro* or *in vivo* except for a few cases with two-cell embryos (McGrath and Solter, 1984; Tsunoda *et al.*, 1987). When nuclei from four- and eight-cell embryos were fused with enucleated blastomeres of two-cell embryos, live mice with normal fertility were obtained (Tsunoda *et al.*, 1987). Nuclei were not fully reprogrammed, however, because compaction of enucleated two-cell embryos receiving the nuclei of eight-cell embryos occurred at the four- to eight-cell stage, but not at the eight- to sixteen-cell stage, as in normal embryos.

As in the frog (DiBerardino, 1997), the nuclei from mammalian preimplantation embryos can be reprogrammed in the cytoplasm of unfertilized oocytes (Willadsen, 1986). Willadsen reported that blastomeres of eight- to sixteen-cell sheep embryos fused with enucleated oocytes, and the fused oocytes were cultured in the sheep oviduct to the morula to blastocyst stages. He obtained live young after transfer of nuclear-transferred embryos to recipients. Since then, a number of mice, rabbits, goats, sheep, bovines, pigs, and rhesus monkeys have been produced after transfer of blastomeres of two-, four-, and eight-cell and morula-stage embryos to enucleated oocytes at the second metaphase (Tsunoda and Kato, 2000). The development of nuclear-transferred embryos to term was not clearly different among developmental stages of donor embryos. Enucleated oocytes receiving inner cell mass cells of sheep (Smith and Wilmut, 1989), bovines (Keefer *et al.*, 1994; Collas and Barnes, 1994), and mice (Tsunoda and Kato, 1998) also developed into living offspring. The development of embryonic stem (ES) cells to term has been limited. Live calves were produced after transfer of short-term cultured inner cell mass cells of blastocysts into enucleated oocytes (Sims and First, 1993). Wakayama *et al.*

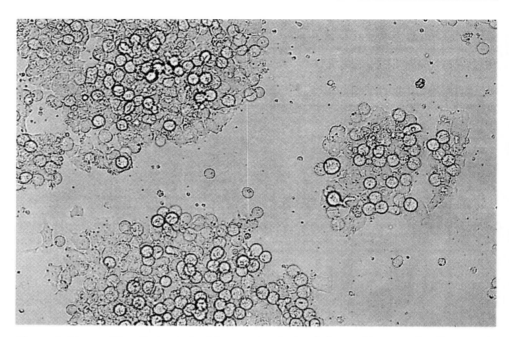

Figure 1 *Mouse ES cells treated with nocodazole for 3 hours. The round cells were in metaphase.*

(1999) and Rideout *et al.* (2000) demonstrated that it was possible to obtain mice from ES cells using piezoelectric-actuated nuclear transfer procedures (see Chapter 16, this volume). Live mice can also be produced following transfer of enucleated oocytes at the second metaphase fused with ES cells by HVJ (Amano *et al.*, 2001a,b). In our reports, the cell cycles of ES cells were synchronized at M phase before nuclear transfer by culturing them with nocodazole for 3 hours. Most of the cells arising from the ES cell colony after nocodazole treatment had metaphase chromosomes (Fig. 1). The development of nuclear-transferred oocytes into blastocysts was relatively high (34–88%) but development into live offspring was very low (1–3%). There was no clear correlation between the developmental potential after nuclear transfer and the chimera formation rate or genetic background of ES cells. Most of the offspring (16/21), however, died soon after birth and had morphologic abnormalities. Three of the remaining mice had normal fertility. On the other hand, Rideout *et al.* (2000) reported that the incidence of postnatal death was different among offspring from various ES cell lines.

The enucleated oocytes receiving day 12.5–16.5 male fetal germ cells, which were at the G0 stage of the cell cycle, developed into blastocysts but did not develop into offspring (Tsunoda *et al.*, 1989). The failures are probably related to the starting point of gamete imprinting for the next generation in male germ cells (Kato *et al.*, 1999a). One live calf was born after nuclear transfer of male fetal germ cells on days 50–57 (Zakhartchenko *et al.*, 1999a), suggesting that the starting point of gamete imprinting might be different.

DEVELOPMENTAL POTENTIAL OF SOMATIC CELL LINE NUCLEI FROM DIFFERENT TISSUES

Mammalian embryos first differentiate into two distinct cell lineages at the blastocyst stage; one is the inner cell mass, which forms the embryo proper, and the other is the trophectoderm (TE), which contributes to form the placenta and fetal membranes but does not participate in the formation of the fetus proper. At least some mouse trophectoderm cell nuclei have the same developmental totipotency as inner cell mass cells, because fertile mice were obtained after TE nuclear transfer into enucleated oocytes (Tsunoda and Kato, 1998). Moreover, Wilmut et al. (1997) reported that nuclei from fetal and adult somatic cells have the potential to develop into offspring. In nuclear transfer experiments, Wilmut and colleagues (1997) obtained four lambs from cultured embryonic disc cells, three lambs from fibroblast cells of a day 26 fetus, and one lamb, named Dolly, from mammary gland cells of a 6-year-old pregnant female.

A large number of cloned female and male sheep (Wilmut et al., 1997; Schnieke et al., 1997), mice (Wakayama et al., 1998; Kato et al., 1999b; Wakayama and Yanagimachi, 1999; Ogura et al., 2000; Ono et al., 2001), calves (Kato et al., 1998, 2000; Cibelli et al., 1998; Vignon et al., 1998; Zakhartchenko et al., 1999b; Wells et al., 1999; Renard et al., 1999; Shiga et al., 1999; Kubota et al., 2000; Lanza et al., 2000), goats (Baguisi et al., 1999), and pigs (Betthauser et al., 2000; Onishi et al., 2000; Polejaeva et al., 2000) have been produced after nuclear transfer of somatic cells cultured from various tissues of fetuses, newborns, and adults.

Cloned animals have been produced after nuclear transfer of somatic cells from organs such as mammary gland, cumulus, oviduct, ear, skin, muscle, liver, tail, and Sertoli cells. It is not clear, however, which cell types or cell origins are most successful for mammalian cloning. We have systematically examined the developmental potential of enucleated oocytes receiving somatic cells from various tissues of adult, newborn, and fetal cows (Kato et al., 2000). Adult cells were obtained from Japanese beef and Holstein cows of unknown age, including a 10-year-old live beef bull. Cells were also obtained from newborn calves and fetuses. For the collection of cumulus cells, cumulus–oocyte complexes obtained from ovarian follicles were cultured in embryonic stem cell/Dulbecco's modified Eagle's medium (ES-DMEM) (Robertson, 1987) supplemented with 10% fetal bovine serum. When cell growth extended over the bottom of the culture dish, cells were dispersed using phosphate-buffered saline (PBS) containing trypsin and ethylenediaminetetraacetic acid (EDTA) and transferred to new dishes. Oviductal cells were obtained by squeezing oviduct epithelial cells with a pair of tweezers and culturing them as for cumulus cells. For the collection from other tissues, each tissue fragment was washed several times in PBS, minced in trypsin–EDTA solution, and incubated for 20 minutes at 37°C. The cell suspension was centrifuged and cells were resuspended, cultured, and passaged several times as for cumulus cells. Before nuclear transfer, cultured somatic cells were induced to the G0/G1 stage of cell cycle by serum starvation. A single donor cell from each tissue was fused electrically with enucleated oocytes matured in vitro. Successfully fused oocytes were stimulated electrically again and cultured with CR1-amino acid medium (Rosenkraus and First, 1991) containing cycloheximide for 5 to 6 hours and then cultured in the same medium without cycloheximide. On day 3 of in vitro culture, the nuclear-transferred embryos were cocultured with inactivated mouse fetal fibroblast cells. On days 7 to 9, morphologically normal embryos at the morula to blastocyst stages were transferred to recipient cows at seven different prefecture livestock stations.

As shown in Table 1, the percentages of blastocysts that developed from donor cells from various tissues were not largely different among donor cells and there was also no difference in the percentages of blastocysts that developed from oocytes containing adult (42%), newborn (37%), or fetal calf (40%) nuclei, or between female (39%) and male (40%) nuclei. The proportions of oocytes receiving somatic cells that developed into blastocysts were higher than those obtained in multiple nuclear transfer of embryonic nuclei (27–37%) (Takano et al., 1997).

As shown in Table 2, 55 of 139 recipients (40%) became pregnant on days 40 to 60 after transfer of cloned blastocysts. The frequency of abortions was 27% by day 100, 38% by day 150, and 49% by day 250. The abortion rate was relatively high compared with that observed in multiple nuclear transfer of embryonic nuclei

Table 1 Developmental Potential of Somatic Nuclear Transplant to Blastocysts in Vitro[a,b]

	Adult		Newborn		Fetus	
Origin	Female	Male	Female	Male	Female	Male
Cumulus	30					
Oviduct	38					
Uterus	50		33			
Skin	52	49	39	27	46	41
Ear	48	41	44	25		
Heart	34	40	41	46		45
Liver	44	53	32	47		44
Kidney	36	33	28	43	47	38
Muscle			36		23	
Lung			40			
Mammary gland				42		
Testis[c]				38		
Epididymis[c]				30		
Gut					41	37
Tongue						42

[a]Source: Reproduced from Tsunoda and Kato (2000), with permission of Japan Academic Association.
[b]Percentage of blastocysts.
[c]Only somatic fibroblast cells were used.

(3/10) (Takano *et al.*, 1997). Moreover, abortions occurring after day 200, which is very unusual, were also observed. The duration of pregnancy for the recipients was 242 to 295 days, which is similar to that of Japanese beef (285–287 days) and Holstein cattle (279 days), except that the gestation period of 242–244 days in calves 1, 2, and 29 was shorter than that of control cows (Table 3). The remaining 28 pregnant recipients produced 32 calves but 14 of them died around or within 7 days after parturition. Significant morphologic abnormalities of the kidney or phocomelia were observed in eight calves (calves 19–24, 28, and 29); most of them originated from male adult skin or ear cells, as shown in Fig. 2. The neck was bent backward, the hind legs were stretched tightly, or the second joints were bent toward the opposite direction from the normal position. Calf 16, derived from female adult skin cells, died 19 days after parturition from colisepticemia. Calf 27, derived from male adult ear cells, and calf 32, derived from male fetus skin cells, died suddenly 180 and 138 days after parturition, respectively, but no gross abnormalities were observed.

Because there are few data on embryo transfer of nuclear-transferred oocytes receiving different somatic cells, precise comparison of the potential to develop to live offspring and peri- or postnatal death of calves is difficult. Two groups in which

Table 2 Summarized Data on Transfer of Bovine Somatic Cell Nuclear-Transferred Embryos

Parameter	Value
No. of blastocysts transferred	182
No. of recipients	139
Pregnant (%)	55 (40)
Calved (%)	28 (20)
No. of calves produced (%)	32 (18)

Table 3 Details of Cloned Calves[a]

| Number | Origin of donor cell | | | Day of birth | Birth weight (kg) | Status at | | | Fertility |
	Sex[b]	Maturity	Derived from			Parturition	Day 7	Present	
1	Female (JB)	Adult	Oviduct 1	242	18.2	Survived	Survived	Living	On April 23, 2001
2	Female (JB)	Adult	Oviduct 1	242	17.3	Survived	Survived	Living	On March 26, 2001
3	Female (JB)	Adult	Cumulus 1	266	32.0	Survived	Dead		
4	Female (JB)	Adult	Cumulus 1	267	17.3	Survived	Dead		
5	Female (JB)	Adult	Cumulus 1	267	34.8	Survived	Dead		
6	Female (JB)	Adult	Cumulus 1	276	23.0	Survived	Survived	Living	On September 17, 2000
7	Female (JB)	Adult	Cumulus 1	276	27.5	Survived	Survived	Living	On July 10, 2000
8	Female (JB)	Adult	Oviduct 1	287	30.1	Survived	Dead		
9	Female (JB)	Adult	Cumulus 1	278	33.0	Survived	Survived	Living	On December 23, 2000
10	Female (JB)	Adult	Cumulus 1	278	33.0	Survived	Survived	Living	On September 10, 2001
11	Female (H)	Adult	Cumulus 2	283	46.0	Survived	Survived	Living	On April 13, 2001
12	Female (JB)	Adult	Oviduct 1	286	34.0	Survived	Survived	Living	On February 20, 2001
13	Female (JB)	Adult	Oviduct 1	279	35.4	Survived	Survived	Living	On April 28, 2001
14	Female (JB)	Adult	Uterine	271	40.7	Dead	Dead		
15	Female (JB)	Adult	Uterine	271	45.9	Dead	Dead		
16	Female (JB)	Adult	Skin	278	39.0	Survived	Survived	Dead	Died on day 19
17	Female (JB)	Adult	Skin	285	32.0	Survived	Survived	Living	On March 22, 2001
18	Female (H)	Adult	Skin	289	65.0	Survived	Survived	Living	On August 24, 2001
19	Male (JB)	Adult	Skin	289	36.0	Dead	Dead		
20	Male (JB)	Adult	Skin	264	15.0	Dead	Dead		
21	Male (JB)	Adult	Skin	249	68.0	Dead	Dead		
22	Male (JB)	Adult	Skin	292	53.0	Survived	Dead		
23	Male (JB)	Adult	Skin	289	46.0	Survived	Dead		
24	Male (JB)	Adult	Ear	289	51.0	Dead	Dead		
25	Male (JB)	Adult	Ear	290	52.0	Survived	Survived	Living	On June 6, 2001
26	Male (JB)	Adult	Ear	292	40.0	Survived	Survived	Living	On November 27, 2001
27	Male (JB)	Adult	Ear	294	66.0	Survived	Survived	Dead	Died on day 180
28	Male (JB)	Adult	Ear	295	55.0	Survived	Dead		
29	Male (H)	Newborn	Liver	244	40.0	Survived	Survived		Killed on day 431
30	Male (H)	Newborn	Liver	293	68.0	Dead	Dead		
31	Male (H)	Newborn	Skin	288	54.0	Survived	Survived		Killed on day 490
32	Male (JB)	Fetus	Skin	288	60.0	Survived	Survived	Dead	Died on day 138

[a]Source: Adapted from Kato *et al.* (2000), with permission of Journals of Reproduction and Fertility Ltd.

[b]JB, Japanese beef cattle; H, Holstein cattle.

a relatively large number of cloned blastocysts were transferred are compared in Table 4. The pregnancy rate after transfer of cloned blastocysts originating from adult female cumulus and oviduct cells (28%) was significantly lower than that obtained for adult male skin and ear cells (50%). The proportions of calves produced, however, were not different and significantly more fetuses were aborted in the latter group. Morphologic abnormalities were not observed in calves originat-

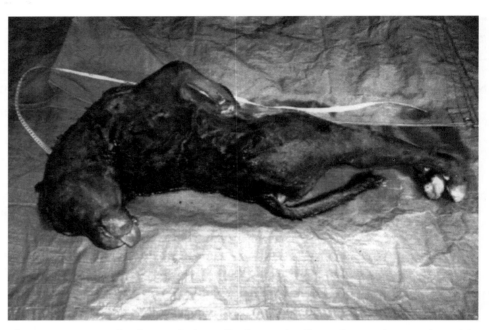

Figure 2 *One example of a morphologically abnormal calf (number 19) that was derived from skin cells of a mature male calf.*

ing from cumulus and oviduct cells whose body weights were within the normal range (17.3–35.4 kg in Table 3). When male skin and ear donor cells were used, 4 of 10 calves exceeded the mean body weight of control calves by more than 40% (Table 3), and the peri- or postnatal death of these calves was significantly high (80%). Morphologic abnormalities were observed in calves originating from skin and ear cells that died around parturition (one example is shown in Fig. 2). As shown in Table 5, differences in developmental potential were also detected even if donor cells from the same origin were used. When cumulus cells and oviduct cells established from the first cow (line 1) were used, the pregnancy rate was high (63%) and the abortion rate was low (10%). The pregnancy rate was low (17%) and the abortion rate was high (80%), however, when cell line 2 was used. When cumulus and oviduct cells established from the third cow (line 3) were used, none of eight recipients became pregnant.

The remaining 15 calves, 11 females and 4 males, grew normally to adulthood. All females, including eight cloned calves, that originated from cumulus and oviduct cell line 1 (Fig. 3) became pregnant after artificial insemination and produced normal calves, the first one on 10 July 2000. Two of four males were killed 431 and 490 days after birth, respectively, and ejaculated or epididymal spermatozoa were col-

Table 4 Comparison of the Developmental Potential of Cloned Blastocysts Originating from Different Donor Cells

Donor cells	No. of pregnant/recipients (%)	No. of calves/blastocysts (%)	No. of dead calves at or after birth (%)
Adult female cumulus and oviduct	15/53 (28)[a]	13/68 (19)	4/13 (31)[a]
Adult male skin and ear	21/42 (50)[a]	10/53 (20)	8/10 (80)[a]

[a]$P < 0.05$.

Table 5 Effects of Cell Lines on the Developmental Potential of Somatic Cell Nuclear-Transferred Embryos

Line	Origin	No. of embryos transferred	Recipients	Pregnancies	Abortions	Calves	Peri- or postnatal death (%)
				Number of (%)			
1	Cumulus	20	10	5 (50)	0 (0)	7 (35)	3 (43)
	Oviduct	9	6	5 (83)	1 (20)	5 (56)	1 (20)
	Total	29	16	10 (63)	1 (10)	12 (41)	4 (33)
2	Cumulus	21	21	4 (19)	3 (75)	1 (5)	0 (0)
	Oviduct	8	8	1 (13)	1 (100)	0 (0)	
	Total	29	29	5 (17)	4 (80)	1 (3)	0 (0)
3	Cumulus	6	6	0			
	Oviduct	4	2	0			
	Total	10	8	0			

lected and frozen in liquid nitrogen. *In vitro* fertilization of frozen-thawed spermatozoa revealed that both of them had normal fertilizing ability because 5–17% of inseminated oocytes developed into blastocysts 8 days after *in vitro* culture. After artificial insemination of frozen-thawed ejaculated spermatozoa obtained from two other males (Fig. 4) that originated from adult male ear cells, nine females became pregnant and produced calves, the first calf on 17 November 2000.

Our (Kato *et al.*, 1998, 2000) and other (Wells *et al.*, 1999) studies demonstrated that cell lines from cumulus and oviduct cells are suitable donor cells for cloning calves. The reasons for the differences in developmental potential among cell origin or cell lines are unknown. There was no correlation between pregnancy success or perinatal death of young and the number of passages (two to seven passages) of donor cells through cell culture. In addition, the chromosome numbers in all cultured cell lines were apparently normal, but this level of analysis does not eliminate the possibility that there might be submicroscopic genetic changes.

Figure 3 *All calves (numbers 1, 2, 6, 7, and 12) were derived from cumulus or oviduct cells of an adult female (line 1) at the Ishikawa prefecture. All of them produced calves after artificial insemination.*

Figure 4 *Both calves (numbers 25 and 26) were derived from ear cells of an adult male at the Kumamoto prefecture.*

There are several possible reasons for the high abortion rate, peri- and postnatal death of young, and abnormalities of young. The most feasible reason is insufficient reprogramming of donor nuclei in recipient oocytes. In contrast to embryonic nuclei, reprogramming of somatic cell nuclei occurs in nonactivated oocyte cytoplasm, but not in activated oocytes (Tani *et al.*, 2001). We examined the *in vitro* developmental potential of nonactivated and activated oocytes receiving cumulus cells synchronized at the G0/G1, G1, G2, S, and M phases. Enucleated oocytes activated 6 hours before receiving cumulus cells at any cell cycle stage stopped developing at the eight-cell stage, but 21 to 50% of nonactivated oocytes receiving donor cells at G0/G1, G1, G2, and M phases developed into blastocysts (Tani *et al.*, 2001). One normal calf was born after transfer of five blastocysts to recipients that developed from oocytes receiving cumulus cells at the M phase. These results indicate that "reprogramming factor(s)" for somatic cell nuclei are involved in nonactivated oocytes. Abortion at a late pregnancy stage or morphologic abnormalities of calves could occur after transfer of nuclear-transferred embryos when reprogramming of donor nuclei is insufficient.

DNA damage by ultraviolet radiation might be another possibility. Cells from the ear and skin tissue are especially likely to suffer sun damage from ultraviolet irradiation (Bernstein *et al.*, 1996).

Another possibility is epigenetic alterations of donor nuclei. This might be true in the case of mouse fetal germ cells and ES cells. When nuclear-transferred blastocysts that developed from male fetal germ cells on days 14.5–16.5 were transferred to recipients, no young were obtained due to the fact that gametic imprints for the next generation started in the germ cells at this stage (Kato *et al.*, 1999a). Postnatal death of offspring was also observed in the ES cell-derived mice that were produced by nuclear transfer (Rideout *et al.*, 2000; Amano *et al.*, 2001a,b), aggregation with tetraploid embryos (Nagy *et al.*, 1990, 1993), and injection into tetraploid (Wang *et al.*, 1997; Amano *et al.*, 2001c) and heat-treated blastocysts (Amano *et al.*, 2000). As shown in Table 6, 62–85% of ES cell-derived mice obtained by three different methods died soon after birth (Amano *et al.*, 2000, 2001a–c). The most feasible reason for these postnatal deaths might be that epigenetic modifications of imprinted genes, such as *Igf2r*, *H19*, *Igf2*, or *V2af1-rS* (Dean *et al.*, 1998), reduced the developmental potential of the ES cell nuclei, as in the somatic cell nuclei.

Table 6 A Comparison of the Incidence of Postnatal Death in ES Cell-Derived Mice

Method	No. of blastocysts transferred	No. of newborns (%)	Postnatal death/no. of ES cell-derived mice (%)
Heat treated[a]	1011	26(3)	10/15 (67)
Tetraploid[b]	1037	48(5)	21/34 (62)
Nuclear transfer	1976	20(1)	17/20 (85)

[a]ES cells were injected into blastocysts treated at 45°C for 20 minutes.
[b]ES cells were injected into tetraploid blastocysts.

It is now possible to produce mammals by nuclear transfer of a variety of somatic cells and ES cells. So far, the most effective donor cells for cloning appear to be cultured cumulus and oviductal cells. The exact origin of donor cells so far used, however, is not clear and most of the somatic cells obtained from various tissues might be fibroblasts, except for cumulus and oviductal cells, which are epithelial cells. Although viable calves are produced by nuclear transfer of senescent somatic cells (Lanza *et al.*, 2000), further studies on the developmental potential of terminally differentiated somatic cells are required.

ACKNOWLEDGMENTS

A part of our work was supported by a grant from the Program for Promotion of Basic Research Activities for Innovative Biosciences (PROBRAIN), from the Ministry of Education, Science, Sports, and Culture (Nos. 11480250, 11876061, 12358014, and 12878161), and the Special Coordination Funds for Promoting Science and Technology from the Ministry of Science and Technology. We thank staff members of Ishikawa, Tochigi, Kumamoto, Aichi, Yamaguchi, Fukui, and Mie prefecture Livestock Station and Animal Public Health Center for embryo transfer, assistance, and management of recipient animals, and assistance in postmortem analysis.

REFERENCES

Amano, T., Nakamura, K., Tani, T., Kato, Y., and Tsunoda, Y. (2000). Production of mice derived entirely from embryonic stem cells after injecting the cells into heat treated blastocysts. *Theriogenology* **56,** 1449–1458.

Amano, T., Tani, T., Kato, Y., and Tsunoda, Y. (2001a). Mouse cloned from embryonic stem (ES) cells synchronized in metaphase with nocodazole. *J. Exp. Zool.* **288,** 139–145.

Amano, T., Kato, Y., and Tsunoda, Y. (2001b). The full-term development of enucleated mouse oocytes fused with ES cells from different cell lines. *Reproduction* **121,** 729–733.

Amano, T., Kato, Y., and Tsunoda, Y. (2001c). Comparison of heat-treated and tetraploid blastocysts for the production of completely ES cell-derived mice. *Zygote* **9,** 153–157.

Baguisi, A., Behboodi, E., Melican, D. T., Pollock, J. S., Destrempes, M. M., Cammuso, C., Williams, J. L., Nins, S. D., Porter, C. A., Midura, P., Palacios, M. J., Ayres, S. L., Denniston, R. S., Hayes, M. L., Ziomek, C. A., Meade, H. M., Godke, R. A., Gavin, W. G., Overstom, E. W., and Echelard, Y. (1999). Production of goats by somatic cell nuclear transfer. *Nature Biotechnol.* **17,** 456–461.

Bernstein, E. F., Chen, Y. Q., Kopp, J. B., Fisher, L., Brown, D. B., Hahn, P. J., Robey, F. A., Lakkakorpi, J., and Uitto, J. (1996). Long-term sun exposure alters the collagen of the papillary dermis. *J. Am. Acad. Dermatol.* **34,** 209–218.

Betthauser, J., Forsberg, E., Augenstein, M., Childs, L., Eilertsen, K., Enos, J., Forsythe, T., Golueke, P., Jurgella, G., Koppang, R., Lesmeister, T., Mallon, K., Mell, G., Misica, P., Pace, M., Pfister-Genskow, M., Strelchenko, N., Voelker, G., Watt, S., Thompson, S., and Bishop, M. (2000). Production of cloned pigs from *in vitro* systems. *Nature Genet.* **18,** 1055–1059.

Cibelli, J. B., Stice, S. L., Golueke, P. J., Kane, J. J., Jerry, J., Blackwell, C., Ponce de Leon, F. A., and Robl, J. M. (1998). Cloned transgenic calves produced from nonquiescent fetal fibroblasts. *Science* **280,** 1256–1258.

Collas, P., and Barnes, F. L. (1994). Nuclear transplantation by microinjection of inner cell mass and granulosa cell nuclei. *Mol. Reprod. Dev.* **38**, 264–267.

Dean, W., Bowden, L., Aitchison, A., Klose, J., Moore, T., Meneses, J. J., Reik, W., and Feil, R. (1998). Altered imprinted gene methylation and expression in completely ES cell-derived mouse fetuses: Association with aberrant phenotypes. *Development* **125**, 2273–2282.

DiBerardino, M. A. (1997). Nuclear potential of differentiated amphibian tissues and cells. *In* "Genomic Potential of Differentiated Cells" (M. A. Diberardino, ed.), pp. 68–82. Columbia University Press, New York.

Kato, Y., Tani, T., Sotomaru, Y., Kurokawa, K., Kato, J., Doguchi, H., Yasue, H., and Tsunoda, Y. (1998). Eight calves cloned from somatic cells of a single adult. *Science* **282**, 2095–2098.

Kato, Y., Rideout, W. M. III., Hilton, K., Barton, S. C., Tsunoda, Y., and Surani, M. A. (1999a). Developmental potential of mouse primordial germ cells. *Development* **126**, 1823–1832.

Kato, Y., Yabuuchi, A., Motosugi, N., Kato, J., and Tsunoda, Y. (1999b). Developmental potential of mouse follicular epithelial cells and cumulus cells after nuclear transfer. *Biol. Reprod.* **61**, 1110–1114.

Kato, Y., Tani, T., and Tsunoda, Y. (2000). Cloning of calves from various somatic cell types of male and female adult, newborn and fetuses. *J. Reprod. Fertil.* **120**, 231–237.

Keefer, C. L., Stice, S. L., and Matthews, D. L. (1994). Bovine inner cell mass cells as donor nuclei in the production of nuclear transfer embryos and calves. *Biol. Reprod.* **50**, 935–939.

Kubota, C., Yamakuchi, H., Todoroki, J., Mizoshita, K., Tabara, N., Barber, M., and Yang, X. (2000). Six cloned calves produced from adult fibroblast cells after long-term culture. *Proc. Natl. Acad. Sci. U.S.A.* **97**, 990–995.

Lanza, R. P., Cibelli, J. B., Blackwell, C., Cristofalo, V. J., Francis, M. K., Baerlocher, G. M., Mak, J., Schertzer, M., Chavez, E. A., Sawyer, N., Lansdorp, P. M., and West, M. D. (2000). Extension of cell life-span and telomere length in animals cloned from senescent somatic cells. *Science* **288**, 665–669.

McGrath, J., and Solter, D. (1983). Nuclear transplantation in the mouse embryo by microsurgery and cell fusion. *Science* **220**, 1300–1302.

McGrath, J., and Solter, D. (1984). Inability of mouse blastomere nuclei transferred to enucleated zygotes to support development *in vitro*. *Science* **226**, 1317–1319.

Nagy, A., Gocza, E., Diaz, E. M., Prideaux, V. R., Ivanyi, E., Markkula, M., and Rossant, J. (1990). Embryonic stem cells alone are able to support fetal development in the mouse. *Development* **110**, 815–821.

Nagy, A., Rossant, J., Nagy, R., Abramow-Newerly, W., and Roder, J. C. (1993). Derivation of completely cell culture-derived mice from early-passage embryonic stem cells. *Proc. Natl. Acad. Sci. U.S.A.* **90**, 8424–8428.

Ogura, A., Inoue, K., Ogonuki, N., Noguchi, A., Takano, K., Nagano, R., Suzuki, O., Lee, J., Ishino, F., and Matsuda, J. (2000). Production of male cloned mice from fresh, cultured, and cryopreserved immature sertoli cells. *Biol. Reprod.* **62**, 1579–1584.

Onishi, A., Iwamoto, M., Akita, T., Mikawa, S., Takeda, K., Awata, T., Hanada, H., and Peryy, A. C. F. (2000). Pig cloning by microinjection of fetal firoblast nuclei. *Science* **18**, 1188–1190.

Ono, Y., Shimozawa, N., Ito, M., and Kono, T. (2001). Cloned mice from fetal fibroblast cells arrested metaphase by a serial nuclear transfer. *Biol. Reprod.* **64**, 44–50.

Polejaeva, L. A., Chen, S. H., Vaught, T. D., Page, R. L., Mullins, J., Ball, S., Walker, S., Ayares, D. L., Colman, A., and Campbell, K. H. S. (2000). Cloning pigs produced by nuclear transfer from adult somatic cell. *Nature* **407**, 505–509.

Renard, J. P., Chastant, S., Chesne, P., Richard, C., Marchal, J., Cordonnier, N., Chavatte, P., and Vignon, X. (1999). Lymphoid hypoplasia and somatic cloning. *Lancet* **353**, 1489–1491.

Rideout, W. M. III., Wakayama, T., Wutz, A., Eggan, K., Jackson-Grusby, L., Dausman, J., Yanagimachi, R., and Jaenisch, R. (2000). Generation of mice from wild-type and targeted ES cells by nuclear cloning. *Nature Genet.* **24**, 109–110.

Robertson, E. J. (1987). Embryo-derived stem cell lines. *In* "Teratocarcinomas and Embryonic Stem Cells: A Practical Approach," pp. 71–112. IRL Press, Oxford.

Rosenkraus, C. F., Jr., and First, N. L. (1991). Culture of bovine zygotes to the blastocyst stage: Effect of amino acids and vitamins. *Theriogenology* **35**, 266.

Schnieke, A. E., Kind, A. J., Ritchie, W. A., Mycock, K., Scott, A. R., Ritchie, M., Wilmut, I., Colman, A., and Campbell, K. H. S. (1997). Human factor IX transgenic sheep produced by transfer of nuclei from transfected fetal fibroblasts. *Science* **278**, 2130–2133.

Shiga, K., Fujita, T., Hirose, K., Sasae, Y., and Nagai, T. (1999). Production of calves by transfer of nuclei from cultured somatic cells obtained from Japanese black bulls. *Theriogenology* **52**, 527–535.

Smith, L. C., and Wilmut, I. (1989). Influence of nuclear and cytoplasmic activity on the development *in vivo* of sheep embryos after nuclear transplantation. *Biol. Reprod.* **40**, 1027–1035.

Sims, M., and First, N. L. (1993). Production of calves by transfer of nuclei from cultured inner cell mass cells. *Proc. Natl. Acad. Sci. U.S.A.* **90**, 6143–6147.

Takano, H., Koyama, K., Kozai, C., Shimizu, S., Kato, Y., and Tsunoda, Y. (1997). Cloning of bovine embryos by multiple nuclear transfer. *Theriogenology* **47**, 1365–1373.

Tani, T., Kato, Y., and Tsunoda, Y. (2001). Direct exposure of chromosomes to nonactivated ovum cytoplasm is effective for bovine somatic cell nucleus reprogramming. *Biol. Reprod.* **64**, 324–330.

Tsunoda, Y., and Kato, Y. (1998). Not only inner cell mass cell nuclei but also trophectoderm nuclei of mouse blastocysts have a developmental totipotency. *J. Reprod. Fertil.* **113**, 181–184.

Tsunoda, Y., and Kato, Y. (2000). The recent progress on nuclear transfer in mammals. *Zool. Sci.* **17**, 1177–1184.

Tsunoda, Y., Yasui, T., Shinoda, Y., Nakamura, K., Uchida, T., and Sugie, T. (1987). Full-term development of mouse blastomere nuclei transplanted into enucleated two-cell embryos. *J. Exp. Zool.* **242**, 147–151.

Tsunoda, Y., Tokunaga, T., Imai, H., and Uchida, T. (1989). Nuclear transplantation of male primordial germ cells in the mouse. *Development* **107**, 407–411.

Vignon, X., Chesne, P., Le Bourhis, D., Flechon, J. E., Heyman, Y., and Renard, J. P. (1998). Developmental potential of bovine embryos reconstructed from enucleated matured oocytes fused with cultured somatic cells. *C.R. Acad. Sci.* **321**, 735–745.

Wakayama, T., and Yanagimachi, R. (1999). Cloning of male mice from adult tail-tip cells. *Nature Genet.* **22**, 127–128.

Wakayama, T., Perry, A. C., Zuccoti, M., Johnson, K. R., and Yanagimachi, R. (1998). Full-term development of mice from enucleated oocytes injected with cumulus cell nuclei. *Nature* **394**, 369–374.

Wakayama, T., Rodriguez, I., Perry, A. C., Yanagimachi, R., and Mombaerts, P. (1999). Mice cloned form embryonic stem cells. *Proc. Natl. Acad. Sci. U.S.A.* **96**, 14984–14989.

Wang, Z. Q., Kiefer, F., Urbanek, P., Wagner, E. F. (1997). Generation of completely embryonic stem cell- derived mutant mice using tetraploid blastocyst injection. *Mech. Dev.* **62**, 137–145.

Wells, D. N., Misica, P. M., and Tervit, H. R. (1999). Production of cloned calves following nuclear transfer with cultured adult mural granulose cells. *Biol. Reprod.* **60**, 996–1005.

Willadsen, S. M. (1986). Nuclear transplantation in sheep embryos. *Nature* **320**, 63–65.

Wilmut, I., Schnieke, A. E., McWhir, J., Kind, A. J., and Campbell, K. H. (1997). Viable offspring derived from fetal and adult mammalia cells. *Nature* **385**, 810–813.

Zakhartchenko, V., Durcova-Hills, G., Schernthaner, W., Stojkovic, M., Reichenbach, H. D., Mueller, S., Steinborn, R., Mueller, M., Wenigerkind, H., Prelle, K., Wolf, E., and Brem, G. (1999a). Potential of fetal germ cells for nuclear transfer in cattle. *Mol. Reprod. Dev.* **52**, 421–426.

Zakhartchenko, V., Durcova-Hills, G., Stojkovic, M., Schernthaner, W., Prelle, K., Steinborn, R., Muller, M., Brem, G., and Wolf, E. (1999b). Effects of serum starvation and re-cloning on the efficiency of nuclear transfer using bovine fetal fibroblasts. *J. Reprod. Fertil.* **115**, 325–331.

PART III
CLONING BY SPECIES

CHAPTER 14

CLONING OF AMPHIBIANS

J. B. Gurdon and J. A. Byrne

INTRODUCTION

The first success in cloning any vertebrate was achieved in 1952 by Briggs and King. The principle behind what used to be called nuclear transplantation is the removal or destruction of the chromosomes of an unfertilized egg, and the injection, in their place, of a nucleus from a somatic cell. It is important to point out that the aim of the early amphibian work had nothing to do with the reproductive cloning of animals. A fundamental question in developmental biology at that time asked whether the genome of somatic cells is complete, or whether the process of cell differentiation involves the loss or stable inactivation of genes no longer needed for a particular direction of cell specialization. If the nucleus of a specialized cell could functionally replace the zygote nucleus by inducing an enucleated egg to develop into a normal animal, then genes no longer required in a specialized cell must nevertheless be present and able to be reactivated.

Using the American species of frog *Rana pipiens*, Briggs and King (1952) found that about 50% of transplanted blastula nuclei generated normal tadpoles, thereby showing that their methodology was successful, at least for embryo cell nuclei. They found, however, that, though technically no more demanding, the nuclei of gastrula endoderm cells could not support normal development (Briggs and King, 1957). At about this time, success had been achieved with nuclear transplantation in *Xenopus laevis*, the South African clawed frog (Fischberg *et al.*, 1958; Gurdon *et al.*, 1958). Compared to members of the genus *Rana*, *Xenopus* has two major advantages. First, eggs and embryos can be obtained throughout the year by injection of commercial preparations of mammalian hormone. *Rana* species can be induced to lay eggs only in some months of the year and then only by injection of amphibian pituitary gland extracts. Second, embryos can be grown to sexual maturity in less than 1 year, compared to about 4 years for most *Rana* species, thereby making it realistic to make use of genetic mutants, most notably of a nuclear marker to prove the success of enucleation (Elsdale *et al.*, 1960). Early experiments with *Xenopus* showed, just as in *Rana*, that the ability to obtain normal nuclear transplant tadpoles decreased the older and more specialized were the embryos from which donor nuclei were taken (Gurdon, 1960). Nevertheless, even the nuclei of fully differentiated cells, such as the intestinal epithelium of feeding larvae, were able in a few cases to generate entirely normal sexually mature male and female adult frogs (Gurdon, 1962; Gurdon and Uehlinger, 1966). These results with *Xenopus* and later ones with *Rana* DiBerardino *et al.*, 1984) elicited the general principle, which appears now to be widely accepted, that the process of cell differentiation does not depend on any change, loss or stable inactivation of genes. It is this principle that makes reproductive and therapeutic cloning from somatic cells theoretically achievable.

The aim of reproductive cloning is to produce normal adult animals from the nuclei of somatic cells. The first adult vertebrate to be obtained by nuclear transplantation was described by Gurdon *et al.* in 1958 (Gurdon *et al.*, 1958), and a clone of adult frogs was reported by Gurdon in 1962 (Gurdon, 1962b). These clones were obtained from the nuclei of embryos. In amphibia, as in much more recent work with mammals, it is always observed that the success of nuclear transfer—that is, the proportion of enucleated eggs receiving an injected nucleus and reaching adulthood—goes down sharply as older and more differentiated donor cells are used. The reasons for this probably include the slow replication and division rate of older and more differentiated cells, the condensed state of their chromatin, and incomplete reprogramming. However, it has been appreciated from a very early time that the number of nuclear transplant embryos can be amplified, in principle indefinitely, by serial nuclear transfer. Even a partial, and hence nonviable, blastula obtained from a specialized cell can be used to generate many more embryos. It seems that once a nucleus has been reprogrammed successfully, cells of the resulting blastula have acquired a relatively high transplantability success rate, characteristic of a blastula, and no longer have the low rate of a specialized cell. A key to cloning success is the reprogramming of a somatic cell nucleus by egg cytoplasm, the major area of future work in this field. Recently, it was observed that most of the cells from nonviable cloned *Xenopus* embryos are capable of normal differentiation when transplanted to normal host embryos, even though those cloned cells expressed a very aberrant gene expression pattern (Byrne *et al.*, 2002). This data supports the theory that vertebrate development is surprisingly tolerant of aberrant gene expression (Humpherys *et al.*, 2001).

EGG FORMATION AND EMBRYO DEVELOPMENT

In amphibia, eggs are formed by a process called oogenesis and meiotic maturation. External fertilization takes place, eggs and sperm being activated by exposure to the low-salt conditions of pond water. Thereafter embryonic development is rapid and is temperature related.

Oogenesis, the formation of a fully grown oocyte from a germ cell, is undergone in the female's ovary over many months, or even years. The growth of a female germ cell, about $20\,\mu m$ in diameter, to the full-grown oocyte, of $\geq 1\,mm$ in diameter, occurs mainly during the diplotene stage of meiotic prophase, when the oocyte chromosomes are tetraploid. Full-grown oocytes remain in the ovary indefinitely until stimulation by a pituitary hormone, such as progesterone, causes completion of meiosis; this entails the rupture of the oocyte's nucleus, the germinal vesicle, and the first meiotic division, with release of the first polar body. During this time, which lasts about 8 hours in *Xenopus*, the oocyte is ovulated from the ovary, surrounded by jelly as it passes down the oviduct, and extruded into the medium in which the frog lays its eggs. This maturation and oviposition process can be stimulated by the subcutaneous injection into a mature female of rather large doses of mammalian luteinizing hormone. Eggs laid by females are arrested in the metaphase of the second meiotic division. Only after receipt of a transplanted nucleus, or fertilization, is this meiotic division completed, with the emission of a second polar body and the formation of a haploid egg pronucleus.

Hormone-injected *Xenopus* females will lay eggs in a medium with a high salt concentration, which fails to activate the eggs, and these can be accumulated for several hours from a laying female. Such nonactivated eggs can be used as recipients for nuclear transfer or they can have a sperm suspension added to them, and caused to start development as soon as they are transferred to a low-salt (tap water) medium.

As soon as eggs are transferred to low salt, or are activated by penetration with a pipette during nuclear transfer, embryo development proceeds very rapidly. It may be noted, in passing, that the eggs of different amphibian species have very different requirements for activation. In *Xenopus* penetration by a nuclear transfer pipette is sufficient; in other species, electrical or other treatments are also needed (Signoret *et al.*, 1962; Gallien *et al.*, 1973). *Xenopus laevis* embryos develop fast, but, conveniently, will tolerate a temperature range from 14–26°C; development at 14°C is 2.5 times slower than at 23°C, the temperature used to prepare the standard normal table for this species (Nieuwkoop and Faber, 1956). At 23°C, the first cell cycle and division of the egg into the two-cell stage takes only 90 minutes. It is within this time that a transplanted somatic cell nucleus has to initiate and complete replication of its chromosomes if it is to participate in normal development. After the first mitosis, the next 11 cell divisions take place even more rapidly, at about 30-minute intervals. At the midblastula stage (4000 cells, 5 hours at 23°C, in *Xenopus*), gene transcription first starts, and the cell cycle lengthens. Thereafter cell division occurs at different rates in different parts of the embryo; gastrulation starts at 9 hours (30,000 cells), muscular response at 16 hours (50,000 cells), and heartbeat at 26 hours (100,000 cells). Tadpoles hatch at 2 days and begin to feed at 4 days.

METHODOLOGY

The cloning procedure for amphibia has three components, namely, the preparation of recipient eggs, the preparation of donor cells, and nuclear transfer.

Xenopus females injected with hormone (see above) will usually lay eggs continuously for about 8 hours. Eggs laid in a high-salt medium (above) can be used as recipients for an hour or more after they have been laid. For cloning experiments, the egg chromosomes need to be removed or destroyed. In *Xenopus* this is best achieved by irradiating eggs with ultraviolet light at 254 nm. Details have been described before (Elsdale *et al.*, 1960). An important point to remember is that ultraviolet light has low penetration through water, and unfertilized eggs must be placed on a dry slide, with the position of the egg chromosomes (conveniently located on the surface of the egg and marked by a pale patch in the pigmented hemisphere) pointing upward towards the source of the light. In other species of amphibia, it is usual to remove manually, with a needle, a small amount of cytoplasm containing the meiotic spindle.

To prepare donor cells, it is necessary to dissociate these from solid tissue unless naturally separate cells, such as blood, or cultured cells, easily trypsinized, are used. This can be difficult, and it is easier, when the design of experiment permits this, to use those cells that grow out in culture from an explant of adult tissue. In most cases, such cells are described as fibroblasts, and may not show the specialized pattern of gene expression of the donor tissue. If the aim of the cloning experiment is just to obtain the most normal development from any source, donor cell characterization is unimportant. Donor cells are best kept in isotonic, Ca^{2+}-deficient medium containing 0.5% serum albumin, and can be used for nuclear transfer for at least 1 hour after preparation.

Finally, there is the nuclear transfer procedure. Until recently this required the dexterous sucking up of a donor cell into a specially microforged micropipette, the shape of which is critical. Success by this method requires that the donor cell is slightly squeezed so that its plasma membrane is broken, but its nucleus is not dislodged from its surrounding cytoplasm. In addition, the pipette must have a sharp point so as to penetrate the recipient egg without damage. To make such micropipettes is particularly challenging when large embryonic donor cells are to be used. In 1976, Gurdon introduced the use of lysolecithin to permeabilize donor cells for use in oocyte nuclear transfer experiments (see below). Using donor cells treated this way, with serum albumin to limit the activity of the lysolecithin, has greatly simplified the nuclear transfer procedure, because donor cells no longer need to be physically ruptured, and the nuclear transfer pipette is fairly easy to make. This procedure was used by Kroll and Amaya (1996) for their transgenesis work in which sperm nuclei were transplanted into nonenucleated eggs. A further improvement was introduced by Chan and Gurdon (1996). This uses streptolycin O as a permeabilizing agent for donor cells; this reagent binds to cells at 0°C, but does not make them permeable until cells are warmed to room temperature, thereby enabling the time for which donor cells remain in a permeabilized, and potentially deteriorating, condition to be kept to a minimum.

In all cloning experiments, it is highly desirable to use genetically marked donor nuclei, thereby checking that the resulting embryos have received no contribution from a failure to enucleate the recipient egg.

NUCLEAR TRANSFER SUCCESS

Figure 1 (Gurdon, 1963) summarizes the results obtained with *R. pipiens* and *X. laevis*. The success rate can be judged by the frequency with which eggs receiving transplanted nuclei reach the muscular response stage (functional muscle and nerve), the swimming tadpole stage (most differentiated cell types formed), or the adult stage. By all these criteria, experiments with *R. pipiens* are less successful than those with *Xenopus*, and the success rate drops rapidly with postembryonic donor cells. Although fertile adults have been obtained from the nuclei of differentiated intestine cells of feeding larvae, they have not been obtained from any cells derived from tissues of an adult frog, as has been done with mammals. In both amphibia and mammals, the success rate for differentiated and adult cells is very low, and only 1% form feeding tadpoles or appear normal at birth. The remaining 99% become abnormal and die at various stages from early to late development.

This raises the question of whether only 1% of the cells of differentiated tissue are genetically multipotent. Other than nuclear transfer, there are few if any experiments that test the genetic integrity and ability of individual cells of differentiated tissue to be reprogrammed. It is known, especially from early experiments of DiBerardino and Hoffner (1970), that many abnormal nuclear transplant embryos contain visibly defective chromosome complements. It is known that these defects are not present in normal somatic tissue to anything like the same extent as in nuclear transplant embryos, and it is widely believed that the nuclear transfer procedure, notably the extremely rapid replication of transplanted nuclei immediately

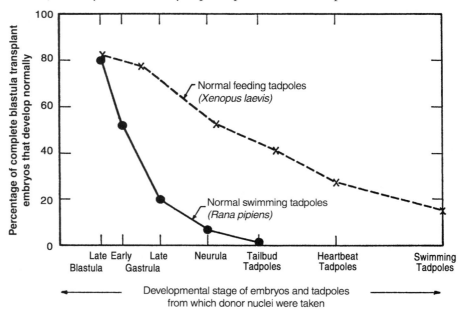

Figure 1 *Graph showing that as cells differentiate a progressively smaller proportion of them contain nuclei able to promote the development of a normal tadpole. From Gurdon (1963). Quarterly Review of Biology. Copyright 1963 The University of Chicago Press, with permission.*

after transfer, is the cause of genetic damage due to incomplete chromosome replication in the time available. Indeed the relatively very long time of nearly 24 hours between nuclear transfer and the first cytoplasmic cell division in mice and other mammals may increase the likelihood of complete chromosome replication, and hence of normal nuclear transfer development.

NUCLEAR REPROGRAMMING

Because, in *Xenopus*, a somatic cell nucleus has to start and finish chromosome replication in 1–2 hours, according to whether it was in the G1 or G2 phase of the cell cycle, it is amazing that this can ever happen successfully. Egg cytoplasm evidently has remarkable "reprogramming" powers. So far attempts to discover the basis of this have been of three kinds.

At least we know that reprogramming can be achieved on the actual DNA of a somatic nucleus, rather than only on the replicated daughter DNA of such a nucleus. This conclusion came from experiments in which multiple somatic nuclei were injected with growing (meiotic prophase) oocytes. These cells do not divide nor do they induce DNA replication of injected nuclei. Many hundred somatic nuclei can be injected into an oocyte, which can be cultured with its contained somatic nuclei for a few days. During this time, the nuclei become increasingly active in RNA synthesis and begin to express oocyte-specific genes (De Robertis and Gurdon, 1977). Thus it is seen that nuclear reprogramming can take place in the absence of replication, and involves changed transcription of the actual chromosomes that were present in a somatic cell.

Turning to the reprogramming of somatic nuclei transplanted to eggs (as opposed to oocytes), we can ask how soon after exposure to egg cytoplasm a reprogramming of nuclear activity can be detected. DNA replication is induced in over 90% of all somatic nuclei injected into eggs, even if such nuclei come from adult tissues such as brain, with a negligible replication rate (Graham *et al.*, 1966). Most such nuclei, however, fail to complete replication by the time of the first cytoplasmic division, in contrast to sperm nuclei, which always do so well before this time. RNA synthesis is normally dormant until the midblastula transition, and this is observed in nuclear transplant embryos. Indeed the pattern of synthesis of major classes of RNA, such as ribosomal RNA and transfer RNA, are the same in nuclear transplant embryos as in normal embryos at the equivalent stages of development. This is also true for some essential early zygotic regulatory genes (Chan and Gurdon, 1996), although quantitatively aberrant expression of such genes is observed in abnormal nuclear transplant embryos (Byrne *et al.*, 2002). We conclude that those transplanted somatic cell nuclei that can promote development to the blastula stage (9 hours) seem to be already reprogrammed by egg cytoplasm. It is, however, possible that nuclear transplant embryos that fail to reach the normal blastula stage may be incompletely reprogrammed, and this could relate to their defective development.

As far as the mechanisms of nuclear reprogramming are concerned, we know only that a massive nuclear swelling and chromatin dispersal take place within 1 hour of transplantation, and that this is accompanied by a two-way exchange of proteins between nucleus and cytoplasm (Gurdon and Woodland, 1970). It has yet to be shown that the chromosomal protein exchanges observed are related to the survival and normality of development of nuclear transplant embryos (Kikyo and Wolffe, 2000).

Nuclear transplantation has been achieved in some other amphibian species, notably in the axolotl (Signoret *et al.*, 1962) and the *Pleurodeles* (Gallien *et al.*, 1973).

More detailed information can be obtained from a number of reviews of nuclear transplantation in amphibia (Gurdon, 1963, 1964, 1977, 1986, 1991; Gurdon and Woodland, 1968; De Robertis and Gurdon, 1979; DiBerardino, 1997).

REFERENCES

Briggs, R., and King, T. J. (1952). Transplantation of living nuclei from blastula cells into enucleated frogs' eggs. *Proc. Natl. Acad. Sci. U.S.A.* **38**, 455–463.

Briggs, R., and King, T. J. (1957). Changes in the nuclei of differentiating endoderm cells as revealed by nuclear transplantation. *J. Morphol.* **100**, 269–312.

Byrne, J. A., Simonsson, S., and Gurdon, J. B. (2002). From intestine to muscle: Nuclear reprogramming through defective cloned embryos. *Proc. Natl. Acad. Sci. U.S.A.* **99**, 6059–6063.

Chan, A. P., and Gurdon, J. B. (1996). Nuclear transplantation from stably transfected cultured cells of *Xenopus*. *Int. J. Dev. Biol.* **40**, 441–451.

DiBerardino, M. (1997). "Genomic potential of differentiated cells." Columbia University Press, New York.

DiBerardino, M. A., and Hoffner, N. (1970). Origin of chromosomal abnormalities in nuclear transplants—A reevaluation of nuclear differentiation and nuclear equivalence in amphibians. *Dev. Biol.* **23**, 185–209.

DiBerardino, M. A., Hoffner, N. J., and Etkin, L. D. (1984). Activation of dormant genes in specialized cells. *Science* **224**, 946–952.

De Robertis, E. M., and Gurdon, J. B. (1977). Gene activation in somatic nuclei after injection into amphibian oocytes. *Proc. Natl. Acad. Sci. U.S.A.* **74**, 2470–2474.

De Robertis, E. M., and Gurdon, J. B. (1979). Gene transplantation and the analysis of development. *Sci. Am.* **241**(6), 74–82.

Elsdale, T. R., Gurdon, J. B., and Fischberg, M. (1960). A description of the technique for nuclear transplantation in *Xenopus laevis*. *J. Embryol. Exp. Morphol.* **8**, 437–444.

Fischberg, M., Gurdon, J. B., and Elsdale, T. R. (1958). Nuclear transplantation in *Xenopus laevis*. *Nature* **181**, 424.

Gallien, C.-L., Aimar, C., and Guillet, F. (1973). Nucleocytoplasmic interactions during ontogenesis in individuals obtained by intra- and interspecific nuclear transplantation in the genus *Pleurodeles* (urodele amphibian). *Dev. Biol.* **33**, 154–170.

Graham, C. F., Arms, K., and Gurdon, J. B. (1966). The induction of DNA synthesis by frog egg cytoplasm. *Dev. Biol.* **14**, 349–381.

Gurdon, J. B. (1960). The developmental capacity of nuclei taken from differentiating endoderm cells of *Xenopus laevis*. *J. Embryol. Exp. Morphol.* **8**, 505–526.

Gurdon, J. B. (1962a). The developmental capacity of nuclei taken from intestinal epithelium cells of feeding tadpoles. *J. Embryol. Exp. Morphol.* **10**, 622–640.

Gurdon, J. B. (1962b). Multiple genetically-identical frogs. *J. Hered.* **53**, 4–9.

Gurdon, J. B. (1963). Nuclear transplantation in Amphibia and the importance of stable nuclear changes in cellular differentiation. *Q. Rev. Biol.* **38**, 54–78.

Gurdon, J. B. (1964). The transplantation of living cell nuclei. *Adv. Morphogen.* **4**, 1–43.

Gurdon, J. B. (1976). Injected nuclei in frog oocytes: Fate, enlargement, and chromatin dispersal. *J. Embryol. Exp. Morphol.* **36**, 523–540.

Gurdon, J. B. (1977). Methods for nuclear transplantation in Amphibia. *In* "Methods in Cell Biology," Vol. 16 (G. Stein, J. Stein, and L. J. Kleinsmith, eds.), pp. 125–139. Academic Press, New York.

Gurdon, J. B. (1986). Nuclear transplantation in eggs and oocytes. *J. Cell Sci.* (Suppl.) **4**, 287–318.

Gurdon, J. B. (1991). Nuclear transplantation in *Xenopus*. *Methods Cell Biol.* **36**, 299–309.

Gurdon, J. B., and Uehlinger, V. (1966). "Fertile" intestine nuclei. *Nature* **210**, 1240–1241.

Gurdon, J. B., and Woodland, H. R. (1968). The cytoplasmic control of nuclear activity in animal development. *Biol. Rev.* **43**, 233–267.

Gurdon, J. B., and Woodland, H. R. (1970). On the long term control of nuclear activity during cell differentiation. *Curr. Topics Dev. Biol.* **5**, 39–70.

Gurdon, J. B., Elsdale, T. R., and Fischberg, M. (1958). Sexually mature individuals of *Xenopus laevis* from the transplantation of single somatic nuclei. *Nature* **182**, 64–65.

Humpherys, D., Eggan, K., Akutsu, H., Hochedlinger, K., Rideout III, W. M., Biniszkiewicz, D., Yanagimachi, R., and Jaenisch, R. (2001). Epigenetic instability in ES cells and cloned mice. *Science* **293**, 95–97.

Kikyo, N., and Wolffe, A. P. (2000). Reprogramming nuclei: Insights from cloning, nuclear transfer and heterokaryons. *J. Cell Sci.* **113**, 11–20.

King, T. J., and Briggs, R. (1956). *Cold Spring Harbor Symp. Quant. Biol.* **21**, 271–290.

Kroll, K. L., and Amaya, E. (1996). Transgenic *Xenopus* embryos from sperm nuclear transplantation reveal FGF signalling requirements during gastrulation. *Development* **122**, 3173–3183.

Nieuwkoop, P. D., and Faber, J. (eds.) (1956). "Normal Table of *Xenopus laevis* (Daudin)." Issued by Hubrecht Laboratory, Utrecht.

Signoret, J., Briggs, R., and Humphrey, R. R. (1962). Nuclear transplantation in the Axolotl. *Dev. Biol.* **4**, 134–164.

CHAPTER 15

CLONING OF FISH

Yuko Wakamatsu[1] and Kenjiro Ozato

Cloning of fish has been studied using a small laboratory fish, medaka (*Oryzias latipes*). In the first step of the study, basic techniques of nuclear transplantation were developed. Single-blastula nuclei of an inbred strain with the wild-type body color were transplanted into nonenucleated unfertilized eggs of an outbred orange-red strain. The nuclear transplants grew to the adult stage, expressing genetic markers of both donor and recipient nuclei; they were triploids and sterile. Foreign genes introduced into donor nuclei could be expressed normally in nuclear transplants when blastula nuclei derived from transgenic fish carrying the green fluorescent protein (*GFP*) gene, driven by the promoter of the medaka elongation factor gene, *EF-1α-A*, were transplanted into nonenucleated eggs. The gene expression pattern was the same as that in the transgenic fish. In the second step of the study, embryonic cell nuclei from transgenic fish carrying the *GFP* gene were transplanted into unfertilized eggs enucleated by X-ray irradiation. Fertile and diploid nuclear transplants were successfully generated, and the natural and GFP markers of the donor nuclei were transmitted to the F_1 and F_2 offspring in a mendelian fashion. This is the first successful nuclear transplantation using embryonic cells. Nuclear transplantation using differentiated cells or cultured cells remains to be realized.

INTRODUCTION

Fish are an important and extensively used animal resource, not only for biomedical research, but also for food production. However, the usefulness of fish as an animal resource is limited to some extent, because modern techniques of genetic modification and cloning, although established in mammalian species (Campbell *et al.*, 1996; Wilmut *et al.*, 1997), are not yet available for fish except for transgenesis by introduction of foreign genes into embryos (Ozato and Wakamatsu, 1994). Nuclear transplantation is a key technique in genetic modification and cloning; thus a nuclear transplantation technique should be established for efficient utilization of fish as an animal resource.

In fish, nuclear transplantation was first reported in the 1960s by Chinese researchers, although they obtained no conclusive results (reviewed by DiBerardino, 1997). Gasaryan *et al.* (1979) transplanted embryonic nuclei into nonenucleated and enucleated eggs of loach and obtained nuclear transplants that developed into feeding larvae, but they did not obtain any nuclear transplants that grew to adults. Since the 1970s, Yan and colleagues have conducted extensive studies of nuclear transplantation, mainly in cyprinid fish, and have produced nucleo–cytoplasmic hybrids by transplanting the nuclei of one species into enucleated eggs of another species (reviewed by Yan 1989, 1998). However, the results of these studies in loach and cyprinid fish have not been successfully reproduced by other groups; hence the

---

[1]Author to whom correspondence should be addressed.

technique of nuclear transplantation remains to be established in fish. Considering this, we aim to develop basic techniques of nuclear transplantation in fish using a small laboratory fish, medaka (*Oryzias latipes*), as a model. Here, we review our studies on nuclear transplantation of embryonic cells into unfertilized eggs.

DEVELOPMENT OF BASIC TECHNIQUES OF TRANSPLANTATION OF EMBRYONIC CELLS TO NONENUCLEATED EGGS

In order to develop basic techniques of nuclear transplantation, nuclei from blastula-stage embryos were transplanted into nonenucleated unfertilized eggs (Niwa *et al.*, 1999). It is known that nuclei obtained from blastula- or gastrula-stage embryos exhibit a higher potential for normal development in nuclear transplants than do those obtained from older embryos in all the amphibian species tested (reviewed by Gurdon, 1986; DiBerardino, 1987). The presence of haploid pronuclei in recipient eggs is considered to have almost no effect on the development of resultant triploid embryos (Subtelny, 1965a,b), or to improve the progress of development from the embryonic to later stages (Kroll and Gerhart, 1994).

MEDAKA

The medaka is an egg-laying freshwater teleost that is widely used as an experimental animal. Its body size (3 cm in total length) makes it the smallest of known vertebrates. Its generation time is short, 2 months. It spawns daily all year round under artificial conditions, and the timing of spawning can be controlled by adjusting light conditions during a 24-hour period. The distinct transparency of its eggs affords an advantage for embryological manipulation. Many mutants, including developmental mutants, have been collected from natural populations and maintained in laboratories (Tomita, 1992). Several inbred strains have been established (Hyodo-Taguchi and Egami, 1985).

DONOR CELLS

An inbred strain with a wild-type body color (HNI-I) (Hyodo-Taguchi, 1996) was used as the donor strain for nuclear transplantation. This strain is homozygous for the dark color allele (*B/B*). The skin is darkly pigmented (Fig. 1A). Blastoderms were collected from 15 to 20 blastula embryos and dissociated into single cells by pipetting in Ca^{2+}- and Mg^{2+}-free phosphate-buffered saline (CMF-PBS) (Wakamatsu *et al.*, 1994). The single cells were suspended in a buffer solution prepared by modifying a transplantation buffer for *Xenopus* (Gurdon, 1976). The solution contained 0.25 M sucrose, 120 mM NaCl, 0.5 mM spermidine trihydrochloride, 0.15 mM spermine tetrahydrochloride, and 15 mM HEPES (pH 7.3). For easier identification of cells during microinjection, the cells were stained with a neutral red solution. The cells were kept at 4°C until use.

RECIPIENT EGGS

An outbred orange-red strain (OR) was used as the recipient strain (Fig. 1B). OR is a recessive-color mutant involving the *B* locus, *b/b*; this mutant exhibits the

Figure 1 Medaka strains used as the donor and recipient for nuclear transplantation. (A) The donor, an inbred strain (HNI-I) with the wild-type body color. (B) The recipient, an outbred strain with the orange-red body color (OR). Bar, 10 mm. From Niwa, K. et al., "Transplantation of blastula nuclei to non-enucleated eggs in the medaka, Oryzias latipes," *Development Growth & Differentiation*, (1999) Vol. 41: 163–172, reproduced with permission.

Figure 1

orange-red body color because it lacks dark pigmentation in the skin. With this combination of the donor and recipient, the participation of donor nuclei in the development of nuclear transplants is confirmed by the appearance of densely pigmented melanophores in the skin. In earlier studies of loach (Gasaryan *et al.*, 1979) and nucleo–cytoplasmic hybrid teleost (reviewed by Yan, 1989, 1998), donor nuclei were transplanted into the blastodisc of artificially activated unfertilized eggs. Many fish eggs, including medaka eggs, have a thick chorion that hardens after activation; this makes pricking by micropipettes difficult during transplantation. For easier microinjection, we used as recipients intact unfertilized eggs before activation. In addition, unfertilized medaka eggs maintain their potential for embryonic development in a balanced salt solution (BSS) for medaka for several hours after being removed from ovaries (Iwamatsu, 1983), allowing us to perform experiments for a longer period. In addition to these technical advantages, the condition in the egg cytoplasm prior to activation is considered to be much better than that after activation for reprogramming of transplanted nuclei in various amphibian and mammalian species (Gurdon, 1986; Prather and First, 1990; First and Prather, 1991; Solter, 1996). Mature oocytes that had been ovulated in the ovarian cavity were collected by dissecting female fish; these oocytes were kept in a 35-mm plastic dish containing BSS supplemented with 100 U/ml penicillin and 100 µg/ml streptomycin at 18°C until use.

In addition to the outbred orange-red strain, an inbred orange-red strain (HO5) (Hyodo-Taguchi and Egami, 1985) was used for experiments to examine the expression of genetic markers and reproductive potentials in nuclear transplants, and inheritance of genetic markers in the offspring.

TRANSPLANTATION

Single donor cells were aspirated into micropipettes and were transplanted into the cytoplasm of the recipient eggs under the micropyle at the animal pole. Sperms penetrate the egg cytoplasm through the micropyle and are then brought to the position at which fusion of the male and female pronuclei occurs in the eggs during fertilization. This process is also thought to be operative for transplanted nuclei when the nuclei are deposited in the animal hemisphere of the eggs (Gurdon, 1991). The inner diameter of the opening of the micropipette was approximately 20–30 µm, which was similar to or slightly smaller than the diameter of the donor cells (Iwamatsu, 1994). The cells usually appear to be slightly ruptured when they are sucked through the small opening of the micropipette. It is known that unfertilized medaka eggs are activated by pricking them with a fine glass needle, but most of the eggs cannot develop further and remain at the one-cell stage (Yamamoto, 1944). Thus, the activation of operated eggs might be caused by pricking with micropipettes at the time of transplantation. To induce further embryonic development, cytoplasmic factors introduced with the ruptured cells are presumed to be essential. The cytoplasmic factors probably act as triggers of cleavage induction (Brun, 1978; Gasaryan *et al.*, 1979). Moreover, the possible contact between the donor nuclei and the recipient egg cytoplasm may contribute to the development of nuclear transplants when the donor cells are ruptured, but their nuclei remain intact and are surrounded by the cytoplasm of the donor cells (Briggs and King, 1953; Gurdon, 1977). Likewise, in medaka, a slight rupture of donor cells may be essential in inducing the development of operated eggs.

The cytoplasm of unfertilized eggs of medaka forms a thin cortical layer surrounding the large yolk sphere, which is commonly observed in many fish species. It became apparent in our study that transplantation of the donor nuclei into this thin cytoplasm of the unfertilized eggs is possible. This method could be applied to eggs of other fish species that have a hard chorion.

Table 1 Development of Nuclear Transplants Generated by Transplantation of Blastula Nuclei into Nonenucleated Unfertilized Eggs[a]

Total no. transplanted	Activated (%)	Cleaved (%)	Blastula[b] (%)	Late gastrula to nine somites (%)	Hatched (%)	Adult (%)
845	779 (92.2)	616 (72.9)	580 (68.6)	372 (44.0)	52 (6.1)	29 (3.4)
			Melanophores +	228	45	27
			Melanophores −	144[c]	7	2

[a]Numbers in parentheses represent the percentage of the total number of transplants. The total numbers in 11 experiments are shown. From K. Niwa, T. Ladydina, M. Kinoshita, K. Ozato, Y. Wakamatsu, "Transplantation of blastula nuclei to non-enucleated eggs in the medaka, *Oryzias latipes*," *Development Growth & Differentiation*, (1999) Vol. 41: 163–172. Table 1, reproduced with permission.

[b]The numbers of nuclear transplants with or without melanophores are indicated separately. +, Nuclear transplants with densely pigmented melanophores; −, nuclear transplants without melanophores or with a small number of weakly pigmented melanophores.

[c]Number includes a few embryos in which appearance of melanophores was not examined. They died before the nine-somite stage, at which stage melanophores appear in the wild-type strain.

NUCLEAR TRANSPLANTATION OF EMBRYONIC CELLS TO NONENUCLEATED EGGS

The results of nuclear transplantation are summarized in Table 1. Of the 845 operated eggs, 372 (44.0%) developed to the stage of embryonic body formation (the late-gastrula stage) or the nine-somite stage. Fifty-two (6.1%) eggs hatched and the fry actively swam; 29 (3.4%) of the fry grew to the adult stage.

EXPRESSION OF GENETIC MARKERS OF DONOR NUCLEI

Melanophores

Melanophores, a genetic marker of the donor nuclei, are first observed in embryos at stage 22 in the wild-type strain (Iwamatsu, 1994). In the nuclear transplant fish, melanophores were also observed at this stage. Of the 372 embryos that developed to the late-gastrula or the nine-somite stage, 228 exhibited densely pigmented melanophores, mainly on the yolk sphere. Of the 52 embryos that developed to the hatching stage, 45 exhibited a large number of densely pigmented melanophores in the skin. Of the 29 individuals that grew to the adult stage, 27 exhibited the wild-type body color (Table 1 and Fig. 2) and two exhibited the orange-red body color.

Allozyme Markers

Medaka has been known to exhibit strain-specific electrophoretic patterns of some allozymes (Sakaizumi *et al.*, 1983). Phosphoglucomutase (PGM) is an enzyme found in the metabolic pathway of glycogen. HNI-I exhibits electrophoretic bands of PGM^a, whereas OR and HO5 exhibit PGM^b (Wakamatsu *et al.*, 1993). All of the examined nuclear transplants with the wild-type body color exhibited both PGM^a and PGM^b bands, suggesting that both donor and recipient nuclei participated in the development (Fig. 3). These results of melanophore and PGM allozyme expression indicate that the donor nuclei participate in the development of the nuclear transplants, appropriately expressing genes originating from them. The percentage of nuclear transplants exhibiting genetic markers of donor nuclei was comparable with that in a previous report on loach, although nuclear transplants of loach did not grow beyond the stage of actively feeding larvae (Gasaryan *et al.*, 1979). On the other hand, the two fish with the orange-red body color exhibited only the PGM^b band, suggesting that only the recipient nuclei participated in the development.

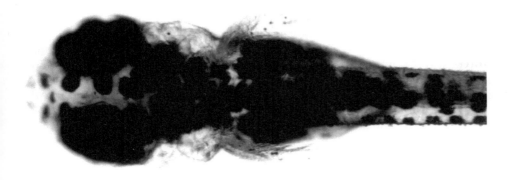

Figure 2 *A nuclear transplant fry immediately after being hatched. The fry has numerous melanophores in the skin, which is characteristic of the donor strain. Bar, 0.5 mm. From Niwa, K. et al., "Transplantation of blastula nuclei to non-enucleated eggs in the medaka, Oryzias latipes," Development Growth & Differentiation, (1999) Vol. 41: 163–172, reproduced with permission.*

PLOIDY STATUS

Relative DNA Content

The ploidy status of the nuclear transplants was assayed based on the relative DNA content of their erythrocytes. The blood obtained from the caudal veins was fixed with 100 µl of Carnoy solution, smeared on a glass slide, and air dried. The blood smears were stained with 4′,6-diamidino-2-phenylindole (DAPI) (Hamada and Fujita, 1983), and the DNA content of the erythrocytes was measured by micro-fluorometry (Naruse et al., 1985; Komaru et al., 1988). The relative DNA content was determined by comparing the mean DNA fluorescence intensity in the nuclear transplants with that in the diploid control fish. The mean fluorescence intensity in the nuclear transplant fish was 1.364–1.563 times higher than that in the diploid control, which fell in the range expected for a triploid.

Chromosome Number

Two- to five-month-old nuclear transplants were colchicinized by intraperi-toneal injection of 0.1 ml of 0.04% colchicine dissolved in CMF-PBS 5–6 hours prior

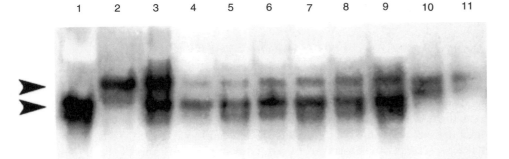

Figure 3 *Electrophoretic patterns of phosphoglucomutase (PGM) allozymes. The upper and lower bands represent PGM^b and PGM^a, respectively. Lane 1, HNI-I; lane 2, inbred orange-red strain (HO5); lane 3, F$_1$ hybrid between HNI-I and HO5; lanes 4–6, male nuclear transplants with the wild-type body color; lanes 7–9, female nuclear transplants with the wild-type body color; lanes 10–11, female individuals with the orange-red body color. All of the six nuclear transplants with the wild-type body color exhibit both PGM^a and PGM^b bands, whereas both the individuals with the orange-red body color exhibit the PGM^b band alone. From Niwa, K. et al., "Transplantation of blastula nuclei to non-enucleated eggs in the medaka, Oryzias latipes," Development Growth & Differentiation, (1999) Vol. 41: 163–172, reproduced with permission.*

to sacrifice. The spleens and gills were subsequently removed, minced finely with scissors, and mixed together. Chromosomal preparations were made according to standard techniques. The chromosome number in all of the nuclear transplants examined was 72, which is a triploid chromosome number. The diploid chromosome number of medaka is 48. Taken together, these results of the relative DNA content and chromosomal count determination suggest that nuclear transplants with the wild-type body color are triploids resulting from the fusion of the diploid transplanted nuclei with the haploid recipient pronuclei.

REPRODUCTIVE POTENTIAL OF NUCLEAR TRANSPLANTS

Twenty-seven nuclear transplants with the wild-type body color exhibited secondary sexual characteristics 2 months after hatching. Fourteen and 13 of these fish were identified to be females and males, respectively, which were crossed individually with HO5 to determine their reproductive potential. Most of the male nuclear transplants were sterile. One male was poorly fertile, but produced a viable offspring. It is unclear why a viable offspring was produced from this triploid male nuclear transplant. A similar phenomenon was reported in which grass carp obtained by crossing of triploid males and diploid females rarely survived up to even the juvenile stage (Van Eenennaam *et al.*, 1990). All of the female nuclear transplants were sterile. The testes appeared normal on gross examination, and had motile spermatozoa. On the other hand, the ovaries showed a degenerated morphology, although small immature oocytes were found in some females.

Two individuals with an orange-red body color were females and fertile. When these fish were crossed with males of HO5, all the offspring had the orange-red body color. These observations, coupled with the results of PGM allozyme analysis, indicate that the fish with the body color of the recipient are diploids in which the donor nuclei do not contribute to the development.

From these studies, it is considered that basic techniques of nuclear transplantation are achieved in medaka.

EXPRESSION OF FOREIGN GENES IN NUCLEAR TRANSPLANTS

It is important to determine whether foreign genes in the donor nuclei are expressed and are therefore useful as genetic markers in nuclear transplants. Furthermore, foreign gene expression in nuclear transplants is an essential condition for genetic modification by nuclear transplantation.

DONOR TRANSGENIC FISH CARRYING THE *GFP* GENE

EF-1α-A/GFP Transgenic Fish

The transgenic HNI-I medaka carrying a mutant *GFP*, driven by the medaka elongation factor 1α-A gene promoter (*EF-1α-A/GFP*) (Kinoshita *et al.*, 2000), was used as the donor for nuclear transplantation. EF-1α is a ubiquitous protein present abundantly in all living cells (Negrutskii and Elskaya, 1998). It is known that medaka possesses two isoforms of EF-1α, namely, EF-1α-A and EF-1α-B (Kinoshita *et al.*, 1999). In this transgenic fish, zygotic expression of GFP is observed in the entire body from the early gastrula stage to the 5-day-old embryo stage. With embryonic development, GFP expression is gradually lost in the trunk and tail regions, in which the skeletal muscle is the most prevalent tissue, and is restricted to the head region and lens after the 5-day-old embryo stage. In adults, GFP is expressed in many organs, except in the skeletal muscle (Fig. 4A). Thus, the *EF-1α-A/GFP* transgenic fish carries genetic markers of GFP, wild-type body color, and PGM[a] allozyme.

β-Act/GFP-N Transgenic Fish

The transgenic fish of the OR carries the *GFP* gene with the nuclear localization signal, driven by the medaka β-actin gene promoter (*β-Act/GFP-N*). The transgenic fish homozygous for *β-Act/GFP-N* was used for nuclear transplantation to enucleated eggs (as described on p. 294). In this transgenic fish, the weak GFP expression was first observed in the somites at the 2-day-old embryo stage. An intense fluorescence was restricted to the skeletal muscle throughout the subsequent developmental stages until the adult stage (Fig. 4B) (Yamauchi *et al.*, 2000).

FOREIGN GENE EXPRESSION

Here, the expression of foreign genes is demonstrated by two experiments using *EF-1α-A/GFP* (Niwa *et al.*, 2000). In the first experiment, blastula embryos produced by mating of HNI-I females with transgenic males heterozygous for the *GFP* gene were used as a source of donor nuclei for nuclear transplantation. These blastula embryos were used without any preselection for the presence of the *GFP* gene, because the GFP fluorescence from the zygotic expression of the *GFP* gene was not observed at the blastula stage. Therefore, individuals with or without the *GFP* gene were expected to segregate at a 1:1 ratio. Of the 655 transplanted eggs, 35 (5.3% of the transplanted eggs) hatched (Table 2), 32 of which exhibited the wild-type body color. One-half (16) of these exhibited GFP fluorescence and the other half (16) did not. Fourteen individuals with the wild-type body color grew to the adult stage. One-half of these (7) exhibited GFP fluorescence and the other half (7) did not. Thus, nuclear transplants at the hatching and adult stages that exhibited the wild-type body color were segregated into GFP-expressing and GFP-nonexpressing individuals at a 1:1 ratio. This may reflect the fact that the donor eggs were obtained by crossing of transgenic fish heterozygous for the *GFP* gene with nontransgenic fish, and were expected to include *GFP* gene carriers and non-carriers at a 1:1 ratio.

Table 2 Expression of the Wild-Type Body Color and GFP Gene in Nuclear Transplants[a]

Experiment/parameter	Hatched	Adult
First experiment		
Survived	35	15
Wild-type body color		
GFP(+)	16	7
GFP(−)	16	7
Orange-red body color		
GFP(+)	0	0
GFP(−)	3	1
Second experiment		
Survived	6	5
Wild-type body color		
GFP(+)	5	4
GFP(−)	1	1
Orange-red body color		
GFP(+)	0	0
GFP(−)	0	0

[a]The number of nuclear transplants that hatched into fry and developed into adults. They were examined based on the body color and green fluorescent protein (GFP) expression: GFP(+), exhibiting GFP fluorescence; GFP(−), not exhibiting GFP fluorescence.

In the second experiment, blastula embryos produced by mating of individuals heterozygous for the *GFP* gene were used as the source of donor nuclei. From these embryos, individuals with or without the *GFP* gene were expected to segregate at a 3:1 ratio. Of the 305 transplanted eggs, 6 (2.0%) hatched and exhibited the wild-type body color, 5 of which exhibited GFP fluorescence. Five of the hatched fry reached the adult stage, four of which exhibited GFP fluorescence. This increased number of GFP-positive fish might reflect the fact that the *GFP* gene-carrying and -noncarrying cells might be in a 3:1 ratio in the donor cell population. Four of the five adult fish were females and one was male.

The presence of the *GFP* gene was confirmed by polymerase chain reaction (PCR) in all of the adult nuclear transplants exhibiting GFP fluorescence. All of the examined adult nuclear transplants with the wild-type body color possessed both PGM^a and PGM^b with or without the GFP expression and were determined to be triploid based on the relative DNA content. None of the nuclear transplants was fertile. This finding suggests that both of the donor and recipient nuclei participated in the development of the nuclear transplants as mentioned above.

Expression of GFP during Development

GFP fluorescence was first observed in the entire blastoderm at the early gastrula stage, and in the entire body of the embryo until the 2-day-old embryo stage (Fig. 5A). As embryonic development progressed, the observed fluorescence became weak in the trunk and tail regions and was then restricted to the head region and lens after the 5-day-old embryo stage (Fig. 5B). A weak GFP expression was observed in the skin. In the adult nuclear transplants, the fluorescence was observed in the lens, the skin, and the fin rays when they were observed from the outside of the body (Fig. 5C).

Expression of GFP in Adult Internal Organs

In the dissected adult nuclear transplants, the fluorescence was observed in various organs. It was intense in the gills and pseudobranchia, moderate in the pharyngeal teeth, kidney, and spleen, and weak in the liver, gut, heart, and brain. Fluorescence was not observed in the muscle. The ovary and testis of the nuclear transplants were poorly developed and smaller than those of normal fish, and exhibited weak or no fluorescence, although a strong GFP expression in these tissues has been reported in transgenic medaka used as a donor fish (Kinoshita *et al.*, 2000).

These expression patterns of GFP in embryos and adult tissues of nuclear transplants were the same as those in the transgenic fish. Thus, it was shown that the expression of the *GFP* gene by donor nuclei is faithful to the property of the promoter in the nuclear transplants. These results indicate that the foreign *GFP* gene is expressed in nuclear transplants and is useful as a donor genetic marker in nuclear transplantation in medaka.

FERTILE AND DIPLOID NUCLEAR TRANSPLANTS

Nuclear transplantation using nonenucleated eggs is realized in medaka as mentioned above. Here, generation of fertile and diploid nuclear transplants by transplantation of embryonic cells into enucleated eggs is described (Wakamatsu *et al.*, 2001).

Enucleation by X-Ray Irradiation

Unfertilized eggs collected from females of OR were placed in BSS. X-Ray irradiation was performed in an 80-kV and 8-mA X-ray facility. Eggs were exposed to a total X-ray dose of 100 or 200 Gy at an exposure rate of 8.51 Gy/min through a 0.2-mm-thick aluminum filter.

Table 3 Development of Nuclear Transplants Generated by Transplantation of Blastula Nuclei into Enucleated Eggs[a]

Experiment	No. of eggs operated	No. of individuals[b]					
		Activated	Cleaved	Blastula	Embryonic body formation	Hatched	Adult
First experiment	588	573 (97.4)	409 (69.6)	386 (65.6)	29 (4.9)	2 (0.3)	1 (0.2)
Second experiment	298	291 (97.7)	209 (70.1)	203 (68.1)	48 (16.1)	7 (2.3)	5 (1.7)

[a]Total numbers in 17 and 10 experiments within the first and second experiments, respectively, are shown. *From Wakamatsu et al. (2001). Copyright 2001 National Academy of Sciences, U.S.A.*
[b]Numbers in parentheses represent the percentage of the total number of transplants.

TRANSPLANTATION OF EMBRYONIC CELLS INTO ENUCLEATED EGGS

In the first experiment, a transgenic fish homozygous for *EF-1α-A/GFP* was used as the donor (Fig. 4A). Of the 588 operated eggs, two (1NT1 and 1NT2) (0.3%) hatched (Table 3). One (1NT2) of the fry, however, exhibited an abnormal swelling of the intestine caused by indigestion 50 days after hatching, i.e., before sexual maturation. 1NT2 was confirmed to be female by the presence of an immature ovary found by dissection immediately after its death. 1NT1 matured sexually within 2.5 months of hatching, as in the case of the donor transgenic fish. It was female and spawned daily with a normal brood size of about 30 eggs per spawning when it was crossed with a male HO5. The survival rate of the nuclear transplants after the blastula stage was lower than that in the case of using nonenucleated eggs as recipients, although the same combination of donor and recipient fish was used (Niwa *et al.*, 2000). Similar results were reported in nuclear transplantation of *Xenopus* (Kroll and Gerhart, 1994). As mentioned above, female pronuclei in recipient eggs may have some ability to promote the embryonic development of nuclear transplants; in addition, X-ray irradiation applied for enucleation may result in some harmful effects on maternal factors of the recipient eggs.

1NT1 and 1NT2 exhibited the wild-type body color, which is a natural genetic marker of the donor. In the nuclear transplants, melanophores appeared on the yolk sphere at the anticipated developmental stage—that is, the nine-somite stage. The PGM[a] allozyme specific to the donor fish was detected in both 1NT1 and 1NT2, but the PGM[b] allozyme specific to the recipient fish was not detected (Table 4). Both of the transplants expressed GFP. The GFP expression pattern was the same as that

Table 4 Characterization of Nuclear Transplants[a]

Experiment	Nuclear transplant	Body color	PGM allozyme	GFP fluorescence	GFP gene	Fertility	Sex	No. of chromosomes	Ploidy
First experiment	1NT1	Wild type	PGM[a]	+	+	+	F*	48	Diploid
	1NT2	Wild type	PGM[a]	+	+	ND	F	ND	ND
Second experiment	2NT1	Orange-red	ND	+	+	+	F	48	Diploid
	2NT2	Orange-red	ND	+	+	+	F	48	Diploid
	2NT3	Orange-red	ND	+	+	+	F	48	Diploid
	2NT4	Orange-red	ND	+	+	+	F	48	Diploid
	2NT5	Orange-red	ND	+	+	+	F	48	Diploid

[a]Abbreviations: PGM, phosphoglucomutase; GFP, green fluorescent protein; ND, not determined; F, female. *From Wakamatsu et al. (2001). Copyright 2001 National Academy of Sciences, U.S.A.*

in the donor fish throughout the course of the embryonic development and its growth to the adult stage (Fig. 6A).

In the second experiment, the transgenic fish homozygous for β-Act/GFP-N were used as donor fish (Fig. 4B). Of the 298 operated eggs, 7 (2.3%) reached the hatching stage; of these, 5 (1.7%) matured sexually within 1.5 months, as in the case of the donor transgenic fish, and all were females that spawned daily with a normal brood size when they were crossed with a male of OR. They exhibited the orange-red body color and exhibited GFP fluorescence with the same expression pattern as that in the donor fish throughout the embryonic and subsequent developmental stages to the adult stage (Table 4 and Fig. 6D).

The transgenes were detected in all of the seven nuclear transplants in both experiments. The diploid chromosome number in all of the six surviving nuclear transplants was determined to be 48 (Table 4).

INHERITANCE OF GENETIC MARKERS

The inheritance patterns of the genetic markers in the F_1 and F_2 offspring derived from the nuclear transplants were analyzed by crossing nuclear transplants with HO5. F_1 offspring derived from 1NT1 carrying *EF-1α-A/GFP* exhibited the wild-type body color and expressed GFP. The expression pattern of GFP was the same as that in the donor fish throughout the course of the embryonic development and growth to the adult stage (Fig. 6B). The F_1 individuals examined expressed both PGM^a and PGM^b. For further analyses of the F_2 generation, both male and female F_1 individuals, derived from 1NT1, were crossed with HO5 or OR by pair mating. F_2 offspring obtained in each crossing were examined with respect to their body color and GFP fluorescence throughout the embryonic development up to the hatching stage. They were segregated into individuals with melanophores and those without melanophores, based on their body color, at a 1:1 ratio. Similarly, they were segregated based on GFP expression into positive and negative at a 1:1 ratio. The body color and GFP expression were inherited independently. The expression pattern of GFP in the offspring was the same as that in the donor fish (Fig. 6C).

Figure 4 *GFP transgenic medaka used as sources of donor nuclei. (A) A 1.5-month-old donor transgenic fish carrying EF-1α-A/GFP. An intense fluorescence in the eyes and weak fluorescence throughout the skin are observed. Fluorescence in the belly and dotlike fluorescence in the trunk are due to autofluorescence of pigment cells. (B) A 3-week-old donor transgenic fish carrying β-Act/GFP-N. An intense fluorescence is observed in the muscle tissue. From Wakamatsu et al. (2001). Copyright 2001 National Academy of Sciences, U.S.A.*

Figure 5 *Expression of EF-1α-A/GFP in nuclear transplants. (A) A 2-day-old embryo. Melanophores are observed in the embryonic body and yolk sac. Yellow spots in the head represent autofluorescence of leucophores. The yolk also exhibits weak autofluorescence. (B) A 5-day-old embryo. GFP fluorescence is hardly seen in the trunk and tail (arrowhead). Leucophores are observed on the dorsal midline and in the head. (C) An adult fish. GFP expression is observed in the skin and lens. The fluorescence in the antero-ventral portion is an artifact due to reflection of light from iridophores in the skin. The bars represent 0.3 mm in A and B, and 10 mm in C.*

Figure 6 *Expression of EF-1α-A/GFP and β-Act/GFP-N in nuclear transplants and their offspring. (A–C) Expression of EF-1α-A/GFP. (A) A 7-month-old nuclear transplant (1NT1). The arrowhead indicates an artifact. (B) A 1-month-old F_1 offspring derived from 1NT1. (C) A 1.5-day-old embryo from an F_2 offspring derived from 1NT1, with the wild-type body color showing densely pigmented melanophores in the embryonic body (arrowheads). (D–F) Expression of β-Act/GFP-N. (D) A 3-week-old nuclear transplant. (E) A 1-month-old F_1 offspring derived from 2NT1. (F) A 5-day-old embryo from an F_2 offspring derived from 2NT4. Yellow spots in the head and on the dorsal midline are autofluorescence of leucophores (arrowheads). Bars in A, B, D, and E represent 5 mm. Bars in C and F represent 0.3 mm. From Wakamatsu et al. (2001). Copyright 2001 National Academy of Sciences, U.S.A.*

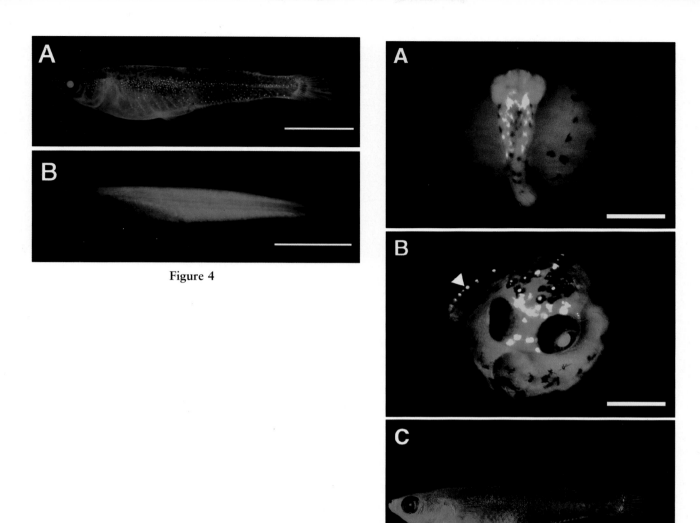

Figure 4

Figure 5

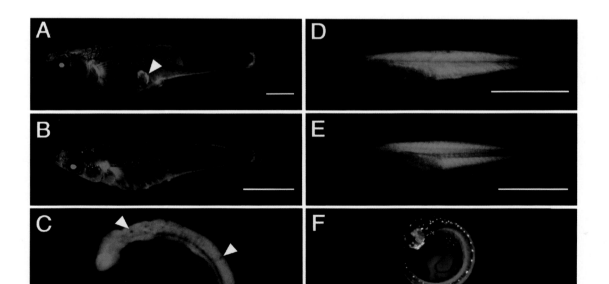

Figure 6

In the second experiment, all the F_1 progenies derived from crossing five nuclear transplants carrying *β-Act/GFP-N* (2NT1–2NT5 in Table 4) with OR by pair mating exhibited the orange-red body color and expressed GFP. The GFP expression pattern in the F_1 offspring was the same as that in the donor transgenic fish from the embryonic to the postembryonic development and the adult stage (Fig. 6E). The F_2 offspring were generated by crossing both male and female F_1 individuals, derived from each of the five nuclear transplants, with OR by pair mating. All of the F_2 offspring lacked densely pigmented melanophores in their skin, as expected of embryos with the orange-red body color. One-half of them were GFP-positive and the other half were GFP-negative. The GFP expression pattern during the course of embryonic development was the same as that in the donor transgenic fish (Fig. 6F).

The transgenes of *EF-1α-A/GFP* or *β-Act/GFP-N* were specifically detected in individuals expressing GFP in F_1 and F_2 generations. The chromosome number in F_1 individuals derived from each of the six surviving nuclear transplants in both experiments was 48, a diploid number. Genetic analyses of the F_1 and F_2 generations revealed that all of the surviving nuclear transplants are homozygous in terms of the natural and introduced marker genes, which are inherited in a mendelian fashion.

CONCLUSIONS AND PERSPECTIVES

Basic techniques for fish nuclear transplantation using medaka embryonic cells were established in this study. The nuclear transplants generated by transplantation of embryonic cells into enucleated eggs were diploid and fertile, and homozygous for marker genes from the donor nuclei. These genetic markers were transmitted to subsequent generations in a mendelian fashion. Foreign genes in the donor nuclei were expressed, faithful to the properties of the promoter.

It is indicated that X-ray irradiation is useful for inactivation of the genome of recipient eggs; the PGM allozyme marker of the recipient eggs was not detected in the nuclear transplants. Furthermore, triploid individuals and individuals with genetic markers of the recipient fish were not obtained, whereas such individuals were obtained by transplantation of embryonic cell nuclei into nonenucleated eggs (Niwa *et al.*, 1999, 2000).

Thus, cloning of fish from embryonic cells has become possible, at least in one fish species, medaka. Cloning from differentiated cells or cultured cells remains to be achieved in fish.

REFERENCES

Briggs, R., and King, T. J. (1953). Factors affecting the transplantability of nuclei of frog embryonic cells. *J. Exp. Zool.* **122**, 485–505.

Brun, R. B. (1978). Developmental capacities of *Xenopus* eggs, provided with erythrocyte or erythroblast nuclei from adults. *Dev. Biol.* **65**, 271–284.

Campbell, K. H. S., McWhir, J., Ritchie, W. A., and Wilmut, I. (1996). Sheep cloned by nuclear transfer from a cultured cell line. *Nature* **380**, 64–66.

DiBerardino, M. A. (1987). Genomic potential of differentiated cells analyzed by nuclear transplantation. *Am. Zool.* **27**, 623–644.

DiBerardino, M. A. (1997). "Genomic Potential of Differentiated Cells." Columbia University Press, New York.

First, N. L., and Prather, R. S. (1991). Genomic potential in mammals. *Differentiation* **48**, 1–8.

Gasaryan, K. G., Hung, N. M., Neyfakh, A. A., and Ivanenkov, V. V. (1979). Nuclear transplantation in teleost *Misgurnus fossilis* L. *Nature* **280**, 585–587.

Gurdon, J. B. (1976). Injected nuclei in frog oocytes: Fate, enlargement, and chromatin dispersal. *J. Embryol. Exp. Morphol.* **36**, 523–540.

Gurdon, J. B. (1977). Methods for nuclear transplantation in Amphibia. *In* "Methods in Cell Biology," Vol. 16 (G. Stein, J. Stein, and L. J. Kleinsmith, eds.), pp. 125–139. Academic Press, San Diego.

Gurdon, J. B. (1986). Nuclear transplantation in eggs and oocytes. *J. Cell Sci.* (Suppl.) **4**, 287–318.

Gurdon, J. B. (1991). Nuclear transplantation in *Xenopus*. In "Methods in Cell Biology," Vol. 36 (B. K. Kay and H. B. Peng, eds.), pp. 299–309. Academic Press, San Diego.

Hamada, S., and Fujita, S. (1983). DAPI staining improved for quantitative cytofluorometry. *Histochemistry* **79**, 219–226.

Hyodo-Taguchi, Y. (1996). Inbred strains of the medaka, *Oryzias latipes*. Fish Biol. J. Medaka **8**, 11–14.

Hyodo-Taguchi, Y., and Egami, N. (1985). Establishment of inbred strains of the medaka *Oryzias latipes* and the usefulness of the strains for biomedical research. *Zool. Sci.* **2**, 305–316.

Iwamatsu, T. (1983). A new technique for dechorionation and observations on the development of the naked egg in *Oryzias latipes*. *J. Exp. Zool.* **228**, 83–89.

Iwamatsu, T. (1994). Stages of normal development in the medaka *Oryzias latipes*. *Zool. Sci.* **11**, 825–839.

Kinoshita, M., Nakata, T., Yabe, T., Adachi, K., Yokoyama, Y., Hirata, T., Takayama, E., Mikawa, S., Kioka, N., Takahashi, M., Toyohara, H., and Sakaguchi, M. (1999). Structure and transcription of the gene coding for polypeptide chain elongation factor 1α of medaka *Oryzias latipes*. *Fish. Sci.* **65**, 765–771.

Kinoshita, M., Kani, S., Ozato, K., and Wakamatsu, Y. (2000). Activity of the medaka fish translation elongation factor 1α-A promoter examined using the *GFP* gene as a reporter. *Dev. Growth Differ.* **42**, 469–478.

Komaru, A., Uchimura, Y., Ikeyama, H., and Wada, K. T. (1988). Detection of induced triploid scallops, *Chlamys nobilis*, by DNA microfluorometry with DAPI staining. *Aquaculture* **69**, 201–209.

Kroll, K. L., and Gerhart, J. C. (1994). Transgenic *X. laevis* embryos from eggs transplanted with nuclei of transfected cultured cells. *Science* **266**, 650–653.

Naruse, K., Ijiri, K., Shima, A., and Egami, N. (1985). The production of cloned fish in the medaka (*Oryzias latipes*). *J. Exp. Zool.* **236**, 335–341.

Negrutskii, B. S., and Elskaya, A. B. (1998). Eukaryotic translation elongation factor 1α: Structure, expression, functions, and possible role in aminoacyl-tRNA channeling. *Prog. Nucleic Acids Res.* **60**, 47–78.

Niwa, K., Ladydina, T., Kinoshita, M., Ozato, K., and Wakamatsu, Y. (1999). Transplantation of blastula nuclei to nonenucleated eggs in the medaka, *Oryzias latipes*. *Dev. Growth Differ.* **41**, 163–172.

Niwa, K., Kani, S., Kinoshita, M., Ozato, K., and Wakamatsu, Y. (2000). *Cloning* **2**, 23–34.

Ozato, K., and Wakamatsu, Y. (1994). Developmental genetics of medaka. *Dev. Growth Differ.* **36**, 437–443.

Prather, R. S., and First, N. L. (1990). Nuclear transfer in mammalian embryos. *Int. Rev. Cytol.* **120**, 169–190.

Sakaizumi, M., Moriwaki, K., and Egami, N. (1983). Allozymic variation and regional differentiation in wild populations of the fish *Oryzias latipes*. *Copeia* **2**, 311–318.

Solter, D. (1996). Lambing by nuclear transfer. *Nature* **380**, 24–25.

Subtelny, S. (1965a). Single transfers of nuclei from differentiating endoderm cells into enucleated and nucleated *Rana pipiens* eggs. *J. Exp. Zool.* **159**, 47–58.

Subtelny, S. (1965b). On the nature of the restricted differentiation-promoting ability of transplanted *Rana pipiens* nuclei from differentiating endoderm cells. *J. Exp. Zool.* **159**, 59–92.

Tomita, H. (1992). The lists of the mutants and strains of the medaka, common gambusia, silver carp, goldfish, and golden venus fish maintained in the Laboratory of Freshwater Fish Stocks, Nagoya University. *Fish Biol. J. Medaka* **4**, 45–47.

Van Eenennaam, J. P., Stocker, R. K., Thiery, R. G., Hagstrom, N. T., and Doroshov, S. I. (1990). Egg fertility, early development and survival from crosses of diploid female × triploid male grass carp (*Ctenopharyngodon idella*). *Aquaculture* **86**, 111–125.

Wakamatsu, Y., Hashimoto, H., Kinoshita, M., Sakaguchi, M., Iwamatsu, T., Hyodo-Taguchi, Y., Tomita, H., and Ozato, K. (1993). Generation of germ-line chimeras in medaka (*Oryzias latipes*). *Mol. Mar. Biol. Biotechnol.* **2**, 325–332.

Wakamatsu, Y., Sasado, T., and Ozato, K. (1994). Establishment of a pluripotent cell line derived from a medaka (*Oryzias latipes*) blastula embryo. *Mol. Mar. Biol. Biotechnol.* **3**, 185–191.

Wakamatsu, Y., Ju, B., Pristyaznhyuk, I., Niwa, K., Ladygina, T., Kinoshita, M., Araki, K., and Ozato, K. (2001). Fertile and diploid nuclear transplants derived from embryonic cells of a small laboratory fish medaka (*Oryzias latipes*). *Proc. Natl. Acad. Sci. U.S.A.* **98**, 1071–1076.

Wilmut, I., Schnieke, A. E., McWhir, J., Kind, A. J., and Campbell, K. H. S. (1997). Viable offspring derived from fetal and adult mammalian cells. *Nature* **385**, 810–813.

Yamamoto, T. (1944). Physiological studies on fertilization and activation of fish eggs. I. Response of the cortical layer of the egg of *Oryzias latipes* to insemination and to artificial stimulation. *Annot. Zool. Jpn.* **22**, 109–125.

Yamauchi, M., Kinoshita, M., Sasanuma, M., Tsuji, S., Terada, M., Morimyo, M., and Ishikawa, Y. (2000). Introduction of a foreign gene into medakafish using the particle gun method. *J. Exp. Zool.* **287**, 285–293.

Yan, S. Y. (1989). The nucleo-cytoplasmic interactions as revealed by nuclear transplantation in fish. *In* "Cytoplasmic Organization Systems: A Primer in Developmental Biology (G. M. Malacinski, ed.), pp. 61–81. McGraw-Hill Publ. Co., New York.

Yan, S. Y. (1998). "Cloning in Fish: Nucleocytoplasmic Hybrids." Educational and Cultural Press Ltd., Hong Kong.

CHAPTER 16

CLONING OF MICE

Teruhiko Wakayama and Anthony C. F. Perry

Mice can now be cloned from cultured and noncultured adult-, fetus-, male-, or female-derived cells. The efficiency is typically 1–2%; an average of one to two offspring will be produced for every 100 starting one-cell embryos. Cloned embryos usually undergo developmental arrest prior to or soon after implantation and offspring often develop cloning-associated phenotypes. In the cryptic causes of these anomalies may lie clues to fundamental mechanisms in biology, including cancer and both cellular and organismal aging. The mouse thus represents a uniquely powerful biological model. We here review recent advances in mouse cloning to illustrate its strengths and promise in the study of mammalian biology and biomedicine.

INTRODUCTION

Cloning here refers to the combination of all or part of a nucleus donor cell—including its chromosomes—with an enucleated oocyte or embryonic cell to produce an embryo and/or offspring. The first cloned mammals were sheep generated by a cell fusion method (Willadsen, 1986). Subsequently, a microinjection method was developed to clone mice by selective nuclear transfer (NT) (Wakayama et al., 1998a). Species cloned by these methods (as opposed to embryo fission) include sheep (Wilmut et al., 1997), cows (Kato et al., 1998; Cibelli et al., 1998; Wells et al., 1999), goats (Baguisi et al., 1999), pigs (Onishi et al., 2000; Polejaeva et al., 2000), monkeys (Meng et al., 1997), cats (Shin et al., 2002), and mice (Wakayama et al., 1998a, 1999). The range of cloning efficiencies is 0 to ~20%, but a rate of ~1–2% (i.e., one or two live offspring per 100 embryos that start developing) is typical. Although most cloning studies have used farm animals, these species are expensive and highly out-bred (genetically heterogeneous) relative to strains of laboratory mice. By contrast, the short generation period (~2 months, including a gestation period of 19.5 days), low cost, and small size of the mouse, coupled to a relatively good understanding of mouse gamete biology, embryology, development, husbandry, and genetics, today make it the archetypal mammalian model species.

Mice have been cloned using a piezo-actuated NT method with cumulus cells (Wakayama et al., 1998a), tail-tip cells (Wakayama and Yanagimachi, 1999), Sertoli cells (Ogura et al., 2000a), and embryonic stem (ES) cells (Wakayama et al., 1999) as nucleus donors. These studies have revealed abnormalities in the development of some, but not all, cloned embryos. But why not all? Why do *any* of the cloned embryos develop within the normal range when the vast majority fail? Here we seek to explore this question, making comparisons to cloning in other species and addressing the relative importance of technical limitations and the biology inherent to cloning. We begin our description with a historical perspective of mouse cloning.

MOUSE CLONING: A HISTORICAL PERSPECTIVE

No historical account of cloning is complete without reference to the first report of cloned mice (Illmensee and Hoppe, 1981) (Table 1). The report described production of three mice (one male and two females) by the coordinated microinjection of inner cell mass (ICM) cell nuclei into zygotes and immediate removal of the preexisting pronuclei. However, the report triggered controversy and disbelief as other investigators failed to repeat or corroborate the method (McGrath and Solter, 1983, 1984a). By the end of the 1980s, most reports of full mouse development involved transfer of the nuclei of one- or two-cell embryos into an enucleated zygote (McGrath and Solter, 1983, 1984a; Howlett *et al.*, 1987). Perhaps the most notable advance (Tsunoda *et al.*, 1987) (Table 1) came with an approach postulated earlier for cattle (Robl *et al.*, 1986). In this method, two-cell (not one-cell) embryos provided recipient cytoplasm, with blastomeres from eight-cell embryos furnishing donor nuclei.

Reports of mouse cloning in the late 1980s and early 1990s were sparse, revealing no success using nuclei from differentiated cells, suggesting that the approaches being taken were flawed. Moreover, the first cloned mammals (sheep) had been produced by transferring donor nuclei into recipient, *unfertilized* eggs (oocytes), not one- or two-cell embryos (Willadsen *et al.*, 1986). These factors may have contributed to the change of technique, to using oocytes as recipient cells in mouse NT (Kono *et al.*, 1991; Cheong *et al.*, 1993). Notwithstanding the relative ease of mouse embryo culture and manipulation, mouse metaphase II oocytes (in contrast to those of most other species) are so sensitive that conventional microinjection through pipette tips wider than $2\,\mu m$ is impracticable and generally results in cell lysis. (Most nuclei require tips wider than $5\,\mu m$.) The intractability of microinjecting mouse metaphase II (MII) oocytes through such wider pipettes was eventually overcome with the application of piezo-actuated micromanipulation (Kimura and Yanagimachi, 1995). This permitted the use of larger microinjection needles, although the realization that they could be utilized in NT to clone mice came later (Wakayama *et al.*, 1998a).

Although mouse cloning remained at an impasse, progress in cloning large animals (whose eggs can be injected without piezo actuation) accelerated rapidly, and in the first half of the 1990s larger animals became the models of choice for NT (Sun and Moor, 1995). Success in larger animals was also potentiated by the view that the zygotic switch occurred relatively late in these species, but at the early two-cell stage in the mouse (Flach *et al.*, 1982; Latham *et al.*, 1991; Ram and Schultz, 1993). The early onset of zygotic genome activation—such as that believed to occur in the mouse—would imply less time for a newly transferred nuclear genome to undergo the reprogramming (defined below) mandatory for development. Mice were cloned following fusion of enucleated oocytes to the blastomeres of late two-cell embryos (Kono *et al.*, 1991), and later those of early eight-cell embryos (Cheong *et al.*, 1993), yet these isolated successes had been equaled or surpassed in the first report of sheep cloning (Willadsen, 1986). Not until later was it possible to clone mice by the serial transfer of nuclei from morula-derived or blastocyst (ICM)-derived cell nuclei first into enucleated oocytes and then the blastomeres of two-cell embryos (Tsunoda and Kato, 1997, 1998). Such serial NT marks a key departure from the earlier method of Illmensee and Hoppe (1981) and may have facilitated reprogramming of the donor cell nuclei (discussed below).

MOUSE CLONING FROM THE CELLS OF ADULTS

Cloning by transfer of somatic cell nuclei from adult mice was finally achieved using a single transfer step in 1997 (Wakayama *et al.*, 1998a), marking the beginning of a new chapter in mouse cloning (Table 1). In this development, mouse NT biology was brought alongside that of the larger species (Wilmut *et al.*, 1997; Kato *et al.*,

Table 1 Key Developments in Mouse Cloning

Date/nucleus donor[a]	Recipient cytoplast	Development to	Researchers
1981			
ICM	Zygote	—	Illmensee and Hoppe
TE	Zygote	—	Illmensee and Hoppe
1983			
Zygote	Zygote	Term	McGrath and Solter
1984			
Two cell to ICM	Zygote	≤Blastocyst	McGrath and Solter
1986			
Two cell	Two cell	Term	Robl et al.
Eight cell	Two cell	Fetus	Robl et al.
1987			
Two cell	Zygote	Term	Tsunoda et al.
Four to eight cell	Zygote	≤Four cell	Tsunoda et al.
Eight cell	Two cell	Term	Tsunoda et al.
ICM	Two cell	≤Four cell	Tsunoda et al.
1989			
Male PGC	Oocyte	Implant	Tsunoda et al.
1990			
EC	Oocyte	≤Blastocyst	Modlinski et al.
1991			
Late two cell	Oocyte	Term	Kono et al.
Eight cell to ICM	Oocyte	≤Blastocyst	Kono et al.
1992			
Four cell	Two-cell blastomere	Term	Kono et al.
1993			
ES cell	Oocyte	Implant	Tsunoda and Kato
Thymus	Oocyte	≤Blastocyst	Kono et al.
Early four cell	Oocyte	Term	Cheong et al.
Early eight cell	Oocyte	Term	Cheong et al.
1995			
Male PGC	Oocyte → two cell	Fetus	Kato and Tsunoda
1996			
Late four cell	Oocyte → one cell	Term	Kwon and Kono
1997			
Early morulae	Oocyte → two cell	Term	Tsunoda and Kato
1998			
Early ICM	Oocyte → two cell	Term	Tsunoda and Kato
TE	Oocyte → two cell	Term	Tsunoda and Kato
Sertoli	Oocyte	Fetus	Wakayama et al.
Brain cell	Oocyte	Fetus	Wakayama et al.
Cumulus	Oocyte	Term	Wakayama et al.
1999			
Tail cell	Oocyte	Term	Wakayama and Yanagimachi
ES cell	Oocyte	Term	Wakayama et al.
Oviduct cell	Oocyte → two cell	Late fetus	Kato et al.
2000			
ES/Tail cells (X^{GFP})	Oocyte	Fetus	Eggan et al.
Sertoli cell	Oocyte	Term	Ogura et al.
Cloned cumulus cell	Oocyte	Term	Wakayama et al.
2001			
ES cell (epigenetics)	Oocyte	Term	Humpherys et al.
Fetal fibroblast	Oocyte → one cell	Term	Ono et al.
Fetal gonadal cell	Oocyte	Term	Wakayama and Yanagimachi
ntES cell	Oocyte	Term	Wakayama et al.
2002			
Mainly Sertoli cell (imprinted genes)	Oocyte	Term	Inoue et al.
Sertoli cell (early death)	Oocyte	Term	Ogonuki et al.

[a]Abbreviations: ICM, inner cell mass; GFP, green fluorescent protein; EC, embryonal carcinoma; ES, embryonic stem; PGC, primoridal germ cell; TE, trophectoderm.

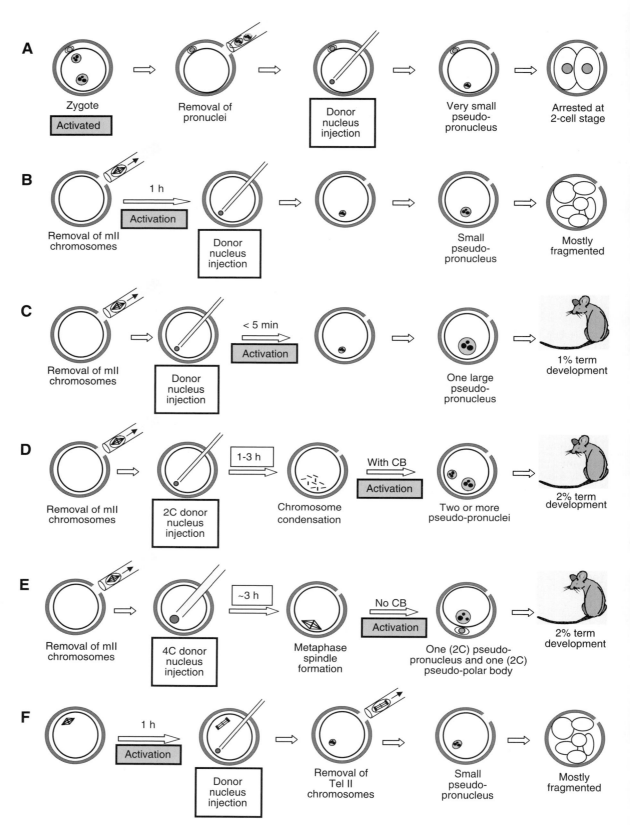

Figure 1 Diagrammatic representation of six mouse nuclear transfer protocols. (A) Zygotes are enucleated ~5–6 hours after fertilization (fertilization is in vivo) followed by diploid (2C) donor nucleus injection. This method results in embryo degeneration at the two- to four-cell stage. (B) Metaphase II (MII) oocytes are enucleated and then activated 1 hour before injection of diploid (2C) donor nuclei. Most of these embryos fragment within 24 hours.

1998; Wakayama *et al.*, 1998). Unless stated otherwise, mouse cloning data and phenomena described here were obtained essentially using the microinjection method of Wakayama *et al.* (1998a). The birth of Dolly the sheep following a fusion-based cloning method (Willadsen, 1986) had been reported in 1997; the first surviving cloned mouse was also born in that year, on October 3, 1997; she was named Cumulina, after the cumulus cell nucleus from which she was derived.

The birth of Cumulina paved the way for a surge in the application of NT to key questions in mammalian biology (e.g., Wakayama *et al.*, 1999, 2000a, 2001; Eggan *et al.*, 2000; Ohgane *et al.*, 2001; Humpherys *et al.*, 2001; Tanaka *et al.*, 2001; Inoue *et al.*, 2002; Ogonuki *et al.*, 2002; Tamashiro *et al.*, 2002). Work on Cumulina and contemporaneously cloned mice suggested rapidly and for the first time that mammalian clones were fertile and produced normal offspring (Wakayama *et al.*, 1998a), that clones could be obtained from adult males using other cell types such as tail-tip (probably fibroblast) cells, that cloning was not female specific (Wakayama and Yanagimachi, 1999), and that clones could be used to generate more clones (Wakayama *et al.*, 1998a, 2000a). Moreover, cloning was demonstrated with cultures from established ES cell lines that had completed G1 phase and were at G2 or M phase (Wakayama *et al.*, 1999). This observation was remarkable because it challenged the prevailing dogma that donor nuclei had to be at G0 in order for them to program full development following transfer (e.g., Campbell *et al.*, 1996b; Wilmut *et al.*, 1997). The question of cell cycle stage specificity in cloning is clearly an important one, but its historical origins lie in the domain of large animal cloning and it is therefore discussed elsewhere in this volume (Chapters 4 and 20). We now outline further key areas of work in mouse NT.

CONTRIBUTION OF THE RECIPIENT CYTOPLASM

OOCYTES

Oocytes (sometimes referred to as "eggs") are female gametes. In many species, including the mouse, mature oocytes are arrested at metaphase of the second meiotic division (MII) and are the cells normally encountered by fertilizing spermatozoa. Mature MII oocytes (attempts to clone using immature ones have not been reported) are biochemically characterized by a cytoplasmic activity termed maturation-promoting factor (MPF; also known as mitosis-, meiosis-, or metaphase-promoting factor) (Masui and Markert, 1971; Whitaker, 1996). Active MPF, a heterodimer of Y15-nonphosphorylated p34^{cdc2} and cyclin B, is a kinase with pleiotropic effects that peak in mitotic and meiotic cells at metaphase (Murray and Hunt, 1993). Mammalian oocytes are arrested at MII due in large part to the high steady-state level of MPF activity. This activity persists in MII oocytes following removal of the maternal chromosomes (to produce enucleated oocytes) as part of the NT procedure (Fig. 1B–E). Immediately following donor cell fusion or nuclear microinjection into the enucleated oocyte (cytoplast), the donor nucleus undergoes nuclear envelope break-

Figure 1 *(continued) (C) Enucleated MII oocytes are activated immediately after transfer of diploid (2C) donor nuclei. This method does not require the inhibition of cytokinesis (e.g., with cytochalasin B) during activation. Of the resulting one-cell embryos, ~1% develop to term. (D) Enucleated MII oocytes are activated 1–3 hours after transfer of diploid (2C) donor nuclei. Of these, ~2% develop to term. (E) G2/M-phase donor cell nuclei were injected into enucleated oocytes, with subsequent activation in the absence of a cytokinesis inhibitor. Donor nuclei are tetraploid (4C). Of these one-cell embryos, ~2% develop to term. (F) Intact MII oocytes are activated without prior enucleation and diploid (2C) donor cell nuclei are transferred soon after. Maternal telophase II chromosomes are removed 1 hour post-NT. Preembryos and embryos in this procedure are at no time deprived of a nucleus. None of the resultant embryos develop beyond the four-cell stage. With the exception of (E), all donor nuclei are diploid (2C). Abbreviations: CB, cytochalasin B; MII, metaphase II; Tel II, telophase II.*

down (NEBD) (Czolowska *et al.*, 1984; Szollosi *et al.*, 1988) and its chromosomes condense. Both the absence of a nuclear envelope (usually via NEBD) and chromosome condensation are critical to the reprogramming that resets incoming nuclear function in MII cytoplasm (see below), and both are downstream consequences of MPF activity. The level of MPF (or analogous substituting activities) is therefore likely to be a key cytoplasmic modulator of cloning competence.

ONE- AND TWO-CELL EMBRYOS

At fertilization, arrest of MII is relieved and the level of MPF activity rapidly declines. The resulting resumption of meiosis ("oocyte activation") can be mimicked by chemical and other agents used to activate oocytes parthenogenetically in NT procedures. Experiments can be performed in which oocytes of larger species are either enucleated and then subjected to an activation stimulus or vice versa prior to the insertion of a donor nucleus (Collas and Robl, 1991; Stice *et al.*, 1994; Campbell *et al.*, 1994). The first type of experiment permits careful control of the time that elapsed between activation and subsequent nuclear insertion. In the second class, removal of pronuclei is not practicable without special imaging until pronuclear membranes become visible, about 5 hours after activation. In both cases, MPF activity has declined *prior* to transfer of the donor nucleus, and neither NEBD nor chromosome condensation occur in the first cell cycle.

It was initially thought that one-cell mouse embryos might support development following transfer of donor nuclei from G1-, S-, or G2-phase cells if the phases of both were synchronized. In sheep, rabbits, cattle, and goats such experiments indeed resulted in offspring from early-stage embryo-derived and more differentiated nucleus donor cells (Prather *et al.*, 1989; Stice *et al.*, 1994; Campbell *et al.*, 1994; Wilmut *et al.*, 1997; Baguisi *et al.*, 1999). It was therefore thought that activated oocyte or zygote cytoplasts generally served as "universal recipients" (Campbell *et al.*, 1996a). However, this "universality" turned out not to hold for the mouse: where cytoplasts were derived from zygotes, the donor cell type was restricted to the blastomeres of either one- or two-cell embryos, with donor nuclei from four-cell and later developmental stages failing to yield term development (McGrath and Solter, 1983; Tsunoda *et al.*, 1987; Wakayama *et al.*, 2000b).

The question of the effect of activation timing relative to NT on development has been revisited in greater detail (Wakayama *et al.*, 2000b). Cumulus cell nuclei were microinjected into enucleated pronuclear one-cell embryos, or enucleated preactivated or MII oocytes (Fig. 1A–E, respectively). Gross physical examination of chromosomes in the resultant embryos (Fig. 2) revealed the percentages containing apparently normal complements (Wakayama *et al.*, 2000b). These data show a consistent correlation between activation–NT timing, apparent chromosome damage, and development *in vitro*. Essentially, the greater the duration for which an oocyte/zygote has been activated prior to NT, the greater the chromosomal damage and poorer the development. In the most extreme case embryos reconstituted using recipient zygote cytoplasts all contained highly fragmented chromosomes (Fig. 2C) and universally failed to develop to the blastocyst stage (Fig. 2). This implies a direct, recipient cell cycle stage-dependent cytoplasmic interaction with incoming chromatin involving endonuclease activity (causing the chromosomes to become shredded). These data are also interesting from a historical perspective in that they comprehensively fail to corroborate the first claim of mouse cloning with zygote-derived cytoplasts (Illmensee and Hoppe, 1981). At the opposite extreme, activation of reconstituted cells 1–3 hours *after* NT resulted in more than 50% of embryos with apparently normal chromosome profiles, and a similar percentage developing to the blastocyst stage (Fig. 2). The positive correlation overall between karyotypic integrity and developmental potential indicates that chromosome damage is a key determinant of cloning (in)efficiency in certain protocols.

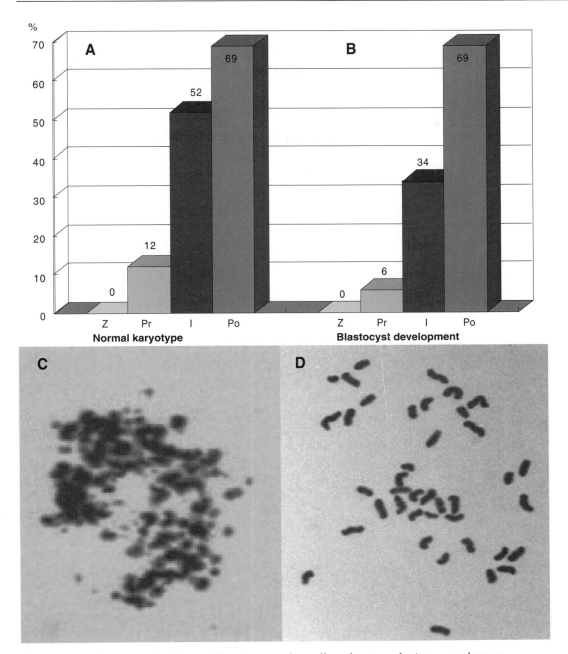

Figure 2 *Development of embryos following cumulus cell nuclear transfer into cytoplasts at different cell cycle stages. (A) The proportion (%) of one-cell embryos harboring an apparently normal karyotype at the first mitotic metaphase. (B) Blastocyst development in vitro at embryonic day 4. Abbreviations: Z, the recipient was a zygote produced by fertilization in vivo (see Fig. 1A); Pr, enucleated MII oocytes were preactivated 1 hour prior to NT (see Fig. 1B); I, enucleated MII oocytes were activated immediately after NT (see Fig. 1C); Po, enucleated MII oocytes were activated 1–3 hours post-NT (see Fig. 1D). (C) Use of zygotes as nucleus recipients (see 1A) invariably produced embryos with extensive chromosome degradation, whereas (D) use of MII oocytes as recipients (see 1B) above) produced karyotypically normal embryos in the majority of cases.*

Finally, cytoplasts derived from the blastomeres of two-cell embryos have also been shown to support development to term by serial NT, starting with donor blastomeres from four- or eight-cell embryos (Tsunoda *et al.*, 1987), as described below.

SERIAL NT

Serial NT involves insertion of a donor nucleus into first one recipient cell (often an enucleated MII oocyte), and its subsequent transfer into a second recipient cell, often an enucleated one-cell embryo (Kwon and Kono, 1996) or blastomere from a two-cell embryo (Tsunoda *et al.*, 1987). This serial NT has been thought to increase the developmental potential of the transferred nucleus (1) by improving donor nucleus reprogramming *via* prolonged exposure of chromatin to factors within the maternal cytoplasm, (2) by increased dilution of donor chromatin transcription and other factors that might interfere with reprogramming, and (3) by permitting the use of cytoplasts derived from *in vivo*-fertilized embryos, presumed to contain the optimal balance of activation pathways. Although serial cloning by NT has been applied to several species in addition to the mouse, there is no evidence for increased efficacy in any, including cattle (Stice and Keefer, 1993), pigs (Onishi *et al.*, 2000; *cf.* Polejaeva *et al.*, 2000), and mice (Wakayama *et al.*, 1998a; *cf.* Kato *et al.*, 1999; Ono *et al.*, 2001a). Indeed, serial NT protocols may produce fewer offspring, are less efficient in terms of labor, and are more likely to induce important subcellular damage *via* increased manipulation.

CONTRIBUTION OF THE NUCLEUS DONOR

CELL TYPE

Until early in 1999, reports of mammalian cloning from adult-derived cells exclusively described the use of female reproductive tissue as the source of nucleus donor (Wilmut *et al.*, 1997; Wakayama *et al.*, 1998a). The first mammalian cloning from a nonreproductive tissue, that of adult males, used mouse tail-tip cells (presumed to be fibroblasts) as nucleus donors (Wakayama and Yanagimachi, 1999).

All adult somatic cell types screened to date can support blastocyst development, albeit with differing efficiencies (Table 2). However, these relative efficiencies should be interpreted with caution because nucleus donor cell populations were not tested with vital markers to assess pre-NT cell viability, and they do not take into account the possibility of differential nuclear damage during NT. Differential developmental potential may thus reflect either an inherent (in)ability of a given nucleus donor cell type to support development or the proportion of cells alive and/or whose (epi)genomes were intact at the time of NT. Either explanation could account for the nuclei of 3% (or fewer) of thymus-derived, and 22% of brain-derived cells being able to support blastocyst development (compared to 53% for cumulus cells) (Table 2). Interpretation is further complicated because these data could be explained if many of the thymus- or brain-derived donor cells had undergone programmed genomic rearrangements that deleted genes important for development (Hozumi and Tonegawa, 1976; Chun and Schatz, 1999). However, cattle have been cloned from peripheral blood leukocytes (Galli *et al.*, 1999) and mouse blastocysts have been cloned from mature B and T lymphocytes (Hochedlinger and Jaenisch, 2002), consistent with the preservation of developmental genes in rearranged genomes.

Mouse offspring have been cloned *via* NT from putative fibroblasts, and from Sertoli and cumulus cells (Wakayama *et al.*, 1998a; Wakayama and Yanagimachi, 1999; Ogura *et al.*, 2000a). Of the remainder of acutely isolated cell types tested, most—possibly all—support embryonic development to the implantation stage; development in these cases is typically arrested at around embryonic days 6–7 (Fig. 3) (Table 2) (T. Wakayama and A. Perry, unpublished observations).

It is formally possible that all clones generated hitherto have been produced from fetus- or adult-derived stem cell subpopulations, with none produced from

Table 2 Nuclear Donor Cell Types Used in Cloning, and Development following Nuclear Transfer at E8.5–12.5[a]

Donor cell type	Sex of cell donor	No. (%) reconstructed oocytes		No. (%) morulae/ blastocysts	No. (%) transferred embryos (recipient)	No. (%) of			Live offspring[b]
		Total	Activated (%)			Implantation sites	Fetuses		
							Total	Alive	
Fetal gonadal cell (E13.5)	Female	63	42 (66.7)	31 (68.8)[c]	22 (3)	15 (68.2)	4 (18.2)	2 (9.1)	Yes (1)
	Male	18	14 (77.8)	12 (85.7)	12 (1)	11 (91.7)	4 (33.3)	2 (16.7)	Yes (1)
Adult cells									
Cumulus	Female	136	120 (88.2)	64 (53.3)	45 (5)	32 (71.1)	7 (15.6)	5 (11.1)	Yes (2)
Sertoli	Male	159	159 (100)	63 (39.6)	59 (8)	41 (69.5)	1 (1.7)	1 (1.7)	Yes (3)
Fibroblast	Female	239	219 (91.6)	85 (38.8)	85 (5)	63 (74.1)	4 (4.7)	0	Yes (4)
	Male	429	306 (71.3)	182 (59.5)	176 (15)	120 (68.2)	13 (7.4)	3 (1.7)	Yes (5)
Thymus	Female	176	168 (95.5)	5 (3.1)	0	—	—	—	No (1)
	Male	96	58 (60.4)	0	0	—	—	—	No (1)
Spleen	Female	80	49 (61.3)	11 (22.4)	11 (2)	10 (90.9)	2 (18.2)	0	No (1)
	Male	52	38 (73.1)	8 (21.1)	8 (1)	6 (75)	0	—	No (1)
Macrophage	Female	308	187 (60.7)	58 (31.0)	52 (5)	26 (50.0)	4 (7.7)	0	No (1)
	Male	205	109 (53.2)	25 (22.9)	25 (3)	19 (76.0)	0	—	No (1)
Brain	Female	228	223 (97.8)	50 (22.4)	46 (5)	25 (54.3)	1 (2.2)	0	No (2)

[a]Does not include ES cell data. Recipient females were euthanized at E8.5–12.5 for examination of developing fetuses.

[b]References (numbers in parentheses) are as follows: (1) Wakayama and Yanagimachi, 2001a,b; (2) Wakayama *et al.*, 1998a,b; (3) Ogura *et al.*, 2000a; (4) Ogura *et al.*, 2000b; (5) Wakayama and Yanagimachi, 1999.

[c]Some embryos were not cultured to morulae/blastocysts but were transferred to recipients at the two-cell stage.

differentiated cell types. Nuclear transfer ES cells have been derived from blastocysts cloned by NT from T or B lymphocytes that had, respectively, undergone programmed T cell receptor or Ig gene locus rearrangements (Hochedlinger and Jaenisch, 2002). Although such rearrangements are indeed indicative of a differentiated state, cloned mice have not been produced by NT from B or T cells, and the question remains as to whether the nuclei of differentiated cells are susceptible to reprogramming sufficient for full development. The conservation of cloning efficiency for different acutely isolated and primary cultured cell types (1–2% of one-cell NT embryos resulting in term development) requires that approximately the same proportion of stem cells would have to inhabit all nucleus donor cell preparations if they were the ones responsible for supporting term development following NT.

There are clearly differences between developmental outcomes following somatic and ES cell NT (Wakayama *et al.*, 1998a, 1999; Eggan *et al.*, 2001; Humpherys *et al.*, 2001; Inoue *et al.*, 2002). Clones derived from ES cell NT often perish around term, apparently due to respiratory failure (Eggan *et al.*, 2001), whereas perinatal mortality is less frequent in somatic cell-derived clones (Inoue *et al.*, 2002). Moreover, NT from ES cells produces neonatal clones of varying size, being reportedly on average smaller than controls (Amano *et al.*, 2001a), the same as controls (Wakayama *et al.*, 1999; Ono *et al.*, 2001b), or approximately 30% larger (Eggan *et al.*, 2001). Offspring that were derived exclusively from ES cells by embryo complementation, rather than by NT, lacked the overgrowth phenotype (Eggan *et al.*, 2001). This suggests that the overgrowth phenotype reflects a peculiar synergy between ES cells and NT.

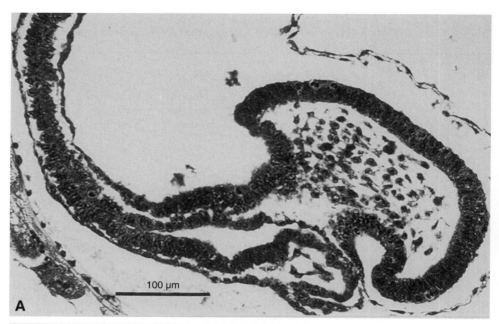

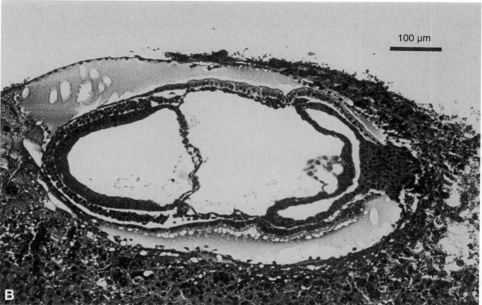

Figure 3 *Development of embryonic day 8.5 fetuses following nuclear transfer from different cell types. (A) Cloned fetus derived from a tail-tip cell (presumptive fibroblast). Fetal size is normal, matching that of a fertilization-derived fetus of the same age. This contrasts with the retarded development programmed by transfer of a spleen cell nucleus into an enucleated MII oocyte (B), in which the developmental stage corresponds to that of normal embryos at ~E6. The NT protocol used here was that of Fig. 1D.*

STRAIN

The ability of nucleus donor cells to support development is influenced by (in addition to cell type) the mouse strain from which the cells originated (Wakayama *et al.*, 1999; Rideout *et al.*, 2000; Wakayama and Yanagimachi, 2001a; Eggan *et al.*, 2001). Cloning by NT from cumulus cells (Table 3) or ES cells (Table 4)

Table 3 Term Development Programmed by Cumulus Cell Nuclei from Inbred and Hybrid Strains

| | Nucleus donor source | Nucleus recipient source | No. oocytes | | | Activated oocytes reaching morulae/ blastocysts[a] (%) | No. of two- to eight-cell embryos or morulae/ blastocysts transferred (recipients) | No. (%) live offspring at E19.5[b] |
			Total	Surviving injection	Activated (%)			
Inbred	C57BL/6	B6D2F$_1$	1098	1045	1006 (96.3)	23.8	413 (24)	0
	C3H/He	B6D2F$_1$	322	305	297 (97.4)	48.4	200 (16)	0
	DBA/2	B6D2F$_1$	382	370	354 (95.7)	59.3	308 (16)	1 (0.3) [0.3]
	DBA/2	DBA/2	57	51	46 (90.2)	—[c]	44[c] (4)	0
	129/SvJ	129/SvJ	51	22	19 (86.4)	0	—	0
	129/SvJ	B6D2F$_1$	341	286	273 (67.5)	58.0	166 (16)	2 (1.2) [0.7]
	129/SvE[d]	B6D2F$_1$	238	217	188 (86.6)	24.4	46 (4)	3 (6.5) [1.4]
Total			**2489**	**2296**	**2183 (95.1)**	**—**	**1177 (80)**	**6 (0.5) [0.3] a**
Hybrid	B6D2F$_1$	B6D2F$_1$	2261	2145	2034 (94.8)	59.5	1294 (88)	38 (2.9) [1.7]
	B6C3F$_1$	B6D2F$_1$	502	473	454 (96.0)	71.1	312 (19)	7 (2.2) [1.5]
	B6D2F$_1$	B6C3F$_1$	381	372	354 (95.2)	49.4	189 (18)	7 (3.7) [1.9]
	B6C3F$_1$	B6C3F$_1$	367	341	307 (90.0)	81.4	267 (20)	5 (1.9) [1.5]
	B6D2F$_1$[d]	B6D2F$_1$[d]	826	779	583 (74.8)	31.3	269 (26)	12 (4.5) [1.5]
	B6C3F$_1$[d]	B6D2F$_1$[d]	764	663	594 (89.6)	29.1	242 (22)	11 (4.5) [1.7]
	Tg F$_3$	B6D2F$_1$	92	78	73 (96.3)	23.1	18 (2)	1 (5.6) [1.3]
Total			**5193**	**4851**	**4399 (90.7)**	**—**	**2591 (195)**	**81 (3.1) [1.7] b**

[a]These data were obtained after 78 hours of embryo culture. Some embryos were transferred at the two- to eight-cell stage.
[b]Expressed as (% of transferred embryos) and [% of surviving oocytes after nucleus injection]. Letters **a** and **b** indicate significant differences ($P < 0.005$) between **a** and **b** when analyzed by Chi-squared tests.
[c]All embryos were transferred to recipient females at the two-cell stage.
[d]Mice were purchased from NCI except where indicated: inbred, Taconic; hybrid, Jackson Laboratory.

of hybrid strains is significantly more efficient than analogous NT from inbred strains.

Microinjection of cumulus cell nuclei from inbred mice into cytoplasts from hybrid strains (heterotypic NT) supported development to morulae/blastocysts in 24–59% of cases after ~4 days of culture. However, these generally failed to develop to term after embryo transfer to surrogate mothers, with the exceptions of embryos derived by NT from inbred strains DBA/2 or 129/Sv into B6D2F$_1$ (hybrid) cytoplasts (Table 3) (Wakayama and Yanagimachi, 2001a).

The most efficient cloning was achieved when enucleated oocytes of hybrid mice (B6D2F$_1$ or B6C3F$_1$) were injected with cumulus cell nuclei also of hybrid mice (B6D2F$_1$ or B6C3F$_1$). Yet why do cumulus cell nuclei from most inbred strains fail to support postimplantation development when those of hybrid strains occasionally succeed? Success with inbred strains 129/Sv and DBA/2 shows that heterotypic NT *per se* is not the cause of the general inability of the nuclei of inbred strains to support cloning. In searching for alternative explanations, it is notable that strain 129/Sv is susceptible to spontaneous testicular and ovarian teratomas and teratocarcinomas and has served as the source of embryonic stem or embryonic carcinoma (EC) cell lines (Stevens, 1973; Simpson *et al.*, 1997). Other inbred strains for which NT cloning has been attempted, such as C57BL/6 and C3H/He, are probably less amenable to the derivation of ES cell lines (Ledermann and Burki, 1991; Kitani *et al.*, 1996). Perhaps the propensity of strain 129/Sv to yield ES cell lines is

Table 4 Cloning Mice with ES or ntES Cells as Nucleus Donors[a]

| | | Nucleus donor | | | | Nuclear transfer and development *in vitro* | | Development *in vivo* | |
| | | | | | | No. reconstructed oocytes | Morula/ blastocyst development (%) | Fetus/ Placenta | Live cloned offspring (%) |
Mouse strain	Source embryo	Sex	Cell line name	Type of strain	Passage				
Inbred	Fertilized embryo	Male	V 18.6	129/SvJ	5–6	155	19 (12.3)	1 (0.6)	4 (2.6)
		Male	J1	129/SvJ	8–11	419	34 (8.1)	6 (1.4)	4 (1.0)
		Male	J1	129/SvJ	35	151	23 (15.2)	10 (6.6)	0
		Male	E14	129/Ola	22–33	1765	323 (18.3)	8 (0.5)	5 (0.3)
		Male	E14-targ.	129/Ola	>30	663	218 (32.9)	10 (1.5)	1 (0.2)
		Male	AB1	129/Sv	nd	103	32 (31.1)	0	0
		Male	AB2.2	129/Sv	nd	193	48 (24.9)	2 (1.0)	0
		Male	TL1	129/Sv	nd	120	34 (28.3)	4 (3.3)	0
	Somatic cell NT embryo	Female	C6	129/SvTac	5–8	181	26 (14.4)	0	0
		Male	C7	129/SvTac	5–8	296	146 (49.3)	22	0
		Male	CN1,2	C57BL/6[nu/nu]	5–8	675	88 (19.0)	2	1 (0.6)
Total						**4721**	**991 (21.0)**	**65 (1.4)**	**15 (0.3)**
Hybrid	Fertilized embryo	Male	R1	129/Sv × 129/Sv-CP	<32	1087	312 (28.7)	20 (1.8)	26 (2.4)
		Male	V 6.5	129SvJ × C57BL/6	5–9	280	27 (9.6)	2 (0.7)	6 (2.1)
		Male	V 6.5-targ.	129SvJ × C57BL/6	8–9	51	7 (13.7)	0	1 (2.0)
		Male	TT2	C57BL/6 × CBA	nd	286	81 (28.3)	1 (0.3)	0
	Somatic cell NT embryo	Female	C1–5,8	C57BL/6 × DBA/2	5–8	933	386 (41.4)	15 (1.6)	11 (1.2)
		Male	C13–17	129/SvTac × B6D2F₁	5–8	712	199 (27.9)	24 (3.4)	8 (1.1)
		Male	CT1	EGFP Tg F6	5–8	168	46 (27.4)	1 (0.6)	0
Total						**3517**	**1058 (30.1)**	**63 (1.8)**	**52 (1.5)**

[a]ntES cells are described in Wakayama *et al.* (2001).

linked to the relative ease with which 129/Sv supports cloning. The mechanism underlying any such general link would, however, have to explain cloning success with the nontumorigenic inbred strain, DBA/2, for which ES cell lines are rare.

Collectively, ES and cumulus cell NT data strongly suggest that cloning hybrid strains is more efficient than cloning inbred ones (Wakayama *et al.*, 1999; Eggan *et al.*, 2001; Wakayama and Yanagimachi, 2001a). Such a clear difference between the NT biology of inbred and hybrid strains could be related to well-established contrasts in developmental phenomena between the two groups. Embryos of inbred strains are more susceptible *in vitro* to arrested development at the two-cell stage (the "two-cell block") and are generally more difficult to culture (Suzuki *et al.*, 1996). Strain-specific differences also exist in androgenetic development between the two groups (Mann and Stewart, 1991; Latham, 1994). Alternatively any hybrid/inbred cloning disparity might imply that, in the mouse at least, heterosis facilitates cloning. Because even hybrid mouse strains contain relatively homogeneous genomes, a correlation between degree of heterosis and cloning seemingly predicates increased efficiency of cloning in outbred species, such as cattle. However, this is apparently not the case; the efficiency of cloning does not vary markedly between species and the role of heterosis in mammalian cloning remains enigmatic.

Perhaps most significantly of all, these studies reveal influences on developmental outcome following NT that are cell-type independent (Wakayama *et al.*, 1999; Eggan *et al.*, 2001). Because it is possible for a single nucleus donor cell type (e.g., the cumulus cell) from different strains to exhibit such disparate abilities to support development, cell type *per se* cannot be the predominant determinant of cloning outcome in these cases. As we shall see (p. 330), analyses of cloning phenotypes suggest that donor cell type may exert a phenotypic influence at some level. However, the cell-type independence indicated here would imply that the transcriptional state of donor cell at the time of NT does not guarantee the ability of that cell to support development thereafter. At one extreme, this argues that at some levels, the state of a donor chromatin at the time of transfer plays no decisive role in reprogramming but interacts passively with oocyte cytoplasmic reprogramming effectors. Reprogramming is considered in greater detail below.

DEVELOPMENTAL STAGE

A comparison of the efficiency of mouse cloning using embryo-, fetus-, and adult-derived cells as nucleus donors is shown in Fig. 4. The efficiency of term development following NT from acutely derived (preimplantation) embryonic cells (2.4%) or fetal cells (1.0 and 2.2%; female and male, respectively) is slightly higher than analogous rates for adult-derived cells (0.5 and 1.7%; female and male, respectively) (Wakayama *et al.*, 1998a, 1999; Wakayama and Yanagimachi, 2001a). These data sets are, like the differences, small, and variables include cell type in addition to the developmental stage of the source. However, the differences are consistent (T. Wakayama and A. Perry, unpublished observations), and if borne out by further experimentation this finding would be consonant with an enhanced ability of embryonic cell nuclei to be reprogrammed compared to their adult-derived counterparts. Alternatively, the difference may reflect an accumulation of (epi)genomic assaults (Dean *et al.*, 1998) in adult-derived cells that preclude development in cloning by NT (Humpherys *et al.*, 2001).

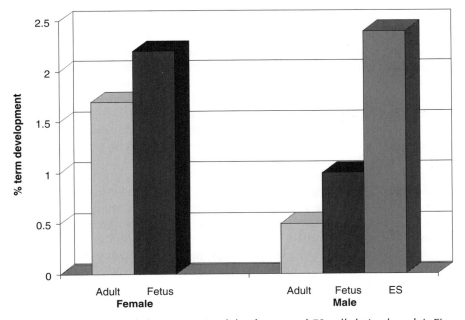

Figure 4 *Efficiencies of cloning using adult-, fetus-, and ES cell-derived nuclei. Figures correspond to the proportion (%) of starting (one-cell) embryos that developed to term. Independent of sex, the cloning success rate diminishes as the developmental stage of the nucleus donor cell source advances, i.e., the efficiency increases in the order adult → fetus → ES (ΨICM) nucleus donor cells.*

CELL CYCLE

Until the late 1990s, dogma held that the cell cycle stage of the nucleus donor was critical in mammalian cloning: specifically, nucleus donor cells had to be quiescent (G0) at the time of transfer (Campbell *et al.*, 1996b; Wilmut *et al.*, 1997). This assertion has since been challenged by work in cattle (Cibelli *et al.*, 1998) and mice, which have (for example) been cloned by NT from somatic cells cultured in serum-containing medium (Wakayama and Yanagimachi, 1999; Ogura *et al.*, 2000b). This does not of itself refute the "G0 dogma", because even rapidly dividing cell cultures may contain a fraction of cells that are quiescent, and these alone may be the ones that are cloning competent. However, a mandatory G0 nucleus donor state would result in a drop in cloning efficiency proportional to this fraction; if 1% of cells in a dividing culture were at G0, compared to 99% in a quiescent culture, then the cloning efficiency using cells from the former would be approximately 0.01× that of the latter. This is not the case; cow and mouse cloning efficiencies are not markedly affected by culturing donor cells in a range of serum concentrations expected to alter the fraction of quiescent cells in each population (Cibelli *et al.*, 1998; Wakayama and Yanagimachi, 1999).

Further NT experiments in the mouse suggest that donor cells at different cell cycle stages (G1, G2, and M phases) are cloning competent. These pivot on the twin observations (described in detail below) that the cloning procedure can be successfully adapted in a manner that would yield cloned offspring (1) if the nucleus donor contained four, but not two, genomic complements (4C but not 2C), and (2) by causing mitotic arrest to enrich for 4C donor cells prior to NT.

In practice, G0/G1- and G2/M-phase cells can be distinguished on the basis of size, with cells in the second group being larger in preparation for cleavage (Fig. 5). Figure 6 shows the orderly alignment of chromosomes on a metaphase spindle following NT from G2/M-phase donor cells into enucleated oocytes. The ordered array (Fig. 6B) presumably reflects mitotic spindle assembly and consequently non-random orientation of bivalent chromosomes, because this order is not seen in chromosomes transferred from presumptively prereplicated G0- or G1-phase donor cells (Fig. 6A).

Activation of 4C reconstructed cells is in the *absence* of cytochalasin B (see Fig. 1E), because in these embryos it is important that the second meiotic division reduces chromosome number (genomic complement) to 2C, thereby preventing the formation of inviable tetraploid embryos (Fig. 6). This method has been used to produce cloned offspring from G2/M-phase cells of the ES cell line, R1 (Wakayama *et al.*, 1999). Cloning efficiencies were approximately equal whether large (4C) or small (2C) cells were used as nucleus donors in appropriately adapted protocols.

In a second class of experiments, nucleus donor cultures were enriched for cells arrested at M phase by exposure to the microtubule-disrupting agent, nocodazole (Wakayama *et al.*, 1999; Amano *et al.*, 2001a,b; Ono *et al.*, 2001a). Such treatments profoundly affect cultured cell morphology, causing cells to round up and often float free of the culture vessel wall. Cells from cultures affected in this way have been used as nucleus donors in protocols adapted to accommodate (and thereby confirm) their presumed 4C status (Johnson *et al.*, 1988; Kato and Tsunoda, 1992). Again, cloning efficiencies in these experiments were comparable to those using protocols that were not adapted for nocodazole treatment.

Whereas these data clearly show that a G0 nucleus donor cell is not required for cloning, they do not suggest that the cell cycle stage is unimportant. Nucleus donor cells at S phase (i.e., actively replicating their DNA) at the time of transfer to enucleated MII oocytes fail to support development in currently available protocols (Kono, 1997). Cytogenetic analysis following transfer of S-phase nuclei into MII oocytes reveals gross DNA damage (Rao and Johnson, 1970; Johnson and Rao,

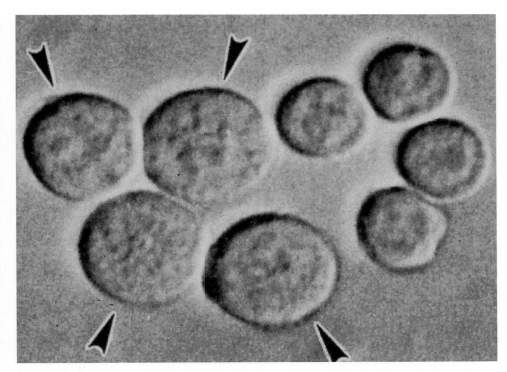

Figure 5 *Selection of E14 ES cells for subsequent nucleus microinjection into enucleated mouse oocytes. Representatives of small (average diameter, 10 μm) and large (arrowheads; average diameter, 18 μm) cells are here shown grouped together to show contrasting sizes between members of different groups. The large cells are presumed to be post-S phase (4C) whereas the smaller cells are in G1 phase (2C).*

1970; Collas *et al.*, 1992), suggesting that S-phase donor nuclei are incompatible with MII oocytes in cloning by NT.

NUCLEAR–CYTOPLASMIC COORDINATION

Cloning by NT generally requires that chromatin performing a set of instructive functions in a given cell type becomes adapted on transfer to its new cellular environment and performs a distinct set of instructive functions: those that initiate the program for full development. This adaptation is known as reprogramming, which is loosely defined both structurally (in terms of the architectural changes accompanying chromatin remodeling) or functionally (in terms of changes in transcription and developmental fate). Reprogramming of chromatin in the donor nucleus occurs in response to the new milieu of the host cytoplast. Clearly, this response is pivotal to determining the outcome of NT and may have consequences far downstream of the initial NT event.

Exposure of donor nuclei to the cytoplasm of MII oocytes is rapidly followed by NEBD and chromosome condensation (Masui and Markert, 1971; Czolowska *et al.*, 1984; Szollosi *et al.*, 1988; Whitaker, 1996). In the first report of mouse cloning from differentiated cells (Wakayama *et al.*, 1998a), 1–6 hours were allowed to elapse between NT and activation (the elapsed time is sometimes referred to as the "latent period"). In other words, the incoming nucleus had at least 1 hour to undergo the reprogramming changes required for subsequent development. Practical constraints in NT protocols mean that the shortest time between nuclear microinjection and exposure to an activating stimulus is approximately 5 minutes, although it should be noted that full activation does not occur immediately on expo-

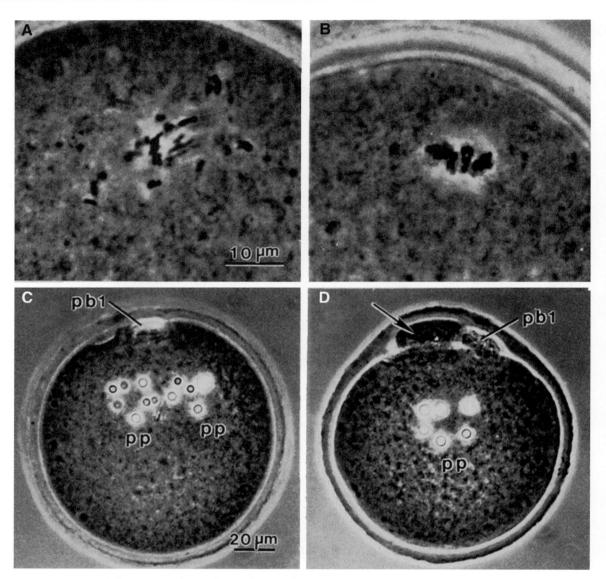

Figure 6 Analysis of preembryos following transfer of E14 ES cell nuclei into MII oocytes. (A) At 3–4 hours after transfer of putatively diploid (2C) small cell (see Fig. 5) nuclei, chromosomes condense to form a disorganized array. (B) At 3–4 hours after transfer of putatively tetraploid (4C) large cell (see Fig. 5) nuclei, chromosomes condense to form an orderly chromosome array more reminiscent of the maternal metaphase plate. (C) A one-cell embryo following transfer of a small (2C) cell nucleus and subsequent activation by exposure to Sr^{2+} for 6 hours in the presence of the cytokinesis inhibitor, cytochalasin B. Two pseudo-pronuclei (pp) are discernible, each containing several nucleoli. Remnants of the first polar body (pb1) are also visible. (D) A one-cell embryo following transfer of a large (4C) cell nucleus and subsequent activation by exposure to Sr^{2+} for 6 hours. In this case, the reductive extrusion of a pseudo-second polar body is desired (to generate a 2C embryo), thus activation is in the absence of a cytokinesis inhibitor. A degenerate pb1, single pp, and pseudo-second polar body (indicated by an arrow) are visible.

sure to the stimulus (see Table 5 and below). Experiments in which the time between NT and activation was carefully controlled suggest moderately fast reprogramming kinetics: 30 minutes was sufficient for full reprogramming following the transfer of nuclei into $B6D2F_1$ recipient cytoplasts (Wakayama and Yanagimachi, 2001b).

Presumably, the more closely the donor nucleus resembles that of a one-cell embryo at the time of transfer, the less extensive the remodeling it must undergo in

Table 5 Effects of Activation Immediately after NT, Dimethylsulfoxide, and Cytochalasin B on Development[a]

Activation medium[b]	No. of enucleated oocytes	No. of oocytes surviving nucleus injection	No. of activated oocytes (%)	No. of oocytes developing to blastocyst (% ± SD)[c]	No. of transferred embryos (recipients)	No. of live offspring (%)
DMSO + CB	172	135	102 (75.6)	59 (43.7 ± 14.8) (a)	59 (5)	0
DMSO, no CB	257	235	193 (82.1)	153 (65.1 ± 17.3) (b)	149 (9)	3 (2.0)
No DMSO, no CB	252	220	168 (76.4)	101 (46.0 ± 24.2) (a)	97 (7)	3 (3.1)

[a]Data represent five replicates.
[b]Dimethylsulfoxide (DMSO) is at 1% (v/v); cytochalasin B (CB) is at 5 µg/ml.
[c]Data analyzed using the Chi-square test show a significant difference between a and b ($P < 0.05$).

order to direct embryonic development. Early successes in mouse cloning may have reflected this; the blastomeres of G2/M-phase four- to eight-cell mouse embryos contain chromosomes that are either condensed or in an advanced state of condensation prior to NT, possibly facilitating reprogramming by removing the requirement for cytoplasmic recondensation following NT (Cheong *et al.*, 1994; Kwon and Kono, 1996).

Consistent with the above ideas, there is little difference in cloning efficiency whether the incoming nucleus is allowed 1 or 6 hours of reprogramming time prior to activation (Wakayama *et al.*, 1998a). Because development typically does not progress far beyond the blastocyst stage, this indicates an inherent nuclear–cytoplasmic inability to achieve nuclear remodeling in the majority of cases, not that remodeling fails to go to completion due to an insufficient length of nuclear exposure to the reprogramming environment. The interval between NT and activation may, however, be important for processes other than reprogramming. Microinjection of G2/M-phase (4C) ES cell nuclei into MII ococytes followed by more than 3 hours of ooplasmic exposure prior to activation enables the formation of a spindle, upon which the incoming chromosomes become aligned in an orderly manner (Fig. 6B) (Wakayama *et al.*, 1999). Complete spindle formation would clearly be a critical prelude to balanced cytokinesis. In these experiments, 68% of reconstructed cells extruded a pseudo-second polar body, with some resultant embryos developing to term (Wakayama *et al.*, 1999). Whereas chromosomal choreography following the transfer of a 2C nucleus is apparently more random than that observed with 4C nuclei (*cf.* Fig. 6, panels A and B), here too, there may have to be sufficient time to coordinate events that target cytoplasmic components to the newly transferred chromatin to tether the chromosomes to a spindle. These putative events are largely undescribed, let alone understood at high resolution.

DOES CYTOPLAST NUCLEAR DEPRIVATION (THE WINDOW OF ENUCLEATION) DEPRESS CLONING EFFICIENCY?

So far, successful NT has used enucleated (or chromosome-free) cells (cytoplasts) as nucleus recipients. When cytoplasts are derived from MII oocytes, it is not known whether the duration for which the recipient lacks a nucleus or chromosomes (the "window of enucleation") affects the cloning efficiency. Windows of enucleation of ~1 to 4 hours do not produce marked differences in cloning outcome (T. Wakayama and A. Perry, unpublished observations). However, it is possible that nuclear/chromosomal absence is more critical in NT to cytoplasts derived from one- or two-cell embryos (i.e., *following* activation). This potential role was investigated by transfer of cumulus cell nuclei into activated oocytes followed by telophase II chromosome removal (Fig. 1F). Such reconstructed embryos never lacked a nucleus, although none developed beyond the four-cell stage. Developmental arrest at this stage also occurs in the analogous achromosomal situation in which nuclei are transferred immediately following activation of oocytes that were already enucleated. These similarities suggest that failure following NT into preactivated cytoplasts is not due to nuclear deprivation. To summarize, there is no evidence that the window of enucleation in NT protocols alters the efficiency of cloning (Wakayama and Yanagimachi, 2001b).

TECHNICAL DETERMINANTS OF DEVELOPMENTAL OUTCOME IN CLONING

The extent to which the outcome of NT reflects technical details of the methodologies used *vs.* properties inherent to the biological material is not known (Perry and Wakayama, 2002). Many of the differences in developmental outcome following NT are likely due to subtle relative differences in operator skills. Speed of manipulation is an obvious variable in this respect, and may have profound consequences for developmental outcome by reducing or prolonging trauma to cells and cell fractions during microsurgery. Workers performing experiments side-by-side with shared samples can produce irreconcilable data (Perry and Wakayama, 2002). Technical considerations should therefore form a major part of data interpretation. Data on NT embryos are biologically meaningless absent a description of the developmental potential of those embryos. The mouse as a model lends itself to determining the contribution of technical constraints to the outcome of this development. We now discuss comparatively what is known of some of the gross technical parameters that have been investigated in mouse cloning protocols.

OOCYTE ACTIVATION

Oocyte activation has been induced following NT in sheep and goats with an electric pulse (Wilmut *et al.*, 1997; Baguisi *et al.*, 1999) and in cattle by an electric current plus the protein synthesis inhibitor, cyclohexamide (Kato *et al.*, 1998), calcium ionophore A23187 plus the protein kinase inhibitor, 6-dimethylaminopurine (DMAP) (Cibelli *et al.*, 1998), and the calcium ionophore, ionomycin, plus DMAP (Wells *et al.*, 1999).

Mouse cloning is typically achieved by activation using strontium (Sr^{2+}) ions (Wakayama *et al.*, 1998a), although activation by ethanol, an electric pulse, or even spermatozoa may result in equivalent cloning efficiencies (Kishikawa *et al.*, 1999; Amano *et al.*, 2001a). Although the biochemical consequences of different activation stimuli are likely to overlap, they are also distinct. For example, exposures of oocytes to calcium ionophore A23187, ethanol, or an electric pulse all generate a single, large Ca^{2+} transient, whereas activation by Sr^{2+} is accompanied by a series of Ca^{2+} oscillations that resemble those induced at fertilization (Cuthbertson *et al.*, 1981; Swann and Ozil, 1994; Kline and Kline, 1992). However, down-regulation of the inositol 1,4,5-trisphosphate receptor that normally follows fertilization is not triggered by parthenogenic activation stimuli (Brind *et al.*, 2000; Jellerette *et al.*, 2000). The functional consequences—if any—of these differences are not known. Convergent research on oocyte activation and cloning by NT will qualitatively describe the role of activation as a determinant of cloning efficacy; perhaps the universally low rate of cloning success (~1–2%) reflects a universally suboptimal feature common to current activation protocols.

AGENTS USED TO PREVENT CYTOKINESIS: THE EFFECT OF CYTOCHALASIN B ON EMBRYONIC DEVELOPMENT

The use of 2C donor cells in NT generally requires the inhibition of cytokinesis to prevent chromosome loss *via* the extrusion of a pseudo-second polar body

(Willadsen, 1986; Kono *et al.*, 1993; Cheong *et al.*, 1994; Wakayama *et al.*, 1998a, 1999). As an exception, activation of MII oocytes immediately after NT does not require the inhibition of cytokinesis because there is insufficient time for spindle formation and thus there is no pseudo-second polar body extrusion (Table 5) (Wakayama and Yanagimachi, 2001b). Indeed, inclusion of the cytokinesis inhibitor cytochalasin B was detrimental to blastocyst and subsequent development in these experiments; offspring were generated when cytochalsin B was omitted, but not when it was present (Table 5). This situation contrasts with preimplantation development after immediate post-NT activation in rabbits (Collas and Robl, 1990) and sheep (Smith and Wilmut, 1989).

If cytochalasin B has a deleterious effect on embryo development, prevention of cytokinesis following NT by some means other than exposure to cytochalasin B might increase the success rate. Like cytochalasin B, the more potent cytokinesis inhibitor, cytochalasin D, is also a microfilament-disrupting reagent and has been used to induce polyploidy in preimplantation embryos (Snow, 1973; Siracusa *et al.*, 1980; Bos-Mikich *et al.*, 1997). Cytochalasin D does not appear to affect mouse development to the blastocyst stage *in vitro* (Siracusa *et al.*, 1980) and produces NT zygotes with two or three pseudo-pronuclei, reflecting microtubular integrity (Wakayama and Yanagimachi, 2001b). The effect of replacing cytochalasin B with cytochalasin D on development of NT embryos beyond the blastocyst stage is not known. Inhibition of cytokinesis by nocodazole (an inhibitor of tubulin polymerization and hence microtubule integrity) following NT produced zygotes with many small pseudo-pronuclei (Fig. 7), presumably reflecting chromosome scattering. Inhibition of cytokinesis by nocodazole neither markedly reduces nor enhances development of fertilization-derived offspring (Kato and Tsunoda, 1992; Samake and Smith, 1996).

Although it may be inferred from data concerning inhibition of cytokinesis by alternative agents (or none) that cytochalasin B does not reduce the cloning efficiency, common to all authenticated reports of cloned offspring to date, enucleation to produce recipient cytoplasts prior to NT was also permitted by exposure to cytochalasin B. Such a universal treatment may exert a subtle yet lingering effect that contributes to consistently poor development of NT embryos. This possibility could be tested using other methods of enucleation, such as "chemical enucleation"

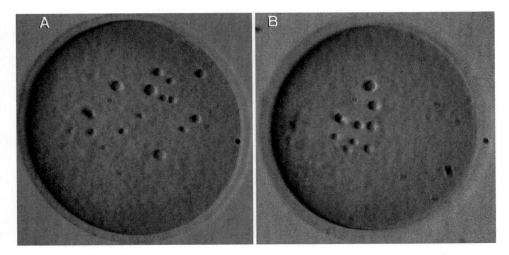

Figure 7 *Pseudo-pronuclei within one-cell embryos 6 hours after activation. Embryos were produced by transferring cumulus cell nuclei into enucleated MII oocytes. (A) Activation in the presence of nocodazole (3 µg/ml), revealing many small pseudo-pronuclei. (B) Activation in the presence of cytochalasin B (5 µg/ml). Activated embryos had two or three pseudo-pronuclei.*

(Fulka and Moor, 1993; Baguisi and Overström, 2000). The precisely timed application of the microtubule inhibitor, colcemid, to anaphase II oocytes has been reported to result in the extrusion of all maternal chromosomes into a pseudo-second polar body and is a candidate for chemical enucleation (Baguisi and Overström, 2000).

DIMETHYLSULFOXIDE AS AN EFFECTOR OF DEVELOPMENT

Dimethylsulfoxide (DMSO) is a commonly used solvent for cytochalasin B and is thus present [at 1% (v/v) in initial experiments] during oocyte activation (Wakayama *et al.*, 1998a). DMSO is not a bystander in this process and can exert a negative influence on blastocyst development, as demonstrated by activating in various DMSO concentrations (Fig. 8). Although DMSO can cause the release of Ca^{2+} from intracellular stores (Morley and Whitfield, 1993) exposure of MII oocytes to 0.05–2.00% DMSO in the absence of Sr^{2+} does not induce oocyte activation (Imahie *et al.*, 1995; T. Wakayama and A. Perry, unpublished observations). Thus, any effect of DMSO is probably not due to a qualitative role in oocyte activation. Instead, the mechanism of action may be linked to its morphogenic properties: DMSO promotes differentiation of EC cells into cardiac and skeletal muscle (McBurney *et al.*, 1982; Rudnicki *et al.*, 1990; Vidricaire *et al.*, 1994). DMSO may thus interfere with reprogramming and/or early embryonic cell proliferation following NT.

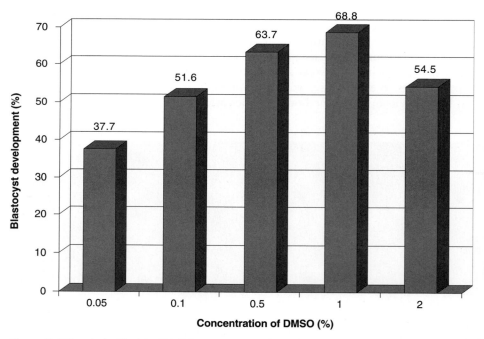

Figure 8 *Dimethylsulfoxide (DMSO) exerts a modest effect on embryo development* in vitro *following nuclear transfer. The proportion (%) of embryos developing to the blastocyst stage is shown following transfer of (diploid) cumulus cell nuclei into enucleated MII oocytes according to the method of Fig. 1D. Subsequent activation was in the presence of cytochalasin B solubilized in different concentrations (v/v) of DMSO (0.05–2.00%). Activation in the presence of 1% DMSO yielded development to the blastocyst stage at the highest rate (69%), 1.8 times that of the lowest rate (38%).*

THE INTERVAL BETWEEN NT AND ACTIVATION: HOW LONG IS LONG ENOUGH?

When reconstructed cells are exposed to an activating stimulus (Sr^{2+}) immediately after NT, the briefest interval is ~5 minutes (Wakayama and Yanagimachi, 2001b). Because Sr^{2+} induces activation of $B6D2F_1$ oocytes within ~25 minutes of exposure (T. Wakayama and A. Perry, unpublished observations), the donor nucleus is exposed to relatively high, but decreasing, MPF activity for ~5–30 minutes. Should a high MPF milieu be critical for reprogramming, this implies an upper limit on the time necessary, of ~30 minutes. For $B6D2F_1$ oocytes, longer exposure of up to ~6 hours prior to activation does not alter the cloning efficiency, with latencies of up to ~10 hours permitting full development to term (Wakayama *et al.*, 1998a; T. Wakayama and A. Perry, unpublished observations).

EMBRYO CULTURE CONDITIONS

The gross contribution of embryo culture conditions to the developmental outcome of NT can be gauged by transferring embryos at either the two-cell or morula/blastocyst stages (respectively, into oviducts or uteri). For $B6D2F_1$ embryos generated by intracytoplasmic sperm injection or *in vitro* fertilization, rates of development *in vitro* can be as good as those *in vivo* (Wakayama *et al.*, 1995, 1998b). Culture of NT embryos resulted in 87 and 53% development to the two-cell or morula/blastocyst stages, respectively (Table 6). Expressed as fractions of the number of starting embryos, similar numbers of two-cell embryos (2.3%) and morulae/blastocysts (2.8%) produced offspring (Table 6). Hence, although different culture conditions affect mouse embryonic gene expression (Doherty *et al.*, 2000), optimal culture of $B6D2F_1$ NT embryos *in vitro* (as opposed to growth *in vivo*) for the first 3–

Table 6 Nuclear Transfer Embryo Development[a]

Stage at embryo transfer	Cytokinesis inhibitor and solvent concentration[b]		No. of surviving oocytes after nucleus injection	No. of activated oocytes (%)	No. of embryos (%)		No. of transferred embryos (recipients)	No. of live pups	Pups (%) per	
	Agent	DMSO			Two cell	Morula/ blastocyst			Transferred embryo	Injected oocyte
Two cell	CB	1%	71	67 (94.4)	59 (84.3)	—	59 (3)	2	3.4	2.8
	CD	0.2%	74	74 (100)	62 (83.8)	—	62 (3)	1	1.6	1.4
	CD	1%	77	71 (92.2)	70 (92.1)	—	70 (4)	2	2.9	2.6
	Nocod	1%	79	72 (91.1)	68 (87.2)	—	68 (4)	2	3.0	2.5
Total			**301**	**284 (94.4)**	**259 (86.9)**	**—**	**259 (14)**	**7**	**2.7**	**2.3**
Morula/ blastocyst	CB	1%	77	73 (94.5)	—	47 (61.0)	47 (4)	3	6.4	3.9
	CD	0.2%	84	83 (98.8)	—	42 (50.0)	42 (4)	1	2.4	1.2
	CD	1%	78	77 (98.7)	—	46 (59.0)	46 (4)	2	4.4	2.6
	Nocod	1%	83	73 (88.0)	—	35 (42.2)	35 (4)	3	8.6	3.6
Total			**322**	**306 (95.0)**	**—**	**170 (52.9)**	**170 (16)**	**9**	**5.3**	**2.8**

[a]Embryos develop to term whether transferred to surrogate mothers at the two-cell stage or the morula/blastocyst stage. These data represent three independent replicates. In addition, success rates in contemporaneous experiments were not significantly altered by transferring two-cell embryos oviductally or morula/blastocysts into uteri (Wakayama *et al.*, 1995, 1997; Tanemura *et al.*, 1997; Wakayama and Yanagimachi 1998a,b,c).

[b]CB, Cytochalasin B at a final concentration of 5 µg/ml; CD, cytochalasin D at a final concentration of 1 µg/ml; Nocod, nocodazole at a final concentration of 3 µg/ml. The solvent used in each case was dimethylsulfoxide (DMSO) at the concentrations shown (v/v).

4 days of development therefore does not apparently depress the cloning efficiency *per se.*

METHODS OF NUCLEUS DELIVERY

Four methods have been reported for donor nucleus delivery in mouse cloning by NT: fusion with Sendai virus (Kato *et al.*, 1999), electrical fusion (Ogura *et al.*, 2000b), conventional microinjection (Zhou *et al.*, 2000), and piezo-actuated microinjection (Wakayama *et al.*, 1998a). The first mouse cloning from adult-derived cells utilized piezo-actuated microinjection (Wakayama *et al.*, 1998a), in which a piezoelectric pulse is harnessed to "fire" a microinjection needle tip first through the zona pellucida and then through the plasma membrane. Piezo-actuated microinjection permits rapid sample processing and is by far the most prevalent method of mouse cloning to date. In perhaps the most comparable studies using different nucleus insertion methods, donor nuclei from the ES cell line R1 were either fused with enucleated MII oocytes using Sendai virus (Amano *et al.*, 2001b) or microinjected with piezo actuation (Wakayama *et al.*, 1999). Efficiencies of cloned offspring production were 2/582 (0.3%) for the fusion protocol *vs.* 16/529 (3%) for piezo-actuated microinjection (Amano *et al.*, 2001b; Wakayama *et al.*, 1999). However, additional variables may have been responsible for this 10-fold difference. Indeed, data are in general too scarce to allow a comprehensive assessment of whether different cell/nucleus delivery methods significantly affect the outcome of cloning by NT.

CLONE DEVELOPMENT

Given that cloning by NT circumvents gametogenesis and places enormous demands on both the donor nucleus and recipient cytoplasm, it is perhaps remarkable that it ever succeeds in potentiating full development. However, in the vast majority of cases—usually more than 98% of them—it fails. This is the case for both of the main species studied to date, with pre- and perinatal fatality documented for cattle (Kato *et al.*, 1998; Wells *et al.*, 1999; Renard *et al.*, 1999; Kubota *et al.*, 2000; Hill *et al.*, 1999, 2000) and mice (Wakayama *et al.*, 1999, 2000a, 2001; Ogura *et al.*, 2000a,b; Ono *et al.*, 2001a; Amano *et al.*, 2001b). Yet these failures should be instructive. We now describe the stages at which mouse cloning fails and their associated pathologies.

PRENATAL DEVELOPMENT

Between 50 and 60% of embryos cloned by NT from mouse cumulus cells develop to the morula/blastocyst stages, with 50–70% of these initiating the implantation process following transfer to surrogate mothers (Wakayama *et al.*, 1998a). The efficiency of blastocyst development apparently varies according to the nucleus donor cell type (Table 2). It is therefore possible that different nucleus donor cell types exhibit different probabilities of generating embryos that fail at given points along the developmental pathway. These hypothetical points of failure ("developmental restriction points") are evidenced by comparing parallel NT experiments with cumulus cells and the ES cell line R1 (both derived from hybrid strains) as nucleus donors (Fig. 9) (Wakayama *et al.*, 1999).

The association of a given cell type with a given profile of restriction points needs to be verified in further experiments. Such an association would be one manifestation—hypothetical at this point—of the anticipated functional link between reprogramming and development. For example, ES cell-derived NT embryos at embryonic days 15–17 (E15–17) were either arrested in fetal development or were composed exclusively of placental tissue (Fig. 10) (Wakayama *et al.*, 1999; Amano *et al.*, 2001a). This is perhaps paradoxical given that ES cells make only a limited

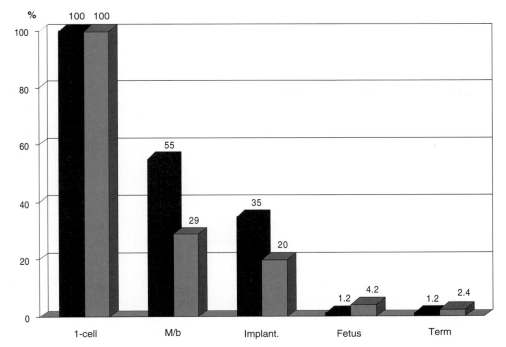

Figure 9 *The fate of cloned embryos at different times after nuclear transfer. The proportion of oocytes alive and activated following NT is set at 100%. Progress to the morula/blastocyst (M/b) stage is recorded after 3.5 days of culture in vitro. After transfer, cesarean section of pregnant surrogates at E19.5 reveals the number of implantation sites, developmentally arrested fetuses, and dead and living term pups. Fetuses are here grouped with pups that had developed to term. Black, cumulus cell nucleus as donor; gray, ES cell nucleus as donor.*

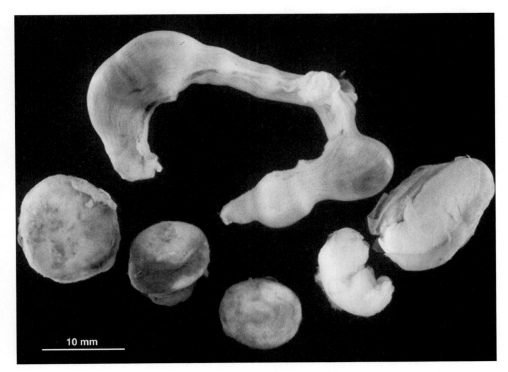

Figure 10 *Cloned fetuses, intrauterine cloned fetuses, and cloning-associated placentas. Nucleus donors were from the F_1 hybrid-derived ES cell line, R1. Dead fetuses and isolated placentas without an embryo proper are commonly observed at E19.5. Clonally derived placentas are enlarged relative to those of nonclones.*

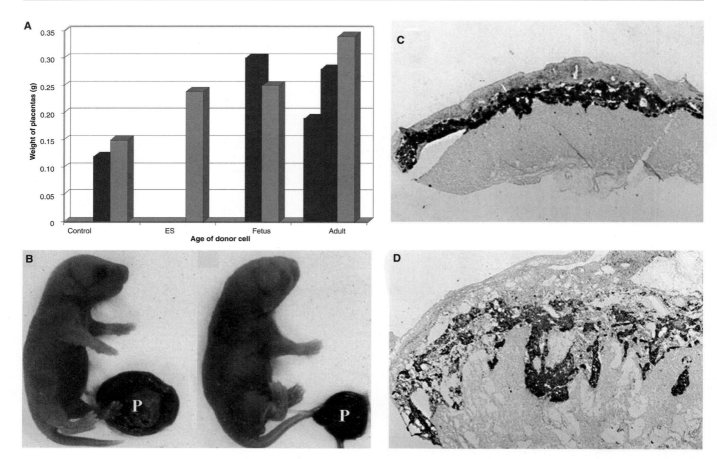

Figure 11 *Cloned fetuses and placentas at E19.5. (A) The relative weights of placentas associated with mice cloned from the nuclei of the ES cell line, R1, and from acutely isolated fetal and adult somatic cells. Black, female cells as donor; gray, male cell as donor. The 129 adult females are also denoted as black. (B) E19.5 fetus cloned from a cumulus nucleus donor cell (left) and fertilization-derived (control) fetus (right) at E19.5. The relatively large placenta (P) of the cloned fetus relative to that of the noncloned one is clearly apparent. In situ hybridization analysis of control (C) and clone-associated (D) placentas using antisense riboprobes corresponding to the trophoblast-specific gene, Tpbp. The prominent expansion of the spongiotrophoblast layer and perturbation of the spongiotrophoblast and labyrinth layers were evident in the cloned mouse placenta relative to controls.*

contribution (if any) to the placenta following classical fertilization-derived blastocyst injection (Nagy *et al.*, 1990), but suggests that ES cell nuclear reprogramming favors a gene expression pattern reminiscent of that of the trophectoderm in some NT embryos.

PLACENTAL ANOMALIES AT TERM

One striking and possibly diagnostic effect of mouse NT cloning is to generate offspring that have placentas enlarged ~two- to threefold relative to those of noncloned counterparts (Fig. 11) (Wakayama and Yanagimachi, 1999, 2001a; Ono *et al.*, 2001a; Tanaka *et al.*, 2001; Inoue *et al.*, 2002). The effect, as far as we can tell, is universal in the mouse, and is thus independent of the nucleus donor source.

Clone-associated placental enlargement has been characterized when somatic cells from F₁ hybrids were used as nucleus donors, and was found to be predominantly due to an exaggeration of the basal layer: spongiotrophoblasts, giant trophoblasts, and especially glycogen cells (Tanaka *et al.*, 2001; Inoue *et al.*, 2002). Moreover, placental zonation is apparently disrupted by the apparent invasion of spongiotrophoblasts and glycogen cells into the labyrinthine layer, in which the fetal capillaries become irregularly branched and dilated (C and D, Fig. 11). These abnormalities presumably reflect dysregulation of gene function in cells of the trophectodermal lineage, implicitly accounted for by aberrant reprogramming. Indeed, the steady-state levels for transcripts of a subset of placental genes are markedly reduced in NT clone-associated placentas (Tanaka *et al.*, 2001; Inoue *et al.*, 2002). This reduction includes mRNAs expressed from the paternal (*Peg1/Mest*) and maternal (*Meg1/Grb10* and *Meg3/Gt12*) alleles, and nonimprinted genes (*Igfbp2*, *Igfbp6*, *Vegfr2/Flk1*, and *Esx1*) (Inoue *et al.*, 2002). It seems that the reduction of imprinted gene expression is not universal, because transcript levels for both paternally and maternally imprinted genes (*H19* and *Igf2*, respectively) are not necessarily affected (Inoue *et al.*, 2002).

The placentas of transferred embryos occasionally developed in the absence of an associated fetus, a phenomenon not observed in the development of embryos generated in other ways, such as intracytoplasmic sperm injection (ICSI) (Fig. 10) (Wakayama *et al.*, 1999; Amano *et al.*, 2001a). Such placentas are also hypertrophic, with histological features similar to those of fetal placentas, although with generally more extensive disruption of the labyrinthine layer. The size of placentas associated with offspring produced by clone × clone crosses is within the normal range (T. Wakayama and A. Perry, unpublished observations). This suggests that the mechanism underlying the cloning-associated phenotype of placental enlargement is not genetic, but that it is a reversible epigenetic anomaly that is corrected as a consequence of germ line transmission. Similarly, cloning-associated obesity (discussed below) is not transmitted through the germ line of clones (Tamashiro *et al.*, 2002).

PERINATAL DEVELOPMENT AND ANOMALIES

Offspring (as opposed to their associated placentas) derived from somatic cell NT are reportedly essentially normal at birth (Wakayama *et al.*, 1998a; Inoue *et al.*, 2002). However, neonatal abnormalities have been reported in restricted cases for NT cloned mice derived from ES cells (Wakayama *et al.*, 1999; Eggan *et al.*, 2001). Anomalous development of the embryo proper at term is manifest as increased or decreased fetal weight (compared to controls) reported for some clones derived from ES cell NT (Eggan *et al.*, 2001; Amano *et al.*, 2001a). In mouse cloning, fetal overgrowth (~30%) is to date exclusively associated with NT from ES cells. The birth weights of neonates derived by somatic cell NT are either within the normal range or smaller than controls (Tanaka *et al.*, 2001), with no indication of large-offspring syndrome exhibited in some cases of cattle cloning (Young *et al.*, 1998). Cloned offspring derived by NT from ES cells have also been described with birth weights within the normal range (Wakayama *et al.*, 1999; Ono *et al.*, 2001b) or smaller (Amano *et al.*, 2001a). This is arguably a reflection of epigenetic instability exhibited by ES cells during culture (Humpherys *et al.*, 2001).

Moreover, clones derived by NT from ES cells exhibit an anomalously high rate (22% for F_1-derived and 100% for inbred-derived lines) of perinatal mortality, often due to respiratory catastrophe (Eggan *et al.*, 2001; Amano *et al.*, 2001a,b). This is in contrast to ~7% mortality of (155) clones derived by NT from a variety of hybrid-derived somatic cells (Inoue *et al.*, 2002). Moreover, these somatic-cell-derived neonates were, at low resolution, anatomically normal and they rapidly became active (Inoue *et al.*, 2002). In an earlier study (Tamashiro *et al.*, 2000), cloned neonates exhibited delays in eye opening and the onset of ear twitching and negative geotaxis.

POSTNATAL DEVELOPMENT

The behavioral parameters so far examined for adult cloned mice (including assessments of home cage activity, the Morris water maze test, the Krushinsky test, and motor tests) show no significant differences from control mice (Tamashiro et al., 2000). This tentative holistic indication that adult mice cloned from somatic cell nuclei are essentially physiologically normal is subject to major caveats.

First, the body mass of NT clones in some cases, and in a strain-dependent manner, becomes significantly greater than that of noncloned controls postpubertally, from approximately 8 to 10 weeks of age (A and B, Fig. 12). Higher body mass reflects increased fat and is not a consequence of low activity or increased dietary intake; corresponding behavioral assays did not show any significant difference between clones and controls (Tamashiro et al., 2000, 2002). Clones exhibited elevated levels of plasma leptin and insulin, but not corticosterone, relative to controls. One interpretation of these data is that the underlying metabolic determinant of obesity in cloned mice is independent of leptin and reflects a hitherto undescribed mechanism.

The clone-associated, epigenetically predisposed phenotype has variable penetrance: 77% of adult females cloned from B6C3F$_1$ nuclei became obese, compared to only 20% of those cloned from B6D2F$_1$ nuclei (Tamashiro et al., 2000; T. Wakayama and A. Perry, unpublished observations). Further, as with placental enlargement, the obese phenotype is not heritable and is absent from the progeny of clone × clone and clone × wild type crosses (Tamashiro et al., 2002). This implies that, like placental enlargement, the obesity of somatic cell-derived NT clones is caused by an imprinting and/or reprogramming anomaly that is corrected during gametogenesis.

A second set of consequences of mouse NT cloning from somatic cells can apparently lead to premature death (Ogonuki et al., 2002). In this study by Ogonuki et al., males cloned from immature Sertoli cells of genotype B6C3F$_1$ exhibited reduced antibody production (suggesting immune impairment) and elevated serum NH$_4^+$ and lactate dehydrogenase (suggesting hepatic dysfunction). The remaining serum parameters tested (total protein, albumin, alanine and aspartate aminotransferases, blood urea nitrogen, glucose, total cholesterol, triglyceride, calcium, alkaline phosphatase, creatinine, creatine phosphokinase, bilirubin, and amylase) were within normal (control) ranges. Premature death accounted for 10/12 of the clones studied (>80%) and necropsies performed on 6 of them corroborated the immune and hepatic phenotypes: all had contracted pneumonia, four suffered necrotic lesions of the liver, and two had tumors (leukaemia and lung cancer).

Collectively, these findings suggest that NT cloning can have subtle and specific sequelae that are often delayed in postnatal clones (Tamashiro et al., 2000, 2002; Ogonuki et al., 2002). Whereas mice cloned from B6C3F$_1$ cumulus cells become obese (Tamashiro et al., 2000, 2002), those cloned from immature Sertoli cells of the same strain do not (Ogonuki et al., 2002), suggestive of cell-type-specific cloning phenotypes.

DNA Methylation and Imprinting in Cloned Offspring

Putative links between methylation and imprinting (Jaenisch, 1997) and imprinting and development (Kaufman et al., 1977; McGrath and Solter, 1984b) have led to studies on methylation of imprinted and other genes in clones and noncloned controls. Pangenomic differences have been reported between cleavage patterns generated by the methylation-sensitive restriction endonuclease NotI in different tissues from NT embryos and controls (Ohgane et al., 2001). The differences suggest that a small subset (0.5% of the total) of genes are differentially methylated in placental and other tissues from clones and controls. These clones had been derived by NT from cumulus (i.e., adult-derived somatic) cell nuclei (Ohgane et al., 2001).

Analysis of selected imprinted genes in offspring derived by NT from ES cells found either no anomalous methylation in clones, or marked methylation anom-

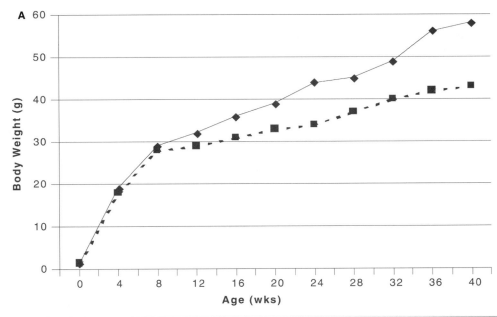

Figure 12 Preliminary analysis of the body weights of B6C3F₁ mice cloned by nuclear transfer. (A) Growth of cloned and control, noncloned mice. Postpubertal growth of some cloned animals is greater than that of their control counterparts. Clone body weight showed greater variance than that of controls; penetrance of the obese phenotype varies within and between strains. ◆, Cloned mice; ■, noncloned control mice. (B) Cloned (left) and noncloned (right), control mice at 1 year, weighing 77.95 and 42.90g, respectively. Both mice are of strain B6C3F₁ produced following in vitro manipulation and single-offspring pregnancies.

alies that were not predictive of developmental outcome (Humpherys *et al.*, 2001). The mammalian developmental program is therefore not exquisitely sensitive to the methylation status or expression levels of all imprinted genes. However, as noted above, the value of ES cell NT in laying the foundation for general conclusions concerning NT biology is questionable. It is perhaps more significant that clones produced by transferring the nuclei of somatic cells did not exhibit a marked departure from controls in steady-state transcript levels either for maternally expressed (*Igf2r*, *H19*, *Meg1/Grb10*, *Meg3/Gtl2*, and *p57^{Kip2}*) or paternally expressed (*Igf2* and *Peg1*) imprinted genes (Inoue *et al.*, 2002). This preservation of imprint memory by somatic cells and following NT cloning hints that the relevance of imprinting to postimplantation clone development may not be great.

REITERATIVE CLONING AND TELOMERE LENGTH

The advent of mouse cloning has facilitated an answer to the question "Can mammalian clones be cloned?" The answer is "Yes"—to at least six generations (Wakayama *et al.*, 1998a, 2000a). This was shown in an experiment in which two clonal female mouse lines were established following cumulus cell NT for successive generations. In both lines, the overall trend of the cloning efficiency was downward (Fig. 13). The numbers in these experiments are small, and it may be that the sixth generation (G6) does not represent the upper limit in reiterative cloning; in fact, the sole G6 pup was cannibalized, precluding an assessment of whether it too was cloning competent. Without having established the point at which it becomes zero, it seems that cloning competence in the mouse decreases along successive clonal generations.

Reiterative cloning generates animals containing cells that have undergone many population doublings without meiotic intervention. These cells are thus not subject to the reparative and screening processes of gametogenesis (Kozik *et al.*, 1998; Surani, 1998; Baarends *et al.*, 2001), and are presumably subject to cumulative aging effects. These might include genetic and epigenetic lesions. At an organismal level, there is no evidence of age-related attrition of learning ability (as judged by the Morris water maze and Krushinsky tests), or of strength, agility, coordina-

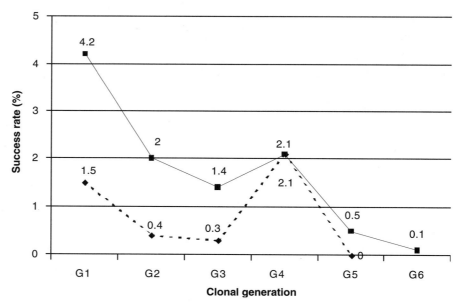

Figure 13 Full-term development after successive rounds of reiterative cloning. Clonal generations arising from two founder mice, A and B, are represented as G1, G2, and so on. Nuclei were taken from acutely isolated cumulus cells according to the method outlined in Fig. 1D. Percentages represent the fraction of one-cell embryos following NT that developed to term.

tion or home cage activity compared to noncloned controls (Tamashiro *et al.*, 2000). Mice cloned from Sertoli cell nuclei die prematurely (Ogonuki *et al.*, 2002) but there is no description of parameters indicative of senescence (lordokyphosis, precocious reduction in body mass, delayed wound healing, impaired hair regrowth, reduced dermal thickness, etc.). It is thus as yet unclear whether mice cloned by somatic cell NT age at rates within the normal range, although clearly a subset will live for at least as long as controls (Ogonuki *et al.*, 2002); Cumulina died aged 2 years and 7 months.

At the cellular level, telomere length in peripheral blood lymphocytes of clones did not reveal evidence of shortened telomeres in the cloned mice (Wakayama *et al.*, 2000a). On the contrary, telomeres may exhibit a modest lengthening with each successive generation of clones. Because the mice in this study were sampled simultaneously, an age-related contribution to this increase—in which younger mice have longer telomeres—cannot be excluded (Fig. 14) (Wakayama *et al.*, 2000a). However, these studies on the mouse suggest that cloned animals have telomeres that are at least as long as those of noncloned counterparts. This concords with work in cattle cloned from fetal cells cultured to near senescence; in these clones telomeres are longer than those of age-matched controls (Lanza *et al.*, 2000). Cloned cattle have also been reported to possess telomeres of the same length as those of age-matched controls (Tian *et al.*, 2000; Betts *et al.*, 2001). The general mechanism (if one exists) by which telomeres are preserved and possibly extended in clones remains to be elucidated; cloning efficiency conceivably reflects a selection for embryos derived from nucleus donor cells containing the longest telomeres.

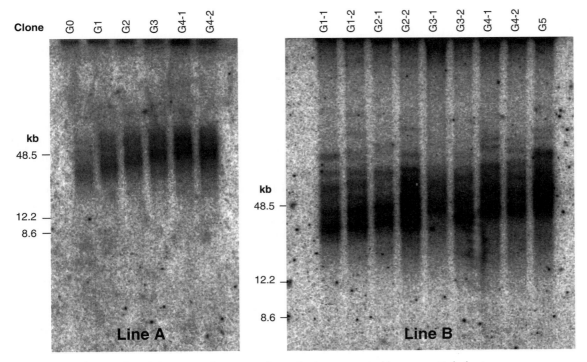

Figure 14 *Telomere length in successive generations of mice generated by sequential cloning from cumulus cells in two lines (A and B). Southern blot analysis of terminal restriction fragments using a telomere-specific probe indicates that telomeres do not exhibit incremental erosion in successive clonal generations. Samples were obtained on the same day, with the age of each mouse (in months) as follows: Line A, G1, 18; G2, 16; G3, 14; G4, 12. Line B, G1, 15.5; G2, 13; G3, 11; G4, 9; G5, 7.*

QUESTIONS AND APPLICATIONS

Many of the unresolved issues in mammalian cloning are wrapped up in the single question: "Why is cloning by nuclear transfer so inefficient?" When hybrid mouse strains provide the somatic donor cell nucleus, an efficiency of ~1–2% (i.e., a failure rate of ~98–99%) is today typical. Because mice have been cloned from a wide variety of cell types (including Sertoli, cumulus, and ES cells) the range of cloning competent cell types is likely to be broad. Studies in the mouse have also put beyond reasonable doubt the lack of a strict correlation between nucleus donor cell cycle stage and cloning competence. This needs rigorous demonstration using cell cycle markers available in the mouse [for example, short-lived enhanced green fluorescent proteins (EGFPs) driven by cell cycle-regulated promoters such as those of cyclins]. Collectively, commonalities in cloning phenomena—the most salient of which is the conserved efficiency—suggest either that the recipient cytoplast is the dominant determinant of cloning efficiency, or that a subset of highly conserved cellular traits shared by nucleus donors is critical for cloning to succeed. Studies are therefore likely to focus on the role of the recipient cytoplasm and its contribution to the interactions that reprogram incoming donor nucleus chromatin to potentiate embryonic gene expression.

HOW IS REPROGRAMMING ACHIEVED?

Elucidating the molecular biology of reprogramming following NT in mammals is confounded by the scarcity of material. Much is at present inferred from work in other systems. The level of methylation at CpG islands correlates inversely with transcriptional activity, and positively with histone hypoacetylation (Issa *et al.*, 1997; Eden *et al.*, 1998). Indeed, evidence suggests that transcriptional repression is directly mediated by histone deacetylases (HDACs) in complex with methyl-CpG-binding domain proteins (reviewed by Rountree *et al.*, 2001). Furthermore, HDACs interact with DNA methyltransferases (DNMTs), which also induce transcriptional repression. For example, repression in a human cell line, HeLa, is mediated via a complex of DNMT1 and HDAC1 (Robertson *et al.*, 2000) and *Dnmt3a* silences transcription in association with HDAC1 (Fuks *et al.*, 2001). In this context, it is noteworthy that bovine oocytes and early embryos contain HDACs 1, 2, and 3 (Segev *et al.*, 2001). In the yeast *Saccharomyces cerevisiae*, the NAD-dependent histone deacetylase Sir2p mediates gene silencing (Guarente, 1999) and calorie-restricted life-span extension (Lin *et al.*, 2000). The mammalian ortholog of Sir2p (mSir2α) could exert a beneficial effect in cloning by eliciting global silencing of gene expression either prior to or soon after NT (Guarente, 1999; Imai *et al.*, 2000); such silencing would be beneficial if it potentiated reprogramming. Evidence supporting this possibility comes from work on *Xenopus* oocyte extracts, which contain a nucleosomal ATPase activity, ISWI, which actively removes TATA-binding protein and possibly the general transcription factor, TFIIB, from somatic cell nuclei to which it is exposed (Kikyo *et al.*, 2000). A mouse functional homolog of ISWI would also be a strong candidate in the search for molecules that remodel chromatin *en route* to reprogramming in mammalian NT. Although the bulk level of core histones in incoming nuclei remains largely unaltered, chromatin–cytoplasm exchange following *Xenopus* NT involves replacement of histone H1 with the oocyte linker-type histone, B4 (Dimitrov and Wolffe, 1996) and the loss of nucleolin (Kikyo *et al.*, 2000). Comprehensively cataloging the factors exchanged on NT between the incoming nucleus and its new cellular environment is a major challenge in mouse cloning, and one that is likely to reveal that the dynamics of successful reprogramming are subtle.

It is possible that NT protocols ensure reprogramming of only a subset of genes—in successful cases, the smallest subset sufficient to guarantee full development. If reprogramming of, say, six master genes is necessary for full development, with a 50% chance of reprogramming in each case, the overall probability that all six undergo the change is $\frac{1}{2}^6 \times 100 = 1.6\%$. This likely oversimplification has the attraction of producing a value that agrees closely with the observed efficiency of ~1–2%. Were such genes to exist, their identification would allow rescue of their activities even when NT did not result in the appropriate chromatin remodeling. If possible, this could pave the way for artificially induced reprogramming not only in oocytes, but using either cytoplasts derived from any cell, or by inducing reprogramming in somatic cells *in situ*.

IMPRINTING

These speculations would need to accommodate the possible contribution of imprinting in determining developmental outcome following NT. The role of genes whose parent of origin prescribes their function (imprinted genes) is described in detail in Chapter 5. The previously posited necessity for balanced imprinted gene expression (Moore and Haig, 1991), coupled with the gradual erasure of imprint marks in cultured cells (Dean *et al.*, 1998), argue that what is most remarkable about cloning is that it succeeds at all, albeit with a low efficiency. However, embryos generated by NT of ES cell nuclei are apparently able to buffer imbalances in imprinted gene expression during development (Humpherys *et al.*, 2001). Neither the extent nor the generality of this tolerance is known. Cloning by NT will allow us to address this issue and to interfere experimentally with imprint marks to evaluate their contribution to different stages of development after fertilization and cloning.

The reported segregation of maternal and paternal genomes in around 5% of adult somatic cells (Mayer *et al.*, 2000) suggests the possibility that these cells constitute the subpopulation that is cloning competent. Moreover, DNA methylation (a candidate for imprint marks) of the paternal genome is reduced on fertilization prior to the onset of S phase (Oswald *et al.*, 2000). It will be interesting to learn whether a similar phenomenon occurs in the DNA of donor nuclei in cloning; determination of the methylation status of the genomes in NT embryos whose developmental potential is known will provide further insights into the role of methylation in embryogenesis. In the mouse, it is formally possible to interfere with DNA methylation experimentally (either genetically or chemically) prior to or during embryonic development to test putative links between methylation and development in NT embryos.

APPLICATIONS TO GENOMICS AND THERAPEUTIC CLONING

Mouse cloning provides the preeminent paradigm for the development of applications of relevance to other species. Completion of the mouse genome project (see http://www.ncbi.nlm.nih.gov/genome/seq/MmHome.html) will enhance the role of the mouse in genome-wide pharmacological screening (pharmacogenomics). This will benefit from cloning, first by reducing uncontrolled variation between mice and second by allowing the propagation of subfertile or sterile strains.

A more direct application of mouse cloning to biomedicine is exemplified by the union of NT and stem cell biology, the ultimate goal of which is therapeutic cloning (Munsie *et al.*, 2000; Kawase *et al.*, 2000; Wakayama *et al.*, 2001). Adult-derived somatic cell NT can be used to produce cloned embryos from which ES cell lines can be derived (Fig. 15; Table 7). These cells, called ntES cells, exhibit full pluripotency in that they can be caused to differentiate along prescribed pathways *in vitro* (to produce, for example, dopaminergic neurons), contribute to the germ line following injection into blastocysts, and support full development following nuclear transfer (Wakayama *et al.*, 2001). Reports of human ES cell-like cell lines

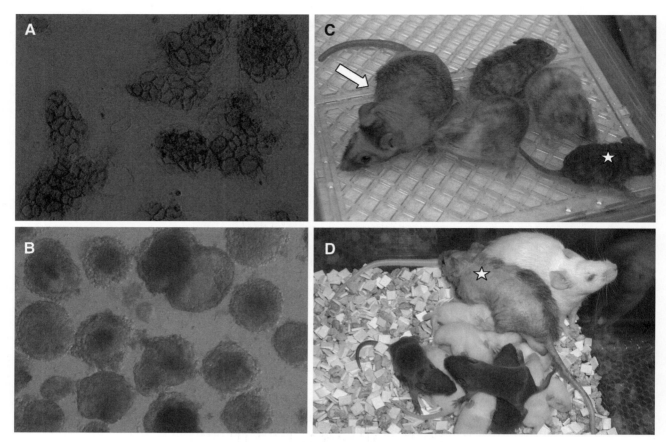

Figure 15 *Characterization of nuclear transfer ES (ntES) cells in vitro. (A) Staining of near-confluent cultures for a marker of nondifferentiated state, alkaline phosphatase, reveals islands of undifferentiated ntES cells. (B) Embryoid bodies are readily formed by ntES cells in vitro. (C) Contribution in chimeric offspring of ntES cells derived from a male (arrow) of the nude strain, C57BL/6^{nu/nu}. Chimeras were generated by injecting the ntES cells into ICR × ICR fertilization-derived blastocysts; the ntES cell contribution in offspring is manifest as a dark coat color, reflecting the C57BL/6^{nu/nu} genotype. (D) One male (indicated with an asterisk in C and D) was crossed at 8 weeks with an albino (ICR) female, producing a litter containing three offspring with uniformly dark coats, confirming the contribution of C57BL/6^{nu/nu} to the germ line of their chimeric father.*

(Thomson *et al.*, 1998; Shamblott *et al.*, 1998) coupled to this work on the mouse raises the hope that ntES cells will provide a source of differentiated cells for human autologous transplant therapy; therapeutic cloning (Gurdon and Colman, 1999).

Furthermore, because ES cells support homologous recombination at a relatively high efficiency, genetic alterations might be introduced in ntES cells by gene targeting or transgenic complementation before they are used to establish germ line chimeras or in cloning. This should facilitate the establishment of germ cells, individuals, and cell lines containing site-directed genomic alterations.

CONCLUDING REMARKS

The advent of mouse cloning from adult-derived cells in July 1998 marked a new departure in the study of key biological problems in NT biology (Table 8). Although tractable in the mouse, many of these experiments are difficult in other species. Since 1998, some 40,000 experimental NT mouse embryos have been reported, a value that massively exceeds what is practicable in sheep, goats, cattle, and pigs within a similar time frame. Yet mouse cloning by NT remains nontrivial. Although it is

Table 7 *Establishment of ntES Cell Lines*[a]

Nucleus donor			Establishment of ntES cell via nuclear transfer			In vivo differentiation after ntES cell injection into blastocysts[c]			
Strain	Sex	Tissue	No. of reconstructed oocytes	Blastocyst development (%)	Established ntES cell line (%)[b]	Per cent normal karyotype[d]	No. of injected blastocysts	No. of chimeras/ offspring	No. of germ line transmitting cell lines (black/pups)[e]
B6D2F1	F	Cumulus	130	57 (43.8)	9 (15.8) [6.9]	67.8 ± 14.1 (6)	129	39/102	1 (5/196)
129/Sv	F	Cumulus	44	13 (29.5)	1 (10.0) [2.3]	51.8 (1)	25	2/15	1 (1/72)
129/Sv	M	Tail tip	88	42 (47.7)	1 (2.4) [1.1]	66.2 (1)	24	17/20	1 (2/127)
129F1	M	Tail tip	182	54 (29.7)	7 (13.0) [3.8]	50.5 ± 16.7 (4)	49	16/25	1 (3/100)
C57BL/6[nu/nu]	F	Tail tip	159	75 (47.2)	5 (6.7) [3.1]	25.8, 31.3 (2)	24	4/22	0
C57BL/6[nu/nu]	M	Tail tip	210	88 (41.9)	4 (4.5) [1.9]	46.1 ± 33.0 (3)	44	16/25	2 (10/119)
EGFP Tg	F	Cumulus	118	50 (42.4)	7 (14.0) [5.9]	10.3 (1)	14	3/13	—
EGFP Tg	M	Tail tip	85	19 (22.4)	1 (5.3) [1.2]	68.8 (1)	39	8/15	1 (8/31)
Total (%) [%]			1016	398 (39.2)	35 (8.8) [3.4]	48.8 ± 20.4 (19)	355	105/237	7 (24/645)

[a]Following nuclear transfer from adult-derived cumulus or tail-tip cells and examination of pluripotency following ntES cell injection into fertilization-derived blastocysts.

[b]Expressed as % of blastocysts () and of reconstructed oocytes [].

[c]Data refer to karyotyped ntES cell lines only.

[d]More than 50 metaphase cells were examined for each ntES cell line. The number of ntES cell lines examined is shown in parentheses.

[e]Data are shown for ntES cell lines that exhibited germ line transmission in chimeras. Data from nontransmitting chimeras have been omitted.

Table 8 Mouse Cloning Reports since July 1998

Year	Nucleus donor			Development	Principal findings[a]	Reference
	ES	Fetus	Adult			
1998			√	Term	**First mouse cloned from adult-derived cells; first vertebrate cloned from a clone; first cloning by piezo-actuated microinjection into an MII ooplast**	Wakayama *et al.* (1998a,b)
1999			√	Late fetus	Serial nuclear transfer; temperature treatment of donor cell	Kato *et al.* (1999)
			√	Term	Comparison of oocyte activation methods. Ethanol, electric current, Sr^{2+}, and sperm gave identical cloning success rates	Kishikawa *et al.* (1999)
			√	Term	**First cloned male mammal; nucleus donor cell culture serum starvation had no effect on cloning efficiency; placental enlargement in cloned mice**	Wakayama and Yanagimachi (1999)
	√			Term	**First cloning by NT with ES cell nuclei; first live offspring cloned from targeted ES cell; clones derived from G1- and G2/M-phase nucleus donors; first indication that F_1 hybrids support more efficient cloning than inbred strains**; analysis of restriction points in clonal development	Wakayama *et al.* (1999)
2000	√		√	E12 fetuses	**Analysis of X-chromosome inactivation; first to show that epigenetic marks can be removed and reestablished during cloning**	Eggan *et al.* (2000)
			√	E14 fetuses	Mouse cloning by nuclear microinjection and fusion	Hosaka *et al.* (2000)
		√		Culture	ES cell-like cells are derived from fetus neuronal cells following nuclear transfer	Kawase *et al.* (2000)
			√	Culture	**First derivation of ES cell-like cells derived from somatic cells following nuclear transfer, from a single cloned blastocyst**	Munsie *et al.* (2000)
			√	Term	**First cloning from Sertoli cells and frozen nucleus donor cells; use of donor cells containing a stably integrated DNA construct**	Ogura *et al.* (2000a,b)
			√	Term	Mouse cloning by nuclear microinjection and fusion occurs at similar efficiencies	Ogura *et al.* (2000a,b)
	√			Term	Offspring cloned from targeted ES cells; corroboration of the notion that hybrid-derived nuclei support cloning more efficiently than those derived from inbred strains	Rideout *et al.* (2000)
	√			E14 fetuses	Cloning by NT with ES cell	Sato *et al.* (2000)
			√	Term	**The behavior of cloned mice is apparently normal; greater body weight of some clones compared to controls**	Tamashiro *et al.* (2000)
			√	Term	Cloned mice from cloned mice; **telomere length does not markedly decrease in successive clonal generations**; behavior is indistinguishable from that of controls	Wakayama *et al.* (2000a)
			√	E3.5 blast	**Donor egg cell cycle is a critical determinant of cloning, with nucleus insertion followed by activation being the most efficient**	Wakayama *et al.* (2000b)
			√	Term	**First clone produced by conventional injection of the donor cell nucleus**	Zhou *et al.* (2000)

(continues)

Table 8 *(continued)*

Year	ES	Fetus	Adult	Development	Principal findings[a]	Reference
2001	√			Term	Cloning by NT with ES cell; most cloned mice dead soon after birth	Amano *et al.* (2001a)
	√			Term	Cloning by NT with ES cell using fusion method	Amano *et al.* (2001b)
	√			Term	Cloning by NT with ES cell nuclei; F$_1$ hybrids support more efficient cloning than inbred strains; clones have large placentas	Eggan *et al.* (2001)
	√			Term	Cloning by NT with ES cell nuclei; epigenetic instability and lack of correlation between methylation of imprinted genes and developmental outcome	Humphreys *et al.* (2001)
			√	Term	**First analysis of DNA methylation in cloned mammalian offspring**	Ohgane *et al.* (2001)
		√		Term	**First offspring generated from nucleus donors acutely derived from fetuses**; serial NT; observation of cloned large placentas	Ono *et al.* (2001a,b)
		√	√	Term	F$_1$ hybrids support more efficient cloning than inbred strains; "inbred" 129/Sv gives a relatively high success rate; **cloning efficiency inversely proportional to developmental stage of the donor (adult < fetus < ES)**	Wakayama and Yanagimachi (2001a)
			√	Term	Cytokinesis inhibitor and timing of embryo transfer has no effect on cloning efficiency; DMSO can affect clonal development; success with immediate activation following NT	Wakayama and Yanagimachi (2001b)
			√	Culture and term	**First comprehensive demonstration of ES cell lines from somatic cell by NT (ntES). ntES cells differentiate into gametes *in vivo* and dopaminergic neurons *in vitro/vivo*; mice cloned from ntES cell lines by NT**	Wakayama *et al.* (2001)
			√	E12 fetuses	Spermin, protamine, and putrescine in nucleus donor cell culture has no effect on cloning efficiency	Yabuuchi *et al.* (2001)
	√			Term	Cloning by NT with ES cell	Zhou *et al.* (2001)
	√			Term	Cloning by NT with ES cell	Ono *et al.* (2001a,b)
		√		Term	Cloning by NT with fetus neuronal cells	Yamazaki *et al.* (2001)
			√	Term	Detailed observation of cloned large placentas and gene expression	Tanaka *et al.* (2001)
2002			√	Term	**Faithful expression of imprinted genes in cloned mice**	Inoue *et al.* (2002)
			√	Term	**Early death of mice cloned from somatic cells**	Ogonuki *et al.* (2002)
			√	Term	**Cloned mice have an abese phenotype that is not transmitted to their offspring**	Tamashiro *et al.* (2002)
			√	E3.5 blast	ES cell cells are derived from B and T cells following nuclear transfer	Hochedlinger and Jaenisch (2002)

The table header spans: Year | Nucleus donor (ES, Fetus, Adult) | Development | Principal findings | Reference

[a]Major "firsts" are in boldface type.

available today to relatively few, the immense power of the technique should ensure that it will be practiced by many more as its pioneers learn how to relay their expertise. With this dissemination will come a comprehensive description of cloning-associated anomalies and a picture of their underlying mechanisms. Such a mechanistic understanding will in turn promise new horizons in the safe and efficient modulation of mammalian genomes *via* nuclear transfer.

REFERENCES

Amano, T., Kato, Y., and Tsunoda, Y. (2001a). Full-term development of enucleated mouse oocytes fused with embryonic stem cells from different cell lines. *Reproduction* 121, 729–733.

Amano, T., Tani, T., Kato, Y., and Tsunoda, Y. (2001b). Mouse cloned from embryonic stem (ES) cells synchronized in metaphase with nocodazole. *J. Exp. Zool.* 289, 139–145.

Baarends, W. M., van Der Laan, R., and Grootegoed, J. A. (2001). DNA repair mechanisms and gametogenesis. *Reproduction* 121, 31–39.

Baguisi, A., and Overström, E. W. (2000). Induced enucleation in nuclear transfer procedures to produce cloned animals. *Theriogenology* 53, 209.

Baguisi, A., Behboodi, E., Melican, D. T., Pollock, J. S., Destrempes, M. M., Cammuso, C., Williams, J. L., Nims, S. D., Porter, C. A., Midura, P., Palacios, M. J., Ayres, S. L., Denniston, R. S., Hayes, M. L., Ziomek, C. A., Meade, H. M., Godke, R. A., Gavin, W. G., Overström, E. W., and Echelard, Y. (1999). Production of goats by somatic cell nuclear transfer. *Nat. Biotechnol.* 17, 456–461.

Betts, D. H., Bordignon, V., Hill, J. R., Winger, Q., Westhusin, M. E., Smith, L. C., and King, W. A. (2001). Reprogramming of telomerase activity and rebuilding of telomere length in cloned cattle. *Proc. Natl. Acad. Sci. U.S.A.* 98, 1077–1082.

Bos-Mikich, A., Whittingham, D. G., and Kones, K. T. (1997). Meiotic and mitotic Ca^{2+} oscillations affect cell composition in resulting blastocysts. *Dev. Biol.* 182, 172–179.

Brind, S., Swann, K., and Carroll, J. (2000). Inositol 1,4,5-trisphosphate receptors are downregulated in mouse oocytes in response to sperm or adenophostin A but not to increases in intracellular Ca^{2+} or egg activation. *Dev. Biol.* 223, 251–265.

Campbell, K. H. S., Loi, P., Cappai, P., and Wilmut, I. (1994). Improved development to blastocyst of ovine nuclear transfer embryos reconstructed during the presumptive S-phase of enucleated activated oocytes. *Biol. Reprod.* 50, 1385–1393.

Campbell, K. H. S., Loi, P., Otaegui, P. J., and Wilmut, I. (1996a). Cell cycle co-ordination in embryo cloning by nuclear transfer. *Rev. Reprod.* 1, 40–45.

Campbell, K. H. S., McWhir, J., Ritchie, W. A., and Wilmut, I. (1996b). Sheep cloned by nuclear transfer from a cultured cell line. *Nature* 380, 64–66.

Cheong, H. T., Takahashi, Y., Kanagawa, H. (1993). Birth of mice after transplantation of early cell-cycle-stage embryonic nuclei into enucleated oocytes. *Biol. Reprod.* 48, 958–963.

Cheong, H. T., Takahashi, Y., and Kanagawa, H. (1994). Relationship between nuclear remodeling and subsequent development of mouse embryonic nuclei transferred to enucleated oocytes. *Mol. Reprod. Dev.* 37, 138–145.

Chun, J., and Schatz, D. G. (1999). Rearranging views on neurogenesis: neuronal death in the absence of DNA end-joining proteins. *Neuron* 22, 7–10.

Cibelli, J. B., Stice, S. L., Golueke, P. J., Kane, J. J., Jerry, J., Blackwell, C., Ponce de Leon, F. A., and Robl, J. M. (1998). Cloned transgenic calves produced from nonquiescent fetal fibroblasts. *Science* 280, 1256–1258.

Collas, P., and Robl, J. M. (1990). Factors affecting the efficiency of nuclear transplantation in the rabbit embryo. *Biol. Reprod.* 43, 877–884.

Collas, P., and Robl, J. M. (1991). Relationship between nuclear remodeling and development in nuclear transplant rabbit embryos. *Biol. Reprod.* 45, 455–465.

Collas, P., Pinto-Correia, C., Ponce de Leon, F. A., and Robl, J. M. (1992). Effect of donor cell cycle stage on chromatin and spindle morphology in nuclear transplant rabbit embryos. *Biol. Reprod.* 46, 501–511.

Cuthbertson, K. S., Whittingham, D. G., and Cobbold, P. H. (1981). Free Ca^{2+} increases in exponential phases during mouse oocyte activation. *Nature* 294, 754–757.

Czolowska, R., Modlinski, J. A., and Tarkowski, A. K. (1984). Behaviour of thymocyte nuclei in non-activated and activated mouse oocytes. *J. Cell Sci.* 69, 19–34.

Dean, W., Bowden, L., Aitchison, A., Klose, J., Moore, T., Meneses, J. J., Reik, W., and Feil, R. (1998). Altered imprinted gene methylation and expression in completely ES cell-derived mouse fetuses: association with aberrant phenotypes. *Development* 125, 2273–2282.

Dimitrov, S., and Wolffe, A. P. (1996). Remodeling somatic nuclei in *Xenopus laevis* egg extracts: Molecular mechanisms for the selective release of histones H1 and H1(0) from chromatin and the acquisition of transcriptional competence. *EMBO J.* 15, 5897–5906.

Doherty, A. S., Mann, M. R., Tremblay, K. D., Bartolomei, M. S., and Schultz, R. M. (2000). Differential effects of culture on imprinted H19 expression in the preimplantation mouse embryo. *Biol. Reprod.* 62, 1526–1535.

Eden, S., Hashimshony, T., Keshet, I., Cedar, H., and Thorne, A. W. (1998). DNA methylation models histone acetylation. *Nature* 394, 842.

Eggan, K., Akutsu, H., Hochedlinger, K., Rideout, W. 3rd., Yanagimachi, R., and Jaenisch, R. (2000). X-Chromosome inactivation in cloned mouse embryos. *Science* **290**, 1578–1581.

Eggan, K., Akutsu, H., Loring, J., Jackson-Grusby, L., Klemm, M., Rideout, W. M. 3rd, Yanagimachi, R., and Jaenisch, R. (2001). Hybrid vigor, fetal overgrowth, and viability of mice derived by nuclear cloning and tetraploid embryo complementation. *Proc. Natl. Acad. Sci. U.S.A.* **98**, 6209–6214.

Epstein, C. J. (1986). "The Consequences of Chromosome Imbalance." Cambridge University Press, Cambridge.

Flach, G., Johnson, M. H., Braude, P. R., Taylor, R. A. S., and Bolton, V. N. (1982). The transition from maternal to embryonic control in the 2-cell mouse embryo. *EMBO J.* **1**, 681–686.

Fuks, F., Burgers, W. A., Godin, N., Kasai, M., and Kouzarides, T. (2001). Dnmt3a binds deacetylases and is recruited by a sequence-specific repressor to silence transcription. *EMBO J.* **20**, 2536–2544.

Fulka, J., Jr., and Moor, R. M. (1993). Noninvasive chemical enucleation of mouse oocytes. *Mol. Reprod. Dev.* **34**, 427–430.

Galli, C., Duchi, R., Moor, R. M., and Lazzari, G. (1999). Mammalian leukocytes contain all the genetic information necessary for the development of a new individual. *Cloning* **1**, 160–175.

Guarente, L. (1999). Diverse and dynamic functions of the Sir silencing complex. *Nat. Genet.* **23**, 281–285.

Gurdon, J. B., and Colman, A. (1999). The future of cloning. *Nature* **402**, 743–746.

Hill, J. R., Roussel, A. J., Cibelli, J. B., Edwards, J. F., Hooper, N. L., Miller, M. W., Thompson, J. A., Looney, C. R., Westhusin, M. E., Robl, J. M., and Stice, S. L. (1999). Clinical and pathologic features of cloned transgenic calves and fetuses (13 case studies). *Theriogenology* **51**, 1451–1465.

Hill, J. R., Burghardt, R. C., Jones, K., Long, C. R., Looney, C. R., Shin, T., Spencer, T. E., Thompson, J. A., Winger, Q. A., and Westhusin. M. E. (2000). Evidence for placental abnormality as the major cause of mortality in first-trimester somatic cell cloned bovine fetuses. *Biol. Reprod.* **63**, 1787–1794.

Hochedlinger, K., and Jaenisch, R. (2002). Monoclonal mice generated by nuclear transfer from mature B and T donor cells. *Nature* **415**.

Howlett, S. K., Barton, S. C., and Surani, M. A. (1987). Nuclear cytoplasmic interactions following nuclear transplantation in mouse embryos. *Development* **101**, 915–923.

Hozumi, N., and Tonegawa, S. (1976). Evidence for somatic rearrangement of immunoglobulin genes coding for variable and constant regions. *Proc. Natl. Acad. Sci. U.S.A.* **73**, 3628–3632.

Humpherys, D., Eggan, K., Akutsu, H., Hochedlinger, K., Rideout III, W. M., Biniszkiewicz, D., Yanagimachi, R., and Jaenisch, R. (2001). Epigenetic instability in ES cells and cloned mice. *Science* **293**, 95–97.

Illmensee, K., and Hoppe, P. C. (1981). Nuclear transplantation in *Mus musculus*: Developmental potential of nuclei from preimplantation embryos. *Cell* **23**, 9–18.

Imahie, H., Sato, E., and Toyoda, Y. (1995). Parthenogenetic activation induced by progesterone in cultured mouse oocytes. *J. Reprod. Dev.* **41**, 7–14.

Imai, S., Armstrong, C. M., Kaeberlein, M., and Guarente, L. (2000). Transcriptional silencing and longevity protein Sir2 is an NAD-dependent histone deacetylase. *Nature* **403**, 795–800.

Inoue, K., Kohda, T., Lee, J., Ogonuki, N., Mochida, K., Noguchi, Y., Tanemura, K., Kaneko-Ishino, T., Ishino, F., and Ogura, A. (2002). Faithful expression of imprinted genes in cloned mice. *Science* **295**, 297.

Issa, J.-P., Baylin, S. B., and Herman, J. G. (1997). DNA methylation changes in hematologic malignancies: Biologic and clinical implications. *Leukemia* **11**, 7–11.

Jaenisch, R. (1997). *Trends Genet.* **13**, 323.

Jellerette, T., He, C. L., We, H., Parys, J. B., and Fissore, R. A. (2000). Down-regulation of the inositol 1,4,5-trisphosphate receptor in mouse eggs following fertilization or parthenogenetic activation. *Dev. Biol.* **223**, 238–250.

Johnson, R. T., and Rao, P. N. (1970). Mammalian cell fusion: Induction of premature chromosome condensation in interphase nuclei. *Nature* **226**, 717–722.

Johnson, M. H., Pickering, S. J., Dhiman, A., Radcliffe, G. S., and Maro B. (1988). Cytocortical organization during natural and prolonged mitosis of mouse 8-cell blastomeres. *Development* **102**, 143–158.

Kato, Y., and Tsunoda, Y. (1992). Nuclear transplantation of mouse fetal germ cells into enucleated two-cell embryos. *Theriogenology* **37**, 769–778.

Kato, Y., and Tsunoda, Y., (1995). Germ cell nuclei of male fetal mice can support development of chimeras to midgestation following serial transplantation. *Development* **121**, 779–783.

Kato, Y., Tani, T., Sotomaru, Y., Kurokawa, K., Kato, J., Doguchi, H., Yasue, H., and Tsunoda, Y. (1998). Eight calves cloned from somatic cells of a single adult. *Science* **282**, 2095–2098.

Kato, Y., Yabuuchi, A., Motosugi, N., Kato, J., and Tsunoda, Y. (1999). Developmental potential of mouse follicular epithelial cells and cumulus cells after nuclear transfer. *Biol. Reprod.* **61**, 1110–1114.

Kaufman, M. H., Barton, S. C., and Surani, M. A. (1977). Normal postimplantation development of mouse parthenogenetic embryos to the forelimb bud stage. *Nature* **265**, 53–55.

Kawase, E., Yamazaki, Y., Yagi, T., Yanagimachi, R., and Pedersen, R. A. (2000). Mouse embryonic stem (ES) cell lines established from neuronal cell-derived cloned blastocysts. *Genesis* **28**, 156–163.

Kikyo, N., Wade, P. A., Guschin, D., Ge, H., and Wolffe, A. P. (2000). Active remodeling of somatic nuclei in egg cytoplasm by the nucleosomal ATPase ISWI. *Science* **289**, 2360–2362.

Kimura, Y., and Yanagimachi, R. (1995). Intracytoplasmic sperm injection in the mouse. *Biol. Reprod.* **52**, 709–720.

Kishikawa, H., Wakayama, T., and Yanagimachi, R. (1999). Comparison of oocyte-activating agents for mouse cloning. *Cloning* **1**, 153–159.

Kitani, H., Takagi, N., Atsumi, T., Kawakura, K., Imamura, K., Goto, S., Kusakabe, M., and Fukuta, K. (1996). Isolation of a germline-transmissible embryonic stem (ES) cell line from C3H/He mice. *Zool. Sci.* **13**, 865–871.

Kline, D., and Kline, J. D. (1992). Repetitive calcium transients and the role of calcium in exocytosis and cell cycle activation in the mouse egg. *Dev. Biol.* **149**, 80–89.

Kono, T. (1997). Nuclear transfer and reprogramming. *Rev. Reprod.* **2**, 74–80.

Kono, T., Tsunoda, Y., and Nakahara, T. (1991). Production of identical twin and triplet mice by nuclear transplantation. *J. Exp. Zool.* **257**, 214–219.

Kono, T., Ogawa, M., and Nakahara, T. (1993). Thymocyte transfer to enucleated oocytes in the mouse. *J. Reprod. Dev.* **39**, 301–307.

Kozik, A., Bradbury, E. M., and Zalensky, A. (1998). Increased telomere size in sperm cells of mammals with long terminal (TTAGGG)n arrays. *Mol. Reprod. Dev.* **51**, 98–104.

Kubota, C., Yamakuchi, H., Todoroki, J., Mizoshita, K., Tabara, N., Barber, M., and Yang, X. (2000). Six cloned calves produced from adult fibroblast cells after long-term culture. *Proc. Natl. Acad. Sci. U.S.A.* **97**, 990–995.

Kwon, O. Y., Kono, T. (1996). Production of identical sextuplet mice by transferring metaphase nuclei from four-cell embryos. *Proc. Natl. Acad. Sci. U.S.A.* **93**, 13010–13013.

Lanza, R. P., Cibelli, J. B., Blackwell, C., Cristofalo, V. J., Francis, M. K., Baerlocher, G. M., Mak, J., Schertzer, M., Chavez, E. A., Sawyer, N., Lansdorp, P. M., and West, M. D. (2000). Extension of cell life-span and telomere length in animals cloned from senescent somatic cells. *Science* **288**, 665–669.

Latham, K. E. (1994). Strain-specific differences in mouse oocytes and their contributions to epigenetic inheritance. *Development* **120**, 3419–3426.

Latham, K. E., Garrels, J. I., Chang, C., and Solter, D. (1991). Quantitative analysis of protein synthesis in mouse embryos. I. Extensive reprogramming at the one- and two-cell stages. *Development* **112**, 921–932.

Ledermann, B., and Burki, K. (1991). Establishment of a germ-line competent C57BL/6 embryonic stem cell line. *Exp. Cell Res.* **197**, 254–258.

Lin, S.-J., Defossez, P.-A., and Guarente, L. (2000). Requirement of NAD and *SIR2* for life-span extension by calorie restriction in *Saccharomyces cerevisiae*. *Science* **289**, 2126–2128.

Mann, J. R., and Stewart, C. L. (1991). Development to term of mouse androgenetic aggregation chimeras. *Development* **113**, 1325–1333.

Masui, Y., and Markert, C. L. (1971). Cytoplasmic control of nuclear behavior during meiotic maturation of frog oocytes. *J. Exp. Zool.* **177**, 129–145.

Mayer, W., Smith, A., Fundele, R., and Haaf, T. (2000). Spatial separation of parental genomes in preimplantation mouse embryos. *J. Cell Biol.* **148**, 629–634.

McBurney, M. W., Jones-Villeneuve, E. M., Edwards, M. K., and Anderson, P. J. (1982). Control of muscle and neuronal differentiation in a cultured embryonal carcinoma cell line. *Nature* **299**, 165–167.

McGrath, J., and Solter, D. (1983). Nuclear transplantation in the mouse embryo by microsurgery and cell fusion. *Science* **220**, 1300–1302.

McGrath, J., and Solter, D. (1984a). Inability of mouse blastomere nuclei transferred to enucleated zygotes to support development *in vitro*. *Science* **226**, 1317–1319.

McGrath, J., and Solter, D. (1984b). Completion of mouse embryogenesis requires both the maternal and paternal genomes. *Cell* **37**, 179–183.

Meng, L., Ely, J. J., Stouffer, R. L., and Wolf, D. P. (1997). Rhesus monkeys produced by nuclear transfer. *Biol. Reprod.* **57**, 454–459.

Modlinski, J. A., Gerhauser, D., Lioi, B., Winking, H., and Illmensee, K. (1990). Nuclear transfer from teratocarcinoma cells into mouse oocytes and eggs. *Development* **108**, 337–348.

Moore, T., and Haig, D. (1991). Genomic imprinting in mammalian development: A parental tug-of-war. *Trends Genet.* **7**, 45–49.

Morley, P., and Whitfield, J. F. (1993). The differentiation inducer, dimethyl sulfoxide, transiently increases the intracellular calcium ion concentration in various cell types. *J. Cell Physiol.* **156**, 219–225.

Munsie, M. J., Michalska, A. E., O'Brien, C. M., Trounson, A. O., Pera, M. F., and Mountford, P. S. (2000). Isolation of pluripotent embryonic stem cells from reprogrammed adult mouse somatic cell nuclei. *Curr. Biol.* **10**, 989–992.

Murray, A., and Hunt, T. (1993). The cell cycle—An introduction. Oxford University Press, Oxford.

Nagy, A., Gocza, E., Diaz, E. M., Prideaux, V. R., Ivanyi, E., Markkula, M., and Rossant, J. (1990). Embryonic stem cells alone are able to support fetal development in the mouse. *Development* **110**, 815–821.

Oronuki, N., Inoue, K., Yamamoto, Y., Noguchi, Y., Tanemura, K., Suzuki, O., Nakayama, H., Doi, K., Ohtomo, Y., Satoh, M., Nishida, A., and Ogura, A. (2002). Early death of mice cloned from somatic cells. *Nat. Genet.* **30**, in press.

Ogura, A., Inoue, K., Ogonuki, N., Noguchi, A., Takano, K., Nagano, R., Suzuki, O., Lee, J., Ishino, F., and Matsuda, J. (2000a). Production of male cloned mice from fresh, cultured, and cryopreserved immature sertoli cells. *Biol. Reprod.* **62**, 1579–1584.

Ogura, A., Inoue, K., Takano, K., Wakayama, T., and Yanagimachi, R. (2000b). Birth of mice after nuclear transfer by electrofusion using tail tip cells. *Mol. Reprod. Dev.* **57**, 55–59.

Ohgane, J., Wakayama, T., Kogo, Y., Senda, S., Hattori, N., Tanaka, S., Yanagimachi, R., and Shiota, K. (2001). DNA methylation variation in cloned mice. *Genesis* **30**, 45–50.

Onishi, A., Iwamoto, M., Akita, T., Mikawa, S., Takeda, K., Awata, T., Hanada, H., and Perry, A. C. F. (2000). Pig cloning by microinjection of fetal fibroblast nuclei. *Science* **289**, 1188–1190.

Ono, Y., Shimozawa, N., Ito, M., and Kono, T. (2001). Cloned mice from fetal fibroblast cells arrested at metaphase by a serial nuclear transfer. *Biol. Reprod.* **64**, 44–50.

Ono, Y., Shimozawa, N., Muguruma, K., Kimoto, S., Hioki, K., Tachibana, M., Shinkai, Y., Ito, M., and Kono, T. (2001b). Production of cloned mice from embryonic stem cells arrested at metaphase. *Reproduction* **122**, 731–736.

Oswald, J., Engemann, S., Lane, N., Mayer, W., Olek, A., Fundele, R., Dean, W., Reik, W., and Walter, J. (2000). Active demethylation of the paternal genome in the mouse zygote. *Curr. Biol.* **10**, 475–478.

Perry, A. C. F., and Wakayama, T. (2002). Untimely ends and new beginnings in mouse cloning. *Nat. Genet.* **30**, 2–3.

Polejaeva, I. A., Chen, S. H., Vaught, T. D., Page, R. L., Mullins, J., Ball, S., Dai, Y., Boone, J., Walker, S., Ayares, D. L., Colman, A., and Campbell, K. H. S. (2000). Cloned pigs produced by nuclear transfer from adult somatic cells. *Nature* **407**, 86–90.

Prather, R. S., Sims, M. M., and First, N. L. (1989). Nuclear transplantation in early pig embryos. *Biol. Reprod.* **41**, 414–418.

Ram, P. T., and Schultz, R. M. (1993). Reporter gene expression in G2 of the 1-cell mouse embryo. *Dev. Biol.* **156**, 552–556.

Rao, P. N., and Johnson, R. T. (1970). Mammalian cell fusion: Studies on the regulation of DNA synthesis and mitosis. *Nature* **225**, 159–164.

Renard, J. P., Chastant, S., Chesne, P., Richard, C., Marchal, J., Cordonnier, N., Chavatte, P., and Vignon, X. (1999). Lymphoid hypoplasia and somatic cloning. *Lancet* **353**, 1489–1491.

Rideout, W. M. 3rd., Wakayama, T., Wutz, A., Eggan, K., Jackson-Grusby, L., Dausman, J., Yanagimachi, R., and Jaenisch, R. (2000). Generation of mice from wild-type and targeted ES cells by nuclear cloning. *Nat. Genet.* **24**, 109–110.

Robertson, K. D., Ait-Si-Ali, S., Yokochi, T., Wade, P. A., Jones, P. L., and Wolffe, A. P. (2000). DNMT1 forms a complex with Rb, E2F1 and HDAC1 and represses transcription from E2F-responsive promoters. *Nat. Genet.* **25**, 338–342.

Robl, J. M., Gilligan, B., Critser, E. S., and First, N. L. (1986). Nuclear transplantation in mouse embryos: Assessment of recipient cell stage. *Biol. Reprod.* **34**, 733–739.

Rountree, M. R., Bachman, K. E., Herman, J. G., and Baylin, S. B. (2001). DNA methylation, chromatin inheritance and cancer. *Oncogene* **20**, 3156–3165.

Rudnicki, M. A., Jackowski, G., Saggin, L., and McBurney, M. W. (1990). Actin and myosin expression during development of cardiac muscle from cultured embryonal carcinoma cells. *Dev. Biol.* **138**, 348–358.

Samake, S., and Smith, L. C. (1996). Effects of cell-cycle-arrest agents on cleavage and development of mouse embryos. *J. Exp. Zool.* **274**, 111–120.

Sato, K., Hosaka, K., Ohi, S., Uchiyama, H., Tokieda, Y., and Ishiwata, I. (2000). Mouse fetuses by nuclear transfer from embryonic stem cells. *Hum. Cell* **13**, 197–202.

Segev, H., Memili, E., and First, N. L. (2001). Expression patterns of histone deacteylases in bovine oocytes and early embryos, and the effect of their inhibition on embryo development. *Zygote* **9**, 123–133.

Shamblott, M. J., Axelman, J., Wang, S., Bugg, E. M., Littlefield, J. W., Donovan, P. J., Blumenthal, P. D., Huggins, G. R., and Gearhart, J. D. (1998). Derivation of pluripotent stem cells from cultured human primordial germ cells. *Proc. Natl. Acad. Sci. U.S.A.* **95**, 13726–13731.

Shin, T., Kraemer, D., Pryor, J., Liu, L., Rugila, J., Howe, L., Buck, S., Murphy, K., Lyons, L., and Weshusin, M. (2002). A cat cloned by nuclear transplantation. *Nature* **415**, 859.

Simpson, E. M., Linder, C. C., Sargent, E. E., Davisson, M. T., Mobraaten, L. E., and Sharp, J. J. (1997). Genetic variation among 129 substrains and its importance for targeted mutagenesis in mice. *Nat. Genet.* **16**, 19–27.

Siracusa, G., Whittingham, D. G., and DeFelici, M. (1980). The effect of microtubule- and microfilament-disrupting drugs on preimplantation mouse embryos. *J. Embryol. Exp. Morphol.* **60**, 71–82.

Smith, L. C., and Wilmut, I. (1989). Influence of nuclear and cytoplasmic activity on the development *in vivo* of sheep embryos after nuclear transplantation. *Biol. Reprod.* **40**, 1027–1035.

Snow, M. H. L. (1973). Tetraploid mouse embryos produced by cytochalasin B during cleavage. *Nature* **244**, 513–515.

Stevens, L. C. (1973). A new inbred subline of mice (129-terSv) with a high incidence of spontaneous congenital testicular teratomas. *J. Natl. Cancer Inst.* **50**, 235–242.

Stice, S. L., and Keefer, C. L. (1993). Multiple generational bovine embryo cloning. *Biol. Reprod.* **48**, 715–719.

Stice, S. L., Keefer, C. L., and Matthews, L. (1994). Bovine nuclear transfer embryos: Oocyte activation prior to blastomere fusion. *Mol. Reprod. Dev.* **38**, 61–68.

Sun, F. Z., and Moor, R. M., (1995). Nuclear transplantation in mammalian eggs and embryos. *Curr. Top. Dev. Biol.* **30**, 147–176.

Surani, M. A. (1998). Imprinting and the initiation of gene silencing in the germ line. *Cell* **93**, 309–312.

Suzuki, O., Asano, T., Yamamoto, Y., Takano, K., and Koura, M. (1996). Development *in vitro* of pre-implantation embryos from 55 mouse strains. *Reprod. Fertil. Dev.* **8**, 975–980.

Swann, K., and Ozil, J. P. (1994). Dynamics of the calcium signal that triggers mammalian egg activation. *Int. Rev. Cytol.* **152**, 183–222.

Szollosi, D., Czolowska, R., Szollosi, M. S., and Tarkowski, A. K. (1988). Remodeling of mouse thymocyte nuclei depends on the time of their transfer into activated, homologous oocytes. *J. Cell Sci.* **91**, 603–613.

Tamashiro, K. L. K., Wakayama, T., Blanchard, R. J., Blanchard, D. C., and Yanagimachi, R. (2000). Postnatal growth and behavioral development of mice cloned from adult cumulus cells. *Biol. Reprod.* **63**, 328–334.

Tamashiro, K. L. K., Wakayama, T., Akutsu, H., Yamazaki, Y., Lachey, J. L., Wortmsan, M. D., Seeley, R. J. D'Alessio, D. A., Woods, S. C., Yanagimachi, R., and Sakai, R. R. (2002). Cloned mice have an obese phenotype that is not transmitted to their offspring. *Nat. Med.* **63**, 328–334.

Tanaka, S., Oda, M., Toyoshima, Y., Wakayama, T., Tanaka, M., Yoshida, N., Hattori, N., Ohgane, J., Yanagimachi, R., and Shiota, K. (2001). Placentomegaly in cloned mouse concepti caused by expansion of the spongiotrophoblast layer. *Biol. Reprod.* **65**, 1813–1821.

Thomson, J. A., Itskovitz-Eldor, J., Shapiro, S. S., Waknitz, M. A., Swiergiel, J. J., Marshall, V. S., and Jones, J. M. (1998). Embryonic stem cell lines derived from human blastocysts. *Science* **282**, 1145–1147.

Tian, X. C., Xu, J., and Yang, X. (2000). Normal telomere lengths found in cloned cattle. *Nat. Genet.* **26**, 272–273.

Tsunoda, Y., and Kato, Y. (1997). Full-term development after transfer of nuclei from 4-cell and compacted morula stage embryos to enucleated oocytes in the mouse. *J. Exp. Zool.* **278**, 250–254.

Tsunoda, Y., and Kato, Y. (1998). Not only inner cell mass cell nuclei but also trophectoderm nuclei of mouse blastocysts have a developmental totipotency. *J. Reprod. Fertil.* **113**, 181–184.

Tsunoda, Y., Yasui, T., Shioda, Y., Nakamura, K., Uchida, T., and Sugie, T. (1987). Full-term development of mouse blastomere nuclei transplanted into enucleated two-cell embryos. *J. Exp. Zool.* **242**, 147–151.

Vidricaire, G., Jardine, K., and McBurney, M. W. (1994). Expression of the *Brachyury* gene during mesoderm development in differentiating embryonal carcinoma cell cultures. *Development* **120**, 115–122.

Wakayama, T., and Yanagimachi, R. (1998a). Fertilizability and developmental ability of mouse oocytes with reduced amount of cytoplasm. *Zygote* **6**, 341–346.

Wakayama, T., and Yanagimachi, R. (1998b). The first polar body can be used the production of normal offspring. *Biol. Reprod.* **59**, 100–104.

Wakayama, T., and Yanagimachi, R. (1998c). Development of normal mice from oocytes injected with freeze-dried spermatozoa. *Nat. Biotechnol.* **16**, 639–641.

Wakayama, T., and Yanagimachi, R. (1999). Cloning of male mice from adult tail-tip cells. *Nat. Genet.* **22**, 127–128.

Wakayama, T., and Yanagimachi, R. (2001a). Mouse cloning with nucleus donor cells of different age and type. *Mol. Reprod. Dev.* **58**, 376–383.

Wakayama, T., and Yanagimachi, R. (2001b). Effect of the timing of oocyte activation, cytokinesis inhibitor and DMSO on mouse cloning using cumulus cell nuclei. *Reproduction* **122**, 49–60.

Wakayama, T., Tanemura, K., Suto, J., Imamura, K., Fukuta, K., Mori, H., Kuramoto, K., Kurohmaru, M., and Hayashi, Y. (1995). Production of term offspring by *in vitro* fertilization using old mouse spermatozoa. *J. Vet. Med. Sci.* **57**, 545–547.

Wakayama, T., Hayashi, Y., and Ogura, A. (1997). Participation of the female pronucleus derived from the second polar body in full embryonic development of mice. *J. Reprod. Fertil.* **110**, 263–266.

Wakayama, T., Perry, A. C. F., Zuccotti, M., Johnson, K. R., and Yanagimachi, R. (1998a). Full-term development of mice from enucleated oocytes injected with cumulus cell nuclei. *Nature* **394**, 369–374.

Wakayama, T., Whittingham, D. G., and Yanagimachi, R. (1998b). Production of normal offspring from mouse oocytes injected with spermatozoa cryopreserved with or without cryoprotection. *J. Reprod. Fertil.* **112**, 11–17.

Wakayama, T., Rodriguez, I., Perry, A. C. F., Yanagimachi, R., and Mombearts, P. (1999). Mice cloned from embryonic stem cells. *Proc. Natl. Acad. Sci. U.S.A.* **96**, 14984–14989.

Wakayama, T., Shinkai, Y., Tamashiro, K. L. K., Niida, H., Blanchard, D. C., Blanchard, R. J., Ogura, A., Tanemura, K., Tachibana, M., Perry, A. C. F., Colgan, D. F., Mombaerts, P., and Yanagimachi, R. (2000a). Cloning of mice to six generations. *Nature* **407**, 318–319.

Wakayama, T., Tateno, H., Mombaerts, P., and Yanagimachi, R. (2000b). Nuclear transfer into mouse zygotes. *Nat. Genet.* **24**, 108–109.

Wakayama, T., Tabar, V., Rodriguez, I., Perry, A. C. F., Studer, L., and Mombaerts, P. (2001). Differentiation of embryonic stem cell lines generated from adult somatic cells by nuclear transfer. *Science* **292**, 740–743.

Wells, D. N., Misica, P. M., and Tervit, H. R. (1999). Production of cloned calves following nuclear transfer with cultured adult mural granulosa cells. *Biol. Reprod.* **60**, 996–1005.

Whitaker, M. (1996). Control of meiotic arrest. *Rev. Reprod.* **1**, 127–135.

Willadsen, S. M. (1986). Nuclear transplantation in sheep embryos. *Nature* **320**, 63–65.

Wilmut, I., Schnieke, A. E., McWhir, J., Kind, A. J., and Campbell, K. H. S. (1997). Viable offspring derived from fetal and adult mammalian cell. *Nature* **385**, 810–813.

Yabuuchi, A., Tani, T., Kato, Y., and Tsunoda, Y. (2001). Nuclear transfer of mouse follicular epithelial cells pretreated with spermine, protamine, or putrescine. *J. Exp. Zool.* **289**, 208–212.

Yamazaki, Y., Makino, H., Hamaguchi-Hamada, K., Hamada, S., Sugino, H., Kawase, E., Miyata, T., Ogawa, M., Yanagimachi, R., and Yagi, T. (2001). Assessment of the developmental totipotency of neural cells in the cerebral cortex of mouse embryo by nuclear transfer. *Proc. Natl. Acad. Sci. U.S.A.* **98**, 14022–14026.

Young, L. E., Sinclair, K. D., and Wilmut, I. (1998). Large offspring syndrome in cattle and sheep. *Rev. Reprod.* **3**, 155–163.

Zhou, Q., Boulanger, L., and Renard, J. P. (2000). A simplified method for the reconstruction of fully competent mouse zygotes from adult somatic donor nuclei. *Cloning* **2**, 35–44.

CHAPTER 17

CLONING OF RABBITS

András Dinnyés, X. Cindy Tian, and Xiangzhong Yang

INTRODUCTION

Nuclear transfer using somatic cells offers new opportunities for genetic engineering, genome preservation, and tissue regeneration. As described in detail in the corresponding chapters of this book, nuclear transfer of somatic cells has succeeded in various species, including sheep (Schnieke *et al.*, 1997; Wilmut *et al.*, 1997), cattle (Cibelli *et al.*, 1998; Kato *et al.*, 1998; Vignon *et al.*, 1998; Wells *et al.*, 1998; Shiga *et al.*, 1999; Zakhartchenko *et al.*, 1999; Kubota *et al.*, 2000), goats (Baguisi *et al.*, 1999), mice (Wakayama *et al.*, 1998), pigs (Polejaeva *et al.*, 2000; Onishi *et al.*, 2000; Betthauser *et al.*, 2000), and gaurs (Lanza *et al.*, 2001). Although the efficiency of the overall cloning process has been low, rapid progress has been made in the improvement and application of this technology.

Rabbits were one of the first species in which blastomere nuclear transfer succeeded; as a model species, rabbits played a central role in developing the micromanipulation technologies in embryos. There are several advantages of using rabbits as an experimental model:

1. The costs related to animal procurement, animal care, and oocyte production in rabbits are relatively low compared to large animals. For example, the cost of a cattle embryo produced *in vivo* is about 30 times that of a rabbit embryo.

2. The developmental biology of rabbit embryos and fetuses resembles more closely that of large farm animals than that of rodent model species (mouse, rat, and hamster), including the transition from maternal to embryonic control of embryo development (see below).

3. The pregnancy of rabbits is relatively short (1 month), allowing rapid evaluation of fetal and postnatal development. In comparison, the gestation length for cattle is nine times longer.

4. The sizeable milk production of rabbits allows their use as test animals for therapeutic protein expression in milk, or as a living bioreactor.

5. Rabbits are induced ovulators. Domesticated rabbits are nonseasonal breeders and produce multiple offspring in one litter. These reproductive patterns make the use of rabbits for reproductive research highly efficient.

Despite the success with blastomere nuclear transfer (Stice and Robl, 1988; Collas and Robl, 1990; Yang *et al.*, 1992) and the efforts of several research teams (Mitalipov *et al.*, 1999; Yin *et al.*, 2000; Dinnyes *et al.*, 2001b) on somatic cell nuclear transfer, for several years no live births have been obtained in rabbit cloning using adult or fetal somatic cells. However, recent success by Chesné *et al.* (2002) show that modifications in the activation method and embryo transfer synchrony have achieved the long-awaited breakthrough in rabbit somatic cell nuclear transfer. There is hope that these results will facilitate progress in improving the cloning efficiency in this as well as other species, and practical applications of the cloning technology will follow in the near future. In this chapter we discuss the state of the art of blastomere and somatic cell nuclear transfer, along with specific issues on development and embryo technologies in rabbits.

BACKGROUND OF RABBIT EMBRYOLOGY

HISTORICAL BACKGROUND OF RABBIT BIOTECHNOLOGY

The first successful embryo transfer, using the rabbit, was reported in 1890 by Heape. Over the years rabbits continued to be an important model species for research in reproductive biology. The rabbit oviduct and uterus provided an environment with easy access and low cost for embryo culture, until *in vitro* culture technologies were developed sufficiently (Chang *et al.*, 1971; Ellington *et al.*, 1990; Totey *et al.*, 1992; Petters and Wells, 1993). Rabbits have also been a good model for the development and improvement of many micromanipulation techniques. These include production of transgenic rabbits by pronuclear microinjection (Hammer *et al.*, 1985), identical twin rabbits by embryo splitting (Yang and Foote, 1987), live young from intracytoplasmic sperm injection (Deng and Yang, 2001), chimera production by injection of inner cell mass (ICM) cells into morulae or blastocysts (Giles *et al.*, 1993), nuclear transfer with embryonic blastomeres (see details below), and recently with somatic cells (Chesné *et al.*, 2002).

EMBRYONIC AND FETAL DEVELOPMENT IN RABBITS

Rabbit embryonic and fetal development provides an important model for other species and has been studied in detail. The main events of the preimplantation development of rabbit embryos are described in Table 1. A good description of embryonic and fetal development and further details on the biology of the laboratory rabbits can be found in the work of Adams *et al.* (1961), Edwards (1968), Hagen (1974), Anderson and Henck (1994), and Harkness and Wagner (1995).

Rabbits are induced ovulators and mature preovulatory follicles are present constantly on the surface of their ovaries. A luteinizing hormone (LH) surge resulting from administration of LH or human chorionic gonadotrophins (hCGs), or stimulated by gonadotrophin-releasing hormone (GnRH), can induce ovulation. Mating,

Table 1 Preimplantation Development in Rabbits[a]

Time postmating (hours)	Embryo stage (cell number)	Doubling time (hours)	Location of embryos	Major events
12–14	Oocyte	—	Oviduct	Fertilization
18–20	Zygote	—	Oviduct	Pronuclear formation (6 hours)
24–26	Two cell	6	Oviduct	—
30–32	Four cell	6	Oviduct	—
38–40	Eight cell	8	Oviduct	Maternal–zygotic transition
46–48	Sixteen cell	8	Oviduct	
54–56	Morula (32 cell)	8	Oviduct	Compaction and transport to uterus
64–66	Compact morula (64 cell)	10	Oviduct/uterus	Morula–blastocyst transition
76–78	Early blastocyst (128 cell)	12	Oviduct/uterus	
84–86	Expanded blastocyst (256 cell)	8	Uterus	Blastocoel expansion
94–96	Hatched blastocyst (512 cell)	8	Uterus	Hatching

[a]Modified from Yang (1991).

which mechanically stimulates the central nervous system and thus surges of GnRH and LH, is the natural mechanism for inducing ovulation. This makes the timing of ovulation as well as the age of the oocytes highly predictable and controllable (Yang, 1991). Ovulation normally occurs 10–12 hours after the onset of hormonal or mechanical stimulus, although superovulation can cause slightly prolonged ovulation periods (Varian *et al.*, 1967).

A unique feature of rabbit embryology is that a mucin coat (Fig. 1), which is a unique glycoprotein covering the rabbit embryos, is accumulated during oviductal transport; the mucin coat can reach a thickness of 110 μm by 72 hours after ovulation (Adams, 1958; Denker and Gerdes, 1979; Leiser and Denker, 1988; Fisher *et al.*, 1991).

Fertilization occurs shortly after ovulation in the ampullar region of the oviducts by sperm already present for 10–12 hours and capacitated in the female genital tract (Chang, 1951). The resulting zygote progresses through the next stage of development rather quickly. Pronuclear formation takes about 6 hours, and cell-doubling time is about 6 hours for the first two divisions, all under maternal or zygotic control. Overlap of cell stages among different embryos is commonly observed, especially following superovulation (Varian *et al.*, 1967).

The switch from maternal to embryonic control of development occurs around the 8- to 16-cell stages, similar to that in cattle and sheep (Barnes and Eyestone, 1990; Telford *et al.*, 1990). It is characterized by the loss or decay of mRNA of maternal origin, activation of transcription of the embryonic genome, developmental arrest of the embryo in the presence of transcription inhibitors, and marked qualitative changes in protein synthesis patterns in the embryos (Telford *et al.*, 1990). Cell-doubling time is about 8 hours after the transition until compaction, at the 32- to 64-cell stages (Ziomek *et al.*, 1990).

Embryos reach the uterus between 60 and 72 hours after ovulation; however, in superovulated rabbits this process can be delayed (Yang, 1991). Blastocoel formation starts at around the 128-cell stage, with an increased cell-doubling time of

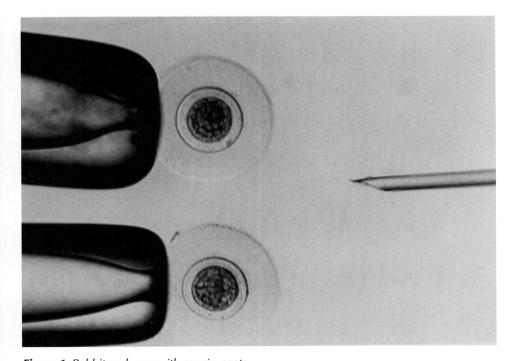

Figure 1 *Rabbit embryos with mucin coats.*

12 hours (Stice and Robl, 1988; Ziomek *et al.*, 1990). By this time, the mucin layer and the zona pellucida are less resistant to enzymatic removal (Ziomek *et al.*, 1990) and mechanical penetration by micropipette (Yang and Foote, 1987). Blastocyst expansion and zona pellucida dissolution start around 84 hours postovulation. In *in vivo*-developed blastocysts, the embryos do not undergo a real hatching process because of the presence of the mucin coat. The disappearance of zona pellucida and mucin coat occurs around 96 and 140 hours postovulation, respectively (Adams, 1958; Fisher *et al.*, 1991). Before implantation, the blastocysts secrete proteins, forming a new layer of coating under the mucin coat. This new layer is termed the neozona and its formation does not take place under the nonphysiological developmental conditions of *in vitro* culture (Fisher *et al.*, 1991). Implantation of rabbit embryos takes place around 7 days following artificial insemination (Orsini, 1962).

The length of the gestation varies slightly by breed, averaging 30–31 days, depending on litter size. The number of corpora lutea required for pregnancy maintenance is minimal and varies depending on the breed (Beatty, 1958; Feussner *et al.*, 1992). There are two critical periods in rabbit pregnancies when most of the postimplantation losses occur. Adams (1960) observed that among the approximate 18% postimplantation losses, 7% happened immediately postimplantation, 66% between days 8 and 17, and 27% between days 17 and 23. The first critical period, around day 13, is related to the placentation changes from the yolk sac to the hemochorial type, and the second period falls between days 22 and 23, when the tense, rounded fetal structures are susceptible to dislodgment. Around day 23 the placenta becomes hemoendothelial, with only one layer separating the maternal and fetal circulation.

STATE OF THE ART

The procedures of rabbit nuclear transfer are similar to those used in other species in regard to oocyte and somatic cell donors (Fig. 2). In the following section, the important steps of the procedures are discussed in detail, with special emphasis on the aspects unique to rabbits.

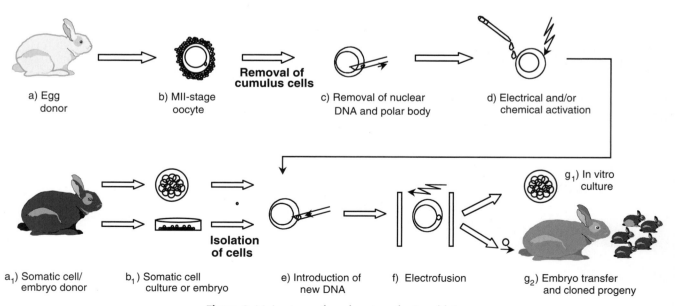

Figure 2 *Main steps of nuclear transfer in rabbits.*

ORIGIN OF RECIPIENT OOCYTES AND THEIR ENUCLEATION

In vitro-matured follicular oocytes were successfully used as recipients for blastomere nuclear transfers, and live pups were produced from such experiments. However, it is not clear whether the malformations observed in some of these progeny are related to the use of such oocytes (Park *et al.*, 1998). Modifying the *in vitro* maturation process in oocytes might make them more suitable to reprogram the donor nuclei (Lonergan *et al.*, 2000; Kasinathan *et al.*, 2001).

Most rabbit nuclear transfer experiments are dependent on the use of *in vivo*-matured and -ovulated oocytes. Ovulated oocytes are usually collected by midventral laparotomy *in situ* and flushing of the oviducts (Maurer, 1978). In order to increase oocyte production, a variety of superovulation protocols can be used and approximately 30–40 ova per doe can be obtained (Kennelly and Foote, 1965; Maurer *et al.*, 1968; Carney and Foote, 1990; Yang *et al.*, 1990a,b; Kauffman *et al.*, 1998). The response to the superovulation treatment in the number and quality of the oocytes varies among donor animals. Superovulation protocols can also affect the quality of the resulting embryos, including their cell numbers, sensitivity to cryopreservation, and the subsequent pregnancy rates (Carney and Foote, 1990; Kauffman *et al.*, 1998). However, very little is known about its effect on the quality of oocytes for nuclear transfer. Potentially, natural cycles or improved superovulation protocols might result in better quality oocytes.

Cryopreserved oocytes represent another resource to produce cytoplasts for nuclear transfer as described for cattle (Dinnyes *et al.*, 2000). Rabbit oocytes are not particularly sensitive to low temperature, and cryopreserved oocytes were successfully used to produce progeny following *in vitro* fertilization (Vincent *et al.*, 1989). However, cryopreserved rabbit oocytes have not been tested in nuclear transfer experiments.

The breed of the oocyte donors may also have a major effect on the outcome of nuclear transfer, as was demonstrated in mice (Wakayama and Yanagimachi, 2001) and sheep (Dinnyes *et al.*, 2001a). Only a few rabbit breeds were tested in earlier nuclear transfer experiments (Mitalipov *et al.*, 1999; Chesné *et al.*, 2001; Dinnyes *et al.*, 2001b). Furthermore, interactions between the genetic origin of the recipient oocytes and that of the donor cells are yet to be investigated. Studies on these variables may shed light on whether this was among the factors contributing to the lack of success in the development of cloned embryos in several studies.

The age of the oocytes relative to their ovulation may also be an important factor for the success of nuclear transfer. In rabbits, activation rates were successfully increased by using aged oocytes. However, their ability to fuse with nuclear donor cells decreased with oocyte aging (Stice and Robl, 1988; Collas and Robl, 1990). Recent observations by Lagutina *et al.* (1999) indicated a potential increase in morula/blastocyst development following the use of older oocytes for blastomere nuclear transfer. Oocyte aging can also result in the migration of the metaphase plate toward the center of the oocyte and the fragmentation of the first polar body, both of which reduce the enucleation efficiency (Collas and Robl, 1990; Mitalipov *et al.*, 1999).

The enucleation of oocytes in rabbits is similar to that in other species, as described in detail in other chapters of this book. The zona pellucida and the cell membrane are neither too hard nor too flexible to create a major difficulty for micromanipulations. Depending on the breed of oocyte donors, Nomarski optics usually allows the visualization of the metaphase chromosomes in most of the oocytes, and chromosome removal can be performed under visual control. However, the variability of rabbit oocyte morphology and quality often reduces the efficiency of the micromanipulation. The coloration and granulation of the cytoplasm can make it impossible to see the chromosomes by standard light microscopy, and may necessitate the use of epifluorescent staining and ultraviolet light exposure of the oocytes

(Dinnyes *et al.*, 2001b). Similarly, Mitalipov *et al.* (1999) also reported that oocytes from Dutch-belted rabbits were more difficult to manipulate than those from New Zealand whites, due to their darkness and opacity. In some oocytes, the metaphase plate and the polar body are often not in proximity; therefore, in many cases blind enucleation is not efficient to remove the nuclear material, and epifluorescent staining is necessary to complete the enucleation. Furthermore, fragmentation of oocytes from some donors can reach as high as 100% shortly after micromanipulation, whereas other donors' oocytes from the same treatment groups showed no (or low) occurrence of such fragmentation (A. Dinnyes and X. Yang, unpublished observation). The enucleation conditions, especially the composition of the media, have been shown to have a major effect on further development (see Collas and Robl, 1990).

OOCYTE ACTIVATION

Activation of the recipient oocytes is of major importance to the outcome of nuclear transfer experiments (Collas and Robl, 1990). The general mechanisms of activation are described in detail in earlier chapters of this book. During fertilization, sperm induces periodic increases in intracellular calcium (Ca^{2+}) concentrations and activates the oocytes (Fissore and Robl, 1993; Swann and Lai, 1997). Activation of the oocytes can also be achieved by injection of rabbit sperm extracts (Stice and Robl, 1990). However, this method is not very efficient. Identification and purification of the molecules responsible for this phenomenon will provide an efficient approach to activation induction.

Various artificial activation methods exist to create the transient Ca^{2+} oscillations, but they usually fail to mimic completely the natural events, and result in a single intracellular Ca^{2+} rise. In rabbits, oocyte activation can be induced by an electric pulse, allowing a transmembrane influx of Ca^{2+} (Collas and Robl, 1990). Multiple electric pulses have resulted in greater activation efficiency (Collas and Robl, 1990, 1991; Ozil, 1990; Escriba and Garcia-Ximenez, 1999, 2000; Ozil and Huneau, 2001). However, this protocol requires either increased handling of the oocytes or special equipment (Ozil, 1990; Ozil and Huneau, 2001). Results of Yang *et al.* (1992) in rabbit studies suggested that a stronger electrical field (2.4 kV/cm) was more efficient to achieve activation, compared to a weaker one (1.8 or 1.2 kV/cm). Furthermore, addition of an alternating current (AC) pulse prior to the activating direct current (DC) pulse was beneficial to activation and subsequent blastocyst development (Yang *et al.*, 1992). Chesné *et al.* (2002) applied 3DC pulses (3.2 kV/cm) for fusion and activation, repeated after 1 hour for an additional activation effect.

Various chemicals are used for oocyte activation based on their ability to increase intracellular Ca^{2+}. Successful activation of rabbit oocytes can be obtained by the use of ionomycin (Mitalipov *et al.*, 1999; Lagutina *et al.*, 2000), inositol 1,4,5-trisphosphate (IP3) (Fissore and Robl, 1993; Mitalipov *et al.*, 1999), thimerosal (which potentiates IP3-sensitive Ca^{2+} release pathways) (Fissore and Robl, 1993), and guanosine 5'-O-(3-thiotriphosphate), which is a G-protein stimulator (Fissore and Robl, 1994). These treatments all result in the inactivation of maturation-promoting factor (MPF), which is a prerequisite to release oocytes from metaphase II arrest.

Targeting the signaling pathway downstream of the Ca^{2+} signal can also induce activation. Protein synthesis inhibitors, such as cycloheximide or puromycin, prevent the production of cyclin B, the regulatory subunit of MPF, and induce a subsequent drop in MPF levels and thus allow the oocyte to progress to interphase (Nussbaum and Prather, 1995). Treatments with protein kinase inhibitors such as 6-dimethylaminopurine (6-DMAP) or staurosporine also trigger meiotic resumption in several species due to their involvement in inactivating MPF and mitogen-activated protein kinase (MAPK) (Prather *et al.*, 1997; Wang *et al.*, 1997). A combined treatment, beginning with the induction of a transient increase in intracellular Ca²⁺, followed by a protein synthesis or kinase inhibitor, resulted in higher oocyte activation and developmental rates in several species (Susko-Parrish *et al.*, 1994; Nussbaum and Prather, 1995; Grocholova *et al.*, 1997; Liu *et al.*, 1998), including rabbits (Mitalipov *et al.*, 1999; Yin *et al.*, 2000; Lagutina *et al.*, 200; Chesné *et al.*, 2001; Dinnyes *et al.*, 2001b) (see Fig. 3). This is exemplified by Chesné *et al.* (2001), who used an activation treatment of two electric pulses followed by a combined cycloheximide and 6-DMAP treatment, and achieved a near 90% parthenogenetic blastocyst rate in rabbits. The duration of this treatment can be crucial for success in rabbit. The shortening of the cycloheximide and 6-DMAP treatment to one hour is considered one of the major contributing factors for the first successful somatic cell cloning in this species (Chesné *et al.*, 2002).

The composition of the solution used for electric activation can have a major effect on the embryos, especially on the frequency of lysis following treatment. In rabbits, mannitol (Stice and Robl, 1988; Collas and Robl, 1990, 1991; Yang *et al.*, 1990b; Du *et al.*, 1995; Chesné *et al.*, 2001; Dinnyes *et al.*, 2001b; Chesné *et al.*, 2002), Zimmerman solution (Yin *et al.*, 2000), inositol (Mitalipov *et al.*, 1999), or glucose solution (Ozil, 1990; Piotrowska *et al.*, 2000) were successfully used for activation or fusion. Reduction in the osmolarity of the pulsing mannitol solution might increase the parthenogenetic activation rates (Escriba and Garcia-Ximenez, 1999).

The activation event has long-lasting effects on subsequent embryo development. Ozil and Huneau (2001) demonstrated that characteristics of Ca²⁺ oscillation, i.e., amplitude, number, and frequency, are linked to the efficiency and quality of postimplantation development. Interestingly, the cell division dynamics were less affected by variations in Ca²⁺ oscillation, yet development of parthenogenetic embryos was altered. These observations may indicate that remodeling of the chromosomes and chromatin structures by the Ca²⁺ changes contributes significantly to determining the functional pattern of gene activity at later stages (Thompson *et al.*, 1995; Schultz *et al.*, 1999).

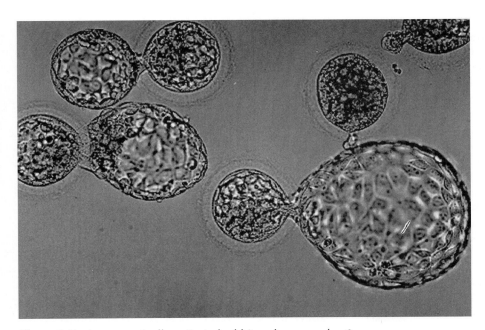

Figure 3 *Parthenogenetically activated rabbit embryos on day 6.*

Most activation treatments affect a broad spectrum of cellular components in addition to the oocyte-activating effect, and therefore the physiology of the oocytes may be altered in various ways. A more specific, noninvasive method that affects only the oocyte activation events would be desirable. In this regard, the rabbit provides a good model for the search for such a method and also to elucidate the fundamental activation mechanisms for advancement of nuclear transfer technologies.

EMBRYONIC BLASTOMERE NUCLEAR TRANSFER IN RABBITS

Embryonic blastomeres can be reprogrammed in an oocyte environment, resulting in the birth of offspring, as demonstrated in sheep by Willadsen (1986) and in rabbits by Stice and Robl (1988). The biological background of this reprogramming process is described in detail in other chapters of this book. Rabbits, as a model species, have played an important role in the development of this research area. Here only some rabbit-specific aspects of the procedures are discussed.

Donor Embryo Production and Blastomere Separation

Embryos as nuclear donors for blastomere nuclear transfer are usually produced following superovulation (see above) and surgical embryo collection by oviduct and/or uterus flushing (described in detail by Maurer, 1978).

In order to prepare blastomeres as cell donors for nuclear transfer, the mucin coat and the zona pellucida must be removed. This can be achieved with acidic phosphate-buffered saline (PBS) (Stice and Robl, 1988; Collas and Robl, 1990), pronase (Piotrowska *et al.*, 2000), or a combination of the two reagents (Yang *et al.*, 1990a,b; Du *et al.*, 1995). The embryonic blastomeres can then be separated mechanically (Stice and Robl, 1988) after treatment with trypsin (Collas and Robl, 1990; Du *et al.*, 1995) or Ca^{2+}- and Mg^{2+}-free medium (Yang *et al.*, 1990; Piotrowska *et al.*, 2000). The blastomere separation method was found to affect the outcome of the experiments (Yang *et al.*, 1990a,b), possibly due to damage to the membrane of the blastomeres.

Developmental Stage of Donor Embryos

In rabbits, the birth rate of young from embryos reconstructed with embryonic blastomeres varied between 3 and 10%, using nuclei before the maternal-to-zygotic transition (MZT) at the 8- to 16-cell stages (Stice and Robl, 1988; Collas and Robl, 1990, 1991), or following this transition at the 32- to 64-cell stages (Collas and Robl, 1990, 1991; Heyman *et al.*, 1990; Yang *et al.*, 1992; Lagutina *et al.*, 1999). Rabbit MZT is rather different from that in the mouse and has many similarities to MZT in cattle and sheep. It includes a long period of "minor transcription activity" prior to MZT and the slow degradation of maternal transcripts (Henrion *et al.*, 2000; Brunet-Simon *et al.*, 2001). The changes in transient gene expressions around the MZT can affect the pattern of gene activity following nuclear transfer and reprogramming (Christians *et al.*, 1994; Kanka *et al.*, 1996).

Polarization, another important biological event, occurs around the 32-cell stage in rabbit embryos (Koyama *et al.*, 1994). It is not clear if early polarization events affect the blastomeres' capability to be reprogrammed by nuclear transfer, because *in vitro* developmental rates of embryos cloned from pre- and postpolarization stage were not different (Tan *et al.*, 1997).

Differentiation of embryonic cells might have an important effect on their potential to be reprogrammed by nuclear transfer. Collas and Robl (1991) obtained a lower developmental rate for nuclear transfer using blastomeres from the inner cell mass and mural trophectoderm compared to that with 8-cell embryo blastomeres (17, 0, and 61% blastocysts, respectively). Du *et al.* (1995) compared the use of blastomeres from 16-cell-stage embryos vs. putative embryonic stem (ES) cells

from two different and passaged cell lines and obtained 31, 29, and 5% of blastocysts, respectively, indicating the potential differences among ES cell lines for nuclear
transfer. The resulting blastocysts contained a similar ICM/trophoblast cell ratio
compared to fertilized control embryos (around 25% ICM cells), although the total
cell numbers were lower. No birth of progeny was reported from the ES-like cell
nuclear transfers.

Cell Cycle Stage Synchronization between Donor Blastomeres and the Recipient Oocytes

The cell cycle stage of the donor blastomeres has a major influence on the
further development of cloned embryos. The cell cycle is composed of two distinct
phases. The "M phase" refers to the period of mitosis, and the "interphase" is composed of three periods, G1 (pre-DNA replication), S (DNA replication), and G2
(post-DNA replication) (Barnes and Eyestone, 1990). When a blastomere is fused
with an enucleated M-phase oocyte, premature condensation of the donor chromosomes occurs. In rabbits, Collas and Robl (1991) demonstrated that when a G1-
or early S-phase blastomere was used, relatively normal embryos could be produced.
When late S-phase nuclei were used, however, gross chromosome and spindle abnormalities in the reconstructed embryos were observed. Synchronization of morula
blastomeres to G1 phase with colcemid did not affect cleavage rates of the reconstructed embryos, although increases in the formation of blastocysts were reported
(Lagutina *et al.*, 1999).

Another possibility is the transfer of G2- or M-phase nuclei into an M-phase
oocyte cytoplasm. In this case the incidence of tetraploid embryos was very high
(Yang, 1991). A related problem is that electric activation of rabbit oocytes can
result in a high rate of polar body retention, compared to that of spontaneously
activated oocytes (Yang *et al.*, 1992).

The use of a "universal recipient," a preactivated recipient oocyte, can solve the
problems related to the cell cycle incompatibility of donor blastomeres and recipient oocytes, because in this environment the normal ploidy is achieved in every combination of donor cell and recipient cytoplasm (Campbell *et al.*, 1996a). When the
donor blastomere is fused to an oocyte after the decline of MPF by activation, the
nuclear membrane is maintained, no premature chromosome condensation occurs,
and DNA synthesis proceeds without interruption (Barnes *et al.*, 1993). In rabbits,
aging of the recipient oocytes before nuclear transfer also results in MPF decline,
and the aged oocytes can be used as the "universal recipients" (Heyman and Renard,
1996). The use of preactivated cytoplasts has proved to be beneficial to nuclear
transfer using blastomeres from 8-cell rabbit embryos and is superior when compared to using metaphase II cytoplasts or blastomeres synchronized in G1 phase
with nocodazole (Piotrowska *et al.*, 2000).

Enucleated zygotes in S phase as recipients resulted in lower developmental rates
compared to metaphase II cytoplasts (Modlinski and Smorag, 1991), suggesting that
a low MPF level is not the only prerequisite for successful reprogramming. The poor
development from enucleated zygotes as recipients might be due to the possibility
that some reprogramming factors closely associated with the pronucleus are
removed during enucleation (Heyman and Renard, 1996).

Fusion of Donor Cells to Recipient Oocytes

Transfer of the donor blastomere adjacent to the cytoplast cell membrane is a
relatively simple micromanipulation step. This procedure can be facilitated by
osmotic shrinkage of the cytoplast (Yang *et al.*, 1990a). A high rate of membrane
fusion between the blastomere and the enucleated oocyte is critical for the efficiency
of the entire nuclear transfer procedure, and can be achieved by several methods.
In the rabbit, electrical pulses combined with mechanical alignment resulted in good
fusion rates (Stice and Robl, 1988; Collas and Robl, 1990; Yang *et al.*, 1990b). In

addition to the DC pulses used for fusion, the application of AC pulses can assure a better alignment of the recipient cytoplast and the donor cells, and has been reported to be beneficial to increase the fusion rates of the relatively small ICM cells (Smith and Wilmut, 1989) and rabbit blastomeres (Yang *et al.*, 1990b, 1992). Importantly, the age of the recipient oocyte may have a major influence on fusion rates. Newly ovulated oocytes fuse better with the donor blastomeres, especially following multiple pulses (Collas and Robl, 1990).

Other treatment of the fused blastomere–cytoplast complexes depends on the activation protocol (see above). However, the use of cytochalasin B postfusion might be beneficial in preserving the normal ploidy of the transferred DNA and has been shown to reduce the fragmentation of the reconstructed embryo when blastomeres were from 16- to 64-cell embryos (Collas and Robl, 1990).

Serial Nuclear Transfer of Cloned Embryos (Recycling)

In order to maximize the number of cloned embryos and progeny with the same genetic origin, embryos generated by nuclear transfer of blastomeres can in turn be used as nuclear donors in further cycles of nuclear transfer experiments. In cattle, despite a large number of embryos created by this method (Willadsen, 1989; Stice and Keefer, 1993), the subsequent round of cloning increased the embryonic and fetal mortality. In rabbits, experiments have demonstrated that the use of metaphase II oocytes drastically reduced embryonic development as early as the second round of nuclear transfers. The fusion of preactivated cytoplast and morula-stage blastomeres, however, resulted in high rates of blastocyst development even in the third round, and normal young were obtained following embryo transfer. The largest number of morulae/blastocysts obtained from an individual donor embryo in this experiment was 27 (Piotrowska *et al.*, 2000).

SOMATIC CELL NUCLEAR TRANSFER IN RABBITS

Successful nuclear transfer from differentiated embryonic cell lines (Campbell *et al.*, 1996b) and from adult cells (Wilmut *et al.*, 1997) provided new impetus for research on cloning. In rabbits, the success of nuclear transfer with embryonic cells and the advanced development of embryo micromanipulation methods, as described above, provided a good starting point for somatic cell nuclear transfer in this species. Cloned blastocysts have been obtained by several research groups. However, developmental failure following embryo transfer was encountered by most rabbit cloning teams. Yin *et al.* (2000) observed a few (2%) implantation sites (no fetuses) by day 14 following embryo transfer. The failure of development of rabbit embryos from somatic cell nuclear transfer beyond blastocyst stage in those cases may have indicated an incomplete reprogramming of the donor nucleus. The authors contributed the success of nuclear transfer with cumulus cells to improvements in activation methods and in the synchrony of the cloned embryos to the recipient (Chesné *et al.*, 2002). These results emphasize the difficulty of optimizing conditions in this complex cloning procedure and the importance of better understanding the unique physiology and requirements of cloned embryos.

Establishment of Donor Cell Culture and the Effects of Cell Type and Length of Culture

Various cell types have been used for somatic cell nuclear transfer. In rabbits, fibroblast cells can be isolated and maintained similarly as described for other species. A method for establishing fibroblast culture using tissue explants is described in detail in the appendix of this chapter.

Previously, Wakayama *et al.* (1998) and Kato *et al.* (1999) demonstrated that donor cell type affects the cloning competence in mouse and cattle. In the rabbit, transfer of adult (Mitalipov *et al.*, 1999; Dinnyes *et al.*, 2001b) and fetal fibroblasts (Galat *et al.*, 1999; Lagutina *et al.*, 2000; Chesné *et al.*, 2001) as well as freshly isolated (Chesné *et al.*, 2001) and cultured cumulus cells (Yin *et al.*, 2000) generated blastocysts. The cell type, however, influences the developmental rates. When gonial cells, immediately following isolation from rabbit fetuses at 18–20 days of pregnancy, were tested for nuclear transfer (Moens *et al.*, 1996), the *in vitro* development to blastocyst from male gonial cells was four times higher (16%) than that of female cells (4%). Additionally, Chesné *et al.* (2001) compared nuclear transfer of fresh cumulus cells with cycling fetal fibroblast cells, and the former resulted in significantly higher blastocyst rates (47 vs. 4%, respectively). Moreover, the first cloned progeny were produced with fresh cumulus cells (Chesné *et al.*, 2002).

In rabbits, cells derived directly from tissue explants (nonpassaged) were less likely to support blastocyst development than were cells from later passages (Dinnyes *et al.*, 2001b). This could be due to the fact that a lower proportion of the nonpassaged cells are in G0/G1 as revealed by fluorescence-activated cell sorting (FACS) analysis. Further, DNA in cells cultured *in vitro* is less methylated and may be easier for reprogramming. This was supported by a study in cattle in which embryos cloned from cells at passages 10 and 15 gave better development than those from cells at passage 5 (Kubota *et al.*, 2000). Extremely long culture of donor cells, however, should be avoided in somatic nuclear transfer because chromosome abnormalities can occur in long-term cultures.

The number of times a cell doubles in culture before reaching senescence is of primary importance for gene-targeting experiments. In this regard, fibroblasts are superior to cumulus cells because of their higher doubling capacity in culture (A. Dinnyes and X. Yang, unpublished observation). Cells passaged 15 times in culture, equivalent to 30 cell doublings, are still competent for cloning (Kubota *et al.*, 2000) and are sufficient for selections of cells undergoing the highly rare gene-targeting events (Capecchi, 2000; Denning *et al.*, 2001).

Donor Cell Cycle

Similar to blastomere nuclear transfer, the cell cycle of the donor somatic cells has a major influence on the reprogramming process and the subsequent blastocyst development (Stice *et al.*, 1993). The effects of somatic cell cycle synchronization on the success of nuclear transfer were described in detail in Chapter 20, with special emphasis on the use of G0-stage cells, which was found beneficial in the first successful somatic cell cloning experiments (Wilmut *et al.*, 1997). Serum starvation or contact inhibition (confluency) will arrest the majority of cells in either G0 or G1 stage, respectively (Boquest *et al.*, 1999; Dinnyes *et al.*, 2001b). However, the accurate evaluation of the cell cycle effect on the development of individual embryos is difficult, because a given donor cell cannot be analyzed. Because of the low efficiency of nuclear transfer at the current state of the art, rare events may lead to the production of the 2–5% cloned progeny. Therefore, statistical probabilities may not provide the best estimates in the analysis of donor cells for somatic cell nuclear transfer.

In rabbits, cloned blastocysts have been obtained from serum-starved G0-stage adult fibroblasts (Mitalipov *et al.*, 1999) and from serum-starved G0/G1-stage cultured cumulus cells (Yin *et al.*, 2000). *In vitro* development to the blastocyst stage was not different in embryos cloned from serum-starved vs. contact-inhibited fibroblast cells (Dinnyes *et al.*, 2001b). Galat *et al.* (1999) compared quiescent and cycling fetal fibroblast cells as nuclear donors and observed similar cleavage development. However, increased blastocyst rates were obtained using cells subjected to serum starvation. Cloned progeny were ultimately obtained using fresh cumulus cells, which are naturally in a resting stage (Chesné *et al.*, 2002).

Fusion of Donor Cells to Recipient Oocytes

Similar to the blastomere nuclear transfer experiments, membrane fusion between the intact somatic cells and enucleated oocytes can be achieved by several methods, including electric pulses (Mitalipov *et al.*, 1999; Yin *et al.*, 2000; Dinnyes *et al.*, 2001b). In rabbits, an AC pulse prior to the DC pulse, combined with mechanical alignment of donor and recipient, resulted in good fusion rates (Dinnyes *et al.*, 2001b). In this case, the AC pulse is not assisting the alignment, because the somatic cells are too small to respond to such an effect. The AC pulse may, however, have a beneficial effect on the cell membranes, promoting fusion. Cell type and size can also influence fusion rates. Smaller cumulus cells fuse less efficiently compared to fibroblast cells (Chesné *et al.*, 2001).

An alternative micromanipulation method, relatively efficient in mouse (Wakayama *et al.*, 1998), is based on the injection of somatic nuclei directly into the oocytes' cytoplasm following removal of the donor cell membrane. In rabbits, this method has not been extensively studied. Although partially successful with ICM cells, it has shown very little promise with cumulus cells (Zhou *et al.*, 1997).

Emerging evidence from several species indicates that somatic cell nuclei, unlike embryonic blastomeres, need a longer exposure to an environment high in MPF in order to be fully reprogrammed. Because the reprogramming procedure is far from being understood, it is difficult to explain the exact mechanism. It might, therefore, be beneficial to use nonactivated cytoplasts for fusion and to delay the activation step for several hours after the nuclear transfer step (Wakayama *et al.*, 1998). In rabbits, however, experiments to test the preactivated vs. simultaneous activation strategies using an electrical pulse and various 6-DMAP treatments revealed no obvious differences in the *in vitro* development by cloned embryos (Dinnyes *et al.*, 2001b). Similarly, Lagutina *et al.* (2000) observed no differences in developmental rates using young or aged oocytes with delayed or simultaneous activation treatments, or with preactivated oocytes.

EMBRYO CULTURE AND CRYOPRESERVATION

Significant progress has been made during recent years in the area of *in vitro* culture of rabbit zygotes (Kane and Foote, 1971; Maurer, 1978). Using either coculture (Carney and Foote, 1990; Carney *et al.*, 1990) or media containing peritoneal fluid (Collas *et al.*, 1991) or vitreous humor of the eyes of female rabbits (Collas and Robl, 1991), or even completely defined protein-free systems (Carney and Foote, 1991; Li *et al.*, 1993; Farrell and Foote, 1995; Li and Foote, 1996; Giles and Foote, 1997), high rates of blastocyst development have been obtained. Menezo B2 medium was successfully used to culture parthenogenetically activated oocytes to the blastocyst stage (Ozil, 1990) and somatic cell nuclear transfer embryos prior to transfer into recipients (Chesné *et al.*, 2002). Both TCM 199 and KSOM media were successfully used for the culture of rabbit embryos from somatic cell nuclear transfer (Dinnyes *et al.*, 2001b; Chesné *et al.*, 2001). Although both media support high rates of *in vitro* blastocyst development, a short culture in KSOM resulted in superior *in vivo* survival of the embryos (B. Wang and X. Yang, unpublished data). Our experience is that only a short period of *in vitro* culture before embryo transfer is advisable to increase the chance of obtaining progeny. Mitalipov *et al.* (1999) found that Earle's balanced salt solution (EBSS) supplemented with minimum essential medium (MEM) nonessential amino acids, basal medium Eagle amino acids, L-glutamine, sodium pyruvate, and fetal bovine serum (FBS) better supported blastocyst development of activated oocytes compared to Dulbecco's modified Eagle's medium (DMEM/RPMI) with FBS or CR1aa and FBS. Using the same medium, they also obtained a good blastocyst development rate (30%) by embryos from somatic cell nuclear transfer. Similar results were also achieved with the EBSS system by Yin *et al.* (2000).

Embryo cryopreservation, another important technique for practical embryology research, is fairly successful in rabbits (Tsunoda *et al.*, 1982; Renard *et al.*, 1984; al-Hasani *et al.*, 1989; Vincent *et al.*, 1989; Kasai *et al.*, 1992). Blastomeres from frozen rabbit morulae were successfully used as nuclear donors, resulting in the birth of young (Heyman *et al.*, 1990). Furthermore, cloned embryos from blastomere nuclear transfer were cryopreserved/thawed before being used as donors in a second round of nuclear transfers (Rao *et al.*, 1998).

EMBRYO TRANSFER

Rabbit embryos accumulate a mucin coat during their passage through the oviduct, and without this mucin layer their viability is limited (Murakami and Imai, 1996). Although successful progeny development was recently reported (Jin *et al.*, 2000) following prolonged *in vitro* embryo culture and blastocyst-stage embryo transfer, rabbit embryos need to be transferred shortly after micromanipulation in order to obtain high number of progeny. Embryos are usually surgically transferred into the oviduct, by midventral laparotomy (Yang *et al.*, 1990a,b). Laparoscopic transfer to the fallopian tubes also offers an efficient and faster alternative for early and late cleavage-stage embryos (Besenfelder and Brem, 1993; Besenfelder *et al.*, 1998). Successful nonsurgical transfer of later stage embryos through the cervix into the uterus has also been reported (Kidder *et al.*, 1999).

The synchrony between the recipient oviduct or uterus and the transferred embryos is an important issue in rabbits (Chang, 1951). Asynchronous transfer can result in various abnormalities and death of the embryos. Rabbit blastocysts are more sensitive to asynchrony than are earlier cleavage-stage embryos (Fisher, 1989). Depending on the age of the embryo, synchronous or ±1-day asynchronous transfers resulted in the highest fetal development (Tsunoda *et al.*, 1982; Fisher, 1989; Yang and Foote, 1990). A minimum of two implantations seems to be necessary to carry a pregnancy to term (Adams, 1970). Conversely, overcrowding of the uterus by the transfer of too many embryos can reduce embryonic and fetal survival (Hafez, 1964). This was demonstrated in a report by Besenfelder and Brem (1993), who obtained a higher rate of progeny survival following transfer of 10–20 embryos as compared to 30–65 embryos. Synchronous embryo transfer with cloned embryos failed to establish pregnancy, however, a relatively large 22-h asynchrony was one of the major factors for the success in producing the first somatic cell cloned rabbits (Chesné *et al.*, 2002).

PREGNANCY MONITORING AND PROGENY PRODUCTION

Rabbit pregnancy can be detected by palpation around days 10–14 after artificial insemination, as described by Harkness and Wagner (1995). During certain periods of pregnancy (around day 23, as discussed earlier), the fetuses are very sensitive to handling, and palpation can cause abortion.

Embryos from blastomere nuclear transfer have resulted in pregnancies and healthy progeny (Stice and Robl, 1988; Collas and Robl, 1990; Heyman *et al.*, 1990; Yang *et al.*, 1992). Full-term development, however, is infrequent due to the high incidence of postimplantation losses (Heyman *et al.*, 1990; Heyman and Renard, 1996). Birth of abnormal young has also been reported (Park *et al.*, 1998).

Somatic cell nuclear transfer can often result in a high rate of pregnancy losses and a variety of fetal and perinatal abnormalities, as has been shown in cattle and sheep (see Chapters 19 and 20, this volume). The frequency of the abnormalities depends on many factors, and species differences are likely among them. Reports on progeny from goat and pig somatic cell nuclear transfer showed that the majority of them are healthy during late pregnancy and after birth (Polejaeva *et al.*, 2000; Betthauser *et al.*, 2000; Baguisi *et al.*, 1999). The first litters of somatic cell cloned rabbits were normal in their morphological appearance, and the loss of 2 of 6 kits can be contributed to failure in the adoptive process from the lactating mother. The four others were developing normally and two of them were proven to be fertile (Chesné *et al.*, 2002).

Rabbits have been extensively used for production of therapeutic proteins in their mammary gland. Milking of the does is relatively simple (Lebas, 1970; Marcus *et al.*, 1990). Milk production varies significantly according to breed, and as much as 200 ml/day can be obtained in New Zealand Whites.

APPLICATIONS OF SOMATIC CELL NUCLEAR TRANSFER IN RABBITS

Rabbit nuclear transfer with embryonic blastomeres represented a major contribution to the development of the field of cloning. The fundamental background learned in rabbit embryology and fetal development has made this species an important model for farm animals and human medicine.

The success of somatic cell nuclear transfer in the mouse makes this species an excellent model for this area of research, especially because of the detailed information on the entire mouse genome. Although information on the genome of rabbits is limited, many aspects of nuclear transfer-related problems in farm animals and primates may still be better studied in a rabbit model than in the mouse because of the higher degree of similarities in embryology and fetal development. Rabbit extraembryonic membranes, for example, more closely resemble those of the human than those of rodents (Foote and Carney, 2000).

The fact that rabbits are good milk producers makes this species a suitable model for mammary gland-related genetic modifications. Transgenic rabbits produced by pronuclear microinjection are used not only as models for sheep and cattle, but as bioreactors for the production of pharmaceutical proteins. The choice of animal species as bioreactors depends on the market demand for the specific protein. For example, to produce enough human serum albumin for the world market (2000 tons/year), dairy cattle would be the best choice. In contrast, human α-glucosidase, which may be used to cure Pompe's disease, is needed in a much smaller amount despite its billion-dollar market value, and approximately 200 rabbits will produce enough of this protein for the entire world (Yang *et al.* 2000). The recent success of rabbit somatic cell nuclear transfer is expected to allow detailed studies on gene knockin and knockout technologies with a major potential impact on pharmaceutical and medical applications.

APPENDIX: PROTOCOL FOR RABBIT SOMATIC-CELL NUCLEAR TRANSFER

A protocol for rabbit somatic cell nuclear transfer is provided below. Many other alternatives exist for almost every step, including ones that are potentially more efficient. In the light of recent success by Chesné *et al.* (2002), the activation and embryo transfer synchrony steps need special attention in this species. This protocol resulted in development of hatched blastocysts with high cell numbers; however, on transfer of cloned zygotes and two-cell-stage embryos, no progeny was obtained (Dinnyes *et al.*, 2001b).

SUPEROVULATION OF OOCYTE DONORS AND OOCYTE COLLECTION

Mature Dutch-belted female rabbits were superovulated (Yang *et al.*, 1990a) with two 0.3-mg and four 0.4-mg subcutaneous injections of follicle-stimulating hormone (FSH) (FSH-P; Schering-Plough Animal Health, Kenilworth, NJ), given 12 hours apart. Twelve hours after the last FSH injection, 100 IU hCG (Sigma, St. Louis, MO) was injected intravenously to induce ovulation. At 13.5 hours after the hCG injection, the animals were laparatomized and ovulated oocytes were flushed from the oviducts with Dulbecco-modified PBS supplemented with 3 mg/ml BSA (Fraction V; Sigma, Cat. No. A9418) (D-PBS). Cumulus cells were removed by short exposure to 10 µg/ml hyaluronidase (Sigma, Cat. No. H3506) in D-PBS solution and subsequent pipetting with a small-bore Pasteur pipette.

ENUCLEATION OF THE OOCYTES

Oocytes, freed of cumulus, were enucleated by micromanipulation. Briefly, oocytes were placed into a small drop of PBS plus 15% fetal bovine serum (FBS; Hyclone, Cat. No. 10099-41) and 5 µg/ml cytochalasin B under oil on a depression slide. An inverted Nikon microscope equipped with Nomarski optics allowed the visualization and removal of the metaphase chromosomes in most of the oocytes, and their removal was under visual control (Collas and Robl, 1990). However, in certain oocytes the coloration and granulation of the cytoplasm made it impossible to see the chromosomes by standard light microscopy. In these cases, oocytes were subjected to Hoechst 33342 staining. The metaphase plate was removed with about 10% of the adjacent cytoplasm, preferably together with the first polar body. Successful enucleation was confirmed by a short (2–3 seconds) ultraviolet light exposure of the presumed karyoplasts removed under an epifluorescent microscope.

DONOR CELL PREPARATION

Fibroblast cells were collected from an ear skin biopsy of an adult male rabbit. The biopsy area was shaved of fur and the surface was cleaned with 70% ethanol. A small piece of tissue (~1 cm^2) was cut from the ear and washed several times in PBS with 10× antibiotic–antimycotic solution (Gibco BRL, Paisley, Scotland; Cat. No. 15240-062), then cut into 1-mm cubes, and placed into a petri dish to culture as explants with DMEM medium plus 10% FBS. After approximately 14 days, the fibroblast cells, growing out of these explants, were washed twice with Ca^{2+}- and Mg^{2+}-free PBS then trypsinized. These cells were washed by centrifugation, resuspended, then used for nuclear transfer (referred to as non passaged, or 0 passage, cells) or cultured further for up to 15 passages in DMEM plus 10% FBS. Nuclear donor cells were isolated from non-serum-starved culture drops supporting a fully confluent cell monolayer for 2–5 days prior to the experiment (Figs. 4 and 5). Alternatively, serum-starved cells were obtained by exposing fully confluent cell cultures to 0.5% FBS in DMEM for 3–5 days. Cell monolayers were trypsinized, and cells were washed by centrifugation in DMEM plus 10% or 0.5% FBS, respectively, then incubated at 39°C in drops and used within 1 hour. Small (10–15 µm in diameter), smooth membrane-surfaced cells were selected for nuclear transfer (Fig. 6).

NUCLEAR TRANSFER

Single donor cells were introduced into the perivitelline space of enucleated oocytes. For electrofusion and activation, the oocyte–fibroblast complexes were manually oriented in a 3.5-mm gap chamber of a BTX 200 Electro Cell Manipulator (San Diego, CA) in 0.3 M mannitol solution containing 0.1 mM calcium chloride and 0.1 mM magnesium chloride, then exposed to a short (2–3 seconds) 0.1-kV AC pulse followed immediately by a 2.4-kV/cm, 60-µsec DC pulse. The fused oocyte–fibroblast complexes were cultured for 2 hours with 2.6 mM DMAP in TCM 199 containing 5.0 µg/ml cytochalasin B (Fig. 7).

EMBRYO CULTURE

Two methods were used: (1) Embryos were cultured for 4.5 days in 100-µl drops of TCM 199 (Gibco, Cat. No. 041 01150H) plus 15% FBS under oil in a humidified atmosphere of 5% CO$_2$ in air at 39°C. (2) Alternatively, embryos were cultured in 100-µl drops of KSOM plus 0.1% BSA for 2 days, and then the medium was replaced with KSOM plus 1% BSA for the remaining culture; all culture took place in a humidified atmosphere of 5% O$_2$, 5% CO$_2$, and 90% N$_2$ at 39°C.

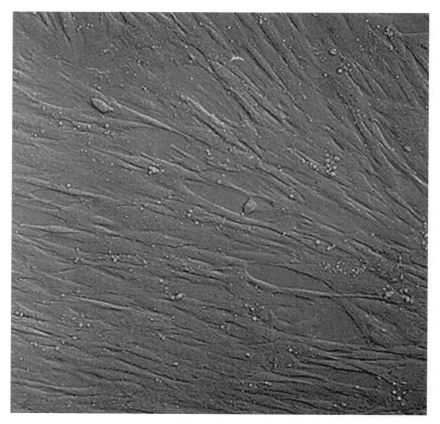

Figure 4 *Adult rabbit fibroblast cell culture.*

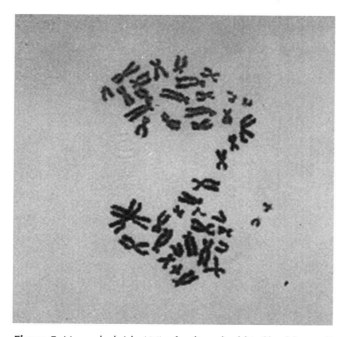

Figure 5 *Normal ploidy (44) of cultured rabbit fibroblast cells.*

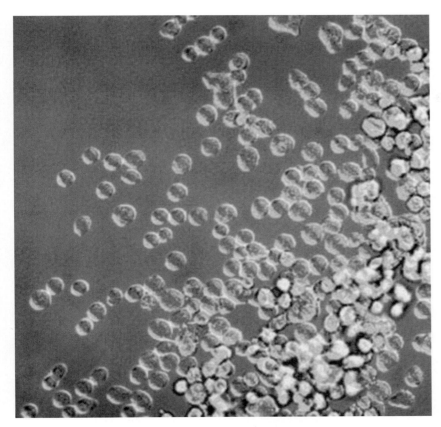

Figure 6 Rabbit fibroblast cells isolated for nuclear transfer.

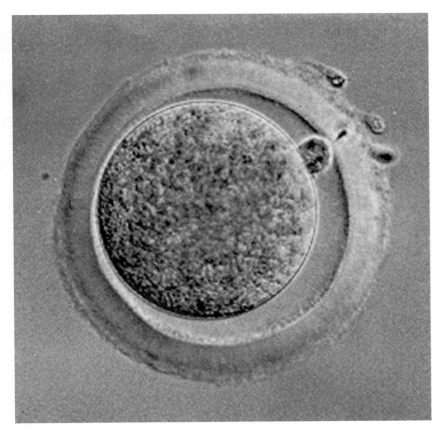

Figure 7 Enucleated cytoplast and fibroblast cell in the process of fusion in rabbit.

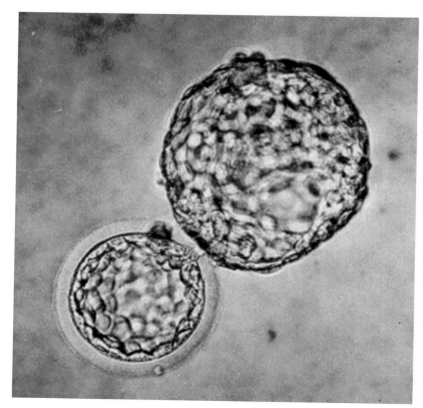

Figure 8 *Hatching day 5 rabbit blastocyst produced by nuclear transfer with an adult fibroblast cell.*

Blastocyst development was recorded and blastocyst cell numbers were counted following Hoechst 33342 epifluorescein staining (Fig. 8).

EMBRYO TRANSFER

Embryos were transferred at the presumed zygote stage (after fusion evaluation) or at the two-cell stage after overnight culture in KSOM medium, as described above. All recipient does were administered 1.2 µg GnRH analog (Cystorelin; Abbot Labs, Athens, GA) to induce ovulation for synchronization with oocyte donors. Embryos were transferred into the oviduct via median laparotomy.

CHARACTERIZATION OF DONOR CELLS

Cell-Specific Marker Staining

Immunocytochemical confirmation of fibroblast phenotype in the cultured cells used for nuclear transfer was performed by staining with monoclonal antibodies directed against the cytoskeletal filament vimentin (fibroblast-specific cell marker) or against cytokeratin (epithelial cell-specific cell marker) (Franke *et al.*, 1978, 1979). Briefly, cells were grown to confluency in Lab-Tek chamber slides (Nalgen Nunc International, Naperville, IL). Cells were washed three times with PBS and fixed in methanol at 4°C for 20 minutes. After fixation the cells were washed three times in PBS and blocked with 3% BSA in PBS for 15 minutes at 37°C. The block was removed and 100 µl of either a 1:40 dilution of antivimentin (Vimentin clone V9, Sigma) or a 1:400 dilution of anticytokeratin (Pan-cytokeratin clone c-11, Sigma) was added. Slides were incubated for 1 hour at 37°C. Cells were washed three times with PBS and incubated for 1 hour with 100 µl of a 1:300 dilution of fluoroscein isothiocyanate (FITC)-labeled antimouse IgG. Cells were washed three times with PBS, coverslipped with 50% glycerol in PBS, and observed under a fluorescent microscope. Appropriate controls for autofluorescence and secondary antibodies were included.

Cell Cycle Determination by Flow Cytometry

Flow cytometry to determine the cell cycle stage profile was performed as described previously (Boquest *et al.*, 1999). Briefly, cells were trypsinized, resuspended in DMEM with 10% FBS at a concentration of approximately 5×10^5 cells/tube. Cells were pelleted and resuspended in 1 ml of 4°C "saline GM" (6.1 mM glucose, 137 mM NaCl, 5.4 mM KCl, 1.5 mM $Na_2HPO_4 \cdot 7H_2O$, 0.9 mM KH_2PO_4, 0.5 mM EDTA). Cells were fixed by slowly adding 4°C ethanol while gently vortexing and were incubated overnight at 4°C. Cells were then washed with PBS and 0.5 mM EDTA and were pelleted. Cell pellets were stained for 1 hour at room temperature with 30 µg/ml of propidium iodide and filtered through a 30-µm mesh (Spectrum, Los Angeles, CA) prior to flow cytometry. Cells were analyzed on a FACs Calibur (Becton Dickinson, San Jose, CA). Ten thousand cells were collected for further cell cycle analysis using the Cell Quest program (Becton Dickinson). The single-parameter histogram of DNA allows for the discrimination of cell populations existing in G0/G1 (2C DNA content), S (between 2C and 4C), and G2 + M (4C) phases of the cell cycle. Percentages were calculated based on the gated cells displaying fluorescence, which correlated to a cell cycle stage.

ACKNOWLEDGMENTS

This manuscript is a scientific contribution (#2089) of the Storrs Experiment Station at the University of Connecticut. A.D. was supported by OTKA Grant No.T030698. The authors wish to thank Marina Julian for careful editing of this manuscript.

REFERENCES

Adams, C. E. (1958). Egg development in the rabbit. The influence of post-coital ligation on the uterine tube and ovariectomy. *J. Endocrinol.* **16**, 283–294.

Adams, C. E. (1960). Prenatal mortality in the rabbit *Oryctolagus cuniculus. J. Reprod. Fertil.* **1**, 36–44.

Adams, C. E. (1970). Maintenance of pregnancy relative to the presence of few embryos in the rabbit. *J. Endocrinol.* **48**, 243–249.

Adams, C. E., Hay, M. F., and Lutwak-Mann, C. (1961). The action of various agents upon the rabbit embryo. *J. Embryol. Exp. Morphol.* **9**, 468–491.

al-Hasani, S., Kirsch, J., Diedrich, K., Blanke, S., van der Ven, H., and Krebs, D. (1989). Successful embryo transfer of cryopreserved and *in-vitro* fertilized rabbit oocytes. *Hum. Reprod.* **4**, 77–79.

Anderson, J., and Henck, J. W. (1994). Toxicity and safety testing. *In "The Biology of the Laboratory Rabbit"* (P. J. Manning, D. H. Ringler and C. E. Newcomer, eds.), pp. 455–466. Academic Press, San Diego.

Baguisi, A., Behboodi, E., Melican, D. T., Pollock, J. S., Destrempes, M. M., Cammuso, C., Williams, J. L., Nims, S. D., Porter, C. A., Midura, P., Palacios, M. J., Ayres, S. L., Denniston, R. S., Hayes, M. L., Ziomek, C. A., Meade, H. M., Godke, R. A., Gavin, W. G., Overstrom, E. W., and Echelard, Y. (1999). Production of goats by somatic cell nuclear transfer. *Nat. Biotechnol.* **17**, 456–461.

Barnes, F. L., and Eyestone, W. H. (1990). Early cleavage and the maternal zygotic transition in bovine embryos. *Theriogenology* **33**, 141–152.

Barnes, F. L., Collas, P., Powell, R., King, W. A., Westhusin, M., and Shepherd, D. (1993). Influence of recipient oocyte cell cycle stage on DNA synthesis, nuclear envelope breakdown, chromosome constitution, and development in nuclear transplant bovine embryos. *Mol. Reprod. Dev.* **36**, 33–41.

Beatty, R. A. (1958). Variation in the number of corpora lutea and in the number and size of 6-day blastocysts in rabbits subjected to superovulation treatment. *J. Endocrinol.* **17**, 248–260.

Besenfelder, U., and Brem, G. (1993). Laparoscopic embryo transfer in rabbits. *J. Reprod. Fertil.* **99**, 53–56.

Besenfelder, U., Strouhal, C., and Brem, G. (1998). A method for endoscopic embryo collection and transfer in the rabbit. *Zentralbl. Veterinarmed. A* **45**, 577–579.

Betthauser, J., Forsberg, E., Augenstein, M., Childs, L., Eilertsen, K., Enos, J., Forsythe, T., Golueke, P., Jurgella, G., Koppang, R., Lesmeister, T., Mallon, K., Mell, G., Misica, P., Pace, M., Pfister-Genskow,

M., Strelchenko, N., Voelker, G., Watt, S., Thompson, S., and Bishop, M. (2000). Production of cloned pigs from *in vitro* systems. *Nat. Biotechnol.* **18**, 1055–1059.

Boquest, A. C., Day, B. N., and Prather, R. S. (1999). Flow cytometric cell cycle analysis of cultured porcine fetal fibroblast cells. *Biol. Reprod.* **60**, 1013–1019.

Brunet-Simon, A., Henrion, G., Renard, J. P., and Duranthon, V. (2001). Onset of zygotic transcription and maternal transcript legacy in the rabbit embryo. *Mol. Reprod. Dev.* **58**, 127–136.

Campbell, K. H. S., Loi, P., Otaegui, P. J., and Wilmut, I. (1996a). Cell cycle co-ordination in embryo cloning by nuclear transfer. *Rev. Reprod.* **1**, 40–46.

Campbell, K. H. S., McWhir, J., Ritchie, W. A., and Wilmut, I. (1996b). Sheep cloned by nuclear transfer from a cultured cell line. *Nature* **380**, 64–66.

Capecchi, M. (2000). How close are we to implementing gene targeting in animals other than the mouse. *Proc. Natl. Acad. Sci. U.S.A.* **97**, 956–957.

Carney, E. W., and Foote, R. H. (1990). Effects of superovulation, embryo recovery, culture system and embryo transfer on development of rabbit embryos *in vivo* and *in vitro*. *J. Reprod. Fertil.* **89**, 543–551.

Carney, E. W., and Foote, R. H. (1991). Improved development of rabbit one-cell embryos to the hatching blastocyst stage by culture in a defined, protein-free culture medium. *J. Reprod. Fertil.* **91**, 113–123.

Carney, E. W., Tobback, C., Ellington, J. E., and Foote, R. H. (1990). Co-culture of rabbit 2-cell embryos with rabbit oviduct epithelial cells and other somatic cells. *Mol. Reprod. Dev.* **27**, 209–215.

Chang, M. C. (1951). Fertilization capacity of sperm deposited in the fallopian tube. *Nature (London)* **168**, 697.

Chang, M. C., Casas, J. H., and Hunt, D. M. (1971). Development of ferret eggs after 2 to 3 days in the rabbit fallopian tube. *J. Reprod. Fertil.* **25**, 129–131.

Chesné, P., Adenot, P., Boulanger, L., and Renard, J. P. (2001). Somatic nuclear transfer in the rabbit. *Theriogenology* **55**, 260 (abstr.).

Chesné, P., Adenot, P. G., Viglietta, C., Baratte, M., Boulanger, L., and Renard, J.-P. (2002). Cloned rabbits produced by nuclear transfer from adult somatic cells. *Nat. Biotechnol.* **20**, 366–369.

Christians, E., Rao, V. H., and Renard, J. P. (1994). Sequential acquisition of transcriptional control during early embryonic development in the rabbit. *Dev. Biol.* **164**, 160–172.

Cibelli, J. B., Stice, S. L., Golueke, P. J., Kane, J. J., Jerry, J., Blackwell, C., de Leon, F. A. P., and Robl, J. M. (1998). Cloned transgenic calves produced from nonquiescent fetal fibroblasts. *Science* **280**, 1256–1258.

Collas, P., and Robl, J. M. (1990). Factors affecting the efficiency of nuclear transplantation in the rabbit embryo. *Biol. Reprod.* **43**, 877–884.

Collas, P., and Robl, J. M. (1991). Relationship between nuclear remodeling and development in nuclear transplant rabbit embryos. *Biol. Reprod.* **45**, 455–465.

Collas, P., Duby, R. T., and Robl, J. M. (1991). *In vitro* development of rabbit pronuclear embryos in rabbit peritoneal fluid. *Biol. Reprod.* **44**, 1100–1107.

Deng, M., and Yang, X. J. (2001). Full term development of rabbit oocytes fertilized by intracytoplasmic sperm injection. *Mol. Reprod. Dev.* **59**, 38–43.

Denker, H. W., and Gerdes, H. J. (1979). The dynamic structure of rabbit blastocyst coverings. I. Transformation during regular preimplantation development. *Anat. Embryol. (Berlin)* **157**, 15–34.

Denning, C., Burl, S., Ainslie, A., Bracken, J., Dinnyes, A., Fletcher J., King, T., Ritchie, M., Ritchie, W. R., Rollo, M., De Sousa, P., Travers, A., Wilmut, I., and Clark, A. J. (2001). Deletion of the α(1,3)galactosyl transferase (GGTA1) gene and the prion protein (PrP) gene in sheep. *Nat. Biotechnol.* **19**, 559–562.

Dinnyes, A., Dai, Y. P., Jiang, S., and Yang, X. Z. (2000). High developmental rates of vitrified bovine oocytes following parthenogenetic activation, *in vitro* fertilization, and somatic cell nuclear transfer. *Biol. Reprod.* **63**, 513–518.

Dinnyes, A., King, T., Wilmut, I., and De Sousa, P. A. (2001a). Sheep somatic cell nuclear transfer: Effect of breed and culture system on embryonic and fetal development. *Theriogenology* **55**, 264.

Dinnyes, A., Dai, Y., Barber, M., Liu, L., Xu, J., Zhou, P., and Yang, X. (2001b). Development of cloned embryos from adult rabbit fibroblasts: Effect of activation treatment and donor cell preparation. *Biol. Reprod.* **64**, 257–263.

Du, F., Giles, J. R., Foote, R. H., Graves, K. H., Yang, X., and Moreadith, R. W. (1995). Nuclear transfer of putative rabbit embryonic stem cells leads to normal blastocyst development. *J. Reprod. Fertil.* **104**, 219–223.

Edwards, J. A. (1968). The external development of the rabbit and rat embryo. *In* "Advances in Teratology" (D. H. M. Woollam, ed.), pp. 239–262. Academic Press, New York.

Ellington, J. E., Farrell, P. B., Simkin, M. E., Foote, R. H., Goldman, E. E., and McGrath, A. B. (1990). Development and survival after transfer of cow embryos cultured from 1–2 cells to morulae or blastocysts in rabbit oviducts or in a simple medium with bovine oviduct epithelial cells. *J. Reprod. Fertil.* **89**, 293–299.

Escriba, M. J., and Garcia-Ximenez, F. (1999). Electroactivation of rabbit oocytes in an hypotonic pulsing medium and parthenogenetic *in vitro* development without cytochalasin B-diploidizing pretreatment. *Theriogenology* **51**, 963–973.

Escriba, M. J., and Garcia-Ximenez, F. (2000). Influence of sequence duration and number of electrical pulses upon rabbit oocyte activation and parthenogenetic *in vitro* development. *Anim. Reprod. Sci.* **59**, 99–107.

Farrell, P. B., and Foote, R. H. (1995). Beneficial effects of culturing rabbit zygotes to blastocysts in 5% oxygen and 10% carbon dioxide. *J. Reprod. Fertil.* **103**, 127–130.

Feussner, E. L., Lightkep, G. E., Hennesy, R. A., Hoberman, A. M., and Christian, M. S. (1992). A decade of rabbit fertility data: study of historical control animals. *Teratology* **46**, 349–365.

Fischer, B. (1989). Effects of asynchrony on rabbit blastocyst development. *J. Reprod. Fertil.* **86**, 479–491.

Fischer, B., Mootz, U., Denker, H. W., Lambertz, M., and Beier, H. M. (1991). The dynamic structure of rabbit blastocyst coverings. III. Transformation of coverings under non-physiological developmental conditions. *Anat. Embryol. (Berlin)* **183**, 17–27.

Fissore, R. A., and Robl, J. M. (1993). Sperm, inositol trisphosphate, and thimerosal-induced intracellular Ca^{2+} elevations in rabbit eggs. *Dev. Biol.* **159**, 122–130.

Fissore, R. A., and Robl, J. M. (1994). Mechanism of calcium oscillations in fertilized rabbit eggs. *Dev. Biol.* **166**, 634–642.

Foote, R. H., and Carney, E. W. (2000). The rabbit as a model for reproductive and developmental toxicity studies. *Reprod. Toxicol.* **14**, 477–493.

Franke, W. W., Schmid, E., Osborn, M., and Weber, K. (1978). Different intermediate-sized filaments distinguished by immunofluorescence microscopy. *Proc. Natl. Acad. Sci. U.S.A.* **75**, 5034–5038.

Franke, W. W., Schmid, E., Winter, S., Osborn, M., and Weber, K. (1979). Widespread occurrence of intermediate-sized filaments of the vimentin- type in cultured cells from diverse vertebrates. *Exp. Cell. Res.* **123**, 25–46.

Galat, V. V., Lagutina, I. S., Mesina, M. N., Chernich, V. J., and Prokofiev, M. I. (1999). Developmental potential of rabbit nuclear transfer embryos derived from donor fetal fibroblast. *Theriogenology* **51**, 203 (abstr.).

Giles, J. R., and Foote, R. H. (1997). Effects of gas atmosphere, platelet-derived growth factor and leukemia inhibitory factor on cell numbers of rabbit embryos cultured in a protein-free medium. *Reprod. Nutr. Dev.* **37**, 97–104.

Giles, J. R., Yang, X., Mark, W., and Foote, R. H. (1993). Pluripotency of cultured rabbit inner cell mass cells detected by isozyme analysis and eye pigmentation of fetuses following injection into blastocysts or morulae. *Mol. Reprod. Dev.* **36**, 130–138.

Grocholova, R., Petr, J., Rozinek, J., and Jilek, F. (1997). The protein phosphatase inhibitor okadaic acid inhibits exit from metaphase II in parthenogenetically activated pig oocytes. *J. Exp. Zool.* **277**, 49–56.

Hafez, E. S. E. (1964). Effects of over-crowding *in utero* on implantation and fetal development in the rabbit. *J. Exp. Zool.* **156**, 269–288.

Hagen, K. W. (1974). Colony husbandry. In *"The Biology of the Laboratory Rabbit"* (S. H. Weisbroth, R. E. Flatt, and A. L. Kraus, eds.), pp. 23–47. Academic Press, New York.

Hammer, R. E., Pursel, V. G., Rexroad, C. E., Wall, R. J., Bolt, D. J., Ebert, K. M., Palmiter, R. D., and Brinster, R. L. (1985). Production of transgenic rabbits, sheep and pigs by microinjection. *Nature* **315**, 680–683.

Harkness, J. E., and Wagner, J. E. (1995). *"The Biology and Medicine of Rabbits and Rodents."* William and Wilkins, Baltimore.

Heape, W. (1890). Preliminary note on the transplantation and growth of mammalian ova within a uterine foster-mother. *Proc. R. Soc. Lond. B Biol. Sci.* **48**, 457–459.

Henrion, G., Renard, J. P., Chesné, P., Oudin, J. F., Maniey, D., Brunet, A., Osborne, H. B., and Duranthon, V. (2000). Differential regulation of the translation and the stability of two maternal transcripts in preimplantation rabbit embryos. *Mol. Reprod. Dev.* **56**, 12–25.

Heyman, Y., and Renard, J. P. (1996). Cloning of domestic species. *Anim. Reprod. Sci.* **42**, 427–436.

Heyman, Y., Chesné, P., and Renard, J. P. (1990). Reprogrammation complete de noyaux embryonnaires congeles apres transfert nucleaire chez le lapin. *C.R. Acad. Sci. (Paris)* **311**, 321–326.

Jin, D. I., Kim, D. K., Im, K. S., and Choi, W. S. (2000). Successful pregnancy after transfer of rabbit blastocysts grown in vitro from single-cell zygotes. *Theriogenology* **54**, 1109–1116.

Kane, M. T., and Foote, R. H. (1971). Factors affecting blastocyst expansion of rabbit zygotes and young embryos in defined media. *Biol. Reprod.* **4**, 41–47.

Kanka, J., Hozak, P., Heyman, Y., Chesné, P., Degrolard, J., Renard, J. P., and Flechon, J. E. (1996). Transcriptional activity and nucleolar ultrastructure of embryonic rabbit nuclei after transplantation to enucleated oocytes. *Mol. Reprod. Dev.* **43**, 135–144.

Kasai, M., Hamaguchi, Y., Zhu, S. E., Miyake, T., Sakurai, T., and Machida, T. (1992). High survival of rabbit morulae after vitrification in an ethylene glycol-based solution by a simple method. *Biol. Reprod.* **46**, 1042–1046.

Kasinathan, P., Knott, J. G., Moreira, P. N., Burnside, A. S., Joseph, J. D., and Robl, J. M. (2001). Effect of fibroblast donor cell age and cell cycle on development of bovine nuclear transfer embryos *in vitro*. *Biol. Reprod.* **64**, 1487–1493.

Kato, Y., Tani, T., Sotomaru, Y., Kurokawa, K., Kato, J., Doguchi, H., Yasue, H., and Tsunoda, Y. (1998). Eight calves cloned from somatic cells of a single adult. *Science* **282**, 2095–2098.

Kato, Y., Yabuuchi, A., Motosugi, N., Kato, J., and Tsunoda, Y. (1999). Developmental potential of mouse follicular epithelial cells and cumulus cells after nuclear transfer. *Biol. Reprod.* **61**, 1110–1114.

Kauffman, R. D., Schmidt, P. M., Rall, W. F., and Hoeg, J. M. (1998). Superovulation of rabbits with FSH alters *in vivo* development of vitrified morulae. *Theriogenology* **50**, 1081–1092.

Kenelly, J. J., and Foote, R. H. (1965). Superovulatory response of pre- and post-puberal rabbits to commercially available gonadotrophins. *J. Reprod. Fertil.* **9**, 131–145.

Kidder, J. D., Roberts, P. J., Simkin, M. E., Foote, R. H., and Richmond, M. E. (1999). Nonsurgical collection and nonsurgical transfer of preimplantation embryos in the domestic rabbit (*Oryctolagus cuniculus*) and domestic ferret (*Mustela putorius furo*). *J. Reprod. Fertil.* **116**, 235–242.

Koyama, H., Suzuki, H., Yang, X., Jiang, S., and Foote, R. H. (1994). Analysis of polarity of bovine and rabbit embryos by scanning electron microscopy. *Biol. Reprod.* **50**, 163–170.

Kubota, C., Yamakuchi, H., Todoroki, J., Mizoshita, K., Tabara, N., Barber, M., and Yang, X. Z. (2000). Six cloned calves produced from adult fibroblast cells after long-term culture. *Proc. Natl. Acad. Sci. U.S.A.* **97**, 990–995.

Lagutina, I. S., Zakhartchenko, V. I., and Prokofiev, M. I. (1999). Nuclear transfer in rabbits and factors affecting it efficiency. *Theriogenology* **51**, 207 (abstr.).

Lagutina, I. S., Mezina, M. N., Chernikh, V. J., Prokofiev, M. I., and Galat, V. V. (2000). Developmental potential of rabbit nuclear transfer embryos produced by various fusion/activation protocols. *Theriogenology* **53**, 230 (abstr.).

Lanza, R. P., Cibelli, J. B., Diaz, F., *et al.* (2000). Cloning of an endangered species (*Bos gaurus*) using interspecies nuclear transfer. *Cloning* **2**, 79–91.

Lebas, F. (1970). Description d'une machine a traire les lapines. *Ann. Zootech.* **19**, 223–228.

Leiser, R., and Denker, H. W. (1988). The dynamic structure of rabbit blastocyst coverings. II. Ultrastructural evidence for a role of the trophoblast in neozona formation. *Anat. Embryol. (Berlin)* **179**, 129–134.

Li, J., and Foote, R. H. (1996). Differential sensitivity of one-cell and two-cell rabbit embryos to sodium chloride and total osmolarity during culture into blastocysts. *J. Reprod. Fertil.* **108**, 307–312.

Li, J., Foote, R. H., and Simkin, M. (1993). Development of rabbit zygotes cultured in protein-free medium with catalase, taurine, or superoxide dismutase. *Biol. Reprod.* **49**, 33–37.

Liu, L., Ju, J. C., and Yang, X. (1998). Differential inactivation of maturation-promoting factor and mitogen-activated protein kinase following parthenogenetic activation of bovine oocytes. *Biol. Reprod.* **59**, 537–545.

Lonergan, P., Dinnyes, A., Fair, T., Yang, X., and Boland, M. (2000). Bovine oocyte and embryo development following meiotic inhibition with butyrolactone I. *Mol. Reprod. Dev.* **57**, 204–209.

Marcus, G. E., Shum, F. T., and Goldman, S. L. (1990). A device for collecting milk from rabbits. *Lab. Anim. Sci.* **40**, 219–221.

Maurer, R. R. (1978). Advances in rabbit embryo culture. In *"Methods in Mammalian Reproduction"* (J. C. Daniel, ed.), pp. 259–272. Academic Press, New York.

Maurer, R. R., Hunt, W. L., Van Vleck, L. D., and Foote, R. H. (1968). Developmental potential of superovulated rabbit ova. *J. Reprod. Fertil.* **15**, 171–175.

Mitalipov, S. M., White, K. L., Farrar, V. R., Morrey, J., and Reed, W. A. (1999). Development of nuclear transfer and parthenogenetic rabbit embryos activated with inositol 1,4,5-trisphosphate. *Biol. Reprod.* **60**, 821–827.

Modlinski, J. A., and Smorag, Z. (1991). Preimplantation development of rabbit embryos after transfer of embryonic nuclei into different cytoplasmic environment. *Mol. Reprod. Dev.* **28**, 361–372.

Moens, A., Chastant, S., Chesné, P., Flechon, J. E., Betteridge, K. J., and Renard, J. P. (1996). Differential ability of male and female rabbit fetal germ cell nuclei to be reprogrammed by nuclear transfer. *Differentiation* **60**, 339–345.

Murakami, H., and Imai, H. (1996). Successful implantation of *in vitro* cultured rabbit embryos after uterine transfer: A role for mucin. *Mol. Reprod. Dev.* **43**, 167–170.

Nussbaum, D. J., and Prather, R. S. (1995). Differential effects of protein synthesis inhibitors on porcine oocyte activation. *Mol. Reprod. Dev.* **41**, 70–75.

Onishi, A., Iwamoto, M., Akita, T., Mikawa, S., Takeda, K., Awata, T., Hanada, H., and Perry, A. C. F. (2000). Pig cloning by microinjection of fetal fibroblast nuclei. *Science* **289**, 1188–1190.

Orsini, M. W. (1962). Study of ovo-implantation in the hamster, rat, mouse, guinea-pig, and rabbit in cleared uterine tracts. *J. Reprod. Fertil.* **3**, 288–293.

Ozil, J. P. (1990). The parthenogenetic development of rabbit oocytes after repetitive pulsatile electrical stimulation. *Development* **109**, 117–127.

Ozil, J. P., and Huneau, D. (2001). Activation of rabbit oocytes: the impact of the Ca^{2+} signal regime on development. *Development* **128**, 917–928.

Park, C. S., Jeon, B. G., Lee, K. M., Yin, X. J., Cho, S. K., Kong, I. K., Lee, H. J., and Choe, S. Y. (1998). Production of cloned rabbit embryos and offsprings by nuclear transplantation using *in vitro* matured oocytes. *Theriogenology* **49**, 325 (abst).

Petters, R. M., and Wells, K. D. (1993). Culture of pig embryos. *J. Reprod. Fertil. (Suppl.)* **48**, 61–73.

Piotrowska, K., Modlinski, J. A., Korwin-Kossakowski, M., and Karasiewicz, J. (2000). Effects of pre-activation of ooplasts or synchronization of blastomere nuclei in G1 on preimplantation development of rabbit serial nuclear transfer embryos. *Biol. Reprod.* **63**, 677–682.

Polejaeva, I. A., Chen, S. H., Vaught, T. D., Page, R. L., Mullins, J., Ball, S., Dai, Y. F., Boone, J., Walker, S., Ayares, D. L., Colman, A., and Campbell, K. H. S. (2000). Cloned pigs produced by nuclear transfer from adult somatic cells. *Nature* **407**, 86–90.

Prather, R. S., Mayes, M. A., and Murphy, C. N. (1997). Parthenogenetic activation of pig eggs by exposure to protein kinase inhibitors. *Reprod. Fertil. Dev.* **9**, 539–544.

Rao, V. H., Heyman, Y., Chesné, P., and Renard, J. P. (1998). Freezing and recloning of nuclear transfer embryos in rabbit. *Theriogenology* **49**, 328 (abst).

Renard, J. P., Bui, X. N., and Garnier, V. (1984). Two-step freezing of two-cell rabbit embryos after partial dehydration at room temperature. *J. Reprod. Fertil.* **71**, 573–580.

Schnieke, A. E., Kind, A. J., Ritchie, W. A., Mycock, K., Scott, A. R., Ritchie, M., Wilmut, I., Colman, A., and Campbell, K. H. S. (1997). Human factor IX transgenic sheep produced by transfer of nuclei from transfected fetal fibroblasts. *Science* **278**, 2130–2133.

Schultz, R. M., Davis, W., Jr., Stein, P., and Svoboda, P. (1999). Reprogramming of gene expression during preimplantation development. *J. Exp. Zool.* **285**, 276–282.

Shiga, K., Fujita, T., Hirose, K., Sasae, Y., and Nagai, T. (1999). Production of calves by transfer of nuclei from cultured somatic cells obtained from Japanese black bulls. *Theriogenology* **52**, 527–535.

Smith, L. C., and Wilmut, I. (1989). Influence of nuclear and cytoplasmic activity on the development *in vivo* of sheep embryos after nuclear transplantation. *Biol. Reprod.* **40**, 1027–1035.

Stice, S. L., and Keefer, C. L. (1993). Multiple generational bovine embryo cloning. *Biol. Reprod.* **48**, 715–719.

Stice, S. L., and Robl, J. M. (1988). Nuclear reprogramming in nuclear transplant rabbit embryos. *Biol. Reprod.* **39**, 657–664.

Stice, S. L., and Robl, J. M. (1990). Activation of mammalian oocytes by a factor obtained from rabbit sperm. *Mol. Reprod. Dev.* **25**, 272–280.

Stice, S. L., Keefer, C. L., Maki-Laurila, M., and Matthews, L. (1993). Donor blastomere cell cycle stage affects developmental competence of bovine nuclear transfer embryos. *Theriogenology* **39**, 318 (abst).

Susko-Parrish, J. L., Leibfried-Rutledge, M. L., Northey, D. L., Schutzkus, V., and First, N. L. (1994). Inhibition of protein kinases after an induced calcium transient causes transition of bovine oocytes to embryonic cycles without meiotic completion. *Dev. Biol.* **166**, 729–739.

Swann, K., and Lai, F. A. (1997). A novel signalling mechanism for generating Ca^{2+} oscillations at fertilization in mammals. *BioEssays* **19**, 371–378.

Tan, J. H., Zhou, Q., Li, Z. Y., Sun, X. S., Liu, Z. H., and He, G. X. (1997). Embryo cloning by nuclear transplantation in rabbits. *Theriogenology* **47**, 237 (abst).

Telford, N. A., Watson, A. J., and Schultz, G. A. (1990). Transition from maternal to embryonic control in early mammalian development: a comparison of several species. *Mol. Reprod. Dev.* **26**, 90–100.

Thompson, E. M., Legouy, E., Christians, E., and Renard, J. P. (1995). Progressive maturation of chromatin structure regulates HSP70.1 gene expression in the preimplantation mouse embryo. *Development* **121**, 3425–3437.

Totey, S. M., Singh, G., Taneja, M., Pawshe, C. H., and Talwar, G. P. (1992). *In vitro* maturation, fertilization and development of follicular oocytes from buffalo (*Bubalus bubalis*). *J. Reprod. Fertil.* **95**, 597–607.

Tsunoda, Y., Soma, T., and Sugie, T. (1982). Effect of post-ovulatory age of recipient on survival of frozen-thawed rabbit morulae. *J. Reprod. Fertil.* **65**, 483–487.

Varian, N. B., Maurer, R. R., and Foote, R. H. (1967). Ovarian response and cleavage rate of ova in control and FSH-primed rabbits receiving varying levels of luteinizing hormone. *J. Reprod. Fertil.* **13**, 67–73.

Vignon, X., Chesné, P., Le Bourhis, D., Flechon, J. E., Heyman, Y., and Renard, J. P. (1998). Developmental potential of bovine embryos reconstructed from enucleated matured oocytes fused with cultured somatic cells. *C.R. Acad. Sci. Paris* **321**, 735–745.

Vincent, C., Garnier, V., Heyman, Y., and Renard, J. P. (1989). Solvent effects on cytoskeletal organization and *in-vivo* survival after freezing of rabbit oocytes. *J. Reprod. Fertil.* **87**, 809–820.

Wakayama, T., and Yanagimachi, R. (2001). Mouse cloning with nucleus donor cells of different age and type. *Mol. Reprod. Dev.* **58**, 376–383.

Wakayama, T., Perry, A. C. F., Zuccotti, M., Johnson, K. R., and Yanagimachi, R. (1998). Full-term development of mice from enucleated oocytes injected with cumulus cell nuclei. *Nature* **394**, 369–374.

Wang, W., Sun, Q., Hosoe, M., and Shioya, Y. (1997). Calcium- and meiotic-spindle-independent activation of pig oocytes by the inhibition of staurosporine-sensitive protein kinases. *Zygote* **5**, 75–82.

Wells, D. N., Misica, P. M., McMillan, W. H., and Tervit, H. R. (1998). Production of cloned bovine fetuses following nuclear transfer using cells from a fetal fibroblast cell line. *Theriogenology* **49**, 330.

Willadsen, S. M. (1986). Nuclear transplantation in sheep embryos. *Nature* **320**, 63–65.

Willadsen, S. M. (1989). Cloning of sheep and cow embryos. *Genome* **31**, 956–962.

Wilmut, I., Schnieke, A. E., McWhir, J., Kind, A. J., and Campbell, K. H. S. (1997). Viable offspring derived from fetal and adult mammalian cells. *Nature* **385**, 810–813.

Yang, X. (1991). Featured article: Embryo cloning by nuclear transfer in cattle and rabbits. *Embryo Transf. Newslett.* **9**, 10–22.

Yang, X. Z., and Foote, R. H. (1987). Production of identical twin rabbits by micromanipulation of embryos. *Biol. Reprod.* **37**, 1007–1014.

Yang, X. Z., and Foote, R. H. (1990). Survival of bisected rabbit morulae transferred to synchronous and asynchronous recipients. *Mol. Reprod. Dev.* **26,** 6–11.

Yang, X., Zhang, L., Kovacs, A., Tobback, C., and Foote, R. H. (1990a). Potential of hypertonic medium treatment for embryo micromanipulation: II. Assessment of nuclear transplantation methodology, isolation, subzona insertion, and electrofusion of blastomeres to intact or functionally enucleated oocytes in rabbits. *Mol. Reprod. Dev.* **27,** 118–129.

Yang, X., Chen, Y., Chen, J., and Foote, R. H. (1990b). Potential of hypertonic medium treatment for embryo micromanipulation: I. Survival of rabbit embryos *in vitro* and *in vivo* following sucrose treatment. *Mol. Reprod. Dev.* **27,** 110–117.

Yang, X., Jiang, S., Kovacs, A., and Foote, R. H. (1992). Nuclear totipotency of cultured rabbit morulae to support full-term development following nuclear transfer. *Biol. Reprod.* **47,** 636–643.

Yang, X., Tian, X. C., Dai, Y., and Wang, B. (2000). Transgenic farm animals: Applications in agriculture and biomedicine. *In* "Biotechnology Annual Review" (M. R. El-Gewely, ed.), Vol. 5, pp. 269–292. Elsevier Science, New York.

Yin, X. J., Tani, T., Kato, Y., and Tsunoda, Y. (2000). Development of rabbit parthenogenetic oocytes and nuclear-transferred oocytes receiving cultured cumulus cells. *Theriogenology* **54,** 1469–1476.

Zakhartchenko, V., Alberio, R., Stojkovic, M., Prelle, K., Schernthaner, W., Stojkovic, P., Wenigerkind, H., Wanke, R., Duchler, M., Steinborn, R., Mueller, M., Brem, G., and Wolf, E. (1999). Adult cloning in cattle: Potential of nuclei from a permanent cell line and from primary cultures. *Mol. Reprod. Dev.* **54,** 264–272.

Zhou, Q., Li, Z. Y., Liu, Z. H., Sun, X. S., He, G. X., and Tan, J. H. (1997). Nuclear transplantation by microinjection in rabbits. *Theriogenology* **47,** 239 (abstr.).

Ziomek, C. A., Chatot, C. L., and Manes, C. (1990). Polarization of blastomeres in the cleaving rabbit embryo. *J. Exp. Zool.* **256,** 84–91.

CHAPTER 18

NUCLEAR TRANSFER IN SWINE

Randall S. Prather

INTRODUCTION

Cloning technologies have been applied to swine. This has resulted in a number of cloned offspring. The goal is to produce animals with specific genetic modifications. In this section a history of the development of the techniques is provided. In the next section a review of the information regarding reprogramming in swine after nuclear transfer is discussed. Finally, the reports of offspring derived from cultured cells are examined.

The cloning and production of transgenic swine is an important endeavor. Swine have become important in biomedical research because they are excellent models for cardiovascular disease, cutaneous pharmacology and toxicology research, lipoprotein metabolism, and pathobiology of intestinal transport, injury, and repair, as well as being considered a potential source of organs for xenotransplantation. Development and improvements in the system of cloning pig embryos will have enormous implications throughout the biomedical field. The process of cloning could lead to developments in creating models for cystic fibrosis, osteoporosis, or diabetes. Cloning could have additional applications in improving the efficiency of meat production and could have a large impact on feeding the growing world population. Addition or removal of specific genes responsible for reproduction, disease resistance, or production characteristics could make production of pork more efficient. Thus, because the pig is the species of choice for understanding many human diseases and pork plays an important role in feeding the world population, we have chosen to work to improve the systems for cloning by nuclear transplantation. It is anticipated that this review will provide history and direction for future studies to further improve and understand the cloning process in swine for biomedical purposes as well as agricultural purposes.

INITIAL STUDIES

The first report of nuclear transfer in the pig was in 1985 by Robl and First (1986) at the Third International Congress of Pig Reproduction held in Columbia, Missouri. The authors simply reported that nuclei could be transferred between two two-cell-stage pig embryos, but their published paper showed (1) that the process of nuclear transfer could be accomplished and (2) how to go about performing a nuclear transfer in the pig. This was followed by reports in sheep (Willadsen, 1986), cattle (Prather *et al.*, 1987) and rabbits (Stice and Robl, 1988), and finally nuclear transfer of a four-cell-stage donor nucleus in the pig was shown to result in term development (Prather *et al.*, 1989a). The production of offspring in the pig from a four-cell-stage donor was not repeated until 2000 (Li *et al.*, 2000). With this background, many studies were initiated to evaluate the changes in nuclear structure and function after nuclear transfer, as well as to attempt to produce offspring. The focus here is first on the changes that occur to nuclei when they are transferred to oocytes, and then their development *in vitro* and finally the production of offspring.

MOLECULAR EVENTS OF REPROGRAMMING AS REVEALED BY STUDIES IN PIGS

Swine embryos have served as an important model for many of the studies that have described the nuclear remodeling that occurs after nuclear transfer. The premise is that if the donor nucleus is to be reprogrammed, it first must be remodeled to be similar to that of a pronucleus. Generally the pronuclei are considered to be transcriptionally inactive. At a species-specific cell stage the embryo begins producing significant amounts of mRNA. In the pig this transition occurs during the four-cell stage (Jarrell *et al.*, 1991; Schoenbeck *et al.*, 1992). This cell stage is correlated with a number of morphological changes to the nuclei. Thus the initial discussion here focuses on morphological changes after nuclear transfer. To set the stage for this discussion a number of morphological differences between a donor nucleus and a pronucleus must be identified. Some of these changes have been identified in the mouse (Prather and Schatten, 1992).

The most obvious difference in morphology between a pronucleus and nuclei from other cell types is the size of the pronucleus. In the fertilized egg the pronuclei are large—larger than many somatic cells. During cleavage the size of the nuclei diminishes as the size of the cell diminishes. In pigs the size of a nucleus from a 2-cell-stage embryo is over 18 μm, and this diminishes to 14.3 and 13.0 μm for nuclei from 8- and 16-cell embryos, respectively (Prather *et al.*, 1990). When nuclei from these 8- and 16-cell embryos are transferred to enucleated (and subsequently activated) meiotic metaphase II oocytes, the nuclei swelled to 27.3 or 27.2 μm, respectively. Interestingly, the degree of swelling was independent of the type of oocyte, i.e., a one-half oocyte versus a relatively intact oocyte. Thus the amount of cytoplasm, in this case, is not limiting the growth of the "pronuclei." Others have clearly shown that this swelling is an important aspect of the remodeling process (Liu *et al.*, 1995) because it indirectly indicates the exchange of proteins that occurs when the donor nucleus is transferred into the cytoplasm of the oocyte.

Nuclear swelling is an indirect indicator of the exchange of proteins that occurs after nuclear transfer. One specific class of proteins that is known to change in composition during early cleavage is the nuclear lamins, and as such they can be used to illustrate this exchange of protein. In mammals there are at least two families of these proteins, broadly categorized into nuclear lamins B and A/C. Lamins A and C are identical, except for an 82-amino acid tail on lamin A. The nuclear lamins are intermediate filament-type proteins that polymerize into a sphere surrounding the chromatin within the nuclear envelope. These proteins depolymerize during metaphase and are thus depolymerized in the oocyte. On fertilization these cytoplasmic lamins are recruited into the newly forming pronuclei. The composition of the nuclear lamins changes after the species-specific transition to embryonic control of development, e.g., while the B type lamins remain, the A/C type disappear. The A/C type lamins reappear later in development in a tissue- and time-dependent manner. So the question is, "if a nucleus is really remodeled after transfer to an oocyte, shouldn't the composition of the nuclear lamins be altered to be the same as the pronuclei?" When nuclei from immediately beyond the species-specific transition are transferred to an oocyte, they do indeed acquire the pronuclear type of lamins, in this case the A/C type (Prather *et al.*, 1989b). Furthermore, in the mouse it has been shown that the A/C lamins are not acquired by the donor nucleus if the recipient oocyte has been either fertilized or preactivated (Prather *et al.*, 1991). Thus we have another example of remodeling of a donor nucleus such that it is similar to a pronucleus.

Another morphological change that occurs is in the structure of the nucleoli. The nucleoli are the sites of rRNA synthesis. By using electron microscopy of sections through the nuclei it has been determined that during the pronuclear stage the nucleoli have a tight, compact appearance with no reticulations and a smooth surface. In contrast, in a cell that is undergoing active rRNA synthesis, at and beyond the four-cell stage in the pig, the nucleoli have a very reticulated or vacuolated appearance with a very rough or granular periphery. So the question is, "if a nucleus is really remodeled after transfer to an oocyte, shouldn't the morphologies of the nucleoli and the pronuclei be similar?" Indeed, when nuclei with nucleoli that have an "active" nucleolar morphology are transferred to an enucleated oocyte and subsequently activated, the nucleolar morphology is modified such that it is similar to the nucleoli in pronuclei, i.e., compact, agranular, and without reticulations or vacuoles (Mayes *et al.*, 1994; Ouhibi *et al.*, 1996), which indirecly indicates that rRNA synthesis has ceased. Similar to the nuclear lamins, this modification is not complete if the oocyte is fertilized or preactivated (see review by Kühholzer and Prather, 2000).

A third method of evaluating the degree of remodeling is to evaluate another structure located in the nucleus that indirectly indicates the activity of RNA synthesis, i.e., the small nuclear ribonuclear proteins (snRNPs). The snRNPs are responsible for the processing of pre-mRNA before it leaves the nucleus and enters the cytoplasm. The B and D core protein can be identified by a monoclonal antibody, Y12. This antibody does not localize to the nucleus between germinal vesicle breakdown and the late four-cell stage in the pig (Prather and Rickords, 1992). The appearance of the Y12 antigen is α-amanitin sensitive. So the question is, "if a nucleus is really remodeled after transfer to an oocyte, shouldn't the distribution and appearance of the Y12 antigen be similar to that of the pronuclei?" When nuclei with readily detectable Y12 antigen are transferred to an enucleated oocyte and the oocyte is activated, then the nuclei lose their reactivity to the Y12 antigen (Prather and Rickords, 1992). This is an indirect indication that there is no mRNA processing and hence no mRNA synthesis. Thus, similar to the nuclear size, nuclear lamin composition, and nucleolar morphology, the transferred nucleus has a reactivity to the Y12 antibody that is similar to that of a pronucleus.

A final method of evaluating RNA synthesis is to look at the incorporation of [³H]uridine in cells before nuclear transfer and then to evaluate the change in incorporation after nuclear transfer. There is little if any RNA synthesis in the pronuclear-stage embryo and significant RNA synthesis and hence incorporation beyond the eight-cell stage. So the question is, "if a nucleus is really remodeled after transfer to an oocyte, shouldn't there be no incorporation of [³H]uridine into the pronuclei?" Hyttel *et al.* (1993) performed such experiments on *in vitro*-matured oocytes and embryos and found that, although there was a significant decrease in [³H]uridine incorporation after nuclear transfer, there still was some incorporation. This suggested that complete remodeling and reprogramming had not occurred.

An extension of the above experiments is to look at specific genes rather than at morphology and classes of RNA. Winger *et al.* (2000) evaluated message for a number of key regulatory enzymes in bovine embryos. They found that lactate dehydrogenase, citrate synthase, and phosphofructokinase were all correctly reprogrammed. Park *et al.* (2001a) found that roughly only half of the nuclei were reprogrammed, as measured by expression of enhanced green fluorescent protein (EGFP). Further studies with transgenic ear-derived fibroblasts revealed that EGFP expression in nuclear transfer embryos can be mosaic (Park *et al.*, 2002), suggesting that reprogramming in all blastomeres is not uniform. Because there are no other studies in the pig that specifically address the issue of nuclear reprogramming, examples from another farm species, the cow, will be used. In an initial study, DeSousa *et al.* (1999) showed by differential display technology that 95% of the transcripts in nuclear transfer blastocysts were similar to the control *in vitro*-produced embryo. However, that also means that 5% of the transcripts were different. Apparently, Daniels *et al.* (2000) have identified some of this 5%. They show that interleukin 6 (IL-6), fibroblast growth factor 4 (FGF4), and FGFr2 are not expressed correctly after nuclear transfer. Thus, although much remodeling occurs normally after nuclear transfer, in some cases it is not complete. (For further discussion of nuclear remodeling and reprogramming, see Chapter 4.)

OOCYTE ACTIVATION

CELL CYCLE SYNCHRONY

As discussed in Chapter 5, the correct ploidy will also need to be maintained. If the donor cell is late in the cell cycle, i.e., S or G2, then donor cell nuclei, on transfer to an oocyte, can complete another round of DNA synthesis after transfer (Stumpf *et al.*, 1992). This can have catastrophic effects on the developmental potential of the embryos. If donor cells in S or G2 are used as donors then the recipient egg must be preactivated (Prather, 2000).

MECHANISMS OF OOCYTE ACTIVATION

After the donor nucleus is transferred into the cytoplasm of the oocyte, the oocyte must be stimulated to initiate the resumption of meiosis. Ideally this would be accomplished via a mechanism similar to that which a sperm would use at fertilization (Macháty *et al.*, 1999). The debated theories for the mechanism of oocyte activation at fertilization include (1) that the sperm binds with a receptor on the surface of the oocyte and transduces a signal across the membrane to signal the egg to initiate development, and (2) that the sperm deposits into the cytoplasm of the egg a factor that is responsible for signaling the egg to begin development. Some of the likely candidates include integrin–disintegrin binding (Campbell *et al.*, 2000) and a sperm factor (Macháty *et al.*, 2000), respectively. It is likely that the sperm may use more than one mechanism to signal the egg to initiate development.

Although a number of methods are available for parthenogenetic activation, in many cases they are used in combination to attack two or more pathways that will result in activation. Activation that dealt specifically with nuclear transfer was reported by Tao *et al.* (1999, 2000b), who removed calcium from the medium after parthenogenetic activation and found that it reduced the percentage of blastocyst-stage but increased the percentage of morula-stage embryos. And in another report, Tao *et al.* (2000a) used cell fusion and found that thimerosal/dithiothreitol treatment could also result in blastocyst formation. Grupen *et al.* (1999) included a second set of electrical pulses 30 minutes after the first set administered to *in vivo*-matured oocytes, which resulted in embryos that developed better to the blastocyst stage (51%), compared to those receiving only a single set of pulses (34%). Three sets of pulses were detrimental to development. However, for *in vitro*-matured oocytes one set of pulses resulted in better development (31%) than did two sets of pulses (16%).

IN VITRO DEVELOPMENT OF NUCLEAR TRANSFER EMBRYOS

There are a number of reports of development to the blastocyst stage after nuclear transfer. Chen *et al.* (1999) produced nuclear transfer-derived blastocysts from an embryonic stem cell line. In contrast, Tao *et al.* (1999) used microinjection, rather than cell fusion, to transfer fetal fibroblast nuclei into *in vitro*-matured oocytes and obtained 5% blastocysts. Verma *et al.* (2000) showed that cleavage was higher for cycling cells (79%) than for serum-starved cells (56%). Development to the blasto-

Figure 1 *Cloned piglets expressing (left) and not expressing (right) the enhanced green fluorescent protein (Reprinted from Park et al., 2001b, by courtesy of Marcel Dekker, Inc.).*

cyst stage was between 1 and 7%. Koo *et al.* (2000) obtained 6 to 11% blastocyst stage. The number of nuclei was higher in blastocysts derived from activation at 2 versus 6 hours after the nuclear transfer. Kühholzer *et al.* (2000) reported the production of transgenic blastocysts derived from the transfer of fetal fibroblasts to *in vitro*-matured oocytes. These oocytes were activated with thimerosal/dithiothreitol as described in Macháty *et al.* (1997). In most cases blastocyst development was lower than that of control parthenogenotes (Kühholzer *et al.*, 2000; Koo *et al.*, 2000). Uhm *et al.* (2000) showed that fetal fibroblast cells could direct the development to morula or blastocyst stage at higher rates (24%) than could cumulus-derived cells (8%). Koo *et al.* (2001) showed that they could get blastocyst development and that the embryo expressed GFP. Kühholzer *et al.* (2001) found that some clonally derived cell lines result in better development than do their identically treated clones.

PRODUCTION OF OFFSPRING RESULTING FROM NUCLEAR TRANSFER

In 2000, three laboratories reported the birth of pigs following nuclear transfer. They all used different techniques to accomplish the transfer. Betthauser *et al.* (2000) used *in vitro*-matured oocytes (from sows); both Polejaeva *et al.* (2000) and Onishi *et al.* (2000) used *in vivo*-matured oocytes. Onishi *et al.* (2000) used microinjection and the other two groups used cell fusion to facilitate the transfer. Polejaeva *et al.* (2000) used two rounds of nuclear transfer (the first to enucleated oocytes at metaphase II, and the second to enucleated pronuclear stage eggs). The methods of activation initially were electrical, but in some cases this was followed by additional treatments of ionophore and dimethylaminopurine (DMAP) (Betthauser *et al.*,

2000). Polejaeva *et al.* (2000) used granulose-derived cells, whereas Onishi *et al.* (2000) and Betthauser *et al.* (2000) used fetal-derived cells. All three groups transferred large numbers of eggs to the synchronized recipient gilts early in the cycle.

Park *et al.* (2001b) showed that the transfer of transgenic nuclei to enucleated oocytes can result in the production of transgenic pigs that clearly express the enhanced green fluorescent protein (Fig. 1). Interestingly, nuclei in G2/M can be used as donors and result in cloned offspring (Lai *et al.*, 2002a); fibroblasts derived from a 4-day-old transgenic neonate (Park *et al.*, 2002) and an adult animal (Bondioli *et al.*, 2001) can be reprogrammed to produce cloned transgenic offspring.

Lai *et al.* (2002b) showed that the gene encoding α-(1,3)-galactosyltransferase can be knocked out of miniature pig fetal-derived fibroblasts. By using these fibroblasts and the nuclear transfer technology, viable offspring can be produced. In addition to having knocked out the gene, the miniature pig line used for these studies does not transmit the porcine endogenous retroviruses to human cells *in vitro*. Thus descendants from these pigs may be very valuable for xenotransplantation.

CONCLUSIONS

The progress over the past decade in the area of nuclear transfer, cloning, and transgenic animal production has been exciting. Development to term has increased from the four-cell stage as donor nuclei to early gestation fetal fibroblasts and adult granulosa cells. Future developments will result in additional knockout animals for both biomedical and production agriculture applications.

ACKNOWLEDGMENTS

This manuscript was prepared while funded by the National Institutes of Health under agreement RR13438 R01 and with the technical assistance of Nicole Whyte.

REFERENCES

Betthauser, J., Forsberg, E., Augenstein, M., Childs, L., Eilertsen, K., Enos, J., Forsythe, T., Golueke, P., Jurgella, G., Koppang, R., Lesmeister, T., Mallon, K., Mell, G., Misica, P., Pace, M., Pfister-Genskow, M., Strelchenko, N., Voelker, G., Watt, S., Thompson, S., and Bishop, M. (2000). Production of cloned pigs from *in vitro* systems. *Nature Biotechnology* **18**, 1055–1059.

Bondioli, K., Ramsoondar, J., Williams, B., Costa, C., and Fodor, W. (2001). Cloned pigs generated from cultured skin fibroblasts derived from a H-transferase transgenic boar. *Mol. Reprod. Dev.* **60**, 189–195.

Campbell, K. D., Reed, W. A., and White, K. L. (2000). Ability of integrins to mediate fertilization, intracellular calcium release, and parthenogenetic development in bovine oocytes. *Biol. Reprod.* **62**, 1702–1709.

Chen, L. R., Shiue, Y. L., Bertolini, L., Medarano, J. F., BonDurant, R. H., and Anderson, G. B. (1999). Establishment of pluripotent cell lines from porcine preimplantation embryos. *Theriogenology* **52**, 195–212.

Daniels, R. V., Hall, A. O., and Trounson, A. (2000). Analysis of gene transcription in bovine nuclear transfer embryos reconstructed with granulosa cell nuclei. *Biol. Reprod.* **63**, 1034–1040.

DeSousa, P. A., Winger, Q., Hill, J. R., Jones, K., Watson, A. J., and Westhusin, M. E. (1999). Reprogramming of fibroblast nuclei after transfer into bovine oocytes. *Cloning* **1**, 63–69.

Grupen, C. G., Verma, P. J., Du, Z. T., McIlfatrick, S. M., Ashman, R. J., and Nottle, M. B. (1999). Activation of *in vivo* and *in vitro* derived porcine oocytes by using multiple electrical pulses. *Reprod. Fertil. Dev.* **11**, 457–462.

Hyttle, P., Prochazka, R., Smith, S., Kanka, J., and Greve, T. (1993). RNA synthesis in porcine blastomere nuclei introduced into *in vitro* matured ooplasm. *Acta Vet. Scand.* **34**, 159–167.

Jarrell, V. L., Day, B. N., and Prather, R. S. (1991). The transition from maternal control of development to zygotic control of development occurs during the 4-cell stage in the domestic pig, *Sus scrofa*: Quantitative and qualitative aspects of protein synthesis. *Biol. Reprod.* **44**, 62–68.

Koo, D. B., Kang, Y. K., Choi, Y. H., Park, J. S., Han, S. K., Park, I. Y., Kim, S. U., Lee, K. K., Son, D. S., Chang, W. K., and Han, Y. M. (2000). *In vitro* development of reconstructed porcine oocytes after somatic cell nuclear transfer. *Biol. Reprod.* **63**, 986–992.

Koo, D.-B., Kang, Y.-K., Choi, Y.-H., Park, J. S., Kim, H.-N., Kim, T., Lee, K.-K., and Han, Y. M. (2001). Developmental potential and transgene expression of porcine nuclear transfer embryos using somatic cells. *Mol. Reprod. Dev.* **58**, 15–21.

Kühholzer, B., and Prather, R. S. (2000). Advances in live stock nuclear transfer. *Proc. Soc. Exp. Biol. Med.* **224**, 240–245.

Kühholzer, B., Tao, T., Macháty, Z., Hawley, R. J., Greenstein, J. L., Day, B. N., and Prather, R. S. (2000). Production of transgenic porcine blastocysts by nuclear transfer. *Mol. Reprod. Dev.* **56**, 145–148.

Kühholzer, B., Hawley, R. J., Lai, L., Kolber-Simonds, D., and Prather, R. S. (2001). Nuclear transfer using different sub-clones of porcine fetal fibroblast cells result in different *in vitro* development. *Biol. Reprod.* **64**, 1695–1698.

Lai, L., Park, K.-W., Cheong, H.-T., Kühholzer, B., Samuel, M., Bonk, A., Im, G.-S., Rieke, A., Day, B. N., Murphy, C. N., Carter, D. B., and Prather, R. S. (2002a). A transgenic pig expressing the green fluorescent protein produced by nuclear transfer using colchicines-treated fibroblasts as donor cells. *Mol. Reprod. Dev.* (in press).

Lai, L., Kolber-Simonds, D., Park, K.-W., Cheong, H.-T., Greenstein, J. L., Im, G.-S., Samuel, M., Bonk, A., Rieke, A., Day, B. N., Murphy, C. N., Carter, D. B., Hawley, R. J., and Prather, R. S. (2002b). Production of α-1,3-galactysyltransferase-knockout inbred miniature swine by nuclear transfer cloning. *Science* **295**, 1089–1092 (*ScienceExpress* 1/4/02).

Li, G. P., Tan, J. H., Sun, Q. Y., Meng, Q. G., Yue, K. Z., Sun, X. S., Lu, Z. Y., Wang, H. B., and Xu, L. B. (2000). Cloned piglets born after nuclear transplantation of embryonic blastomeres into porcine oocytes matured *in vivo*. *Cloning* **2**, 45–52.

Liu, L., Moor, R. M., Laurie, S., and Notarianni, E. (1995). Nuclear remodeling and early development in cryopreserved, porcine primordial germ cells following nuclear transfer into *in vitro* matured oocytes. *Int. J. Dev. Biol.* **39**, 639–644.

Macháty, Z., Wang, W.-H., Day, B. N., and Prather, R. S. (1997). Complete activation of porcine oocytes induced by the sulfhydryl reagent, thimerosal. *Biol. Reprod.* **57**, 1123–1127.

Macháty, Z., Rickords, L. F., and Prather, R. S. (1999). Parthenogenetic activation of porcine oocytes following nuclear transfer. *Cloning* **1**, 101–109.

Macháty, Z., Bonk, A., Kühholzer, B., and Prather, R. S. (2000). Porcine oocyte activation induced by a cytosolic sperm factor. *Mol. Reprod. Dev.* **57**, 290–295.

Mayes, M. A., Stogsdill, P. L., Parry, T. W., Kinden, D. A., and Prather, R. S. (1994). Reprogramming of nucleoli after nuclear transfer of pig blastomeres into enucleated oocytes. *Dev. Biol.* **163**, 542.

Onishi, A., Iwamoto, M., Akita, T., Mikawa, S., Takeda, K., Awata, T., Hanada, H., and Perry, A. C. F. (2000). Pig cloning by microinjection of fetal fibroblast nuclei. *Science* **289**, 1188–1190.

Ouhibi, N., Fulka, J., Kanka, J., and Moor, R. M. (1996). Nuclear transplantation of ectodermal cells in pig oocytes-ultrastructure and radiography. *Mol. Reprod. Dev.* **44**, 533–539.

Park, K.-W., Kuhholzer, B., Lai, L., Macháty, Z., Sun, Q. Y., Day, B. N., and Prather, R. S. (2001a). Development of porcine transgenic embryos derived from nuclear transfer of granulosa-derived cells transfected with the green fluorescent protein gene. *Anim. Repro. Sci.* **68**, 111–120.

Park, K.-W., Cheong, H.-T., Lai, L., Im, G.-S., Kühholzer, B., Bonk, B., Samuel, M., Rieke, A., Day, B. N., Murphy, C. N., Carter, D. B., and Prather, R. S. (2001b). Production of nuclear transfer-derived swine that express the enhanced green fluorescent protein. *Anim. Biotechnol.* **12**, 171–181.

Park, K.-W., Lai, L., Cheong, H.-T., Cabot, R., Sun, Q.-Y., Wu, G., Rucker, E. B., Durtschi, D., Bonk, A., Samuel, M., Rieke, A., Day, B. N., Murphy, C. N., Carter, D. B., and Prather, R. S. (2002). Mosaic gene expression in nuclear transfer-derived embryos and the production of cloned transgenic pigs from ear-derived fibroblasts. *Biol. Reprod.* **66**, 667–674.

Polejaeva, I. A., Chen, S. H., Vaught, T. D., Page, R. L., Mullins, J., Ball, S., Dai, Y., Boone, J., Walker, S., Ayares, D. L., Colman, A., and Campbell, K. H. S. (2000). Cloned pigs produced by nuclear transfer from adult somatic cells. *Nature* **407**, 505–507.

Prather, R. S. (2000). Perspectives: Cloning. Pigs is pigs. *Science* **289**, 1886–1887.

Prather, R. S., and Rickords, L. F. (1992). Developmental regulation of a snRNP core protein epitope during pig embryogenesis and after nuclear transfer for cloning. *Mol. Reprod. Dev.* **33**, 119–123.

Prather, R. S., and Schatten, G. (1992). Construction of the nuclear matrix at the transition from maternal to zygotic control of development in the mouse: An immunocytochemical study. *Mol. Reprod. Dev.* **32**, 203–208.

Prather, R. S., Barnes, F. L., Sims, M. L., Robl, J. M., Eyestone, W. H., and First, N. L. (1987). Nuclear transplantation in the bovine embryo: Assessment of donor nuclei and recipient oocyte. *Biol. Reprod.* **37**, 859–866.

Prather, R. S., Sims, M. M., and First, N. L. (1989a). Nuclear transplantation in early pig embryos. *Biol. Reprod.* **41**, 414–418.

Prather, R. S., Sims, M. M., Maul, G. G., First, N. L., and Schatten, G. (1989b). Nuclear lamin antigens are developmentally regulated during porcine and bovine embryogenesis. *Biol. Reprod.* **41**, 123–132.

Prather, R. S., Sims, M. M., and First, N. L. (1990). Nuclear transplantation in the pig embryo: Nuclear swelling. *J. Exp. Zool.* **255**, 355–358.

Prather, R. S., Kubiak, J., Maul, G. G., First, N. L., and Schatten, G. (1991). The association of nuclear lamins A and C is regulated by the developmental stage of the mouse oocyte or embryonic cytoplasm. *J. Exp. Zool.* **257**, 110–114.

Robl, J. M., and First, N. L. (1986). Manipulation of gamets and embryos in the pig. *J. Reprod. Fertil.* (Suppl.) **33**, 101–114.

Schoenbeck, R. A., Peters, M. S., Rickords, L. F., Stumpf, T. T., and Prather, R. S. (1992). Characterization of DNA synthesis and the transition from maternal to embryonic control in the 4-cell porcine embryo. *Biol. Reprod.* **47**, 1118–1125.

Stice, S. L., and Robl, J. M. (1988). Nuclear reprogramming in nuclear transplant rabbit embryos. *Biol. Reprod.* **39**, 657–664.

Stumpf, T. T., Schoenbeck, R. A., and Prather, R. S. (1992). DNA synthesis during the porcine embryo two-cell stage. *Biol. Reprod.* (Suppl.) **1**, 71.

Tao, T., Machaty, Z., Boquest, A. C., Day, B. N., and Prather, R. S. (1999). Development of pig embryos reconstructed by microinjection of cultured fetal fibroblast cells into *in vitro* matured oocytes. *Anim. Reprod. Sci.* **56**, 133–141.

Tao, T., Boquest, A. C., Machaty, Z., Peterson, A. L., Day, B. N., and Prather, R. S. (2000a). Development of pig embryos by nuclear transfer of cultured fibroblast cells. *Cloning* **1**, 55–62.

Tao, T., Machaty, Z., Abeydeera, L. R., Day, B. N., and Prather, R. S. (2000b). Optimisation of porcine oocyte activation following nuclear transfer. *Zygote* **8**, 69–77.

Uhm, S. J., Kim, N. H., Kim, T., Chung, H. M., Chung, K. H., Lee, H. T., and Chung, K. S. (2000). Expression of enhanced green fluorescent protein and neomycin resistant genes in porcine embryos following nuclear transfer with porcine fetal fibroblasts transfected by retrovirus vector. *Mol. Reprod. Dev.* **57**, 331–337.

Verma, P. J., Du, Z. T., Crocker, L., Faast, R., Grupen, C. G., McIlfatrick, S. M., Ashman, R. J., Lyons, I. G., and Nottle, M. B. (2000). *In vitro* development of porcine nuclear transfer embryos constructed using fetal fibroblasts. *Mol. Reprod. Dev.* **57**, 262–269.

Willadsen, S. M. (1986). Nuclear transplantation in sheep embryos. *Nature* **320**, 63–65.

Winger, Q. A., Hill, J. R., Shin, T., Watson, A. J., Kraemer, D. C., and Westhusin, M. E. (2000). Genetic reprogramming of lactate dehydrogenase, citrate synthase, and phosphofructokinase mRNA in bovine nuclear transfer embryos produced using bovine fibroblast nuclei. *Mol. Reprod. Dev.* **56**, 458–464.

CHAPTER 19

CLONING OF CATTLE

Neal L. First, Zeki Beyhan, and Jennifer D. Ambroggio

INTRODUCTION

Cattle were the second mammalian species cloned by nuclear transplantation. The initial objective was multiplication of high-milk-performance dairy cattle by transfer of totipotent embryonic cells from embryos expected to be of high performance into enucleated bovine oocytes. This required a generation of testing milk production from cows of several clonal lines to select the clonal line to be propagated. At least three companies pursued this approach but abandoned it when the possibility of directly cloning the highest performance cows from their somatic cells became apparent. The cloning of high-performance dairy and beef cattle for the purpose of genetic improvement is presently occurring. Because the efficiency of cloning has been low, only a few clones have been made for agricultural purposes. The major effort has been production of clones from cell lines genetically engineered to produce specific products of extremely high value in milk. These products for human use include pharmaceuticals, blood proteins, antigens for antibody production, and perhaps someday specialty foods or neutraceuticals through milk.

The principal companies producing genetically engineered milk-borne products include Advanced Cell Technology (Worcester, MA), Gala Design (Prairie du Sac, WI), Genzyme Transgenics (Framingham MA), Infigen (Deforest, WI), and Protein Products Ltd. (Edinburgh, Scotland, and Blacksburg, VA). At present, each of these companies has produced a few to more than 100 cloned cattle. As the efficiency of cloning improves, the cloning of cattle is expected to impact human medicine and agricultural milk and meat production. Additionally, identical animals can help resolve issues of genome and environment interaction for specific traits, providing valuable information about the genetics of cattle breeding. The present ability to make clones from cell lines cultured to large populations of cells will also allow gene deletion and gene expression experiments capable of elucidating the genomic control of specific traits and physiological functions involved in normal and abnormal physiology and pathology of cattle. This chapter focuses on the history of cloning of cattle, the methodology, successes and failures, possible causes of failures, and applications of cloning of cattle.

HISTORY

The cloning of cattle began in our laboratory in 1986 with introduction of embryonic cleavage-stage nuclear donor cells into enucleated metaphase II oocytes, resulting in pregnancy and birth (Prather *et al.*, 1987). Our success and success at Granada Genetics near the same time were made possible by several factors. Studies in amphibians showed more advanced development occurring from transfer of nuclei into enucleated metaphase oocytes rather than enucleated pronuclear zygotes (Hoffner and DiBerardino, 1980). Evidence from mouse studies demonstrated that the frequency of development after nuclear transfer was much higher when nuclei were fused into mouse oocytes rather than microinjected (Mcgrath and Solter, 1983). The studies of Willadsen (1986) in sheep showed that lambs could be produced by electrofusion of embryonic cells into enucleated metaphase II oocytes. The emergence of methods for producing embryos *in vitro* and for culturing embryos to the blastocyst stage also facilitated the advent of cattle cloning. Bovine embryos are commonly transferred into cows at the blastocyst stage to produce offspring (Gordon, 1994; First *et al.*, 1999).

Cleavage-stage embryonic cells were used for transfer into enucleated metaphase oocytes in these early studies because early amphibian literature suggested that embryonic cells were totipotent, but that somatic or fetal differentiated cells were not (DiBerardino, 1997). This was confirmed in mammals by studies such as that of Navara *et al.* (1992). In this study, the use of polarized and trophoblast lineage committed cells as nuclear donors resulted in very few blastocyst-stage embryos (7%; $n = 158$), whereas nonpolarized and nondifferentiated cells as nuclear donors yielded a high frequency of blastocysts (47%; $n = 184$). In comparison, 30% of 139 unselected embryos developed to the blastocyst stage. The conclusion in the 1980s that only nondifferentiated cell types were developmentally effective as nuclear donors was due to the major technical limitation that artificial means were unable to activate oocytes as quickly and effectively as with sperm (Ware *et al.*, 1989). This forced the use of aged oocytes, which have lost the ability to reprogram a differentiated nucleus. A comparison of young and aged oocytes is shown in Table 1. In fact, with many early protocols the best success was achieved when the oocyte was so aged that the donor cell nuclear envelope and its contents were retained *in toto* without nuclear envelope breakdown (Leibfried-Rutledge *et al.*, 1992; Poccia and Collas, 1997). Oocytes that were moderately aged would allow nuclear envelope breakdown, but would rapidly reassemble the nuclear envelope and exclude some of the chromatin from the nucleus.

The ability to activate young bovine oocytes came about through development of procedures that more completely and continuously suppressed the activity of metaphase-promoting factor (MPF) and cytostatic factor (CSF), compared to Ca^{2+} ionophores (Liu and Yang, 1999). MPF is assayed by H1 kinase activity and CSF is assayed by mitogen-activated protein kinase (MAPK) activity; together they maintain metaphase oocytes in meiotic arrest. This ability to activate young oocytes provided the foundation for harnessing the ability of metaphase II oocytes to reprogram fetal or somatic nuclear donor cells in nuclear transfer. One such procedure is the activation of young metaphase II oocytes with a calcium ionophore, followed by a 4- to 6-hour exposure to 6-dimethylaminopurine (6-DMAP) to interfere with phosphorylation of serine and threonine residues on MPF and MAPK (Susko-Parrish *et al.*, 1994). Another method uses butyrolactone for the same purpose (Motlik *et al.*, 1998). Activation can also be achieved by inhibition of protein synthesis with

Table 1 Use of Young versus Aged Oocytes in Nuclear Transfer

Process	24 hours	48 hours
Artificial activation	Poor	Excellent
Nuclear envelope breakdown and reassembly	Yes	No
Synchrony of nucleus and oocyte	Essential	Nonessential
Cell and genomic reprogramming of donor cell	Yes	No
Sequence and timing of activation	Critical	Not critical
Successful donor cells	Differentiated adult or fetal	Totipotent embryonic

cycloheximide after an ionophore treatment, followed by incubation with cytochalasin B to prevent chromatin extrusion from the cell and thereby maintain a diploid state (Liu and Yang, 1999).

Blastocyst development after *in vitro* fertilization or nuclear transfer was further enhanced by the discovery that the first oocytes to reach metaphase II at approximately 16 hours after follicle removal, when inseminated 8 hours later, produced a much higher frequency of blastocysts than did embryos obtained from oocytes mature at 24 hours (Dominko and First, 1997). The ability to produce highly competent young oocytes that could be activated early allowed us to produce blastocysts, pregnancies, and offspring from blastocyst inner cell mass cells that had been cultured, multiplied, and passaged for four to six passages (Sims and First, 1994). The efficiency was not high; approximately 15% of the nuclear transfers produced blastocysts and four calves resulted from 34 embryo transfers. The efficiency of nuclear transfer was higher when cleavage-stage blastomeres or morula cells were used in the aged oocyte model.

To date several hundred calves have been produced from nuclear transfer using embryonic cells as nuclear donors. Recloning has been accomplished (Stice and Keefer, 1993), and frozen donor cells and *in vitro*-matured oocytes have been used (Barnes *et al.*, 1993). At best, 20–50% of the nuclear transfers have resulted in blastocysts for transfer into cows, of which approximately 50% become pregnancies. However, the pregnancy losses are greater than normal, with as many as 20% failing to deliver and requiring induction of labor, with larger than normal calves often resulting.

Commercial companies such as Alta Genetics, American Breeders Service, and Granada Genetics have attempted to improve the efficiency of embryonic cloning to a level and cost suitable for commercial cattle production. Unfortunately a sufficient level of efficiency has not been reached. In dairy cattle the production of embryonic clones of high-milk-producing cattle required rearing to milk production and selection of high-milk-producing embryonic clonal lines. Commercial interest in embryonic cloning disappeared with the advent of cloning from primordial germ cells and fetal or adult somatic cell lines. There are still attempts to culture bovine embryonic stem cells to numbers large enough for gene transfer and transgenic colony selection before nuclear transfer (Mitalipova *et al.*, 2001).

CLONING FROM FETAL CELLS

Fetal cells have been especially robust in terms of long-term cell culture, abundance of cells for genetic engineering, efficiency of genetic engineering, and efficiency of offspring production after use as donor cells in nuclear transfer. The principal origins of fetal cells have been germ line cells of the genital ridge or fetal somatic cells such as fibroblasts.

FETAL GERM LINE-DERIVED CELLS

Most of the data on use of fetal germ line-derived cells as nuclear donors come from the company Infigen, which cloned the first calf from a fetal cell in 1997. This bull calf was named Gene. The germ line cells are derived from the genital ridge of fetuses of the sex desired at 40 to 60 days of gestational age. The germinal cells are isolated, and long-term cell cultures with large numbers of cells are created. For most applications, the cells are genetically engineered to contain a gene of commercial product interest. Transgenic cell colonies are selected and propagated to provide transgenic cells for nuclear donors in nuclear transfer into enucleated bovine oocytes.

Some of the highest developmental efficiencies reported thus far have been from the use of long-term cultured fetal germ line-derived cells as nuclear donors in nuclear transfer (Brink *et al.*, 2000). Totipotent embryonic germ (EG) cells were derived from the genital ridge of female bovine fetuses and selectively cultured prior to being electrically fused with enucleated metaphase II (MII) oocytes. Approximately 20% of the transgenic nuclear transfer embryos reached the blastocyst stage, comparable to the developmental efficiency achieved with nontransfected nuclear transfer embryos. Of 73 nontransgenic nuclear transfer blastocysts transferred into cows, 19% completed pregnancy. Of 44 transgenic nuclear transfer blastocysts transferred into cows, 25% completed pregnancy and all were confirmed transgenic calves. In another study, 42% of 13 nontransgenic embryo transfers and 26% of 44 transgenic embryo transfers became delivered offspring (Forsberg *et al.*, 2001). In a later study by the same group, 25% of 2170 NT embryos transferred into cows resulted in pregnancy of which 20% resulted in live births, of these 77% remained alive and healthy for a year or more. Data comparing fetal germ line-, fetal somatic cell-, and adult somatic cell-derived nuclear donor cell lines and their use in nuclear transfer by the same laboratory are not presently available.

FETAL SOMATIC CELLS

The first calves cloned from somatic cells were produced by nuclear transfer of fibroblasts from a day 55 male fetus (Cibelli *et al.*, 1998). Donor cells from actively dividing cultures were used at 70–80% confluence. Within 2–4 hours of nuclear transfer, reconstructed oocytes were activated by a 4-minute incubation with a calcium ionophore followed by a 3-hour treatment with 6-DMAP. Embryos were cocultured with mouse embryonic fibroblasts for 6.5 days prior to transfer. Fetal fibroblasts have an inherently long G1 phase, so over half of the cells are in this quiescent point of the cell cycle at any given time. Consequently, no serum starvation was required to synchronize donor cells. Cloned fetuses can be reused as a source of fetal fibroblasts for a succession of nuclear transfers, enabling further genetic manipulation. Cibelli's group subsequently cloned six calves from a female fetus, and performed an analysis of telomere length to explore the relationship between nuclear transfer and cellular aging (Lanza *et al.*, 2000) [see below, and also Betts *et al.* (2001), for a discussion of telomere length in fetal fibroblast clones].

Another group found that fibroblasts of a 37-day-old male fetus that were serum starved for 8 days in culture prior to nuclear transfer exhibited significantly better preimplantation development than did nonstarved fibroblasts (Zakhartchenko *et al.*, 1999) (see Table 2). Oocytes were denuded 18–20 hours after maturation and enucleated within 2 hours. Nuclear transfer was performed with fibroblast cells at 20–22 hours postmaturation, and reconstructed oocytes were electrofused in

Table 2 Relative Efficiency of Bovine Fetal Cell Nuclear Transfer Protocols

Research group/year	Fused[a]	Developed to blast	Transferred (% fused)	Born (% transferred)	Overall efficiency (live births/fused)
Cibelli/1998	276	33	28 (10.1)	4 (14.3)	1.45%
Zakhartchenko/1999	Nss, 174; Ss, 205	35; 80	7 (4.0); 16 (7.8)	2 (28.6); 0	1.15%; 0%
Lanza/2000	1896	87	79 (4.2)	6 (7.6)	0.32%
Kato/2000	Skin, 26; liver, 36	15; 13	7 (26.9); 6 (16.7)	1 (14.3); 0	3.85%; 0%
Brink/2000 (EG cells)	Non-tg, ?; Tg, ?	?; ?	73; 44	14 (19.2); 11 (25.0)	?; ?
Forsberg/2001 (EG cells)	Gfs, 465; no Gfs, 575;	103; 153; 156	13 (2.8); 37 (6.4); 44 (5.7)	5 (38.5); 4 (10.8); 11 (25.0)	1.08%; 0.70%; 1.42%
Pace/2001 (EG cells)	Tg, 775		2170	106 (5%)	

[a]Abbreviations: Nss, not serum starved; Ss, serum starved; Non-tg, nontransgenic; Tg, transgenic; Gfs, growth factors.

Zimmerman fusion medium. At 24 hours postmaturation, fused couplets were activated by a 5-minute culture with ethanol followed by a 5-hour exposure to cycloheximide and cytochalasin B. Although nuclear transfer embryos derived from starved fibroblasts exhibited superior *in vitro* development, offspring were obtained only from embryos derived from nonstarved fibroblasts. Furthermore, recloning into preactivated enucleated oocytes with cells from nuclear transfer-derived morulae resulted in significantly higher rates of blastocyst development than from the initial round of fibroblast cloning. This observation suggests the potential for increasing the efficiency of nuclear transfer.

Fibroblasts are not the only fetal cells that can support development after nuclear transfer. Kato and colleagues (2000) compared the efficiency of nuclear transfer with skin, kidney, gut, and muscle cells from female bovine fetuses, as well as skin, heart, liver, kidney, gut, and tongue cells from male bovine fetuses. Of the female fetal cell types, kidney cells supported the greatest development to the blastocyst stage ($n = 28$; 47%), whereas only 23% ($n = 15$) of muscle cell-derived embryos became blastocysts. The range for male cell types was smaller; 45% of nuclei from heart cells vs. 37% from gut cells supported blastocyst development. Only embryos derived from male fetal skin and ear cells were transferred to recipient cows, resulting in the birth of one calf. Interestingly, this study found no significant differences between the developmental rates of nuclear transfer embryos derived from fetal, newborn, or adult cells.

CLONING FROM ADULT SOMATIC CELLS

The birth of Dolly the lamb, cloned from a cultured adult mammary cell line, was enabled by serum starvation of the donor cells to induce a state of quiescence (Wilmut *et al.*, 1997). Dolly did not remain the only mammal cloned from an adult somatic cell for long; within a year several "litters" of mice cloned from cumulus cells joined her in the spotlight (Wakayama *et al.*, 1998). Sertoli, neuronal, and cumulus cells are naturally in the G0 phase, negating the need for serum starvation or *in vitro* culture. Nuclei were directly injected into enucleated MII oocytes with a piezo micromanipulator. This innovation bypasses the need for electrofusion, and may minimize trauma to both the donor nucleus and the recipient cell. Subsequently, bovine adult somatic cell cloning has come into its own. A comparison of the experiments that have led to the birth of live calves from adult somatic cell nuclear transfer may help to determine an optimal protocol (Table 3).

Cows were the next animals to join the list of adult somatic cell clones (Kato *et al.*, 1998). Cumulus and oviduct cells were cultured for several passages and exposed to serum starvation for 3–4 days. Quiescent nuclei were transferred into enucleated oocytes, electrofused in Zimmerman fusion medium, activated by treatment with cycloheximide, and cultured *in vitro* until the blastocyst stage. Embryos reconstructed with cumulus cell nuclei exhibited greater developmental potential during the preimplantation period, compared to embryos cloned from oviduct cell nuclei. Four of the eight offspring died shortly after birth.

Another cow clone met with an early death from lymphoid hypoplasia (Renard *et al.*, 1999). This cow was cloned using skin cells grown from an ear biopsy taken from a 15-day-old calf. The progenitor calf survived and became an adult, and was in fact produced by nuclear transfer using an embryonic blastomere. The somatic clone was obtained from frozen/thawed skin cells after eight passages during *in vitro* culture. Recipient oocytes were matured for 24 hours, then aged for 10 hours at 10°C in TCM 199 prior to enucleation. An hour after electrofusion and activation, the reconstructed embryos were moved into B2 medium with 10% fetal calf serum (FCS) and cocultured with Vero cells for 7 days. The death of the resultant offspring is not extraordinary in light of the neonatal mortality noted in experiments yielding higher numbers of live progeny.

Table 3 Comparison of Techniques Used for Bovine Adult Somatic Cell Nuclear Transfer

Research group/year (cell type)	Donor cell state	Fusion	Activation
Kato/1998 (cumulus/ oviduct)	Cultured several passages, serum starved 3–4 days, quiescent	Electric pulses in sucrose + K_2PO_4, BSA, Ca^{2+}, and Mg^{2+}	Another pulse to activate, 5–6 hours with cycloheximide
Renard/1999 (ear skin from clone)	Frozen/thawed, 8 passages/4 weeks	Electric pulses in mannitol + Ca^{2+} and Mg^{2+}	Simultaneous
Wells/1999 (mural granulosa)	Frozen/thawed, ≥9 passages, serum starved, quiescent	Electric pulses in mannitol + BEPES, BSA, Ca^{2+}, and Mg^{2+}	6–8 hours postfusion: 4 minutes with ionomycin/4 hours with 6-DMAP, vs. recloning into preactivated cytoplasts
Kubota/2000 (fibroblast from a 17-year-old bull)	Long-term culture, ≤3 months, 10–15 passages, serum starved	Electrofusion, media not specified	5-hour culture with cycloheximide
Kato/2000 (cumulus, skin, oviduct, ear, uterine)	1–16 passages, serum starved, quiescent	Electric pulses in sucrose + K_2PO_4, BSA, Ca^{2+}, and Mg^{2+}	Another pulse to activate, 5–6 hours with cycloheximide

High rates of neonatal mortality resulted from a study examining the suitability of a variety of cell types for nuclear transfer (Kato *et al.*, 2000). A total of 24 calves were cloned from both male and female adult, newborn, and fetal somatic cells, 9 of which weighed 40% above average at birth. Nineteen calves were cloned from adult cell types, either cumulus, oviduct, uterus, skin, or ear cells; however, 6 died at parturition. The surviving 6 females include clones from the following cell types: 3 cumulus, 2 oviduct, and 1 skin, along with 3 males cloned from ear cells. Cultured donor cells were passaged 1–16 times and serum starved into quiescence. *In vitro* matured MII oocytes were enucleated 22–24 hours after initiation of maturation. Donor cells were fused with two electric pulses in Zimmerman fusion medium, and this procedure was repeated every 15 minutes until fusion was confirmed. Activation was initiated by a final electric stimulation, followed by a 5- to 6-hour incubation with cycloheximide.

To enhance *in vitro* development, reconstructed embryos were cocultured with mouse fetal fibroblasts starting on day 3. Prior to examining *in vivo* development, the range of cell types were assessed for *in vitro* developmental potential. Male adult skin cells had the highest fusion efficiency of 89% (65/73), whereas female adult oviduct cells came in last place with a fusion efficiency of 55% (93/170). Of the somatic cell types tested, male liver cells exhibited the highest rate of development to the blastocyst stage (53% of the successful fusions, 34/66), whereas female cumulus cells had the lowest rate of 30% (19/73). Regardless of the lower preimplantation development rate of cumulus cell clones, these embryos contributed significantly to the number of surviving calves. It is important to note that cumulus clones exhibited normal birth weight. The calves that died at birth or shortly thereafter suffered from a wide range of abnormalities. Near the time of parturition most of the recipient cows failed to experience labor or mammary development, indicating a lack of fetal–maternal communication.

On the other hand, all of 10 calves cloned from cultured, frozen/thawed mural granulosa cells developed normally after birth (Wells *et al.*, 1999). Oocytes were obtained from slaughterhouse ovaries, matured within the cumulus–oocyte complexes for 20 hours at 39°C. MII oocytes were stained in HSOF medium with 10% FCS, Hoechst 33342, and CCB for 20 minutes. After enucleation the recipient cytoplasts were dehydrated in HSOF medium with 10% FCS and 5% sucrose. Donor cells made quiescent by serum starvation were inserted into the perivitelline space. After rehydration, reconstructed embryos were electrically fused in mannitol solution with HEPES, BSA, Ca^{2+}, and Mg^{2+}. Chemical activation by a 4-minute exposure to ionomycin took place 4–6 hours after fusion. Embryos were then washed before a 4-hour incubation with 6-DMAP.

In an attempt to allow for a longer period of nuclear reprogramming, blastomeres from cloned morulae were fused to S-phase cytoplasts 6–8 hours after preactivation with ionomycin and 6-DMAP. Although over half of these recloned embryos reached the morula/blastocyst stage during *in vitro* culture, none that were transferred survived to day 100 of gestation. Transcription of the embryonic genome may have already started in the blastomeres used for recloning, so cells from earlier cleavage-stage embryos might yield better long-term results. In comparison, approximately 70% of the embryos from the fusion-before-activation group completed preimplantation development, resulting in 10 term pregnancies. The authors (Wells *et al.*, 1999) attribute these relatively high rates of development to the use of 6-DMAP, a protein kinase inhibitor that may prevent phosphorylations required for function of the spindle apparatus, thereby delaying micronuclei formation until after activation.

Cattle have also been cloned from fibroblast cells of a 17-year-old bull after long-term culture (Kubota *et al.*, 2000). Nuclear donor cells from early-, middle-, and late-passage cultures were either serum starved or harvested when confluent. Enucleation of recipient oocytes was confirmed with Hoechst 33342 staining, then donor cells 10–15 µm in diameter were inserted subzonally. Electrofusion was said to activate the couplets simultaneously, but all reconstructed embryos were further activated by a 5-hour incubation with cycloheximide. Cleaving embryos were cocultured with cumulus cells until transfer at the blastocyst stage. No offspring were obtained from early-passage cells. Four of the six cloned offspring have survived, demonstrating that prolonged culture does not thwart the competence of nuclear donor cells. This group also cloned 10 calves from fibroblast and cumulus cells of a 13-year-old dairy cow to analyze telomere length after adult somatic cell nuclear transfer (Tian *et al.*, 2000) [see below, and also Betts *et al.* (2001)].

Infigen (Deforest, WI), reported a total of 75 calves cloned from somatic cells (Pace *et al.*, 2001). Donor nuclei were obtained from six different cell types, representing 18 different genetic lines. The report does not mention what percentage of these clones is from adult somatic cell lines versus fetal cell lines. Seven of the calves were dead at birth, and 14 more died within 5 days of birth due to either digestive dysfunction or "multiple systematic dysfunctions." The 54 remaining calves are exhibiting normal growth rates. Table 4 compares efficiencies of bovine adult somatic cell nuclear transfer protocols.

TELOMERES

Several reports support hopes that nuclear transfer is capable of resetting the life span of somatic cells. Ten calves were cloned from fibroblast and cumulus cells of a 13-year-old cow, of which four survived (Tian *et al.*, 2000). Telomeres of the clones were longer than those of the donor cow, and comparable to those of age-matched controls. Terminal restriction fragment analysis of size of the clones that died shortly after birth was not different from that of the survivors. Telomere elongation takes place during gametogenesis, but because nuclear donor cells do not experience this process telomere restoration would have to occur during embryogenesis. Telomerase activity was detected at gradually increasing levels in successive developmental stages of cloned embryos, indicating active reversal of senescence. Although telomerase is naturally active in embryonic cells, reactivation of this enzyme in most

Table 4 Relative Efficiency of Bovine Adult Somatic Cell Nuclear Transfer Protocols

Research group/year	Fused[a]	Developed to blast	Transferred (% fused)	Born (% transferred)	Overall efficiency (live births/fused)
Kato/1998	47 cumulus, 94 oviduct	18, 20	6 (12.77), 4 (4.26)	5 (83.33), 3 (75.0)	10.64%, 3.19%
Renard/1999	175	Not given	6 (3.43)	1 (16.67)	0.57%
Wells/1999	552 FBA, 146 recloned	383, 84	100 (18.12), 16 (10.96)	10 (10.0), 0	1.81%, 0%
Kubota/2000	Passg 5 Nss, 102; passg 5 Ss, 114; Passg 10 Ss, 115; Passg 15 Ss, 109	28, 24, 43, 36	? (10 recip.), 15 (13.16), 22 (19.13), 17 (15.60)	0, 0, 4 (18.18), 2 (11.76)	0%, 0%, 3.48%, 1.83%
Kato/2000	Female cells: 319 cumulus, 250 oviduct, 98 uterus, 51 skin, 40 ear	173, 84, 51, 25, 13	41 (12.85), 17 (6.80), 14 (14.29), 4 (7.84), 2 (5.0)	3 (7.32), 2 (11.76), 0, 2 (50.0), 0	Female cells (0.92%), 0.94%, 0.8%, 0%, 3.92%, 0%
	Male cells: 248 skin, 90 ear	69, 34	23 (9.27), 30 (33.33)	2 (8.70), 4 (13.33)	Male cells (1.78%), 0.81%, 4.44%
Lanza			496	30	

[a]Abbreviations: FBA, fusion before activation; Nss, not serum starved; Ss, serum starved.

somatic cells is associated with oncogenic transformation (Brenner, 1999). It remains to be determined whether telomerase activity lingers in the somatic cells of nuclear transfer clones, increasing a predisposition for cancer development.

An additional six healthy calves were cloned from a senescent somatic cell line, derived from a 45-day-old female fetus (Lanza *et al.*, 2000). These clones exhibited telomeres *longer* than those of age-matched controls. Fetal fibroblasts were passaged until completion of greater than 95% of their life span, at which point cells demonstrated morphological features of replicative senescence. The cloned calves may have fared differently than Dolly because of species differences, or because different donor cell types were used. The bovine embryos may have overcompensated in telomere reprogramming in reaction to the donor cells being near the limit of telomere shortening.

Telomerase activity appears to be fully reprogrammable by the blastocyst stage of bovine development, which is slightly delayed in comparison to embryos derived from fertilization (Betts *et al.*, 2001). Calves cloned from adult fibroblasts, fetal fibroblasts, and granulosa cells did not have telomeres significantly different in length from those of age-matched controls. Initially during donor cell culture, telomerase activity decreased 42% in 24th-passage fetal fibroblasts when compared to early-passage cells. Prolonged culture may lead to cumulative oxidative stress-induced telomeric damage. Additionally, telomerase activity decreased 51.4% in first-passage fetal fibroblasts after 10 days of serum starvation. Cell cycle exit into a quiescent state, induced by serum starvation, may be responsible for this decrease in telomerase, consequently increasing vulnerability to telomeric damage. Other regulatory mechanisms that have not yet been analyzed in cloned animals may contribute to the determination of telomere length, which could explain the current controversy in the literature as to the effect of nuclear transfer on telomere length. Various procedures may restore telomere length in certain animal systems yet not in others. A more complete discussion of cell aging and telomere lengths is found in Chapter 4, this volume.

LARGE-OFFSPRING SYNDROME

As illustrated in Table 5, fetal survival, parturition, live birth, and postnatal survival are compromised in a small way by *in vitro* production and culture of bovine

Table 5 *Development of Bovine Embryos Produced in Vivo, in Vitro, and by Nuclear Transfer*

Development	*In vivo*	*In vitro*	Nuclear transfer
To blastocyst	60–80%	30–50%	20–30%
Pregnancy	55–60%	50–55%	0–25%
Fetal loss	5–20%	30–35%	20–80%
Large-calf syndrome	Rare	Moderate +/–	Frequent
Postnatal survival	Most (95%+)	Slightly reduced	50–90%
Reduced life span? (telomeres)	Normal	Normal	Normal

embryos, and to a greater extent for bovine nuclear transfer embryos. This problem, commonly known as the large-calf or large-offspring syndrome (LOS), imposes a restriction on the efficiency and potential applications of nuclear transfer. The etiology as summarized across the work of several investigators, beginning around day 35 of gestation, begins with placental cotyledons, which are larger and fewer than normal, and with a partial deficiency in mesodermal-derived allantoic tissue. There is placental hydrops, and considerable loss of fetuses after day 35. Gestation is prolonged, and parturition does not occur. Calves at predelivery, or delivered by cesarean section or induced labor, are larger than normal and often die at the time of delivery or soon afterward. Fetal lung and maternal mammary development are retarded. Intensive postnatal care is required because neonatal mortality is high for clones (9–50%). Suckling reflex is poor, and hence colostrum intake is reduced and delayed. Calves born with this syndrome often have metabolic and respiratory acidosis, greatly elevated plasma insulin levels, hypothyroidism but enlarged thyroid gland, hypoxemia, hypoglycemia, lung fibrosis, and cirrhosis of the liver (Sinclair *et al.*, 2000; Stice *et al.*, 1996; Garry *et al.*, 1996; Wilson *et al.*, 1995; Behboodi *et al.*, 1995; Farin and Farin, 1995; Kruip and den Daas, 1997; McEvoy *et al.*, 1998; Peterson and McMillan, 1998; Schmidt *et al.*, 1996; van Wagtendonk-deLeeuw *et al.*, 1998; Wells *et al.*, 1999; Young *et al.*, 1998).

For *in vitro*-produced embryos, at least a part of failed development has been attributed to embryo cultures containing high serum concentrations or feeder cells (Walker *et al.*, 1996; Thompson *et al.*, 1995; Sinclair *et al.*, 2000). There is some evidence that exposure of embryos to high serum concentration may affect their lipid composition and gene expression (Sinclair *et al.*, 2000). Whether the causative effects after nuclear transfer are the same as after *in vitro* culture is unknown. At least four hypotheses are being pursued, regarding cause of or mechanisms causing LOS.

The first says that the etiology of the nuclear transfer LOS is similar to that occurring after *in vitro* production of bovine embryos. This hypothesis may be true, but similarity with increased frequency does not explain the cause for either *in vitro*-produced or nuclear transfer-produced embryos. Causative culture conditions and potential media complications are known for *in vitro*-produced embryos (Sinclair *et al.*, 2000). Critical experiments comparing *in vivo* embryo culture to *in vitro* maturation and culture need to be done.

The second hypothesis is that LOS is caused by failed reprogramming of the genome, manifest either as expression of fetal or adult genes in the embryo, or failed expression of embryonic genes. A limited comparison of genes expressed after nuclear transfer versus *in vitro* embryo production is shown in Table 6.

A difficulty with these experiments is that the variables of levels of expression and precise timing of expression are missed in many of the studies to date. The

Table 6 Differences between Embryos Derived from Nuclear Transfer and from in Vitro Fertilization and Culture in Expression of Specific Genes

Gene	Expression level NT vs. IVF	Analyzed stages	Quantification method	Reference
PFK	Similar	Eight cell, blastocyst	Qualitative	Winger et al., 2000
CS	Similar			
LDH	Similar			
Oct4	Similar	Two, four, and eight cell, morula, blastocyst	Qualitative	Daniels et al., 2000, 2001
FGF-2	Similar			
Gp130	Similar			
PolyAP	Similar			
IL-6	Similar or reduced			
FGF-4	Reduced			
FGF-2r	Similar or reduced			
DNMT	Similar or reduced	Blastocyst	Semiquantitative	Wrenzycki et al., 2001
HSP	Similar or reduced			
Hash-2	Similar or elevated			
GluT	Similar			
Dcll	Similar			
E-cadherin	Similar			
Igf2r	Similar			
IF-tau	Similar or elevated			
Igf2	Similar or elevated	Blastocyst, fetal liver	Semiquantitative	Winger et al., 2001

importance of level of gene expression has been illustrated in a study by Niwa *et al.* (2000) in the mouse. They studied a member of the Pou transcription factor family, Pou5f1, and its product Oct4. Oct4 was known to be expressed in totipotent and pluripotent germ line lineages, and to recognize an 8-bp DNA sequence of promoter and enhancer regions of developmentally important expressed genes. Niwa *et al.* (2000) showed that usual levels of Oct4 message maintain embryonic cells in the embryonic stem cell or germ line lineage. A mouse knockout of Pou5f1 (Oct4) caused the embryonic cells to become trophectoderm. Overexpression of Oct4 caused formation of extraembryonic endoderm and mesoderm (Niwa *et al.*, 2000). If the important variable affecting development is level of gene expression for Oct4 or several other genes, the search for a genomic failure would provide false-negative answers. Sensitive quantitative methods for assaying gene expression need to be developed and used.

A third hypothesis contends that LOS is caused by failed or inappropriate expression of imprinted genes. Favorite known imprinted genes used to test this hypothesis have been the *Igf2–H19* complex and the *Igf2r* (receptor) or the *Xist* gene, and X-chromosome inactivation. Regarding X-chromosome inactivation, Eggan *et al.* (2000) have shown for day 12.5 mouse fetuses derived from somatic cell nuclear transfer that both X chromosomes were active during cleavage of cloned embryos, and were followed by random X inactivation in the embryo proper, whereas in the trophectoderm X inactivation was nonrandom with the inactivated X of the nuclear donor cell chosen for inactivation. Thus the imprint of the somatic cell was retained. When mouse embryonic stem cells that are not yet X chromosome inactivated were used as donor cells, random X inactivation occurred in both trophectoderm and embryo. It is not known whether the retention and expression of the donor cell inactive X as the inactive X of the new trophectoderm are affecting normal trophectoderm and placental development. The etiology of LOS is partially similar to failed expression of a mouse and human imprinted gene, the *Igf2–H19* group and the *Igf2r*. This similarity has led several investigators to examine the effect of nuclear transfer on the *Igf2–H19* complex (Sinclair *et al.*, 2000; Young *et al.*, 2001). The relation of the *Igf2–H19* genes to LOS is still unclear for embryos derived from *in vitro* culture or nuclear transfer. However, the loss of both mouse and sheep fetuses derived from embryo culture has been associated with decreased expression of the *Igf2* receptor gene and increased expression of the gene for IGF binding protein 2 (Young *et al.*, 2001). Whether this holds true for embryos or fetuses derived from nuclear transfer, or for bovine embryos or fetuses derived from *in vitro* culture, remains unknown. It does illustrate the importance for examining several players in control of a physiological event (IGF) rather than drawing conclusions from change in expression of one single gene at one specific moment.

The fourth hypothesis states that LOS is caused by failed chromatin organization or failed initiation of transcription after nuclear transfer. (The substance of this hypothesis is reviewed more extensively in Chapters 3 and 12, this volume.) However, two lines of reasoning are noteworthy here. The first pertains to initiation of polymerase I-driven transcription from the nucleolar organizing regions (NORs) associated with chromosomes. In bovines, polymerase I functional NORs develop at the embryonic eight-cell stage, which is the beginning of the major turn on of the embryonic genome. Kanka *et al.* (1999) have shown that functional NORs develop in bovine embryos derived from nuclear transfer, but at one cell cycle earlier than for *in vitro*-produced embryos. Whether development of NORs is related to LOS has not been tested.

The second line of reasoning pertains to a variable other than Oct4, as discussed earlier, which may also affect mesoderm and hence placental formation. This variable is a nuclear histone known as linker histone H1, which is highly mobile in the nucleus. It localizes to the nucleosome regions and prevents access of transcription factors and chromatin remodeling complexes to DNA (Misteli, 2001). The active linker form of H1 is not detectable in MII oocytes, and in bovine *in vitro*-produced embryos it reassembles onto chromatin during the fourth cell cycle at the time of embryonic genome activation. Its loss in bovine nuclear transfer embryos is most rapid when fusion is close to the time of oocyte activation (Bordignon *et al.*, 1999). Activation by 6-DMAP or cycloheximide delays linker H1 loss from transplanted donor nuclei.

Linker histone H1 is involved in regulation of gene expression in amphibian early embryonic development (Wolffe *et al.*, 1997). In *Xenopus*, the time at which mesoderm-inducing genes are transcribed is altered by experimental manipulation of the time when somatic linker H1 becomes detectable in embryos (Steinbach *et al.*, 1997). As noted earlier, overexpression of the mesoderm-inducing gene *Oct4* leads to endoderm–mesoderm formation. Inappropriate timing might cause abnormal placental development. Conditions such as activation, which affect loss of linker H1 activity in reconstructed oocytes and the time or extent of renewed linker H1 activity, need to be associated with the frequency of LOS in cloned cattle. Although we are left with defendable hypotheses, none has yet been tested sufficiently to explain precisely the cause, in whole or in part, of LOS in cattle or other cloned species.

APPLICATIONS OF NUCLEAR TRANSFER

Several of the applications for cloning of cattle were indicated in the introduction to this chapter. The realization of each is dependent on the efficiency of the cloning process. Some applications require the ability to culture donor cells to large numbers in order to be made transgenic for the production of transgenic offspring. An example is the production of pharmaceutical proteins and other highly valuable proteins in milk. A procedure for producing genetically engineered products in milk is shown in Fig. 1.

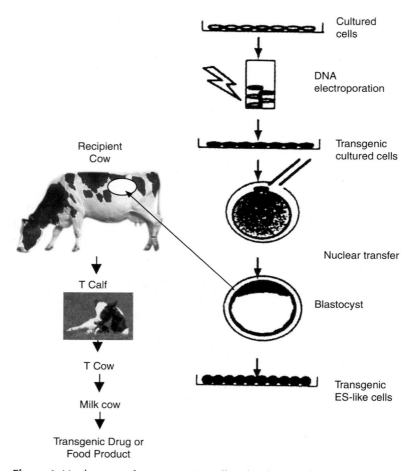

Figure 1 *Nuclear transfer: transgenic cell technology and end products.*

Thus far, at least three companies are producing cloned cows designed to produce pharmaceutical proteins. In spite of the losses during pregnancy after nuclear transfer, the efficiency of producing transgenic offspring from nuclear transfer derived blastocysts (25%) is still much greater than from the traditional microinjection of DNA into pronuclear bovine zygotes (less than 5%). This higher efficiency is because essentially 100% of the nuclear transfer-derived offspring is indeed transgenic (Brink *et al.*, 2000). This application does not require making a copy of a particular cow or a cow of established high-level milk production. Therefore, fetal germ line or fetal somatic cells with more robust culture and developmental capabilities are the most common nuclear donor cells.

Another application is the use of somatic cells in nuclear transfer for the purpose of cloning proven high-performance dairy or beef cattle. Here the intent is to broaden the peak of a genetic pyramid of performance, wherein a few elite animals represent the peak and most cattle of the breed are between the middle and base of the pyramid. Whereas the average dairy cow produces over 15,000 pounds of milk per year, a very few cows have produced over 60,000 pounds and only two cows have produced over 70,000 pounds of milk per year. The making of as few as 100 clones from one of these cows would, through their descendants, have a large impact on the average milk production per cow in the United States. Geneticists estimate that clones will be approximately 70% like the founding cow (G. Shook, personal communication). Until the year 2001 the low efficiency and hence high cost of cloning was considered too expensive for production of milk- or meat-producing cattle. However, three companies are now producing clones for milk or for meat production improvement. This agricultural application will likely be followed by the use of nuclear transfer to produce cattle transgenic for performance, disease resistance, or product traits. Even new food products such as a unique and tasty cheese, artificial sweeteners, more flavorful or tender meat, or industrial proteins might be produced by transgenic cloned cattle.

An alternative application of nuclear transfer concerns interspecies nuclear transfer. The mechanisms regulating early embryonic development appear to be compatible between a variety of mammalian species (Dominko *et al.*, 1999). Bovine metaphase oocyte cytoplasm has successfully supported development of embryos reconstructed with skin fibroblast nuclei from a sheep, pig, cynomolgus monkey, and rat. Oocytes that extruded the first polar body after 16 hours of culture were selected as recipients rather than later maturing oocytes. Donor cells were subjected to serum starvation for 3–10 days prior to transfer by whole cell fusion. Prior to activation with ionomycin and 6-DMAP, fibroblasts were exposed to the cytoplast environment for at least 6 hours to enhance modification of the somatic nuclei. The recipient cytoplast seemed to control timing of the first two cleavage divisions. Subsequent development steered the embryos toward a developmental schedule characteristic of the donor species, indicating that embryonic transcription had successfully begun.

High proportions of reconstructed embryos developed beyond the 16-cell stage, a subset of which compacted into morulae. The observation that many of the late cleavage-stage embryos failed to compact suggests that the products synthesized by the donor nucleus were incompatible with residual elements from the recipient cytoplasm. Only two cow–sheep embryos resulted in implantation, but no fetal heartbeats were detected. Bovine ooplasm can also support development of nuclei transferred from human lymphocytes and oral mucosal epithelial cells (Robl *et al.*, 1998). A calf was produced by fusion of a nucleus from a closely related subspecies of cattle, the gaur, into a *Bos taurus* oocyte; however this clone lived only a short time after birth (Cibelli *et al.*, 2000). Mitochondrial DNA within the recipient cytoplasm must be interacting with the nuclear genome in order to produce protein systems in the mitochondria. These experiments indicate that such interactions are possible between different species; however, the lack of live offspring from all but closely related intraspecies nuclear transfer embryos demonstrates that their effectiveness is limited. Further study of cytoplasm or mitochondria–nucleus compatibility is needed.

One potential problem with transspecific nuclear transfer concerns the appropriate expression of imprinted genes. The oocyte cytoplasm may contain species-specific factors that would fail to activate appropriately genes encoded within the nucleus of another species, leading to aberrant expression and compromised development (Colman and Kind, 2000). Although transspecific nuclear transfer is not yet practical for the production of cloned offspring, species in which oocytes are limited or hard to obtain may someday benefit from this technique. If interspecies nuclear transfer were perfected, it would be possible to use enucleated bovine oocytes to rescue endangered species from only a few donor cells of that species. It might also be possible to produce cell lines for human cell therapies from human nuclear donor somatic cells. The latter application raises ethical concerns about whether it should be done, and this will be discussed in a later chapter.

REFERENCES

Barnes, F., Collas, P., Powell, R., King, W., Westhusin, M., and Shepherd, D. (1993). Influence of recipient oocyte cell cycle stage on DNA synthesis, nuclear envelope breakdown, chromosome constitution and development in nuclear transplant embryos. *Mol. Reprod. Dev.* **36**, 33–41.

Behboodi, E., Anderson, G. B., BonDurant, R. H., Cargill, S. L., Kreuscher, B. R., Medrano, J. F., and Murray, J. D. (1995). Birth of large calves that developed from *in-vitro*–derived bovine embryos. *Theriogenology* **44**, 227–232.

Betts, D. H., Bordignon, V., Hill, J. R., Winger, Q., Westhusin, M. E., Smith, L. C., and King, W. A. (2001). Reprogramming of telomerase activity and rebuilding of telomere length in cloned cattle. *Proc. Natl. Acad. Sci. U.S.A.* **98**, 1077–1082.

Bordignon, V., Clarke, H. J., and Smith, L. C. (1999). Developmentally regulated loss and reappearance of immunoreactive somatic histone H1 on chromatin of bovine morula-stage nuclei following transplantation into oocytes. *Biol. Reprod.* **61**, 22–30.

Brenner, C. A. (1999). Telomeres, Telomerase. *Alpha Newsletter* **January**.

Brink, M. F., Bishop, M. D., and Pieper, F. R. (2000). Developing efficient strategies for the generation of transgenic cattle which produce biopharmaceuticals in milk. *Theriogenology* **53**, 139–148.

Cibelli, J. B., Stice, S. L., Golueke, P. J., Kane, J. J., Jerry, J., Blackwell, C., Ponce de Leon, F. A. P., and Robl, J. M. (1998). Cloned transgenic calves produced from non quiescent fetal fibroblasts. *Science* **280**, 1256–1258.

Colman, A., and Kind, A. (2000). Therapeutic cloning: concepts and practicalities. *Trends Biotechnol.* **18**, 192–196

Daniels, R., Hall, V., and Trounson, A. O. (2000). Analysis of gene transcription in bovine nuclear transfer embryos reconstructed with granulosa cell nuclei. *Biol. Reprod.* **63**, 1034–1040.

Daniels, R., Hall, V. J., French, A. J., Korfiatis, N. A., and Trounson, A. O. (2001). Comparison of gene transcription incloned bovine embryos produced by different nuclear transfer techniques. *Mol. Reprod. Dev.* **60**, 281–288.

DiBerardino, M. (1997). "Genomic Potential of Differentiated Cells." Columbia University Press, New York.

Dominko, T., and First, N. L. (1997). Timing of meiotic progression in bovine oocytes and its effect on early embryo development. *Mol. Reprod. Dev.* **47**, 456–467.

Dominko, T., Mitalipova, M., Haley, B., Behyan, Z., Memili, E., McKusick, B., and First, N. (1999). Bovine oocyte cytoplasm supports development of embryos produced by nuclear transfer of somatic cell nuclei from various mammalian species. *Biol. Reprod.* **60**, 1496–1502.

Eggan, K., Akutsu, H., Hochedlinger, K., Rideout, W., Yanagimacho, R., and Jaenisch, R. (2000). X-Chromosome inactivation in cloned mouse embryos. *Science* **290**, 1578–1581.

Farin, P. W., and Farin, C. E. (1995). Transfer of bovine embryos produced *in-vivo* or *in-vitro*: Survival and fetal development. *Biol. Reprod.* **52**, 676–682.

First, N., Mitalipova, M., and Kent-First, M. (1999). Reproductive technologies and transgenics. *In* "The Genetics of Cattle" (R. Fries and A. Ruvinsky, eds.), pp. 411–435. CAB Int., Wallingford, Oxford, UK.

Forsberg, E. J., Betthauser, J., Strelchenko, N., Golueke, P., Childs, L., Jurgella, G., Koppang, R., Mell, G., Pace, M. M., and Bishop, M. (2001). Cloning non-transgenic and transgenic cattle. *Theriogenology* **55**, 269 (abstr.).

Garry, F. B., Adams, R., McCann, J. P., and Odde, K. G. (1996). Postnatal characteristics of calves produced by nuclear transfer cloning. *Theriogenology* **45**, 141–152.

Gordon, I. (1994). "Laboratory Production of Cattle Embryos." CAB Int., Wallingford, Oxford, UK.

Hoffner, N. J., and DiBerardino, M. A. (1980). Developmental potential of somatic nuclei transplanted into meiotic oocytes of *Rana pipiens*. *Science* **209**, 517–519.

Kanka, J., Smith, S. D., Soloy, E., Holm, P., and Callesen, H. (1999). Nucleolar ultrastructure in bovine nuclear transfer embryos. *Mol. Reprod. Dev.* **52**, 253–263.

Kato, Y., Tani, T., Sotomaru, Y., Kurokawa, K., Kato, J., Doguchi, H., Yasue, H., and Tsunoda, Y. (1998). Eight calves cloned from somatic cells of a single adult. *Science* **282**, 2095–2098.

Kato, Y., Tani, T., and Tsunoda, Y. (2000). Cloning of calves from various somatic cell types of male and female adult, newborn, and fetal cows. *J. Reprod. Fertil.* **120**, 231–237.

Kruip, T. A. M., and den Daas, J. H. G. (1997). *In-vitro* produced and cloned embryos: Effects on pregnancy, parturition, and offspring. *Theriogenology* **47**, 43–52.

Kubota, C., Yamakuchi, H., Todoroki, J., Mizoshita, K., Tabara, N., Barber, M., and Yang, X. (2000). Six cloned calves produced from adult fibroblast cells after long-term culture. *Proc. Natl. Acad. Sci. U.S.A.* **97**, 990–995.

Lanza, R. P., Cibelli, J. B., Blackwell, C., Cristofalo, V. J., Francis, M. K., Baerlocher, G. M., Mak, J., Schertzer, M., Chavez, E. A., Sawyer, N., Lansdorp, P. M., and West, M. D. (2000). Extension of cell life-span and telomere length in animals cloned from senescent somatic cells. *Science* **288**, 665–669.

Leibfried-Rutledge, M. L., Northey, D. L., Nuttleman, P. R., and First, N. L. (1992). Processing of donated nucleus and timing of post-activation events differ between recipient oocytes 24 or 42 hr. age. *Theriogenology* **37**, 244.

Liu, L., and Yang, X. (1999). Interplay of maturation-promoting factor and mitogen-activated protein kinase inactivation during metaphase-to-interphase transition of activated bovine oocytes. *Biol. Reprod.* **61**, 1–7.

McEvoy, T. G., Sinclair, K. D., Broadbent, P. J., Goodhand, K. L., and Robinson, J. J. (1998). Post-natal growth and development of Simmental calves derived from *in vivo* or *in vitro* embryos. *Reprod. Fertil. Dev.* **10**, 459–464.

McGrath, J., and Solter, D. (1983). Nuclear transplantation in the mouse embryo by microsurgery and cell fusion. *Science* **200**, 1300–1320.

Misteli, T. (2001). Protein dynamics: implications for nuclear architecture and gene expression. *Science* **291**, 843–847.

Mitalipova, M., Beyhan, Z., and First, N. L. (2001). Pluripotency of bovine embryonic cell lines derived from precompacting embryos. *Cloning* 60(2), 59–67.

Motlik, J., Pavlok, A., Kubelka, M., Kaous, J., and Kalab, P. (1998). Interplay between cdc2 kinase and MAP kinase pathway during maturation of mammalian oocytes. *Theriogenology* **49**, 461–469.

Navara, C. S., Sims, M. M., and First, N. L. (1992). Timing of polarization in bovine embryos and developmental potential of polarized blastomeres. *Biol. Reprod.* **46**(Suppl. 1).

Niwa, H., Miyazaki, J., and Smith, A. G. (2000). Quantitative expression of Oct-3/4 defines differentation, dedifferentiation or self-renewal of embryonic stem cells. *Nat. Genet.* **24**, 372–376.

Pace, M. M., Mell, G., Forsberg, E. J., Betthauser, J., Strelchenko, N., Golueke, P., Childs, L., Jurgella, G., Koppang, R., and Bishop, M. (2001). Cloning using somatic cells: an analysis of 75 calves. *Theriogenology* **55**, 281 (abstr.).

Pace, M., Augenstein, M., Betthauser, J., Childs, L., Eilertsen, K., Enos, J., Forsberg, E., Golueke, P., Graber, D., Kemper, J., Koppang, R., Lange, G., Lesmeister, T., Mallon, K., Mell, G., Misica, P., Pfister-Genskow, M., Strelchenko, N., Voelker, G., Watt, S., and Bishop, M. D. (2002). Ontogeny of cloned cattle to lactation. *Biol. Reprod.*, in press.

Peterson, A. J., and McMillan, W. H. (1998). Allantoic aplasia—a consequence of *in vitro* production of bovine embryos and the main cause of late gestation embryo loss. [Abstr. No. 23] *In* "Proceedings of the Annual Conference Aust. Soc. Reprod. Biol.," Perth, Australia.

Poccia, D., and Collas, P. (1997). Nuclear envelope dynamics during male pronuclear development. *Dev. Growth Differ.* **39**, 541–550.

Prather, R. S., Barnes, F. L., Sims, M. M., Robl, J. M., Eyestone, W. H., and First, N. L. (1987). Nuclear transplantation in the bovine embryo: Assessment of donor nuclei and recipient oocytes. *Biol. Reprod.* **37**, 859–866.

Renard, J. P., Chastant, S., Chesne, P., Richard, C., Marchal, J., Cordonnier, N., and Chavatte, X. V. (1999). Lymphoid hypoplasia and somatic cloning. *Lancet* **353**, 1489–1491.

Robl, J., Cibelli, J., and Stice, S. L. (1998). Embryonic or stem-like cell lines produced by cross species nuclear transplantation. International patent application number WO 98/07841. World Intellectual Property Organization.

Schmidt, M., Greve, T., Avery, B., Beckers, J. F., Sulon, J., and Hansen, H. B. (1996). Pregnancies, calves and calf viability after transfer of *in vitro* produced bovine embryos. *Theriogenology* **46**, 527–539.

Sims, M., and First, N. L. (1994). Production of calves by transfer of nuclei from cultured inner cell mass cells. *Proc. Natil. Acad. Sci. U.S.A.* **91**, 6143–6147.

Sinclair, K., Young, L., Wilmut, I., and McEvoy, T. (2000). *In-utero* overgrowth in ruminants following embryo culture: Lessons from mice and a warning to men. *Hum. Reprod.* **15**(Suppl. 5), 68–86.

Steinbach, O. C., Wolffe, A. P., and Rupp, R. A. (1997). Somatic linker histones cause loss of mesodermal competence in *Xenopus*. *Nature* **389**, 395–399.

Stice, S. L., and Keefer, C. L. (1993). Multiple generational bovine embryo cloning. *Biol. Reprod.* **48**, 715–719.

Stice, S. L., Strelchenko, N. S., Keefer, C. L., and Matthews, L. (1996). Pluripotent bovine embryonic cell lines direct embryonic development following nuclear transfer. *Biol. Reprod.* **54**, 100–110.

Susko-Parrish, J. L., Leibfried-Rutledge, M. L., Northey, D. L., Schutzkus, V., and First, N. L. (1994). Inhibition of protein kinases after an induced calcium transient causes transition of bovine oocytes to embryonic cycles, without meiotic completion. *Dev. Biol.* **166**, 729–739.

Thompson, J. G., Gardner, D. K., Pugh, P. A., McMillan, W. H., and Tervit, H. R. (1995). Lamb birth weight is affected by culture system utilized during *in vitro* pre-elongation development of ovine embryos. *Biol. Reprod.* **53**, 1385–1391.

Tian, X. C., Xu, J., and Yang, X. (2000). Normal telomere lengths found in cloned cattle. *Nat. Genet.* **26**, 272–273.

van Wagtendonk-de Leeuw, A. M., Aerts, B. J. G., and den Daas, J. H. G. (1998). Abnormal offspring following *in vitro* production of bovine preimplantation embryos: A field study. *Theriogenology* **49**, 883–894.

Wakayama, T., Perry, A. C. F., Zuccotti, M., Johnson, K. R., and Yanagimachi, R. (1998). Full term development of mice from enucleated oocytes injected with cumulus cell nuclei. *Nature* **394**, 369–374.

Walker, S. K., Hartwich, K. M., and Seamark, R. F. (1996). The production of unusually large offspring following embryo manipulation: Concepts and challenges. *Theriogenology* **45**, 111–120.

Ware, C. B., Barnes, F. I., Malki-Laurilla, M., and First, N. L. (1989). Age dependence of bovine oocyte activation. *Gamete Res.* **22**, 265–275.

Wells, D., Misica, P., Day, A., and Tervit, H. R. (1999). Production of cloned calves following nuclear transfer with cultured adult mural granulosa cells. *Biol. Reprod.* **60**, 996–1005.

Willadsen, S. M. (1986). Nuclear transplantation in sheep embryos. *Nature* **320**, 63–65.

Wilmut, I., Schnieke, A. E., McWhir, J., Kind, A. J., and Campbell, K. H. S. (1997). Viable offspring derived from fetal and adult mammalian cells. *Nature* **385**, 810–813.

Wilson, J. M., Williams, J. D., Bondiols, K. R., *et al.* (1995). Comparison of birth weight and growth characteristics of bovine calves produced by nuclear transfer (cloning), embryo transfer and natural mating. *Anim. Reprod. Sci.* **68**, 73–83.

Winger, Q. A., Hill, J. R., Shin, T., Watson, A. J., Kraemer, D. C., and Westhusin, M. E. (2000). Genetic reprogramming of lactate dehydrogenase, citrate synthase, and phosphofructokinase mRNA in bovine nuclear transfer embryos produced using bovine fibroblast cell nuclei. *Mol. Reprod. Dev.* **56**, 458–464.

Winger, Q. A., Hill, J. R., Shin, T., Watson, A. J., and Westhusin, M. E. (2001). Reprogramming of IGF-II mRNA in bovine somatic cell nuclear transfer embryos and fetal liver. *Theriogenology* **55**, 296.

Wolffe, A. P., Khochbin, S., and Dimitrov, S. (1997). What do linker histones do in chromatin? *BioEssays* **19**, 249–255.

Wrenzycki, C., Wells, D., Herrmann, D., Miller, A., Oliver, J., Tervit, R., and Niemann, H. (2001). Nuclear transfer protocol affect messenger RNA expression patterns in cloned bovine blastocysts. *Biol. Reprod.* **65**, 309–317.

Yang, X., Kubota, C., Suzuki, H., Taneja, M., Bols, P. E. J., and Presice, G. A. (1998). Control of oocyte maturation in cows—Biological factors. *Theriogenology* **49**, 471–482.

Young, L. E., Sinclair, K. D., and Wilmut, I. (1998). Large offspring syndrome in cattle and sheep. *Rev. Reprod.* **3**, 155–163.

Young, L., Fernandes, K., McEvoy, T., Butterworth, S., Gutierrez, G., Carolan, C., Broadbent, P., Robinson, J., Wilmut, I., and Sinclair, K. (2001). Epigenetic change in IGF2R is associated with fetal overgrowth after sheep embryo culture. *Nat. Genet.* **27**, 153–154.

Zakhartchenko, V., Durcova-Hills, G., Stojkovic, M., Schernthaner, W., Prelle, K., Steinborn, R., Muller, M., Brem, G., and Wolf, E. (1999). Effects of serum starvation and re-cloning on the efficiency of nuclear transfer using bovine fetal fibroblasts. *J. Reprod. Fert.* **115**, 325–331.

CHAPTER 20

CELL CYCLE REGULATION IN CLONING

Keith H. S. Campbell

INTRODUCTION

"The cell cycle of a growing cell is the period between the formation of the cell by the division of its mother cell and the time when the cell itself divides to form two daughters. It is a fundamental unit of time at the cellular level since it defines the life cycle of a cell" (Mitchison, 2002).

During the past three decades significant progress has been made toward understanding the molecular events that occur during the cell cycle and identifying many of the genes and proteins involved in this complex series of events. The objectives here are not to review the extensive literature that exists relating to these mechanisms, but rather to relate the cell cycle to the production of embryos by nuclear transfer. In particular, the focus is on interactions between the donor nucleus and components of the recipient cytoplasm and the impact on subsequent development.

The development of live offspring by nuclear transfer was originally restricted to the use of donor nuclei derived directly from the cells of early embryos, in general blastomeres; however, in 1986 offspring were derived from embryonic cells that had differentiated in culture (Campbell *et al.*, 1996b). Since this time offspring have been reported from a range of cell types, including those derived from embryonic, fetal, and adult populations. Although the underlying mechanisms observed in different species and different donor cell types (embryonic vs. somatic) are similar, some differences are observed. The major cell cycle effects observed across all species are summarized in the following discussions; where relevant, differences are examined in detail.

INITIAL EVENTS

During a single cell cycle, a cell must double all of its components, segregate its genetic material equally to the two daughter cells, and undergo cell division. Early studies of the cell cycle were largely devoted to defining the temporal sequence of events within the cycle. The most obvious of these events are those concerned with the replication and partition of the genetic material to the two daughters—namely, DNA replication, DNA segregation, and cell division. In eukaryotic cells, which contain a defined nuclear structure, DNA segregation is achieved by mitosis and nuclear division. The advent of autoradiographic techniques demonstrated that DNA synthesis is a periodic event during the cell cycle of most eukaryotes. This periodicity led Howard and Pelc (1953) to divide the cell cycle of eukaryotes into four distinct phases: G1, S, G2 and M. The discrete period of DNA synthesis is preceded by G1, the pre-DNA synthetic period, and followed by G2, the post-

DNA synthetic or premitotic period. The M phase denotes the mitotic segregation of the duplicated genetic material and is followed by cell division.

Early biochemical and morphological studies on cells in culture generated a temporal map of many stage-specific processes during the cell cycle. These were referred to as landmark events. In addition, many other stage-specific functions were identified by the use of inhibitors of specific phases or by the isolation of mutants defective (or altered) in specific phases of the cell cycle. Cell cycle mutants have been isolated from prokaryotic and both higher and lower eukaryotic cell types. The majority of these mutants are temperature-sensitive lethals that are unable to complete the cell cycle at the restricted temperature. Analysis of data generated from these studies demonstrated that, in general, the cell cycle is composed of a series of dependent sequences of events. However, there may be more than a single sequence, and cell cycle events may occur on a series of parallel pathways (for reviews see Hartwell, 1991; Hartwell and Weinert, 1989; Jacobs, 1992; Kirschner, 1992). One such dependent sequence of events is the "nuclear division cycle," now termed the "cell division cycle."

The nuclear division cycle involves two major events, DNA replication (S phase) and segregation of the duplicated genetic material (M phase, or mitosis). During a single cell cycle all of the genetic material must be replicated once and only once and segregated equally to the two daughter cells. Failure to replicate, re-replication of a portion of the genetic material, or unequal segregation will lead to daughter cells having an incorrect content of DNA (aneuploidy). The mechanisms that control DNA replication are complex and involve a variety of proteins (for review see Coverley and Laskey, 1994); however, central to this control is the maintenance of an intact nuclear envelope (Blow and Laskey, 1988). Segregation of replicated DNA occurs by mitosis; following nuclear envelope breakdown (NEBD), the DNA condenses into chromosomes and becomes arranged on the mitotic spindle and the chromosomes are then segregated equally to the two new daughters. Early studies using cell fusion experiments demonstrated that the onset of M phase was controlled by a cytoplasmic factor. Fusion of late G2- or M-phase cells to cells in G1 caused premature entry into mitosis (Johnson and Rao, 1970a; Johnson *et al.*, 1970). These studies also demonstrated the role of the nuclear envelope, in that fusion of S- or G2-phase cells to G1 cells did not result in DNA re-replication (Johnson and Rao, 1970b). Studies using frog eggs had also demonstrated the role of a cytoplasmic factor in controlling the onset of meiosis during oocyte maturation (Masui and Markert, 1971); this factor was termed maturation-promoting actor (MPF). MPF is now known to be a protein kinase cdc2, the activity of which is caused by association with cyclins, a group of proteins that are temporally expressed during the cell cycle (Gautier *et al.*, 1990), and phosphorylation at specific sites by other cell cycle-specific proteins (for reviews see Coleman and Dunphy, 1994; Nurse, 1993). The cdc2 gene is ubiquitous to all eukaryotic cells and the kinase activity controls both meiotic and mitotic divisions (Nurse, 1990).

Nuclear transfer involves the transfer of genetic material from a donor cell to a recipient cell from which the genetic material has been removed (for review see Campbell, 1999b). In general, a mature oocyte or unfertilized egg has been the recipient of choice; however, activated oocytes (Campbell *et al.*, 1994), fertilized zygotes (McGrath and Solter, 1983), and two-cell embryos (Landa, 1989) have also been used. The cytoplasmic states of each of these recipients will vary and affect development of the reconstructed embryo.

OOCYTE DEVELOPMENT AND THE CELL CYCLE

In mammals, mitotically active oocyte precursor cells, the oogonia, populate the primitive ovary during prenatal development. At about the time of birth, meiosis is initiated, and these cells, now called "primary oocytes," progress through most of prophase I and become arrested at the diplotene stage. The recruitment signal for initiation of follicle development is unknown, but in response to it, dramatic changes occur within the follicle, leading to growth and differentiation. The diplotene- or prophase-arrested oocyte is characterized by a prominent nucleus, termed the "germinal vesicle," containing decondensed chromatin that is transcriptionally active. Critical developmental changes occur in germinal vesicle (GV)-stage mammalian oocytes as they near completion of their growth phase. These changes pertain to the acquisition of competence to undergo both nuclear and cytoplasmic maturation and are essential for the formation of a competent oocyte, which has the capacity for fertilization and development to live offspring. During the growth phase, the GV-stage oocyte prepares for later development by stockpiling mRNA, proteins, and other macromolecules needed for the completion of meiosis, fertilization, and early embryonic development. Nuclear maturation encompasses the processes inducing resumption of meiotic division and subsequent arrest at metaphase II of the second meiotic division (MII). The mature oocyte remains at MII until fertilization occurs.

The resumption and control of meiotic division requires MPF (for review see Parrish *et al.*, 1992), MPF kinase activity is maximal at metaphase of the meiotic division; at MI, MPF activity declines and then rises until MII, at which it remains high. MPF activity is stabilized at MII by cytostatic factor (CSF), the product of the c-*mos* protooncogene (Hunt, 1992). On fertilization or activation meiosis resumes, MPF activity declines, the second polar body is extruded, and pronuclei are formed. In different species the rates of MPF decay and pronuclear formation vary with time [e.g., cattle (Campbell *et al.*, 1993), sheep (Cmpbell *et al.*, 1994)], in addition, MPF activity in MII-arrested oocytes decays with time and can affect both the ease of activation and the rate of pronuclear formation [e.g., pigs (Kikuchi *et al.*, 2000)]. Following pronuclear formation, DNA synthesis occurs. The replicated DNA is then segregated to the two daughter cells by mitosis; the onset of M phase is controlled by MPF activity, which rises in G2 following completion of S phase.

The use of MII oocytes as cytoplast recipients therefore involves transferring a nucleus into a cytoplasmic environment containing high levels of MPF activity. By definition, MPF induces entry into M phase. This is characterized by NEBD and chromosome condensation. Unless the transferred nucleus was about to enter M phase, then these events occur prematurely; this has been termed "premature chromosome condensation" (PCC). The effects of NEBD and PCC are dependent on the cell cycle stage of the donor nucleus at the time of transfer (for review see (Campbell *et al.*, 1996a). Nuclei in the G1 phase (diploid, pre-S) form single chromatids and nuclei in the G2 phase (tetraploid, post-S) form double chromatids; in both of these situations there appears to be no damage to the DNA. In contrast, nuclei, which are in S phase at the time of transfer, have a typical pulverized appearance and undergo DNA damage (Collas *et al.*, 1992). When mitotic chromosomes are transferred the chromosomes remain condensed (Alberio *et al.*, 2000).

CELL CYCLE EFFECTS OF OOCYTE-DERIVED CYTOPLAST RECIPIENTS

In general, oocytes are enucleated at MII and are used as cytoplast recipients; however, transfer of the donor nucleus may occur prior to enucleation at MII or enucleation may be performed following activation at telophase (TII) of the second meiotic division (Baguisi *et al.*, 1999; Bordignon and Smith, 1998; Liu *et al.*, 2000). In addition, oocytes may be activated or fertilized and enucleated following pronuclear formation. The cell cycle stage of each of these recipients will therefore differ and in particular the levels of MPF activity will differ. For simplicity we will differentiate these recipients into those with and without MPF activity (for review see Campbell *et al.*, 1996a).

RECIPIENTS WITH MPF ACTIVITY

Nuclei may be transferred into MII oocytes at the time of activation or activation may be induced following transfer; these two methods may have different impacts on subsequent development as a result of chromatin segregation due to cytoskeletal effects or other cytoplasmic effects (see below). Following activation, MPF activity declines and the nuclear envelope reforms; DNA synthesis then occurs regardless of the original cell cycle stage of the nuclear donor. Diploid nuclei, which have not previously replicated their DNA, undergo a single round of DNA synthesis and following mitosis result in the formation of two diploid daughters. G2 or tetraploid nuclei undergo a second round of DNA synthesis and result in tetraploid daughter nuclei; those in S-phase undergo an unknown amount of re-replication and result in aneuploid daughter cells. Nuclei transferred in M phase will also reform the nuclear envelope and undergo DNA synthesis, and the ploidy of the resultant daughter cells will depend on segregation of the mitotic chromosomes due to cytoskeletal effects.

When oocytes are enucleated at TII, the decay of MPF activity has been initiated, thus exposure to MPF will depend on the rate of its decay and the rate of donor cell fusion or introduction of the donor nucleus. If MPF activity is sufficient then NEBD and PCC will occur. However, the extent of PCC may be reduced due to lower levels of MPF activity or the reduced duration of exposure. The occurrence of NEBD will result in the patterns of DNA replication described above.

These observations suggest that only diploid nuclei should be transferred into MII oocytes or oocytes containing MPF activity. However, cytoskeletal effects and the occurrence of pseudomeiotic/mitotic divisions have allowed the production of live offspring following the transfer of donor nuclei in G2 and M phases [e.g., mouse (Cheong et al., 1993), cattle (Tani et al., 2001)]. The role of the cytoskeleton in maintaining ploidy is described below. In reality, the effects of fusion to enucleated MII oocytes on the donor nuclei are extremely variable. Reports using a number of species have shown that not all transferred nuclei undergo NEBD [e.g., cattle (Campbell et al., 1993)], suggesting that MPF activity has declined. Other reports have shown that if NEBD does occur the extent and duration of PCC are variable [e.g., cattle (Barnes et al., 1993)]. Reports on a number of species have shown that the level of activity of MPF kinase in MII oocytes declines as the oocyte ages, as determined by morphological examination of the transferred nucleus [e.g., rabbit (Adenot et al., 1997)], or by kinase activity [e.g., pigs (Kikuchi et al., 2000)]. A second factor that may reduce the level of MPF activity is the enucleation process. Factors associated with MPF and its control are reported to be associated with the chromatin in both mitotic and meiotic cells (Adlakha et al., 1982; Czolowska et al., 1986; Maldonado-Codina and Glover, 1992). Therefore, by removing the chromatin the activity of MPF kinase may be removed or destabilized and its activity subsequently reduced; in fact, enucleation of sheep MII oocytes significantly reduces the level of MPF activity (H.-L. Joon and K. H. S. Campbell, personal communication, 2002). In addition, the kinetics of MPF decay following activation of enucleated oocytes may differ from the unenucleated controls, therefore reducing the level of exposure or duration of exposure of the donor nucleus. Interactions with other cytoplasmic kinases thought to be important for meiotic maturation may also be disrupted i.e., mitogen-activated protein kinases (MAPKs) (Anderson, 1992; Lenormand et al., 1993; Verlhac et al., 1994). The cell cycle stage of the donor cell may also play a role in controlling MPF activity; inhibitors of MPF activity have been reported in the cytoplasm of the G1 phase, but not G0, S, or G2 phases, of somatic cells (Adlakha et al., 1983), and other studies have suggested that MPF activity may be the result of an interplay between metaphase and interphase factors (Balakier and Masui, 1986).

RECIPIENTS WITH NO MPF ACTIVITY

Following activation or fertilization MPF activity declines. When intact nuclei from donors in G0/G1, S, or G2 phase are transferred into such recipients, no NEBD and no PCC occur. DNA damage due to PCC is avoided and DNA replication is co-coordinated to the cell cycle stage of the donor. Diploid nuclei undergo a single round of replication, S-phase nuclei continue replication, and G2 nuclei do not re-replicate. Therefore, all of the resultant daughter cells will be diploid. An exception to coordinated replication expected in the case of diploid nuclei may exist with the use of quiescent, or G0, cells. DNA replication requires the presence of chromosome-bound factors that are thought to attach following mitosis and prior to nuclear assembly. In quiescent cells these factors or their activity are lost with time and initiation of DNA replication requires permeabilization of the nuclear membrane (Leno and Munshi, 1994). In the case of mitotic donors there are no published examples; however, decondensation of the DNA, nuclear reformation, and DNA synthesis may expected, resulting in the production of aneuploid (possibly tetraploid) daughter cells.

In the mitotic cell cycle of growing cells a number of mechanisms exist that coordinate the completion of one event with the onset of another (for reviews see Polymenis and Schmidt, 1999; Russell, 1998). In early embryos these mechanisms may or may not operate. One example of such a mechanism is the completion of S phase and the onset of M phase. In cultured cells delaying the completion of S phase by treatment with inhibitors of DNA replication (i.e., aphidicolin) prevents the onset of mitosis and cell division (for review see Enoch and Nurse, 1991). In early embryos of *Xenopus* (the South African clawed toad) a similar mechanism has been described (Dasso and Newport, 1990); however, in mammalian embryos these mechanisms have been little studied. The existence of such mechanisms may be of great importance in determining the timing of transfer within the cell cycle in order to ensure that replication is completed.

OTHER FACTORS RELATED TO THE RECIPIENT CELL CYCLE PHASE

The recipient cytoplast or the combination of the recipient cytoplast and the nuclear donor cell may have other effects on subsequent development. These can be split into two categories: chromatin fate and nuclear reprogramming.

CHROMATIN FATE

On activation or fertilization of an MII oocyte, the second meiotic division resumes and is completed with expulsion of the second polar body. Segregation of the chromosomes is dependent on an intact organized spindle, and polar body extrusion is dependent on cytoskeletal microtubules. In the mouse, transfer of donor nuclei from blastomeres in the G2 phase of the cell cycle can result in polar body extrusion and live offspring, demonstrating a controlled mitotic event in the donor nucleus. Similarly, in the mouse, transfer of blastomere nuclei arrested in mitosis by treatment with nocodazole to MII oocytes treated with cytochalasin B can form two functional pronuclei on subsequent activation (Kwon and Kono, 1996), again suggesting a controlled mitotic event. However, other reports, particularly with somatic cells, have demonstrated the presence of an unorganized spindle and the formation of multiple pseudopronuclei on activation (Wakayama *et al.*, 1998). In other species, the reports are similarly confused; in cattle, transfer of M-phase nuclei from cumulus cells has resulted in the birth of live offspring, suggesting a coordinated mitotic event and polar body extrusion (Tani *et al.*, 1991). However, other reports using M-phase-arrested blastomere nuclei have shown that spindle organization occurs with time and polar body extrusion is also dependent on the time after activation [e.g., cattle (Alberio *et al.*, 2000)]; however, in this report the ploidy of the pronucleus and extruded polar body is unknown. In sheep, observation of an organized spindle after transfer of M-phase fibroblast nuclei is variable and also dependent on time; in addition, loss of chromosomes from the spindle is frequently observed (H.-L. Joon and K. H. S. Campbell, personal communication, 2002). In cattle and pigs, transfer of somatic diploid nuclei to unactivated MII oocytes results in NEBD and PCC. On subsequent activation multiple pseudopronuclei are formed; this can be prevented with inhibitors of spindle polymerization (i.e., nocodazole), suggesting the presence of an unorganized spindle (Campbell and Wilmut, 1999).

Reported differences in the behavior of the transferred chromatin may be dependent on a number of factors. The first factor may be species differences. In the mouse, spindle organization requires microtubule organizing centers (MTOCs) (Maro *et al.*, 1986), and each oocyte contains multiple MTOCs. In contrast, other species use spindle pole bodies to organize their spindle and these are removed with the meiotic spindle. The second factor may be the level of MPF activity, and the duration of exposure may affect the level of chromosome condensation (see previous discussion). Third, the source of the donor cell and the method of reconstruction may be factors. The function of mitotic spindles may vary when spindles are placed into the meiotic environment, and this may also vary between embryonic and somatic donors. Cell fusion transfers the complete contents of the donor cell; in contrast, nuclear injection can transfer only the nucleus and may not transfer factors or structures necessary for spindle formation.

NUCLEAR REPROGRAMMING

The changes that are required to modify the donor chromatin in order to obtain development of the reconstructed embryo have been termed "nuclear reprogramming." Although the molecular events associated with "reprogramming" are slowly being elucidated, at present the only true measure of reprogramming is development to term. The recipient cytoplasm may have effects on the ploidy of the reconstructed embryo, and it may also affect development and therefore nuclear reprogramming. Early studies in the rabbit suggested that nuclear reprogramming was increased when using unactivated oocytes as cytoplast recipients (Collas and Robl, 1991). Subsequently it was suggested that prolonged condensation of the donor chromatin and its exposure to the oocyte cytoplasm may increase reprogramming (Campbell and Wilmut, 1999). In cattle, experiments have demonstrated an improved frequency of development to the blastocyst stage following prolonged exposure (Wells *et al.*, 1999). In addition, a decrease in the relative abundance of a range of transcripts has been reported in such embryos when compared to embryos produced with a shorter period of exposure (Wrenzycki *et al.*, 2001).

The effects of MPF on reprogramming may be related to the degree and duration of PCC for a number of reasons; first, chromosome condensation displaces transcription factors from the DNA (Martinez-Balbas *et al.*, 1995), and the degree of displacement may be dependent on the extent and duration of condensation. Second, displacement of transcription factors will allow oocyte-specific chromatin-binding factors access to the transferred chromatin. Again, the optimum for these processes to occur may be related to the level of MPF activity and duration of exposure, as demonstrated by the pattern of transcription in the resulting embryo (Wrenzycki *et al.*, 2001).

EFFECTS OF CELL CYCLE COMBINATIONS ON DEVELOPMENT

Coordination of the cell cycles of the donor nucleus and the recipient cytoplasm may be obtained through a variety of combinations. However, do these combinations affect subsequent development? Comparing data from nuclear transfer studies carried out in different species and in different laboratories is difficult and unreliable due to the effects of numerous other factors on development, including culture media, oocyte quality, and methodology. In addition the differentiated status and the cell cycle stage of the donor nucleus may also have effects on development. When using unsynchronized early embryos as nuclear donors, the majority of the nuclei are in S phase of the cell cycle. Transfer of such nuclei to enucleated preactivated oocytes resulted in an increase in development in sheep (Campbell *et al.*, 1994). In the literature, examples can be found whereby offspring have been produced from all cell cycle stages when using embryonic donors. However, the extent of nuclear reprogramming required when using embryonic donors may differ from that using somatic donors. Experiments with somatic cell nuclear donors have produced offspring from a variety of combinations, including MII, TII and pre-activated cytoplasts and G1-, G0-, S-, G2-, and M-phase donor nuclei. Surveying the literature, it appears that the frequency of development is increased by the combination of MII or TII recipients with diploid donor cells (G1/G0). This may reflect the possible effects of MPF on nuclear reprogramming as previously discussed, or effects of the donor cell (see below).

EFFECTS OF THE DONOR CELL CYCLE STAGE ON DEVELOPMENT

The advent of animal production from cultured cell populations (Campbell *et al.*, 1996b) has allowed more detailed analysis of the effects of donor cell cycle stage. Live offspring have now been produced from a range of donor cell cycle stages; however, little evidence has emerged of an increased frequency of development from a particular stage. The relationship between components of the recipient cytoplasm and nuclear reprogramming has resulted in an apparent preference for the transfer of diploid donor cells to MII oocytes. Diploid donor cells may be obtained in a variety of cell cycle stages, early in G1, late in G1, or having exited the cell cycle and arrested in G0. The use of quiescent or G0 cell populations has been suggested to be advantageous for embryo development due to the chromatin state (for discussion see Campbell, 1999a). Although many offspring in a range of species have been produced using quiescent cell populations, other studies have suggested that G1 cells may be more amenable to controlling development (Kasinathan *et al.*, 2001). At the present time the low frequency of development to term of nuclear transfer embryos makes comparisons difficult. Similarly, the precise state of the donor and recipient at the time of reconstruction of successful embryos can only be derived from the remaining populations, and factors other than those described here may be involved.

SUMMARY

The cell cycle stages of the recipient oocyte and the donor nucleus interact to affect subsequent development of nuclear transfer embryos. At the present time the molecular mechanisms controlling nuclear reprogramming and development are unknown, although the cell cycle is intimately linked to the success of nuclear transfer.

REFERENCES

Adenot, P. G., Szollosi, M. S., Chesne, P., Chastant, S., and Renard, J. P. (1997). *In vivo* aging of oocytes influences the behavior of nuclei transferred to enucleated rabbit oocytes. *Mol. Reprod. Dev.* **46**, 325–336.

Adlakha, R. C., Sahasrabuddhe, C. G., Wright, D. A., Lindsey, W. F., and Rao, P. N. (1982). Localization of mitotic factors on metaphase chromosomes. *J. Cell Sci.* **54**, 193–206.

Adlakha, R. C., Sahasrabuddhe, C. G., Wright, D. A., and Rao, P. N. (1983). Evidence for the presence of inhibitors of mitotic factors during G1 period in mammalian cells. *J. Cell Biol.* **97**, 1707–1713.

Alberio, R., Motlik, J., Stojkovic, M., Wolf, E., and Zakhartchenko, V. (2000). Behavior of M-phase synchronized blastomeres after nuclear transfer in cattle. *Mol. Reprod. Dev.* **57**, 37–47.

Anderson, N. G. (1992). MAP kinases—Ubiquitous signal transducers and potentially important components of the cell cycling machinery in eukaryotes. *Cell. Signal.* **4**, 239–246.

Baguisi, A., Behboodi, E., Melican, D. T., Pollock, J. S., Destrempes, M. M., Cammuso, C., Williams, J. L., Nims, S. D., Porter, C. A., Midura, P., Palacios, M. J., Ayres, S. L., Denniston, R. S., Hayes, M. L., Ziomek, C. A., Meade, H. M., Godke, R. A., Gavin, W. G., Overstrom, E. W., and Echelard, Y. (1999). Production of goats by somatic cell nuclear transfer. *Nat. Biotechnol.* **17**, 456–461.

Balakier, H., and Masui, Y. (1986). Interactions between metaphase and interphase factors in heterokaryons produced by fusion of mouse oocytes and zygotes. *Dev. Biol.* **117**, 102–108.

Barnes, F. L., Collas, P., Powell, R., King, W. A., Westhusin, M., and Shepherd, D. (1993). Influence of recipient oocyte cell-cycle stage on DNA synthesis, nuclear envelope breakdown, chromosome constitution, and development in nuclear transplant bovine embryos. *Mol Reprod. Dev.* **36**, 33–41.

Blow, J. J., and Laskey, R. A. (1988). A role for the nuclear envelope in controlling DNA replication within the cell cycle. *Nature* **332**, 546–548.

Bordignon, V., and Smith, L. C. (1998). Telophase enucleation: An improved method to prepare recipient cytoplasts for use in bovine nuclear transfer. *Mol. Reprod. Dev.* **49**, 29–36.

Campbell, K. H. S. (1999a). Nuclear equivalence, nuclear transfer, and the cell cycle. *Cloning* **1**, 3–16.

Campbell, K. H. (1999b). Nuclear transfer in farm animal species. *Semin. Cell Dev. Biol.* **10**, 245–252.

Campbell, K. H. S., and Wilmut, I. (1999). Unactivated oocytes as cytoplast recipients for nuclear transfer. Roslin Institute. Patent WO 97/07668, 1999.

Campbell, K. H., Ritchie, W. A., and Wilmut, I. (1993). Nuclear-cytoplasmic interactions during the first cell cycle of nuclear transfer reconstructed bovine embryos: implications for deoxyribonucleic acid replication and development. *Biol. Reprod.* **49**, 933–942.

Campbell, K. H., Loi, P., Cappai, P., and Wilmut, I. (1994). Improved development to blastocyst of ovine nuclear transfer embryos reconstructed during the presumptive S-phase of enucleated activated oocytes. *Biol. Reprod.* **50**, 1385–1393.

Campbell, K. H., Loi, P., Otaegui, P. J., and Wilmut, I. (1996a). Cloning mammals by nuclear transfer. Co-ordinating nuclear and cytoplasmic events. *Rev. Reprod.* **1**, 40–46.

Campbell, K. H., McWhir, J., Ritchie, W. A., and Wilmut, I. (1996b). Sheep cloned by nuclear transfer from a cultured cell line [see comments]. *Nature* **380**, 64–66.

Cheong, H. T., Takahashi, Y., and Kanagawa, H. (1993). Birth of mice after transplantation of early cell-cycle-stage embryonic nuclei into enucleated oocytes. *Biol Reprod.* **48**, 958–963.

Coleman, T. R., and Dunphy, W. G. (1994). Cdc2 regulatory factors. *Curr. Opin. Cell Biol.* **6**, 877–882.

Collas, P., and Robl, J. M. (1991). Relationship between nuclear remodeling and development in nuclear transplant rabbit embryos. *Biol. Reprod.* **45**, 455–465.

Collas, P., Pinto-Correia, C., Ponce, d. L., and Robl, J. M. (1992). Effect of donor cell cycle stage on chromatin and spindle morphology in nuclear transplant rabbit embryos. *Biol. Reprod.* **46**, 501–511.

Coverley, D., and Laskey, R. A. (1994). Regulation of eukaryotic DNA replication. *Annu. Rev. Biochem.* **63**, 745–776.

Czolowska, R., Waksmundzka, M., Kubiak, J. Z., and Tarkowski, A. K. (1986). Chromosome condensation activity in ovulated metaphase II mouse oocytes assayed by fusion with interphase blastomeres. *J. Cell Sci.* **84**, 129–138.

Dasso, M., and Newport, J. W. (1990). Completion of DNA replication is monitored by a feedback system that controls the initiation of mitosis *in vitro*: Studies in *Xenopus*. *Cell* **61**, 811–823.

Enoch, T., and Nurse, P. (1991). Coupling M phase and S phase: Controls maintaining the dependence of mitosis on chromosome replication. *Cell* **65**, 921–923.

Gautier, J., Minshull, J., Lohka, M., Glotzer, M., Hunt, T., and Maller, J. L. (1990). Cyclin is a component of maturation-promoting factor from *Xenopus*. *Cell* **60**, 487–494.

Hartwell, L. H. (1991). Twenty-five years of cell cycle genetics. *Genetics* **129**, 975–980.

Hartwell, L. H., and Weinert, T. A. (1989). Checkpoints: Controls That ensure the order of cell cycle events. *Science* **246**, 629–634.

Hunt, T. (1992). Cell biology: Cell cycle arrest and c-*mos*. *Nature* **355**, 587–588.

Jacobs, T. (1992). Control of the cell cycle. *Dev. Biol.* **153**, 1–15.

Johnson, R. T., and Rao, P. N. (1970a). Mammalian cell fusion: Induction of premature chromosome condensation in interphase nuclei. *Nature* **226**, 717–722.

Johnson, R. T., and Rao, P. N. (1970b). Mammalian cell fusion: Studies on the regulation of DNA synthesis and mitosis. *Nature* **225**, 159–164.

Johnson, R. T., Rao, P. N., and Hughes, H. D. (1970). Mammalian cell fusion. III. A HeLa cell inducer of premature chromosome condensation active in cells from a variety of species. *J. Cell Physiol.* **76**, 151–158.

Kasinathan, P., Knott, J. G., Wang, Z., Jerry, D. J., and Robl, J. M. (2001). Production of calves from G1 fibroblasts. *Nat. Biotechnol.* **19**, 1176–1178.

Kikuchi, K., Naito, K., Noguchi, J., Shimada, A., Kaneko, H., Yamashita, M., Aoki, F., Tojo, H., and Toyoda, Y. (2000). Maturation/M-phase promoting factor: A regulator of aging in porcine oocytes. *Biol. Reprod.* **63**, 715–722.

Kirschner, M. (1992). The cell cycle then and now. *Trends Biochem. Sci.* **17**, 281–285.

Kwon, O. Y., and Kono, T. (1996). Production of identical sextuplet mice by transferring metaphase nuclei from four-cell embryos. *Proc. Natl. Acad. Sci. U.S.A.* **93**, 13010–13013.

Landa, V. (1989). Transplantation of nuclei from 2- to 16-cell embryos into enucleated blastomeres of 2-cell mouse embryos. *Folia Biol. (Praha)* **35**, 299–305.

Leno, G. H., and Munshi, R. (1994). Initiation of DNA replication in nuclei from quiescent cells requires permeabilization of the nuclear membrane. *J. Cell Biol.* **127**, 5–14.

Lenormand, P., Pages, G., Sardet, C., L'Allemain, G., Meloche, S., and Pouyssegur, J. (1993). Map kinases: Activation, subcellular localization and role in the control of cell proliferation. *Adv. Second Messenger Phosphoprotein Res.* **28**, 237–244.

Liu, J. L., Wang, M. K., Sun, Q. Y., Xu, Z., and Chen, D. Y. (2000). Effect of telophase enucleation on bovine somatic nuclear transfer. *Theriogenology* **54**, 989–998.

Maldonado-Codina, G., and Glover, D. M. (1992). Cyclins A and B associate with chromatin and the polar regions of spindles, respectively, and do not undergo complete degradation at anaphase in syncytial *Drosophila* embryos. *J. Cell Biol.* **116**, 967–976.

Maro, B., Johnson, M. H., Webb, M., and Flach, G. (1986). Mechanism of polar body formation in the mouse oocyte: An interaction between the chromosomes, the cytoskeleton and the plasma membrane. *J. Embryol. Exp. Morphol.* **92**, 11–32.

Martinez-Balbas, M. A., Dey, A., Rabindran, S. K., Ozato, K., and Wu, C. (1995). Displacement of sequence-specific transcription factors from mitotic chromatin. *Cell* **83**, 29–38.

Masui, Y., and Markert, C. L. (1971). Cytoplasmic control of nuclear behaviour during meiotic maturation of frog oocytes. *J. Exp. Zool.* **177**, 129–145.

McGrath, J., and Solter, D. (1983). Nuclear transplantation in mouse embryos. *J. Exp. Zool.* **228**, 355–362.

Mitchison, J. M. (2002). *The Biology of the Cell Cycle.* Cambridge Univ. Press, Cambridge.

Nurse, P. (1990). Universal control mechanism regulating the onset of M-phase. *Nature* **344**, 503–507.

Nurse, P. (1993). The Wellcome lecture, 1992. Cell cycle control. *Philos. Trans. Roy. Soc. Lond. B: Biol. Sci.* **341**, 449–454.

Parrish, J. J., Kim, C. I., and Bae, I. H. (1992). Current concepts of cell-cycle regulation and its relationship to oocyte maturation, fertilization and embryo development. *Theriogenology* **38**, 277–296.

Polymenis, M., and Schmidt, E. V. (1999). Coordination of cell growth with cell division. *Curr. Opin. Genet. Dev.* **9**, 76–80.

Russell, P. (1998). Checkpoints on the road to mitosis. *Trends Biochem. Sci.* **23**, 399–402.

Tani, T., Kato, Y., and Tsunoda, Y. (2001). Direct exposure of chromosomes to nonactivated ovum cytoplasm is effective for bovine somatic cell nucleus reprogramming. *Biol. Reprod.* **64**, 324–330.

Verlhac, M. H., Kubiak, J. Z., Clarke, H. J., and Maro, B. (1994). Microtubule and chromatin behavior follow MAP kinase activity but not MPF activity during meiosis in mouse oocytes. *Development* **120**, 1017–1025.

Wakayama, T., Perry, A. C., Zuccotti, M., Johnson, K. R., and Yanagimachi, R. (1998). Full-term development of mice from enucleated oocytes injected with cumulus cell nuclei [see comments]. *Nature* **394**, 369–374.

Wells, D. N., Misica, P. M., and Tervit, H. R. (1999). Production of cloned calves following nuclear transfer with cultured adult mural granulosa cells. *Biol. Reprod.* **60**, 996–1005.

Wrenzycki, C., Wells, D., Herrmann, D., Miller, A., Oliver, J., Tervit, R., and Niemann, H. (2001). Nuclear transfer protocol affects messenger RNA expression patterns in cloned bovine blastocysts. *Biol. Reprod.* **65**, 309–317.

PART IV

CURRENTLY SOUGHT AFTER SPECIES

CHAPTER 21

CLONING OF RATS

Philip Iannaccone, Michael Bader, and Vasiliy Galat

INTRODUCTION

The relationship between nucleus and cytoplasm is one of the most intriguing and enduring issues in biology. From the first recognition that the nucleus was a critical organelle, a notion popularized by Virchow, to the speculation that the nucleus is another endosymbiont in life's history, the range of interest and effort in this arena is truly grand. The current intellectual pendulum places us at a time when the nucleus is supposed to have preeminence of control of cellular events. But is this a fair picture of reality? It is already clear that mitochondria may have a tighter grip on things in the cell (beyond respiration) than had been previously thought (Wallace *et al.*, 1998). Polarity of the early mouse embryo now seems to occur similarly in embryos of other species in that the site of sperm fusion controls the placement of division planes and the timing of cleavage (Piotrowska and Zernicka-Goetz, 2001). The process of cloning animals in a sense is the logical conclusion of studies aimed at determining if the cytoplasm on its own could direct embryonic development. Normal metameric development cannot occur in the absence of a nucleus, but cleavage without differentiation could occur in the embryos of many species in the absence of a nucleus. Adding back a sequestered nucleus could direct a robust developmental program. Cloning of amphibians (see Chapter 14, this volume) showed that even nuclei from cancer cells could direct normal development at least to the tadpole stage. Clearly the ability of the nucleus to direct a developmental program could be regained after differentiation because of the influence of appropriate cytoplasm. A fascinating historical review of these topics was provided by Bob McKinnell a pioneer in amphibian cloning (McKinnell, 1985). The extent (that is, the range of species and the degree of differentiation) to which differentiated nuclei could be reprogrammed has been pursued vigorously with controversial results. Along the way we learned that both male and female pronuclei were required for normal fetal development and thereby learned of imprinted genes (for a discussion see Fitchev *et al.*, 1999). The successful cloning of several species from adult and even senescent cells has put that issue to rest.

WHY CLONE THE RAT?

At the time of this writing experimentalists using nuclear transfer with adult cells as nuclear donors have successfully cloned several important species. These include sheep, cattle, pigs, mice, and goats. Work is currently underway to clone rats, rabbits, and primates. What is the rationale to undertake the effort of working out the methods required to clone another species, such as the rat?

The rat is an extremely important experimental model of human disease. It has been used for over a century as a model of choice for a wide variety of physiological and behavioral studies with direct relevance to the human condition. We have seen a tremendous increase in available data from this experimental system in recent years, with genomic techniques applied to the rat in order to ensure uniformity of experimental models and to establish the expressed genes in the species. Radiation hybrid maps are available, high-density markers cover nearly 99% of the rat genome for many strains, there are over 100,000 National Center for Biotechnology Information (NCBI) UniGene clusters in the expressed sequence tag data base (http://www.ncbi.nlm.nih.gov/UniGene/Rn_DATA/lib_report.html), and the rat has been chosen for complete genome sequencing. The National Center for Research Resources of the National Institutes of Health (NIH) has established a National Repository to maintain and ensure the genetic and microbiological integrity of the most frequently used strains of rat. Importantly, modern methods of phenotyping now allow the simultaneous tracking of many hundreds of quantitative traits and genomic markers (Cowley et al., 2000; Stoll et al., 2000).

An important deficiency in the use of this model, however, is the current inability to make targeted mutations in the rat. This procedure is now routine in the mouse and in general relies on the ability to use homologous recombination of DNA in order to take a normal allele of a gene and replace it with specific desired mutant forms. Chimeras are produced from ES cells bearing the mutation. Animals derived from the cultured embryonic stem (ES) cells with a mutant allele, identified by a combination of coat color, Southern analysis, and RNA analysis, are bred to homozygosity by mating heterozygous mutant ES cell-derived offspring of the chimera with each other. The offspring of this mating can be back-crossed on another strain of mouse. As powerful as this approach is, it has not worked in species other than the mouse because so far it has not been possible to produce ES cells from the blastocysts of nonmouse species that have been proved to produce germ line chimeras. Nuclei from cells with targeted mutations used as nuclear donors, though, can generate knockout animals by cloning (Lai et al., 2002) in other species.

Beyond the value the rat offers as a model species of human disease, it is important to remember that in many gene knockout experiments in the mouse no observable phenotype results. It is usually assumed that this indicates redundant gene activity whereby some gene other than the one targeted assumes the function of the gene that is knocked out. In some cases this has been established by identifying the redundant genes and knocking them out as well revealing phenotypes related to the overlapping functions of the genes. However, in many instances this has not been achieved and in some of those instances it has been established that the phenotype depends on the genetic background of the mouse. When, with some considerable effort, the knockout is put into a new genetic background the phenotype becomes manifest or is significantly altered. This is extremely important to the study of gene function in development and disease because it indicates that there are modifying genes capable of strongly influencing the function of the targeted gene (LeCouter et al., 1998; Bonyadi et al., 1997). The ability to do knockouts in other species will greatly enhance our ability to understand the range of these modifying genes by offering phenotypic comparison across large phylogenetic differences while retaining the underlying sequence and syntenic homologies.

STEPS IN CLONING

Cloning an intact animal requires the obtainment of oocytes, enucleation of the oocyte or removal of chromosomal DNA, insertion of a nucleus from the species to be cloned with or without genetic modification, activation of the reconstituted egg, and development to term in a suitable surrogate mother (Fig. 1). Each of these steps has species-specific peculiarities that are not easily predicted and for which solutions are only obtained empirically. This places a burden on the development of cloning procedures. In those instances in which technical variations need to be tested in a systematic way, the use of intermediate surrogates has proved useful.

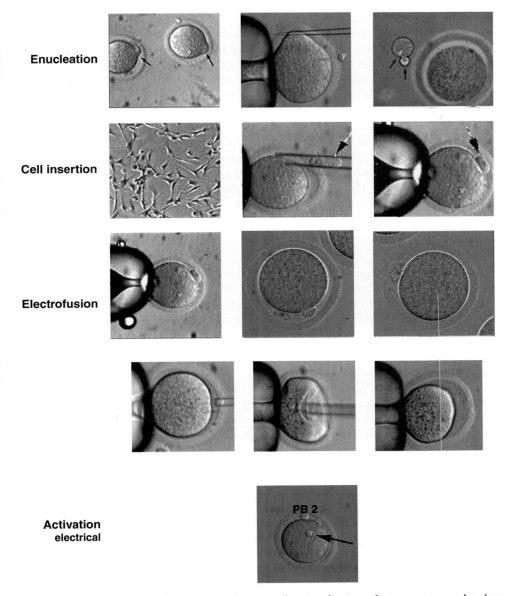

Enucleation

Cell insertion

Electrofusion

Activation
electrical

PB 2

Figure 1 *Steps in cloning the rat by nuclear transfer. Enucleation of rat oocytes can be done in various ways. Illustrated here is a method of removing the chromosomes from a cytoplasmic bulge containing the metaphase II plate. A slit is made in the zona pellucida, then the cytoplasmic bulge is extruded with mild suction; finally, the chromosomal material pinches off into a small cellular fragment. The DNA is highlighted (right panel) in ultraviolet light illumination because of staining with Hoechst 33342 (arrows). Cell insertion is shown here as the placement of cultured β-galactosidase-marked embryonic fibroblasts (first panel and arrows) in the perivitelline space through the slit in the zona pellucida. In the electrofusion process, the first panel shows the fibroblast placed in the perivitelline space, the middle panel shows the embryo immediately after the electrical pulse, and the right panel shows the fused embryo. Alternatively, the donor nucleus can be directly injected with the aid of a piezoelectric device. Electrical activation is illustrated with an unmanipulated oocyte that is developing nuclear structure.*

That is, one places a reconstituted egg into a surrogate mother, which is then sacrificed a few days later to determine the developmental potential of the embryo following nuclear transfer, or to pass on developed embryos to a second surrogate mother for development to term. This is a highly cumbersome and inefficient method. Studies done with mice eliminated this problem because it was routine to culture mouse fertilized embryos from one cell to blastocyst stage in chemically defined media. This allowed reasonable tests of technical variations to be completed before committing to an experimental approach with development to term. Moreover, culture *in vitro* after activation of reconstituted eggs could be used with impunity to obtain a selected group of nuclear transfer embryos for final transfer. The development of these *in vitro* culture systems for the mouse required very intensive and largely thankless efforts of dedicated labs (Lawitts and Biggers, 1993; Quinn *et al.*, 1982; Erbach *et al.*, 1994).

This is the desired approach for developing methods for cloning the rat as well. Unfortunately, the *in vitro* culture of fertilized rat embryos from one cell to blastocyst stage has been an elusive goal. Nevertheless, we have utilized several systems that are in the published scientific literature to develop an approach to testing cloning variables.

EMBRYO CULTURE

Utilizing a culture medium (mR1ECM) developed by Niwa and colleagues (Miyoshi *et al.*, 1997; Miyoshi and Niwa, 1997) we were able to obtain blastocyst-stage embryos from 76% of SD strain rat fertilized one-cell embryos after 5 days in culture (Iannaccone *et al.*, 2001). A small percentage of the embryos (6%) progressed to blastocysts after 4 days in culture. Inbred strains performed much more poorly (Table 1).

OOCYTES

Most protocols for obtaining oocytes utilize hormone treatments to superovulate the donor females. This enhances the yield of oocytes from each animal, increases the number of animals from which oocytes can be obtained, and allows precise determination of the ovular age of oocytes. Superovulation of rat oocytes and embryos has been well described (Mukumoto *et al.*, 1995; Charreau *et al.*, 1996, 1997).

In our lab we have used several approaches to superovulation. In general, we have successfully employed pregnant mare serum (PMS) (G-4877; Sigma Chemical Co., St. Louis, MO) as a source of follicle-stimulating activity followed 47–51 hours later by human chorionic gonadotropin (HCG) (CG-10; Sigma Chemical Co.) in young (4-week old) or adult (>8-week old) rats maintained on a 12-hour light/12-hour dark light cycle (1800–0600 hours is the dark period). Following the second injection, the females were placed with vasectomized males and examined for

Table 1 In Vitro Culture of One-Cell Fertilized Embryos from Various Strains of Rat

Rat strain[a]	Zygote development to stage (%)		
	Two cell (after 2 days)	Four cell (after 3 days)	Blastocyst (after 5 days)
WF (23)	56	35	0
LEW (53)	13	13	9
F344 (43)	72	42	12
PVG (16)	88	31	6
SD (113)	96	94	76

[a]Fertilized eggs were obtained at the one-cell stage and cultured as described. Numbers in parentheses are the total number of embryos cultured. Embryos are removed from the oviduct the day following mating and cultured in mR1ECM (Miyoshi *et al.*, 1997; Miyoshi and Niwa, 1997) and observed every 24 hours. From Iannaccone *et al.* (2001), with permission.

copulatory plugs the following morning, and eggs were flushed from the oviduct. There are strain-dependent variations in yield of one-cell eggs from mating with vasectomized males. Our usual protocol for young animals is as follows: 15 IU PMS is injected intraperitoneally between 1200 and 1500 hours followed by 15 IU of HCG injected intraperitoneally 47–50 hours later into female rats 28–30 days old. This results in an average of 25 eggs per female in inbred PVG strain rats but much poorer results in the outbred SD strain. For adult animals (>8 weeks old) we inject 20 IU of PMS subcutaneously between 1000 and 1130 hours. This is followed 4–6 hours later with 30 IU of PMS subcutaneously and then 50 IU of HCG intraperitoneally 47–49 hours later. This yielded an average of 25 eggs per inbred PVG female and an average of 17 eggs per outbred SD female. We have also used subcutaneous osmotic minipumps (Hamilton and Armstrong, 1991) to deliver folltropin (follicle-stimulating hormone; FSH). These pumps are inserted in 28- to 31-day-old female rats between 1000 and 1200 hours followed by 11 IU of HCG injected intraperitoneally 49–51 hours later. This procedure yielded an average of 54 eggs per animal.

More consistent results were obtained in our lab with hormones from a different source and utilizing different light cycles. Intergonan (PMS; Intervet GmbH), 20 IU injected intramuscularly, was followed 48–50 hours later with Ovogest (HCG; Intervet GmbH), 20 IU injected intraperitoneally into 23- to 31-day-old females. The second injection was made 2–3 hours before the release of endogenous luteinizing hormone (LH). Ovulation occurred 12–13 hours after the HCG injection and resulted in 40–70 eggs per female.

ENUCLEATION

Removal of the nuclear material from the unfertilized rat oocyte is a straightforward surgical procedure that can be successfully performed by several microsurgical methods and at several different ovular ages. The first polar body (PB1) in the rat egg is visible only in about 1% of the ovulated oocytes. This prevents the use of PB1 as a marker in locating the metaphase II (MII) plate during enucleation. The egg remains in the MII phase for about 30 minutes after isolation. This limits the time during which enucleation (removal of the MII plate) can be performed. In eggs that have progressed through metaphase II (MII) and are approaching the second polar body separation, a cytoplasmic bulge that will become the second polar body contains the second metaphase plate. This is easily removed by mild suction through a slit in the zona pellucida. The oocytes are manipulated in 10-cm dishes with microdrops of M2 culture medium (Quinn *et al.*, 1982; Sturm and Tam, 1993) (Specialty Media, Phillipsburg, NJ). Glass instruments are coated with 10% polyvinylpyrrolidone (PVP) (Specialty Media, Phillipsburg, NJ) in M2. The second metaphase plate is removed by slitting the zona pellucida in the region of the nascent polar body using a microneedle, with subsequent suction of the nascent PB2 through the slit with the egg-holding pipette (Fig. 1). Cytochalasin B is not used in this procedure.

Oocytes can be enucleated prior to MII, when the first metaphase plate can be removed using a beveled glass pipette or piezo-actuated pipettes as in intracytoplasmic sperm injection (Kimura and Yanagimachi, 1995; Wakayama *et al.*, 1998). Oocytes must be removed from the ovary in order to remove chromosomal DNA at this stage. The chromosomes are not as well localized at this stage of development and their removal is enhanced with the use of DNA dyes such as Hoechst 33342 and epifluorescent illumination on the micromanipulator. It is important to remember, though, that ultraviolet light illumination is very cytotoxic and will kill the oocytes even after brief exposures (Tsunoda *et al.*, 1988; Hockberger *et al.*, 1999). Therefore, observation of fluorescent stained DNA must be made very briefly using a shutter, or after the procedure, to demonstrate that the DNA is in the extracted fragment.

RECONSTITUTION

Nuclei are injected into enucleated oocytes using glass pipettes that are drawn to an approximate inner diameter of 10 µm, acid etched with 30% hydrofluoric acid (HF) (Aldrich catalog no. 33,926-1; Aldrich, Milwaukee, WI), and washed extensively with distilled water filtered through 0.2-µm millipore filters. These pipettes are fitted to piezo actuators and coated with PVP prior to use. Nuclei are released from the donor cells with shear stress induced by piezoelectric actuation immediately before insertion. The nucleus is then injected into the cytoplasm of the oocyte. This procedure was enhanced by back-loading the pipette with mercury, but piezo devices now available no longer require this. Alternatively, mercury can be replaced with Fluorinert FC-77 (Sigma F4758).

FUSION OF NUCLEAR DONORS WITH OOCYTES

Alternatively, the donor nucleus can be supplied to the enucleated oocyte by electrofusion. A single donor cell is placed in the perivitelline space with a pipette through the same slit that is used for enucleation (Fig. 1). Karyoplast–cytoplast complexes (KCCs) are examined for the presence of tight contact between the oolema and cell membrane. Cell adhesion to the zona pellucida should be avoided if possible. Up to 15 KCCs are washed in fusion media (0.3 M mannitol, 0.1 mM magnesium, 0.05 mM calcium, and 0.05 mg/ml bovine serum albumin or 0.1% PVP-360) and placed in a fusion chamber away from the electrodes. The KCCs are placed between electrodes (gap distance, 350 µm) one at a time and positioned by applying an alternating current (AC) pulse (3–5 seconds), with some manual assistance to aid the alignment. Cell fusion is induced with two direct current (DC) pulses followed immediately by an AC pulse. DC pulses of 2.0 kV/cm for 20 µsec each were delivered by a XPONOΣ-2 apparatus. Stimulated KCCs are cultured 10–15 minutes in M16 medium then are checked for fusion by microscopic examination. A second round of fusion is performed if needed. Overall efficiency of fusion is greater than 90% when primary embryonic fibroblast cells are used.

DONOR CELLS

We have used several cell types as nuclear donors with successful production of cloned embryos. These include adult rat female cumulus (ovarian follicle) cells (Schuetz *et al.*, 1989), primary rat embryonic fibroblasts, and genetically modified primary rat embryonic fibroblasts.

The cumulus cells used as a source of nuclei are removed from the cumulus mass surrounding oocytes isolated following spontaneous ovulation by brief treatment with hyaluronidase in M2. The cells, maintained at 4°C in M2, were used within 1 hour and nuclei were isolated just before injection.

Primary embryonic fibroblasts are obtained from SD strain rat fetuses. To prepare the cells as nuclear donors, subconfluent cultures are grown in 0.5% fetal bovine serum (FBS) for 5 days prior to their use. The serum-deprived cultures are trypsinized for 5 minutes, then washed in M2 twice. The washed cell pellet is resuspended in 0.5 ml of M2 and is maintained at 4°C. Some of the rat embryonic fibroblasts were genetically modified by transfection with *pcUBI-hgfp* (green fluorescent protein regulated with the human ubiquitin C promoter) or with *pCX-EGFP* (enhanced green fluorescent protein regulated with a cytomegalovirus enhancer and chicken β-actin promoter (Okabe *et al.*, 1997) using a Bio-Rad Gene Pulser II electroporator. The DNA (30 µg) was added to 10^7 cells in electroporation buffer (ES-003-D; Specialty Media, Phillipsburg, NJ) at 4°C. Following electroporation (280 V, 500 µF), the cells were cultured in Dulbecco's minimal essential medium (DMEM) with 10% FBS for 24 hours and then with G418 (200 µg/ml) for 5–6 days. Surviving colonies were selected, expanded, and used as described above.

ACTIVATION

A critical aspect of reprogramming the transferred nucleus and starting embryonic development is activation of the reconstituted oocyte to begin progression through cleavage stages. Activation is a poorly understood sequence of events leading to completion of meiotic division and first cleavage following fertilization. Activation can be induced by electrical current or by treatment with chemical ionophores. Most activation protocols can be modified for use in the rat, in which successful activation scored as parthenogenetic development must be balanced against toxicity.

The critical events may in some way be related to normal gene imprinting, critical to normal postgastrulation development. It is believed that the normal postimplantation development requires that the donor nucleus be inserted into the oocyte before activation (Rideout *et al.*, 2000). However, the rat ovulated nonfertilized egg is unusual insofar as it does not remain in the metaphase II arrest after isolation, but undergoes spontaneous activation (Zernika-Goetz, 1991). This, however, is incomplete activation and leads to the formation of a second polar body (PB2) but not of pronuclei. If the egg is artificially activated late (e.g., 3 hours after spontaneous activation), micronuclei rather than normal pronuclei are formed. Spontaneous activation in the unfertilized ovulated rat oocyte can be triggered by delayed egg isolation (5 minutes or more after the death of the animal) or by exposure to room temperature during a standard egg isolation procedure.

In our lab, activation is achieved by incubating renucleated oocytes in CZB (Lawitts and Biggers, 1993) activation medium (Specialty Media, Phillipsburg, NJ) for 2.5–3 hours (1.5–2.0 hours for genetically modified fibroblasts). The activation medium is CZB as supplied, without HEPES or $CaCl_2$, to which 5 mM $SrCl_2$ (2.5 mM for genetically modified fibroblasts) and cytochalasin B (5 µg/ml) are added. Following this activation period embryos are incubated in mR1ECM (Miyoshi *et al.*, 1997) at 37°C.

Unmanipulated nonfertilized oocytes can be activated by culture in $SrCl_2$. If oocytes are incubated in the medium described above several hours after isolation, development does not proceed. In four experiments, 35 of 94 Sr^{2+}-activated oocytes developed to the two-cell stage, but none of the 94 Sr^{2+}-activated oocytes developed to the four-cell, eight-cell, or blastocyst stage. Oocytes isolated and placed into mR1ECM culture without other treatment developed poorly. Two of 26 unactivated oocytes developed to the two-cell stage, and none of these 26 developed to the four-cell, eight-cell, or blastocyst stage. However, if oocytes were chemically activated immediately following isolation, the efficiency of parthenogenetic development to the blastocyst stage was greatly increased, both for spontaneously ovulated eggs and for superovulated eggs. Of 72 spontaneously ovulated SD oocytes that were activated for 1.5 hours immediately following isolation, 37 (51%) developed to blastocyst stage in culture. Of 54 oocytes obtained from superovulation, 30 (56%) developed to blastocyst stage in culture (Iannaccone *et al.*, 2001).

Serial electric pulses trigger calcium oscillations that have an effect similar to that of Sr^{2+}. Such assisted activation stimuli begin before (30–180 minutes) the time that second polar body separation would occur in the absence of cytochalasin B. This time depends on the ovular age of the oocytes. Electroactivation is done with a fusion apparatus using three serial pulses of 1.3 kV/cm for 20 µsec each, applied at intervals of 30 minutes. This leads to 100% activation of oocytes cultured in M16 with of 5 µg/ml cytochalasin B. However, parthenogenetic embryos rarely develop even to the morula stage.

Alternatively, 20-hour post-hCG eggs are activated by an inhibitor of methylaminopurine kinase 6, dimethylaminopurine (DMAP) (Sigma, 1.9 mM). A 3-hour activation results in development of 45% (38/85) of parthenogenetic embryos to the blastocyst stage. Freshly isolated oocytes, which are in the process of spontaneous activation, do not require a triggering stimulus such as incubation with an

Table 2 Development following Nuclear Transfer

	Nuclear donor[b]		
Stage[a]	SD cumulus cell	SD embryonic fibroblast	Genetically altered fibroblast
Two cell	40/98 (41%)	31/65 (47%)	85/168 (51%)
			25/64 (39%)
Four cell	19/98 (19%)	16/65 (25%)	58/168 (34%)
			16/64 (25%)
Eight cell	11/98 (11%)	9/65 (13%)	40/168 (24%)
			10/64 (16%)
Blastocyst	8/98 (8%)	5/65 (8%)	21/168 (12%)
			6/64 (9%)

[a]Observations from days 2, 3, 4, and 5 in culture.

[b]Number of SD rat strain embryos that develop to the given stage/total number of embryos surviving manipulation; 98/141 survived manipulation in the cumulus cell experiments, 65/107 survived manipulation in the embryonic fibroblast experiments, 168/242 survived manipulation in the genetically altered fibroblast experiments using *pcUBI-hgfp* for transfection (first line), and 64/108 survived manipulation in the genetically altered fibroblasts using *pCX-EGFP* for transfection (second line).

ionophor or electric stimulation that usually precedes DMAP treatment. Total inhibition of protein synthesis by puromycine also leads to activation of the eggs (Zernika-Goetz, 1991); however, development of such embryos is restricted before compaction.

OUTCOMES

A total of 107 embryos from four experiments were reconstructed by injecting fibroblast cell nuclei and activation with Sr^{2+}. From these, 65 survived to be cultured. After 24 hours in culture, 32% developed to the two-cell stage and after 48 hours in culture, 42% developed to the two-cell stage (Table 2). Following 3 days in culture, 25% developed to the four-cell stage and after 4 days in culture 14% of the zygotes had developed to the eight-cell stage. A total of 8% of the manipulated cultured embryos developed to the blastocyst stage after 5 days in culture (Fig. 2).

The ability of the embryonic fibroblast nuclei to support advanced preimplantation development following nuclear transfer is not diminished following transfection and selection in antibiotic; 12% of embryos reconstructed by injecting fibroblast nuclei and activating with Sr^{2+} developed to the blastocyst stage. Similar results were obtained with a second transgene, *pCX-EGFP*, which codes for enhanced green fluorescent protein (see Fig. 3). Blastocysts developed from nuclear transfer zygotes using fibroblasts selected for stable integration of this transgene as a nuclear donor at about the same frequency as from *pcUBI-hgfp* (Table 2). No sham operated zygotes developed.

TRANSFER OF RECONSTITUTED EMBRYOS TO SURROGATE MOTHERS

In the middle afternoon following overnight culture, the reconstructed embryos are transferred to the oviduct of surrogate mothers on day 0 of their pseudopregnancy. SD rats are used as surrogate mothers. They are mated with vasectomized male rats and on the following morning vaginal smears are performed. The presence of a copulatory plug is taken as evidence of mating and the female is considered to be at day 0 of pregnancy.

Cloned rat embryos were cultured overnight and then transferred to surrogate mothers; 21/86 (24%) of these embryos implanted with nuclear transfer embryos

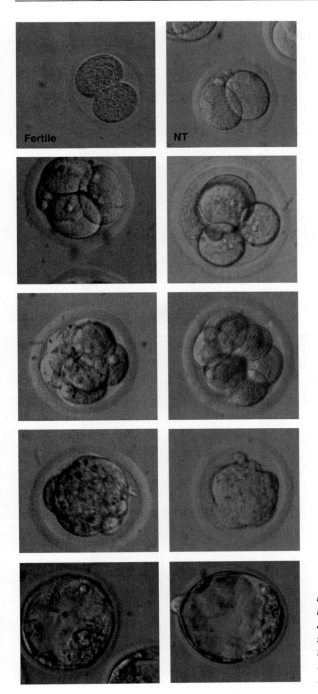

Figure 2 *Preimplantation rat embryos. Left panels from the top show normally fertilized embryos at the two-cell, four-cell, eight-cell, sixteen-cell, and blastocyst stages in culture. The right panels show the same culture system at the same time points with embryos generated with nuclear transfer (NT). The embryos are approximately 100 μm in diameter. Adapted from Iannaccone et al. (2001), with permission from Cambridge University Press.*

developed robust decidual swellings. None of the nuclear transfer embryos developed postgastrulation structures. Of the 35 nuclear transfer embryos that were allowed to go to term in surrogates, none gave birth, as opposed to 59/166 (36%) live pups/transferred fertilized oocytes cultured to the two-cell stage overnight before transfer (Iannaccone *et al.*, 2001). Similar results with rat embryos cloned by NT have been reported recently (Hayes *et al.*, 2001).

LIMITATIONS

At this point we have been able to develop cloned rat preimplantation embryos that are capable of implanting. We do not yet have evidence of postimplantation devel-

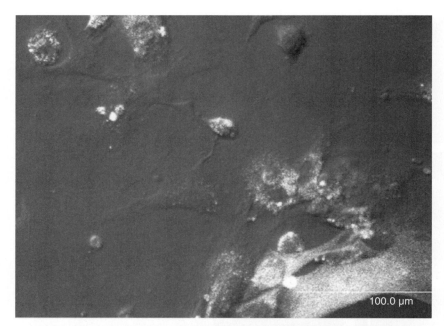

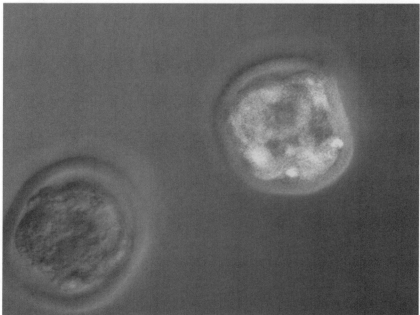

Figure 3 *Top panel: Rat embryonic fibroblasts transfected with* pcUBI-hgfp *(green fluorescent protein). Bar = 100 μm. Bottom panel: Cloned rat embryos generated by nuclear transfer using the cells from fibroblast cultures illustrated in the top panel as nuclear donors. The blastocyst on the right has a strong green fluorescent protein signal. The embryos are approximately 100 μm in diameter. Embryos and cells were illuminated in an Olympus confocal microscope (excitation, 488 nm; emission, 515–540 nm) and superimposed on the Nomarski Differential Interference Contrast image. Adapted from Iannaccone et al. (2001), with permission from Cambridge University Press.*

opment. Because so few embryos have been transferred so far, it may simply be a matter of not having tried hard enough. On the other hand, it is possible that the approaches outlined above will not permit normal postimplantation development. If this turns out to be true, the most likely explanation is that the eggs are too mature at the time of nuclear transfer to support reprogramming of the nucleus. We are currently working on methods to rectify this by working with oocytes of younger ovular age and increasing the number of nuclear transfer embryos transferred.

Rat oocytes will spontaneously activate following isolation from the animal (Zernika-Goetz, 1991). Our results establish that although unfertilized rat oocytes will develop to the two-cell stage, they do so at a low frequency and may represent a subset of activated eggs. This frequency of two-cell stage development can be significantly increased by chemical activation, and subsequent development of such eggs to later preimplantation stages may be affected by the timing of chemical activation, because the efficiency of parthenogenetic development of oocytes is greatly increased when chemical activation occurs immediately following isolation. This is probably important to further development when a donor nucleus is inserted in the egg. It has been argued that activation must occur following the insertion of a donor nucleus, for reprogramming and subsequent development to occur (Rideout *et al.*, 2000; Wakayama *et al.*, 2000). It is possible that spontaneous activation of the rat oocyte is either incomplete or that it can be reinduced by chemical treatment subsequent to the insertion of a donor nucleus in enucleated eggs. On the other hand, we have used this culture system to support growth *in vitro* of fertilized SD strain embryos to the two-cell stage, with subsequent transfer to surrogate mothers, and live birth rates are high, although development following implantation of nuclear transfer embryos has not so far occurred. The failure of nuclear transfer embryos to develop following implantation may be a result of transferring the nuclei to oocytes that have activated spontaneously prior to the transfer. This result is consistent with the notion that reprogramming of the transferred nucleus involves resetting the genomic imprint. If the imprinting is incomplete or inaccurate, development to the blastocyst stage with poor or no postimplantation development might be expected. This may indicate that methods preventing spontaneous activation will be required to obtain further development of nuclear transfer embryos.

QUESTIONS OF INTEREST IN CLONING

There are a number of important and compelling issues raised by animal cloning beyond the purely technical aspects. Several of these are discussed at length elsewhere in this book, but it might be useful to reiterate at least a partial list here.

We began with a discussion of nucleus and cytoplasm. Cloning by nuclear transfer will offer an experimental approach to the big question of what the cytoplasm does to limit the developmental potential of the nucleus. We will be able to address the question of what factors reside in cytoplasmic compartments that have influence over the early developmental stages.

Because the cells used as nuclear donors are older than embryos derived from them, cloning by nuclear transfer poses and helps to answer the question of what is meant by aging. Is there a different age for the donor cell and the animal cloned from it? Can the donor nucleus impart its aging machinery to the clone?

What do the mitochondria contribute to the phenotypes in the cloned animal (including age)? Because the mitochondria are normally maternal, does the inclusion of "adult" mitochondria from the donor cell result in different cloning outcomes when cell fusion is used rather than injection of the nucleus into the oocyte? Is the age of the clone determined by the age of the mitochondria?

How does the nuclear transfer embryo undergo early patterning events when sperm entry point seems determinant in the process of establishing division planes in preimplantation stages (Piotrowska and Zernicka-Goetz, 2001)?

To date, a very low percentage of the oocytes that are reconstituted will actually develop into live offspring. This is true regardless of the source of the nucleus, the species being cloned, or the exact technical details of the nuclear transfer. There are many abnormalities in the fetuses and most clones seem to have placental abnormalities. Overcoming these problems will challenge the field and undoubtedly uncover more interesting biological issues.

The common use of nuclei from cells in G0 arrest in successful cloning suggests that cell cycle is important to cytoplasm's role in reprogramming the nucleus. Understanding this effect, if generally true, will provide important insights to the relationship between nucleus and cytoplasm. Indeed, precisely what reprogramming at a molecular level means will be an issue rightly occupying the attention of those working in this field.

Similarly, activation is a poorly characterized series of events that lead to the division competence of the zygote. It is a critically important step in cloning and yet although we understand some empirical rules to effect this step, we know very little about it. The relationship of the insertion of the nucleus to activation, the poor postimplantation outcomes, and the placental abnormalities in cloning all seem to suggest that gene imprinting mechanisms may inform the question of what nuclear reprogramming involves. How does resetting the imprint need to occur in the ooplasm? Are the enzymes that regulate setting gene imprints the cytoplasmic factors so critical to successful nuclear reprogramming?

Taken together, it is clear that, completely aside from the joy of sifting through myriad technical details of cloning various species, this field will provide all who choose to work in it many happy hours of pondering fundamental biological problems.

REFERENCES

Bonyadi, M., Rusholme, S. A., Cousins, F. M., Su, H. C., Biron, C. A., Farrall, M., and Akhurst, R. J. (1997). Mapping of a major genetic modifier of embryonic lethality in TGF beta 1 knockout mice. *Nat. Genet.* **15**, 207–211.

Charreau, B., Tesson, L., Soulillou, J. P., Pourcel, C., and Anegon, I. (1996). Transgenesis in rats: Technical aspects and models. *Transgen. Res.* **5**, 223–234.

Charreau, B., Tesson, L., Menoret, S., Buscail, J., Soulillou, J. P., and Anegon, I. (1997). Production of transgenic rats for human regulators of complement activation. *Transplant. Proc.* **29**, 1770.

Cowley, A. W., Jr., Stoll, M., Greene, A. S., Kaldunski, M. L., Roman, R. J., Tonellato, P. J., Schork, N. J., Dumas, P., and Jacob, H. J. (2000). Genetically defined risk of salt sensitivity in an intercross of Brown Norway and Dahl S rats. *Physiol. Genom.* **2**, 107–115.

Erbach, G. T., Lawitts, J. A., Papaioannou, V. E., and Biggers, J. D. (1994). Differential growth of the mouse preimplantation embryo in chemically defined media. *Biol. Reprod.* **50**, 1027–1033.

Fitchev, P., Taborn, G., Garton, R., and Iannaccone, P. (1999). Nuclear transfer in the rat: Potential access to the germline. *Transplant. Proc.* **31**, 1525–1530.

Hamilton, G. S., and Armstrong, D. T. (1991). The superovulation of synchronous adult rats using follicle-stimulating hormone delivered by continuous infusion. *Biol. Reprod.* **44**, 851–856.

Hayes, E., Galea, S., Verkuylen, A., Pera, M., Morrison, J., Lacham-Kaplan, D., and Trounson, A. (2001). Nuclear transfer of adult and genetically modified fetal cells of the rat. *Physiol. Genomics* **5**, 193–204.

Hockberger, P. E., Skimina, T. A., Centonze, V. E., Lavin, C., Chu, S., Dadras, S., Reddy, J. K., and White, J. G. (1999). Activation of flavin-containing oxidases underlies light-induced production of H2O2 in mammalian cells. *Proc. Natl. Acad. Sci. U.S.A.* **96**, 6255–6260.

Iannaccone, P. M., Taborn, G., and Garton, R. (2001). Preimplantation and postimplantation development of rat embryos cloned with cumulus cells and fibroblasts. *Zygote* **9**, 135–143.

Kimura, Y., and Yanagimachi, R. (1995). Intracytoplasmic sperm injection in the mouse. *Biol. Reprod.* **52**, 709–720.

Lai, L., Tolber-Simands, D., Park, K. W., Cheong, H. T., Greenstein, J. L., Im, G. S., Samuel, M., Borik, A., Rieke, A. Day, B. N., Murphy, C. N., Carter, D. B., Hawleg, R. J., and Prather, R. S. (2002). Production of alpha-1,3-galactosyltransterase knockout pigs by nuclear transfer cloning. *Science* **295**, 1089–1092.

Lawitts, J. A., and Biggers, J. D. (1993). *In* "Guide to Techniques in Mouse Development," Vol. 225 (P. M. Wassarman and M. L. DePamphilis eds.), pp. 153–164. Academic Press, San Diego.

LeCouter, J. E., Kablar, B., Whyte, P. F., Ying, C., and Rudnicki, M. A. (1998). Strain-dependent embryonic lethality in mice lacking the retinoblastoma-related p130 gene. *Development* **125**, 4669–4679.

McKinnell, R. G. (1985). "Cloning of Frogs, Mice, and Other Animals." University of Minnesota Press, Minneapolis.

Miyoshi, K., and Niwa, K. (1997). Stage-specific requirement of phosphate for development of rat 1-cell embryos in a chemically defined medium. *Zygote* **5**, 67–73.

Miyoshi, K., Kono, T., and Niwa, K. (1997). Stage-dependent development of rat 1-cell embryos in a chemically defined medium after fertilization *in vivo* and *in vitro*. *Biol. Reprod.* **56**, 180–185.

Mukumoto, S., Mori, K., and Ishikawa, H. (1995). Efficient induction of superovulation in adult rats by PMSG and hCG. *Exp. Anim.* **44**, 111–118.

Okabe, M., Ikawa, M., Kominami, K., Nakanishi, T., and Nishimune, Y. (1997). "Green mice" as a source of ubiquitous green cells. *FEBS Lett.* **407**, 313–319.

Piotrowska, K., and Zernicka-Goetz, M. (2001). Role for sperm in spatial patterning of the early mouse embryo. *Nature* **409**, 517–521.

Quinn, P., Barros, C., and Whittingham, D. G. (1982). Preservation of hamster oocytes to assay the fertilizing capacity of human spermatozoa. *J. Reprod. Fertil.* **66**, 161–168.

Rideout, W. M., 3rd, Wakayama, T., Wutz, A., Eggan, K., Jackson-Grusby, L., Dausman, J., Yanagimachi, R., and Jaenisch, R. (2000). Generation of mice from wild-type and targeted ES cells by nuclear cloning. *Nat. Genet.* **24**, 109–110.

Schuetz, A. W., Whittingham, D. G., and Legg, R. F. (1989). Alterations in the cell cycle characteristics of granulosa cells during the periovulatory period: Evidence of ovarian and oviductal influences. *J. Exp. Zool.* **249**, 105–110.

Stoll, M., Kwitek-Black, A. E., Cowley, A. W., Harris, E. L., Harrap, S. B., Krieger, J. E., Printz, M. P., Provoost, A. P., Sassard, J., and Jacob, H. J. (2000). New target regions for human hypertension via comparative genomics. *Genome Res.* **10**, 473–482.

Sturm, K., and Tam, P. L. (eds.) (1993). "Isolation and Culture of Whole Postimplantation Embryos and Germ Layer Derivitives." Academic Press, San Diego.

Tsunoda, Y., Shioda, Y., Onodera, M., Nakamura, K., and Uchida, T. (1988). Differential sensitivity of mouse pronuclei and zygote cytoplasm to Hoechst staining and ultraviolet irradiation. *J. Reprod. Fertil.* **82**, 173–178.

Wakayama, T., Perry, A. C. F., Zuccotti, M., Johnson, K. R., and Yanagimachi, R. (1998). Full-term development of mice from enucleated oocytes injected with cumulus cell nuclei. *Nature* **394**, 369–374.

Wakayama, T., Tateno, H., Mombaerts, P., and Yanagimachi, R. (2000). Nuclear transfer into mouse zygotes. *Nat. Genet.* **24**, 108–109.

Wallace, D. C., Brown, M. D., Melov, S., Graham, B., and Lott, M. (1998). Mitochondrial biology, degenerative diseases and aging. *Biofactors* **7**, 187–190.

Zernika-Goetz, M. (1991). Spontaneous and induced activation of rat oocytes. *Mol. Reprod. Dev.* **28**, 169–176.

PART V

NUCLEAR TRANSFER IN PRIMATES

CHAPTER 22

CLONING IN NONHUMAN PRIMATES

Tanja Dominko, Calvin Simerly, Crista Martinovich, and Gerald Schatten

INTRODUCTION

The genetic similarity between nonhuman primates and humans makes nonhuman primates, especially the rhesus monkey, a uniquely suited model for biomedical research. Bridging the genetic gap between mice (the species commonly used as a model to study human disease) and humans, nonhuman primates are proving to be indispensable. Unlike mice, however, nonhuman primates are a very costly experimental model proposition. To reduce the number of animals that have to be used in controlled experimental situations, an identical genetic background of experimental subjects would be required—a prospect that became feasible for the first time in 1996. Campbell and co-workers (1996) published the birth of two genetically identical sheep (Megan and Morag), mammals born from embryos reconstructed by nuclear transfer of nuclei from embryonic cells grown in culture. Similar experiments have been attempted in nonhuman primates and despite the continuous successes published in other mammalian species (reviewed by Pennisi and Vogel, 2000), the generation of genetically identical primates remains an elusive goal. This chapter reviews the ongoing research efforts to derive genetically identical nonhuman primates.

WHY GENETICALLY IDENTICAL NONHUMAN PRIMATES?

There is no doubt that nonhuman primates provide the ultimate experimental model for investigating human disease (reviewed by Sibal and Samson, 2001). Their close evolutionary relationship offers a very high degree of genotypic and phenotypic homologies that cannot be found between humans and other mammals (Stone *et al.*, 1987). Similarities between nonhuman and human genetics (VandeBerg, 1987), immunology (Schramm *et al.*, 2001), susceptibility to disease and disease manifestation (VandeBerg and Williams-Blangero, 1996), and aging-induced degenerative disorders (Lane, 2000; Roth *et al.*, 2000; Laman *et al.*, 2001; Black *et al.*, 2001), to name a few, are remarkable. Because of these similarities, nonhuman primates are ideally suited for research of not only traditional human maladies, but also offer a unique system for developing and testing new therapeutic approaches in human medicine. Therapeutic applications of gene therapy (Heim and Dunbar, 2000), cell, tissue, and organ transplantation (Starr *et al.*, 1999; Vons *et al.*, 2001; Kirk *et al.*, 2001), stem cell therapies (Tarantal *et al.*, 2000; Heim and Dunbar, 2000; Rosenzweig *et al.*, 2001), and vaccines and drugs for unique primate conditions can only be reliably studied in nonhuman primates. Conditions specific to primates include asthma (Hart *et al.*, 2001), HIV (Kent and Lewis, 1998), hepatitis (Purcell and Emerson, 2001), leishmaniasis (Kenney *et al.*, 1999), parainfluenza (Durbin *et al.*, 2000), tuberculosis (McMurray, 2000), cancer (Brewer *et al.*, 2001), addiction (Schneider *et al.*, 2001), obesity (Winegar *et al.*, 2001), and mitochondrial diseases (Pulkes and Hanna, 2001).

Genetic diversity between animals imposes an experimental constraint that can only be overcome either by larger groups of experimental animals or by the availability of genetically identical animals. Eliminating genetic variability among experimental animals would allow for much fewer animals to be used while obtaining clinically significant results. Invariable genetics improve accuracy, reducing the number of animals needed while providing perfect controls. Identicals with different birth dates would have numerous applications: phenotypic analysis of the offspring before propagating its siblings, serial transfer of stem cells to address cellular aging beyond life expectancy, and investigation of maternally induced epigenetic effects on offspring carried sequentially in the same surrogate (Schneider *et al.*, 2001). Because numerous embryos could be produced by cloning technology, embryonic stem cells (Thomson *et al.*, 1995) and stem cell-derived differentiated cells produced *in vitro* could be used to evaluate transplantation outcomes after allo- and autologous cell transplantation (Lanza *et al.*, 1998, 2000; Rosenzweig *et al.*, 2001). Rare animals affected with heritable diseases (retinitis pigmentosa, congenital hypothyroidism) could be reproduced by somatic cell cloning and used as models for developing new therapeutic approaches. On the other hand, animals with specific genotypes of interest could be derived from cells (stem cells or somatic cells) in which genes have been knocked out or knocked in, changed by site-directed mutagenesis or other cell transfection approaches.

The importance of nonhuman primates in biomedical research cannot be underscored enough, and availability of identical nonhuman primates would bring current biomedical research a step closer to therapeutic realization.

FACTORS INFLUENCING RESEARCH IN ASSISTED REPRODUCTIVE TECHNOLOGIES

Even though desired as a penultimate animal model, the expense associated with acquisition and maintenance of primates, the complexity of facilities that support intricate social structure within and between groups, the need for highly skilled personnel, and their ever-increasing limited availability make nonhuman primates a precious commodity. In addition to a logistically demanding undertaking, several reproductive characteristics of nonhuman primates prevent swift progress in assisted reproduction research. Animals reach sexual maturity between 4 and 5 years of age, and oocytes collected from prepubertal animals have very low developmental competence unless stimulated (Zheng *et al.*, 2001). Protocols for *in vitro* maturation of nonhuman primate oocytes undergoing development during the past 10 years (Boatman, 1987; Wolf *et al.*, 1996) have finally allowed production of developmentally competent mature oocytes that, after *in vitro* fertilization, form normal embryos that can be carried to term (Schramm and Paprocki, 2000). Good quality oocytes can be obtained only after hormonal stimulation for oocyte growth and development (superovulation) using exogenous gonadotropins. In the absence of rhesus-specific gonadotropins, human recombinant hormones are used (Zelinski-Wooten *et al.*, 1995). Induction of immune response over repeated stimulation protocols (Bavister *et al.*, 1986) in the same animal results in decreased superovulation response—lower numbers of oocytes with decreased developmental potential are observed after the third stimulation (Zelinski-Wooten, 1996). Reproductive seasonality in rhesus monkey further limits the availability of oocytes (Zheng *et al.*, 2001). Between May and September, the animals have reduced ovulatory and mating sequences and this seasonality is independent of the environment [the same in free-ranging animals and those housed in controlled environment (Hutz *et al.*, 1985)].

In addition, the very similarities with humans that make them a desired model impose experimental constraints that are difficult to overcome. Some of the limitations are listed below for rhesus monkeys:

1. Rhesus monkey and human menstrual cycles are very similar. On average, one oocyte is ovulated every 28 days and uterine lining is shed by menstrual bleeding approximately 2 weeks after ovulation in nonpregnant animals. Response to superovulation protocols varies extensively between animals and is difficult to predict. During the past few years, ultrasonography for imaging of ovaries was employed successfully and improved evaluation of stimulation protocols. Size of the ovaries and the number and size of developing follicles can be measured in order to evaluate the response. This has resulted not only in improved yields of good quality oocytes but has also reduced the number of animals undergoing surgical recovery attempts in the absence of a satisfactory ovarian stimulation.

2. Superovulation yields inconsistent and unpredictable numbers of good quality oocytes. Average yield from a first-time superovulated animal ranges between 0 and 156 (Wolf *et al.*, 1999; Zheng *et al.*, 2001; T. Dominko, personal observation), averaging between 15 and 20 per collection. As in other species, increase in the number of obtained oocytes invariably signifies their lower developmental potential.

3. Recovery of oocytes relies on surgical procedures. Laparoscopic follicular aspiration (Dierschke and Clark, 1976) has become the procedure of choice and allows for multiple retrievals in the same animal. It is less traumatic for animals, compared to laparotomy, reduces the appearance of postoperative peritoneal adhesions, and allows faster recovery. Under optimal conditions, retrievals can be performed during every second menstrual cycle; however, the number and quality of retrieved oocytes decrease after the third stimulation (Wolf *et al.*, 1999; Zelinski-Wooten *et al.*, 1996).

4. To achieve significant pregnancy rates embryo transfer has to be performed surgically. This is dictated by two sets of experimental observations. First, transfer of early preimplantation embryos (two- to eight-cell stage) into oviducts of recipient females assures optimal embryo culture environment. Second, anatomical structure of the cervix in rhesus monkey does not allow for easy access to the uterus. Even though rare successes have been achieved by nonsurgical transcervical embryo transfer of blastocysts, the method of choice remains laparoscopy.

5. There are no protocols for synchronization of menstrual cycles in rhesus monkeys. Recipient animals have to be monitored for the levels and dynamics of luteinizing hormone (LH), estradiol, and progesterone in peripheral blood (either by radioimmunoassay or enzyme-linked immunosorbent assay in order to determine time of ovulation. Embryos are usually transferred into oviducts 2 or 3 days after the preovulatory LH surge.

6. Difficulty in carrying twin pregnancies to term, which reduces the number of embryos that must be transferred into recipients.

7. High incidence of stillbirths when animals are allowed to carry pregnancies to term. It appears most deliveries occur at night, when intervention and immediate assistance by veterinary staff are not available.

8. High neonatal and perinatal mortality rates in offspring rejected by their birth mothers due to cesarean delivery. Neonatology and perinatalogy are in a developmental phase for nonhuman primates.

9. Routine monitoring of ongoing pregnancies (hormone tests, ultrasound) has to be performed under sedation to avoid stress-induced elevation of cortisol, which is the physiological inducer of labor.

Only a handful of laboratories conduct research in assisted reproductive technologies in nonhuman primates and consequently the progress has been relatively slow. Only recently have techniques that are commonly offered by human fertility clinics been successfully adopted in nonhuman primates: superovulation (Zelinski-Wooten et al., 1995; Weston et al., 1996), offspring derived after transfer of embryos produced by in vitro fertilization (IVF) (Bavister et al., 1984), intracytoplasmic sperm injection (ICSI) (Hewitson et al., 1999), elongated spermatid injection (ELSI) (L. Hewitson et al., unpublished observations), embryo cryopreservation (Wolf et al., 1989; Lanzendorf et al., 1990), and artificial insemination with fresh and frozen semen (Sanchez-Partida et al., 2000).

Development of these techniques allows for enhanced and controlled propagation of the species, and at the same time offers a unique opportunity to investigate the effect of assisted reproductive technologies on normalcy of the created embryos and offspring and extrapolation to human reproductive medicine (Hewitson et al., 1998, 1999; Sanchez-Partida et al., 2000). In-depth understanding of primate gametogenesis and embryogenesis is expected to enable development of protocols for generation of genetically identical animals.

METHODS FOR PRODUCTION OF GENETICALLY IDENTICAL ANIMALS

NUCLEAR TRANSFER

The main advantage offered by nuclear transfer (NT) technology is the possibility of creating multiple animals with identical nuclear genotype. Blastomeres from preimplantation embryos, when used as nuclear donors, could satisfy most of the requirements for generation of small numbers of identical primate offspring, but this source of genetics is inherently limited by the number of blastomeres available in donor embryos. The reported proportion of NT embryos that will produce live offspring ranges between 1 and 3% across all the mammals produced by this technique. Consequently, attempts to produce identical offspring by NT of nuclei in early preimplantation embryos (4-cell to 16-cell stages) inevitably carry a high risk of failure from the very beginning. To approach a successful outcome statistically, postcompaction preimplantation embryonic stages, such as blastocysts or blastocyst-derived embryonic stem cells and somatic cells, could assure increased numbers of identical donor cells. The situation in primates is further complicated by the need for simultaneous availability of good quality donor embryos (produced by either IVF or ICSI) and mature recipient oocytes. Due to the limitations described above, one is often faced with a situation wherein, on the day when embryos are available for NT, mature oocytes are in short supply or simply unavailable, and vice versa. Even when both recipients and donors are available and NT embryos can be reconstructed under optimal biological conditions, it is very likely that surrogate females at the desired stage of the menstrual cycle will not be at a researcher's disposal for embryo transfer. Despite these difficulties, preimplantation embryos in nonhuman primates have been produced by nuclear transfer of embryonic, embryonic stem cell, and differentiated somatic cell nuclei (Meng et al., 1997; Wolf et al., 1999; Simerly et al., 2002; Mitalipov et al., 2001) (Table 1). Because only one study published to date reports birth of monkeys from embryos produced by nuclear transfer of embryonic blastomeres into enucleated oocytes (Meng et al., 1997), it is prudent to examine the approach utilized by Meng and co-workers and compare it with approaches from later unsuccessful attempts (Wolf et al., 1999; Simerly et al., 2002). Experimental steps employed in these studies are summarized in Table 2.

The oocytes are obtained from stimulated females exhibiting normal menstrual cycles (Zelinski-Wooten et al., 1995). Donor IVF embryos are produced by oocyte insemination with freshly ejaculated sperm (Bavister et al., 1983) or by

Table 1 Preimplantation Development of Nonhuman Primate Embryos Produced by Nuclear Transfer

Recipient Stage	Donor Stage	Enucleated N	Fused N (%)	Cleaved N (%)[a]	<16 cell N (%)[a]	Reference
MII	Embryonic blastomere	97	78	59	NR[b]	Meng et al., 1997
MII	Embryonic blastomere	31	23	14	8	Simerly et al., 2002
MII	Fetal fibroblast	136	100	57	NR	Wolf et al., 1999
MII	Adult fibroblast	25	9	4	NR	Wolf et al., 1999
MII	Embryonic stem cell	51	35	18	NR	Wolf et al., 1999
Old MII	Embryonic blastomere	30	27	10	NR	Meng et al., 1997
Old MII	Embryonic blastomere	75	62	46	21	Simerly et al., 2002
Activated MII	Embryonic blastomere	79	63	47	14	Simerly et al., 2002
Zygote	Embryonic blastomere	11	10	9	NR	Meng et al., 1997
Zygote	Embryonic blastomere	5	3	3	0	Simerly et al., 2002

[a]Percentage of fused.
[b]NR, Not reported.

Table 2 Comparison of Protocols Used in Primate Nuclear Transfer[a]

Variable	Meng et al. (1997)	Wolf et al. (1999)	Simerly et al. (2002)
Age of MII oocytes at enucleation (hours post-hCG)	27–39	27–34	27–34
Hoechst 33342	0	NR	1 µg/ml
Cytochalasin	7.5 µg/ml CB	NR	7.5 µg/ml CB
Donor	From 4- to 32-cell unsynchronized blastomeres	Fetal fibroblasts (G0), adult fibroblasts (G0), ES cells	From 4- to 32-cell unsynchronized blastomeres; SPIT[b]
Donor	IVF	Somatic	IVF, ICSI
Donor	Fresh and frozen	—	Fresh
Fusion parameters	2.2 kV/cm, 50 µsec, 2 pulses	NR	1.2 kV/cm, 30 µsec, 2 pulses; same for SPIT[b]
Fusion medium	0.3 M mannitol, 100 µM Ca and Mg	NR	0.3 M D-sorbitol, 50 µM Ca and Mg
Cell dissociation	Trypsin in Ca-, Mg-free PBS	Trypsin EDTA	Ca-, Mg-free TALP, EDTA, EGTA
Activation time	1 hour prefusion to 2 hours postfusion continuously	NR	2 hours postfusion, 2 hours prefusion, at the same time
Activation protocol	7.5 µg/ml cycloheximide, 7.5 µg/ml CB	NR	Sperm factor; ICSI for SPIT[b]
Embryo culture	BRL monolayers, CMRL 1066	Sequential (S1, S2)	BRL monolayers, CMRL 1066
Embryos transferred	Frozen	NR	Fresh
Pregnancies/surrogates (embryos transferred)	2/9 (29), 2/17 (53)[c]	0/6 (12) somatic cells, 0/8 (24)[c] blastomeres	0/14 (29), 1/8 (16) SPIT[b]
Offspring	2	0	0

[a]Abbreviations: NR, Not reported; CB, cytochalasin B; ES, embryonic stem; IVF, *in vitro* fertilization; ICSI, intracytoplasmic sperm injection; SPIT, spindle transfer.
[b]Instead of discarding the removed maternal chromatin, the MII-containing karyoplast was placed back under the zona and an MII oocyte was reconstructed by electrical fusion. These oocytes were fertilized 2–3 hours after spindle fusion by ICSI.
[c]Values were calculated based on the difference between those reported by Wolf *et al.* (1999) and Meng *et al.* (1997).

intracytoplasmic sperm injection (Hewitson *et al.*, 1999). Embryos generated by both IVF and ICSI have been shown to be developmentally normal and reproducibly result in live offspring (Hewitson *et al.*, 1999). Removal of maternal chromatin is most commonly aided by labeling of maternal DNA with Hoechst 33342/bisbenzimide—an intercalating DNA dye that is excited in ultraviolet range of the spectrum—in order to confirm complete DNA removal. A single blastomere is placed under the zona of an enucleated oocyte (A and B, Fig. 1). Electric fusion is used to reconstitute the enucleated recipient cytoplast with donor cell nucleus. Mannitol- or sorbitol-based fusion medium (0.25–0.3 M) is used, supplemented with 0.05–0.1 mM Ca and Mg and 0.1% bovine serum albumin (BSA). Activation protocols employed in NT of mice and domestic species vary and births have been obtained after combination of Ca ionophores and kinase inhibitors (ionomycin/dimethylaminopurine), protein synthesis inhibitors and electric current (cycloheximide/electric), strontium alone, and electric current alone. The cell cycle stage of the recipient cytoplasm can be tightly controlled with these protocols and both high-concentration metaphase-promoting factor (MPF) oocytes (fresh metaphase II) and low-concentration MPF oocytes (preactivated or aged oocytes), when used as recipients, result in live offspring in several mammalian species (Campbell *et al.*, 1993; Campbell, 1999). Several differences can be noted between the protocols summarized in Table 1:

1. *Labeling of maternal DNA.* Meng *et al.* (1997) never employed Hoechst 33342 to visualize maternal DNA under ultraviolet (UV) light. It is well known that exposure of biological material to UV light is damaging, and even though this damage may not manifest itself during early development, the effects may become detrimental during pregnancy.

2. *Fusion medium.* In mammalian species, offspring have been produced from embryos fused in either mannitol- or sorbitol-based media. Mannitol may be better suited for monkey embryos, although the exact reasons for the better results in nonhuman primates are presently unknown.

3. *Activation.* A unique approach was employed by Meng *et al.* (1997). The recipient oocytes were preactivated (an hour before fusion) by cycloheximide; the nucleus was introduced by fusion and the activation continued in the presence of cycloheximide and electric pulsing for an additional 2 hours. During the entire procedure, the oocytes were exposed to cytochalasin B to prevent expulsion of donor chromatin.

Nevertheless, the two monkeys born showed that the NT techniques that result in births of healthy offspring in several mammalian species (sheep, cattle, rabbits, mice, and pigs) can also be used in nonhuman primates. In addition, it appeared that the biology of nucleocytoplasmic interactions required for reprogramming and development after NT was well conserved among various species of mammals.

To examine in more detail the behavior of the introduced nucleus, we monitored the progression of donor chromatin through the first four embryonic cell cycles (Simerly *et al.*, 2002). Within hours of the embryonic pronucleus being formed, additional small nuclei were frequently observed and increased in number throughout zygotic interphase (C and G, Fig. 1). DNA and microtubule imaging revealed abnormalities in NT embryos, with multiple microtubule organizing centers (centrosomes) and spindle poles (H and I, Fig. 1), extensive DNA fragmentation, unequal chromosome segregation (J and K, Fig. 1), and the presence of both condensed and decondensed chromatin within the same cytoplasm (Simerly *et al.*, 2002). At 14 hours after activation, 42% of embryos with one pronucleus (PN) and 16% of the embryos with two or more pronuclei were observed [the remaining zygotes (24%) underwent nuclear envelope breakdown]. When the same embryos were examined 16 hours after activation, 25% remained with one PN and 34% contained two or more pronuclei [81 (41%) underwent nuclear envelope

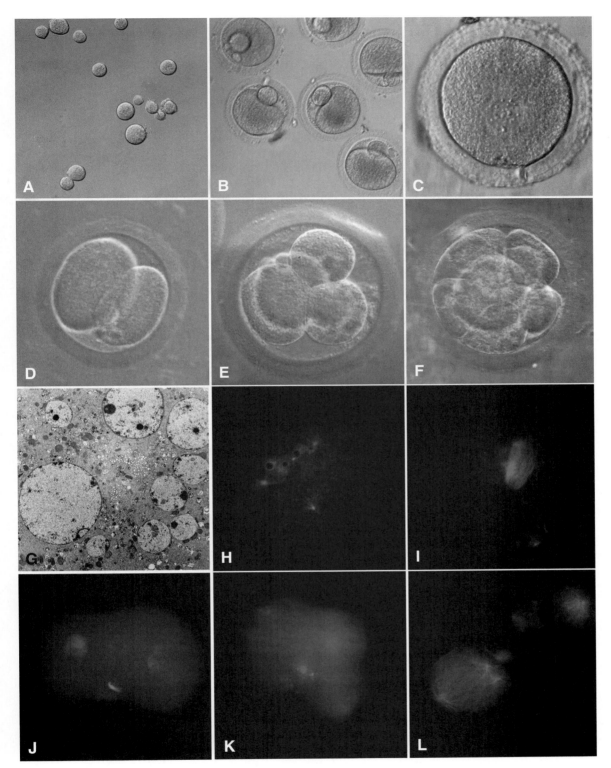

Figure 1 Development of nuclear transfer-derived nonhuman primate embryos. Donor embryonic blastomeres (A) are dissociated and inserted under the zonae of enucleated oocytes (B). Reconstructed and fused units are placed into culture medium and observed for pronuclear formation (C). Embryos progress through first embryonic divisions and appear morphologically normal (D–F). Commonly, however, zygotes develop several pronuclei (G; transmission electron microscope image of nine karyomeres in a zygote) and instead of two microtubule-organizing centers, several foci can be seen by immunocytochemistry (H). Alignment of zygotic chromatin on the first mitotic spindle is incorrect (I) and this results in abnormal cleavage (J; a two-cell embryo). Incorrect ploidy is propagated with each subsequent cell cycle (K and L; four-cell and 6-cell embryos, respectively). Microtubules, red; DNA, blue.

breakdown]. (Simerly *et al.*, 2002). A significant proportion of multinucleated NT embryos had been previously reported after the introduction of interphase somatic cell nuclei into mouse cytoplasts (Wakayama *et al.*, 1998), and in monkey NT embryos reconstructed with unsynchronized embryonic blastomeres (42%) (Mitalipov *et al.*, 2001). This was attributed to MPF-induced premature chromatin condensation (PCC) and pulverization of replicating DNA, but this has not been observed with activated cytoplasts (Campbell *et al.*, 1993). In rhesus NT embryos, multinucleation seems to be the consequence of the activation protocol (ionomycin/DMAP) and not the cell cycle stage of the donor nucleus (Mitalipov *et al.*, 2001). Using ionomycin/roscovitin (cdc2 inhibitor), multinucleation was not observed; however, extrusion of an undetermined amount of nuclear DNA in the form of one or two "polar bodies" occurred (Mitalipov *et al.*, 2001). In our studies, parthenogenetic embryos produced by activation of nucleated MII oocytes with rhesus sperm factor did not show these abnormalities (Simerly *et al.*, 2002). First mitosis started earlier in NT embryos (16–18 hours postactivation) when compared to IVF or ICSI embryos (24 hours) (Hewitson *et al.*, 1998, 1999). As in amphibians (DiBerardino, 1979), shortened DNA replication may be forcing S-phase chromatin into premature condensation and incorrect spindle alignment (Fig. 1I). Fragmentation of the chromatin did not affect the progression of the embryos through the first or subsequent cell cycles. These embryos cleaved, resulting in unequal chromosome segregation during subsequent cell divisions (J–L, Fig. 1). The appearance of these embryos using Hoffman Modulation Contrast (HMC) optics was indistinguishable from parthenogenetic, IVF, or ICSI embryos (D–F, Fig. 1). These observations highlight cytoskeletal differences in the dynamics of the first cell cycle between NT embryos from domestic species and rhesus monkeys, and may be a consequence of one or more of the manipulation steps.

The consequences of the NT procedures on normal fertilization were investigated using a combination of enucleation and fertilization procedures (Simerly *et al.*, 2002). MII oocytes were first enucleated, fused back with the karyoplast containing the MII spindle, and then fertilized by ICSI, i.e., <u>sp</u>indle <u>t</u>ransfer (SPIT). Although the development of these embryos was significantly better than that of NT embryos, a high proportion of SPIT embryos eventually arrested in development and displayed similar chromosomal abnormalities during the first cell cycle. This suggests that exposure to Hoechst 33342, fusion medium, electric fusion, and/or a combination of these procedures has deleterious effects in nonhuman primates. Furthermore, apoptosis in both the trophectoderm (TE) and inner cell mass (ICM) cells of SPIT blastocysts was almost 40%, whereas the cells of control ICSI embryos evaluated on the same day of development were between 0 and 0.3% (Simerly *et al.*, 2002). The transfer of 16 SPIT embryos into 8 recipient females resulted in a single pregnancy, which was later diagnosed as a "blighted pregnancy" by ultrasound on day 25 after transfer. The well-developed placenta contained no fetal tissue, suggesting abnormalities in early differentiation and the absence of a viable ICM. Because control ICSI offspring are routinely produced (Hewitson *et al.*, 1999), the observed abnormality is attributed to factors other than the ICSI procedure, the time oocytes spend outside of the incubator, or the use of manipulation medium. Arrested NT embryos show a high degree of apoptosis and the transfer of 29 morphologically normal two- to eight-cell NT embryos with single nucleated blastomeres into 14 recipient animals produced no pregnancies. The successful initiation of the SPIT pregnancy now permits analysis of the consequences of each NT manipulation by comparison with control ICSI embryos. Several alternative procedures for presumably less damaging visualization of maternal DNA, fusion, and activation have been described for various mammals, including primates (Dominko *et al.*, 1999).

Our spindle transfer experiment suggests that the difference in activation protocols was not the cause of failure in recent experiments. SPIT oocytes were activated by fertilization via intracytoplasmic sperm injection—a procedure that repeatedly results in live offspring in monkeys (Hewitson *et al.*, 1999). If so, the remaining differences observed may very well be responsible for continuous failures. Labeling of maternal DNA, electrical fusion in sorbitol-based medium, or a combination of both are then likely candidates that may need to be eliminated or replaced before NT can be successful in primates.

EMBRYO SPLITTING

An alternative to producing identical offspring is embryo splitting. It has been shown in mice, sheep, goats, bovines, and the rhesus monkey (Chan *et al.*, 2000) that separated blastomeres from early preimplantation embryos retain their full developmental potential and can result in healthy offspring. Using embryo splitting results in animals that not only have identical nuclear genomes, but a mitochondrial DNA complement as well, which has to be considered when the animals are to be used for testing responses that depend on mitochondria-regulated metabolic pathways.

Nonhuman primate identical embryos that can develop to the blastocyst stage *in vitro* have been produced successfully. It has been shown that two blastomeres from an eight-cell embryo (quadruplets) contain sufficient information to support not only development to the blastocyst stage, but also development to term. "Tetra," born in 1999, was the only offspring born from such an embryo (Chan *et al.*, 2000). The remaining three quadruplets miscarried or formed only placental tissue (Chan *et al.*, 2000). A much higher rate of offspring can be produced from twin embryos; however, inevitably one of the twins is lost either before implantation, during early pregnancy (A. Chan *et al.*, unpublished), or due to late-pregnancy abortion (T. Dominko *et al.*, unpublished observations).

Genetically identical embryos occur naturally in mammals by separation of embryonic inner cell mass (cells) prior to implantation. Artificially, identical embryos can be created *in vitro* by manual separation of embryonic blastomeres during the 2- to 16-cell stages of embryonic development. Mechanically separated blastomeres are inserted into host zonae pellucidae, obtained from unfertilized oocytes or arrested embryos. Depending on the developmental stage of the donor embryo, two or more identical embryos can be created. This technique was successfully applied to monkey embryos and could theoretically result in production of identical animals (Chan *et al.*, 2000). The steps were as follows:

1. A donor embryo at the 4- to 16-cell stage is placed into 0.5% pronase to remove the zona pellucida. The zona-free embryo is washed thoroughly through several washes of Tyrode's albumin/lactate/pyruvate–hydroxyethylpiperazinesulfonic acid (TALP–HEPES), and is placed into a drop of Ca- and Mg-free TALP–HEPES, supplemented with 1 mM each EDTA and EGTA (Chan *et al.*, 2000; Mitalipov *et al.*, 2001). After 15 minutes, embryonic blastomeres are dissociated by manual pipetting through a narrow fire-polished glass pipette under a dissecting microscope. Alternatively, blastomeres are separated in a manipulation drop by the transfer pipette under an inverted microscope equipped with micromanipulators.

2. Depending on the developmental stage of the donor embryo and consequently the size of individual blastomeres, a blunt or angled transfer pipette is used for picking up blastomeres and depositing them into empty host zonae.

3. Host zonae are prepared from any source: unfertilized oocytes, developmentally arrested embryos, immature oocytes, dead oocytes, or even oocytes of a different species (Chan *et al.*, 2000). Zonae can be emptied by aspiration of the contents using micromanipulation or alternatively by manually aspirating oocytes/embryos through a sharp-edged pipette that has a slightly smaller diameter than the oocytes/embryos. This procedure will likely result in more

zonae that have been "shredded" and have more than one zona opening due to the harshness of manual aspiration. For developmental purposes, this does not present a problem because the zona retains its spherical shape. When the number of host zonae is limited, aspiration of the material using micromanipulation is preferred. Zonae are washed of cellular debris several times and placed into a manipulation drop. Empty zonae can be collected over time and stored in PBS at 4°C (T. Dominko *et al.*, unpublished observations).

4. Manipulation medium used for reconstruction can be supplemented with 5–7.5 µg/ml cytochalasin B in order to make blastomere plasma membranes more plastic. This is especially important when blastomeres from four-cell embryos are used. Their size requires a large transfer pipette and on many occasions the size of the transfer pipette prevents successful passage through the zona opening. Empty zonae collapse easy when pressure is applied to their surface and one may have to use a smaller diameter transfer pipette. To avoid lysis of blastomeres during their aspiration into the transfer pipette, cytochalasin B or another microfilament inhibitor is used. After transfer, blastomeres return to their normal round shape within 15 minutes.

5. Reconstructed embryos are removed from manipulation medium, washed free of cytochalasin B, and placed into *in vitro* culture.

Embryos that contain between 3 and 16 blastomeres are the best candidates for splitting, and as many as six identical embryos that retained the ability to form blastocysts have been produced. No differences were observed during *in vitro* development of embryos reconstructed from fresh or frozen donor embryos created by either IVF or ICSI. Timing and quality of reconstructed embryos differed from IVF and ICSI controls (Chan *et al.*, 2000). Reconstructed embryos contained fewer cells at the blastocyst stage. Reduction in the cell number was observed in both ICM cells and trophectoderm. Reconstructed embryos contained more apoptotic cells compared to controls, and these cells were mainly located in the ICM. This would indicate that despite their normal morphological appearance, these embryos are disadvantaged at the time of implantation. Insufficient ICM may compromise normal fetal development and reduced trophectoderm may not be able to develop into a functional placenta. Even though multiple animals have been born from one-half of a split embryo, a set of identical animals has yet to be carried successfully to term using these techniques.

CONCLUSIONS

Development of nonhuman primate models for human disease through nuclear transplantation or embryo splitting would greatly benefit ongoing research on drug testing, vaccine development, gene therapy, and many other areas of importance. Genetically identical animals would allow use of smaller numbers of animals in the research while dramatically increasing the accuracy and interpretation of results due to reduced or eliminated genetic variability between animals.

To accomplish these objectives requires a better understanding of cellular events that take place during reprogramming of the somatic nucleus in oocyte cytoplasm. Failed attempts as well as the successful ones to create these embryos and pregnancies have to be critically analyzed. Availability of only a few identical animals cannot satisfy the needs of the biomedical research community, and the ultimate goal has to be the development of a reliable, efficient, and safe technology. In order to justify research programs focused on this costly technology, potential scientific merit has to be very clearly defined and the benefits need to outweigh the cost significantly if the results are to be cost-effective, reliable, repeatable, and safe. An evaluation of the various steps involved in nuclear transfer through the use of SPIT embryos may provide insight into the causes of NT failures, perhaps leading to an enabling NT technology for nonhuman primates.

ACKNOWLEDGMENTS

We would like to thank The Oregon Regional Primate Research Center (ORPRC) in Beaverton, OR, and the NIH (NCRR, NICHD) for supporting this research; we also thank members of Dr. Schatten's lab (Laura Hewitson, Diana Takahashi, Brian McVay, Evelyn Neuber, Tonya Swanson, Kevin Muller, and John Bassir), members of the ORPRC's Department of Animal Resources (John Fanton, Darla Jacob, Mike Cook, and Kevin Grund), and all at the ORPRC for assistance, especially all the personnel involved with animal care.

REFERENCES

Bavister, B. D., Boatman, D. E., Leibfried, L., Loose, M., and Vernon, M. W. (1983). Fertilization and cleavage of rhesus monkey oocytes *in vitro. Biol. Reprod.* **28**, 983–999.

Bavister, B. D., Boatman, D. E., Collins, K., Dierschke, D. J., and Eisele, S. G. (1984). Birth of rhesus monkey infant after *in vitro* fertilization and non surgical embryo transfer. *Proc. Natl. Acad. Sci. U.S.A.* **81**, 2218–2222.

Bavister, B. D., Dees, H. C., and Schultz, R. D. (1986). Refractoriness of Rhesus monkeys to repeated ovarian stimulation by exogenous gonadotropins is caused by formation of nonprecipitating antibodies. *Am. J. Reprod. Immunol. Microbiol.* **11**, 11–16.

Black, A., Tilmont, E. M., Handy, A. M., Scott, W. W., Shapses, S. A., Ingram, D. K., Roth, G. S., and Lane, M. A. (2001). A nonhuman primate model of age-related bone loss: A longitudinal study in male and premenopausal female rhesus monkeys. *Bone* **28**, 295–302.

Boatman, D. E. (1987). *In vitro* growth of nonhuman primate pre- and peri-implantation embryos. *In* "The Mammalian Preimplantation Embryo: Regulation of Growth and Differentiation *In Vitro*" (B. D. Bavister, ed.), pp. 273–308. Plenum Press, New York.

Brewer, M., Utzinger, U., Satterfield, W., Hill, L., Gershenson, D., Bast, R., Wharton, J. T., Richards-Kortum, R., and Follen, M. (2001). Biomarker modulation in a nonhuman rhesus primate model for ovarian cancer chemoprevention. *Cancer Epidemiol. Biomarkers Prev.* **8**, 889–893.

Campbell, K. H. S. (1999). Nuclear equivalence, nuclear transfer, and the cell cycle. *Cloning* **1**, 3–15.

Campbell, K. H. S., Ritchie, W. A., and Wilmut, I. (1993). Nuclear-cytoplasmic interactions during the first cycle of nuclear transfer reconstructed bovine embryos: implications for deoxyribonucleic acid replication and development. *Biol. Reprod.* **49**, 933–942.

Campbell, K. H. S., McWhir, J., Ritchie, W. A., and Wilmut, I. (1996). Sheep cloned by nuclear transfer from a cultured cell line. *Nature* **380**, 64–66.

Chan, A. W., Dominko, T., Luetjens, C. M., Neuber, E., Martinovich, C., Hewitson, L., Simerly, C. R., and Schatten, G. P. (2000). Clonal propagation of primate offspring by embryo splitting. *Science* **287**, 317–319.

DiBerardino, M. A. (1979). Nuclear and chromosomal behavior in amphibian nuclear transplants. *Int. Rev. Cytol.* (Suppl.) **9**, 129–160.

Dierschke, D. J., and Clark, J. R. (1976). Laparoscopy in *Macaca mulatta*: Specialized equipment employed and initial observations. *J. Med. Primatol.* **5**, 100–110.

Dominko, T., Ramalho-Santos, J., Chan, A. W. S., Moreno, R., Luetjens, C. M., Simerly, C., Hewitson, L., Takahashi, D., Martinovich, C., White, J. M., and Schatten, G. (1999). Optimization strategies for production of mammalian embryos by NT. *Cloning* **3**, 143–152.

Durbin, A. P., Elkins, W. R., and Murphy, B. R. (2000). African green monkeys provide a useful nonhuman primate model for the study of human parainfluenza virus types-1, -2, and -3 infection. *Vaccine* **18**, 2462–2469.

Hart, T. K., Cook, R. M., Zia-Amirhosseini, P., Minthoru, E., Sellers, T. S., Maleeff, B. E., Eustis, S., Schwartz, L. W., Tsui, P., Appelbaum, E. R., Martin, E. C., Bugelski, P. J., and Herzyk, D. J. (2001). Preclinical efficacy and safety of mepolizumab (SB-240563), a humanized monoclonal antibody to IL-5, in cynomolgus monkeys. *J. Allergy Clin. Immunol.* **108**, 250–257.

Heim, D. A., and Dunbar, C. E. (2000). Hematopoietic stem cell gene therapy: Towards clinically significant gene transfer efficiency. *Immunol. Rev.* **178**, 29–38.

Hewitson, L., Takahashi, D., Dominko, T., Simerly, C., and Schatten, G. (1998). Fertilization and embryo development to blastocysts after intracytoplasmic sperm injection in the rhesus monkey. *Hum. Reprod.* **13**, 3449–3455.

Hewitson, L., Dominko, T., Takahashi, D., Martinovich, C., Ramalho-Santos, J., Sutovsky, P., Fanton, J., Jacob, D., Monteith, D., Neuringer, M., Battaglia, D., Simerly, C., and Schatten, G. (1999). Unique checkpoints during the first cell cycle of fertilization after intracytoplasmic sperm injection in rhesus monkeys. *Nature Med.* **5**, 431–433.

Hutz, R. J., Dierschke, D. J., and Wolf, R. C. (1985). Seasonal effect on ovarian folliculogenesis in rhesus monkeys. *Biol. Reprod.* **33**, 653–659.

Kenney, R. T., Sacks, D. L., Sypek, J. P., Vilela, L., Gam, A. A., and Evans-Davis, K. (1999). Protective immunity using recombinant human IL-2 and alum as adjuvants in a primate model of cutaneous leishmaniasis. *J. Immunol.* **163**, 4481–4488.

Kent, S. J., and Lewis, I. M. (1998). Genetically identical primate modeling systems for HIV vaccines. *Reprod. Fertil. Dev.* **10**, 651–657.

Kirk, A. D., Tadaki, D. K., Celniker, A., Batty, D. S., Berning, J. D., Colonna, J. O., Cruzata, F., Elster, E. A., Gray, G. S., Kampern, R. L., Patterson, N. B., Szklut, P., Swanson, J., Xu, H., and Harlan, D. M. (2001). Induction therapy with monoclonal antibodies specific for CD80 and CD86 delays the onset of acute renal allograft rejection in non-human primates. *Transplantation* **72**, 377–384.

Laman, J. D., Visser, L., Massena, C. B., de Groot, C. J., de Jong, L. A., Hart, B. A., vaan Meurs, M., and Schellekens, M. M. (2001). *J. Neuroimmunol.* **119**, 124–130.

Lane, M. A. (2000). Nonhuman primate models in biogerontology. *Exp. Gerontol.* **35**, 533–541.

Lanza, R. P., Cibelli, J. B., and West, M. D. (1998). Human therapeutic cloning. *Nature Med.* **280**, 1256–1258.

Lanza, R. P., Caplan, A. L., Silver, L. M., Cibelli, J. B., West, M. D., and Green, R. M. (2000). The ethical validity of using nuclear transfer in human transplantation. *JAMA* **284**, 3175–3179.

Lanzendorf, S. E., Zelinski-Wooten, M. B., Stouffer, R. L., and Wolf, D. P. (1990). Maturity at collection and the developmental potential of rhesus monkey oocytes. *Biol. Reprod.* **42**, 703–711.

McMurray, D. N. (2000). A nonhuman primate model for preclinical testing of new tuberculosis vaccines. *Clin. Infect. Dis.* (Suppl.) **3**, S210–S212.

Meng, L., Ely, J. J., Stouffer, R. L., and Wolf, D. P. (1997). Rhesus monkeys produced by nuclear transfer. *Biol. Reprod.* **57**, 454–459.

Mitalipov, M. M., Nusser, K. D., and Wolf, D. P. (2001). Parthenogenetic activation of rhesus monkey oocytes and reconstituted embryos. *Biol. Reprod.* **65**, 253–259.

Pennisi, E., and Vogel, G. (2000). Animal cloning. Clones: A hard act to follow. *Science* **288**, 1722–1727.

Pulkes, T., and Hanna, M. G. (2001). Human mitochondrial DNA diseases. *Adv. Drug Deliv. Rev.* **49**, 27–43.

Purcell, R. H., and Emerson, S. U. (2001). Animal models of hepatitis A and E. *ILAR J.* **42**, 161–177.

Rosenzweig, M., Marks, D. F., DeMaria, M. A., Connole, M., and Johnson, R. P. (2001). Identification of primitive hematopoietic progenitor cells in the rhesus macaque. *J. Med. Primatol.* **30**, 36–45.

Roth, G. S., Ingram, D. K., Black, A., and Lane, M. A. (2000). Effects of reduced energy intake on the biology of aging: the primate model. *Eur. J. Clin. Nutr.* **54** Suppl 3, S15–20.

Sanchez-Partida, L. G., Maginnis, G., Dominko, T., Martinovich, C., McVay, B., Fanton, J., and Schatten, G. (2000). Live rhesus offspring by artificial insemination using fresh sperm and cryopreserved sperm. *Biol. Reprod.* **63**, 1092–1097.

Schneider, M. L., Moore, C. F., and Becker, E. F. (2001). Timing of moderate alcohol exposure during pregnancy and neonatal outcome in rhesus monkeys (*Macaca mulatta*). *Alcohol Clin. Exp. Res.* **25**, 1238–1245.

Schramm, R. D., and Paprocki, A. M. (2000). Birth of rhesus monkey infant after transfer of embryos derived from *in-vitro* matured oocytes: short communication. *Hum. Reprod.* **15**, 2411–2414.

Schramm, R. D., Paprocki, A. M., and Watkins, D. I. (2001). Birth of MHC-defined rhesus monkeys produced by assisted reproductive technology. *Vaccine* **20**, 603–607.

Sibal, L. R., and Samson, K. J. (2001). Nonhuman primates: A critical role in current disease research. *ILAR J.* **42**, 74–84.

Simerly, C., Dominko, T., St. John, J., Payne, C., Jacoby, E., Neuber, E., Takahashi, D., Martinovich, C., Mountain, V., Compton, D., Hewitson, L., and Schatten, G. (2002). Primate cloning inheritance defects: Deviation in centrosome and mitochondria transmissions. *Science* (in review).

Starr, P. A., Wichmann, T., van Horne, C., and Bakay, R. A. (1999). Intranigral transplantation of fetal substantia nigra allograft in the hemiparkinsonian rhesus monkey. *Cell Transplant.* **8**, 37–45.

Stone, W. H., Treichel, R. C., and VandeBerg, J. L. (1987). Genetic significance of some common primate models in biomedical research. *Prog. Clin. Biol. Res.* **229**, 73–93.

Tarantal, A. F., Goldstein, O., Barley, F., and Cowan, M. J. (2000). Transplantation of human peripheral blood stem cells into fetal rhesus monkey (*Macaca mulatta*). *Transplantation* **69**, 1818–1823.

Thomson, J. A., Kalishman, J., Golos, T. G., Durning, M., Harris, C. P., Becker, R. A., and Hearn, J. P. (1995). Isolation of primate embryonic stem cell line. *Proc. Natl. Acad. Sci. U.S.A.* **92**, 7844–7848.

VendeBerg, J. L. (1987). Historical perspective of genetic research with nonhuman primates. *Genetica* **73**, 7–14.

VandeBerg, J. L., and Williams-Blangero, S. (1996). Strategies for using nonhuman primates in genetic research on multifactorial diseases. *Lab. Anim. Sci.* **46**, 146–151.

Vons, C., Loux, N., Simon, L., Mahieu-Caputo, D., Dagher, I., Andreoletti, M., Borgnon, J., Di Rico, V., Bargy, F., Capron, F., Weber, A., and Franco, D. (2001). Transplantation of hepatocytes in nonhuman primates: A preclinical model for the treatment of hepatic metabolic diseases. *Transplantation* **72,** 811–818.

Wakayama, T., Perry, A. C., Zuccotti, M., Johnson, K. R., and Yanagimachi, R. (1998). Full-term development of mice from enucleated oocytes injected with cumulus cell nuclei. *Nature* **394,** 369–374.

Winegar, D. A., Brown, P. J., Wilkinson, W. O., Lewis, M. C., Ott, R. J., Tiong, W. Q., Brown, H. R., Lehmann, J. M., Kliewer, S. A., Plunket, K. D., Waay, J. M., Bodkin, N. L., and Hansen, B. C. (2001). Effects of fenofibrate on lipid parameters in obese rhesus monkeys. *J. Lipid Res.* **42,** 1543–1551.

Weston, A. M., Zelinski-Wooten, M. B., Hutchinson, J. S., Stouffer, P. L., and Wolf, P. (1996). Developmental potential of embryos produced by *in-vitro* fertilization from gonadotropin-releasing hormone antagonist-treated macaques stimulated with recombinant human follicle stimulating hormone alone or in combination with luteinizing hormone. *Hum. Reprod.* **11,** 608–613.

Wolf, D. P., VandeVoort, C. A., Meyer-Haas, G. R., Zelinski-Wooten, M. B., Hess, D. L., Gaughman, W. L., and Stouffer, R. L. (1989). *In vitro* fertilization and embryo transfer in the rhesus monkey. *Biol. Reprod.* **41,** 335–346.

Wolf, D. P., Alexander, M., Zelinski-Wooten, M., and Stouffer, R. L. (1996). Maturity and fertility of rhesus monkey oocytes collected at different intervals after an ovulatory stimulus (Human chorionic gonadotropin) in *in vitro* fertilization cycles. *Mol. Reprod. Dev.* **43,** 76–81.

Wolf, D. P., Meng, L., Ouhibi, N., and Zelinski-Wooten, M. (1999). Nuclear transfer in the Rhesus monkey: Practical and basic implications. *Biol. Reprod.* **60,** 199–204.

Zelinski-Wooten, M. B., Hutchison, J. S., Hess, D. L., Wolf, D. P., and Stouffer, R. L. (1995). Follicle-stimulating hormone alone supports follicle growth and oocyte development in gonadotropin-releasing hormone antagonist-treated monkeys. *Hum. Reprod.* **10,** 1658–1666.

Zelinski-Wooten, M. B., Alexander, M., Molskness, T. A., Stouffer, R. L., and Wolf, D. P. (1996). Use of recombinant human gonadotropins for repeated follicular stimulation in rhesus monkeys. *In* "Program of the XVIth Congress of the International Primatological Society and XIXth Conference of American Society of Primatologists," Abstr. 133. Madison, WI.

Zheng, P., Si, W., Wang, H., Zou, R., Bavister, B. D., and Ji, W. (2001). Effect of age and breeding season on the developmental capacity of oocytes from unstimulated and follicle-stimulating hormone-stimulated rhesus monkeys. *Biol. Reprod.* **64,** 1417–1421.

PART VI

APPLICATIONS

CHAPTER 23

NUCLEAR TRANSFER FOR STEM CELLS

Alan Trounson

INTRODUCTION

The demonstration that pluripotent embryonic stem (ES) cells can be produced from the culture of inner cell mass (ICM) cells of the human blastocyst (Reubinoff *et al.*, 2000; Thomson *et al.*, 1998) has evoked interest in the potential of ES cells for transplantation for cell therapies in human medicine (Trounson, 2001; Trounson and Pera, 2001). Only a small number of human ES cell lines are available at the present time, and although they do not express the major histocompatibility complex (MHC), this antigen appears during ES cell differentiation and would ultimately result in rejection of transplants derived from incompatible ES cell derivatives.

There are a number of strategies for preventing rejection of cell and tissue transplants derived from ES cells. It is possible to generate a very large panel of ES cells with a wide range of human leukocyte antigen (HLA) subtypes. The production of 200,000 or more different ES cell lines of HLA types suitable for tissue matching with patients may be possible by utilizing the large numbers of human embryos disposed of by infertility clinics worldwide. If there were a range of ES cells with HLA types, immune therapies may enable establishment of new stem cell populations, for recovery of function of the tissues of interest. It is also possible that hematopoietic derivatives of ES cells could be used to induce tolerance in patients to allow transplantation of the cell and tissue types that would correct pathologies, diseases, or injuries.

Given the beneficial clinical potential for stem cell therapies, there is a need to consider the option of developing a bank of ES cells with an appropriate range of HLA subtypes, or of customizing ES cells for transplantation compatibility by the technique of nuclear transfer for stem cells (NTSC), using somatic cells derived from the patients. This has been termed "therapeutic cloning," but this term does not adequately describe the manipulations or intentions involved, and there is a strong preference to use the term NTSC.

NUCLEAR TRANSFER USING SOMATIC CELLS TO PRODUCE ES CELLS

PROOF OF CONCEPT STUDIES

Embryonic stem cell lines have been established from mouse blastocysts (Table 1) generated by nuclear transfer of mouse somatic cell nuclei. The somatic cell types were ovarian cumulus cells (Munsie *et al.*, 2000; Wakayama *et al.*, 2001), tail-tip fibroblasts (Wakayama *et al.*, 2001), and fetal neuronal cells (Kawase *et al.*, 2000).

Table 1 Efficiency of NTSC Blastocysts Producing ES Cell Lines

No. of nuclear transfer blastocysts derived from somatic cells	Number of ES cell lines produced	Reference
10	1 (10%)	Munsie *et al.*, 2000
101	5 (5%)	Kawase *et al.*, 2000
398	35 (9%)	Wakayama *et al.*, 2001

The ES cells produced in these experiments were shown to differentiate in culture and produced teratomas when injected under the testis capsule of severe combined immune-deficient (SCID) mice. The three primary embryonic lineages were all represented in these teratomas (Kawase *et al.*, 2000; Munsie *et al.*, 2000) and normal chimeric progeny were obtained when the ES cells were injected into host blastocysts (Munsie, 2000; Wakayama *et al.*, 2001). The ES cells contributed to all tissues of the chimeric mice (Fig. 1), including germ line transmission.

The primary difficulty for nuclear transfer in mice is the efficiency of insertion of somatic cell nuclei into oocytes (Munsie, 2000). It is difficult to electrofuse cells that are very different sizes, and despite attempts to vary the techniques of electrofusion, reduced size differences, and other parameters, nuclear injection has

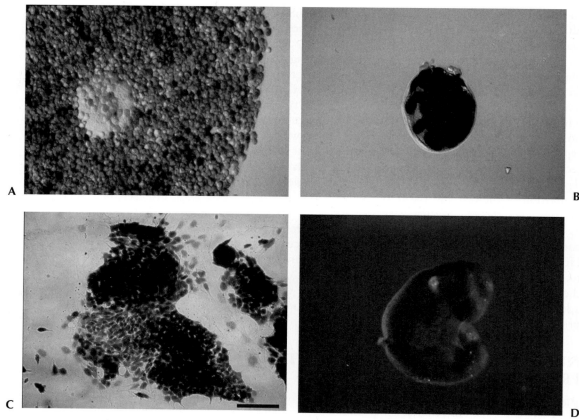

Figure 1 *LacZ expression in ZIN40 transgenic nuclei of (A) cumulus cells recovered from ZIN40 females–somatic cells used for NTSC, (B) a nuclear transfer morula, (C) ES cells produced by NTSC, and (D) chimeric fetus composed of NTSC and wild-type blastocyst cells (Munsie, 2000).*

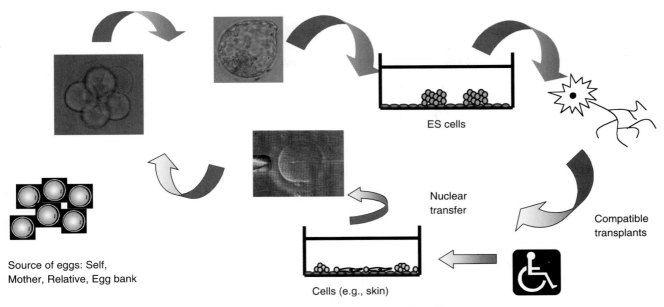

ES cells

Nuclear
transfer

Compatible
transplants

Source of eggs: Self,
Mother, Relative, Egg bank

Cells (e.g., skin)

Figure 2 *Nuclear transfer for stem cells. Cells provided by a patient are introduced into enucleated oocytes, and the inner cell mass cells of the nuclear transfer blastocyst produced are used to derive transplant-compatible ES cells. Directed differentiation of the ES cells into the desired lineage (e.g., neuronal cells) (Reubinoff et al., 2000) may be used to correct the existing pathology.*

remained the method of choice (Trounson *et al.*, 1998). The use of piezo electronic impact micromanipulation (Prime Tech Ltd or Burleigh Instruments) enables very improved efficiencies of nuclear transfer to produce viable embryos (Wakayama *et al.*, 1998). This method was used to produce nuclear transfer embryos for ES cell production in all the studies reported to date.

There were no obvious differences between strains of mice used as donor cells nor between cumulus cells or fibroblasts, in the development to blastocysts, establishment of ES cell lines, and number of chimeric offspring produced by combining NTSC cells with wild-type blastocysts (Wakayama *et al.*, 2001). This would suggest that the donor cell origin or tissue type has no effect on the ability to generate ES cells, and this would be an attractive strategy to explore for generation of ES cells for human cell therapies (Fig. 2). The concern that remains for NTSC in the human is the potential for tissue developmental abnormality that exists for nuclear transfer embryos. These embryos develop to blastocysts *in vitro* at rates equivalent to fertilized embryos, but have a dramatically reduced implantation rate, increased fetal growth, and increased structural abnormalities and losses, birth defects, and postnatal losses (Fig. 3) (Lewis *et al.*, 2001). Nuclear transfer embryos have been shown to have aberrant epigenetic methylation of CpG-rich islands in their genome, which may result in abnormal gene expression (Kang *et al.*, 2001; Dean *et al.*, 2001). This may explain the aberrant expression of key genes involved in implantation in bovines and the high rate of implantation failure of cloned embryos (Daniels *et al.*, 2000, 2001).

NORMALITY OF NUCLEAR TRANSFER EMBRYONIC STEM CELLS

Embryonic stem cells are frequently used for determining gene function, chimeric mice are used to produce transgenic animals with genetic alteration to the germ line. It is of interest that mouse ES cells produced from fertilized oocytes are extremely unstable in their epigenetic control of some imprinted genes (e.g., *Peg1* and *H19*), with variations in subclones for methylation and expression of these genes

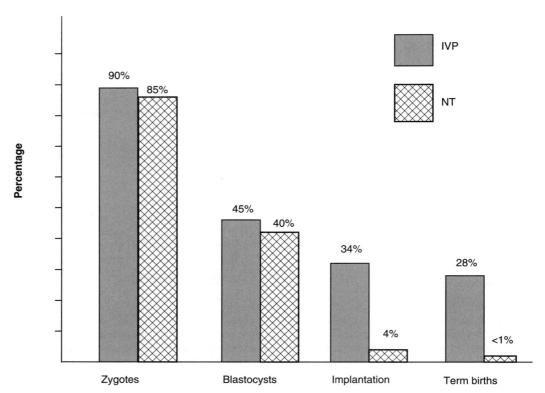

Figure 3 *Survival of in vitro-produced (IVP) and nuclear transfer (NT) bovine embryos (data from the Monash IVP and NT programs).*

(Humpherys *et al.*, 2001). However, this does not impair development of chimeric embryos, fetuses, or offspring created with ES cells and wild-type blastocysts. The epigenetic variants among ES cells are transmitted to offspring when these cells are used for nuclear transfer (Humpherys *et al.*, 2001), which explains some of the developmental abnormalities observed in cloned animals.

Given the epigenetic instability in ES cells and the abnormalities in development of cloned embryos, it is of interest that no abnormalities have been reported in nuclear transfer embryonic stem (NTES) cell lines and chimeric fetuses or offspring. The same cells used for cloning showed that only 50% of the live-born young survived to adulthood (Wakayama *et al.*, 2001), which is the usual postnatal loss observed for cloned embryos. This was not substantially altered by creating nuclear transfer embryos with tetraploid cells (Humpherys *et al.*, 2001). This combination of cell types allows developmentally competent transgenic ES cell lines to form the fetal tissues and relegate the tetraploid components to extraembryonic membranes (Nagy *et al.*, 1993).

A detailed analysis of embryo production and normality of development of cloning of mice from inbred and F_1 hybrid ES cell lines was reported by Eggan *et al.* (2001). It was apparent that the rate of blastocyst formation was the same for inbred and F_1 ES cell donor nuclei, but more pups were born from F_1 ES cell nuclei (17%) than from inbred nuclei (8%). None of the inbred pups survived for long after birth, whereas 78% of the F_1 pups survived to adulthood. These data strongly argue for a role for increased heterozygosity for survival of nuclear transfer embryos. There was no benefit for constructing embryos of the inbred strain ES cells with tetraploid cells. The majority (14/15) of pups were unable to sustain respiration after birth. Hence the losses and phenotypes were similar to cloned inbred mouse strains and tetraploid inbred ES cell composites, confirming the developmental incompetence of these cloned ES cell types. The situation for F_1 hybrid ES cell lines is different, and although embryonic losses are high, those that develop to term have a high expectation of development to adulthood. It is again interesting to compare these results with those of creating ES cells with fertilized wild-type blastocysts. Their rates of development to term and survival to adulthood are comparable to those of fertilized embryos.

It is possible that normal developing tissue is capable of reprogramming epigenetically unstable ES cells and even NTES cells that have aberrant epigenetic signatures remaining from their somatic origins. This may be the same mechanism that enables transdifferentiation of some adult stem cells after experimental transplantation (Blau *et al.*, 2001). Further experiments are needed to explore this hypothesis.

OOCYTE AVAILABILITY FOR NTSC

Presently nuclear transfer requires a mature oocyte for reprogramming a somatic cell nucleus. It is possible that half or less of the oocyte's cytoplasm is necessary for somatic cell reprogramming (Trounson *et al.*, 1998), and it is apparent that the proteins involved in erasure of epigenetic regulatory factors are conserved across species, at least for reprogramming early development (Dominko *et al.*, 1999). It has been suggested that xenotransfer of human somatic cells to enucleated animal oocytes could be utilized for NTSC, but this would be strongly opposed. There would also be concerns about potential developmental problems that occur because of nuclear–cytoplasmic (including mitochondrial) incompatibility, and about epigenetic aberrations that could occur because of this incompatibility.

The availability of human oocytes is a major problem for NTSC because it is likely that only 1 of 10 oocyte fusions will develop to a blastocyst and only 1 of 10 blastocysts will develop to an ES cell line (Table 1). As a result, 100 oocytes or more are needed to establish each ES cell line. Even if substantially improved, it is likely that 20–50 oocytes will be needed in the future for each NTSC line. Large numbers of human oocytes will need to be harvested for NTSC, and it is unlikely that such numbers will readily be available for "customizing" ES cell lines for individual patients.

Studies of the reprogramming factors have been initiated. In the frog, demethylation occurs as a consequence of chromatin remodeling due to maternal activity of the nucleosome-dependent ATPase ISWI protein, in cloned somatic cells (Kikyo *et al.*, 2000). Chromatin remodeling occurs after nuclear transfer and cell hybridization, and alterations to the basal transcriptional complex may facilitate loose chromatin and loss of somatic epigenetic memory (Tada *et al.*, 2001).

If the cytoplasmic nuclear reprogramming factors can be identified, these could be synthesized and used to erase epigenetic determinants that characterize somatic cells. Although factors have not as yet been identified in mammals, it is likely that they are conserved across species. These factors may need to be introduced into somatic cells to be effective in producing a pluripotential cell line with desired genomic compatibility for transplantation. The benefit of using reprogramming factors to reconstitute a pluripotential cell line for a patient, rather than the complete oocyte, is the additional homology of mitochondrial DNA (mtDNA). There are concerns that heterogeneous mtDNA produced by nuclear transfer of somatic cells into donor oocytes could produce unpredictable outcomes for mitochondrial function, because of nuclear protein interactions with mitochondria.

STEM CELL HYBRIDIZATION

It has been shown that mouse embryonic germ (EG) cells and mouse ES cells can be used to reprogram somatic cells by cell hybridization (Tada *et al.*, 1997, 2001).

The hybrid cells are able to contribute to the formation of endoderm, mesoderm, and ectoderm when injected into wild-type blastocysts. This pluripotent plasticity is also shared with neuronal stem cells (Clarke *et al.*, 2000). EG cells are able to reprogram somatic methylation patterns, but ES cells are unable to reprogram parental imprints, when used to hybridize with somatic cells. Given the apparent dominance of EG and ES cells in their ability to reprogram the epigenetic status of somatic nuclei, this could be used to induce pluripotency in adult somatic cells of patients, without the need to resort to the use of oocytes for NTSC.

It is known that the quantitative expression of the POU transcription factor Oct4 determines pluripotency in the mouse (Niwa *et al.*, 2000). In mouse ES cells, a two-fold increase in expression causes differentiation into endoderm and mesoderm, whereas repression induces a loss of pluripotency. In the majority of ES cell–somatic cell hybrids, Oct4, coupled with green fluorescent protein (Oct4-GFP), was expressed (Tada *et al.*, 2001) and equates with the pluripotency observed for the hybrid cells in chimeric embryos. The primary methylation imprints for *H19* and *Igf2r* genes were maintained. In this study the full epigenetic profile of developmental genes was not determined, so it is not possible to predict the developmental competence of the stem cell hybrids. It will be necessary to remove the ES cell nucleus to determine developmental normality of ES cell–somatic cell hybrids in due course.

CONCLUSIONS

NTSC, potentially a very important strategy for the production of pluripotent stem cells, can be directed into a wide range of tissue types (Trounson, 2001). The proof-of-concept studies for NTSC have been done in mice, and it is of interest to note that the ES cell lines produced contribute normally to all the tissues of chimeric offspring, including the germ line. The developmental abnormalities observed in cloned embryos, fetuses, and offspring are not observed in the chimeric offspring or in their NTSC genomic progeny. This raises the hypothesis that the normal tissues of wild-type embryos may reprogram epigenetic instability or epigenetic remnants of the original somatic cells. This may be analogous to the epigenetic reprogramming observed in transdifferentiation.

The primary problem for NTSC is the need for large numbers of mature oocytes to serve as recipients of somatic cell nuclei. This may be solved by epigenetic engineering using reprogramming factors isolated from oocytes or synthesized by molecular techniques. It is also likely that hybridization of somatic cells with stem cell cytoplasm will enable pluripotent stem cells to be derived for individual patients in the future. These techniques may be the basis for cell therapies in the future for a wide range of degenerative disorders, serious pathologies, and tissue damage.

REFERENCES

Blau, H. M., Brazelton, T. R., and Weimann, J. M. (2001). The evolving concept of a stem cell: Entity or function? *Cell* **105**(7), 829–841.

Clarke, D. L., Johansson, C. B., Wilbertz, J., Veress, B., Nilsson, E., Karlstrom, H., Lendahl, U., and Frisen, J. (2000). Generalized potential of adult neural stem cells. *Science* **288**(5471), 1660–1663.

Daniels, R., Hall, V., and Trounson, A. O. (2000). Analysis of gene transcription in bovine nuclear transfer embryos reconstructed with granulosa cell nuclei. *Biol. Reprod.* **63**(4), 1034–1040.

Daniels, R., Hall, V. J., French, A. J., Korfiatis, N. A., and Trounson, A. O. (2001). Comparison of gene transcription in cloned bovine embryos produced by different nuclear transfer techniques. *Mol. Reprod. Dev.* **60**(3), 281–288.

Dean, W., Santos, F., Stojlcovic, M., Zakhartchenko, V., Walter, J., and Reik, W. (2001). Conservation of methylation reprogramming in mammalian development: Aberrant reprogramming in cloned embryos. *Proc. Natl. Acad. Sci. U.S.A.* **98**(24), 13734–13738.

Dominko, T., Mitalipova, M., Haley, B., Beyhan, Z., Memili, E., McKusick, B., and First, N. L. (1999). Bovine oocyte cytoplasm supports development of embryos produced by nuclear transfer of somatic cell nuclei from various mammalian species. *Biol. Reprod.* **60**(6), 1496–1502.

Eggan, K., Akutsu, H., Loring, J., Jackson-Grusby, L., Klemm, M., Rideout, W. M., 3rd, Yanagimachi, R., and Jaenisch, R. (2001). Hybrid vigor, fetal overgrowth, and viability of mice derived by nuclear cloning and tetraploid embryo complementation. *Proc. Natl. Acad. Sci. U.S.A.* **98**(11), 6209–6214.

Humpherys, D., Eggan, K., Akutsu, H., Hochedlinger, K., Rideout, W. M., 3rd, Biniszkiewicz, D., Yanagimachi, R., and Jaenisch, R. (2001). Epigenetic instability in ES cells and cloned mice. *Science* **293**(5527), 95–97.

Kang, Y.-K., Koo, D.-B., Park, J.-S., Choi, Y.-M., Chung, A.-S., Lee, K.-K., and Han, Y.-M. (2001). Aberrant methylation of donor genome in cloned bovine embryos. *Nature Gen.* **28**, 173–177.

Kawase, E., Yamazaki, Y., Yagi, T., Yanagimachi, R., and Pedersen, R. A. (2000). Mouse embryonic stem (ES) cell lines established from neuronal cell-derived cloned blastocysts. *Genesis* **28**(3–4), 156–163.

Kikyo, N., Wade, P. A., Guschin, D., Ge, H., and Wolffe, A. P. (2000). Active remodeling of somatic nuclei in egg cytoplasm by the nucleosomal ATPase ISWI. *Science* **289**(5488), 2360–2362.

Lewis, I. M., Munsie, M. J., French, A. J., Daniels, R., and Trounson, A. O. (2001). The cloning cycle: From amphibia to mammals and back. *Reprod. Med. Rev.* **9**, 3–33.

Munsie, M. J. (2000). Nuclear transfer in the mouse: Reprogramming to pluripotency. *In* "Institute of Reproduction and Development." Monash University, Melbourne.

Munsie, M. J., Michalska, A. E., O'Brien, C. M., Trounson, A. O., Pera, M. F., and Mountford, P. S. (2000). Isolation of pluripotent embryonic stem cells from reprogrammed adult mouse somatic cell nuclei. *Curr. Biol.* **10**(16), 989–992.

Nagy, A., Rossant, J., Nagy, R., Abramow-Newerly, W., and Roder, J. C. (1993). Derivation of completely cell culture-derived mice from early-passage embryonic stem cells. *Proc. Natl. Acad. Sci. U.S.A.* **90**(18), 8424–8428.

Niwa, H., Miyazaki, J., and Smith, A. G. (2000). Quantitative expression of Oct-3/4 defines differentiation, dedifferentiation or self-renewal of ES cells. *Nat. Genet.* **24**(4), 372–376.

Reubinoff, B. E., Pera, M. F., Fong, C. Y., Trounson, A., and Bongso, A. (2000). Embryonic stem cell lines from human blastocysts: Somatic differentiation in vitro. *Nat. Biotechnol.* **18**(4), 399–404.

Tada, M., Tada, T., Lefebvre, L., Barton, S. C., and Surani, M. A. (1997). Embryonic germ cells induce epigenetic reprogramming of somatic nucleus in hybrid cells. *EMBO J.* **16**(21), 6510–6520.

Tada, M., Takahama, Y., Abe, K., Nakatsuji, N., and Tada, T. (2001). Nuclear reprogramming of somatic cells by *in vitro* hybridization with ES cells. *Curr. Biol.* **11**(19), 1553–1558.

Thomson, J. A., Itskovitz-Eldor, J., Shapiro, S. S., Waknitz, M. A., Swiergiel, J. J., Marshall, V. S., and Jones, J. M. (1998). Embryonic stem cell lines derived from human blastocysts. *Science* **282**(5391), 1145–1147.

Trounson, A. (2001). The derivation and potential use of human embryonic stem cells. *Reprod. Fertil. Dev.* **13**, 523–532.

Trounson, A., and Pera, M. (2001). Human embryonic stem cells. *Fertil. Steril.* **76**(4), 660–661.

Trounson, A., Lacham-Kaplan, O., Diamente, M., and Gougoulidis, T. (1998). Reprogramming cattle somatic cells by isolated nuclear injection. *Reprod. Fertil. Dev.* **10**(7–8), 645–650.

Wakayama, T., Perry, A. C., Zuccotti, M., Johnson, K. R., and Yanagimachi, R. (1998). Full-term development of mice from enucleated oocytes injected with cumulus cell nuclei. *Nature* **394**(6691), 369–374.

Wakayama, T., Tabar, V., Rodriguez, I., Perry, A. C., Studer, L., and Mombaerts, P. (2001). Differentiation of embryonic stem cell lines generated from adult somatic cells by nuclear transfer. *Science* **292**(5517), 740–743.

CHAPTER 24

CURRENT RESEARCH AND COMMERCIAL APPLICATIONS OF CLONING TECHNOLOGY

Steven L. Stice

Cloning technology has made several steps forward in recent years, most notably in somatic cell cloning and in combining cloning with transgenic technologies. This chapter explores some of the immediate uses of cloning technology in medicine and agriculture and focuses on the application that may not require an extensive regulatory approval process before being used. First, cloning valuable research animals may alleviate a limited availability of highly informative genotypes while reducing the number of animals used to obtain informative results. Second, cloning and transgenics are now available for producing phenotypes in large animals to study devastating disease; rodent models have been inadequate for this purpose. Finally, cloned parent stock in the cattle and hog industries can add value, reduce waste, and produce a more uniform product, resulting in a better profits for the producer and better products for the consumer. Having a better understanding of cloning technology's potential benefits and limitations as they relate to the needs of researchers, animal producers, and consumers will facilitate implementation of the technology.

INTRODUCTION

Cloning technology promises to change our lives in the future, but what are some of the immediate uses of the technology? For the past 15 years many review articles on cloning have discussed improvements in cloning procedures and outlined some of the potential applications. Some of these review articles were written as far back as the late 1980s, and unfortunately there are still discussions of some of the same "future" uses of cloning. So have we made progress and are we using the technology more extensively today than we did back in the 1980s? Yes, we have made progress in developing the technology, but when it comes to implementing the technology, the gains are not dramatic. Today, cloning plays a relatively small role in research and commercial endeavors. In order to move the applications and implementation of cloning ahead, we must understand what the technology can do and how it can be integrated into the processes and procedures already being used in research and commercial settings.

Other chapters in this volume have described the use of cloning technology to produce therapeutics in milk and for cell replacement or therapy. These areas, along with xenotransplantation and animal agriculture transgenics, require regulatory approval in the form of clinical trials or safety studies. Stringent regulatory approval suggests that these areas will not have an immediate impact on the biomedical industry or society, but they surely will in the future. Regulatory trials or testing can last at least 3–5 years and often times longer. To date, no group has publicly disclosed regulatory trials on products derived from cloned animals.

The purpose of this chapter is to address how cloning technology can be used in shorter time horizons, or, said another way, to describe cloned animal products or uses that are not likely to require stringent regulatory approval. Two areas are to be discussed: cloned animals with or without genetic manipulation to be used in research and cloned animals used in animal agriculture. There will be examples in each of these areas to demonstrate how the technology can be implemented relatively soon to address issues in the research and commercial areas. Clear advantages of cloning technology are highlighted as well as reasons why some potential uses of cloning will lag behind in biomedicine and agriculture.

CLONING RESEARCH ANIMALS

Animal research is expensive, not only monetarily, but also in the time and effort that are obviously necessary to justify the use of any animal in research. Therefore, it is in the best interest of everyone, from the investigator to the public, to eliminate animal studies when possible. When animal studies are needed, however, we should reduce the number of animals used, if possible. It is estimated that large lab animals such as rabbits, dogs, and primates cost $2000 to more than $5000 per animal per year to maintain. The cost of maintaining these animals over their lifetime often exceeds the purchase price of the animal 10-fold or more, thus often putting researchers contemplating long-term studies in difficult situations. The ability to produce animals that can provide answers to important biomedical studies, in less time and with fewer animals, will benefit all. This can be accomplished in a number of ways, but one very appealing option is to eliminate genetic variation as a variable.

In 1986, Biggers demonstrated the statistical power of using genetically identical animals in research. In this study, uniformity among animals was measured and assigned estimates of the interclass correlation coefficients. Traits such as body mass, muscle deposition, and milk production, to list a few, are highly heritable and have coefficients often greater than 0.90. Using these coefficients, one can determine the number of randomly assigned animals potentially replaced with sets of clones. For example, two clones could replace 40 randomly chosen animals in experiments on traits with greater than a 0.90 coefficient, without loss of statistical significance. The number of clones needed for experimentation varies depending on the coefficient. Obviously, cloned animals can replace more genetically dissimilar animals in studies when the trait being investigated is highly heritable. This reduction in experimental animal use is accomplished without losing statistical significance.

An efficient means of producing nonhuman primates for research would expedite progress toward cure and prevention of many human diseases; production of vaccines against pathogens such as the human immunodeficiency virus (HIV) could greatly benefit. Nonhuman primates are valuable models for testing the efficacy of candidate vaccines and treatment for HIV and also for fertility-regulating agents (immunocontraceptives).

It is known that susceptibility to certain diseases differs within subpopulations, both in human and in nonhuman primates. In human immunodeficiency virus type 1 (HIV-1)-infected individuals, disease progression varies considerably. These same phenomena were observed after experimental infection of macaques with simian immunodeficiency virus (SIV). Major histocompatibility complex (MHC) genes may influence disease progression in primate species. Homozygosity for *Mhc-Mamu (Macaca mulatta)-DQB1*0601* is associated with rapid disease progression in SIV-infected macaques. Six unrelated monkeys homozygous for *Mamu-DQB1*0601* and *DRB1*0309-DRB*W201* and six heterozygous monkeys were infected with

SIVmac. Five of the homozygous and only one of the heterozygous monkeys died soon after infection, with manifestations of AIDS. These results were validated by a retrospective survival analysis of 71 SIV-infected monkeys (Sauermann *et al.*, 2000). Now, the *Mhc-Mamu* homozygous animal is a highly desirable animal model for vaccine development because more information can be obtained in less time than with heterozygous animals. However, there are reports of 10-year waiting lists for some of these animals (Don Wolf, personal communication). Obviously, the waiting list eliminates many of the advantages gained with this animal. Efficient means of cloning monkeys homozygous for *Mhc-Mamu (Macaca mulatta)-DQB1*0601* would not only reduce the backlog but would also reduce variation among cohorts, as demonstrated in a study based on the statistical model of Biggers (1986). This could mean further reductions in the total number of animals need in a study to obtain statistical significance.

Cloning of adult animals of known phenotype is also an important advantage on which to capitalize in order to advance our understanding of disease states. For example, there are a number of animal models for familial hypercholesterolemia. In swine and other animals, including humans, this phenotype is characterized by elevated levels of total plasma cholesterol and apolipoprotein B (apoB), and reduced levels of high-density lipoprotein (HDL) cholesterol and apoA-I are associated with the development of spontaneous atherosclerotic lesions. In one study, concentrations of plasma lipids and apolipoproteins B, C-III, and E in six parental animals of two cholesterol concentration phenotypes and their 32 offspring were segregated into high, intermediate, and normal cholesterol phenotypes (Hasler-Rapacz *et al.*, 1995). The extent of atherosclerotic lesion development in coronary arteries was compared to the concentrations of plasma lipids and apolipoproteins in the parents and two offspring per family. The histological analysis of the major coronary arteries from members of the three families showed considerable variation in the severity of lesions, ranging from foci of adaptive intimal thickening consisting of two to six layers of smooth muscle cells to advanced lesions containing necrotic cores, cholesterol clefts, calcification, and hemorrhage (type V). This variation within a family indicates a polygenetic trait and possible environmental interactions. Therefore, in order to use this informative pig model via traditional breeding for propagation, individual offspring must be screened before becoming eligible for a particular study. The cloning of specific phenotypes may reduce the number of animals initially produced, and depending on how highly heritable the trait, reduce the number of animals needed per study. The potential cost savings could be immediate, reducing costs by over $2000 or more (animals, food, and housing). In addition to familial hypercholesterolemia, other desirable animal models that have multiple gene interactions could be enhanced by cloning now and in the future. The role of gene product interactions in phenotype is becoming more evident as we learn more about the genome. It appears that we have fewer genes than previously believed, and the larger effect is how gene products interact to produce the end phenotype. By starting with a uniform and stable genotype (clones), much will be discovered faster and easier. Cloning also allows researchers to study environmental and genetic interactions, an increasing interest of scientists everywhere.

CLONED TRANSGENIC LARGE ANIMALS AS RESEARCH MODELS

As mentioned in earlier chapters, the use of cloning in combination with gene insertion and/or knockout is a powerful tool. Live cloned calves produced as a result of nuclear transfer with cultured transgenic fetal cell lines demonstrate the possibility of producing cloned transgenic cattle (Cibelli *et al.*, 1998a). Gene targeting, using homologous recombination in fetal ovine fibroblast cells and cloning, is now a reality (McCreath *et al.*, 2000). The advantage of using nuclear transfer is that genetically transformed cells can be selected *in vitro*, and only cells with stably integrated transgenes are used as donor cells. All of the animals created from such cells should be transgenic.

Today, work is underway to produce through cloning and transgenics both a nonhuman primate and a farm animal model for ataxia telangiectasia (A-T). Ataxia telangiectasia is a genetic disease that causes degeneration of neurons in the cerebellum, dilated blood vessels in the eyes and face, and an increased risk of cancer. One or more mutations in the ATM (for ataxia telangiectasia mutated) gene cause A-T (Savitsky *et al.*, 1995). Affected heterozygotes exhibit an increased risk of cancer, particularly breast cancer, whereas homozygotes exhibit the full range of symptoms (Kastan, 1995).

The ability to understand and combat a human disease such as A-T is often dependent on finding or developing an informative animal model. Mice are the species of choice because of the wealth of information known about the mouse genome, and a multiple homologous recombination strategy for disrupting genes has already been demonstrated in mice. Also, generation times are short in mice, and animal availability and maintenance are not problems, thus reducing cost compared to other laboratory animals. Unfortunately for A-T sufferers and researchers, and those with a number of other human diseases (cystic fibrosis, etc.), mice may not be good animal models, because transgenic mice do not always have the same phenotype observed in humans. Three groups investigating A-T have mutated different exons of the mouse ATM gene (Barlow *et al.*, 1996; Elson *et al.*, 1996; Xu *et al.*, 1996), producing null mutant mice that exhibited some of the same symptoms observed in A-T sufferers. The most common symptom is the development of cancer, which kills most of the mice between 3 and 10 months of age (Rotman and Shiloh, 1998). The degeneration of Purkinje cells of the cerebellum and the profound ataxia seen in patients were not present in most of the mouse mutants, despite some abnormalities of movement (Barlow *et al.*, 1996) and subtle signs of degeneration in the cerebellum (Kuljis *et al.*, 1997). The reason ATM knockout mice may not exhibit the same severity of cerebellar degeneration seen in humans with mutations in this gene is that patients do not begin to exhibit ataxia until 2 years of age, and the mutant mice are often dead within 3 months.

No one presently knows why the A-T mouse does not present all the phenotype of the human disease. In comparative mapping of the genome it appears that the organization of the genome is more similar among cattle, pigs, and humans than between mice and humans (Kappes, 1999). Another reason may be the large disparity in the temporal and spatial developmental rates between mice and humans. Cattle and humans have the same gestation length (approximately 9 months), whereas for mice, it is only 21 days. Ideally, a nonhuman primate model would be desirable, and this work is progressing in Don Wolf's laboratory. Another alternative is to produce a pig and/or cattle model for A-T. Because pigs and cattle have temporal and spatial embryonic development similar to that of humans and a longer life span than mice have, the expectation is that they will exhibit the neurodegenerative phenotype, severe cerebellar degeneration, and ataxia, characteristic of A-T patients. The hope is that if produced, the cloned ATM knockout nonhuman primate, cow, or pig will provide insights into the disease that are not currently possible with mutant ATM knockout mice.

Other factors in considering which species to use are gestation length and how fast animals of a specific genotype can be multiplied (Table 1). Clearly, the most advantageous species when considering reproductive efficiency is the pig. A short gestation length and large litters make pigs ideal candidates. Also, miniature breeds of pigs can be back-crossed to achieve animal sizes that are more easily managed in animal care facilities. Initially, sheep may be a viable animal model. However, a major disadvantage is that sheep are seasonal breeders. Therefore, research on sheep cloning and transgenics is limited to 4–6 months a year unless unusual and costly management practices are employed. Although cattle have long reproductive cycles,

Table 1 Reproductive Efficiency of Farm Animals

Species	Gestation	Number of offspring	Sexual maturity	Seasonal breeding
Pig	4 months	7–10	7 months	No
Cattle	9 months	1	10 months	No
Sheep	5 months	1–2	7 months	Yes

the research can be conducted just about any time of year. As mentioned above, the gestation length of cattle also more closely mimics that of humans, and cattle produce just one offspring per pregnancy.

The other factor to be considered besides reproductive efficiency is the ease in making the gene-targeted farm animal. Table 2 summarizes the state of cloning technology today. A problem currently encountered in making the A-T model is that most transgenic farm animals have been produced using somatic cells to provide the donor nucleus rather than embryonic stem (ES) cells. Intuitively, ES cells would be better candidate cells, because A-T is an autosomal recessive disease requiring disruption of both copies of the ATM gene to produce the desired phenotype in cattle and/or pigs. Stem cells by definition are immortal, and therefore are more easily genetically manipulated so that both genes are knocked out initially. In contrast, when using primary somatic cells, the cells in culture may enter senescence prior to selection of those having a mutated second copy.

Mouse ES cells have been used to produce cloned mice. However, we have extensive experience in cattle and pig ES-like cells in both gene modification and cloning (Stice *et al.*, 1996; Cibelli *et al.*, 1998b), and after extensive nuclear transfer studies, we were never able to produce a cloned offspring derived from our ES-like cell lines (Stice *et al.*, 1996). Second, ES-like cells in cattle (Cibelli *et al.*, 1998b) and nonhuman primates (Thomson and Marshall, 1998) are not highly amenable to gene modifications because of the laborious culture systems required to maintain them as undifferentiated cells and the inability to propagate these cells clonally *in vitro*. Therefore, the use of somatic cells, given their past (Cibelli *et al.*, 1998a) and recent successes in gene targeting (McCreath *et al.*, 2000), is currently the best alternative to obtaining a large-animal knockout model for A-T disease and other diseases.

Although somatic cells appear to be the best cell type for use in farm species for cloning and transgenics, what cell type to use and whether somatic cells from any one species are inherently easier to modify are still unknowns. However, within a species, researchers have compared the ease of genetically modifying several cell types. In cattle, sheep, pigs, and goats, fetal cells have been used for transgenic livestock animal production because of their rapid growth and potential for many cell divisions before senescence in culture (Cibelli *et al.*, 1998a; Baguisi *et al.*, 1999; Schnieke *et al.*, 1997; Kuhholzer *et al.*, 2000).

Table 2 Status of Gene Modification Using Cloning Technology in Farm Animals

Species	Clones	Transgenic clones	Knockout clones	*In vitro* or *in vivo* systems?
Cattle	Yes	Yes	Yes	*In vitro*
Sheep	Yes	Yes	Yes	*In vivo*
Pig	Yes	Yes	Yes	*In vivo*

Arat *et al.* (2001) reported that genetically modified granulosa cells from late passages could produce viable transgenic embryos at an acceptable rate. Bovine granulosa cells were transfected with a plasmid containing the enhanced green fluorescence protein gene. Both transfected and nontransfected cells were used for cloning between passages 10 and 15. There was no significant difference in cleavage rates or development to the blastocyst stage for cloned embryos from transfected or nontransfected cells. On day 7, total cell numbers in blastocyst-stage embryos derived from transfected and nontransfected donor cells were higher than those reported by Akagi *et al.* (2000), who used cumulus cells as donor material, and similar to results with adult fibroblasts on day 8 (Knott *et al.*, 2000). When compared with other cloning studies, the results from the present study indicate that transfection and selection of transgenes did not impair the developmental potential of cloned transgenic embryos. Transgenic granulosa cells might also have similar advantages in other species. Nontransgenic porcine granulosa cells were a satisfactory donor cell type for producing nuclear transfer pigs (Polejaeva *et al.*, 2000). If porcine granulosa cells behave like the bovine granulosa cells in the study by Arat and co-workers (2001), then using genetically manipulated granulosa cells to produce cloned transgenic pigs may be preferable compared to using fibroblast cells. Granulosa cells may provide a viable alternative cell type for gene modification, especially when female offspring may be more desirable than males. In addition, Kubota and co-workers (2000) demonstrated that adult somatic cells remained totipotent after long-term culture; thus, it may be possible to produce nuclear transfer embryos and offspring derived from genetically manipulated donor cells of late passage number.

Last, ease of cloning the individual species must also be considered. Of the farm species, studies using sheep and cattle cells have advanced the farthest. To date, more groups have cloned cattle than sheep, probably due to the ease in obtaining material. In cattle, both the oocytes used to produce offspring and the culture methods are all standard *in vitro* techniques. In contrast, most sheep cloning work was accomplished using *in vivo* oocytes, *in vivo* culture techniques, and surgical embryo transfer, which usually requires surgical procedures on large numbers of sheep. In contrast, all the cattle cloning procedures are carried out *in vitro* and nonsurgical embryo transfer is used. Several groups have now produced cloned pigs (Onishi *et al.*, 2000; Polejaeva *et al.*, 2000; Betthauser *et al.*, 2000), but in some cases, the methods described are more laborious than those for cattle and sheep. *In vitro* material has been used to produce cloned pigs, but in order to achieve this feat, over 100 embryos were transferred to a recipient to establish a pregnancy. Therefore, the pig is farther behind in development, but improvements are being made.

The above analysis indicates that pigs, sheep, and cattle have distinct advantages and disadvantages for animal models. Once produced, viable pig offspring with ATM gene knockout can be propagated faster than counterparts in cattle, and therefore move faster into studies to test treatments for A-T. However, cloning in cattle is more advanced than in pigs and more is known about the embryology and genome of cattle than any other farm animal (Kappes, 1999). Therefore, the most expedient course of action to obtain gene-targeted large animals may be to pursue more than one species simultaneously. Currently, research is progressing in making both knockout cattle and pigs for A-T disease.

PREVIOUS AND PRESENT OBSTACLES TO COMMERCIAL CLONING IN ANIMAL AGRICULTURE

Cloning promises to improve the quality of genetic stock available to the market, by developing better systems for large-scale animal breeding. It can compress the time to develop and distribute elite genotypes, enabling more rapid genetic improvements. This promise was first made back in the late 1980s, when the first commercial cattle embryo cloning companies were established. The three main companies at that time were Granada Biosciences, American Breeders Service, and Alta Genetics. All three companies made similar commercialization forecasts of cloning technology. However, none of them has been able to fulfill these promises to the extent that they or their customers desired.

There were three basic commercialization obstacles that became apparent in embryo cloning research and development. The first was low cloning efficiencies. There are multiple steps in the cloning process, and in order to produce a large number of embryos and or offspring economically, one must be able to do that efficiently. There were some improvements in enucleation, fusion, and embryo development rates (Stice and First, 1993); however, pregnancy rates did not surpass the 25% level, making it cost prohibitive for the average dairy or beef producer. Anytime a cow is not pregnant, the producer is incurring operational costs without hope of recouping them. For a commodity industry, such as beef or dairy cattle production, this is disastrous. Second, embryo cloning meant taking 16- to 32-cell-stage embryos and making a limitless number of embryos. This of course sounds impossible to start with, because at best one might get 32 clones from one 32-cell embryo. However, the hypothesis was that a cloned embryo could be grown to the 32-cell stage and then used again in the cloning procedure. This was dubbed "multiple generational cloning," or "recloning" for short. If cloning efficiencies are high, embryos could be multiplied exponentially with each generation. It was envisioned that thousands of embryos could be produced in relatively few generations. This proved impossible. As mentioned previously, the cloning process was not very efficient, and the second problem was that these efficiencies went down with each subsequent generation (Stice and Keefer, 1993). None of the three companies could produce viable pregnancies from greater than three generations of cloning. The final and most problematic obstacle to date is the health issues encountered with some of the cloned calves. Large birth weights and cardiovascular, pulmonary, liver, and kidney anomalies were present with embryo cloning (Willadsen *et al.*, 1991). These major obstacles could not be overcome soon enough to make embryo cloning commercially viable.

APPLICATION OF EMBRYO CLONING IN THE DAIRY INDUSTRY

Although embryo cloning technology could not produce large numbers of viable cloned embryos, the commercialization strategies that were developed while the research was carried out showed how cloning of embryos and, now, adult animals, can be implemented. The dairy industry relies heavily on progeny testing to determine the genetically superior bulls in the bull stud operations. In its most simplistic form, the bulls' daughters' milking records are compiled and averaged and a predictive index is calculated to forecast how well the offspring of the bull will perform. As the number of daughters increases, the reliability of the index also increases. Basically, 1 in 10 bulls that already have a good pedigree will be good enough to reach the bull lineup for wide distribution of their germplasm. In embryo cloning, the commercial bull stud operations envisioned production testing the female clonal lines. With this procedure, female embryos are cloned, producing 5–10 offspring. These animals would then go through a production-testing regime while additional cloned embryos would remain in cryoprotective storage. Many female clonal lines would be tested, possibly hundreds, and a certain percentage of these would produce superior milking records on average. The embryos from those superior females could then be thawed and used in the recloning process to produce thousands of clones to sell to dairy producers. The genetic improvement achieved by such a scenario could mean hundreds of dollars in increased milk production per lactation. But the production testing of embryo clones was never implemented because of the cloning difficulties mentioned previously.

Today, if efficient somatic cell cloning rates are achieved, a proved phenotype (adult cells) could be used as the initial donor material, and the strategy used for embryo cloning could be implemented in the dairy industry with even greater genetic gain. Several clonal lines would still be tested to gain the statistical significance needed, because environmental variation can never be eliminated. Thus, for every female cloned, a set of females would enter production testing in order to determine the genetically superior genotypes. Adult cloning is advantageous because instead of cloning embryos obtained from genetically superior parent stock, the parent stock, with its own production records, is used to produce additional animals for production testing, thus generating a more accurate predictive index. This provides better predictability of marketplace performance of the clonal lines.

So why are cloned dairy females not entering production-testing programs today? The reason is that two of the three problems that plagued the embryo cloning commercialization are still unsolved. The problems associated with recloning have been eliminated because with somatic cell cloning, there is a limitless number of cells that can be collected or propagated from one genotype, thus providing the possibility of an unlimited number of clones without requiring recloning. However, the inefficiency and prenatal and neonatal mortality problems persist today, and in some cases are more pronounced. In the dairy industry, the high price of low pregnancy rates is dramatic. Maintaining an "open" (nonpregnant) cow can cost approximately $5 a day. So in order to place cloned embryos in production dairy cattle, the pregnancy rates and outcome are very important. Therefore, unless the remaining two problems are solved, it is unlikely that cloning will be used on a large scale in the dairy industry.

LIMITED CLONING OF RARE OR HIGH-VALUE CATTLE

There have been several news reports of commercial cloning of high-value animals, including some from our laboratory (Walker, 2001). This is feasible because some elite cattle breeders place a high price on these genetics. In the popular press, cloned cattle have been sold for prices ranging from $63,000 to $100,000 (InfiGen, 2000; press release). These high prices are not limited to cloned cattle. Registered elite animals have values of up to $1 million per animal, and their offspring sell for up to $200,000.

The value of a registered animal is not clear until several offspring have been produced, and in some cases the full genetic value of these animals is not realized until late in their reproductive life. Cattle are fertile only until approximately 12 years old. This makes a short interval between the time when it is determined that an individual animal produces high-value offspring and the end of its fertility, thus limiting the potential to produce high-value offspring, especially in females. In addition, some of these females do not respond well to superovulation and embryo transfer. This timing issue is less important for bulls, but is still important. Cloning can potentially extend the fertile life of an elite cow by producing a clone of that high-value animal, given that the clone has a full reproductive life. Today, a few commercial companies provide a cloning service to breeders of registered animals, and later employ genetic engineering to enhance rapidly the existing elite breeds with desirable traits. Additionally, companies will store tissue (i.e., the genetics) of elite cattle in preparation for cloning services. Breeders send skin, saliva, or other cell samples of elite cattle to the lab to be stored for use when effective cloning has been established in cattle. This is an economically important service to owners of live high-value animals that are reaching the end of their lives.

CLONING BEEF CATTLE

Pregnancy rates and outcomes with cloned embryos are problematic in the beef industry, but less so than in the dairy industry because of how the industry might envision using cloning technology. Unlike in the dairy industry, clones or their products will not be produced to enter the market directly. There are several reasons for this, and the reasoning requires an understanding of how the beef industry operates. First, the beef industry does not rely on artificial insemination to the great extent that the dairy industry does. This is a management issue. Because dairy animals are handled each day for milking, it is easier to implement a technology that is labor intensive, such as artificial insemination. Not only must the cow be inseminated twice, but she must also be observed twice daily to know when to inseminate. In the beef industry cows are not managed on a daily basis, and if cows must be handled often for estrus synchronization, heat detection, and insemination, the payback in better offspring has to be dramatic. Therefore, less than 5% of beef cows in the United States are artificially inseminated. The cost of this management is usually only justified to produce replacement females for breeding stock. In other words, artificial insemination is used primarily to produce parent-breeding stock.

Cloning bulls, instead of artificial insemination, may be cost-effective. The ability to clone and distribute elite bulls, rather than the germplasm from elite bulls, seems counterintuitive. However, a cloned bull instead of one derived from artificial insemination may be cost-effective because it eliminates the management costs (estrus synchronization, heat detection, and insemination) associated with obtaining high-value germplasm through artificial insemination. The expectation is that in addition to the added value through lower labor cost, using a cloned elite bull as the sire for the generation going into the feedlot can generate an additional $100 revenue per animal. This is primarily due to the tremendous selection pressure on the elite animal that will be cloned and the market animal's improved uniformity. Much like production testing of dairy females, bull stud operations and producers already collect production data that may have predictive values for many other traits, including reproductive traits, and having more clones of these animals adds reliability to the indexes. In addition, the technology may allow not only existing elite bulls to be cloned but steers as well. Steers' performance through the production environment may be monitored as well as their end carcass characteristics. The top steer having cells in storage could then be cloned. This may provide additional strength to predicting how the offspring from the cloned animal will perform in various production environments.

Cloning elite beef bulls or steers will be perhaps the most widespread use of cloning technology in the next few years. Initially, high-value bulls can be cloned and distributed to replace artificial insemination processes for producing heifer replacements, the next generation of genetic providers. This approach has multiple steps that encourage broader market coverage as efficiencies increase and costs decrease. In the next phase, when economically feasible, cloned bulls or steers will service females to produce the next generation of feedlot steers. When this occurs, it will be important to have developed alliances across the marketplace so that animal breeders can realize some of the downstream value of cloning, both in the feedlot and when the animals go to market.

BEEF INDUSTRY

As it develops, cloning technology will encourage further alliances and consolidation within the beef industry. Besides advances in research and development, the industry must overcome the impediment of a highly fragmented industry and market. In the United States, there are over 800,000 beef cattle producers, with the average farm producing about 30 animals (Table 3). It is anticipated that some

Table 3 Beef Operations in the United States

Herd size (heads)	Number of herds	Percentage of total
1–49	635,112	78.9
50–99	96,592	12.0
100–199	45,439	5.6
200–499	22,119	2.7
500–1,000	3,942	0.5
<1,000	1,391	0.2
Total	804,595	

consolidation in this industry will occur over time, but it will never follow the consolidation process that has occurred in the poultry and hog production industries. Large producers' consolidation efforts are limited because land is such a major cost factor. No producer can afford all the land it would take to control large portions of the market. Table 4 illustrates the distribution of the global beef cattle market for 1996. Although production methods do vary among countries and regions of the world, consolidation and/or alliances worldwide will also be the most effective way to realize fully the advantages of cloning technology for the beef industry.

CLONING HOGS FOR ANIMAL AGRICULTURE

Unlike cattle cloning, pig cloning using somatic cells has not been as widely achieved (Onishi *et al.*, 2000; Polejaeva *et al.*, 2000; Betthauser *et al.*, 2000), and the full potential for cloning in the hog industry is just now being investigated. Most, if not all, of the research and development support for pig cloning has been to develop xenotransplantation applications. Novartis, a major pharmaceutical company, supports some of this research. They suggest that the xenotransplant market could someday reach $6 billion. However, as mentioned previously, this application will require extensive and rigorous regulatory approval before commercialization. So, as with cattle, can advances made in pig cloning be utilized by the pig industry more immediately? In recent years the hog industry has placed great value on genetics, and most modern pig operations utilize artificial insemination. These two technical developments combined with consolidation within the industry suggest that pig cloning may indeed be of value to the hog industry.

HOG INDUSTRY

Approximately 44% of the global meat supply comes from pigs. In 1997, in the United States alone, 92 million hogs were sold for $11 billion, resulting in a retail

Table 4 1996 World Beef Cattle Slaughtered[a]

Region	Beef cattle (thousands)	Percentage of world cattle
Brazil	149,228	14.3
China	132,058	12.7
United States	103,487	10.0
Other	653,220	62.9
Total world	1,037,993	

[a]Source: USDA Foreign Agricultural Service.

Table 5 1997 World Hog Production[a]

Region	Hogs (thousands)	Percentage of world hogs
China	560,000	54
European Union	187,589	18
United States	91,961	9
Other	193,165	19
Total world	1,032,715	100

[a]U.S. National Pork Producers Council, Pork Facts 98.

market of $30 billion. Globally, more than 1 billion hogs were slaughtered in 1998. In terms of volume, little annual growth in the United States is expected, but perhaps 1.5% globally is anticipated in the period 1998 to 2010, driven by the increased pork consumption predicted for the Asian region. The value of the world hog market is expected to exceed $100 billion in 2010. Table 5 illustrates the distribution of the global hog market. China is the largest market, Europe is second, and the United States is third. No other market holds more than 2% of world production. The hog market in the United States has consolidated considerably in the past 20 years, reflecting customer demand for lean meat, available only through elite breeds supplied by breeders, and the cost advantage created by the use of feeding and breeding technology that is most effective at large scale. It is anticipated that this trend will continue and that further consolidation will occur in the hog production market.

Hog production involves a number of stages, each successive stage or level of parent stock involving less technology and adding less value. Hogs in the first stage, often called the great-great-grandparent stock, are of great value, because these are the breeding pigs with elite genetics. The largest player in elite pig genetics is the Pig Improvement Company (PIC), whose market includes approximately 20 million hogs, generating $374 million in revenues. Table 6 is a summary of the largest genetics supply companies in the hog industry. Genetics suppliers can affect a producer's profitability and provide a strategic input to production. The access to breeding stock also affects expansion of production and repopulation efforts for large producers. If there are disease problems in the genetic providers' lines, production is

Table 6 Major Global Hog Breeders, 1997

Company	United States share (%)	Worldwide share (%)	Worldwide sales (× $1000)
PIC (independent)	20–25	7	363
NPD (owned by Smithfield Foods and Carrolls Foods)	3–5	≤2	NA[a]
DeKalb	6	≤2	56
Newsham	6	≤2	<50
Cotswald	≤2	≤2	<50
Danbred	≤2	≤2	<50
Other	4	8	NA
Bred on farms	50	73	NA

[a]NA, Not available.

jeopardized. Large producers might find cloning their own animals more efficient, because those animals are more likely to be resistant to infectious agents that may arise from time to time in that environment. It may also be a better alternative to purchasing breeding animals when expanding or repopulating herds.

Elite pigs are used to generate far larger quantities of pigs through cross-breeding. It takes 4 years or more to disseminate elite genetics broadly by breeding a large number of boars through generations, for large-scale artificial insemination of sows. Current genetics companies provide the genetics via live animals or semen and the cost of genetics in the final hog in the United States market can range from $2.00 to $5.00 per market animal produced, depending on the efficiency and scale of the operation. The largest producers (top 20) receive substantial discounts from their genetics providers for large quantities and pay genetics fees at the lower end of the range. Large producers complete in-house much of the cross-breeding from the elite great-great-grandparent stock to parents of production-grade hogs. They also almost exclusively use artificial insemination as a means to disseminate the elite genetics. Still, genetics costs for large producers are in the range of $1.50 to $2.50. Smaller producers buy grandparent stock from their genetics providers after some scale-up by the producer, and may or may not use artificial insemination. Genetics supply costs for small producers are in the range of $2.50 to $5.00 per market pig. This can be up to 2–5% of the gross revenues from the pigs marketed, depending on the market conditions. This cost is high for a commodity.

CONSOLIDATED PIG PRODUCTION CAN CAPITALIZE ON CLONING

The large hog producers are highly vertically integrated, and the same company typically runs all of the stages, from genetics to the market place to the meat case. The hog market in the United States has consolidated considerably in the past 20 years, reflecting customer demand for lean meat, available only through elite breeds supplied by breeders, and the cost advantage created by the use of feeding and breeding technology that is most effective at large scale. It is anticipated that this trend will continue, and that further consolidation will occur in the hog production market. In the United States market, cloning could offer large-scale producers an alternative to existing genetics providers by supplying large numbers of cloned elite hogs (of the parent generation). The following points emphasize the value of pig cloning to the hog industry:

1. More valuable animals or animal products to market. This, of course, is an obvious benefit. For example, an animal that can produce high-quality pork chops efficiently can be cloned, and the clone reproduces these traits.

2. Decreased cost of production. If the high-quality pork chop is only achievable with any one factor—increased feed costs, lowered reproductive efficiencies, or slower growth rates—then these costs must be passed on to the consumer. However, cloning may permit both high quality and decreased production costs through rigorous production testing of clonal lines.

3. Elimination of costs to the existing genetics provider. This benefit offers a strategic alternative to breeding companies for supply of elite genetics, because new agricultural cloning companies have been established.

4. Value of the genetic enhancements. Because cloning can replicate the very best genetics at the top of the genetic pyramid, levels of multiplier herds can be eliminated. Therefore, instead of requiring 4 years or more to move new genetics into the market place, this period may be cut in half, allowing producers to more rapidly respond to new consumer demands.

5. Lower disease risk. Because a producer could clone more of his own animals' elite genetics, the introduction of new outside animals into the breeding herd would be reduced.

6. Ability to respond rapidly to customer and market demands. Today in the pork industry, there is a high demand for pork bellies, but tomorrow it may be hams. The ability to clone animals having specific genotypes will allow change to occur faster.

7. Ability to customize genetics to reflect specific customer production conditions. Specific genotypes are known to perform differently in different environments. Heat tolerance is one example.

8. Ability to customize genetics to produce a differentiable product. Today pork is a commodity. Cloning enables producers to differentiate their products in the marketplace. Branded products of the future may include lower cholesterol or desirable tenderness or some other characteristic deemed important.

INBREEDING AND REDUCED GENETIC VARIABILITY

An issue often raised is the rate of inbreeding or reduced genetic variability when cloning is used. With decreased genetic variability, animal production traits, including reproduction, would likely decline; therefore, cloning is a management tool that producers will need to monitor. Because inbreeding often occurs when family information is used in making breeding selections, however, the ability to make breeding choices based on other information, such as environmental and genetic interaction in market animals, will reduce the tendency toward making mating selection based on pedigree alone. Therefore, cloning is expected potentially to find new gene combinations and reduce inbreeding. As always, producers will guard against inbreeding by how they manage their animals. Some methods might include using multiple cloned sires and monitoring inbreeding levels. There are quantitative measures that producers can use to monitor and reduce inbreeding (Woolliams and Wilmut, 1989). Inbreeding will be monitored because it is in the producer's best interest to do so. The consequence of inbreeding is lower production rates, gravely affecting the thin profit margins in animal agriculture.

SUMMARY

Understanding cloning technology potential and limitations as they relate to the needs of both researchers and animal producers will facilitate implementation of the technology. Two areas in which cloning can be used immediately on a small scale and grow as the technology advances are in producing and multiplying valuable animals for research and agriculture. Cloning can reduce genetic variation and in turn reduce animal use, in addition to multiplying specific genotypes that are more informative than are random-bred animals. Cloning can greatly enhance animal agriculture today. The consolidating industry can now more fully realize the benefits of multiplying superior genotypes, and genetics can be used to produce a more consistent and higher quality product for consumers.

REFERENCES

Akagi, S., Takahashi, S., Noguchi, T., Hasegawa, K., Shimizu, M., Hosoe, M., and Izaike, Y. (2000). Development of embryos using cumulus cells for nuclear transfer. *Theriogenology* 53(1), 208.

Arat, S., Rzucidlo, S. J., Gibbons, J., Miyoshi, K., and Stice, S. L. (2001). Production of transgenic bovine embryos by transfer of transfected granulosa cells into enucleated oocytes. *Mol. Reprod. Dev.* 60, 20–26.

Baguisi, A., Behboodi, E., Melican, D. T., Pollock, J. S., Desrempes, M. M., Cammuso, C., Williams, J. L., Nims, S. D., Porter, C. A., Midura, P., Palacios, M. J., Ayres, S. L., Denniston, R. S., Hayes, M. L., Ziomek, C. A., Meade, H. M., Godke, R. A., Gavin, W. G., Overstrom, E. W., and Echelard, Y. (1999). Production of goats by somatic cell nuclear transfer. *Nat. Biotechnol.* 17, 456–461.

Barlow, C., Hirotsune, S., Paylor, R., Liyanage, M., Eckhaus, M., Collins, F., Shiloh, Y., Crawley, J. N., Ried, T., Tagle, D., and Wynshaw-Boris, A. (1996). Atm-deficient mice: A paradigm of ataxia telangiectasia. *Cell* 86, 159–171.

Betthauser, J., Forsberg, E., Augenstein, M., Childs, L., Eilertsen, K., Enos, J., Forsythe, T., Golueke, P., Jurgella, G., Koppang, R., Lesmeister, T., Mallon, K., Mell, G., Misica, P., Pace, M., Pfister-Genskow, M., Strelchenko, N., Voelker, G., Watt, S., Thompson, S., and Bishop, M. (2000). Production of cloned pigs from *in vitro* systems. *Nat. Biotechnol.* **18**, 1055–1059.

Biggers, J. D. (1986). The potential use of artificially produced monozygotic twins for comparative experiments. *Theriogenology* **26**, 1–25.

Cibelli, J. B., Stice, S. L., Golueke, P. J., Kane, J. J., Jerry, J., Blackwell, C., Ponce de Leon, F. A., and Robl, J. M., (1998a). Cloned transgenic calves produced from non-quiescent fetal fibroblasts. *Science* **280**, 1256–1258.

Cibelli, J. B., Stice, S. L., Golueke, P. J., Kane, J. J., Jerry, J., Blackwell, C., de Leon, F. A., and Robl, J. M. (1998b). Transgenic bovine chimeric offspring produced from somatic cell-derived stem-like cells. *Nat. Biotechnol.* **16**, 642–646.

Elson, A., Wang, Y., Daugherty, C. J., Morton, C. C., Zhou, F., Campos-Torres, J., and Leder, P. (1996). Pleiotropic defects in ataxia-telangiectasia protein-deficient mice. *Proc. Natl. Acad. Sci. U.S.A.* **93**, 13084–13089.

Hasler-Rapacz, J., Prescott, M. F., Von Linden-Reed, J., Rapacz, J. M., Jr., Hu, Z., and Rapacz, J. (1995). Elevated concentrations of plasma lipids and apolipoproteins B, C-III, and E are associated with the progression of coronary artery disease in familial hypercholesterolemic swine. *Arterioscler. Thromb. Vasc. Biol.* **15**(5), 583–592.

InfiGen (2000). First clone of elite cow offered for sale at world dairy expo's world classic, August 15, DeForest, WI.

Kappes, S. M. (1999). Utilization of gene mapping information in livestock animals. *Theriogenology* **51**, 135–148.

Kastan, M. (1995). Ataxia-telangiectasia—Broad implications for rare disorders. *New Engl. J. Med.* **333**, 662–663.

Knott, J. G., Kasinathan, P., and Robl, J. M. (2000). Development of nuclear transfer embryos from fetal and adult bovine fibroblasts. *Theriogenology* **53**(1), 227.

Kubota, C., Yamakuchi, H., Todoroki, J., Mizoshita, K., Tabara, N., Barber, M., and Yang, X. (2000). Six cloned calves produced from adult fibroblast cells after long-term culture. *Proc. Natl. Acad. Sci. U.S.A.* **97**, 990–995.

Kuhholzer, B., Tao, T., Machaty, Z., Hawley, R. J., Greenstein, J. L., and Day, B. N. (2000). Production of transgenic porcine blastocysts by nuclear transfer. *Mol. Reprod. Dev.* **56**, 145–148.

Kuljis, R. O., Xu, Y., Aguila, M. C., and Baltimore, D. (1997). Degeneration of neurons, synapses, and neuropil and glial activation in a murine Atm knockout model of ataxia-telangiectasia. *Proc. Natl. Acad. Sci. U.S.A.* **94**, 12688–12693.

McCreath, K. J., Howcroft, J., Campbell, K. H. S., Colman, A., Schnieke, A. E., and Kind A. J. (2000). Production of gene-targeted sheep by nuclear transfer from cultured somatic cells. *Nature* **405**, 1066–1069.

Onishi, A., Iwamoto, M., Akita, T., Mikawa, S., Takeda, K., Awata, T., Hanada, H., and Perry, A. C. F. (2000). Pig cloning by microinjection of fetal fibroblast nuclei. *Science* **289**, 1188–1190.

Polejaeva, I. A., Chen, S. H., Vaught, T. D., Page, R. L., Mullins, J., Ball, S., Dai, Y., Boone, J., Walker, S., Ayares, D. L., Colman, A., and Campbell, K. H. S. (2000). Cloned pigs produced by nuclear transfer from adult somatic cells. *Nature* **407**, 86–90.

Rotman, G., and Shiloh, Y. (1998). ATM: From gene to function. *Hum. Mol. Genet.* **7**, 1555–1563.

Sauermann, U., Stahl-Hennig, C., Stolte, N., Muhl, T., Krawczak, M., Spring, M., Fuchs, D., Kaup, F. J., Hunsmann, G., and Sopper, S. (2000). Homozygosity for a conserved Mhc class II DQ-DRB haplotype is associated with rapid disease progression in simian immunodeficiency virus-infected macaques: Results from a prospective study. *J. Infect. Dis.* **182**(3), 716–724.

Savitsky, K., Bar-Shira, A., Gilad, S., Rotman, G., Ziv, Y., Vanagaite, L., Sfez, S., Ashkenazi, M., Pecker, I., Frydman, M., Harnik, R., Patanjali, S. R., Simmons, A., Clines, G. A., Sartiel, A., Gatti, R. A., Chessa, L., Sanal, O., Lavin, M. F., Jaspers, N. G. J., Taylor, A. M. R., Arlett, C. F., Miki, T., Weissman, S. M., Lovett, M., Collins, F. S., and Shiloh, Y. (1995). A single ataxia telangiectasia gene with a product similar to PI-3 kinase. *Science* **268**, 1749–1753.

Schnieke, A. E., Kind, A. J., Ritchie, W. A., Mycock, K., Scott, A. R., Ritchie, M., Wilmut, I., Colman, A., and Campbell, K. H. (1997). Human factor IX transgenic sheep production by transfer of nuclei from transfected fatal fibroblasts. *Science* **278**, 2130–2133.

Stice, S. L., and First, N. L. (1993). Progress towards efficient commercial embryo cloning. *Anim. Reprod. Sci.* **33**, 83–98.

Stice, S. L., and Keefer, C. L. (1993). Multiple generational bovine embryo cloning. *Biol. Reprod.* **48**, 715.

Stice, S. L., Strelchenko, N. S., Keefer, C. L., and Matthews, L. (1996). Pluripotent bovine embryonic cell lines direct embryonic development following nuclear transfer. *Biol. Reprod.* **54**, 100–107.

Thomson, J. A., and Marshall, V. S. (1998). Primate embryonic stem cells. *Curr. Top. Dev. Biol.* **38**, 133–165.

Walker, A. (2001). New technique increases survival of cloned cows. At *CNN.com*, June 26, 2001. Posted at 5:03 PM EDT (2103 GMT).

Willadsen, S. M., Jangen, R. E., McAlister, R. J., and Shea, B. F. (1991). The viability of late morulae and blastocysts produced by nuclear transfers in cattle. *Theriogenology* 35, 161–170.

Woolliams, J. A., and Wilmut, I. (1989). Embryo manipulation in cattle breeding and production. *Anim. Prod.* 48, 3–30.

Xu, Y., Ashley, T., Brainerd, E. E., Bronson, R. T., Meyn, M. S., and Baltimore, D. (1996). Targeted disruption of ATM leads to growth retardation, chromosomal fragmentation during meiosis, immune defects, and thymic lymphoma. *Genes Dev.* 10, 2411–2422.

CHAPTER 25

TRANSGENIC CLONED GOATS AND THE PRODUCTION OF THERAPEUTIC PROTEINS

Esmail Behboodi, LiHow Chen, Margaret M. Destrempes,
Harry M. Meade, and Yann Echelard[1]

The use of recombinant therapeutic proteins has increased during the past two decades. Clinical applications often require large amounts of highly purified molecules, sometimes for multiple treatments. The development of very efficient expression systems is essential to the full exploitation of recombinant technology. Production of recombinant protein in the milk of transgenic goats is currently being tested for the production of a number of therapeutic antibodies as well as an alternative to plasma fractionation for the manufacture of human antithrombin. In this chapter the focus is on the potential of somatic cell nuclear transfer to generate transgenic goats used in the production of recombinant therapeutics.

INTRODUCTION

One of the challenges created by the biotechnology revolution is the development of methods for the economical production of highly purified proteins at large scale. The expression of recombinant proteins in the milk of transgenic dairy animals appears particularly well suited for the economical production of complex polypeptides (reviewed in Houdebine, 1994; Clark, 1998; Meade *et al.*, 1998). To that end, transgenic sheep, cows, goats, and even pigs (hardly dairy animals!) have been generated. Dairy goats offer a nice compromise. Goats have a relatively short generation interval (when compared with cattle) and a relatively large average milk output (compared with pigs and to a lesser extent sheep). These properties allow the generation of small herds of transgenic goats able to yield hundreds of kilograms of unpurified product per year, within a timeline compatible with the clinical and regulatory framework of drug development.

To express a recombinant protein in the milk of a transgenic animal, first the gene encoding the protein of interest is fused to milk-specific regulatory elements to generate the transgene. These DNA constructs have traditionally been introduced in the germ line of dairy animals by pronuclear microinjection of one-cell embryos (Hammer *et al.*, 1985; Bondioli *et al.*, 1991; Ebert *et al.*, 1991; Wright *et al.*, 1991). Microinjected embryos, at various stages of development depending on the species, were then transferred to a surrogate mother. Up to (and often less than) 5–10% of offspring resulting from pronuclear microinjections in large animals carried the transgene. Following integration into the germ line, the mammary gland-specific

---

[1]To whom correspondence should be addressed.

transgene, if expressed, becomes a dominant genetic characteristic that will be predictably inherited by offspring of the founder animal, depending on its degree of mosaicism. Often, transgenic animals express the protein(s) of interest in gram-perliter quantities; this can depend on the mammary-specific regulatory sequences, the gene to be expressed, and the integration site of the transgene.

However, the introduction of transgenes in the germ line of large animals has often proved challenging and very labor intensive. Although successful and widely used, the pronuclear microinjection approach has had limited efficiency. Transgene integration into the genome of founder animals is low and the frequent generation of mosaics (Wilkie *et al.*, 1986; Burdon and Wall, 1992; Whitelaw *et al.*, 1993) has sometimes complicated the expansion of transgenic herds (Williams *et al.*, 1998, 2000). Transgenic founders often carry multiple integration sites, frequently with various degrees of mosaicism. Moreover, in cases in which the cointegration of multiple transgenes is necessary—for example, for the expression of recombinant antibodies (Pollock *et al.*, 1999)—generation of animals carrying only one of the transgenes, or only one of the transgene within a specific chromosomal integration site, further decreases the frequency of "useful" founders.

The discovery that cultured cell lines can efficiently function as karyoplast donors for nuclear transfer has expanded the range of possibilities for germ line modification in large animals. First sheep (Campbell *et al.*, 1996; Wilmut *et al.*, 1997), then cattle (Cibelli *et al.*, 1998), goats (Baguisi *et al.*, 1999; Keefer *et al.*, 2001), and pigs (Onishi *et al.*, 2000; Polejaeva *et al.*, 2000; Betthauser *et al.*, 2000) have successively been generated by this technique. Nuclear transfer with transfected somatic cells allows a more controlled introduction of transgenes and, in some circumstances, can reduce the number of animals (egg donors and recipients) used during the foundering process. It also overcomes the problem of founder mosaicism. The ability to preselect transgenic cell lines before the generation of cloned transgenic embryos by analyzing transgene integration sites is also valuable. It is particularly important for the transgenic production of recombinant monoclonal antibodies in milk, because often several transgenes have to be expressed in the same secretory cells of the mammary epithelium at equivalent levels. Cointegration of the transgenes in the same chromosomal locus, to avoid segregation of heavy-chain and light-chain genes during herd propagation, is also desirable.

EARLY APPLICATIONS OF SOMATIC CELL NUCLEAR TRANSFER FOR THE GENERATION OF TRANSGENIC GOATS

The application of nuclear transfer technology using blastomeres of early caprine embryos has been reported (Yong and Yuqiang, 1998). Using serial nuclear transfer, six generations of cloned goat embryos were produced. Although a large number of kids were obtained, this approach was limited by the inability to introduce foreign genetic material into blastomeres and by their limited availability.

Our group, in collaboration with laboratories at Tufts University and Louisiana State University, tested the feasibility of somatic cell nuclear transfer in goats by using primary cells derived from a 35-day transgenic female fetus (Baguisi *et al.*, 1999). The fetus from which the primary donor cell line originated was produced by artificially inseminating a nontransgenic doe with fresh-collected semen from a transgenic founder buck carrying a transgene targeting the expression of recombinant human antithrombin III (AT-III) to the mammary gland. This transgenic goat line was chosen as a source of donor cells because its genetics and the characteristics of the recombinant protein that it produces have been extensively studied (Edmunds *et al.*, 1998; Williams *et al.*, 1998; van Patten *et al.*, 1999; Minnema *et al.*, 2000; Lu *et al.*, 2000; Nevière *et al.*, 2001).

As proof of concept, three transgenic female goats (Fig. 1A) were first generated by nuclear transfer of the fetal cell lines to *in vivo*-derived oocytes (Baguisi *et*

Figure 1 *Transgenic goats derived from nuclear transfer using a cultured fetal cell line. (A) Three kids derived from oocytes matured in vivo. (B) Five kids derived from oocytes matured in vitro.*

al., 1999). A variety of genotyping approaches (Southern blotting, fluorescence *in situ* hybridization, and polymerase chain reaction with restriction fragment length polymorphism) demonstrated that these animals were clones of the female AT-III cell line. All three kids were healthy and did not exhibit weight, placentation, and other health abnormalities often encountered with sheep or cows derived from nuclear transfer, or, to a lesser extent, from *in vitro* maturation, *in vitro* fertilization, and *in vitro* culture (Behboodi *et al.*, 1995; Wilson *et al.*, 1995; Garry *et al.*, 1996; Holm *et al.*, 1996; Walker *et al.*, 1996; Kruip and den Daas, 1997; Wilmut *et al.*, 1997; Hill *et al.*, 1999; Renard *et al.*, 1999; Young and Fairburn, 2000). Hormonal induction protocols were initially used to study the characteristics of recombinant AT-III produced by these cloned females. These studies showed no differences between these and other transgenic does from the same line obtained by natural breeding. Furthermore, these cloned goats appear to have normal reproductive and natural lactation properties.

The AT-III fetal cell line was also used to study the production of cloned goats from *in vitro*-matured oocytes (Reggio *et al.*, 2001). Five cloned transgenic goats (Fig. 1B) were produced using the same cell line and oocytes aspirated from either follicle-stimulating hormone (FSH)-stimulated or nonstimulated (slaughterhouse-derived) ovaries. All five kids had birth weights ranging from 3.4 to 4.3 kg, which is within the normal range for the breed. In these experiments with the AT-III fetal cell line, all initiated pregnancies resulted in the births of healthy kids, with no fetal loss occurring during gestation and no postnatal morbidity. There was no statistical difference in the percentage of matured oocytes successfully enucleated and reconstructed according to the source (FSH-stimulated ovaries vs. nonstimulated ovaries). Oocyte source also had no effect on embryo development and overall pregnancy rate in the two treatment groups.

These early experiments demonstrated the feasibility of somatic cell nuclear transfer in goats. However, because these cells were directly generated from a transgenic fetus, more experiments were needed to demonstrate the applicability of nuclear transfer to transgenic foundering.

MAMMARY GLAND-SPECIFIC TRANSGENES FOR SOMATIC CELL NUCLEAR TRANSFER

Transgenes targeting the expression of recombinant proteins to the mammary gland are usually chimeric, being derived from the fusion of the gene encoding the protein(s) of interest with mammary-specific regulatory sequences. Regulatory sequences from several milk-specific genes were isolated and characterized first in transgenic mice, then in large animals, as reviewed by Maga and Murray (1995), Echelard (1996), and Meade *et al.* (1998). These groups and others have used regulatory sequences from the caprine β-casein gene to target expression of heterologous proteins to the lactating mammary gland of transgenic mice (DiTullio *et al.*, 1992; Roberts *et al.*, 1992; Persuy *et al.* 1992, 1995; Wilburn *et al.*, 1997; Litscher *et al.*, 1999; Newton *et al.*, 1999), transgenic goats (Ebert *et al.*, 1991, 1994; Gavin *et al.*, 1997; Edmunds *et al.*, 1998; Pollock *et al.*, 1999), and even transgenic cows (Behboodi *et al.*, 2001a,b; Echelard *et al.*, 2002). The intron/exon organization of the caprine gene is similar to that of other homolog β-casein genes in other species and its expression is principally limited to the mammary gland during lactation. High-level expression was first observed in goats transgenic for a construct containing 6.2 kb of 5′ and 7.1 kb of 3′ goat β-casein flanking sequence fused to a variant of the human tissue plasminogen activator cDNA (Ebert *et al.*, 1991), then for human AT-III (Edmunds *et al.*, 1998) as well as several antibodies (Gavin *et al.*, 1997; reviewed in Pollock *et al.*, 1999).

The engineering of mammary gland-specific transgenes for the creation of somatic cell nuclear transfer cell lines is complicated by the need to use a selection marker (for example, neomycin resistance) to establish the transfected cell lines. Expression of the selection marker in the transfected primary cells must be robust enough so that clonal colonies carrying the transgenes can be efficiently isolated and purified. In addition, the regulatory sequences driving the cellular expression of the marker must not interfere with the proper regulation of the mammary gland-specific expression unit. Examples of promoter interference have been encountered in a number of gene-targeting experiments (Kim *et al.*, 1992; Fiering *et al.*, 1995; Hug *et al.*, 1996; Olson *et al.*, 1996; McDevitt *et al.*, 1997) in which the introduction of a transcriptionally active element, such as a neomycin resistance cassette, into a chromatin domain have led to endogenous gene silencing or dysregulation. Hence, not only must the mammary gland-specific transcription unit expressing the heterologous gene in the mammary gland be protected from the interference of the

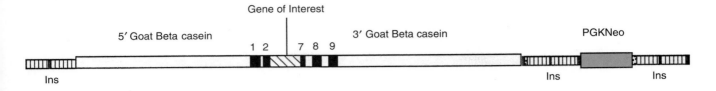

Figure 2 *Schematic representation of a mammary gland expression transgene. Tandem repeats of the 5'CHS4 insulator element (Ins, vertically striped boxes) flank 5' and 3' goat β-casein sequences (open boxes indicate nontranscribed sequences; closed boxes indicate noncoding exon sequences) and the linked PGKNeo (G418 resistance cassette, gray box). The gene of interest (represented by the diagonally striped box) can be inserted in a restriction site located where the translation start site of the goat β-casein gene is normally situated.*

surrounding chromatin at the integration site, as in any transgenic experiment, but, it must be protected from the potential interference of the selection marker.

To address these issues, we have used the chicken β-globin insulator sequence (5'CHS4) (Chung *et al.*, 1993) to shield the mammary expression cassette from both the selection cassette and the surrounding chromatin (Fig. 2). The 5'CHS4 element is a DNA sequence isolated from the 5' boundary of the chicken β-globin domain. It was originally shown to insulate a reporter gene from the activating effects of the β-globin locus control region when assayed in a human erythroid cell line (Chung *et al.*, 1993). Its role *in vivo* appears to separate two differentially regulated loci: a folate receptor gene and the chicken β-globin gene (Prioleau *et al.*, 1999). The 5'CHS4 β-globin insulator is hypothesized to be important for the proper regulation of these two adjacent domains activated at two different stages during differentiation. Experiments in transgenic *Drosophila* (Chung *et al.*, 1993), in K562 cells (Walters *et al.*, 1999), in adeno-associated virus- or retoviruse-infected MEL cells and bone marrow (Inoue *et al.*, 1999; Emery *et al.*, 2000; Rivella *et al.*, 2000), as well as in transgenic mice (Taboit-Dameron *et al.*, 1999; Potts *et al.*, 2000) have shown that flanking a transgene with the chicken β-globin 5'CHS4 can significantly reduce silencing due to chromosomal position effects and promoter interference.

As depicted in Fig. 2, we have used direct repeats of the 1.2-kb 5'HS4 element to flank both the mammary gland expression cassette and the selection marker cassette (in this case neomycin resistance). Instead of linking the two cassettes it has also been possible to cotransfect the selection marker, flanked on both sides with insulator sequences, with one or several mammary-specific (goat β-casein vector) expression constructs. In our experience (Meade *et al.*, 2002), addition of the insulator element to transgenic constructs has significantly reduced the sensitivity of mammary gland expression cassettes to position effect. Furthermore, expression analysis of lines carrying insulator-containing bipromoter constructs (such as in Fig. 2) containing both a mammary gland expression cassette and a selection cassette showed no evidence of promoter interference in both transgenic mice and cows (Echelard *et al.*, 2002).

GENERATION OF TRANSGENIC FOUNDERS BY TRANSFECTED SOMATIC CELL NUCLEAR TRANSFER

In the context of recombinant production, transgenic foundering is the most important application of somatic cell nuclear transfer. Because only primary cells can be used as nuclear donors, one challenge has been to select the transfected clones before

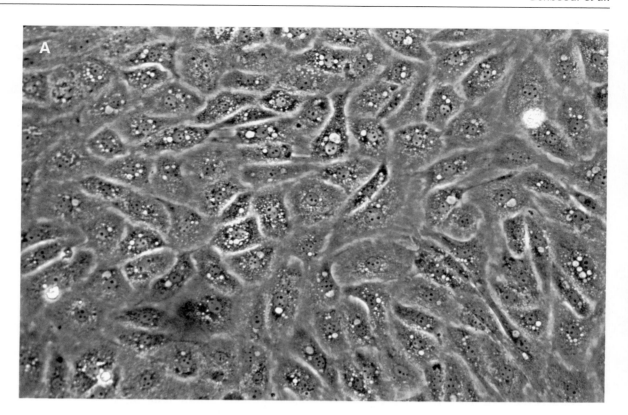

Figure 3 Generation of transgenic goats through nuclear transfer with transfected cells. (A) In vitro-cultured transfected cells used in this study; magnification 400×. (B) Two transgenic offspring resulting from somatic cell nuclear transfer with a transfected cell line. (C, D, E) Interphase FISH showing the transgene integration (green signal) in, respectively, the donor cell line and leukocytes from each of the cloned offspring; magnification 400×. (F, G, H) Metaphase FISH showing the identical transgene chromosomal integration sites (green signal) in, respectively, the donor cell line and leukocytes from each of the cloned offspring; magnification 1000×.

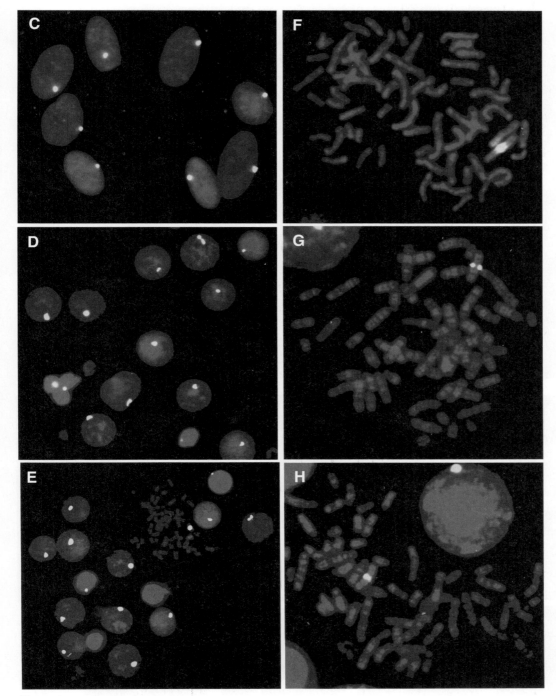

Figure 3 (continued)

the onset of growth arrest brought on by tissue culture-induced senescence. Other issues relate to the impact of the selection process on the ability of the recombinant cell lines to function effectively as cell donors and possible effects of these treatments on the health of the resulting clones. The cell population to be transfected is also in question. For example, fetal cell populations have been widely used for somatic cell nuclear transfer and fetal cells are believed to be able to undergo more cell divisions than most cell types derived from juvenile or adult animals before experiencing growth arrest. However, fetal cells are limited in numbers, more diffi-

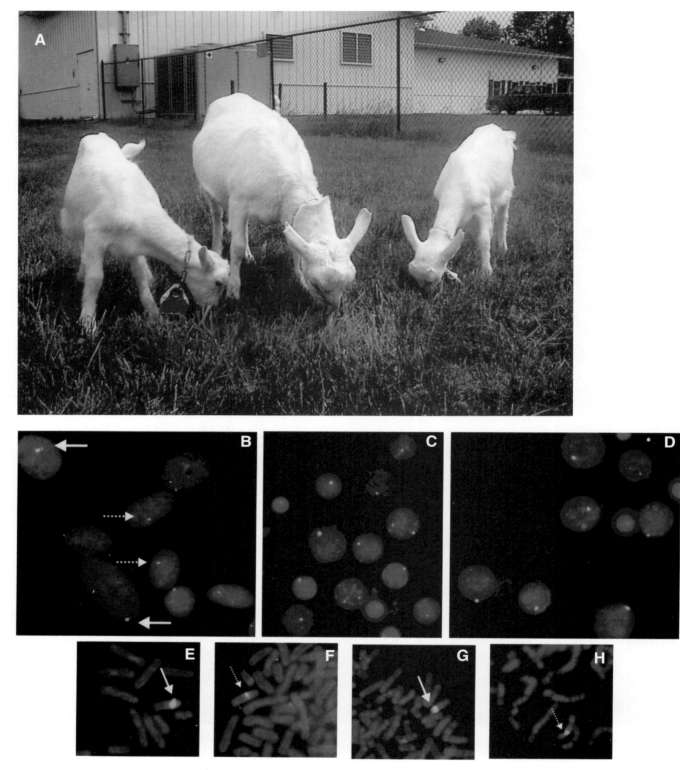

Figure 4 Generation of transgenic goats through nuclear transfer with adult cell isolates. (A) Two transgenic offspring (D108, left, and D119, right) resulting from somatic cell nuclear transfer with isolates from the female transgenic founder (C690, center); this founder was generated by pronuclear microinjection. (B, C, D) Interphase FISH showing the transgene integration site (green signal) in, respectively, C690 fibroblast cells, D108 leukocytes, and D119 leukocytes; magnification 400×. The solid arrows point to the high-copy integration sites. The punctuated arrows point to the low-copy integration site. (E–H) Leukocyte metaphase FISH showing the transgene chromosomal integration sites (green signal), respectively, C690 high-copy signal, C690 low-copy signal, D119 high-copy signal, and D108 low-copy signal; magnification 1000×.

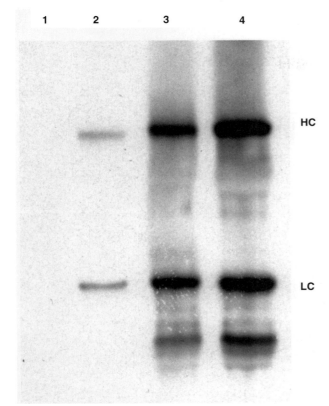

Figure 5 Western blot of an antibody/protein fusion expressed in the milk of transgenic goats. Lane 1, nontransgenic goat milk; lane 2, C690 milk; lane 3, D108 milk; lane 4, D119 milk. HC, Heavy chain; LC, light chain.

cult to harvest, and the characteristics of the resulting offspring are less predictable than if the nuclear transfer donors are derived from a juvenile or adult animal.

Chen *et al.* (2001) and Keefer *et al.* (2001) reported the generation of transgenic cloned goats from transfected somatic cells. In both reports, clones were obtained by introducing the transgene into fetal cells using a lipid-mediated transfection protocol. In both studies, offspring were apparently normal at birth and no perinatal loss or morbidity was observed. Although the total number of nuclear transfer attempts in these studies was low, the live-birth frequency per attempt was not very different than that observed with nontransfected fetal cells. In aggregate, these data suggest that the transfection and selection treatment do not appear to harm the ability of the cell line to serve as nuclear transfer donors. However, note that as for any transgene integration, there is always the possibility of creating a dominant mutation at the site of chromosomal insertion. These mutations may not necessarily affect the cells in culture but could affect the viability of the nuclear transfer offspring.

The generation of transgenic offspring through transfected somatic cell nuclear transfer is summarized in Fig. 3. In this experiment, fetal cells (Fig. 3A) were cotransfected with two mammary gland expression vectors. One vector targeted the expression of an antibody light-chain protein fusion; the other one expressed the corresponding heavy-chain protein fusion. Following transfection and under selection, aliquots of colonies were frozen as soon as possible to limit the number of cell divisions prior to nuclear transfer. Characterization and genotyping were performed on the remainder of the cells. Polymerase chain reaction (PCR) and Southern blotting analyses were used to ascertain the presence and integrity of both transgenes. Fluorescence *in situ* hybridization (FISH) analyses (C and F, Fig. 3) were performed to verify the presence of only one chromosomal integration site. In these experiments (Chen *et al.*, 2001), two healthy female offspring were obtained (Fig. 3B),

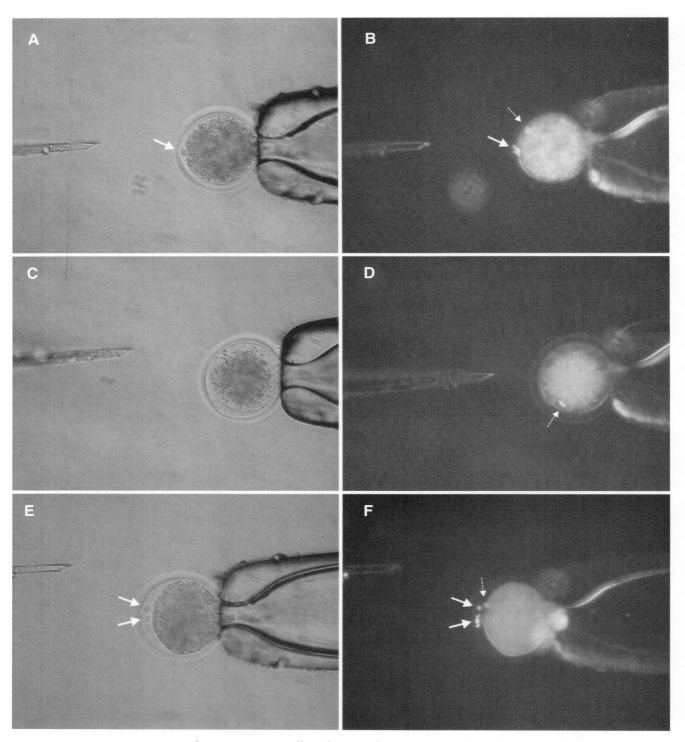

Figure 6 *Somatic cell nuclear transfer using in vivo-derived oocytes. Phase contrast (A) and fluorescence (B) microscopy pictures of metaphase II oocytes. Phase contrast (C) and fluorescence (D) microscopy pictures of late metaphase II oocytes. Phase contrast (E) and fluorescence (F) microscopy pictures of telophase II oocytes. (A–F) The solid arrows point to polar bodies and the punctuated arrows point to metaphase plates. (G) Trypsinized donor cells just prior to reconstruction with enucleated oocytes. (H) Reconstructed oocytes; the arrows point to donor cells fusing with the cytoplasts. Phase contrast (I) and fluorescence (J) microscopy pictures of nuclear transfer goat embryos after a 24-hour culture period. The two-cell embryo (a) appears to contain normal nuclei; the four-cell embryo (b) does not appear to contain normal nuclear material and may have resulted from fragmentation. Magnification for all panels, 400×.*

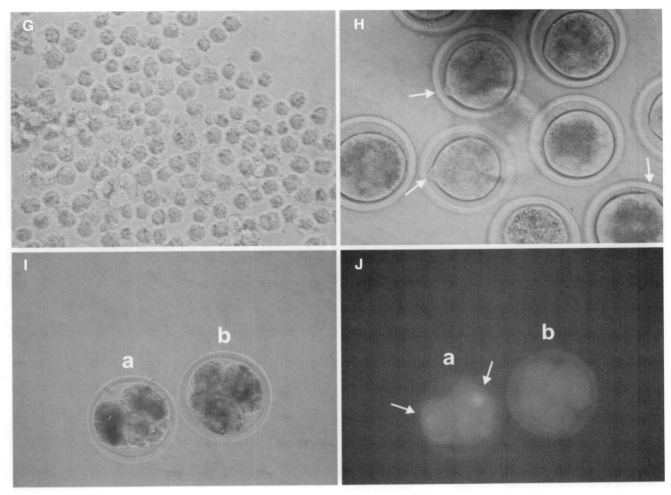

Figure 6 (continued)

both carrying the same transgenic integration as the transfected cell line (D, E, G, and H, Fig. 3).

PRODUCTION OF TRANSGENIC GOATS BY ADULT SOMATIC CELL NUCLEAR TRANSFER

The ability to clone a juvenile or an adult animal, instead of using fetal somatic cells, could be desirable. One may want to reproduce the exact characteristics of an animal without going through the "genetic lottery" of creating a fetal cell line by mating the animal to be reproduced. Adult somatic cell nuclear transfer is also an option to facilitate the propagation of, or even an attempt to "regenerate," an endangered species or subspecies.

In the specific case of transgenic production, adult cloning could have at least two applications: (1) generation of a genetically homogeneous producing herd and (2) separation of multiple integration sites in a mosaic transgenic founder generated by DNA microinjection. The first situation could possibly apply to animals producing a therapeutic protein with a complex glycosylation pattern, for which very specific characteristics (for example, degree of sialylation) are necessary. One could imagine producing a homogeneous herd through cloning of a founder that produces

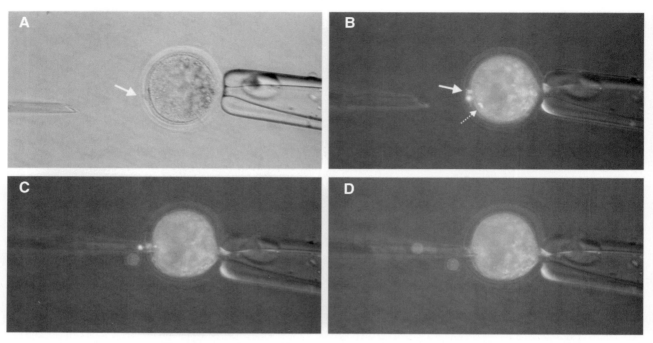

Figure 7 *Enucleation of a metaphase II goat oocyte. Phase contrast (A) and fluorescence (B, C, D) microcopy; magnification 400×. The solid arrows show the polar body and the punctuated arrow shows the metaphase plate.*

proteins with the preferred specifications. On the other hand, although dairy goats are very outbred, it is not clear that glycosylation of a heterologous protein expressed in milk will be more homogeneous in cloned transgenic siblings than in fraternal offspring carrying the same transgene integration.

We have employed somatic cell nuclear transfer to help in the propagation of a complex mosaic transgenic founder (Behboodi *et al.*, 2001). This founder (C690, Fig. 4A) was generated by pronuclear comicroinjection of two antibody fusion protein transgenes and, as it is often the case with this technique, carried several integrations in a complex mosaic pattern (Fig. 4B). Two main integration sites, one high copy and one low copy, were found in different cells (B, E, and F, Fig. 4). Because this founder is a female, separating these transgene integration sites through standard breeding would be haphazard and at best time-consuming. Instead, a skin biopsy from the transgenic founder female was cultured and fibroblasts were used as the nuclear transfer donor. Two healthy transgenic offspring, D108 and D119, were obtained (Fig. 4A). As expected, these does were phenotypically identical to C690. Genetic analyses showed that each of these kids was nonmosaic for one of the two main transgene integration sites (C, D, G, and H, Fig. 4). Examination of induced lactation samples from C690, D108, and D119 (Fig. 5) indicate that the cloned offspring appear to express higher levels of the recombinant antibody fusion protein, compared to the founder. This is most likely due to the fact that they are not mosaic for the transgene integration sites.

IN VITRO-DERIVED OOCYTES VS. *IN VIVO*-DERIVED OOCYTES

Recombinant therapeutic protein production is subject to a number of regulatory requirements. One issue of importance is the minimization of risk of contamination with adventitious agents. In the case of recombinant production from the milk of transgenic goats, strategies have been devised to ensure that starting material is of the highest quality (reviewed in Ziomek, 1996, 1998). Maintenance of closed goat herds of known origin is a convenient step to preclude contamination with adventitious agents. This means that recipient oocytes used in the generation of transgenic founders should ideally be collected from closed herds rather than from abattoir material. The use of closed herds as the source of recipient oocytes dictates that nuclear transfer has to be fairly efficient, because donor animals are limited in number.

Another consideration is the method of collection used to obtain recipient oocytes. Should the oocytes be flushed from the oviduct of the donor goats (hence matured *in vivo*), or should oocytes be aspirated and matured *in vitro*? The developmental stage of aspirated and *in vitro*-matured oocytes can be better controlled, leading to a more homogeneous population of recipient oocytes. As reported earlier (Baguisi *et al.*, 1999) and illustrated in Fig. 6, ovulated oocytes are often heterogeneous. We have used three types of oocytes as nuclear transfer recipients, metaphase II (A and B, Fig. 6; Fig. 7), late metaphase II (C and D, Fig. 6), as well as oocytes spontaneously obtained in telophase II, and have obtained live birth with all three stages of development (Baguisi *et al.*, 1999; Behboodi *et al.*, 2001). It has also been possible to obtain healthy cloned goats from nuclear transfer protocols using abattoir-derived oocytes (Chen *et al.*, 2001) and aspirated oocytes (Keefer *et al.*, 2001; Reggio *et al.*, 2001). Use of *in vivo*-matured oocytes may also lead to healthier offspring. Although we do not have enough data yet to support such claims for goats, there are numerous data from cattle and sheep studies suggesting that *in vitro* maturation of oocytes can have deleterious effects on offspring. For example, when we compared ovulated *in vivo*-matured embryos to *in vitro*-matured/*in vitro*-fertilized embryos in a cow microinjection program, we found that calves derived from *in vitro* oocytes had a significant increased incidence of large-offspring syndrome (Behboodi *et al.*, 2001). It could very well be that use of minimally manipulated oocytes as recipients and avoidance of *in vitro* culture following nuclear transfer are the reasons for the relative absence of nuclear transfer-associated morbidity observed so far with goats.

CONCLUSION

Over the past several years, nuclear transfer protocols have evolved that make it possible to generate transgenic goats efficiently. These methods offer several advantages over the use of pronuclear microinjection: preselection of transgenic integration sites, nonmosaic founder, and apparent reduction in the number of surgical procedures necessary to generate founders. Moreover, it seems that goats generated by somatic cell nuclear transfer do not suffer from the health problems (large birth weight, placentation defects, hematological dysfunction) sometimes reported with cloned sheep and cattle. However, very few data are yet available on the long-term health or reproductive and lactation performance of these cloned goats. Some of the research emphasis will now move from the development of nuclear transfer protocols to the careful study of the cloned offspring.

ACKNOWLEDGMENTS

The authors thank Robert Godke, Carol Ziomek, Sandy Ayres, and William Gavin as well as David Melican, Robin Butler, Michael O'Coin, Brett Reggio, Richard Denniston, and Jennifer Williams for their encouragement and outstanding contributions at numerous stages of these projects. We also wish to thank Alexander Baguisi and Eric Overstrom for their involvement in the early phases of the goat cloning work.

REFERENCES

Baguisi, A., Behboodi, E., Melican, D. T., Pollock, J. S., Destrempes, M. M., Cammuso, C., Williams, J. L., Nims, S. D., Porter, C. A., Midura, P., Palacios, M. J., Ayres, S. L., Denniston, R. S., Hayes, M. L., Ziomek, C. A., Meade, H. M., Godke, R. A., Gavin, W. G., Overström, E. W., and Echelard, Y. (1999). Production of transgenic goats by somatic cell nuclear transfer. *Nat. Biotechnol.* **17**, 456–461.

Behboodi, E., Anderson, G. B., BonDurant, R. H., Cargill, S. L., Kreusher, B. R., Medrano, J. F., and Murray, J. D. (1995). Birth of large calves that developed from *in vitro*-derived bovine embryos. *Theriogenology* **44**, 227–232.

Behboodi, E., Melican, D. T., Liem, H., Chen, L. H., Destrempes, M. M., Williams, J. L., Flanagan, P. A., Butler, R. E., Meade, H. M., Gavin, W. G., and Echelard, Y. (2001a). Production of transgenic goats by adult somatic cell nuclear transfer. *Theriogenology* **55**, 254.

Behboodi, E., Groen, W., Destrempes, M. M., Williams, J. L., Olhrichs, C., Gavin, W. G., Broek, D. M., Ziomek, C. A., Faber, D. C., Meade, H. M., and Echelard, Y. (2001b). Transgenic production from *in vivo*-derived embryos: Effect on calf birth weight and sex ratio. *Mol. Reprod. Dev.* **60**, 27–37.

Betthauser, J., Forsberg, E., Augenstein, M., Childs, L., Eilertsen, K., Enos, J., Forsythe, T., Golueke, P., Jurgella, G., Koppang, R., Lesmeister, T., Mallon, K., Mell, G., Misica, P., Pace, M., Pfister-Genskow, M., Strelchenko, N., Voelker, G., Watt, S., Thompson, S., and Bishop, M. (2000). Production of cloned pigs from *in vitro* systems. *Nat. Biotechnol.* **18**, 1055–1059.

Bondioli, K. R., Biery, K. A., Hill, K. G., Jones, K. B., and De Mayo, F. J., (1991). Production of transgenic cattle by pronuclear injection. *Biotechnology* **16**, 265–273.

Burdon, T. G., and Wall, R. J. (1992). Fate of microinjected genes in preimplantation mouse embryos. *Mol. Reprod. Dev.* **33**, 436–442.

Campbell, K. H., S., McWhir, J., Ritchie, W. A., and Wilmut, I. (1996). Sheep cloned by nuclear transfer from a cultured cell line. *Nature* **380**, 64–66.

Chen, L. H., Behboodi, E., Reggio, B. C., Liem, H., Destrempes, M. M., Green, H. L., Ziomek, C. A., Denniston, R. A., Echelard, Y., Godke, R. A., and Meade, H. M. (2001). Production of transgenic goats from a transfected cell line. *Theriogenology* **55**, 259.

Chung, J. H., Whiteley, M., and Felsenfeld, G. (1993). A 5′ element of the chicken beta-globin domain serves as an insulator in human erythroid cells and protects against position effect in *Drosophila*. *Cell* **74**, 505–514.

Cibelli, J. B., Stice, S. L., Golueke, P. J., Kane, J. J., Jerry, J., Blackwell, C., Ponce de Leon, F. A., and Robl, J. M. (1998). Cloned transgenic calves produced from nonquiescent fetal fibroblasts. *Science* **280**, 1256–1258.

Clark, A. J. (1998). The mammary gland as a bioreactor: Expression, processing, and production of recombinant proteins. *J. Mammary Gland Biol. Neoplasia* **3**, 337–350.

DiTullio, P., Cheng, S. H., Marshall, J., Gregory, R. J., Ebert, K. M., Meade, H. M., and Smith, A. E. (1992). Production of cystic fibrosis transmembrane conductance regulator in the milk of transgenic mice. *Biotechnology (N.Y.)* **10**, 74–77.

Ebert, K. M., Selgrath, J. P., DiTullio, P., Denman, J., Smith, T. E., Memon, M. A., Schindler, J. E., Monastersky, G. M., Vitale, J. A., and Gordon, K. (1991). Transgenic production of a variant of human tissue-type plasminogen activator in goat milk: Generation of transgenic goats and analysis of expression. *Biotechnology (N.Y.)* **9**, 835–838.

Ebert, K. M., DiTullio, P., Barry, C. A., Schindler, J. E., Ayres, S. L., Smith, T. E., Pellerin, L. J., Meade, H. M., Denman, J., and Roberts, B. (1994). Induction of human tissue plasminogen activator in the mammary gland of transgenic goats. *Biotechnology (N.Y.)* **12**, 699–702.

Echelard, Y. (1996). Recombinant protein production in the milk of transgenic animals. *Curr. Opin. Biotechnol.* **7**, 536–540.

Echelard, Y., Destrempes, M. M., Koster, J. A., Blackwell, C., Groen, W., Pollock, D., Williams, J. L., Behboodi, E., Pommer, J., and Meade, H. M. (2002). Production of recombinant human serum albumin in the milk of transgenic cows. *Theriogenology* **57**, 779.

Edmunds, T., Van Patten, S. M., Pollock, J., Hanson, E., Bernasconi, R., Higgins, E., Manavalan, P., Ziomek, C., Meade, H., McPherson, J. M., and Cole, E. S. (1998). Transgenically produced human antithrombin: structural and functional comparison to human plasma-derived antithrombin. *Blood* **91**, 4561–4571.

Emery, D. W., Yannaki, E., Tubb, J., and Stamatoyannopoulos, G. (2000). A chromatin insulator protects retrovirus vectors from chromosomal position effects. *Proc. Natl. Acad. Sci. U.S.A.* **97**, 9150–9155.

Fiering, S., Epner, E., Robinson, K., Zhuang, Y., Telling, A., Hu, M., Martin, D. I. K., Enver, T., Ley, T. J., and Groudine, M. (1995). Targeted deletion of 5′HS2 of the murine beta-globin LCR reveals that it is not essential for proper regulation of the beta-globin locus. *Genes Dev.* **9**, 2203–2213.

Garry, F. B., Adams, R., McCann, J. P., and Odde, K. G. (1996). Postnatal characteristics of calves produced by nuclear transfer cloning. *Theriogenology* **45**, 141–152.

Gavin, W. G., Pollock, D., Fell, P., Yelton, D., Cammuso, C., Harrington, M., Lewis-Williams, J., Midura, P., Oliver, A., Smith, T. E., Wilburn, B., Echelard, Y., and Meade, H. (1997). Expression of the antibody hBR96 in the milk of transgenic mice and production of hBR96 transgenic goats. *Theriogenology* **47**, 214.

Hammer, R. E., Pursel, V. G., Rexroad, C. E. jr, Wall, R. J., Bolt, D. J., Ebert, K. M., Palmiter, R. D., and Brinster, R. L. (1985). Production of transgenic rabbits, sheep and pigs by microinjection. *Nature* **315**, 680–683.

Hill, J. R, Roussel, A. J., Cibelli, J. B., Hooper, N. L., Miller, M. W., Thompson, J. A., Looney, C. R., Westhusin, M. E., Robl, J. M., and Stice, S. L. (1999). Clinical and pathological features of cloned transgenic calves and fetuses (13 case studies). *Theriogenology* **51**, 1451–1465.

Holm, P., Walker, S. K., and Seamark, R. F. (1996). Embryo viability, duration of gestation and birth weight in sheep after transfer of *in vitro* matured and *in vitro* fertilized zygotes cultured *in vitro* or *in vivo*. *J. Reprod. Fertil.* **107**, 175–181.

Houdebine, L. M. (1994). The production of pharmaceutical proteins from the milk of transgenic animals. *J. Biotechnol.* **34**, 269–287.

Hug, B. A., Wessleschmidt, R. L., Fiering, S., Bender, M. A., Epner, E., Groudine, M., and Ley, T. J. (1996). Analysis of mice containing a targeted deletion of beta-globin locus control region 5′ hypersensitive site 3. *Mol. Cell. Biol.* **16**, 2906–2912.

Inoue, T., Yamaza, H., Sakai, Y., Mizuno, S., Ohno, M., Hamasaki, N., and Fukumaki, Y. (1999). Position-independent human beta-globin gene expression mediated by a recombinant adeno-associated virus vector carrying the chicken beta-globin insulator. *J. Hum. Genet.* **44**, 152–162.

Keefer, C. L., Baldassarre, H., Keyston, R., Wang, B., Bhatia, B., Bilodeau, A. S., Zhou, J. F., Leduc, M., Downey, B. R., Lazaris, A., and Karatzas, C. N. (2001). Generation of dwarf goat (*Capra hircus*) clones following nuclear transfer with transfected and nontransfected fetal fibroblasts and *in vitro*-matured oocytes. *Biol. Reprod.* **64**, 849–856.

Kim, C. G., Epner, E. M., Forrester, W. C., and Groudine, M. (1992). Inactivation of the human beta-globin gene by targeted insertion into the beta-globin locus control region. *Genes Dev.* **6**, 928–938.

Kruip, T. A. M., den Daas, J. H. G. (1997). *In vitro* produced and cloned embryos: Effects on pregnancy, parturition and offspring. *Theriogenology* **47**, 43–52.

Litscher, E. S., Liu, C., Echelard, Y., and Wassarman, P. M. (1999). Zona pellucida glycoprotein mZP3 produced in milk of transgenic mice is active as a sperm receptor, but can be lethal to newborns. *Transgenic Res.* **8**, 361–369.

Lu, W., Mant, T., Levy, J. H., and Bailey, J. M. (2000). Pharmacokinetics of recombinant transgenic antithrombin in volunteers. *Anesth. Analg.* **90**, 531–534.

Maga, E. A., and Murray, J. D. (1995). Mammary gland expression of transgenes and the potential for altering the properties of milk. *Biotechnology (N.Y.)* **13**, 1452–1457.

McDevitt, M. A., Shivdasani, R. A., Fujiwara, Y., Yang, H., and Orkin, S. H. (1997). A "knockdown" mutation created by cis-element gene targeting reveals the dependence of erythroid cell maturation on the level of transcription factor GATA-1. *Proc. Natl. Acad. Sci. U.S.A.* **94**, 6781–6785.

Meade, H. M., Echelard, Y., Ziomek, C. A., Young, M. W., Harvey, M., Cole, E. S., Groet, S., Smith, T. E., and Curling, J. M. (1998). Expression of recombinant proteins in the milk of transgenic animals. "Gene Expression Systems: Using Nature For The Art of Expression" (J. M. Fernandez and J. P. Hoeffler, eds.), Academic Press, San Diego.

Meade, H. M., *et al.* (2002). In preparation.

Minnema, M. C., Chang, A. C., Jansen, P. M., Lubbers, Y. T., Pratt, B. M., Whittaker, B. G., Taylor, F. B., Hack, C. E., and Friedman, B. (2000). Recombinant human antithrombin III improves survival and attenuates inflammatory responses in baboons lethally challenged with *Escherichia coli. Blood* **95**, 1117–1123.

Nevière, R., Tournoys, A., Mordon, S., Marechal, X., Song, F. L., Jourdain, M., and Fourrier, F. (2001). Antithrombin reduces mesenteric venular leukocyte interactions and small intestine injury in endotoxemic rats. *Shock* **15**, 220–225.

Newton, D. L., Pollock, D., DiTullio, P., Echelard, Y., Harvey, M., Wilburn, B., Williams, J., Hoogenboom, H. R., Raus, J. C. M., Meade, H. M., and Rybak, S. M. (1999). Antitransferrin receptor antibody-RNase fusion protein expressed in the mammary gland of transgenic mice. *J. Immunol. Methods* **231**, 159–167.

Olson, E. N., Arnold, H. H., Rigby, P. W. J., and Wold, B. J. (1996). Know your neighbors: Three phenotypes in null mutants of the myogenic bHLH gene *MRF4. Cell* **85**, 1–4.

Onishi, A., Iwamoto, M., Akita, T., Mikawa, S., Takeda, K., Awata, T., Hanada, H., and Perry, A. C. (2000). Pig cloning by microinjection of fetal fibroblast nuclei. *Science* **289**, 1188–1190.

Persuy, M. A., Stinnakre, M. G., Printz, C., Mahe, M. F., and Mercier, J. C. (1992). High expression of the caprine beta-casein gene in transgenic mice. *Eur. J. Biochem.* **205**, 887–893.

Persuy, M. A., Legrain, S., Printz, C., Stinnakre, M. G., Lepourry, L., Brignon, G., and Mercier, J. C. (1995). High-level, stage- and mammary-tissue-specific expression of a caprine kappa-casein-encoding minigene driven by a beta-casein promoter in transgenic mice. *Gene* **165**, 291–296.

Polejaeva, I. A., Chen, S. H., Vaught, T. D., Page, R. L., Mullins, J., Ball, S., Dai, Y., Boone, J., Walker, S., Ayares, D. L., Colman, A., and Campbell, K. H. (2000). Cloned pigs produced by nuclear transfer from adult somatic cells. *Nature* **407**, 86–90.

Pollock, D. P., Kutzko, J. P., Birck-Wilson, E., Williams, J. L., Echelard, Y., and Meade, H. M. (1999). Transgenic milk as a method for the production of recombinant antibodies. *J. Immunol. Methods* **231**, 147–157.

Potts, W., Tucker, D., Wood, H., and Martin, C. (2000). Chicken beta-globin 5′HS4 insulators function to reduce variability in transgenic founder mice. *Biochem. Biophys. Res. Commun.* **273**, 1015–1018.

Prioleau, M. N., Nony, P., Simpson, M., and Felsenfeld, G. (1999). An insulator element and condensed chromatin region separate the chicken beta-globin locus from an independently regulated erythroid-specific folate receptor gene. *EMBO J.* **18**, 4035–4048.

Reggio, B. C., James, A. N., Green, H. L., Gavin, W. G., Behboodi, E., Echelard, Y., and Godke, R. A. (2001). Cloned transgenic offspring resulting from somatic cell nuclear transfer in the goat: Oocytes derived from both FSH-stimulated and nonstimulated ovaries. *Biol. Reprod.* **65**, 1528–1533.

Renard, J. P, Chastant, S., Chesne, P., Richard, C., Marchal, J., Cordonnier, N., Chavatte, P., and Vignon, X. (1999). Lymphoid hypoplasia and somatic cloning. *Lancet* **353**, 1489–1491.

Rivella, S., Callegari, J. A., May, C., Tan, C. W., and Sadelain, M. (2000). The cHS4 insulator increases the probability of retroviral expression at random chromosomal integration sites. *J. Virol.* **74**, 4679–4687.

Roberts, B., DiTullio, P., Vitale, J., Hehir, K., and Gordon, K. (1992). Cloning of the goat beta-casein gene and expression in transgenic mice. *Gene* **121**, 255–262.

Taboit-Dameron, F., Malassagne, B., Viglietta, C., Puissant, C., Leroux-Coyau, M., Chereau, C., Attal, J., Weill, B., and Houdebine, L. M. (1999). Association of the 5′HS4 sequence of the chicken beta-globin locus control region with human EF1 alpha gene promoter induces ubiquitous and high expression of human CD55 and CD59 cDNAs in transgenic rabbits. *Transgenic Res.* **8**, 223–235.

van Patten, S. M., Hanson, E., Bernasconi, R., Zhang, K., Manavalan, P., Cole, E. S., McPherson, J. M., and Edmunds, T. (1999). Oxidation of methionine residues in antithrombin. Effects on biological activity and heparin binding. *J. Biol. Chem.* **274**, 10268–10276.

Walker, S. K., Hartwich, K. M., and Seamark, R. F. (1996). The production of unusually large offspring following embryo manipulation: Concepts and challenges. *Theriogenology* **45**, 111–120.

Walters, M. C., Fiering, S., Bouhassira, E. E., Scalzo, D., Goeke, S., Magis, W., Garrick, D., Whitelaw, E., and Martin, D. I. (1999). The chicken beta-globin 5′HS4 boundary element blocks enhancer-mediated suppression of silencing. *Mol. Cell. Biol.* **19**, 3714–37126.

Whitelaw, C. B., Springbett, A. J., Webster, J., and Clark, A. J. (1993). The majority of G0 transgenic mice are derived from mosaic embryos. *Transgenic Res.* **2**, 29–32.

Wilburn, B., Woodworth, L., Gronbeck, A., Lewis-Williams, J., Harrington, M., Pollock, D., Richards, S. M., Meade, H., and Echelard, Y. (1997). High-level expression of recombinant human prolactin in the milk of transgenic mice. *Theriogenology* **47**, 219.

Wilkie, T. M., Brinster, R. L., and Palmiter, R. D. (1986). Germline mosaicism and somatic in transgenic mice. *Dev. Biol.* **118**, 9–18.

Williams, J. L., Ponce de Leon, F. A., Midura, P., Harrington, M., Meade, H., and Echelard, Y. (1998). Analysis of multiple transgene integration sites in a beta casein antithrombin III goat line. *Theriogenology* **49**, 398.

Williams, J. L., Pollock, D. P., Midura, P., Nims, S. D., Hawkins, N., Smith, T. E., Gavin, W. G., Meade, H. M., and Echelard, Y. (2000). FISH analysis of transgenic goats carrying two transgenes. *Theriogenology* **53**, 522.

Wilmut, I., Schnieke, A. E., McWhir, J., Kind, A. J., and Campbell, K. H. S. (1997). Viable offspring derived from fetal and adult mammalian cells. *Nature* **385**, 810–813.

Wilson, J. M., Williams, J. D, Bondioli, K. R., Looney, C. R., Westhusin, M. E., and McCalla, D. F. (1995). Comparison of birth weight and growth characteristics of bovine calves produced by nuclear transfer (cloning), embryo transfer and natural mating. *Anim. Reprod. Sci.* **38**, 73–83.

Wright, G., Carver, A., Cottom, D., Reeves, D., Scott, A., Simons, P., Wilmut, I., Garner, I., and Colman, A. (1991). High level expression of active human alpha-1-antitrypsin in the milk of transgenic sheep. *Biotechnology (N.Y.)* **9**, 830–834.

Yong, Z., and Yuqiang, L. (1998). Nuclear-cytoplasmic interaction and development of goat embryos reconstructed by nuclear transplantation: Production of goats by serially cloning embryos. *Biol. Reprod.* **58**, 266–269.

Young, L. E., and Fairburn, H. R. (2000). Improving the safety of embryo technologies: Possible role of genomic imprinting. *Theriogenology* **53**, 627–648.

Ziomek, C. A. (1996). Minimization of viral contamination in human pharmaceuticals produced in the milk of transgenic goats. *Dev. Biol. Stand.* **88**, 265–268.

Ziomek, C. A. (1998). Commercialization of proteins produced in the mammary gland. *Theriogenology* **49**, 139–144.

PART VII

ETHICAL AND LEGAL AFFAIRS

CHAPTER 26

ETHICAL IMPLICATIONS OF CLONING

Ronald M. Green

The science of reproductive and therapeutic cloning has stirred heated ethical debates. Some of the fears and apprehensions are based on misconceptions of the science, but others are real and substantial. It may be ethically possible to undertake reproductive cloning once the direct physiological risks of cloning are reduced to the level of current assisted reproductive technologies, but this should be done only in a research context with careful monitoring and followup of children and families. Therapeutic cloning raises several different but related ethical issues. Among them are the issue of the destruction of nascent human life, the tie to reproductive cloning, and the treatment of egg and cell donors. Although none of these issues justifies prohibiting therapeutic cloning research, scientists engaged in this research should be sensitive to the ethically controversial nature of their work.

INTRODUCTION

No biomedical technology has produced more ethical controversy or fear than has human cloning. Long before any type of cloning was technically possible, novels such as Huxley's "Brave New World" and Levin's "Boys from Brazil" associated cloning with totalitarianism. Following the announcement in February 1997 of the cloning of a lamb named Dolly, by veterinary researchers in Scotland, many religious groups and bioethicists issued statements opposing human cloning. Politicians rushed to pass laws that would restrict cloning research.

These fears and apprehensions have many cultural sources. Most fundamentally, they stem from the threat to the sanctity of human life that cloning is seen to represent. The possibility of producing multiple copies of a person infers that each individual is a replaceable unit, like the machine parts in the industrial system. Many fear that in a world in which individuals are easily duplicated, life itself will become cheap. Cloning is thus seen as opening the way to murder, slavery, and other forms of human degradation.

In contrast to these fears, some people look to cloning as a source of technological "immortality" (Talbot, 2001). They see cloning as a way for their own or a loved one's existence to be perpetuated in a new, cloned body. Such hopes rest on the mistaken assumption that individuals with similar genotypes will be the same person.

An ethical analysis of human cloning cannot ignore these fears and exaggerated hopes. Nevertheless, such an analysis must be based not on fiction but on scientific and social facts. What follows here is a critical examination of some of the leading concerns that have been voiced about human cloning. In the first part of this discussion, the focus is on reproductive cloning, the effort to produce a human being by means of cloning technology. The second part explores some of the issues raised by *therapeutic cloning*, the use of cloning technology to produce isogenic (immunologically compatible) stem cell lines for the treatment of disease or trauma.

REPRODUCTIVE CLONING

KEY DISTINCTIONS

In approaching reproductive cloning, it is helpful to make several important distinctions. One is between cloning by blastomere separation (embryo splitting) and cloning by somatic cell nuclear transfer (SCNT) technology. The former involves the deliberate creation of multiple copies of a genome by separating or multiplying the individual cells of an early embryo. The latter involves the use of a somatic cell nucleus from an existing (or even a deceased) individual to replicate the genome of that individual. Embryo splitting is an object of concern because it makes possible the production of multiple identical genotypes. SCNT cloning raises the issue of reproducing the genome of a living being. In fact, the distinction between these two types of cloning is not so sharp. Using blastomere separation and embryo freezing it is possible to produce the identical twin of an individual born years before (Macklin, 1994). SCNT cloning can also be used to mass-produce identical genomes by starting with multiple copies of a somatic cell (Wilmut *et al.*, 2000). Because SCNT cloning permits the deliberate reproduction of the genome of an existing offspring or adult and can also be used for mass reproduction, it raises the sharpest ethical questions. The following discussion thus focuses on SCNT.

When speaking about reproductive cloning, it is also useful to distinguish the specifically *procreative* uses of this technology from its deliberately *replicative* uses.[1] The former, for example, could involve the use of SCNT by people who cannot otherwise have a biologically related child. This might occur in the case of a heterosexual couple, both of whom lack gametes for sexual reproduction, or when only one partner has gametes and the couple does not wish to introduce a third party into their relationship through the use of artificial insemination or egg donation (Orentlicher, 1999). This kind of procreative SCNT technology may be attractive to lesbians or gay men. Some lesbian couples, for example, are hesitant to employ donor insemination because they fear that the sperm donor may later assert parenting claims. Procreative SCNT may also appeal to individuals who wish to avoid transmitting a serious genetic disease through sexual reproduction. This might be true when a disorder is involved for which not all the mutations are known and which cannot be effectively avoided by prenatal screening. In all these cases, the aim is not so much to replicate an existing genotype but to have a healthy child that is biologically related to the parents.

Replicative reproductive cloning, in contrast, deliberately aims at the duplication or multiplication of an existing genome. For example, couples seeking to replace a deceased child or to produce a "twin" suitable as a tissue donor for an existing child might use replicative cloning. Replicative cloning might also be used commercially in connection with the sale of "celebrity" genomes for reproductive purposes. Finally, governments or other organizations might use it to produce many desired genotypes for military or other purposes. Because replicative SCNT cloning aims directly at producing an individual with desired genetic characteristics, it raises more ethical questions than does procreative SCNT cloning.

NIGHTMARE SCENARIOS

Fictional literature, film, and popular treatments of cloning often begin with a series of highly imaginative but often scientifically dubious scenarios. In discussing the

---

[1]Davis (1997) makes a conceptually similar but terminologically different distinction when she speaks of logistical versus duplicative cloning.

ethics of human cloning, it is useful to begin by subjecting these scenarios to critical scrutiny. Human cloning raises serious ethical questions, but for a variety of reasons, the concerns raised by these scenarios are not foremost among them.

One such scenario, as was depicted in the movies "Multiplicity" and "The Sixth Day," portrays existing individuals as losing their individual identity or rights through the instant (and possibly unconsented and surreptitious) production of cloned "copies" of them. All these scenarios miss the point that cloning produces at most a genotypically similar human embryo that must still be raised through all the stages of gestation, infancy, childhood, and adolescence. Even if we grant what is not the case, that "genotype equals phenotype" (see below), it would still take decades to produce the cloned counterpart of any existing adult. Because these fictional scenarios rely on purely imaginative technologies of instant replication of body and mind, we do not have to take them seriously.

A related scenario envisions cloning as being used by despots to mass-produce obedient armies of "superwarriors." This vision, too, is flawed on several counts. For one thing, it presumes that a militarily desirable phenotype can be identified and mass-produced by cloning technology. As recent experience in the Gulf and Balkan conflicts illustrates, however, military technology can change dramatically in the several decades it takes a human being to mature. As a result, after years of investment of time and energy, a dictator might see his cloned superwarriors decimated on the battlefield by a handful of frail but adept computer technicians sitting miles away from the battlefield. Beyond this, it should be obvious here that what is really troubling about this scenario is the existence of a dictator willing to use biotechnology for aggression and reproductive slavery. Referring to the specter of a future Hitler "making an army of ten million identical robotic killers," Stephen Jay Gould (1998) observes, "If our society ever reaches a state in which such an outcome might be realized, we are probably already lost."

It is worth noting in this connection that we should also not invoke as arguments against cloning those scenarios that see cloned individuals being produced singly or *en masse* as organ donors, slaves, or otherwise disposable entities to be used at will by other noncloned human beings. Our basic ethical principle of respect for persons condemns such behavior. No less than existing natural clones—monozygotic twins—cloned individuals would be persons with all the rights and legal protections possessed by any human being. A cloned biological origin would in no way reduce status. Theoretically, if such abuses occurred, the problem would lie not with cloning, but with the unjust social and legal systems that permitted them to happen. The same basic ethical principle of respect for persons also condemns the production of "organ bank individuals" in the form of headless fetuses, children, or adults. Although this idea has been mentioned by some scientists (Zorpette, 1998) and has been treated in fictional literature (Egan, 1999), it runs counter to the respect for the human body that is nearly a universal feature of human culture. The considerations that make it wrong to desecrate the body of a deceased individual also prohibit producing and then scavenging a whole body via cloning. The ethical route to this goal is through therapeutic cloning technology with the aim of producing cells and tissues that could be incorporated into scaffolds to produce "neo-organs" (Lanza *et al.*, 1999).

Finally, some scenarios focus on the massive replacement of traditional sexual reproduction and parenting with clonal reproduction. The fear here is that many individuals or couples will be attracted to having children with identifiable and desirable genotypes and will use commercially available cloned embryos for this purpose. The result would be a race of Michael Jordan athletes or Kim Basinger look-alikes. Not only would this have a cultural impact, in the reduced diversity of human talents and abilities, but also a small number of "monocultures" could replace existing human genetic diversity. But these fears are largely without foundation. There is substantial evidence that people—even those suffering from infertility, who must use donor gametes—basically want to have children like themselves.

The "Noble Laureates" sperm bank opened some years ago has done little business (Brock, 2001). It is true that couples using donor gametes frequently select a sperm or egg donor with desirable qualities, such as evidence of advanced education or physical attractiveness. But as the director of one of the nation's largest fertility registries observes, these couples typically want a child whose intelligence is comparable to theirs and with whom they will be able to relate. "They're not looking for any Mensa applicant," she says (Barlow, 1999).

In the absence of profound social change or coercion, there is no reason to believe that many human beings would choose to build their family from commercially available cloned embryos. Furthermore, if a small number of people used cloning for procreative reproductive purposes, there would be no net impact on human genetic diversity. To further prevent this from happening, and to prevent the possible confusions caused by the existence of many individuals with the same genotype and physical appearance, it would be reasonable for society to set a limit on the number of individuals that could be produced from a single genome (Macklin, 1994; Brock, 2001).

SERIOUS ETHICAL CONCERNS

That reproductive cloning may benefit some people is an important consideration in its moral evaluation. Most ethical theories agree that we should strive to protect human liberty and the pursuit of happiness unless such exercise threatens disproportionate harm to individuals and society. Like other assisted technologies, cloning is likely to be desired and used by a small number of people who could not otherwise have a child (Strong, 1998). From an ethical standpoint, therefore, the relevant question is whether this modest benefit is likely to be outweighed by harms created by the practice of cloning. Debates about cloning typically take the form of identifying and assessing a series of alleged harms.

Physiological Harms to Offspring

The most immediate set of ethical questions revolves around the possible risks to any offspring produced by the cloning procedure. Foremost among these are physiological risks. There is widespread agreement that we should try to avoid deliberately and knowingly imposing birth defects, lifelong morbidity, or premature mortality on our offspring. The focus here is on the born offspring who might result from cloning. The substantial loss of early reconstructed embryos that has been common in cloning procedures is a direct source of concern only to those who believe these embryos are moral persons (a matter that is addressed below in connection with therapeutic cloning). For people who are less concerned with the fate of the early embryo, the high prenatal rate of loss is worrisome because it points to enduring physiological problems that cloning may produce.

There is significant scientific evidence that SCNT cloning as an assisted reproductive procedure is not yet safe enough for clinical implementation. The high ratios of transferred embryos to live births in most animal species experimented on to date, and the frequent occurrence of perinatal deaths and birth defects, indicate that we do not yet fully understand the process of epigenetic reprogramming that goes on in a reconstructed cloned embryo and cannot assure that the process will be accomplished successfully in most cases (Wolffe and Matzke, 1999; Pennisi and Vogel, 2000; Kolata, 2001). Still unresolved are complex questions concerning the risk of passing on somatic cell mutations or shortened telomeres to the resulting offspring (Wilmut *et al.*, 2000). Some believe that long-term postnatal survivors are likely to have subtle epigenetic defects that, although below the threshold that threatens viability, can nevertheless pose serious health problems later in life (Jaenisch and Wilmut, 2001). Chance successes such as Dolly, although promising from a long-term standpoint, cannot currently justify clinical implementation of this technology.

The National Bioethics Advisory Commission arrived at this same conclusion in its 1997 review of the issues. Putting aside most other moral questions that had been raised, the Commissioners concluded that at this time cloning would be a "premature experiment that would expose the fetus and the developing child to unacceptable risks" (National Bioethics Advisory Commission, 1997). These risks led the Commission to call for a legal ban on efforts to clone a human child. A proposed "sunset" clause in the ban would have allowed it to be periodically reviewed as scientific research went forward.

Although most people would agree with these conclusions, there are some bioethicists who would justify cloning even at this time. These thinkers base their views on the writings of philosophers such as Derek Parfit (1984), David Heyd (1992), and others.[2] In the cloning debate, the lawyer–bioethicist John Robertson (1994, 1997, 1998) has most frequently taken this position. According to this view, a child born with serious birth defects caused by a reproductive technology that brought it into being cannot be said to be harmed so long as the child would not otherwise have been born without the technology and so long as its life is "worth living." This conclusion stems from the fact that judgments of harm usually involve the comparison of an individual's condition before the alleged injury was done with the individual's condition after the alleged injury. Harm is done if someone is made *worse off* than they were, as a result of another's actions. Robertson acknowledges that when cloning is used to assist some people to have a child that they could not otherwise have, it may cause a child to be born with serious physical and psychological injuries. However, because the child would not exist without the use of the cloning procedure, we cannot say that it has been made "worse off." The condition of the child "before" cloning was not to exist at all. Because even individuals born with birth defects are usually glad to be alive, says Robertson, it cannot be said that even a seriously damaged cloned child would have preferred never to have lived. In Robertson's view, clinicians or parents clearly harm a child by cloning only in rare cases in which the child's suffering is so great that most individuals with the same problems would rather die than continue to live with their impairments.

Robertson's view has many problems. For one thing, it confuses the state of never coming into being with dying. Most of us are averse to dying, but it is by no means clear that we care about never having come into being in the first place. It is also doubtful that coming into being should be regarded as a benefit against which possible harms should be weighed. People routinely avoid the conception of children, and do not believe that they are denying anyone a benefit in doing so. If we were to conclude that production of life is an inherently positive benefit for the child born, we would have to review seriously our practices of birth limitation and population control. Finally, the Parfit–Robertson view assumes that harm always involves making someone worse off than they were before. But there are cases, usually encountered in the area of investment law and fraud (Kelly, 1991), in which people are viewed as harmed when they are made materially better off but *less* well off than they might have been. This shows that we can also harm someone by knowingly disappointing his or her reasonable expectations. Similar logic may apply to cloned children. A child produced by cloning who faces a lifetime of serious illness or disability as a result of this procedure would have reason to feel that his or her reasonable expectations had been diminished. This tells us that, despite the philosophical challenges, our common sense intuition is correct that it would be morally wrong to try to produce a child given the current state of cloning technology.

A further question is how much we must reduce the risks in cloning before we can ethically employ it. What level of risk may a reproductive technology impose on the child it helps bring into being? A standard of "no risk at all" is certainly too

---

[2]See Woodward (1986), Hanser (1990), Steinbock and McClamrock (1994), Brock (1995), Heller (1996), and Roberts (1996).

high. Even normal conception and birth bring with them risks of congenital defects and premature death. It is tempting to make this normal level of risks the baseline for cloning attempts, but even this level may be too demanding.[3] Most accepted assisted reproductive technologies have a heightened level of risk over normal conception and birth (Wilcox *et al.*, 1996; Saunders *et al.*, 1996; Bernasko *et al.*, 1997). In the case of *in vitro* fertilization (IVF), for example, this is partly due to the practice of transferring multiple embryos in order to enhance success rates. The resulting multiple births are often accompanied by prematurity and associated perinatal problems. We do not have to approve of all these practices or the level of risk currently allowed in some areas of reproductive medicine to see that some degree of increased risk over normal conception and birth is permissible here. The reason for this is that infertile parents desperately want a child and these legitimate claims must be balanced against the health needs of the child (Green, 1997). If current opinion and practice are a guide, it is reasonable to permit modest, but not extraordinary, increments of risk over the normal baseline in order to help people have a wanted child. Human cloning procedures will meet this physiological risk standard when substantial animal research and human embryo research indicate that cloning is likely to be no less safe than IVF and intracytoplasmic sperm injection are today. In addition, such risks must be measured not just at birth, but across the child's whole lifetime. For example, the fear that cloning may lead to premature aging or heightened risk of adult-onset cancers must be put to rest before cloning can judged to meet this safety standard.

Psychological Harms

The issue of possible psychological risks to children produced by cloning is weighted with assertions that are often highly speculative and difficult to assess. This does not diminish their force as reasons for concern, but it does urge caution. It also recommends additional research before these assertions are made the basis for ethical or policy conclusions.

Procreative reproductive cloning seems to pose fewer problems in this regard than does replicative cloning. Although the parents may wish some degree of genetic connectedness to their offspring, they are not trying to produce a child with specific genetic traits. This reduces the risks associated with imposing expectations on the child. However, even procreative cloning raises novel ethical questions. Although sibling identical twins are common, we have never before had families with the kind of parent–child identical twins that cloning makes possible. In cases in which the genetic (somatic cell donor) parent is also the social or rearing parent, will there be more similarity and intimacy between the child and parent than is good for the child's healthy development? Bioethicist Laurie Zoloth (2001) voices the concerns of many people when she states the belief that cloning violates the essence of parenting as "the act of encounter with the other who is both not-you and of-you."

There is some reason for believing that these worries may be overstated. For one thing, there are already varying degrees of similar intimacy between parents and their sexually produced children, and this is not commonly viewed as a source of moral concern. As a result, these concerns may not rise to the level of justifying a ban on the first cloning attempts. Nevertheless, they recommend substantial preparatory counseling of people using this technology for procreative purposes and ongoing counseling of both the parents and children. They also suggest that any first attempts at cloning should be undertaken in a research context with careful monitoring and recording of physiological and psychological developments. To reduce risks to the child created by notoriety and overexposure to publicity, it is

---

[3]Pence (1998) argues that the appropriate level of risk is "the normal range of risk that is accepted by people in sexual reproduction." However, he does not make clear whether "normal" here includes assisted reproductive technologies.

also important that parents and caregivers work to protect the child from excessive media attention. Although some fear that this attention would be unavoidable and excessively harmful for the first cloned children, the experience with Louise Brown, the world's first "test tube baby," suggests that the risks are manageable.

Some bioethicists have argued that even when undertaken only for procreative purposes, cloning threatens to disrupt family relationships and through this inflicts unknown injuries on the child. In the words of Leon Kass (1997):

> Troubled psychic identity (distinctiveness), based on all-too-evident genetic identity (sameness), will be made much worse by the utter confusion of social identity and kinship ties. For, as already noted, cloning radically confounds lineage and social relations, for "offspring" as for "parents." As bioethicist James Nelson has pointed out, a female child cloned from her "mother" might develop a desire for a relationship to her "father," and might understandably seek out the father of her "mother," who after all is also her biological twin sister. Would "grandpa," who thought his paternal duties concluded, be pleased to discover that the clonant looked to him for paternal attention and support?

Kass does observe that such confusions are not unique to the area of cloning. Adoption and assisted reproductive practices such as intrafamilial sperm and egg donation already raise similar issues. In Kass's view, however, there is no good reason to augment these difficulties by permitting procreative cloning. In opposition, those who conclude that there is value to developing this additional assisted reproductive option would point to existing practices as evidence that these psychological difficulties are not likely to be overwhelming or insurmountable. They also argue that these risks call not so much for a ban as for caution in proceeding and the provision of effective counseling and followup.

Replicative reproductive cloning raises a host of additional concerns. Foremost among these is the problem of imposed expectations. For example, it appears to be the case that many people today who say they would like to avail themselves of cloning do so because they wish to "replace" a child who has died. However, in such cases it is reasonable to worry that the "replacement" child may be forced to live its life under the shadow of the deceased sibling. In addition, some fear for the emotional maturity and even psychological health of parents who would seek to respond to the death of a child in this way. In testimony before Congress on a bill to ban reproductive cloning, bioethicist Thomas Murray (2001), who lost a 20-year-old daughter in senseless act of violence, sought to address a parent who looked to cloning as a way of responding to the death of his 11-month-old son:

> Massive waves of sorrow knock us down, breathless; we must learn to live with them. When our strength returns we stagger to our feet, summon whatever will we can, and do what needs to be done. Most of all we try to hold each other up. We can no more wish our grief away than King Canute could stem the ocean's tide. . . . There are no technological fixes for grief; cloning your dear dead son will not repair the jagged hole ripped out of the tapestry of your life. Your letter fills me with sadness for you and your wife, not just for the loss of your child but also for the fruitless quest to quench your grief in a genetic replica of the son you lost. It would be fruitless even—especially—if you succeeded in creating a healthy biological duplicate. But there is little chance of that.

There is a great deal of wisdom in Murray's remarks. Surely, researchers and clinicians working with families seeking cloning for these reasons must be aware that the requests may evidence emotional problems and failures of adjustment that are not only a source of worry in their own right, but that could ultimately imperil the cloned child. Nevertheless, we might ask whether all such requests are on a par. Some might argue that the use of cloning to replace a deceased infant should be assessed differently than a request to replace an older child. Given the limited influence of genes on phenotype, it is unrealistic to expect that a cloned child will be substantially the same as the teenager it is supposed to replace. False expectations in this regard could disappoint the parents and harm the child. However, most of the observable features of the clone of an infant who has died will be the same as its forerunner. It may be, as Zoloth (2001) insists, that all such requests stem from a "fear of mortality that lurks always at the corners of birth and fecundity." But cloning in such cases may also be a reasonable way for some parents to cope with grief. Whether cloning can be used in these ways by some people without serious harm to themselves or their children is a question that ongoing research, monitoring, and followup may help us resolve.

Replicative cloning might also be used to produce a matching tissue donor for a dying sibling. Some who object to this argue that such practices have the effect of making the individual nothing but an instrument to the well-being of another. They see this as violating the Kantian precept that we should never treat other human beings "merely as means" but always as "ends in themselves" (Kitcher, 1998). However, the precise meaning and applicability of the Kantian precept are far from clear (Green, 2001). The precept does not rule out using persons as "means" but only "merely as means." Because a child conceived (or cloned) as a tissue donor for a sibling will almost certainly be loved and respected in its own right, we can ask whether this would not satisfy the Kantian requirement that the child also be regarded and treated as "an end in itself." In any case, this issue is again not unique to cloning. Several cases in United States law have addressed parents who conceived a child as a bone marrow donor for a dying sibling. Reflecting the ethical balancing of parental and sibling interests, courts and medical practitioners have usually permitted this practice (Stewart, 1991; Robertson, 1994; Nelson and Nelson, 1995). There is no reason to think that the conclusion would be different for cloning.

This leaves for consideration other intentional uses of cloning to replicate a known genome. This could happen if infertile people choose to have a child using a "celebrity" genome, perhaps in the hopes of having their own Albert Einstein or Tiger Woods. I have already argued that we probably do not have to fear the larger social effects of such choices, because they are not likely to be numerous. But it is still important to consider the impact of such choices on the psychological welfare of the resulting child. What does it mean to be brought into being in a family in which one faces such intense expectations? Would it not be a "no-win" situation for such a child? To the extent that the child fulfilled its parents' expectations and was in every way as successful as its somatic cell donor prototype, its achievements would be chalked up to "good genes." However, if it failed to live up to expectations, the child could easily be faulted for insufficient effort or diligence in developing its genetic gifts. Whatever the outcome, the child would live its life under pressure to meet the expectations that brought it into being, expectations that may not correspond significantly with the child's own interests or talents. Yearning to write poetry, the child might be incessantly driven to become a golf champion.

Over 25 years ago, well before cloning was possible, these worries led the philosopher Hans Jonas to articulate a novel human right to "ignorance" about the implications of one's own genome. According to Jonas (1974), such ignorance is "a condition for the possibility of authentic action" and human freedom. In a related vein, the philosopher Joel Feinberg (1992) has defended each child's "right to an open future." Some have employed this concept to argue against the legitimacy of cloning (Davis, 2001).

All these moral concerns make sense if we assume that there is a high degree of correlation between genotype and phenotype—that is, if we assume that one's genes substantially contribute to and dictate one's expressed physical and mental characteristics and abilities. In that case, parental expectations of a cloned child would be understandable if not entirely justifiable morally. However, although there is little doubt that some traits, such as skin or eye color, are highly heritable, many other complex traits, especially behavioral traits, are far less so (McGuffin *et al.*, 2001). This leads the molecular biologist Richard Lewontin (1997) to speak of what he calls "the fallacy of genetic determinism," which he defines as the supposition that genes "make" the organism. However, says Lewontin,

> It is a basic principle of developmental biology that organisms undergo a continuous development from conception to death, a development that is the unique consequence of the interaction of the genes in their cells, the temporal sequence of environments through which the organisms pass, and random cellular processes that determine the life, death, and transformations of cells. As a result, even the fingerprints of identical twins are not identical. Their temperaments, mental processes, abilities, life choices, disease histories, and deaths certainly differ despite the determined efforts of many parents to enforce as great a similarity as possible.

This understanding has several implications for human cloning. First, it tells us that parents who attempt to use cloning to replicate the excellences of some known person are likely to be disappointed. In order to prevent their disappointment from becoming a burden to the child, they must be made aware of these facts. Information and education, not the banning of this technology, are the most effective way to prevent these harms. Second, it tells us that in the longer run we have less to fear than is believed from the use of cloning. Despite fictional representations, human clones will not be soulless, "cookie cutter" replicants. Like identical twins, they will be discreet individuals with different physical characteristics and personalities. Indeed, the differences between cloned offspring are likely to be greater than those between monozygotic twins. Unlike cloned children, monozygotic twins also share the 3% of genetic material constituted by the mitochondria and experience the same environment from conception onward. In short, the real worry here is not cloning, but misconceptions that are best eliminated by adequate parental selection and by a thorough informed-consent process.

Social Harms

Ethicists debate whether above and beyond the harms inflicted directly or indirectly on individuals there are also harms that can be done to society as a whole. Without settling this debate, it is clear that damaging changes in cultural patterns or institutions caused by cloning should be taken into account because, ultimately, such changes could erode the quality of life for many individuals. Of course, not every change to existing institutions or cultural values is necessarily harmful. The challenge is to separate out reasonable concerns about changes that could harm people from concerns based merely on the fear of change.

Many who object to the prospect of reproductive cloning do so because of what they see as its negative impact on the institutions of family and marriage. These institutions are indisputably important to the well-being of those nurtured by them. Speaking before the National Bioethics Advisory Commission, the ethicist Lisa Sowle Cahill (1997) observes, that "families in all their variety of cultural forms have been key institutions for the structuring of societies and transmission of knowledge, values, and practices, as well as training us in moral dispositions such as empathy, fidelity, honesty, and altruism." Part of the reason for this, Cahill believes, is because traditional procreation links the individual to two different lineages. In her view, human cloning represents a fundamental threat to the dynamics that make family life so important:

> It is not too strong to say that cloning is a violation of the essential reality of human family and of the nature of the socially related individual within it.

We all take part of our identity, both material or biological and social, from combined ancestral kinship networks. The existing practice of "donating" gametes when the donors have no intention to parent the resulting child is already an affront to this order of things. But in such cases, as in cases of adoption where the rearing of a child within its original combined-family network is impossible or undesirable, the child can still in fact claim the dual-lineage origin that characterizes every other human being. Whether socially recognized or not, this kind of ancestry is an important part of the human sense of self (as witnessed by searches for "biological" parents and families), as well a [*sic*] foundation of important human relationships. Cloning would create an unprecedented rupture in those biological dimensions of embodied humanity which have been most important for social cooperation.

The ethical literature on cloning contains many similar statements. In some cases, religious appeals are made to biblical or other texts to ground the claim that departures such as cloning, from sexual reproduction, somehow violate the naturally or divinely constituted order of human procreation, and as such, represent a hazardous departure from one of our most important biological and social realities.

Again, all these claims are very difficult to assess. As Cahill indicates, many current reproductive and family practices, from egg donation to adoption to single-parent families, already depart in various ways from aspects of the standard patterns of "dual-lineage" procreation. There is no evidence to suggest that children raised in such families fare worse than others, at least when such things as socioeconomic levels are kept constant. Furthermore, studies have shown that the offspring of lesbian couples, some of whom result from anonymous-donor artificial insemination, are well adjusted (Sultan, 1995). Nor is it entirely correct to say that a cloned child would lack dual lineages. Such lineages would be present up to the point of the child's genetic, cell-donor parent.

One cannot help thinking that these family-based objections rest more on a fear of changing patterns than on solid evidence or even reasonably held apprehensions. Because standard reproductive and familial practices have been largely successful in facilitating the continuance of society, it is easy to conjure up the many ways that departures from them might damage children or society. Nevertheless, it is appropriate to bring critical sensibilities to such ways of thinking. For example, it is clear that the picture of the heterosexual family offered here is highly idealized. In reality, there have been a vast number of dysfunctional heterosexual families. It is also true that the surest way to achieve effective socialization of children and ensure their well-being is to ensure their birth as wanted children, to parents with the wisdom and the material and emotional resources to rear them. There is no reason to believe that the parents of cloned children would not meet this description.

Also lacking here is almost any sense that cloning might bring new and positive dimensions to parenthood or child rearing. Of course, it is hard to imagine relational possibilities that have never existed before. Nevertheless, studies suggest that monozygotic twins appear to have fewer psychological problems than do singletons (Macklin, 1994). Such twins often report satisfaction with the opportunities for enhanced communication afforded by their shared qualities. Some fictional literature, such as Ursula LeGuin's story "Nine Lives" (1998), portrays intense intimacy between cloned siblings. It is not unimaginable that parents and their cloned children will also experience new and positive degrees of communication and bonding (Chin, 1998).

In summary, all these concerns over the various psychological and social harms created by cloning do not appear to rise to a level sufficient to justify an outright ban on this technology when it becomes physiologically safe enough to employ. They tell us, however, that if cloning is to be used for procreative or other desirable and ethically justifiable purposes, we should proceed cautiously. At least initially, cloning should take place only in research contexts, where it is possible to ensure appropriate selection and counseling of prospective parents and to maintain followup of each cloned child in its family setting. In this way, if individual or familial problems arise, they can be mitigated and the experience gained can be used to revise, or even halt, further cloning procedures.

THERAPEUTIC CLONING[4]

Therapeutic cloning (or egg activation by means of nuclear transfer) is a new biomedical technology that has the potential to revolutionize medicine. It involves the transfer of the nucleus from one of the patient's cells into an enucleated donor oocyte for the purpose of making medically useful and immunologically compatible cells and tissues (Lanza *et al.*, 1999). Because it requires the creation and dissection *ex utero* of blastocyst-stage embryos, it also raises complex ethical questions (McGee and Caplan, 1999; Shapiro, 1999).[5]

ETHICAL OBJECTIONS

Ethical objections to these technologies fall into three categories. The first pertains to all cell replacement technologies using stem cells derived from human embryos [human embryonic stem (hES) cells] or from the gonadal ridge of aborted fetuses (hEG cells). This objection has to do the way in which these technologies depend on the destruction of nascent human life. A second objection has to do with the fear that any cloning of human embryos opens the way to the eventual cloning of a human being through the reproductive uses of this technology. A third set of objections focuses on perceived threat to women's health and freedom, from the potential market in human eggs that therapeutic cloning may create.

Objections Based on the Threat to Nascent Human Life

Many who oppose therapeutic cloning research also oppose human stem cell research using hES or hEG cells. They base their position on the view that human life, in a moral sense, begins at conception. Those holding this position believe that from conception onward, the embryo or fetus is the moral equal of any human child or adult (Mitchell, 2000; Doerflinger, 1999; Pellegrino, 1999). This means that it cannot ethically be used in research that risks its healthy survival. Obviously, hES cell research, which depends on the destruction of a human embryo, cannot meet this test. Because of its close association with abortion, many holding this view also oppose hEG cell research.

To many of those who hold this view, it does not change things that, where stem cell research is concerned, the embryos used to produce the cell lines are almost certain to be destroyed. Opponents of hES cell research liken the frozen or extra (unused) embryo to a dying child or adult and feel that its circumstances call for enhanced, not reduced, research protections (Doerflinger, 1999; Pellegrino, 1999). They prefer research that aims at the development of adult stem cell lines and point to the promise of some recent research in this direction (Bjornson *et al.*, 1999; Johansson *et al.*, 1999; McKay, 1999). They are also willing to accept delays in the progress of stem cell research rather than permit the utilization of cell line sources that they regard as morally objectionable.

Some who believe that human life begins at conception or who otherwise oppose the destruction of embryos for scientific research are willing to permit the

---

[4]The following discussion draws heavily on the ethical arguments I developed in an article I co-authored with R. P. Lanza *et al.* (2000).

[5]I do not consider here the many complex ethical questions that will arise once it is possible to develop isogenic stem cell lines *in vitro* and research is begun on their clinical utilization. Then issues of the safety of transplanting these cell lines into healthy or ill research volunteers will move to the fore.

use of spare embryos, unused in infertility procedures, for hES cell research.[6] They reason that no useful purpose is served by refusing to use the cells or tissues made available in these ways. They also believe that it is unlikely that the use of these cells or tissues in research will encourage either the creation of spare embryos in infertility medicine or the practice of abortion, because there are independent reasons why these go on. For example, in 1996, 3600 embryos unwanted by their progenitors were destroyed in compliance with British regulations (Ibrahim, 1996). Until the efficiency of infertility procedures is greatly increased, couples will routinely produce more embryos than they can successfully transfer or donate for adoption. Regulations in the United States prohibiting women from benefiting from fetal tissue donations also appear to have reassured many people that the permission for such donations does not encourage abortion.

This limited acceptance of hES or hEG cell research vanishes, however, when an embryo must be created *de novo* for a stem cell research protocol, as in the case of therapeutic cloning research. This makes this research particularly objectionable to all those who believe that life begins at conception. It might be argued that an egg activated by nuclear transfer is not a human "embryo" in the traditional sense of that term, because it is not the result of fertilization. Nevertheless, most who believe that life begins at conception can be expected to extend their view to this entity as well, on the grounds that its developmental potential is the same as a naturally fertilized egg. It is a sign of this way of thinking that existing U.S. federal regulations prohibiting federal funding of embryo research define the embryo as "any organism . . . that is derived by fertilization, parthenogenesis, cloning, or any other means from one or more human gametes" (Public Law, 1996).

In addition to this large body of opponents, some who do not share the view that life begins at conception oppose any research in which embryos are deliberately created and destroyed (Washington Post Editorial Board, 1999). These opponents fear the symbolic implications of the deliberate creation and destruction of a form of human life. They worry that these practices could start us down a precarious path toward the use of other classes of research subjects in a harmful way, without their consent.

Replies to These Objections

Many people do not agree with the claim that human life in a moral sense begins at conception. Some hold a "developmental view" that sees prenatal life as growing in moral weight over the course of a pregnancy and reaching full equality with other human beings only very late in pregnancy or at birth. Where the very early embryo is concerned, many considerations support this view and weaken the claim that it should be given substantial moral weight. For example, almost all views holding that human life begins at conception maintain that this is the moment when a new and unique human individual comes into being. However, because twinning and chimerism are still possible during the early stages of development (Grobstein, 1988; Strain *et al.*, 1998), it is doubtful that one can speak of human individuality at this time (Ford, 1988; Shannon and Woltor, 1990; McCormick, 1991). Developmental individuality is not attained until the body axis (primitive streak) begins to form at gastrulation.

Because the early embryo lacks organs, it is not capable of thinking, feeling, or having experiences. It cannot suffer pain, or regret the loss of its opportunity to develop. This leaves its *potential* for development into a human being as the sole consideration that might justify granting it significant moral weight. However, it is not clear how much this potential should count in justifying its protection. Most entities with potential to develop are not valued or treated in the same way as their

[6]A *Washington Post* (1999) editorial reveals a position generally opposed to embryo destruction for research purposes, but permitting of the use of spare embryos.

developed form (Warren, 1973). Also relevant here is the very high rate of early embryo loss. Some estimates place this as high as 80% of all conceptions (Roberts and Lowe, 1975; O'Rahilly and Müller, 1992). In most cases, the great majority of embryos will not go on to develop into a human being. This loss rate reduces the force of the potentiality argument. All these considerations support a developmental view that accords significantly less weight to the pregastrulation embryo and that justifies its use in research that could greatly benefit children and adults. Indeed, where research promises sufficient therapeutic benefit, this view may even morally require such research.

Some who accept these conclusions concerning the embryo nevertheless resist the deliberate creation of embryos for research on stem cells or egg activation through nuclear transfer. They believe that such practices could lead to the "instrumentalization" or "commodification" of human life generally (National Bioethics Advisory Commission, 1999; Annas *et al.*, 1996). Those holding this view often advance some kind of "slippery slope" argument involving the claim that a practice that is not inherently objectionable may nevertheless lead to others that are clearly wrong (Williams, 1985). This can occur because the line between pregastrulation and postgastrulation embryo is not clear enough to anticipate that it will be long respected. Or, the slide can happen because the attitudes and practices established by such research habituate people and prepare them psychologically or socially for other, more worrisome practices.

On neither count, however, is there reason to believe that a permission to create and use hES cells in research, or the development of therapeutic cloning, will lead to the predicted harms. The line established by gastrulation and the appearance of the primitive streak is a clear one. There is also no evidence that the use of embryos in research leads to abuse of human research subjects. Since 1990, Great Britain has permitted the use of embryos in research, including research involving the deliberate creation of embryos, and no such abuses have been recorded. On the contrary, it is reasonable to believe that where embryo research is permitted and monitored under carefully defined regulations, it is less likely that poor or ethically irresponsible research will occur.

Critics of hES cell and therapeutic cloning research point out that it may someday be possible to achieve the direct reprogramming of adult somatic body cells or adult stem cells for therapeutic purposes. On this ground, they recommend abandoning or postponing current research that involves the creation or destruction of embryos (Doerflinger, 1999). However, this conclusion does not follow. Even if therapeutic cloning research proves to be only "transitional," it is worthwhile to pursue it now in order to hasten the day that other cell replacement therapies become practicable (Chief Medical Officer's Expert Group, 2000). All these matters support the conclusion that when its nature and purposes are understood, therapeutic cloning research can be ethically justified.

The Tie to Reproductive Cloning

Some object to therapeutic cloning research on the grounds that it will hasten the day when cloning becomes a reproductive option (Beale, 2001). This objection rests on the premise that it is wrong to clone a human being. However, as the discussion in the first part of this chapter shows, it is by no means clear that this will always be the case. Indeed, if the principal objections to cloning are the physiological risks to any child produced in this way, then it might be argued that extensive embryo research in the context of therapeutic cloning could contribute to improving our understanding of these risks and reducing them.

Therapeutic cloning research may make possible the premature reproductive cloning of a human being. This could occur if reconstructed embryos produced for cell research are wrongly diverted to reproductive purposes. To prevent this, researchers performing therapeutic cloning research must give special attention to security issues. Eggs and embryos should be photographed, counted, and carefully monitored throughout research procedures. They should be maintained in a locked incubator under constant video surveillance. To reduce the risks of diversion, no unauthorized individuals should have access to eggs and embryos, and the presence of at least two members of the research team should be required for all embryo manipulations. Finally, steps should be taken to ensure that all cloned embryos are destroyed by 14 days of development and that their destruction is documented. A period of 14 days permits blastocyst development and the removal of hES cells from the embryonic disk, but it halts research before gastrulation occurs *in vitro*. This period of 14 days also conforms to an existing international consensus on the reasonable limits of human embryo research (National Institutes of Health Human Embryo Research Panel, 1995).

It should also be said here that informed consent by both somatic cell donors and egg donors is a requirement of *all* nuclear transfer research, whether reproductive or therapeutic. Just as no one should have their gametes used to create an embryo or child without their consent, no one should have their body cells or eggs transformed into an embryonic human being or born child without their knowledge or permission (Amer, 1996). It follows from this that children should not ordinarily be used as somatic cell donors in therapeutic cloning research. The exception to this would be cases in which the transfer procedure is designed to provide direct medical benefit for the child whose cell is being used. In such cases, the parents of the child can consent to the procedure and, if the child is judged to be sufficiently mature, its assent should also be secured (Department of Health and Human Services, 1991).

The Issue of a Market in Eggs

In the foreseeable future, therapeutic cloning research will require a substantial supply of human oocytes. Some fear that this will expose many women to the medical risks associated with hyperstimulation syndrome (liver damage, kidney failure, or stroke) as well as the longer term risks of ovulation-stimulating drugs, including a perhaps heightened risk of ovarian cancer, as has been suggested by some studies (Rossing *et al.*, 1994). Because few women can be expected to donate eggs without compensation, there is also concern that payment for eggs will compromise extraeconomic values of deep importance or even foster the commodification and commercialization of human body parts generally (Cohen, 2001; Dresser, 2001).

The fact is, however, that a substantial market in human eggs for reproductive purposes already exists. A statement by the Ethics Committee of the American Society for Reproductive Medicine (2000) permits such compensation so long as it is fair and not so substantial that it becomes an undue inducement that leads egg donors to discount potential risks. It is not clear that a willingness to contribute to health-related research is any less valid a motivation for egg donation.

Nevertheless, it remains true that the ethical treatment of egg donors is one of the most important responsibilities of therapeutic cloning researchers. Researchers must strive to ensure full and informed consent on the part of donors. This includes informing them of all the medical risks, immediate and long term, associated with ovulation induction. Donors should understand that their eggs will be used to produce cloned embryos that will be dissected to develop immortalized hES cell lines. They should be notified that there might be commercial benefits to this research that may accrue to others. Although it can be ethical to compensate donors, this is best thought of as reimbursement for a volunteer's time and effort. This means that payments should not be so substantial that they distort a donor's ability to assess the risks of the study. Payments may be prorated, depending on the extent of a donor's completion of the induction regimen, but it is not advisable that they be conditioned on or proportioned to the number of eggs produced. This reduces the incentives that might lead the donor into overstimulation. To prevent this, as well, special care should be taken to maintain drugs at safe levels and to monitor closely donor response. Finally, because researchers understandably wish to secure the largest number of eggs for their research, they have a conflict of interest in such cases. For this reason, it is advisable that an egg donor program be in the hands of a separate team and that an independent study monitor is used to ensure that all procedures have been properly developed and implemented.

CONCLUSION

Future research on animals and human embryos will reveal whether cloning is ever likely to be safe enough to be used in human reproduction. The benchmark of safety is set by current reproductive techniques such as IVF. No efforts at reproductive cloning can ethically be undertaken until there is reasonable confidence within the research community that this standard can be met. If and when this occurs, the initial efforts at cloning should take place in research contexts and be subject to careful monitoring and followup.

Therapeutic cloning raises a different but related set of moral issues. These include the appropriateness of creating embryos with the intention of destroying them, of hastening the day when reproductive cloning might become feasible, and of fostering a market in human oocytes for research purposes. It is argued here that none of these issues constitutes a valid reason for ethically prohibiting therapeutic cloning research.

These conclusions contrast with the extraordinary degree of popular antipathy to the very idea of cloning. Much of this opposition is based on misunderstanding of the science involved. Some is attributable to the novelty of this whole area of research and its perceived threat to traditional cultural and family values. As has been true of other new assisted reproductive technologies, we might anticipate that some of this opposition will dwindle if reproductive cloning proves to be physiologically and psychologically safe. Nevertheless, scientists involved in this area should be constantly aware of the controversial nature of their work. They can help to reduce public apprehension by making special efforts to ensure that all aspects of their work conform to the highest standards of research on human subjects.

REFERENCES

Amer, M. S. (1996). Breaking the mold: Human embryo cloning and its implications for a right to individuality. *UCLA Law Rev.* **43**, 1659–1688.

Annas, G. J., Caplan, A. L., and Elias, S. (1996). The politics of human-embryo research—Avoiding ethical gridlock. *New Engl. J. Med.* **334**, 1329–1332.

Barlow, R. (1999). Hoping for extra-brainy offspring, some seek eggs of ivy women. *Valley News* (W. Lebanon, NH), January 24, p. A1.

Beale, R. (2001). Don't clone human embryos! (2 April 2001) dEbate response to Rudolf Jaenisch and Ian Wilmut (Don't clone humans! *Science* **291**, 2552) [http://www.sciencemag.org/cgi/eletters/291/5513/2552#EL1].

Bernasko, J., Lynch, L., Lapinski, R., and Berkowitz R. L. (1997). Twin pregnancies conceived by assisted reproductive techniques: maternal and neonatal outcomes. *Obstet. Gynecol.* **89**, 368–372.

Bjornson, C. R., Rietze, R. L., Reynolds, B. A., Magli, M. C., and Vescovi, A. L. (1999). Turning brain into blood: A hematopoietic fate adopted by adult neural stem cells *in vivo*. *Science* **283**, 534–537.

Brock, D. (1995). The non-identity problem and genetic harms—The case of wrongful handicaps. *Bioethics* **9**, 269–275.

Brock, D. (2001). Cloning human beings: An assessment of the ethical issues pro and con. *In* "Cloning and the Future of Human Embryo Research" (P. Lauritzen, ed.), pp. 93–113. Oxford University Press, New York.

Cahill, L. S. (1997). National Bioethics Advisory Commission, "Hearings on Cloning: Religion-based Perspectives," statement by Lisa Sowle Cahill (Boston College), Watergate Hotel, Washington, D.C., March 13, pp. 6–7.

Chief Medical Officer's Expert Group (2000). "A Report from the Chief Medical Officer's Expert Group Reviewing the Potential of Developments in Stem Cell Research and Cell Nuclear Replacement to Benefit Human Health." UK Department of Health [http://www.doh.gov.uk/cegc/stemcellreport.htm].

Chin, V. (1998). Human cloning and the issue of multiple clones. *Princeton J. Bioethics* **1**, 75–81.

Cohen, C. B. (2001). Letter to the editor, *JAMA* **285**, 1439.

Davis, D. S. (1997). What is wrong with cloning? *Jurimetrics* **38**, 83–89.

Davis, D. S. (2001). "Genetic Dilemmas: Reproductive Technology, Parental Choices, and Children's Futures." Routledge, New York.

Department of Health and Human Services (1991). National Institutes of Health Office for Protection from Research Risks. "Protection of Human Subjects, Code of Federal Regulations, Title 45, Part 46, Subpart D: Additional DHHS Protections for Children Involved as Subjects in Research."

Doerflinger, R. (1999). The ethics of funding embryonic stem cell research: A Catholic viewpoint. *Kennedy Inst. Ethics J.* **9**, 137–150.

Dresser, R. (2001). Letter to the editor. *JAMA* **285**, 1439.

Egan, G. (1998). The extra. *In* "Clones" (J. Dann and G. Dozois, eds.), pp. 55–73. Ace Books, New York.

Ethics Committee of the American Society for Reproductive Medicine (2000). Financial incentives in recruitment of oocyte donors. *Fertil. Steril.* **74**, 116–120.

Feinberg, J. (1992). The child's right to an open future. *In* "Freedom and Fulfillment: Philosophical Essays," pp. 76–97. Princeton University Press, Princeton.

Ford, N. M. (1988). "When Did I Begin?" Cambridge University Press, Cambridge.

Gould, S. J. (1998). Dolly's fashion and Louis's passion. *In* "Flesh of My Flesh: The Ethics of Cloning Human Beings" (Gregory E. Pence, ed.), pp. 101–110. Rowan and Littlefield, Lanham, MD.

Green, R. M. (1997). Parental autonomy and the obligation not to genetically harm one's child: Implications for clinical genetics. *J. Law Med. Ethics* **25**, 5–15.

Green, R. M. (2001). What does it mean to use someone "a means only": Rereading Kant. *Kennedy Inst. Ethics J.* **11**, 249–263.

Grobstein, C. (1988). "Science and the Unborn." Basic Books, New York.

Hanser, M. (1990). Harming future people. *Philos. Public Affairs* **19**, 47–70.

Heller, J. C. (1996). "Human Genome Research & The Challenge of Contingent Future Persons." Creighton University Press, Omaha.

Heyd, D. (1992). "Genethics: Moral Issues in the Creation of People." University of California Press, Berkeley, CA.

Ibrahim, Y. N. (1996) Ethical furor erupts in Britain: Should embryos be destroyed? *New York Times*, August 1, late edition, p. A1.

Jaenisch, R., and Wilmut I. (2001). Don't clone humans! *Science* **291**, 2552–2552.

Johansson, C. B., Momma, S., Clarke, D. L., Risling, M., Lendahl, U., and Frisen, J. (1999). Identification of a neural stem cell in the adult mammalian central nervous system. *Cell* **96**, 25–34.

Jonas, H. (1974). "Philosophical Essays: From Ancient Creed to Technological Man." Prentice-Hall, Englewood Cliffs, NJ.

Kass, L. (1997). The wisdom of repugnance. *New Republic*, June 2, pp. 17–26.

Kelly, M. B. (1991). The rightful position in "wrongful life" actions. *Hasting Law J.* **42**, 505–589.

Kitcher, P. (1998). Whose self is it anyway? *In* "Flesh of My Flesh: The Ethics of Cloning Human Beings" (G. E. Pence, ed.), pp. 65–75. Rowan and Littlefield, Lanham, MD.

Kolata, G. (2001). Researchers find big risk of defect in cloning animals. *New York Times*, March 25, 2001 [online edition: http://www.nytimes.com/2001/03/25/science/25CLON.html?pagewanted=all].

Lanza, R. P., Cibelli, J. B., and West, M. D. (1999). Human therapeutic cloning. *Nat. Med.* **5**, 975–977.

Lanza, R. P., Caplan, A. L., Silver, L. M. Cibelli, J. B., West M. D., and Green, R. M. (2000). The ethical validity of using nuclear transfer in human transplantation. *JAMA* **284**, 3175–3179.

LeGuin, U. (1998). Nine lives. *In* "Cloning" (J. Dann and G. Dozois, eds.), pp. 1–30. Ace Books, New York.

Lewontin, R. (1997). The confusion over cloning. *New York Review of Books*, October 23, pp. 18–23.

Macklin, R. (1994). Splitting embryos on the slippery slope. *Kennedy Inst. Ethics J.* **4**, 209–225.

McCormick, R. A. (1991). Who or what is the preembryo?" *Kennedy Inst. Ethics J.* **1**, 1–15.

McGee, G., and Caplan A. L. (1999). What's in the dish? *Hastings Ctr. Rep.* **29**, 36–38.

McGuffin, P., Riley, B., and Plomin, R. (2001). Toward behavioral genomics. *Science* **291**, 1232–1249.

McKay, R. D. (1999). Brain stem cells change their identity. *Nat. Med.* **5**, 261–262.

Mitchell, C. B. (2000). "Commentary: NIH, Stem Cells, and Moral Guilt." The Center for Bioethics and Human Dignity. Retrieved from http://www.bioethix.org/resources/aps/cbmcomment7.htm.

Murray, T. (2001). Even if it worked, cloning wouldn't bring her back. *Washington Post*, Sunday, April 8, p. B01.

National Bioethics Advisory Commission (1997). "Cloning Human Beings: Report and Recommendations of the National Bioethics Advisory Commission." Rockville, Maryland.

National Institutes of Health Embryo Research Panel (1994). Report of the Human Embryo Research Panel. National Institutes of Health, Office of Science Policy, Washington, D.C.

Nelson, H. L., and Nelson, J. L. (1995). "The Patient in the Family: An Ethics of Medicine and Families." Routledge, New York.

O'Rahilly, R., and Müller, F. (1992). "Human Embryology & Teratology," 2nd Ed. Wiley-Liss, New York.

Orentlicher, D. (1999). Cloning and the preservation of family integrity. *Louisiana Law Rev.* 59, 1019–1040.

Parfit, D. (1984). "Reasons and Persons." Clarendon Press, Oxford.

Pellegrino, E. (2000). Testimony of Edmund D. Pellgrino, M.D. *In* "Human Stem Cell Research, Volume III, Religious Perspectives," pp. F1–F5. National Bioethics Advisory Commission, Rockville, MD.

Pence, G. E. (1998). Will cloning harm people?" *In* "Flesh of My Flesh: The Ethics of Cloning Human Beings" (G. E. Pence, ed.), pp. 115–127. Rowan and Littlefield, Lanham, MD.

Pennisi, E., and Vogel, G. (2000). Animal cloning: Clones: A hard act to follow. *Science* 288, 1722–1727.

Public Law (1996). Public Law 104–99, Section 128, 110 Stat. 34 (01-26-96).

Roberts, M. (1996). Human cloning: A case of no harm? *J. Med. Philos.* 21, 337–554.

Roberts, C. J., and Lowe, C. R. (1975). Where have all the conceptions gone? *Lancet* 1, 498–499.

Robertson, J. (1994). "The Children of Choice." Princeton University Press, Princeton, NJ.

Robertson, J. (1997). Wrongful life, federalism, and procreative liberty: A critique of the NBAC cloning report. *Jurimetrics* 38, 69–82.

Robertson, J. (1998). Liberty, identity, and human cloning. *Texas Law Rev.* 76, 1371–1456.

Rossing, M. A., Daling, J. R., Weiss, N. S., Moore, D. E., and Self, S. E. (1994). Ovarian tumors in a cohort or infertile women. *New Engl. J. Med.* 33, 771–776.

Saunders, K., Spensley, J., Munro, J., and Halasz, G. (1996). Growth and physical outcome of children conceived by *in vitro* fertilization. *Pediatrics* 97, 688–692.

Shannon, T. A., and Woltor, A. B. (1990). Reflections on the moral status of the pre-embryo. *Theolog. Stud.* 51, 603–626.

Shapiro, H. (1999). Ethical dilemmas and stem cell research. *Science* 285, 2065.

Steinbock, B., and McClamrock, R. (1994). When is birth unfair to the child? *Hastings Ctr. Rep.* 24, 15–21.

Stewart, S. A. (1991). Toddler may be sister's lifesaver, *USA Today*, June 4, 1991, p. A3.

Strain, L., Dea, J. C. S., Hamilton M. P. R., and Bonthron, D. T. (1998). A true hermaphrodite chimera resulting from embryo amalgamation after *in vitro* fertilization. *New Engl. J. Med.* 338, 166–169.

Strong (1998). Cloning and infertility. *Cambridge Q. Healthcare Ethics* 7, 279–293.

Sultan, S. L. (1995). The rights of homosexuals to adopt: Changing legal interpretations of "parent" and "family". *J. Suffolk Acad. Law* 10, notes 329 and 330 [www.tourolaw.edu/publications/Suffolk/vol10/part3_txt.htm].

Talbot, M. (2001). A desire to duplicate. *New York Times Magazine*, Feb. 4, pp. 39–63 (inclusive).

Warren. M. A. (1973). On the moral and legal status of abortion. *Monist* 57, 43–61.

Washington Post Editorial Board (1999). "Miracle Cells," *Washington Post*, October 7, p. A34.

Wilcox, L. S., Kiely, J. L., Melvin, C. L., and Martin, M. C. (1996). Assisted reproductive technologies: Estimates of their contribution to multiple births and newborn hospital days in the United States. *Fertil. Steril.* 65, 361–66.

Williams, B. (1985). Which slopes are slippery. *In* "Moral Dilemmas in Modern Medicine" (Michael P. Lockwood, ed.), pp. 126–137. Oxford University Press, New York.

Wilmut, I., Campbell, K., and Tudge, C. (2000). "The Second Creation." Farrar, Straus, and Giroux, New York.

Wolffe, A. P., and Matzke, M. A. (1999). Epigenetics: Regulation through repression. *Science* 286, 481–486.

Woodward, J. (1986). The non-identity problem. *Ethics* 96, 804–831.

Zoloth, L. (2001). Born again faith and yearning in the cloning controversy. *In* "Cloning and the Future of Human Embryo Research" (P. Lauritzen, ed.), pp. 132–141. Oxford University Press, New York.

Zorpette, G. (1998). Off with its head! Headless frog embryos are here. "So what?" biologists say. *Scientific American*, January, p. 41.

FINAL REMARKS

CHAPTER 27

MAMMALIAN CLONING: CHALLENGES FOR THE FUTURE

R. L. Gardner

INTRODUCTION

The number of laboratories for which mammalian reproductive cloning is a major focus of research activity has increased dramatically in the past few years. Despite this, the proportion of eggs or zygotes with a genome that has been transferred from another cell that develops normally to adulthood remains obstinately at a few percent. Although most attrition occurs around the time of implantation, late-gestation and early-postpartum deaths are, regardless of species, unquestionably encountered more commonly than normal in clones. An optimistic view would be that this is simply a matter of the limitations of existing techniques and that once these have been addressed, higher success rates will be achieved routinely. It would be very surprising if the introduction of further technical refinements did not result in some improvement in success rate, but it is still far from clear how substantial this is likely to be. For some studies, a successful outcome may simply entail obtaining just a few viable fertile adults following nuclear replacement using genetically modified donor cells. Concern about efficiency is greater in relation to what has become known as therapeutic cloning, for which the first steps using human cells have been reported (Cibelli *et al.*, 2001). This is because oocytes will remain a scarce resource unless their efficient *in vitro* maturation from the gonads of abortuses or from ovarian tissue made available as a by-product of surgery can be secured (Picton and Gosden, 2000; Picton *et al.*, 2000). It might be argued that the supply of oocytes will not be seriously limiting, providing that rates of development to the blastocyst stage, from which embryonic stem (ES) cells can be derived, are good.

However, unlike in other species, reproductive cloning is not an option for critically testing the effectiveness with which donor nuclei have become reprogrammed in the human. Moreover, it is clear from work in other species that morphologically normal development to the blastocyst stage offers no guarantee that the integrity of the genome has not been compromised (Ford, 1975). Hence, for therapeutic cloning to be realized in our species, an important outstanding issue is how to ensure that the differentiated progeny of ES cells derived in this way are safe for treating patients.

ACTIVATION OF DEVELOPMENT

Aside from integrity of the donor genome, there is the further important matter of ensuring effective activation of development of the host oocyte. It appears that the sperm, which serves to activate the oocyte normally, does so more effectively than any of the many alternative agents that have been employed so far (Loi *et al.*, 1998). In the mouse, activation by the fertilizing sperm results in a succession of transient increases in intracellular Ca^{2+} concentration, of which only the first is initiated at the point of attachment of the sperm to the surface of the vitellus. Successive transients differ in both their site of origin and other characteristics and, regardless of the site of sperm entry, eventually become focused in the zygote's vegetal hemisphere (Deguchi et al., 2000).

None of the artificial activators induce patterns of calcium transients that mimic precisely those due to the fertilizing sperm. Moreover, artificial activation of oocytes may not only perturb the inner cell mass (ICM)/trophectoderm cell ratio at the blastocyst stage (Bos-Mikich et al., 1997), but also may result in anomalies that are not apparent until later, when postimplantation development is well underway (Ozil and Huneau, 2001). Is the difference in efficacy of the sperm versus all artificial modes of activation attributable to something biochemically specific about the natural sperm factor, or to the fact that the sperm activates focally whereas all other agents do so in a delocalized way? The work of Ozil and Huneau (2001) suggests that even delocalized activation may be effective, providing the parameters of the Ca^{2+} transients that are induced closely match those occurring during normal fertilization. A further question, given that artificial activation is difficult to study other than in the context of parthenogenesis (Lawrence et al., 1998), is whether problems relate entirely to inadequacy of this process rather than that the fertilizing sperm does more than contribute a paternal genome and activate development of the oocyte. Regarding the latter possibility, one study claimed to have demonstrated that the sperm entry point dictates the plane of first cleavage in the mouse (Piotrowska and Zernicka-Goetz, 2001). However, an investigation of the distribution of components of the fertilizing sperm rather than surrogate markers of its entry point provided no support for this contention (Davies and Gardner, 2002). Rather, the relationship of the cleavage plane to the sperm entry point was found to be very variable, with the two being more often orthogonal to, than aligned with, each other.

The technique employed to clone the pig avoids possible deficiencies resulting from artificial activation by employing a two-stage process whereby the donor nucleus is ultimately transplanted into normally fertilized eggs deprived of their pronuclei (Polejaeva et al., 2000). This approach, by ensuring physiological activation of development, should help to disentangle the contribution that deficiencies in activation, versus other factors such as anomalies in genomic imprinting (Surani, 2001), make to the relatively poor rate of development of cloned embryos. Additional factors that might affect the outcome are perturbation of the organization of the zygote (Gardner, 2000) and use of induced ovulation, which continues to be linked to an increased incidence of anomalous development (Ertzeid and Storeng, 2001).

THERAPEUTIC CLONING VERSUS DIRECT REPROGRAMMING OF ADULT CELLS

In view of the persisting uncertainties and concerns, it is hardly surprising that opinion is divided within the biomedical research community as to whether therapeutic cloning will offer a viable approach for circumventing rejection in the harnessing of ES cells for tissue repair. There is wider agreement about the value of transplantation of nuclei into oocytes as a tool for gaining insight into how the genome of differentiated cells may be reprogrammed. The latter has become a topic of considerable interest because of a spate of studies attesting to the persisting lability of cells in a wide variety of tissues and organs in adult mammals. Although the cells in question are often referred to as "adult stem cells," this may be a misnomer. Except possibly in the case of the hematopoietic system, there is at present no evidence to suggest that the cells from adults implicated in lineage crossing are restricted to those having the characteristics of stem cells in vivo . Indeed, for most tissues there is nothing to counter the view that, if left in their natural environment, the cells in question would remain strictly postmitotic, and that it is the stimulation to reenter a very active proliferative state under the very unnatural conditions obtaining *in vitro* that is instrumental in reprogramming their gene expression. It is noteworthy in this context that a period of culture has been employed in the great majority of cases in which changes in type have been reported for cells from adults. Among the more notable exceptions is the formation of hepatocytes from bone marrow in the human (Alison *et al.*, 2000; Theise *et al.*, 2000).

The notion that the differentiation of adult cells is susceptible to reprogramming is not new. Aside from the various natural examples of transdifferentiation or metaplasia (Okada, 1991), the results of numerous cell fusion experiments support the view that the patterns of gene expression that serve to distinguish different types of differentiated cells are subject to continuous active regulation (Blau, 1989; Blau and Baltimore, 1991; Gardner, 1993). However, for much of the contemporary work on plasticity of adult cells, the identity of the cells involved is uncertain. Indeed, in the light of the evidence that many hematopoietic stem cells enter the blood every day but remain therein for a very limited time (Wright *et al.*, 2001), the possibility that plasticity is conferred on at least some tissues through the presence of these cells has to be entertained.

REPRODUCTIVE CLONING—QUESTIONS, MYTHS, AND CONCERNS

A very intriguing question that is touched on by Wakayama and Perry in their contribution to this volume (see Chapter 16), but for which a definite answer cannot yet be given, is whether mammals can be propagated indefinitely by cloning without recourse to sexual reproduction (Wakayama *et al.*, 2000). This is particularly pertinent when there might be a need to perpetuate a specific transgenic genotype unchanged through several generations. Also relevant here is the still unsettled matter of whether aging phenomena are completely reversed when nuclei from adult cells are transplanted into the cytoplasm of oocytes or zygotes.

In their contributions, both Seidel (Chapter 10) and Green (Chapter 26) draw attention to the important point that does not seem to have been grasped fully by the public so far. This is that the link between genotype and phenotype is much less rigid than is widely supposed, especially for those characteristics on which our uniqueness as individuals so largely depend (Gardner, 1998). So long as this is not appreciated, and the photocopy analogy continues to enjoy such wide currency, unrealistic expectations about the outcome of reproductive cloning will continue to confuse the debate. One only has to consider the degree of disparity between bilaterally symmetrical components of the body such as the patterns in the irides of the two eyes of the same individual to appreciate that morphogenetic processes are not entirely programmed genetically (Daugman and Downing, 2001). Nevertheless, a question remains about the entitlement of future individuals to the unique shuffling of parental genes that sexual reproduction entails, and which is, at present, enjoyed by all individuals except those resulting from accidental subdivision of the early conceptus.

The very widespread present support for a moratorium on human reproductive cloning, both within the scientific community and outside it, is rooted in two types of concerns. Outside, ethical considerations have been dominant, whereas within the community, concerns about safety have featured most prominently in the debate. There are opposing views on whether worries about the safety of human reproductive cloning are warranted (cf. Jaensich and Wilmut, 2001; Lanza *et al.*, 2001). In looking at this issue it is important not to neglect the existence of important species differences among eutherian mammals relating to viviparity. That the trophoblast lineage may be perturbed in cloned embryos is indicated by the occurrence of disturbed placental morphogenesis, which, notwithstanding their very different modes of placentation, is common to both cloned ungulates and mice (Hill *et al.*, 2000; De Sousa *et al.*, 2001; Tanaka *et al.*, 2001). In the mouse, the main defect seems to be reduction in the amount of labyrinthine relative to spongiotrophoblast tissue, which is not accompanied by any marked disturbance in the expression of placenta-specific genes (Tanaka *et al.*, 2001). The consequent compromise in placental exchange may restrict prenatal growth of cloned fetuses without precluding their becoming seemingly normal adults. In considering possible risks of reproductive cloning in humans, it is therefore necessary also to take into account possible consequences for uterine foster mothers of anomalous development of the placenta. This is particularly important because malignant tumors of the trophoblast, called choriocarcinomas, seem to be peculiar to our species (Gardner, 1983).

In conclusion, studies in mammals have shown what those in amphibia failed to do—namely, that development of fertile adults can be supported by the genome of adult cells. Such findings are of potential practical utility and offer further insight into the fundamental process of cellular differentiation. Nonetheless, many observations have yet to be explained. These include obvious differences between species and donor cell types, regarding the ease with which reproductive cloning can be achieved.

ACKNOWLEDGMENTS

I thank Ann Yates for help in preparing the manuscript and the Royal Society and Wellcome Trust for support.

REFERENCES

Alison, M. R., Poulson, R., Jeffery, R., Dhillon, A. P., Quaglia, A., Jacob, J., Novelli, M., Prentice, G., Williamson, J., and Wright, N. A. (2000). Hepatocytes from non-hepatic adult stem cells. *Nature* **406**, 257.

Blau, H. M. (1989). How fixed is the differentiated state? *Trends Genet.* **5**, 268–272.

Blau, H. M., and Baltimore, D. (1991). Differentiation requires continuous regulation. *J. Cell Biol.* **112**, 781–783.

Bos-Mikich, A., Whittingham, D. G., and Jones, K. T. (1997). Meiotic and mitotic Ca^{2+} oscillations affect cell composition in resulting blastocysts. *Dev. Biol.* **182**, 172–179.

Cibelli, J. B., Kiessling, A. A., Cunniff, K., Richards, C., Lanza, R. P., and West, M. D. (2001). Somatic cell nuclear transfer in humans: pronuclear and early embryonic development. *J. Regen. Med.* **2**, 25–31.

Daughman, J., and Downing, C. (2001). Epigenetic randomness, complexity and singularity of human iris patterns. *Proc. R. Soc. Ser. B.* **B268**, 1737–1740.

Davies, T. J., and Gardner, R. L. (2002). The plane of first cleavage is not related to the distribution of sperm components in the mouse. *Hum. Reprod.* In press.

Deguchi, R., Shirakawa, H., Oda, S., Mohri, T., and Miyazaki, S. (2000). Spatiotemporal analysis of Ca^{2+} waves in relation to the sperm entry site and animal-vegetal axis during Ca^{2+} oscillations in fertilized mouse eggs. *Dev. Biol.* **218**, 299–313.

De Sousa, P. A., King, T., Harkness, L., Young, L. E., Walker, S. K., and Wilmut, I. (2001). Evaluation of deficiencies in cloned sheep fetuses and placentae. *Biol. Reprod.* **65**, 23–30.

Ertzeid, G., and Storeng, R. (2001). The impact of ovarian stimulation on implantation and fetal development in mice. *Hum. Reprod.* **16**, 221–225.

Ford, C. E. (1975). The time in development at which gross genome unbalance is expressed. *In "The Early Development of Mammals"* (M. Balls and A. E. Wild, eds.), pp. 285–304. Cambridge University Press, Cambridge.

Gardner, R. L. (1983). Origin and differentiation of extra-embryonic tissues in the mouse. *Int. Rev. Exp. Pathol.* **24**, 63–133.

Gardner, R. L. (1993). Extrinsic factors in cellular differentiation. *Int. J. Dev. Biol.* **37**, 47–50.

Gardner, R. L. (1998). Cloning and individuality. *In "1998 Oxford Amnesty Lectures, The Genetic Revolution and Human Rights"* (J. Birley, ed.), pp. 29–37. University of Chicago Press, Chicago.

Gardner, R. L. (2000). The initial phase of embryonic patterning in mammals. *Int. Rev. Cytol.* **203**, 233–290.

Hill, J. R., Burghardt, R. C., Jones, K., Ling, C. R., Looney, C. R., Shin, T., Spencer, T. E., Thompson, J. A., Winger, Q. A., and Westhusin, M. E. (2000). Evidence for placental abnormality as the major cause of mortality in first trimester somatic cell cloned bovine fetuses. *Biol. Reprod.* **63**, 1787–1794.

Jaenisch, R., and Wilmut, I. (2001). Developmental biology. Don't clone humans! *Science* **292**, 2552.

Lanza, R. P., Cibelli, J. B., Faber, D., Sweeney, R. W., Henderson, B., Nevala, W., West, M. D., and Wettstein, P. J. (2001). Cloned cattle can be healthy and normal. *Science* **294**, 1893–1894.

Lawrence, Y., Ozil, J.-P., and Swann, K. (1998). The effects of a Ca^{2+} chelator and heavy-metal-ion chelators upon Ca^{2+} oscillations and activation at fertilisation in mouse suggest a role for repetitive Ca^{2+} increases. *Biochem. J.* **335**, 335–342.

Loi, P., Ledda, S., Fulka, J., Jr., Cappai, P., and Moor, R. M. (1998). Development of parthenogenetic and cloned ovine embryos: effect of activation protocols. *Biol. Reprod.* **58**, 1177–1187.

Okada, T. S. (1991). "*Transdifferentiation: Flexibility in Cell Differentiation.*" Clarendon Press, Oxford.

Ozil, J.-P., and Huneau, D. (2001). Activation of rabbit oocytes: The impact of the Ca^{2+} signal regime on development. *Development* **128**, 917–929.

Picton, H. M., and Gosden, R. G. (2000). *In vitro* growth of human primordial follicles from frozen-banked ovarian tissue. *Mol. Cell Endocrinol.* **166**, 27–35.

Picton, H. M., Kimm, S. S., and Gosden, R. G. (2000). Cryopreservation of gonadal tissue and cells. *Br. Med. Bull.* **56**, 603–615.

Piotrowska, K., and Zernicka-Goetz, M. (2001). Role for sperm in spatial patterning of the early mouse embryo. *Nature* **409**, 517–521.

Polejaeva, I. A., Chen, S. H., Vaught, T. D., Page, R. L., Mullins, J., Ball, S., Dai, Y., Boone, J., Walker, S., Ayares, D. L., Colman, A., and Campbell, K. H. (2000). Cloned pigs produced by nuclear transfer from adult somatic cells. *Nature* **407**, 86–90.

Surani, M. A. (2001). Reprogramming of genome function through epigenetic inheritance. *Nature* **414**, 122–128.

Tanaka, S., Oda, M., Toyoshima, Y., Wakayama, T., Tanaka, M., Yoshida, N., Hattori, N., Ohgane, J., Yanagimachi, R., and Shiota, K. (2001). Placentomegaly in cloned mouse concepti caused by expansion of the spongiotrophoblast layer. *Biol. Reprod.* **65**, 1813–1821.

Theise, N. D., Nimmakayalu, M., Gardner, R., Illei, P. B., Morgan, G., Teperman, L., Henegariu, O., and Krause, D. (2000). Liver from bone marrow in humans. *Hepatology* **32**, 11–16.

Wakayama, T., Shinkai, Y., Tamashiro, K. L. K., Niida, H., Blanchard, D. C., Blanchard, R. J., Ogura, A., Tanemura, K., Tachibana, M., Perry, A. C. F., Colgan, D. F., Mombaerts, P., and Yanagimachi, R. (2000). Cloning of mice to six generations. *Nature* **407**, 318–319.

Wright, D. E., Wagers, A. J., Gulati, A. P., Johnson, F. L., and Weissman, I. L. (2001). Physiological migration of haematopoietic stem and progenitor cells. *Science* **294**, 1933–1936.